T0264108

Université Joseph Fourier

Les Houches

Session LXXIX

2003

Quantum Entanglement and Information Processing

Intrication quantique et traitement de l'information

Lecturers

R. Blatt
M. Brune
I. Chuang
M. Devoret
N. Gisin
C. Glattli
P. Grangier
S. Haroche
J. Jones
J. Martinis
H. Mooij
R. Schoelkopf
D. Vion
D. Wineland
A. Wernsdorfer
A. Zeilinger
P. Zoller

ÉCOLE D'ÉTÉ DE PHYSIQUE DES HOUCHES

SESSION LXXIX, 30 JUNE – 25 JULY 2003

EURO SUMMER SCHOOL

ÉCOLE THÉMATIQUE DU CNRS

QUANTUM ENTANGLEMENT AND INFORMATION PROCESSING

INTRICATION QUANTIQUE ET TRAITEMENT DE L'INFORMATION

Edited by

Daniel Estève, Jean-Michel Raimond and Jean Dalibard

2004

Amsterdam – Boston – Heidelberg – London – New York – Oxford
Paris – San Diego – San Francisco – Singapore – Sydney – Tokyo

ELSEVIER B.V.
Sara Burgerhartstraat 25
P.O. Box 211, 1000 AE
Amsterdam,
The Netherlands

ELSEVIER Inc.
525 B Street, Suite 1900
San Diego, CA 92101-4495
USA

ELSEVIER Ltd
The Boulevard, Langford Lane
Kidlington, Oxford OX5 1GB
UK

ELSEVIER Ltd
84 Theobalds Road
London WC1X 8RR
UK

© 2004 Elsevier B.V. All rights reserved

This work is protected under copyright by Elsevier B.V., and the following terms and conditions apply to its use:

Photocopying
Single photocopies of single chapters may be made for personal use as allowed by national copyright laws. Permission of the Publisher and payment of a fee is required for all other photocopying, including multiple or systematic copying, copying for advertising or promotional purposes, resale, and all forms of document delivery. Special rates are available for educational institutions that wish to make photocopies for non-profit educational classroom use.

Permissions may be sought directly from Elsevier's Rights Department in Oxford, UK: phone (+44) 1865 843830, fax (+44) 1865 853333, e-mail: permissions@elsevier.com. Requests may also be completed on-line via the Elsevier homepage (http://www.elsevier.com/locate/permissions).

In the USA, users may clear permissions and make payments through the Copyright Clearance Center, Inc., 222 Rosewood Drive, Danvers, MA 01923, USA; phone: (+1) (978) 7508400, fax: (+1) (978) 7504744, and in the UK through the Copyright Licensing Agency Rapid Clearance Service (CLARCS), 90 Tottenham Court Road, London W1P 0LP, UK; phone: (+44) 20 7631 5555, fax: (+44) 20 7631 5500. Other countries may have a local reprographic rights agency for payments.

Derivative Works
Tables of contents may be reproduced for internal circulation, but permission of the Publisher is required for external resale or distribution of such material. Permission of the Publisher is required for all other derivative works, including compilations and translations.

Electronic Storage or Usage
Permission of the Publisher is required to store or use electronically any material contained in this work, including any chapter or part of a chapter.

Except as outlined above, no part of this work may be reproduced, stored in a retrieval system or transmitted in any form or by any means, electronic, mechanical, photocopying, recording or otherwise, without prior written permission of the Publisher.
Address permissions requests to: Elsevier's Rights Department, at the fax and e-mail addresses noted above.

Notice
No responsibility is assumed by the Publisher for any injury and/or damage to persons or property as a matter of products liability, negligence or otherwise, or from any use or operation of any methods, products, instructions or ideas contained in the material herein. Because of rapid advances in the medical sciences, in particular, independent verification of diagnoses and drug dosages should be made.

First edition 2004

Library of Congress Cataloging in Publication Data
A catalog record is available from the Library of Congress.

British Library Cataloguing in Publication Data
A catalogue record is available from the British Library.

ISBN: 0-444-51728-6
ISSN: 0924-8099

Transferred to digital print, 2007
Printed and bound by CPI Antony Rowe, Eastbourne

Working together to grow
libraries in developing countries

www.elsevier.com | www.bookaid.org | www.sabre.org

ELSEVIER BOOK AID International Sabre Foundation

École de Physique des Houches

Service inter-universitaire commun
à l'Université Joseph Fourier de Grenoble
et à l'Institut National Polytechnique de Grenoble

Subventionné par le Ministère de la Jeunesse,
de l'Éducation Nationale et de la Recherche,
le Centre National de la Recherche Scientifique,
le Commissariat à l'Énergie Atomique

Membres du conseil d'administration :
Yannick Vallée (président), Paul Jacquet (vice-président), Cécile DeWitt, Mauricette Dupois, Thérèze Encrenaz, Bertrand Fourcade, Luc Frappat, Jean-François Joanny, Michèle Leduc, Jean-Yves Marzin, Giorgio Parisi, Eva Pebay-Peyroula, Michel Peyrard, Luc Poggioli, Jean-Paul Poirier, Michel Schlenker, François Weiss, Philippe Wisler, Jean Zinn-Justin

Directeur :
Jean Dalibard (Laboratoire Kastler Brossel, Paris, France)

Directeurs scientifiques de la session LXXIX:
Daniel Estève (Quantronics group, SPEC, CEA-Saclay, France)
Jean-Michel Raimond (Laboratoire Kastler Brossel, ENS, Paris, France)

PREVIOUS SESSIONS

I	1951	Quantum mechanics. Quantum field theory
II	1952	Quantum mechanics. Statistical mechanics. Nuclear physics
III	1953	Quantum mechanics. Solid state physics. Statistical mechanics. Elementary particle physics
IV	1954	Quantum mechanics. Collision theory. Nucleon–nucleon interaction. Quantum electrodynamics
V	1955	Quantum mechanics. Non-equilibrium phenomena. Nuclear reactions. Interaction of a nucleus with atomic and molecular fields
VI	1956	Quantum perturbation theory. Low temperature physics. Quantum theory of solids; dislocations and plastic properties Magnesium; ferromagnetism
VII	1957	Scattering theory. Recent developments in field theory. Nuclear interaction; strong interactions. High energy electrons. Experiments in high energy nuclear physics
VIII	1958	The many body problem
IX	1959	The theory of neutral and ionized gases
X	1960	Elementary particles and dispersion relations
XI	1961	Low temperature physics
XII	1962	Geophysics; the earth's environment
XIII	1963	Relativity groups and topology
XIV	1964	Quantum optics and electronics
XV	1965	High energy physics
XVI	1966	High energy astrophysics
XVII	1967	Many body physics
XVIII	1968	Nuclear physics
XIX	1969	Physical problems in biological systems
XX	1970	Statistical mechanics and quantum field theory
XXI	1971	Particle physics
XXII	1972	Plasma physics
XXIII	1972	Black holes
XXIV	1973	Fluids dynamics
XXV	1973	Molecular fluids
XXVI	1974	Atomic and molecular physics and the interstellar matter
XXVII	1975	Frontiers in laser spectroscopy
XXVIII	1975	Methods in field theory
XXIX	1976	Weak and electromagnetic interactions at high energy
XXX	1977	Nuclear physics with heavy ions and mesons
XXXI	1978	III condensed matter
XXXII	1979	Membranes and intercellular communication
XXXIII	1979	Physical cosmology
XXXIV	1980	Laser plasma interaction
XXXV	1980	Physics of defects
XXXVI	1981	Chaotic behaviour of determininistic systems
XXXVII	1981	Gauge theories in high energy physics
XXXVIII	1982	New trends in atomic physics
XXXIX	1982	Recent advances in field theory and statistical mechanics
XL	1983	Relativity, groups and topology
XLI	1983	Birth and infancy of stars

XLII	1984	Cellular and molecular aspects of developmental biology
XLIII	1984	Critical phenomena, random systems, gauge theories
XLIV	1985	Architecture of fundamental interactions at short distances
XLV	1985	Signal processing
XLVI	1986	Chance and matter
XLVII	1986	Astrophysical fluid dynamics
XLVIII	1988	Liquids at interfaces
XLIX	1988	Fields, strings and critical phenomena
L	1988	Oceanographic and geophysical tomography
LI	1989	Liquids, freezing and glass transition
LII	1989	Chaos and quantum physics
LIII	1990	Fundamental systems in quantum optics
LIV	1990	Supernovae
LV	1991	Particles in the nineties
LVI	1991	Strongly interacting fermions and high T, superconductivity
LVII	1992	Gravitation and quantizations
LVIII	1992	Progress in picture processing
LVIX	1993	Computational fluid dynamics
LX	1993	Cosmology and large scale structure
LXI	1994	Mesoscopic quantum physics
LXII	1994	Fluctuating geometries in statistical mechanics and quantum field theory
LXIII	1995	Quantum fluctuations
LXIV	1995	Quantum symmetries
LXV	1996	From cell to brain
LXVI	1996	Trends in nuclear physics, 100 years later
LXVII	1997	Modélisation du climat de la terre et de sa variabilité
LXVIII	1997	Particules et interactions : le modèle standard mis à l'épreuve
LXIX	1998	Topological aspects of low dimensional systems
LX	1998	Infrared space astronomy, today and tomorrow
LXI	1999	The primordial universe
LXII	1999	Coherent atomic matter waves
LXIII	2000	Atomic clusters and nanoparticles
LXIV	2000	New trends in turbulence
LXV	2001	Physics of bio-molecules and cells
LXVI	2001	Unity from duality: Gravity, gauge theory and strings
LXVII	2002	Slow relaxations and nonequilibrium dynamics in condensed matter
LXVII	2002	Accretion discs, jets and high energy phenomena in astrophysics

Publishers:
Session VIII: Dunod, Wiley, Methuen;
Sessions IX and X: Herman, Wiley;
Session XI: Gordon and Breach, Presses Universitaires;
Sessions XII–XXV: Gordon and Breach;
Sessions XXVI–LXVIII: North-Holland;
Sessions LXIX–LXXVIII: EDP Sciences, Springer.

Lecturers

BLATT Rainer, Institut für Experimentalphysik, Universität Innsbruck, Technikerstrasse 25, A-6020 Innsbruck, Austria

BRUNE Michel, LKB/ENS, 24 rue Lhomond, 75231 Paris cedex 05, France

CHUANG Isaac, Center for Bits and Atoms, Media Lab., MIT, 20 Ames street, E 15-424 Cambridge, Ma 02139 USA

DEVORET Michel, Applied Physics Department, Yale University, Po Box 208284, New Haven, CT 06520-8284, USA

GISIN Nicolas, Group of Applied Physics, Université de Genève, 20 rue de l'École de medecine, 1211 Genève, Switzerland

GLATTLI Christian, LPA/ENS, 24 rue Lhomond, 75231 Paris cedex 05, France

GRANGIER Philippe, Institut d'Optique, BP 147, 91403 Orsay cedex, France

HAROCHE Serge, LKB/ENS, 24 rue Lhomond, 75231 Paris cedex 05, France

JONES Jonathan, Center for Quantum Computation, Clarendon laboratory, Parks road, Oxford OX1 3PU, UK

MARTINIS John, NIST, Mail code 814.03, 325 Broadway, Boulder CO 80305-328, USA

MOOIJ Hans, Department of Nanosciences, Delft University of Technology, Po Box 5046, 2600 GA Delft, The Netherlands

SCHOELKOPF Robert J., Department of Applied Physics, Becton Laboratory, Yale University, New haven, CT 06250-8284, USA

VION Denis, SPEC CEA Saclay, 91191 Gif sur Yvette, France

WERNSDORFER Wolfgang, Laboratoire Louis Néel, BP 166, 25 avenue des martyrs, 38042 Grenoble, France

WINELAND David J., NIST, Division 847, 325 Broadway, Boulder CO 80305-3328, USA

ZEILINGER Anton, Institut für Experimentalphysik, Universität Wien, Boltzmanngasse 5, A-1090 Wien, Austria

ZOLLER Peter, Universität Innsbruck, Tecnikerstrasse 25, A-6020 Innsbruck, Austria

Participants

ANDERSSON Mauritz, Physikalisches Institut, Universität Heidelberg, Philosophenweg 12, D-69120 Heidelberg, Germany

AUFFEVES-GARNIER Alexia, LKB/ENS 24 Rue Lhomond, 75231 Paris cedex 05, France

BALL Jonathan, Centre for Quantum Computation, Clarendon Laboratory, Parks Road, Oxford OX1 3PU, UK

BENHELM Jan, University of Innsbruck, Technikerstr. 25, 6020 Insbruck, Austria

BERGAMINI Silvia, IOTA Bât. 503, Centre Scientifique d'Orsay, 91403 Orsay cedex, France

BESCOND Marc, L2MP UMR CNRS 6137, Maison des Technologies, place Georges Pompidou, 83000 Toulon, France

BRAUN Alexander, Institut für Laserphysik, Universität Hamburg, Luruper chaussee 149, Geb 69, 22 761 HH, Germany

BRION Etienne, Laboratoire Aimé Cotton, Bât. 505, 91405 Orsay cedex, France

CELARDO Giuseppe Luca, Dip. Di Matematica e Fisica, Universita cattolica del Sacro Cuore, via Musei 41, Brescia 25100, Italy

CLAUDON Julien, CRTBT CNRS, 25 avenue des martyrs, BP 166, 38042 Grenoble cedex 9, France

COLLIN Eddy, Quantronics Group, SPEC CEA Saclay, 91191 Gif sur Yvette, France

CUBITT Toby, MPI für Quantenoptik, Hans-Kopfermann Strasse 1, D-85748 Garching, Germany

DANTAN Aurélien, LKB Université Pierre et Marie Curie, 4 Place Jussieu, case 74, 75252 Paris cedex 05, France

FATTAL David, Stanford University, dept. of physics, 382 via Pueblo Mail, Stanford, CA 94305, USA

GABELLI Julien, LPMC/ENS, 24 Rue Lhomond, 75231 Paris cedex 05, France

GUTMANN Henryk, LMU, Physics Department, Group Von Delft, Theresienstr. 37, D-80333 Munich, Germany

ITHIER Grégoire, Groupe Quantronique CEA Saclay, 91191 Gif sur Yvette, France

LAMBERT Neill, Department of Physics, UMIST, PO Box 88, Manchester, M 60 1QD, UK

LAMOUREUX Louis-Philippe, ULB, CP 165/59 Avenue F.D Roosevelt, 1050 Bruxelles, Belgique

LARSON Jonas, Royal Institute of Technology, SCFAB, Roslagstullsbacken 21, S-106 91 Stockholm, Sweden

LEVI Benjamin, Laboratoire de Physique Théorique, IRSAMC, Université Paul Sabatier, 118 route de Narbonne, 31062 Toulouse cedex 04, France

LIM Yuan Liang, 53 Sterling place, South Ealing, London W5 4RA, UK

LISOWSKI Caroline, PIIM, Equipe CIML, Université de Provence, Centre de St Jérôme, Case C21, 13397 Marseille cedex 20, France

LIU Ru-Fen, National Cheng Kung university, Physics department, 1 University road, Tainan, Taiwan, R.O.C

LJUNGGREN Daniel, KTH, Royal Institute of Technology, Lab. of Optics Photonics and Quantum Electronics, Dept of Microelectronics and Information Technology, PO Box Electrum 229, SE-16440 Kista, Sweden

MA Zhaoyuan, Clarendon Laboratory, University of Oxford, Parks road, OX1 3 PU, UK

MARQUARDT Christoph, Lehrstuhl für Optik, Universität Erlangen-Nuruberg, Staudtstrasse 7/B2, 91058 Erlangen, Germany

McHUGH Derek, National University of Ireland Maynooth, Maynooth, CO Kildare, Ireland

MEUNIER Tristan, Groupe d'Electrodynamique des systèmes simples, LKB ENS, 24 rue Lhomond, 75231 Paris cedex 05, France

MIN Hyegeun, Department of Physics, Sookmyung Women's University, Chungpa-Dong 2-Ga, Yongsan-Gu, Seoul 140-742, Korea

ORUS Roman, University of Barcelona, Faculty of Physics, Dept. ECM, Diagonal 647, 08028 Barcelona, Spain

PECHEN Alexander, Steklov Mathematical Institute, Gubkin str, 8, 119991, Moscow, Russia

PINEDA Carlos, Instituto de Fisica, Circuito de la Investigacion Cientifica S-n,

Ciudad Universitaria, CP 04510, Coyoacan, Mexico DF

PIORO-LADRIERE Michel, Institute for Microstructural Sciences, National Research Council of Canada, Montreal road campus, Building M23-A, Room 150, Ottawa, Ontario, Canada, K1A 0R6

PIRANDOLA Stefano, Dipartimento di Fisica, Universita di Camerino, via Madonna delle Carceri, I- 62032 Camerino MC, Italy

PRAXMEYER Ludmila, Institute of Theoretical Physics, Warsaw University, ul.Hoza 69, 00-681 Warsaw, Poland

REJEC Tomaz, Department of Theoretical Physics, Josef Stefan Institute, Jamova 39, SI-1000 Ljubljana, Slovenia

RIBEIRO DE CARVALHO André, MPI for Complex Systems, Nöthnitzer str. 38, D-01187 Dresden, Germany

ROMBETTO Sara, Istituto di Cibernetica "Eduardo Caianiello", via campi Flegrei34, Comprensorio "A. Olivetti", Building 70, 80078 Pozzuoli, Italy

RONCAGLIA Augusto, Departamento de Fisica, Facultad de Ciencias Exactas y Naturales, Universidad de Buenos Aires, Pabellon I, Ciudad Universitaria, 1428 Capital Federal, Argentina

SAURET Olivier, LEPES CNRS, 25 avenue des martyrs, 38000 Grenoble, France

SCHAEFFER David, CNRS LCMI, 25 avenue des martyrs, BP 166, 38042 Grenoble cedex 9, France

SCHRIEFL Josef, Laboratoire de Physique ENS Lyon, 46 allée d'Italie, 69007 Lyon, France

SCIARRINO Fabio, Universita di Roma La Sapienza, Dipartimento de Fisica, P.le Aldo Moro 2, Roma 00185, Italy

SERAFINI Alessio, Dipartimento di Fisica E.R Caianiello, Universita di Salerno, Via S. Allende, 84081 Baronissi, Italy

SHIMIZU Ryosuke, Electronic Quantum Devices Laboratory, Research Institute of Electrical Communication, Tohoku University, 2-1-1 Katahira, Aoba-ku, Sendai 980-8577, Japan

TAKEI Nobuyuki, Department of Applied Physics, School of Engineering, University of Tokyo, 7-3-1 Hongo, Bunkyo-ku, Tokyo 113-8656, Japan

TAN Eng Kiang, Department of Physics, University of Strathclyde, John Anderson Building, 107 Rottenrow, Glasgow G4 ONG, UK

TERRACIANO Matthew, 11100 Stonepath Lane, Charlotte, NC 28277, USA

TORRES Elva, Department of Chemistry, Leone Research Group, University of California, Berkeley CA 94720-1460, USA

TROJEK Pavel, MPI für Quantenoptik, Hans kopfermann str 1, D-85748 Garching, Germany

WALLQUIST Margareta, Department of Microtechnology and Nanoscience, Chalmers, S-41296 Göteborg, Sweden

WELLARD Cameron, School of Physics, University of Melbourne, Vic 3010, Australia

WILSON Christopher, Yale University, PO Box 208284, New Haven CT 06520, USA

YU Terri, 11614 Dawson Drive, Los Altos Hills, CA 94024, USA

PREFACE

Quantum optics has been the central subject of quite a few Les Houches schools, since the historical session of 1964. Mesoscopic physics is a newer field, but has also been well represented in this series. The present session is the first one, though, that unites these two fields and aims at creating a link between the two communities, under the general theme of quantum entanglement and quantum information processing.

It has been recognized quite recently that the weirdness of quantum mechanics could be tamed to realize new functions for information transmission or processing. Quantum systems can be used to carry information (using qubits, two-level systems in a logical state superposition, instead of classical bits). The properties of quantum measurement and the impossibility to clone a quantum object make it possible to use qubits to share information in a completely secret and secure way, opening the way to quantum cryptographic key distribution. Quantum non-locality has also been used to transmit a quantum state, realizing quantum teleportation.

More ambitiously, quantum systems could be used to realize a calculation. The pioneering work of Deutsch and Josza triggered a flourishing activity, with a breakthrough in 1994 when P. Shor discovered a quantum algorithm able to factorize efficiently large numbers, with a considerable potential impact on the cryptographic systems. The constraints imposed on a machine able to implement this algorithm are so formidable, though, that it first appeared to be completely out of reach.

Indeed, quantum coherent evolution is required during the whole computation and the unavoidable residual decoherence makes any sizeable computation completely impossible. A way to circumvent this essential difficulty was proposed in 1996, with quantum error correction codes. Once again, their implementation looks extremely difficult, but they lift the complete ban on quantum computers and raise hopes for practical implementations of quantum communication.

These fascinating perspectives resulted in a renewed interest for experiments on fundamental quantum systems and properties. Major efforts are devoted, throughout the world, to the development of the theoretical concepts and of the experimental implementations of quantum information processing. This extremely active subject was mature enough to necessitate a school presenting an in-depth survey of the field.

Quantum information gathers around a common objective, with a common vocabulary, different communities. The most impressive experimental advances so far are in the fields of quantum optics, nuclear magnetic resonance (NMR) and mesoscopic physics. The control of atoms and photons is a well developed field, in which quantum entanglement studies were underway even well before the outburst of quantum information. The exquisite degree of control which is achievable, the simplicity of the systems under study, make this field quite appealing for fundamental quantum mechanics investigations. NMR provides impressively long coherence lifetimes, typical of nuclear spins, and relies on sophisticated techniques developed over many years for chemical analysis purposes. Finally, mesoscopic physics offers the flexibility of designed circuits, of 'man made atoms'. Which of these systems is better suited for large scale information processing is still an open question.

The aim of this school was thus to gather students and lecturers belonging to these different communities, to provide them with a common language, through introductory courses in quantum information, quantum optics, NMR and mesoscopic physics, and to present finally the latest advances in these fields.

The lecture notes in this volume are grouped in a thematic order. Longer introductory chapters are thus interleaved with more advanced focuses on experimental advances. The book opens with a course by I. Chuang on the basic principles of quantum information processing. How is it possible to process information with quantum systems? How can we characterize the amount of information in a quantum state? How can we describe in a synthetic way the decoherence mechanisms and how can we fight this decoherence? All these essential questions are addressed here.

The following chapters (courses 2 to 9) are devoted to the quantum optics approach to quantum information processing. This part opens with a comprehensive course by S. Haroche on mesoscopic superposition states in quantum optics. Such states are obviously at the heart of large quantum information networks. Their production and decoherence also opens very fundamental questions on the boundary between the quantum and classical worlds. This course provides an introduction to the basic tools involved in the following lectures.

Course 3, by M. Brune, presents quantum information processing with cavity quantum electrodynamics, a field already covered as a test ground for quantum mesoscopic superposition states in course 2. Course 4, by P. Zoller, presents a brief survey of three other possible implementations of quantum information processing in quantum optics: trapped ions, atoms in laser traps and qubits stored in atomic ensembles.

Courses 5 and 6, by R. Blatt and D.J. Wineland, give a much more detailed introduction to quantum information processing with trapped ions, one of the most advanced experimental fields. These courses also provide a survey of recent achievements in this field and outline encouraging mid-term perspectives.

Courses 7 and 8, by N. Gisin and P. Grangier, address the very active field of quantum key distribution, either with single photon qubits (and cover thus the important problem of single photon sources) or with macroscopic fields. The closely related course 9, by A. Zeilinger, describes beautiful experiments on key distribution and teleportation.

The nuclear magnetic resonance approach to quantum information is covered in the lectures by J. Jones (course 10). These comprehensive notes both provide an introduction to the basic techniques of NMR and a survey of recent results and perspectives in this field.

The last chapters are devoted to mesoscopic physics approaches. They open by three introductory lectures, by D.C. Glattli (course 11), M.H. Devoret (course 12) and J.M. Martinis (course 13). These chapters provide readers who have a quantum optics background with the essential tools for mesoscopic quantum conductors and superconducting circuits. How to design and operate electronic circuits displaying quantum effects similar to those found in real atoms is presently a challenge: although quantum superpositions have already been demonstrated, quantum coherence of these artificial atoms is still inferior.

Nevertheless, impressive advances in the coherent control of qubit superconducting circuits have been recently obtained. Some of them are covered in the next course, by D. Vion (course 14). Finally, W. Wernsdorfer (course 15) covers the magnetic properties of nano-particles, a beautiful example of a mesoscopic quantum system.

The convergence between quantum optics and mesoscopic physics triggered by quantum information is particularly evidenced by the last contribution to this volume (course 16). It covers a seminar given by R. Schoelkopf, who proposes to combine the cavity QED techniques, related in the courses by S. Haroche and M. Brune, with the kind of mesoscopic circuitry described in the previous chapters.

We hope that this school and the present volume will contribute to further fruitful collaboration between these communities. Whether or not a practical quantum computer could be achieved with the techniques presented here remains (and will probably remain for some time) a subject of debate. However, innovative physics is bound to emerge from the activity in this field and from more exchanges between different communities.

Acknowledgments

The 79th Les Houches summer school and the present volume have been made possible by the financial support of the following institutions, whose contribution is gratefully acknowledged:
– The "High level scientific conference" program of the Research Directorate of the European Commission under grant HPCF-2002-00041.
– The QUIPROCONE Network Of Excellence IST-1999-29064.
– The "Lifelong learning" program of the Centre National de la Recherche Scientifique (France).
– The Université Joseph Fourier, the French Ministry of Research and the Commissariat à l'Energie Atomique, through their constant support to the Physics School.

The permanent staff of the School, especially Brigitte Rousset and Isabelle Lelièvre, have been of invaluable assistance at every stage of the preparation and development of the school, and we would like to thank them warmly on behalf of all students and lecturers.

J. Dalibard, D. Estève
and J.-M. Raimond
March 2004

CONTENTS

Course 4. Quantum optical implementation of quantum information processing, by P. Zoller, J. I. Cirac, Luming Duan and J. J. García-Ripoll *187*

Course 5. Quantum information processing in ion traps I, by R. Blatt, H. Häffner, C. F. Roos, C. Becher and F. Schmidt-Kaler *223*

Course 11. Introduction to quantum conductors, by D. C. Glattli 401

Course 1

PRINCIPLES OF QUANTUM COMPUTATION

Isaac Chuang

Center for Bits and Atoms & Department of Physics
Massachusetts Institute of Technology, Cambridge, MA 02139, USA

D. Estève, J.-M. Raimond and J. Dalibard, eds.
Les Houches, Session LXXIX, 2003
Quantum Entanglement and Information Processing
Intrication quantique et traitement de l'information
© 2004 Elsevier B.V. All rights reserved

HXH=Z

Contents

1. Introduction

This chapter summarizes lectures I gave at the Les Houches school on quantum information, from June 30, 2003 through July 10, 2003. Much of the material is based on the book *Quantum Computation and Quantum Information* [1] ("QCQI"), co-authored with Michael Nielsen, but aside from a quick introduction to the subject, in this write-up the focus is on new topics and approaches, including quantum circuit analysis methods, a survey of entanglement as a physical resource, a review of information theory, examples of open quantum system dynamics, and a summary of ideas from quantum error correction. Two of the lectures, on quantum Fourier transform and quantum search algorithms, followed the book exactly and are omitted entirely here. QCQI is a helpful companion to these notes, and references to it are provided throughout, for further reading.

2. Fundamentals: quantum mechanics and computer science

This lecture reviews the foundations of computer science and of quantum mechanics. We begin with the notion of complexity, turn to a quick summary of the current state of the theory of quantum computation, then a summary of the four fundamental postulates of quantum mechanics, then conclude with a simple mathematical example, superdense coding, illustrating some surprises hidden within quantum mechanics.

2.1. Computer science and complexity

Computer science is the study of algorithms and their complexity. Here, the term "complexity" means the amount of effort (that is, physical resources such as time, space, and energy) required to solve a given mathematical problem, as a function of the size of the problem.

For example, the problem of adding two n-digit numbers together has $O(n)$ complexity, meaning that the time (or space) required to add the two numbers grows linearly with n. There is some cleverness involved in bounding complexity, however. One might think that the problem of multiplying two $n \times n$ matrices together would require $O(n^3)$ time, since there are n^2 numbers to compute, and each takes $O(n)$ multiplications with straightforward matrix multiplication methods. But there is a better way to multiply matrices, discovered by Strassen, which requires only $O(n^{\log_2 7})$ resources. Since $\log_2 7 \approx 2.8$, this is better.

Complexity is defined by the *best possible* time, or *minimum* resources required, and this can often be hard to define. But there are different *classes* of complexity which are easier to identify than specific complexities. For example, consider the question of identifying whether or not two graphs G_1 and G_2, each with n nodes, are equivalent to each other under relabeling of their nodes; this problem of *graph isomorphism* is believed to have complexity $O(2^n)$, but that has never been proven. Likewise, the problem of factoring a number $x = pq$, the product of two prime numbers p and q, is believed to have complexity $O(2^{n^{1/3}})$ (that is roughly the minimum resources with which the best known algorithm can solve the problem today). But there is no proof of this today.

Nevertheless, Computer Science has successfully crafted great insight in to the nature of mathematical problems, by relating the complexities of various mathematical problems to each other. Consider a boolean function $f(x)$ which is the logical OR of many terms, each of which depend on no more than three bits of the n-bit number x; for example, $f(x) = x_1x_2 + x_2\bar{x}_3 + x_1\bar{x}_4x_7 + \cdots$, where x_k is the kth bit of x, $+$ represents OR, and multiplication AND. Does there exist x such that $f(x) = 1$?

The complexity of this *3-SAT* problem is believed to be $O(2^n)$, but more interestingly, the ability to easily solve 3-SAT implies the ability to easily solve *any* other problem in a wide class of problems known as **NP**, short for "nondeterministic polynomial time." **NP** problems are characterized by having answers that are easy to check, but difficult to find, typically requiring time that is exponential in the problem size. These *hard* problems sit in contrast to **P**, or "polynomial" problems like addition, which are considered *easy*; defining the boundary between such problems is an art at the heart of computer science. We also define *efficient* as meaning that no exponential resources are required.

Of course, the boundary between easy and hard problems depends upon the model of computation being adopted. For example, how does one represent a real number (such as π) with error ϵ? Using a unary representation, this is a hard problem, but with a binary (or any digital) representation, this is easy.

The foundation of computer science rests on this concept, as enshrined by the Modern *Church-Turing Thesis*:

> Any algorithmic process can be efficiently simulated by a probabilistic Turing machine.

This statement may be interpreted as saying that a digital computer with a random number generator (and unlimited memory) can execute any known algorithm with overhead which is at most polynomial in the size of the problem, compared with the resources required for another computer.

The importance of quantum computation is that it violates this thesis! Quantum computers can solve certain mathematical problems faster than is possible using classical resources alone, and moreover, classical computers cannot efficiently simulate quantum computations! Thus, physics makes its entreé into computer science, and vice versa. Each of these fields has something to contribute to the other, as we shall see.

2.2. *Perspectives on quantum computation*

How far have quantum computers come today? Theoretically, they can solve the factoring and discrete logarithm problems in time $O(n^3)$, compared with the exponential time required for the best known classical algorithm. They can search an "unsorted database" (that is, for $f(x) : \{0, N\} \rightarrow \{0, 1\}$, find x_0 such that $f(x_0) = 1$) in time $O(\sqrt{N})$, compared with the $O(N)$ time that would be required classically. And they can efficiently simulate other quantum systems (although not necessarily allow efficient determination of any measurement observable on the simulated system).

Experiments have also made enormous progress since the mid 1990's. Since then, trapped ions of ^{9}Be have been used to create entangled states of four qubits, ^{40}Ca ions have implemented the two-qubit Deutsch-Jozsa algorithm, nitrogen vacancies in diamond have shown single qubit behavior, as have superconducting Josephson Junction devices, and quantum dots have shown coupled two-qubit behavior. Many other systems have also been tried. Most successfully to date, nuclear spins of molecules in liquids have been controlled using magnetic resonance techniques to implement 2, 3, 5, and 7 qubit algorithms, including an experimental demonstration of Shor's quantum factoring algorithm in factoring the number 15.

2.3. *Quantum mechanics in four postulates*

Let us turn now to the mathematical foundations for the remainder of this series of lectures. Remarkably, all of (nonrelativistic) quantum mechanics can be understood as arising from just four fundamental postulates, given below (c.f. QCQI Section 2.2).

> **Postulate 1**: Associated to any isolated physical system is a complex vector space $\mathcal{H}$ with an inner product, "the state space," also known as the Hilbert space. The system is completely described by its state vector, a unit vector in $\mathcal{H}$.

Some issues to consider with respect to this postulate are: (a) what is the right space to use for a given system? For atoms, one may have position, energy, magnetic moment, spin, and many other degrees of freedom. Photons may have polarization, position, momentum, and others. (b) The simplest quantum mechanical system is the two-state system, known as a *qubit*. The two basis states for a qubit are

$$|0\rangle = \begin{bmatrix} 1 \\ 0 \end{bmatrix} \quad |1\rangle = \begin{bmatrix} 0 \\ 1 \end{bmatrix}, \tag{2.1}$$

and for example, a single qubit $|\psi\rangle$ in an arbitrary state may be written as

$$|\psi\rangle = a|0\rangle + b|1\rangle = \begin{bmatrix} a \\ b \end{bmatrix}, \tag{2.2}$$

where a and b are complex numbers satisfying $\sqrt{|a|^2 + |b|^2} = 1$. Recall that the complex conjugate transpose state of this vector is denoted as $\langle\psi| = [a^* \; b^*]$, such that $\langle\psi|\psi\rangle = 1$.

Postulate 2: The evolution of a closed quantum system is described by a unitary transform,

$$|\psi(t_2)\rangle = U(t_2, t_1)|\psi(t_1)\rangle, \tag{2.3}$$

where $U(t_2, t_1)$ is unitary. Equivalently, we may postulate that time evolution is described by the differential equation

$$i\hbar\partial_t|\psi\rangle = H|\psi\rangle, \tag{2.4}$$

where H is a positive definite operator known as the Hamiltonian, which generates a unitary transform, giving

$$U = e^{-iHt/\hbar}. \tag{2.5}$$

For example, possible matrices for U include

$$I = \begin{bmatrix} 1 & 0 \\ 0 & 1 \end{bmatrix} \quad X = \begin{bmatrix} 0 & 1 \\ 1 & 0 \end{bmatrix} \tag{2.6}$$

$$Y = \begin{bmatrix} 0 & -i \\ i & 0 \end{bmatrix} \quad Z = \begin{bmatrix} 1 & 0 \\ 0 & -1 \end{bmatrix}, \tag{2.7}$$

the standard Pauli matrices (which also happen to be unitary). An interesting additional unitary matrix is the *Hadamard* transform,

$$H = \frac{1}{\sqrt{2}} \begin{bmatrix} 1 & 1 \\ 1 & -1 \end{bmatrix}, \tag{2.8}$$

which produces $H|0\rangle = \frac{|0\rangle+|1\rangle}{\sqrt{2}}$ and $H|1\rangle = \frac{|0\rangle-|1\rangle}{\sqrt{2}}$.
Exercise for the reader: compute XY and HXH.

A very useful two-qubit unitary transform is

$$U_{cn} = \begin{bmatrix} 1 & 0 & 0 & 0 \\ 0 & 1 & 0 & 0 \\ 0 & 0 & 0 & 1 \\ 0 & 0 & 1 & 0 \end{bmatrix}; \tag{2.9}$$

this is known as the controlled-NOT transform, for reasons that will later become clear.

Postulate 3: Quantum measurements are described by a collection of operators $\{M_k\}$ which act on the Hilbert space $\mathcal{H}$. k refers to the possible measurement outcomes. If the system is in state $|\psi\rangle$ before the measurement, then the probability of observing k is $\langle\psi|M_k^\dagger M_k|\psi\rangle$ and the system becomes

$$\frac{M_k|\psi\rangle}{\sqrt{\langle\psi|M_k^\dagger M_k|\psi\rangle}} \tag{2.10}$$

afterwards. M_k must satisfy

$$\sum_k M_k^\dagger M_k = I\,. \tag{2.11}$$

Note that the post-measurement state is simply $M_k|\psi\rangle$ re-normalized to have unit norm, so it is easy to understand despite the complex-looking denominator.
Example 1: Let

$$M_0 = |0\rangle\langle 0| \qquad M_1 = |1\rangle\langle 1|\,. \tag{2.12}$$

Note that $M_k^2 = M_k = M_k^\dagger$ because these are projectors. These measurement operators give the usual projective measurements onto the $|0\rangle$ and $|1\rangle$ basis states of a qubit. For $|\psi\rangle = a|0\rangle + b|1\rangle$, we find that prob(0) $= \langle\psi|M_0^\dagger M_0|\psi\rangle = \langle\psi|0\rangle\langle 0|\psi\rangle = |\langle\psi|0\rangle|^2 = |a|^2$, and that the post-measurement state when $k = 0$ is observed is just the state $|0\rangle$. Such projective measurements in the natural qubit basis states are known as *computational basis* state measurements.
Example 2: Let

$$M_0 = |00\rangle\langle 00| + |01\rangle\langle 01| \tag{2.13}$$

$$M_1 = |10\rangle\langle 10| + |11\rangle\langle 11|\,, \tag{2.14}$$

and consider $|\psi\rangle = (|00\rangle + |01\rangle + |10\rangle + |11\rangle)/2$. We find that prob(0) $= \langle\psi|M_0^\dagger M_0|\psi\rangle = 1/2$, with the post-measurement state being $(|00\rangle + |01\rangle)/\sqrt{2}$.
Example 3: Let

$$M_0 = \frac{1}{\sqrt{2}}I \qquad M_1 = \frac{1}{\sqrt{2}}X\,, \tag{2.15}$$

and consider $|\psi\rangle = a|0\rangle + b|1\rangle$. What is prob(1) and for this case, and what is the corresponding post-measurement state?

Postulate 4: The state space of a composite system is the tensor product of their component systems,

$$\mathcal{H}_{12} = \mathcal{H}_1 \otimes \mathcal{H}_2\,. \tag{2.16}$$

Moreover, if systems 1 and 2 are in state $|\psi_1\rangle$ and $|\psi_2\rangle$, then $|\psi_{12}\rangle = |\psi_1\rangle \otimes |\psi_2\rangle$.

It is convenient to use the notation $|\psi\rangle^{\otimes n}$ to denote n tensor product copies of the state $|\psi\rangle$, that is, $|\psi\rangle \otimes |\psi\rangle \cdots \otimes |\psi\rangle$.
Example 1: Let

$$|\psi_1\rangle = |0\rangle = \begin{bmatrix} 1 \\ 0 \end{bmatrix} \qquad |\psi_2\rangle = |0\rangle = \begin{bmatrix} 1 \\ 0 \end{bmatrix}. \tag{2.17}$$

Then the tensor product of these two states is

$$|\psi_{12}\rangle = |0\rangle \otimes |0\rangle = \begin{bmatrix} 1 \\ 0 \\ 0 \\ 0 \end{bmatrix} = |0, 0\rangle = |00\rangle \tag{2.18}$$

where the last two equalities give an example of the notation often employed in the literature for such states; when the contents of the ket are known to be binary, often the comma is suppressed.
Example 2: Let

$$|\psi_1\rangle = \frac{|0\rangle + |1\rangle}{\sqrt{2}} = \frac{1}{\sqrt{2}}\begin{bmatrix} 1 \\ 1 \end{bmatrix} \qquad |\psi_2\rangle = \frac{|0\rangle + |1\rangle}{\sqrt{2}} = \frac{1}{\sqrt{2}}\begin{bmatrix} 1 \\ 1 \end{bmatrix}. \tag{2.19}$$

Then the tensor product of these two states is

$$|\psi_{12}\rangle = \frac{1}{2}\begin{bmatrix} 1 \\ 1 \\ 1 \\ 1 \end{bmatrix} = \frac{|00\rangle + |01\rangle + |10\rangle + |11\rangle}{2}\,. \tag{2.20}$$

Example 3: Let

$$|\psi_1\rangle = a|0\rangle + b|1\rangle \qquad |\psi_2\rangle = c|0\rangle + d|1\rangle . \tag{2.21}$$

Then the tensor product of these two states is

$$|\psi_{12}\rangle = ac|00\rangle + ad|01\rangle + bc|10\rangle + bd|11\rangle . \tag{2.22}$$

Example 4: Let

$$|\psi_{12}\rangle = \frac{|00\rangle + |11\rangle}{\sqrt{2}} . \tag{2.23}$$

Do there exist $|\psi_1\rangle$ and $|\psi_2\rangle$ such that $|\psi_{12}\rangle = |\psi_1\rangle \otimes |\psi_2\rangle$? Why or why not?
Example 5: Operators on composite systems are also constructed from the tensor product of operators on the component systems:

$$I \otimes X = \begin{bmatrix} 1 & 0 \\ 0 & 1 \end{bmatrix} \otimes \begin{bmatrix} 0 & 1 \\ 1 & 0 \end{bmatrix} = \begin{bmatrix} 0 & 1 & 0 & 0 \\ 1 & 0 & 0 & 0 \\ 0 & 0 & 0 & 1 \\ 0 & 0 & 1 & 0 \end{bmatrix} . \tag{2.24}$$

Note that this has the form $\begin{bmatrix} X & 0 \\ 0 & X \end{bmatrix}$, because of the rules of tensor products of matrices. Similarly,

$$X \otimes X = \begin{bmatrix} 0 & 0 & 0 & 1 \\ 0 & 0 & 1 & 0 \\ 0 & 1 & 0 & 0 \\ 1 & 0 & 0 & 0 \end{bmatrix} . \tag{2.25}$$

And a useful identity is that

$$[A \otimes B]\big[|\psi\rangle \otimes |\phi\rangle\big] = \big[A|\psi\rangle\big] \otimes [B|\phi\rangle] . \tag{2.26}$$

Example 5: Refer back to Eq.(2.9), where the two-qubit operation U_{cn} was defined. Note that $U_{cn} = (I+Z) \otimes I + (I-Z) \otimes X$. Do there exist U and V such that $U_{cn} = U \otimes V$? Why or why not?

2.4. Example: superdense coding

The postulates of quantum mechanics are exercised by the following problem. Alice and Bob (the two most famous people in quantum computation after Peter Shor) meet Charlie in a bar. Years later, Alice has two bits k_0 and k_1 she wishes to send to Bob. Despite the fact that these bits have nothing to do with anything the three discussed in their prior meeting, is there any physical resource Charlie could have given Alice and Bob to speed her present task? In the purely classical world, the answer is *no*: Alice must send two bits for Bob to receive her message completely. However, it turns out that by using quantum resources, Alice can send Bob her message by sending only *one* qubit.

Let Charlie create the state $|\psi_0\rangle = \frac{|00\rangle+|11\rangle}{\sqrt{2}}$, and distribute the two qubits to Alice and Bob. This state is manifestly independent of k_0 and k_1. Now, when Alice wishes to send her message to Bob, she follows this recipe: if $k_0k_1 = 00$, apply I to her qubit; if 01, apply X, if 10 apply Z, and if 11, apply Y. She then sends her qubit to Bob. Bob then applies U_{cn} to the two qubits he now has in hand, followed by $H \otimes I$, and a projective measurement in the computational basis. We claim that his final measurement result gives exactly the message Alice sent.

Proof of this claim is straightforward; we simply work out exhaustively the sequence of states for all four possible values of Alice's message. Let $|\psi_1\rangle$ be the state of the two qubits held by Alice and Bob, after Alice's encoding operation; $|\psi_2\rangle$ be the state after Bob performs U_{cn}; $|\psi_3\rangle$ be the state Bob obtains before his final measurement, and $\tilde{k}$ be Bob's measurement result. We thus have (suppressing normalization factors):

k_0k_1	$\|\psi_1\rangle$	$\|\psi_2\rangle$	$\|\psi_3\rangle$	$\tilde{k}$
$\|00\rangle$	$\|00\rangle + \|11\rangle$	$\|00\rangle + \|10\rangle = (\|0\rangle + \|1\rangle)\|0\rangle$	$\|00\rangle$	00
$\|01\rangle$	$\|10\rangle + \|01\rangle$	$\|11\rangle + \|01\rangle = (\|0\rangle + \|1\rangle)\|1\rangle$	$\|01\rangle$	01
$\|10\rangle$	$\|00\rangle - \|11\rangle$	$\|00\rangle - \|10\rangle = (\|0\rangle - \|1\rangle)\|0\rangle$	$\|10\rangle$	10
$\|11\rangle$	$i(\|10\rangle - \|01\rangle)$	$i(\|11\rangle - \|01\rangle) = i(\|0\rangle - \|1\rangle)\|1\rangle$	$-i\|11\rangle$	11

Note how tensor products and operations on tensor product states play an important role in this calculation; they are absolutely vital in the mathematics of quantum information, and are one crucial feature which distinguishes this treatment of quantum mechanics, compared with traditional approaches to the subject.

3. Quantum circuits

This lecture introduces a language for transformations of quantum bits which is analogous to that used for electrical circuits. We find that arbitrary unitary transforms can be constructed from certain single and multi-qubit gates, and illustrate this language with an important elementary quantum algorithm. Please refer to Chapter 4 of QCQI for additional material.

3.1. Boolean circuits and reversible computing

The standard set of boolean gates include the AND and NOT gates, which are *universal*, meaning that any boolean function can be constructed from them. As a proof for this claim, consider the function $f(x_1, x_2, \cdots, x_n)$ which maps $\{0, 1\}^n \to \{0, 1\}$ (that is, n bits to one bit), and define

$$f_0(x_2, x_3, \cdots x_n) = f(0, x_2, x_3, \cdots, x_n) \tag{3.1}$$

$$f_1(x_2, x_3, \cdots x_n) = f(1, x_2, x_3, \cdots, x_n). \tag{3.2}$$

Then it follows that

$$f(x_1, x_2, \cdots x_n) = x_1 f_1(x_2, x_3, \cdots x_n) + \bar{x}_1 f_0(x_2, x_3, \cdots x_n), \tag{3.3}$$

where $+$ represents OR (which can easily be constructed from AND and NOT), multiplication represents AND, and $\bar{x}_1$ is the NOT of x_1. The functions f_0 and f_1 can, in turn, be broken down in a similar fashion, recursively, giving an implementation for f solely in terms of AND and NOT gates.

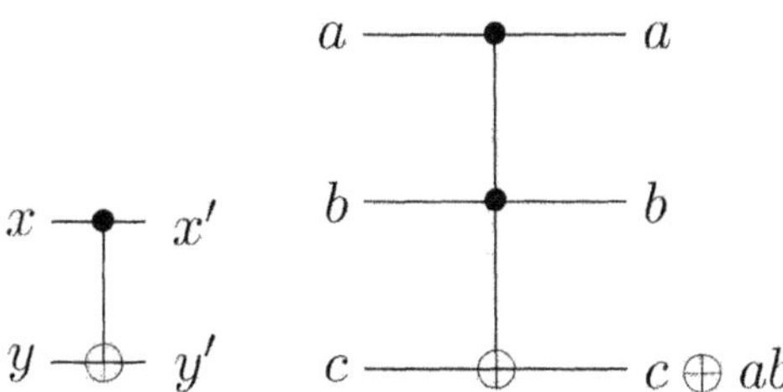

Fig. 1. Controlled-NOT (left) and Toffoli (right) gates.

The AND gate has no direct quantum counterpart because it is irreversible; from the single output bit, it is impossible to determine the two inputs. Reversible circuits are thus constructed from different circuit primitives, two of which are the controlled-NOT ("CNOT") and Toffoli gates, shown in Fig. 1. The CNOT gate outputs $x' = x$ and $y' = x \oplus y$, where $\oplus$ is addition modulo two, while the Toffoli gate has three inputs a, b, and c, and replaces c by $c \oplus ab$. Note that the Toffoli can perform AND and NOT operations easily, by setting inputs appropriately, and thus it is a universal gate.

Any reversible circuit can be constructed from Toffoli gates. Even more interesting is the following; let us define a *garbage-free* reversible computation of $f(x)$ as a circuit which takes as input x (the function argument) and y (a scratchpad for storage of the result) and outputs x and $y \oplus f(x)$. The circuit may, in addition, take as input any number of 0's, as long as it also outputs them unchanged, but there may be no additional inputs our outputs. With this definition in hand, an importatnt theorem can be given:

Theorem: Any boolean circuit can be efficiently simulated by a garbage-free reversible circuit.

Proof: (1) Replace AND and NOT gates with Toffoli gates, obtaining a circuit which transforms $(x, y, 0, 0)$ to $(x, y, f(x), g)$, where g is some undesired output garabage bits. (2) Copy the result, giving $(x, y \oplus f(x), f(x), g)$. (3) Reverse the first computation, obtaining $(x, y \oplus f(x), 0, 0)$. This procedure requires additional space and time which is polynomially larger than the original circuit; a better result is known (see Bennett), in which the overhead is only logarithmic.

3.2. Single qubit gates

Quantum circuits are composed from elementary gates, much like classical circuits. The important quantum gates which act on single qubits are shown in Fig. 2. In addition to those we have already encountered in these lectures, we have the phase gate S and the $\pi/8$ gate T, with unitary transforms

$$S = \begin{bmatrix} 1 & 0 \\ 0 & i \end{bmatrix} \quad T = \begin{bmatrix} 1 & 0 \\ 0 & \sqrt{i} \end{bmatrix}, \tag{3.4}$$

which are $S = \sqrt{Z}$ and $T = \sqrt{S}$, respectively. These gates are important in fault tolerance quantum circuits.

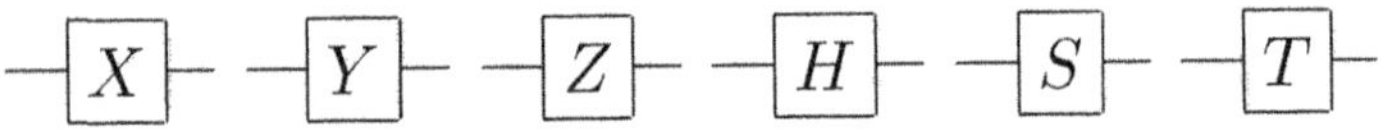

$$-\boxed{X}- \quad -\boxed{Y}- \quad -\boxed{Z}- \quad -\boxed{H}- \quad -\boxed{S}- \quad -\boxed{T}-$$

Fig. 2. Important single qubit gates.

Single qubit gates act upon single qubits, which may be written in general as

$$|\psi\rangle = \cos\frac{\theta}{2}|0\rangle + e^{i\phi}\sin\frac{\theta}{2}|1\rangle, \tag{3.5}$$

where θ and ϕ denote a point on the unit sphere.

More general single qubit operations include rotations

$$R_x(\theta) = e^{-i\frac{\theta}{2}X} \quad (3.6)$$
$$R_y(\theta) = e^{-i\frac{\theta}{2}Y} \quad (3.7)$$
$$R_z(\theta) = e^{-i\frac{\theta}{2}Z} \quad (3.8)$$

about the three principle axes, or the general rotation

$$R_{\hat{n}}(\theta) = e^{-i\frac{\theta}{2}(\hat{n}\cdot\vec{\sigma})} = \cos\frac{\theta}{2} - i(\hat{n}\cdot\vec{\sigma})\sin\frac{\theta}{2}\,. \quad (3.9)$$

Some useful properties these rotations have are, for example,

$$XR_y(\theta)X = R_y(-\theta) \quad (3.10)$$
$$XR_z(\theta)X = R_z(-\theta) \quad (3.11)$$
$$XR_x(\theta)X = R_x(\theta) \quad (3.12)$$
$$XTX = T^\dagger\,. \quad (3.13)$$

Also, it is very useful to keep in mind that $HXH = Z$ and $HZH = X$.

An important result, sometimes known as *Bloch's theorem* or the Bloch decomposition, is that any single qubit gate U can be written as the product of three rotations (and an overall scalar phase that is mostly irrelevant): for all $U \in U(2)$, $\exists\alpha, \beta, \gamma, \delta$ such that $U = e^{i\alpha}R_z(\beta)R_y(\gamma)R_z(\delta)$. It is straightforward to prove this by direct computation.

Exercise for the reader: give the decomposition for X; also show that $T = e^{i\pi/8}e^{-iZ\pi/8}$.

A useful corollary to this is that for all $U \in U(2)$, $\exists\alpha, A, B, C$ such that $ABC = I$ and $U = e^{i\alpha}AXBXC$. The proof for this is to let $A = R_z(\beta)R_y(\gamma/2)$, $B = R_y(-\gamma/2)R_z(-(\delta+\beta)/2)$, and $C = R_z((\delta-\beta)/2)$, with β, γ, δ as in the above Bloch decomposition.

3.3. *Multi-qubit gates*

Let U be a single qubit gate. We define the two-qubit controlled-U gate, shown in Fig. 3, as being the unitary operation

$$\begin{bmatrix} I & 0 \\ 0 & U \end{bmatrix}, \quad (3.14)$$

where this is a matrix of matrices, I being the 2×2 identity operator. The quantum controlled-NOT ("CNOT") gate is an instance of this, with $U = X$.

Claim: the controlled-U gate can be realized by a quantum circit composed of CNOT gates and single qubit gates.

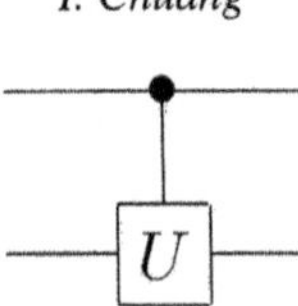

Fig. 3. Controlled-U gate acting on two qubits.

Proof: Let $U = e^{i\alpha} AXBXC$ with $ABC = I$, and construct the circuit shown in Fig. 4. The phase shift matrix

$$\begin{bmatrix} 1 & 0 \\ 0 & e^{i\alpha} \end{bmatrix} \tag{3.15}$$

can be implemented by $R_z(\alpha)$, up to an irrelevant global phase.

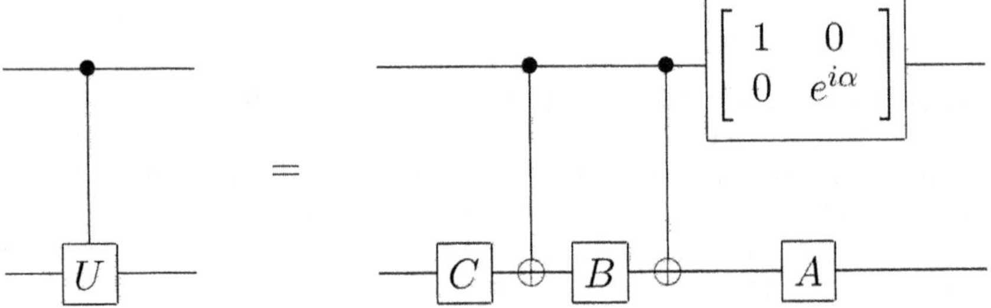

Fig. 4. Implementation of controlled-U gate using CNOT and single qubit gates.

Example: Consider the circit shown in Fig. 5. How is this implemented?

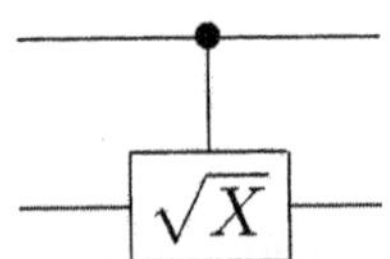

Fig. 5. Circuit for controlled $\sqrt{X}$

Recall that $X = ie^{i\pi/2X}$ by the Bloch decomposition. Thus, $\sqrt{X} = \sqrt{i}e^{i\frac{\pi}{4}X}$. We can therefore compute as follows:

$$\begin{aligned} e^{-i\frac{\pi}{4}X} &= He^{-i\frac{\pi}{4}Z}H && (3.16) \\ &= e^{-i\frac{\pi}{4}}HSH && (3.17) \end{aligned}$$

$$= e^{-i\frac{\pi}{4}} HTTH \tag{3.18}$$
$$= e^{-i\frac{\pi}{4}} (HT)(XT^{\dagger}X)(H)\,, \tag{3.19}$$

so letting $A = HT$, $B = XT^{\dagger}X$, and $C = H$, we find that

$$\sqrt{X} = HT\ XT^{\dagger}X\ H\,, \tag{3.20}$$

giving the circuit construction shown in Fig. 6

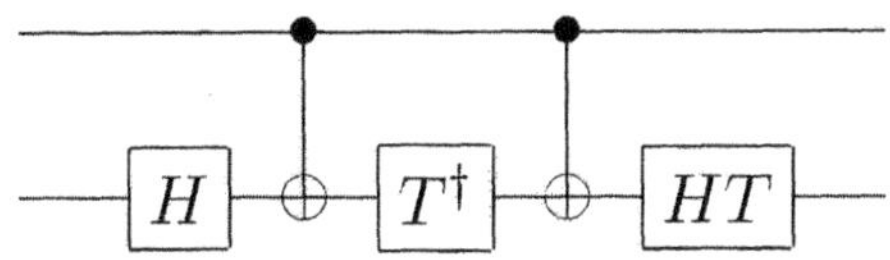

Fig. 6. Circuit implementing controlled $\sqrt{X}$.

Claim: The quantum Toffoli gate can be constructed from controlled-U and CNOT gates.
Proof: Fig. 7

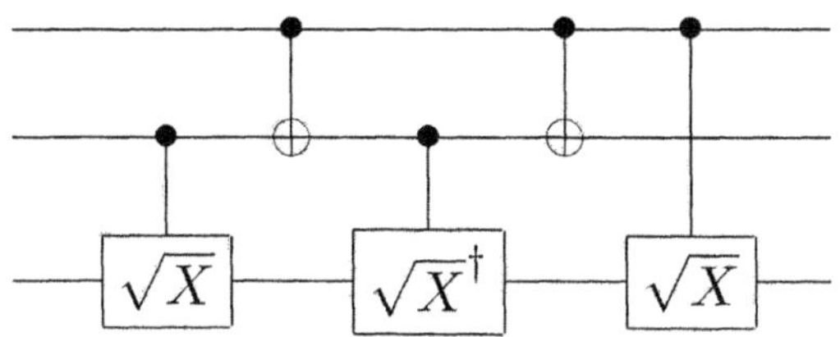

Fig. 7. Circuit implementing Toffoli gate.

Theorem: Any classical circuit may be efficiently simulated using a quantum circuit.
Proof: A classical circuit can be efficiently simulated by a reversible circuit, comprised of Toffoli gates; when these are replaced with quantum gates, a quantum circuit is obtained.

The general idea is classical circuits implement a subset of transformations on input states, which are just permutations. Permutations are a subset of unitary transforms. Thus, quantum computation subsumes classical computation.
Example: Fig. 8 shows a quantum circuit using three CNOT gates to effect a swap between two qubits. Show that this true by multiplying out the unitary transforms.

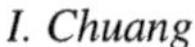

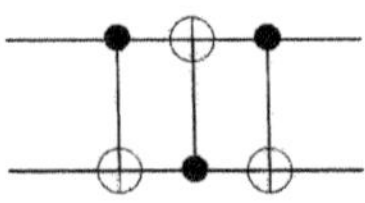

Fig. 8. Quantum circuit using three CNOT gates to swap two qubits.

Example: What is the output of the circuit shown in Fig. 9? This circuit demonstrates the "meter" notation for qubit measurement in the computational basis. The double wires coming out of the meter represent the classical measurement result, which is used to control whether an X operation is performed (or not) in this circuit.

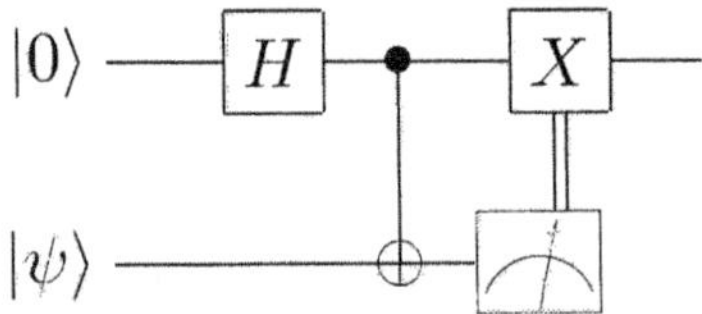

Fig. 9. Sample two-qubit circuit illustrating role of measurement.

3.4. Universality

Theorem: An arbitrary n qubit quantum gate U can be composed from CNOT gates and single qubit gates.
Proof (sketch): U is a product of plane rotations, $U = U_1 U_2 \cdots U_m$, where each U_k acts nontrivially only on a two-dimensional subspace. That subspace may not necessarily be identified with any single qubit, but a permutation can be performed such that the identification follows; each U_k can be expressed as

$$U_k = P_k \begin{bmatrix} 1 & & & & \\ & \ddots & & & \\ & & 1 & & \\ & & & a & b \\ & & & c & d \end{bmatrix} P_k^\dagger = P_k V_k P_k^\dagger \,. \tag{3.21}$$

For each k, the permutation P_k can be performed using CNOT and Toffoli gates, and the operation V_k can be performed by a single qubit gate.

With this theorem in hand, we now have a universal set of gates from which all quantum circuits can be constructed. Unfortunately, this set requires a continum of gates; it is not a discrete set such as in the classical case. On the other hand, we can construct any unitary U *exactly*.

If we are willing to simply *approximate* any transform U with some error ϵ, then it turns out only a discrete set of gates is necessary to be universal.

Def: Let $E(U, V) = \max_{|\psi\rangle} \langle\psi|(U-V)^\dagger(U-V)|\psi\rangle$. This quantity is a measure of the distance (ie error) between two unitary operators U and V.

Theorem: $\forall U \in SU(2)$, for any ϵ, $\exists V$ which is the product of H and T gates such that $E(U, V) \leq \epsilon$.

Proof: Let $R_n(\theta) = THTH$. Up to an irrelevant global phase, this is $e^{-iZ\pi/8} \times e^{-iX\pi/8}$ (using the fact that $HZH = X$). Expanding, we find that $R_n(\theta)$ is a rotation about the axis $\vec{n} = \hat{x}\cos\pi/8 + \hat{y}\sin\pi/8 + \hat{z}\cos\pi/8$, by angle $\theta = \cos^{-1}\left[\cos^2\pi/8\right]$, which is an irrational angle! Thus, it follows that for a desired rotation angle α, and a given ϵ, $\exists k$ integer such that $E(R_n(\alpha), R_n^k(\theta)) \leq \epsilon/3$. And analogous to the Bloch decomposition, we can find α and β such that U is a product of $R_n(\alpha)$ and $R_m(\beta)$, where $R_m(\theta) = HR_n(\theta)H$ is a rotation about an axis $\hat{m}$ which is nonorthogonal to $\vec{n}$. Therefore, $\exists k_1, k_2, k_3$ such that $E(U, R_n^{k_1}(\theta)HR_n^{k_2}(\theta)HR_n^{k_3}(\theta)) \leq \epsilon$.

This construction shows that H and T can be used to approximate any single qubit gate, but it is not very efficient. However, there is a much more powerful result, known as the *Solovay Kitaev Theorem*, according to which any unitary U can be performed with error ϵ using $O(\log^c \frac{1}{\epsilon})$ fixed gates, for example from the set $\{H, T, \text{CNOT}\}$. In this theorem, c is a small integer, below 4. This result, which we do not have room to prove here, shows that any single qubit gate can be *efficiently* approximated by H and T, whereby each additional gate nearly halves the error in the approximation.

3.5. The Deutsch-Jozsa algorithm

The quantum circuits we have studied so far allow us to understand, in a very simple way, one of the earliest quantum algorithms to have been created, known as the Deutsch-Jozsa algorithm. One instance of its modern form, as improved by Cleve, Mosca, Tapp, and others, is as follows. You are given a two-qubit circuit, as shown in Fig. 10, which computes a function $f(x) : \{0, 1\} \to \{0, 1\}$ by taking input bits x and y, and giving as output x and $y \oplus f(x)$ (recall that $\oplus$ is addition modulo two).

There are four possible functions which map one bit to one bit, as shown in Table 1. Analytically, they are the two constant functions $f_1(x) = 0$, $f_2(x) = 1$, and the two balanced functions $f_3(x) = x$ and $f_4(x) = \bar{x}$. It is clear from studying this function that *two* queries of f are necessary to distinguish between constant and balanced functions, classically. This is true even given the quantum oracle of Fig. 10, when limited to classical queries.

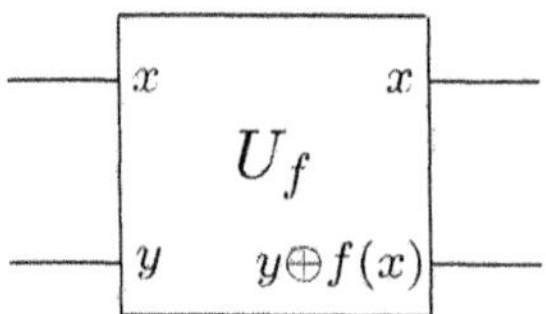

Fig. 10. Quantum circuit for the oracle in the Deutsch-Jozsa problem.

Table 1
The four possible functions which map $\{0, 1\} \to \{0, 1\}$, given as $x, y \to x, y \oplus f(x)$.

xy	f_1	f_2	f_3	f_4
00	00	01	00	01
01	01	00	01	00
10	10	11	11	10
11	11	10	10	11

A quantum circuit for the modern one-bit Deutsch-Jozsa algorithm is shown in Fig. 11. Let the measurement output of the meter be z; we claim that $z = 0$ for balanced functions, and $z = 1$ for unbalanced functions. Usually, this is seen by direct computation, where the quantum state is calculated step-by-step following each gate in the circuit, but here we take another approach, utilizing circuit identities.

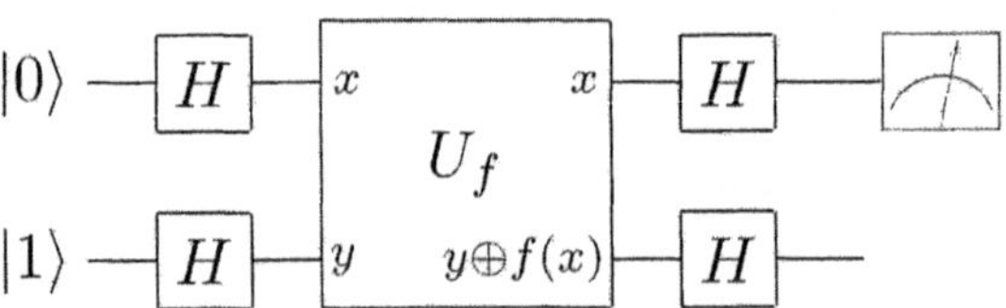

Fig. 11. Quantum circuit for the Deutsch-Jozsa algorithm. Note that the input state is $|0 > |1\rangle$

Specifically, let us consider four circuits used to implement the four functions of Table 1, shown in Fig. 12

Case 1: The unitary quantum circuit implementation of the oracle for function $f_1(x)$ is trivial, since f_1 maps $xy \to xy$ unchanged. Thus, $U_{f_1} = I \otimes I$. The four Hadamard gates thus all cancel out (recall $H^2 = I$), so the final output state is $|0\rangle|1\rangle$.

Case 2: $U_{f_2} = I \otimes X$, since $f_2 : xy \to x\bar{y}$. Now, the two Hadamards on the first qubit (the top wire) cancel, while the second qubit has $HXH = Z$ acting on it, giving the output state $-|1\rangle|1\rangle$.

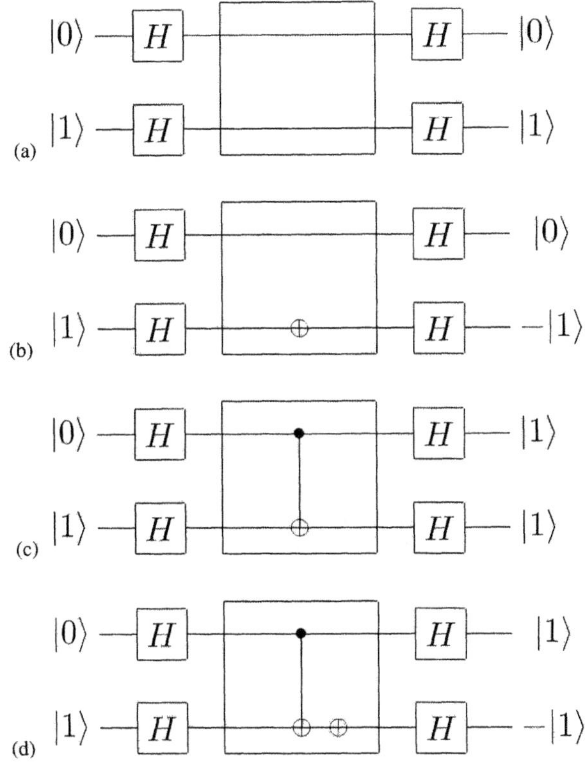

Fig. 12. Explicit quantum circuits for the Deutsch-Jozsa algorithm showing the four cases, with a quantum circuit for the oracle included. Parts (a-d) show cases 1-4, respectively.

Case 3: U_{f_3} is a CNOT with the control on the first qubit, and the target on the second, since $f_3 : x, y \to x, x \oplus y$. The unitary transform performed by a CNOT can be written as $|0\rangle\langle 0| \otimes Z + |1\rangle\langle 1| \otimes X = (I + Z) \otimes Z + (I - Z) \otimes X$. Using the fact that $HZH = X$ and $HXH = Z$, we find that Hadamard gates before and after a CNOT gives us

$$\Big[H \otimes H\Big]\Big[(I + Z) \otimes Z + (I - Z) \otimes X\Big]\Big[H \otimes H\Big] \tag{3.22}$$

$$= (I + X) \otimes X + (I - X) \otimes Z \tag{3.23}$$

$$= I \otimes (I + Z) + X \otimes (I - Z)\,. \tag{3.24}$$

This is a neat result! Four Hadamard gates around a CNOT turn the CNOT gate upside-down, flipping the roles of the control and target qubits. Thus, the output state is $|0\rangle|1\rangle$.

Case 4: U_{f_4} is just like case 3, but with an extra NOT gate on the second qubit, since $f_4 : x, y \rightarrow x, x \oplus \bar{y}$. By inserting $H^2 = I$ on the second qubit between the NOT and the CNOT, it follows immediately from our prior calculations that the output is $-|0\rangle|1\rangle$.

Summarizing the Deutsch Jozsa algorithm: we have found that for f_1 and f_2, the first qubit is output in the state $|0\rangle$, while for f_3 and f_4, the first qubit is output in the state $|1\rangle$. Thus, we have proven the claim that the measurement output z, from measuring the first qubit, distinguishes the balanced functions from the unbalanced ones, in a single oracle query.

4. Entanglement as a physical resource

This lecture introduces the concept of entanglement as being a physical resource. Much like space, time, and energy are physical resources which can be usefully consumed, compared and converted, we show how entangled states are useful for applications such as precision measurement and communication. We show how the amount of entanglement a given state can be quantified, and how entanglement is fungible.

4.1. Mathematical definition of entanglement

Definition: For pure states, a bipartite state $|\psi_{AB}\rangle$ of a composite system is *entangled* if and only if there does not exist $|\psi_A\rangle$, $|\psi_B\rangle$ such that

$$|\psi_{AB}\rangle = |\psi_A\rangle \otimes |\psi_B\rangle . \tag{4.1}$$

Equivalently, we may say that the state is *non-separable*. If it is not entangled, then the state is *separable*.

Definition: For mixed states, the density matrix ρ_{AB} of a bipartite composite system is separable if and only if

$$rho_{AB} = \sum_k p_k \rho_A^k \otimes \rho_B^k , \tag{4.2}$$

for sets of density matrices $\{\rho_A^k\}$ and $\{\rho_b^k\}$, and probabilities $\sum_k p_k = 1$, $\forall k$, $p_k \geq 0$. A separable state is also known as being unentangled.

Example 1: The four Bell states

$$\frac{|00\rangle + |11\rangle}{\sqrt{2}} \quad \frac{|00\rangle - |11\rangle}{\sqrt{2}} \quad \frac{|01\rangle + |10\rangle}{\sqrt{2}} \quad \frac{|01\rangle - |10\rangle}{\sqrt{2}} \tag{4.3}$$

are well known two-qubit entangled states. These are also known as EPR pairs.
Example 2: $|00\rangle + |11\rangle + |22\rangle$ is an entangled state. In fact, this is a common form for writing entangled states, because, for example consider the case when Alice and Bob share two Bell states. They thus have (suppressing normalization factors henceforth):

$$\begin{aligned} (|00\rangle + |11\rangle)^{\otimes 2} &= (|00\rangle + |11\rangle) \otimes (|00\rangle + |11\rangle) && (4.4) \\ &= |0000\rangle + |0011\rangle + |1100\rangle + |1111\rangle\,. && (4.5) \end{aligned}$$

Rearranging the labels by grouping all of Alice's qubits together, followed by Bob's qubits gives us

$$\begin{aligned} & |0000\rangle + |0101\rangle + |1010\rangle + |1111\rangle && (4.6) \\ &= |00\rangle|00\rangle + |01\rangle|01\rangle + |10\rangle|10\rangle + |11\rangle|11\rangle && (4.7) \\ &= \sum_{x=0}^{3} |x\rangle|x\rangle\,, && (4.8) \end{aligned}$$

where x is a two-bit number, conveniently written in decimal form. In fact, n shared EPR pairs may be written as

$$(|00\rangle + |11\rangle)^{\otimes n} = \sum_{x=0}^{2^n-1} |x\rangle|x\rangle \tag{4.9}$$

when appropriately relabeled.
Example 3: $|00\rangle + |01\rangle + |10\rangle$ is entangled.
Example 4: $\sqrt{0.9}|00\rangle + \sqrt{0.1}|11\rangle$ is entangled, but we'd like to think it is somehow *less* entangled than $|00\rangle + |11\rangle$; this is indeed the case, as we will see.
Example 4: $|000\rangle + |111\rangle$ is a three-qubit entangled state known as the GHZ state.

In fact, most multi-qubit states are entangled.

4.2. *Applications of entanglement*

We illustrate the usefulness of entangled states with two applications: precision measurement and teleportation.

4.2.1. *Precision measurement*

The following scenario illustrates the problem of precision measurement. Given

$$-\boxed{\phi}- = \begin{bmatrix} 1 & 0 \\ 0 & e^{i\phi} \end{bmatrix}, \tag{4.10}$$

how precisely can ϕ be determined, as a function of the number of times this gate is used?

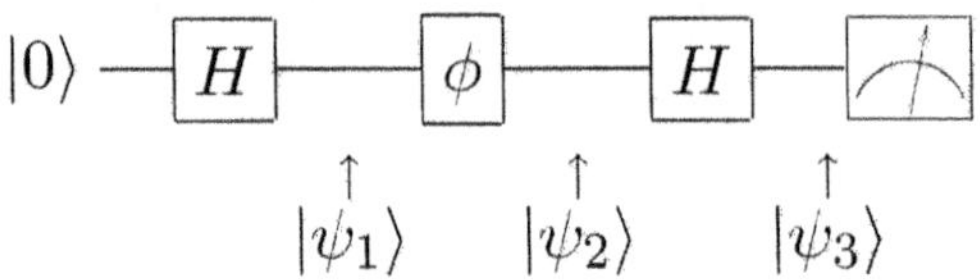

Fig. 13. Quantum circuit for a single qubit Ramsey interferometer.

The standard technique for solving this problem is the method of *Ramsey interferometry*, represented by the quantum circuit shown in Fig. 13. The intermediate states of the circuit are as follows:

$$|\psi_1\rangle = \frac{|0\rangle + |1\rangle}{\sqrt{2}} \tag{4.11}$$

$$|\psi_2\rangle = \frac{|0\rangle + e^{i\phi}|1\rangle}{\sqrt{2}} \tag{4.12}$$

$$|\psi_3\rangle = \frac{1+e^{i\phi}}{2}|0\rangle + \frac{1-e^{i\phi}}{2}|1\rangle\,. \tag{4.13}$$

Thus, the final measurement result is zero with probability

$$p = \text{prob}(0) = \left|\frac{1+e^{i\phi}}{2}\right|^2 = \frac{1+\cos\phi}{2}\,. \tag{4.14}$$

Repeating this cicuit n times gives an estimate of p with standard deviation

$$\Delta p = \sqrt{\frac{p(1-p)}{n}} = \frac{\sin\phi}{2\sqrt{n}}\,, \tag{4.15}$$

such that the uncertainty in our estimate of ϕ, with n uses of the ϕ box, is

$$\Delta\phi = \frac{\Delta p}{|dp/d\phi|} = \frac{1}{\sqrt{n}}\,. \tag{4.16}$$

An alternative strategy, which also uses the ϕ box n times, is to employ entanglement; for example, shown in Fig. 14 is a quantum circuit using a two-qubit entangled state. The intermediate circuit states are:

$$|\psi_1\rangle = \frac{|00\rangle + |10\rangle}{\sqrt{2}} \tag{4.17}$$

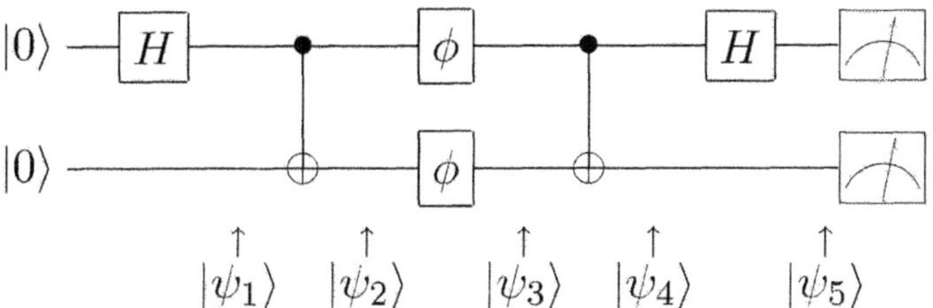

Fig. 14. Quantum circuit for estimating ϕ using an entangled two-qubit state.

$$|\psi_2\rangle = \frac{|00\rangle + |11\rangle}{\sqrt{2}} \tag{4.18}$$

$$|\psi_3\rangle = \frac{|00\rangle + e^{2i\phi}|11\rangle}{\sqrt{2}} \tag{4.19}$$

$$|\psi_4\rangle = \frac{|00\rangle + e^{2i\phi}|10\rangle}{\sqrt{2}} \tag{4.20}$$

$$|\psi_5\rangle = \left[\frac{1+e^{2i\phi}}{2}|0\rangle + \frac{1-e^{2i\phi}}{2}|1\rangle\right]|0\rangle . \tag{4.21}$$

Note how the second qubit is returned to its initial state, and the first qubit is left in a state much like that from the single qubit Ramsey interferometer, Eq.(4.13), but with twice the phase shift angle. With a similar circuit but using n qubits, as shown in Fig. 15, we obtain the output state

$$\left[\frac{1+e^{in\phi}}{2}|0\rangle + \frac{1-e^{in\phi}}{2}|1\rangle\right] \otimes |0\rangle^{\otimes(n-1)} . \tag{4.22}$$

The final measurement output is thus zero with probability

$$p = \text{prob}(0) = \left|\frac{1+e^{in\phi}}{2}\right|^2 = \frac{1+\cos(n\phi)}{2} , \tag{4.23}$$

with uncertainty

$$\Delta p = \sqrt{p(1-p)} = \frac{\sin\phi}{2} , \tag{4.24}$$

giving an uncertainty in the estimate of ϕ to be

$$\Delta\phi = \frac{1}{n} . \tag{4.25}$$

Thus, we see that using entangled states allows the same measurement precision to be obtained using a factor of $\sqrt{n}$ fewer queries, compared to the traditional interferomic technique.

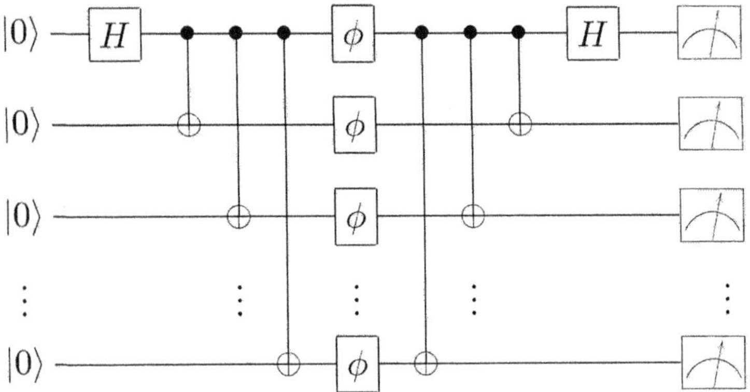

Fig. 15. Quantum circuit for estimating ϕ using an entangled n-qubit state.

4.2.2. Teleportation

As we saw in Section 2.4, prior shared entangled states can allow two parties to communicate two bits of classical information by sending one qubit. A process which is almost the reverse of this is teleportation, which allows one qubit to be sent using two classical bits, given one prior shared EPR pair (known as an "ebit"). An interesting, unusual approach to understanding this process is through an analysis of its quantum circuit implementation, shown in Fig. 16. We claim that the circuit takes in an arbitrary single qubit state $|\psi\rangle$, and outputs the same state on the bottom qubit. The following proves this fact.

The teleportation circuit can be simplified systematically, using some standard circuit identities. First, the X and Z operations controlled by the classical results from the measurements can be replaced by quantum controlled operations. Also, recall that a CNOT is just a controlled-X gate. Second, the quantum circuit (with a Hadamard and CNOT) generating an EPR state can be included explicitly. This gives us the circuit shown in Fig. 17.

A useful classical circuit identity with CNOT gates is given in Fig. 18. We can use that to further simplify our teleportation circuit, by shifting gates along wires where there are no conflicting dependencies, and by adding a CNOT gate near the start which does nothing (because of the state of the input). Specifically, if the target qubit of a CNOT has the state $|0\rangle + |1\rangle$, then the CNOT does nothing, because $|0\rangle + |1\rangle$ is an eigenstate of the X operator. Two other useful steps are to

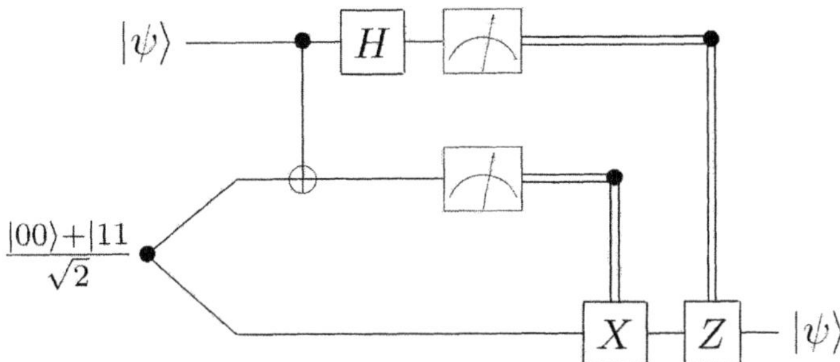

Fig. 16. Quantum circuit for teleporting a single qubit state. The < notation for an EPR state source, used on the bottom left, is a standard way to depict entangled states entering a circuit.

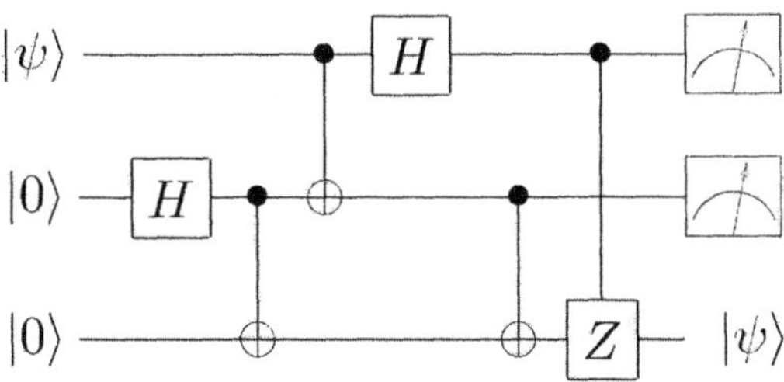

Fig. 17. First set of circuit simplifications applied to the teleportation circuit.

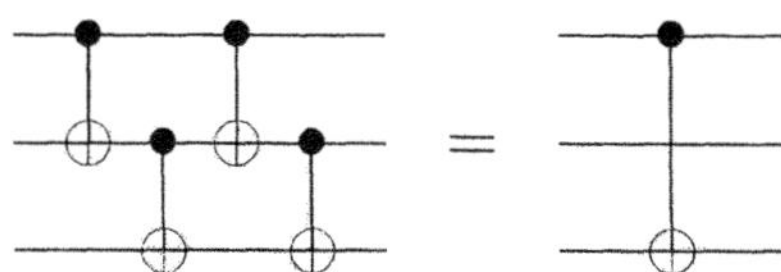

Fig. 18. Useful circuit identity for cascaded CNOT gates.

introduce two hadamard gates which cancel, $H^2 = I$, and to flip the controlled-Z gate upside down. This is possible because the controlled-Z gate performs the operation $(I+Z)\otimes I+(I-Z)\otimes Z = I\otimes(I+Z)+Z\otimes(I-Z)$. The resulting circuit is shown in Fig. 19

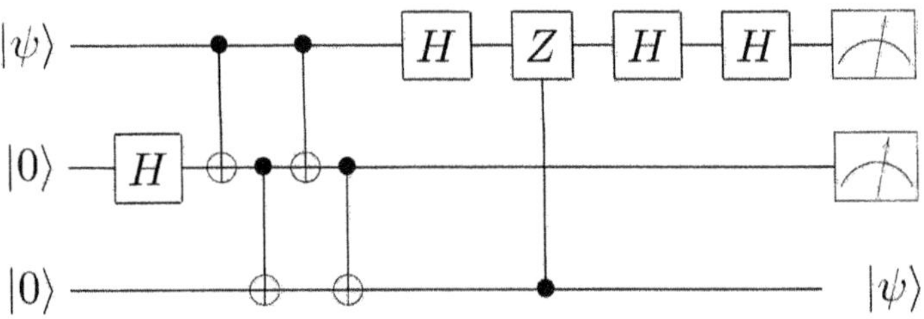

Fig. 19. Second set of circuit simplifications applied to the teleportation circuit.

We can now apply the simplification of Fig. 18, and replace HZH with X. In addition, we can insert another CNOT at the very start of the circuit, controlled on the bottom qubit, since it is in the $|0\rangle$ state (and thus changes nothing). The resulting circuit is shown in Fig. 20.

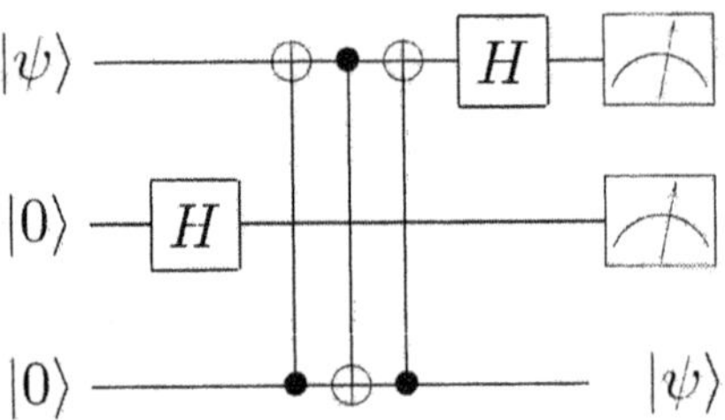

Fig. 20. Third and final set of circuit simplifications applied to the teleportation circuit.

Keep in mind that the circuit of Fig. 20 performs exactly the same transformation, on the given input state, as the orignal circuit of Fig. 16. But now, its function is obvious by inspection. Recall from Fig. 8 that three CNOT gates in this configuration perform a swap operation. This is why the output state (the bottom qubit) is the same as the input state (the initial state of the first qubit).

4.3. *Quantifying entanglement*

How entangled is a given quantum state? One well known test for entanglement is violation of Bell's inequalities, but there are much better meaures known now.

Entropy. Consider a bipartite pure state $|\psi_{AB}\rangle$, and let the partial trace over B be

$$\rho_A = \mathrm{tr}_B|\psi_{AB}\rangle\langle\psi_{AB}| \quad (4.26)$$

$$= \sum_k {}_B\langle k|\psi_{AB}\rangle\langle\psi_{AB}|k\rangle_B\,. \quad (4.27)$$

We define the *entropy of entanglement* of $|\psi_{AB}\rangle$ to be $E(|\psi_{AB}\rangle) = S(\rho_A) = -\mathrm{tr}\,[\rho\log\rho] = -\sum_k \lambda_k \log_2 \lambda_k$, where λ_k is a the kth eigenvalue of ρ_A. This quantity is also known simply as "the entanglement." Note that it applies only to pure states.

Example 1: The von Neumann entropy of a pure state is zero – $S(|\psi\rangle\langle\psi|) = 0$.

Example 2: $|\psi_{AB}\rangle = \frac{|00\rangle+|11\rangle}{\sqrt{2}}$. The corresponding density matrix is

$$\rho_{AB} = \frac{1}{2}\begin{bmatrix} 1 & 0 & 0 & 1 \\ 0 & 0 & 0 & 0 \\ 0 & 0 & 0 & 0 \\ 1 & 0 & 0 & 1 \end{bmatrix}, \quad (4.28)$$

so the state of system A alone is

$$\rho_A = {}_B\langle 0|\rho_{AB}|0\rangle_B + {}_B\langle 1|\rho_{AB}|1\rangle_B \quad (4.29)$$

$$= \frac{1}{2}\begin{bmatrix} 1 & 0 \\ 0 & 1 \end{bmatrix}. \quad (4.30)$$

Thus, $E(|\psi_{AB}\rangle) = S(\rho_A) = 1$.

Schmidt number. Another good measure of entanglement is based on the *Schmidt decomposition*, defined by the following theorem:

Theorem: Let $|\psi_{AB} \in H_A \otimes H_B$. Then $\exists$ orthonormal $|k_A\rangle \in H_A$ and $|k_B\rangle \in H_B$ such that $|\psi_{AB} = \sum_k \lambda_k |k_A\rangle|k_B\rangle$, where $\sum_k \lambda_k^2 = 1$ and $\forall k, \lambda_k \geq 0$. λ_k are known as the *Schmidt coefficients* for $|\psi_{AB}\rangle$.

Proof: In terms of basis vectors $|j\rangle$ and $|l\rangle$ for H_A and H_B, we may write a general $|\psi_{AB}\rangle = \sum_{jl} a_{jl}|j\rangle|l\rangle$, such that by the singular value decomposition, $a = udv$, where a is the matrix with elements a_{jl}, u and v are unitary matrices, and d is diagonal. This gives

$$|\psi_{AB}\rangle = \sum_{jl} u_{jk} d_{kk} v_{kl} |j\rangle|l\rangle \quad (4.31)$$

$$= \sum_k d_{kk}\left(\sum_j u_{jk}|j\rangle\right)\left(\sum_l v_{jl}|l\rangle\right) \quad (4.32)$$

$$= \sum_k d_{kk}|k_A\rangle|k_B\rangle\,, \quad (4.33)$$

where $\left(\sum_j u_{jk}|j\rangle\right)$ is $|k_A\rangle$, and similarly for $|k_B\rangle$, and we may identify λ_k as being d_{kk}.

Definition: The *Schmidt number* of state $|\psi_{AB}\rangle$ is the number of nonzero Schmidt coefficients λ_k in its Schmidt decomposition.

The Schmidt number is known as a good measure of entanglement because it cannot be increased by local operations and classical communication between the two parties. For further reading, see QCQI Section 12.5.

4.4. *Fungibility*

In what way is entanglement a physical resource? If it truly is a resource, then entanglement should not be unique to specific states, but rather, be something that is extractable and convertable between different forms. Resources should have the property that if A and B are equivalent, then the conversions $A \to B$ and $B \to A$ are possible.

Definition: A and B are *asympototically equivalent* if $\exists$ a ratio R such that $\forall \epsilon, \delta > 0, \exists N, \forall n \geq N$

$$A^{\otimes[n(R+\delta)]} \rightarrow B^{\otimes n} \tag{4.34}$$

$$B^{\otimes n} \rightarrow A^{\otimes[n(R-\delta)]}\,. \tag{4.35}$$

Claim: All entangled bipartite pure states are asymptotically equivalent.

Proof idea: Let $|\Psi\rangle = \frac{|00\rangle+|11\rangle}{\sqrt{2}}$; we will consider this EPR pair the "gold standard," to and from which we wish to interconvert all other states. The proof consists of two parts, the first of which is a procedure known as *entanglement concentration*, which performs the conversion

$$|\psi\rangle^{\otimes n} \rightarrow |\Phi\rangle^{\otimes[n(E(|\psi\rangle)-\delta)]} \tag{4.36}$$

with error $< \epsilon$, in the large n limit. Note that the exchange rate, the number of EPR pairs obtainable from each copy of state $|\psi\rangle$, is its entropy of entanglement, $E(|\psi\rangle)$. There is a slight inefficiency δ in the conversion process, but for large enough n, this becomes fractionally negligible. The second part of the proof is the reverse process, *entanglement dilution*, which does

$$|\Phi\rangle^{\otimes[n(E(|\psi\rangle)+\delta)]} \rightarrow |\psi\rangle^{\otimes n}\,. \tag{4.37}$$

This takes a number of EPR pairs and dilutes them to a (generally larger) number of copies of less entangled states $|\psi\rangle$.

We shall leave out details of the dilution procedure, but illustrate the concentration procedure by an example. Let $|\psi\rangle = \sqrt{1-p}|00\rangle + \sqrt{p}|11\rangle$. Note that any two-qubit state can be written in this way, for some p, by the Schmidt decomposition, for appropriate orthonormal $|0\rangle$ and $|1\rangle$ state definitions. The entanglement of this state is

$$E(|\psi\rangle) = -p\log p - (1-p)\log(1-p) \equiv H_2(p)\,, \tag{4.38}$$

the binary entropy of p. By labeling the states appropriately, n copies of this state can be expressed as

$$|\psi\rangle^{\otimes n} = \sum_{x\in\{0,1\}^n} (1-p)^{\frac{n-|x|}{2}} p^{\frac{|x|}{2}} |x\rangle|x\rangle \tag{4.39}$$

$$= \sum_{w=0}^{n} (1-p)^{\frac{n-w}{2}} p^{\frac{w}{2}} \sum_{|x|=w} |x\rangle|x\rangle \tag{4.40}$$

$$= \sum_{w=0}^{n} \sqrt{\binom{n}{w}(1-p)^{n-w} p^w} |S_w\rangle\,, \tag{4.41}$$

where $|x|$ is the Hamming weight of x (the number of one's in its binary representation), and

$$|S_w\rangle = \left[\sqrt{\binom{n}{w}}\right]^{-1} \sum_{|x|=w} |x\rangle|x\rangle\,. \tag{4.42}$$

Each $|S_w\rangle$ is a maximally entangled state of dimension $\binom{n}{w}$, approximately equivalent to $\log\binom{n}{w}$ EPR pairs. Thus, to convert $|\psi\rangle^{\otimes n}$ to a number of EPR pairs, the two parties can both measure the Hamming weight w of their qubits, in the Schmidt basis, collapsing their joint state into $|S_w\rangle$, where

$$\mathrm{prob}(w) = \binom{n}{w} p^w (1-p)^{n-w} \tag{4.43}$$

is an approximately Gaussian probability distribution for $n \to \infty$, with mean np and variance $np(1-p)$. A well known identity in information theory is that

$$\log\binom{n}{np} \approx nH_2(p)\,, \tag{4.44}$$

because of the Stirling approximation, $\log n! \approx n\log n - n$. By use of Eq.(4.38), we have that $nH_2(p) = nE(|\psi\rangle)$, so the number of EPR pairs we obtain, on average, is thus

$$\log\binom{n}{w} \approx \log\binom{n}{np} \approx nE(|\psi\rangle)\,. \tag{4.45}$$

Doing this more carefully (c.f. QCQI Section 12.5.2), one finds that the actual number of EPR pairs obtained is $n[E(|\psi\rangle) - \delta]$, where $n\delta$ grows asymptotically faster than $\sqrt{n}$ as $n \to \infty$.

5. Information theory

This lecture provides an introduction to the classical theory of information. Our goal is to explain fundmental concepts and mathemtical results sufficiently well to provide the necessary background for understanding quantum analogues, in particular quantum error correction and fault tolerance, which are covered in the next lecture. Traditionally, information theory is a semester or year-long course so it is impossible to do due justice to the material in this brief section, but the two fundamental ideas are (1) the definition of information, in terms of entropy, and (2) the ability to and methods for transmitting information robustly through a noisy channel. We describe these ideas below, in sections on entropy, typical sequences, the noiseless coding theorem, mutual information, and the noisy coding theorem. The book by Cover and Thomas [2] is an excellent reference for futher reading.

5.1. Entropy

What is information? In Claude Shannon's 1948 article on the mathematical theory of comminucation, he proposed a model with a message source X that emits a stream of letters $x_1, x_2, x_3, \ldots$, where each letter is drawn from an alphabet $\{1, 2, 3, \ldots, \underline{\mathrm{x}}\}$ of $\underline{\mathrm{x}}$ choices, with a fixed probability distribution P, in which letter k occurs with probability p_k.

Let $H(P)$ be a mathematical function which measures the information from this source. It is desirable for this function to satisfy certain reasonable properties:

1. $H(P)$ is a continuous function of the probabilities p_i.
2. $H(P) \geq 0$ and $H(P) = 0$ iff $p_i = 1$ for some i.
3. $H(P) \leq C(\underline{\mathrm{x}})$ and $\underline{\mathrm{x}}' \geq \underline{\mathrm{x}} \Rightarrow C(\underline{\mathrm{x}}') \geq C(\underline{\mathrm{x}})$, for $C(\underline{\mathrm{x}})$ being some constant function of $\underline{\mathrm{x}}$.
4. $H(P, Q) = H(P) + H(Q)$ if P and Q are independent.

It turns out the these properties are uniquely satisfied, up to scaling factors, by the definition

$$H(P) = -\sum_k p_k \log p_k \,, \tag{5.1}$$

This is known as the *Shannon entropy*. We sometimes write $H(X)$ instead of $H(P)$, refering equivalently to the message source instead of to the probability distribution. Note that also, all logarithms (unless otherwise noted, henceforth) are base 2, such that $\log 2 = 1$.

Example (a): $X = \{0, 1\}$ with $p_0 = p$ and $p_1 = 1 - p$. Then $H(p) = -p \log p - (1 - p)\log(1 - p) \equiv H_2(p)$ is the entropy; this function is also known as the binary entropy. As its plot (Fig. 21) shows, this function is convex and has a maximum value of 1 for $p = 1/2$.

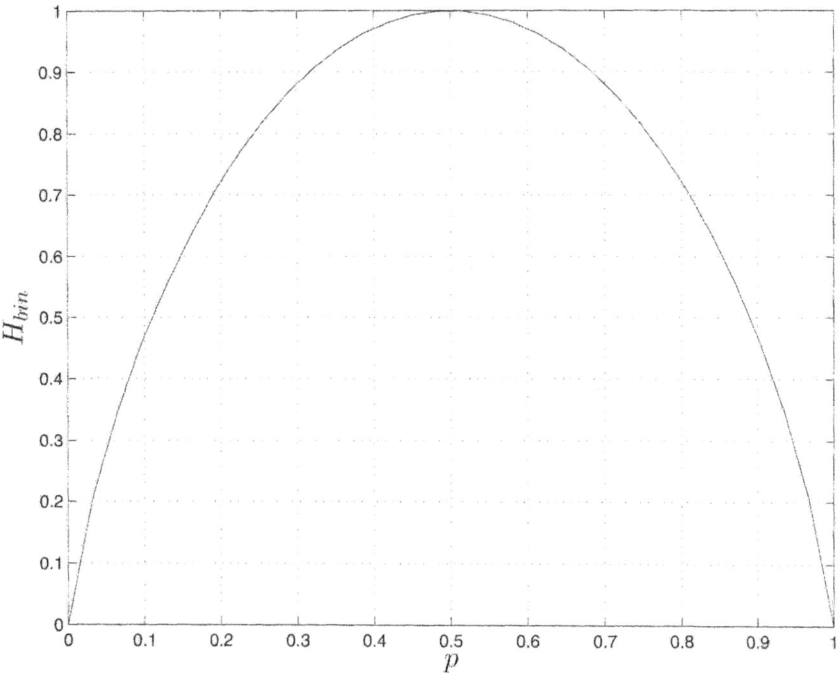

Fig. 21. Plot of the binary entropy function $H_2(p)$.

Example (b): $X = \{0, 1, \ldots, \underline{X}\}$ with $p_k = 1/\underline{x}$. This uniform probability distribution has entropy $H(P) = \log \underline{x}$.

Example (c): $X = \{a, b, c, d\}$ with $p_a = 1/2$, $p_b = 1/4$, $p_c = p_d = 1/8$. This gives $H(P) = 7/4$ bits. As we shall see, this means that the average number of questions needed to guess a symbol is 1.75.

Properties of entropy. It is useful to think of the definition of entropy, Eq.(5.1) as being an expectation value over the probability distribution:

$$H(P) = -\sum_k p_k \log p_k = \mathrm{E}\left[\log \frac{1}{p(x)}\right], \tag{5.2}$$

where $p(x)$ is the probability of symbol x.

5.2. *Typical sequences*

Consider n symbols from an iid (independent, identically distributed) source. The probability of a particular sequence of symbols from that source is

$$\begin{aligned} p(x_1, x_2, \ldots, x_n) &= \prod_{i=1}^{n} p(x_i) && (5.3) \\ &= \prod_{i=1} \underline{\mathrm{x}} p_i^{n_i}, && (5.4) \end{aligned}$$

where n_i is the number of symbols in the sequence equal to letter x_i in the alphabet, and p_i is the probability of that letter. If we take the log of this expression and divide by $-n$, we obtain

$$\begin{aligned} -\frac{1}{n}\log p(x_1, x_2, \ldots, x_n) &= -\frac{1}{n}\log\left[\prod_{i=1} \underline{\mathrm{x}} p_i^{n_i}\right] && (5.5) \\ &= -\sum_{i=1}^{\underline{\mathrm{X}}} \frac{n_i}{n} \log p_i && (5.6) \\ &\approx -\sum_i p_i \log p_i && (5.7) \\ &= H(X). && (5.8) \end{aligned}$$

Thus, $p(x_1, x_2, \ldots, x_n) \approx 2^{-nH(X)}$ as $n \to \infty$, independent of the particular string! This insight is captured by the following statement:

> There are $\underline{\mathrm{x}}^n$ possible strings of length n, and of these, $\sim 2^{nH(X)}$ of these are highly probable strings known as *typical sequences*.

Definition: The set of ϵ-typical strings $A_\epsilon^{(n)}$ of length n is

$$A_\epsilon^{(n)} = \left\{(x_1, \ldots, x_n) \,\middle|\, 2^{-n(H(X)+\epsilon)} \le p(x_1, \ldots, x_n) \le 2^{-n(H(X)-\epsilon)}\right\}. \tag{5.9}$$

The inequality on $p(x_1, x_2, \ldots, x_n)$ can equivalently be written as

$$\left|\frac{1}{n}\log\frac{1}{p(x_1, x_2, \ldots, x_n)} - H(X)\right| \leq \epsilon\,. \tag{5.10}$$

Theorem: Fix $\epsilon > 0$. For any $\delta > 0$ and n sufficiently large, $\mathrm{prob}(A_\epsilon^{(n)}) \geq 1-\delta$.
Proof: By the law of large numbers, for $\epsilon > 0$, for any $\delta > 0$, $\exists n$ sufficiently large such that

$$\mathrm{prob}(|\text{mean} - \text{average}| < \epsilon) \geq 1 - \delta \tag{5.11}$$

when taking n samples from a probability distribution. Suppose we take samples from the distribution $-\log p(x_i)$. This has a mean of

$$\text{mean} = \sum_i \frac{-\log p(x_i)}{n} \tag{5.12}$$

and an average $\mathrm{E}\Big[-\log p(x)\Big]$, so for that distribution, the law of large numbers gives

$$\mathrm{prob}\left(\left|\sum_i \frac{1}{n}\log\frac{1}{p(x_i)} - \mathrm{E}\Big[-\log p(x)\Big]\right| < \epsilon\right) \geq 1 - \delta\,. \tag{5.13}$$

Inserting the expression for entropy, given in Eq.(5.2) into this equation gives Eq.(5.10), the defining relation for ϵ-typical sequences.

The point of defining $A_\epsilon^{(n)}$ is that this set contains nearly all of the important strings that may come from a message source; the relative number of all other messages which may be sent asymptotically goes to zero as the message length n increases, in comparison with the size of the typical set. Nevertheless, at the same time, the size of the typical set is well defined, and is nearly equal to $nH(X)$. These statements can be formalized precisely and mathematically (e.g. QCQI Section 12.2); they form the basis for the operational definition of $H(X)$ as a measure of information.

5.3. Noiseless coding theorem

Shannon's first major result is known as the *noiseless coding theorem*, which relates entropy $H(x)$ to physical costs, establishing information as a kind of resource (compare with our discussion of entanglement as a resource, in Section 4). The scenario is as follows:
Definition: A compression scheme C of rate R (shown in Fig. 22) is *reliable* iff

$$\mathrm{prob}\Big[D(C(x_1, x_2, \ldots, x_n)) = (x_1, x_2, \ldots x_n)\Big] \to 1 \tag{5.14}$$

as $n \to \infty$.

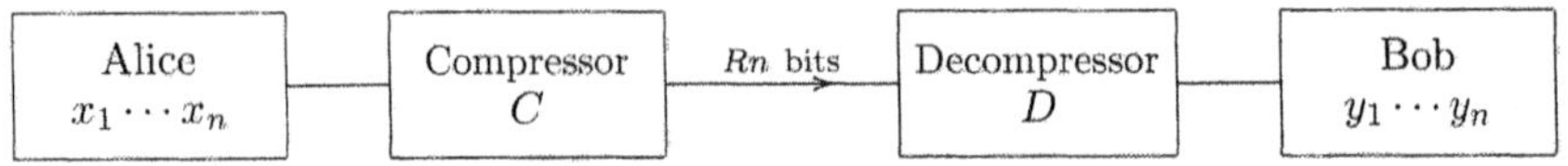

Fig. 22. Noiseless coding communication scenario. C and D maps n symbols to and from Rn bits, respectively.

Theorem: For an iid source of per-symbol entropy $H(X)$, (1) there exists a reliable C if $R > H(X)$, and (2) If $R < H(X)$, any C will be unreliable.
Proof sketch: First, consider (2) – let $S_R = \{D(C(x_1, \ldots, x_n))\}$ be the set of all possible received and decoded messages; by definition, the size of this set $|S_R| < 2^{nR}$, so $|S_R| < 2^{nH(X)}$. However, the *probability* of this set of messages is by and large determined by the typical sequences within S_R, since those are the only messages which occur with high probability (each approximately $2^{-nH(X)}$). Thus,

$$\begin{aligned} \text{prob}(S_R) &\approx \text{prob}(A_\epsilon^{(n)} \cap S_R) && (5.15)\\ &\leq 2^{nR} 2^{-nH(X)} && (5.16)\\ &= 2^{-n(H(X)-R)}\,, && (5.17) \end{aligned}$$

which goes to zero as $n \to \infty$ since $R < H(X)$ by assumption.

For (1), choose $\epsilon < R - H(X)$ so that $\text{prob}(A_\epsilon^{(n)}) > 1-\delta$ goes to 1 as $n \to \infty$, and perform the following procedure for C: enumerate an index $1, 2, \cdots, 2^{nH(X)}$ for each string in $A_\epsilon^{(n)}$. If the message to be transmitted is in $A_\epsilon^{(n)}$, then send the index; else, send a special message FAIL, followed by the original message, unchanged. The compressor performs corresponding operations to recover the orignal message. By the theorem of typical sequences, this compression procedure is reliable.
Example: $X = \{a, b, c, d\}$ with $p_a = 1/2$, $p_b = 1/4$, $p_c = p_d = 1/8$. If we encode $a = 0$, $b = 10$, $c = 110$, and $d = 111$, then the message *abdabaca* can be transmitted losslessly as 01011101001100 and (uniquely) decoded. Note that only 14 bits are used to send the eight symbols, giving a rate of $R = 14/8 = 1.75$ bits per symbol, which achieves the Shannon bound $H(X)$. This rate is less than the 2 bits per symbol (to determine which of four symbols) it would take to send the message with no compression.
Exercise: Suppose that a message source emits 0 with probability p and 1 with probability $1 - p$. Give an encoding scheme which achieves the optimal rate $H_2(p)$.

The noiseless coding theorem (also known as the data compression theorem) essentially *defines* what one bit of information is, because it states that the entropy $H(X)$ (measured in bits) is the minimum rate at which information can be sent from a source X, in order to be able to faithfully reconstruct its message. This equivalence is an asymptotic one, valid in the limit of long messages.

5.4. *Mutual information*

What happens when our goal is not to reconstruct the original source message faithfully, but rather, to reconstruct a good *approximation* of it? Several additional properties of entropy, and new definitions, are necessary to study the noisy scenario.

Definition: The *joint entropy* of X and Y is

$$H(X, Y) = -\sum_{x \in X} \sum_{y \in Y} p(x, y) \log p(x, y) . \tag{5.18}$$

Definition: The *conditional entropy* of Y given X is

$$\begin{aligned} H(Y|X) &= \sum_{x \in X} p(x) H(Y|X = x) && (5.19) \\ &= -\sum_{x \in X} p(x) \sum_{y \in Y} p(y|x) logp(y|x) && (5.20) \\ &= -\sum_{x \in X, y \in Y} p(x, y) \log p(y|x) . && (5.21) \end{aligned}$$

Definition: The *chain rule* for entropies is that

$$H(X, Y) = H(X) + H(Y|X) = H(Y) + H(X|Y) . \tag{5.22}$$

It follows from Bayes' rule, $p(x, y) = p(x)p(y|x)$.
Definition: The *mutual information* between X and Y is

$$\begin{aligned} I(X; Y) &= H(X) + H(Y) - H(X, Y) && (5.23) \\ &= H(X) - H(X|Y) && (5.24) \\ &= H(Y) - H(Y|X) . && (5.25) \end{aligned}$$

5.5. *Noisy coding theorem*

The scenario for the noisy coding theorem is the same as that for the noiseless coding theorem, with the addition of a noisy channel between the compressor (renamed as the encoder) and the decompressor (renamed as the decoder). For many years, it was believed that as the transmission rate through a noisy channel is increased, the error rate also increased, no matter how slow the rate was. However, what Shannon showed, in this second theorem of his, is that as long as the transmission rate is below a certain maximum value, known as the capacity of the channel C, messages could be transmitted with *zero* error, asymptotically in the message length. This capacity is

$$C = \max_{p(X)} I(X; Y) \tag{5.26}$$

We do not have room to include the proof of this theorem here, but the main idea behind the proof can be illustrated as shown in Fig. 23. For inputs X and outputs Y, the number of distinguishable outputs is

$$\frac{2^{nH(Y)}}{2^{nH(Y|X)}} = 2^{n(H(Y)-H(Y|X))} = 2^{nI(X;Y)} . \tag{5.27}$$

The noisy coding theorem states that the rate C is *optimal*, i.e. that rates larger than C are not possible without inducing unrecoverable errors as $n \to \infty$, and *achievable*, meaning a scheme exists which can transmit at any rate less than or equal to C, in the limit as $n \to \infty$.

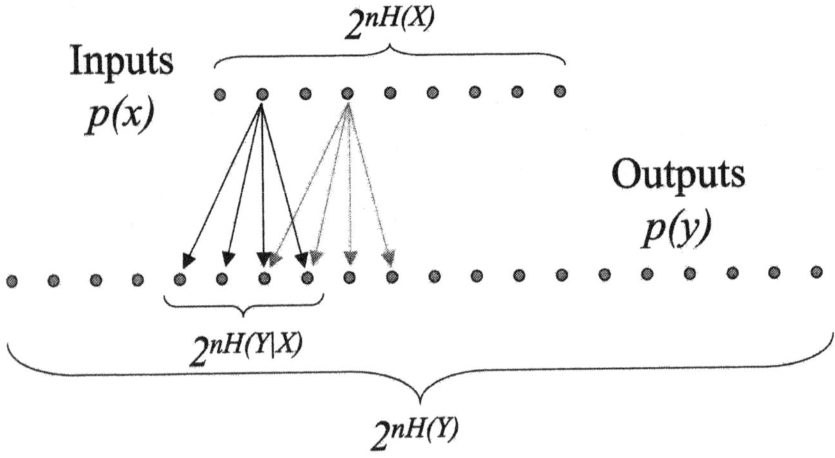

Fig. 23. Diagram illustrating idea behind proof of the noisy coding theorem.

Example: Consider the noisy comminucation channel schematically represented by the diagram in Fig. 24, in which a transmitted 0 is changed to 1 with probability p, or received unchanged with probability $1 - p$, and similarly for a transmitted 1. The mutual information between X and Y is

$$\begin{aligned} I(X;Y) &= H(Y) - H(Y|X) &(5.28)\\ &= H(Y) - \sum_x p(x)H(Y|X=x) &(5.29)\\ &= H(Y) - H_2(p) &(5.30)\\ &\leq 1 - H_2(p), &(5.31) \end{aligned}$$

where, in the last line, the fact that $H(Y) \leq 1$ was used. This establishes $C = 1 - H_2(p)$ as being the capacity of this channel.

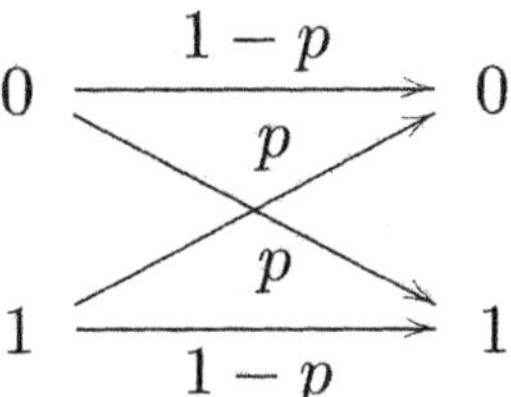

Fig. 24. The binary symmetric channel.

How can data be transmitted error-free through this channel? We can illustrate the basic idea of encoding very simply, but this example will not achieve the channel capacity. Let us represent a message bit 0 by three transmitted bits, 000, and similarly $1 \to 111$. Then the received message has the following probability distribution:

$$000 \quad \to \quad \begin{cases} 000 & (1-p)^3 \\ 001 & p(1-p)^2 \\ 010 & p(1-p)^2 \\ 100 & p(1-p)^2 \\ 011 & p^2(1-p) \\ 101 & p^2(1-p) \\ 110 & p^2(1-p) \\ 111 & p^3 \end{cases} \qquad (5.32)$$

$$111 \quad \rightarrow \quad \left\{ \begin{array}{ll} 000 & p^3 \\ 001 & (1-p)p^2 \\ 010 & (1-p)p^2 \\ 100 & (1-p)p^2 \\ 011 & (1-p)^2 p \\ 101 & (1-p)^2 p \\ 110 & (1-p)^2 p \\ 111 & (1-p)^3 \end{array} \right. . \tag{5.33}$$

If we decode the message by using majority voting, then the probability of error is the probability of having two or more bit flip errors,

$$p(\text{error}) = 3(1-p)p^2 + p^3 = p^2(3 - 3p + p) = O(p^2)\,. \tag{5.34}$$

Generalizing this scheme, if the symbols are repeated $2n+1$ times instead of just three times, then the probability of error becomes $O(p^{n+1})$, achieving a transmission rate of $1/(2n+1)$ bits received per bits sent. As long as $p < 1$, then repetition reduces error exponentially quickly; such error correction is possible is of great interest. By using a recursive error correction scheme (or other equivalent methods), a fix rate with asymptotically zero error as $n \rightarrow \infty$ can be achieved, but we do not have space to explain that procedure here; for more, see e.g. Cover and Thomas [2].

6. Open quantum systems

In this lecture, we return to the subject of quantum systems, and introduce the idea of non-unitary operations which result from interations between the system and an environment. Since the world is fundamentally quantum-mechanical, by including everything in the system we can once again obtain a closed system, and thus unitary dynamics, but it is convenient to make this distinction, so that many undesired degrees of freedom can be relegated to "the environment" and forgotten about. Many mathematical tools have been developed to make this procedure effective, and it is also a useful approach for being able to interpret certain dynamics as being errors that can be corrected, using a quantum extension to classical error correction techniques. We begin by reviewing the theory of density matrices, then introduce a mathematical formalism for quantum operations on open systems, which is illustrated specifically by a form known as the operator sum representation, and a system-environment model. An interesting example concludes.

6.1. Density matrices

A matrix ρ is a density matrix iff (a) $\mathrm{tr}\,(\rho) = 1$, and (b) ρ is positive, that is, for all states $|\phi\rangle$, $\langle\phi|\rho|\phi\rangle$ is real and ≥ 0. Equivalently to (b), we may stipulate that ρ is Hermitean and all eigenvalues are ≥ 0.
Claim: $\sum_k p_k|\psi_k\rangle\langle\psi_k| = \rho$, for p_k probabilities.
Proof: $\mathrm{tr}\,(\rho) = \sum_k p_k \mathrm{tr}\,(|\psi_k\rangle\langle\psi_k|) = \sum_k p_k = 1$. Also, $\langle\phi|\rho|\phi\rangle = \sum_k p_k|\langle\phi|\psi\rangle|^2 \geq 0$.
Claim: Any ρ can be expressed as $\sum_k p_k|\psi_k\rangle\langle\psi_k|$, for some $|\psi_k\rangle$ and probabilities p_k.
Proof: ρ is positive, so it must have spectral decomposition $\rho = \sum_k \lambda_k|k\rangle\langle k|$ where, by $\mathrm{tr}\,(\rho) = 1$, $\sum_k \lambda_k = 1$.
Definition: ρ is *pure* iff $\rho = |\psi\rangle\langle\psi|$ for some $|\psi\rangle$.

An interesting fact about density matrices is that in general, they have an infinite number of *unravelings*: let $\rho = \sum_k p_k|\psi_k\rangle\langle\psi_k|$. Then $\rho = \rho' = \sum_k q_k|\phi_k\rangle\langle\phi_k|$ if

$$\sqrt{p_k}|\psi_k\rangle = \sum_j u_{kj}\sqrt{q_j}|\phi_j\rangle\,, \tag{6.1}$$

for u_{kj} a unitary matrix. Thus, there are an infinite number of ways that a (non-pure) density matrix can be written as a convex combination of pure states; each such description is known as an unraveling.

What is the physical origin of the density matrix? Consider two boxes, the first of which outputs $|0\rangle$ with probability 3/4 or $|1\rangle$ with probability 1/4. The density matrix describing this output state is

$$\rho_1 = \frac{1}{4}\begin{bmatrix} 3 & 0 \\ 0 & 1 \end{bmatrix}. \tag{6.2}$$

The second box puts out states $|a\rangle$ and $|b\rangle$ with equal probability, where $|a\rangle = \sqrt{3/4}|0\rangle + \sqrt{1/4}|1\rangle$, and $|b\rangle = \sqrt{3/4}|0\rangle - \sqrt{1/4}|1\rangle$. Note that $|a\rangle$ and $|b\rangle$ are non-orthogonal. The density matrix describing this output is

$$\rho_2 = \frac{1}{2}\begin{bmatrix} 3/4 & \sqrt{3}/4 \\ \sqrt{3}/4 & 1/4 \end{bmatrix} + \frac{1}{2}\begin{bmatrix} 3/4 & -\sqrt{3}/4 \\ -\sqrt{3}/4 & 1/4 \end{bmatrix} \tag{6.3}$$

$$= \frac{1}{4}\begin{bmatrix} 3 & 0 \\ 0 & 1 \end{bmatrix}. \tag{6.4}$$

So $\rho_1 = \rho_2$. Which description is real? It turns out there is *no possible* experiment which can distinguish between the two boxes; either model is equally real!

Example: Is the matrix

$$\frac{1}{2}\begin{bmatrix} 1 & 0 & 0 & 0 \\ 0 & 0 & 1 & 0 \\ 0 & 1 & 0 & 0 \\ 0 & 0 & 0 & 1 \end{bmatrix} \tag{6.5}$$

a valid density matrix? Why or why not?

6.2. *Quantum operations formalism*

The dyamics of open quantum systems have long been studied in the physics community, primarily with stochastic differential equations and master equations, which describe the continuous evolution of the system in time. In contrast, in the field of quantum information, we shall mostly be interested in the state at discrete points in time, just as we studied how unitary transforms change quantum states in the quantum circuits model.

A closed quantum system in a pure state transforms unitarily as $|\psi\rangle \to U|\psi\rangle$. Similarly, a density matrix transforms as $\rho \to U\rho U^{\dagger}$. In general, however, unitary transforms are not the only legal ways which describe how density matrices may transform.

Example: Let θ be a gaussian random variable with mean 0 and variance 2λ. Then

$$p(\theta) = \frac{1}{\sqrt{4\pi\lambda}} e^{-\theta^2/4\lambda} \tag{6.6}$$

so that if a quantum state $a|0\rangle + b|1\rangle$ is transformed by $R_z(\theta)$, what happens to it *on average*? The answer is given by

$$\rho' = \frac{1}{\sqrt{4\pi\lambda}} \int_{-\infty}^{\infty} R_z(\theta)|\psi\rangle\langle\psi|R_z^{\dagger}(\theta)e^{-\theta^2/4\lambda}\, d\theta \tag{6.7}$$

$$= \frac{1}{\sqrt{4\pi\lambda}} \int_{-\infty}^{\infty} \begin{bmatrix} |a|^2 & ab^* e^{i\theta} \\ a^* b e^{i\theta} & |b|^2 \end{bmatrix} e^{-\theta^2/4\lambda}\, d\theta\,. \tag{6.8}$$

Using the fact that $\langle e^{i\theta}\rangle = e^{-\lambda}$ (where $\langle\cdots\rangle$ denotes an average over the gaussian), we find that

$$\rho' = \begin{bmatrix} |a|^2 & ab^* e^{-\lambda} \\ a^* b e^{-\lambda} & |b|^2 \end{bmatrix}. \tag{6.9}$$

This is a non-unitay transformation known as *phase damping*, which describes, among other things, the effect on the phase of a photon of elastic scattering in a single mode. We will want a more concise mathematical method to describe this, and similar transformations.

Example: Suppose we have a map $\mathcal{E}$ which takes

$$\begin{bmatrix} a & b \\ c & d \end{bmatrix} \stackrel{\mathcal{E}}{\rightarrow} \begin{bmatrix} a & c \\ b & d \end{bmatrix}. \tag{6.10}$$

Is this a legal transformation? Let $|\psi\rangle = \frac{|00\rangle+|11\rangle}{\sqrt{2}}$ such that

$$\rho = |\psi\rangle\langle\psi| = \frac{1}{2}\begin{bmatrix} 1 & 0 & 0 & 1 \\ 0 & 0 & 0 & 0 \\ 0 & 0 & 0 & 0 \\ 1 & 0 & 0 & 1 \end{bmatrix}. \tag{6.11}$$

If we apply $\mathcal{E}$ to the first qubit only, this maps

$$|10\rangle\langle 00| \quad \leftrightarrow \quad |00\rangle\langle 10| \tag{6.12}$$
$$|11\rangle\langle 01| \quad \leftrightarrow \quad |01\rangle\langle 11| \tag{6.13}$$
$$|11\rangle\langle 00| \quad \leftrightarrow \quad |01\rangle\langle 10|, \tag{6.14}$$

giving us the output state

$$\rho' = \mathcal{E}(\rho) = \frac{1}{2}\begin{bmatrix} 1 & 0 & 0 & 0 \\ 0 & 0 & 1 & 0 \\ 0 & 1 & 0 & 0 \\ 0 & 0 & 0 & 1 \end{bmatrix}, \tag{6.15}$$

which is not a legal density matrix.

Definition: A map $\mathcal{E} : \rho \rightarrow \mathcal{E}(\rho)$ is a valid quantum operation if and only if

A1 $\mathrm{tr}\,(\mathcal{E}(\rho)) = 1$

A2 $\mathcal{E}$ is convex and linear: $\mathcal{E}(\sum_k p_k \rho_k) = \sum_k p_k \mathcal{E}(\rho_k)$ for probabilities p_k.

A3 $\mathcal{E}$ is completely positive:

(a) $\mathcal{E}(\rho) > 0$ if $\rho > 0$

(b) $(I_R \otimes \mathcal{E}_Q)(\rho_{RQ})$ must be positive for any bipartite $\rho_{RQ} > 0$.

6.3. Operator sum representation

A convenient representation for a quantum operation, which satisfies the above axiomatic definition, is given by the following theorem.

Theorem: Let $\rho \in H_1$ and $\mathcal{E}(\rho) \in H_2$. $\mathcal{E}$ satisfies A1, A2, and A3 iff

$$\mathcal{E}(\rho) = \sum_k E_k \rho E_k^\dagger, \tag{6.16}$$

for some operators (known as "operation elements") $E_k : H_1 \to H_2$ satisfying

$$\sum_k E_k^\dagger E_k = I \,. \tag{6.17}$$

Proof (←): We prove that a quantum operation of the form in Eq.(6.16) satisfies the axiomatic conditions. First,

$$\begin{aligned} \mathrm{tr}\,(\mathcal{E}(\rho)) &= \sum_k \mathrm{tr}\,(E_k \rho E_k^\dagger) &(6.18)\\ &= \sum_k \mathrm{tr}\,(\rho E_k^\dagger E_k) &(6.19)\\ &= \mathrm{tr}\,(\rho \sum_k E_k^\dagger E_k) &(6.20)\\ &= \mathrm{tr}\,(\rho) = 1\,, &(6.21) \end{aligned}$$

so A1 is satisfied. $\mathcal{E}$ is obviously linear and convex, so A2 is satisfied. Finally, let $\rho_{RQ} \in H_{RQ}$. Then $\forall |\psi\rangle \in H_{RQ}$,

$$\langle\psi|(I_R \otimes E_k)\rho_{RQ}(I_R \otimes E_k^\dagger)|\psi\rangle = \langle\phi_k|\rho_{RQ}|\phi_k\rangle \geq 0 \tag{6.22}$$

where $|\phi_k\rangle = (I_R \otimes E_k^\dagger)|\psi\rangle$. Thus, $\mathcal{E}$ is completely positive, satisfying A3.

6.4. *System-environment model*

Another representation for quantum operations involves the explicit introduction of an environment, according to the following theorem:

Theorem: Any quantum operation $\mathcal{E}$ can be realized by the quantum circuit shown in Fig. 25, with an appropriate choice of Hilbert space for the environment, initial environment state $|e_0\rangle$, and unitary interaction between system and environment U.

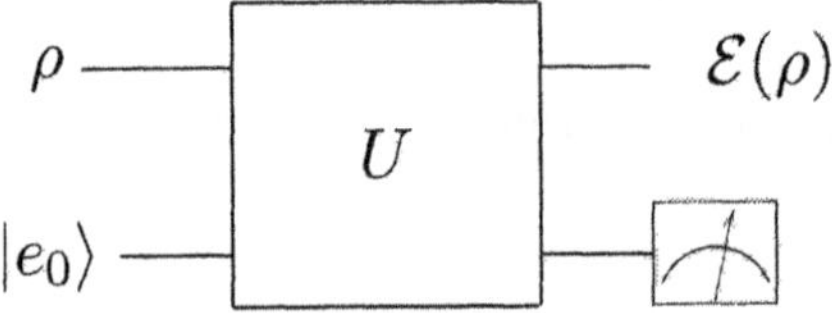

Fig. 25. Quantum circuit representation of a system-environment model for quantum operations.

Proof: Define U such that $E_k = \langle e_k|U|e_0\rangle$, where E_k is an operation element in the operator sum representation of $\mathcal{E}$, and $|e_k\rangle$ provide an orthonormal basis for the Hilbert space of the environment. Thus, $U = \sum_k E_k|e_k\rangle\langle e_0| +$ other terms; essentially, the first column of U is constructed from the E_k, and the size of the environment is chosen to be large enough to accomodate all the E_k terms. The output density matrix after U is then

$$U\Big[\rho \otimes |e_0\rangle\langle e_0|\Big]U^\dagger = \sum_{kk'} E_k \rho E_{k'}^\dagger \otimes |e_k\rangle\langle e_{k'}| \,. \tag{6.23}$$

Measurement of the enviroment in the $|e_k\rangle$ basis then leaves the system in the state

$$\mathcal{E}(\rho) = \sum_k E_k \rho E_k^\dagger \,, \tag{6.24}$$

which is exactly the desired state.

Note that it is simple to use this equivalence in the opposite direction; given any unitary system-environment interaction U, an operator sum representation for the interaction is provided by taking the appropirate matrix elements of U, defined by the initial state of the environment and an orthonormal basis for the environment.

6.5. Examples

Two interesting examples, built on the operator sum representation, and illustrating the interpretative role of the system-environment model, conclude this lecture.

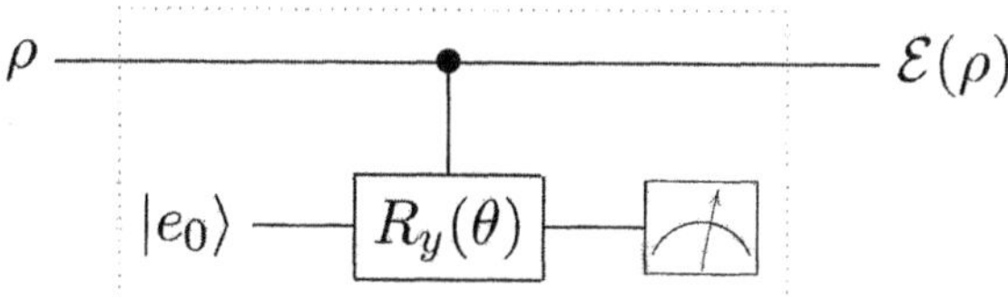

Fig. 26. A system-environment model for a certain quantum operation.

Example 1: Consider the quantum circuit for a system-environment interaction shown in Fig. 26. The unitary operation coupling the system and environment can be expressed as

$$U = \begin{bmatrix} 1 & 0 & \cdot & \cdot \\ 0 & \cos\frac{\theta}{2} & \cdot & \cdot \\ 0 & 0 & \cdot & \cdot \\ 0 & \sin\frac{\theta}{2} & \cdot & \cdot \end{bmatrix}, \tag{6.25}$$

where the first two rows and columns are labeled by $|00\rangle$, $|01\rangle$ with the assignment $|\text{env}, \text{sys}\rangle$; these are the only elements of the matrix we are concerned with, because the environment starts in the $|0\rangle$ state. We can thus read off the operation elements for an operator sum description of this quantum process,

$$E_0 = \begin{bmatrix} 1 & 0 \\ 0 & \sqrt{\gamma} \end{bmatrix} \tag{6.26}$$

$$E_1 = \begin{bmatrix} 0 & 0 \\ 0 & \sqrt{1-\gamma} \end{bmatrix}, \tag{6.27}$$

where we have defined $\sqrt{\gamma} = \cos\frac{\theta}{2}$ just for convenience. The quantum operation is

$$\mathcal{E}(\rho) = E_0 \rho E_0^\dagger + E_1 \rho E_1^\dagger, \tag{6.28}$$

and to understand better what it does, let us apply it to a generic input density matrix $\rho = \begin{bmatrix} a & b \\ c & d \end{bmatrix}$; we find

$$\mathcal{E}(\rho) = \begin{bmatrix} a & \sqrt{\gamma} b \\ \sqrt{\gamma} c & d \end{bmatrix}. \tag{6.29}$$

This is very interesting, because if we identify $\sqrt{\gamma}$ as being $e^{-\lambda}$, then this quantum operation is phase damping – the same as what we saw in the first example of Section 6.2.

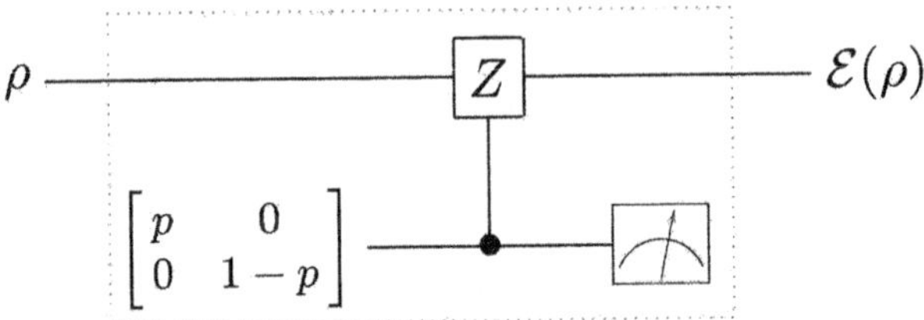

Fig. 27. Another system-environment model for a certain quantum operation.

Example 2: Consider the quantum circuit for another system-environment interaction shown in Fig. 27. In this example, we've allowed the environment to start in a mixed state instead of a pure state; that is valid. Here, it is a diagonal density matrix, and because the circuit is just a controlled-Z operation, it is easy to analyze. With probability p, nothing happens to the system, and with probability $1-p$, the system has a phase flip operation Z applied to it. The quantum operation thus takes an input density matrix $\rho = \begin{bmatrix} a & b \\ c & d \end{bmatrix}$ to

$$\mathcal{E}(\rho) = p\begin{bmatrix} a & b \\ c & d \end{bmatrix} + (1-p)\begin{bmatrix} a & -b \\ -c & d \end{bmatrix} \tag{6.30}$$

$$= \begin{bmatrix} a & (2p-1)b \\ (2p-1)c & d \end{bmatrix}. \tag{6.31}$$

Now, if we identify $2p - 1 = e^{-\lambda}$, this process also describes phase damping, which we've already encountered two other descriptions of!

Interestingly, in this example, we seem to have information from the environment flowing into, and acting upon, the system. Random fluctuations in the environment cause phase flips, damping the off-diagonal coherences of the system's state. In contrast, in the previous example, we seemed to have information flowing out of the system, causing the state of the environment to be rotated by an angle depending on the system qubit! What is the correct interpretation?

It turns out there is no unique way to look at, or model a quantum operation, in general. Much like the infinite unravelings possible for describing a density matrix as a convex combination of pure states, quantum operations also have an infinite number of "unravelings" into different system-environment interaction models, which cannot be distinguished from each other, even in principle. This observation is based upon the following theorem:

Theorem: Let $\mathcal{E}(\rho) = \sum_k E_k \rho E_k^\dagger$, and $\mathcal{F}(\rho) = \sum_k F_k \rho F_k^\dagger$. Then $\mathcal{E} = \mathcal{F}$ iff $\exists$ unitary u_{jk} such that $E_j = \sum_k u_{jk} F_k$.

The theorem is known as that of the unitary degree of freedom in the operator sum representation. It generally allows for multiple physical interpretations of a given quantum operation (those other than unitary transforms), and is conceptually vital in being able to interpret quantum operations, for example, as errors which can be corrected. In the phase damping examples, the first example does not seem to admit a simple procedure for reversing errors; however, the second example makes it clear that phase damping can be identified with random phase flips. It was that identification which allowed the concept of quantum error correction to take birth.

7. Quantum error correction and fault tolerance

This final lecture introduces a topic which is perhaps one of the most surprising and fundamentally interesting, in quantum information: quantum error correction. Before 1995, it was believed that the concept of error correction could not apply to quantum systems, because

- Quantum states collapse when measured
- Errors are continuous
- Quantum states cannot be cloned

However, it turns out there are ways around all of these objections. The part of the Hilbert space containing the information about the state you wish to preserve need not be measured; only the effect of the environment need be determined by a measurement. And using entangled states allow errors to be made orthogonal and distinguishable; entanglement also replaces the role played by redundant copies in classical error correction.

Here, we illustrate the basic ideas behind quantum error correction using a set of specific examples, then a set of algebraic conditions are given which specify when and how quantum error correction is possible. We conclude with a discussion about how this generalizes to provide computation that is robust against errors, and implications this has for scalability of quantum information systems.

7.1. 3-qubit bit-flip code

Suppose errors occur to a qubit as described by the quantum operation $\mathcal{E}(\rho) = (1-p)\rho + pX\rho X$, where p is a probability. This process is known as a "bit-flip" channel, because the qubit is left alone with probability $1-p$, and flipped by X with probability p. We will provide an encoding for a quantum state to allow bit-flip errors to be corrected.

Definition: A $[[n, k]]$ quantum code C is a k qubit subspace of an n qubit Hilbert space.

Example: For $k = 1$, $n = 3$ we can define logical zero as $|0_L\rangle = |000\rangle$ and $|1_L\rangle = |111\rangle$. Much like the three-bit code example in Section 5.5, we find that the effect of the quantum noise process (acting independently and identically on the three qubits in the code) performs the transformation

$$a|0_L\rangle + b|1_L\rangle \stackrel{\mathcal{E}}{\to} \begin{cases} a|000\rangle + b|111\rangle & (1-p)^3 \\ a|001\rangle + b|110\rangle & p(1-p)^2 \\ a|010\rangle + b|101\rangle & p(1-p)^2 \\ a|100\rangle + b|011\rangle & p(1-p)^2 \\ a|011\rangle + b|100\rangle & p^2(1-p) \\ a|101\rangle + b|010\rangle & p^2(1-p) \\ a|110\rangle + b|001\rangle & p^2(1-p) \\ a|111\rangle + b|000\rangle & p^3 \end{cases}, \qquad (7.1)$$

where the probabilities of each pathway are given in the column on the right. How can we determine what error occurred?

Operator measurement: Consider the quantum circuit shown in Fig. 28. Say that U has eigenvalues ± 1 and corresponding eigenvectors $|u_\pm\rangle$. Then a single

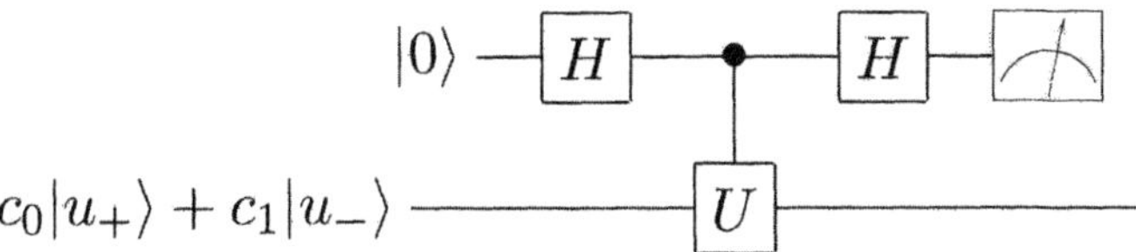

Fig. 28. Quantum circuit for measurement of an operator acting on a single qubit.

qubit state $|\psi\rangle$ can be expressed as $c_0|u_+\rangle + c_1|u_-\rangle$. Plugging this state into the circuit gives the following sequence of transformations, following circuit elements from left to right:

$$
\begin{aligned}
& (|0\rangle)(c_0|u_+\rangle + c_1|u_-\rangle) && (7.2)\\
\rightarrow\quad & \left(\frac{|0\rangle + |1\rangle}{\sqrt{2}}\right)(c_0|u_+\rangle + c_1|u_-\rangle) && (7.3)\\
\rightarrow\quad & \frac{|0\rangle(c_0|u_+\rangle + c_1|u_-\rangle)}{\sqrt{2}} + \frac{|1\rangle(c_0|u_+\rangle - c_1|u_-\rangle)}{\sqrt{2}} && (7.4)\\
\rightarrow\quad & \frac{|0\rangle + |1\rangle}{2}(c_0|u_+\rangle + c_1|u_-\rangle) + \frac{|0\rangle - |1\rangle}{2}(c_0|u_+\rangle - c_1|u_-\rangle) && (7.5)\\
=\quad & c_0|0\rangle|u_+\rangle + c_1|1\rangle|u_-\rangle\,. && (7.6)
\end{aligned}
$$

Thus, when the measurement result is 0, the circuit outputs $|u_+\rangle$ and when 1, outputs $|u_-\rangle$.

We can use this operator measurement procedure to determine the error caused by the environment, without measuring the encoded qubit. Returning to the state in Eq.(7.1), let us measure the two unitary operators

$$U_1 = Z_1Z_2 = ZZI \quad \text{and} \quad U_2 = Z_2Z_3 = IZZ\,. \tag{7.7}$$

These are operators on more than one qubit, but because they both have order 2, they have only two distinct eigenvalues, and can be measured by exactly the same circuit as in Fig. 28, albeit with more qubits entering and leaving the controlled-U operation. Considering just the four high probability output states, we find

State	U_1 result	U_2 result	Recoveryoperation
$a|000\rangle + b|111\rangle$	0	0	I
$a|001\rangle + b|110\rangle$	0	1	X_3
$a|010\rangle + b|101\rangle$	1	1	X_2
$a|100\rangle + b|011\rangle$	1	0	X_1

, (7.8)

where the second and third columns give the measurement results for U_1 and U_2, known as the *syndrome*. Note that none of the measurement results distinguish between a and b, and thus the encoded qubit is left intact. After doing this measurement, we can recover the initial state $a|0_L\rangle + b|1_L\rangle$ by applying the unitary transformation given in the fourth column.

The final output state ρ after performing this correction procedure is not perfect, because of the probability of two or more errors occuring. Quantitatively, the fidelity of the reconstructed state is

$$F(\rho, |\psi\rangle) = \sqrt{\langle\psi|\rho|\psi\rangle} = \sqrt{1 - 3p^2 + 2p^3}\,. \tag{7.9}$$

Interestingly, this scheme also corrects a small rotational error, even though the error is continuous! Let the error be described by the quantum operation

$$\mathcal{E}(\rho) = e^{-i\epsilon X_1} \rho e^{i\epsilon X_1}\,. \tag{7.10}$$

We have from Eq.(3.8) that $|\psi'\rangle = R_{x_1}(2\epsilon)|\psi\rangle = |\psi\rangle - i\epsilon X_1|\psi\rangle$, so the syndrome measurement result collapses this state into either $|\psi\rangle$ or $-i\epsilon X_1|\psi\rangle$. The fidelity of the resulting output, after recovery $\mathcal{R}$, is

$$F(\mathcal{R}(\mathcal{E}(|\psi\rangle)), |\psi\rangle) \approx \sqrt{\infty - \epsilon^{\triangle}} \approx \infty - \epsilon^{\in}\,. \tag{7.11}$$

In comparison, if no encoding were done then the output fidelity would have been $\approx 1 - \epsilon$. This reduction of error from $O(\epsilon)$ to $O(\epsilon^2)$ is a signature of the success of the error correction, even for continuous errors.

7.2. 3-qubit phase-flip code

A similar procedure can be used to correct phase flip errors. Let $\mathcal{E}(\rho) = (1 - p)\rho + pZ\rho Z$. Recall that $HZH = X$ and $HXH = Z$. Thus, $H\mathcal{E}(H\rho H)H = \mathcal{E}_{\text{bit flip}}$. Thus, we can correct phase flip errors by the encoding $|0_L\rangle = |{+}{+}{+}\rangle$, $|1_L\rangle = |{-}{-}{-}\rangle$, where $|\pm\rangle \equiv (|0\rangle \pm |1\rangle)/\sqrt{2}$, and we perform the syndrome measurements XXI and IXX.

The reason phase flip errors are significant and interesting is because of the following.

Claim: Arbitrary qubit erros are a combination of bit-flip (X), phase-flip (Z), and bit-phase flip (XZ) errors.

Proof sketch: Recall that a general quantum operation can be written as $\mathcal{E}(\rho) = \sum_k E_k \rho E_k^\dagger$. Let ρ be a single qubit, and define $E_k = \sum_j c_{jk}\sigma_j$, where $\sigma_j \in \{I, X, Y, Z\}$. Then

$$\begin{aligned}\mathcal{E}(\rho) &= \sum_{kjj'} c_{kj} x^*_{kj'} \sigma_j \rho \sigma_{j'} &(7.12)\\ &= \sum_{kl} \chi_{jl} \sigma_j \rho \sigma_l\,, &(7.13)\end{aligned}$$

where $\chi_{jl} = \sum_k c_{kj} c^* kj'$ is a matrix of c-number coefficients.

For example, for $R_x(2\epsilon)|\psi\rangle \approx |\psi\rangle - i\epsilon X|\psi\rangle$, we have $\mathcal{E}(\rho) = \rho - i\epsilon(X\rho - \rho X) + \epsilon^2 X\rho X$. The syndrome measurement then projects the environment into definite error states, removing the off-diagonal terms (proportional to ϵ instead of to ϵ^2) in this expression.

7.3. *Nine qubit code*

A combination of the bit-flip and phase-flip codes allows correction of *any* single qubit error. This code encodes a single qubit as

$$|0_L\rangle = \frac{(|000\rangle + |111\rangle)^{\otimes 3}}{\sqrt{8}} \tag{7.14}$$

$$|1_L\rangle = \frac{(|000\rangle - |111\rangle)^{\otimes 3}}{\sqrt{8}}. \tag{7.15}$$

The syndrome measurement operators are

$$Z_1Z_2 \;\; Z_2, Z_3 \;\; Z_4, Z_5 \;\; Z_5, Z_6 \;\; Z_7, Z_8 \;\; Z_8, Z_9 \tag{7.16}$$

for bit flip errors, and

$$X_1X_2X_3X_4X_5X_6 \qquad X_4X_5X_6X_7X_8X_9 \tag{7.17}$$

for phase flip errors. It is simple to see that any single qubit bit flip or phase flip or bit-phase flip error gives a unique syndrome measurement result, by direct examination.

7.4. *QEC conditions*

So far, we have given examples of specific quantum codes constructed by analogy from classical codes. More generally, there is a set of criteria which all quantum codes must satisfy, in order to be able to correct for errors caused by a given quantum operation. Let the error (e.g. in a quantum channel) be described by $\mathcal{E}(\rho) = \sum_k E_k \rho E_k^\dagger$.

Theorem: Let C be a quantum code defined by the orthonormal states $\{|\psi_\ell\rangle\}$. There exists a quantum operation $\mathcal{R}$ correcting $\mathcal{E}$ on C iff the following two conditions are satisfied

orthogonality $\langle\psi_\ell|E_j^\dagger E_k|\psi_\ell\rangle = 0, \forall j \neq k, \forall \ell$.

non-deformation $\langle\psi_\ell|E_k^\dagger E_k|\psi_\ell\rangle = d_k, \forall \ell$.

The first condition stipulates that orthogonal basis staes remain orthogonal, and the second, that codespace basis vectors stay of equal length. The important observation is that these are a set of algebraic conditions which allow a quantum code to be found, in principle, for any error process. Of course, the size of the code space required may be nontrivial, so it is not efficient to search randomly for solutions.

Proof: Let $P = \sum_\ell |\psi_\ell\rangle\langle\psi_\ell|$ be the projector onto the codespace C. Note that

$$PE_k^\dagger E_j P = d_k \delta_{jk} P\,. \tag{7.18}$$

By the polar decomposition, $E_k P = U_k\sqrt{PE_k^\dagger E_k P} = \sqrt{d_k} U_k P$, where U_k is unitary.

First, construct the syndrome measurement. Let

$$P_k = U_k P U_k^\dagger = \frac{E_k P U_k^\dagger}{\sqrt{d_k}} = \frac{U_k P E_k^\dagger}{\sqrt{d_k}}\,. \tag{7.19}$$

By Eq.(7.18), the P_k are orthogonal, for $k \neq j$

$$P_k P_j \propto U_k P E_k^\dagger E_j P U_j^\dagger = 0\,. \tag{7.20}$$

So let us measure P_k and obtain k as the syndrome measurement result.

Second, perform a recovery operation to undo the error, using $U_k^\dagger$. Let

$$\mathcal{R}(\rho) = \sum_k U_k^\dagger P_k \rho P_k U_k \tag{7.21}$$

be the recovery operation. Note that for $|\psi\rangle \in C$,

$$U_k^\dagger P_k E_j|\psi\rangle = \frac{1}{\sqrt{d_k}} U_k^\dagger U_k P E_k^\dagger E_j P|\psi\rangle \tag{7.22}$$

$$= \sqrt{d_k}\delta_{jk}|\psi\rangle\,. \tag{7.23}$$

Thus,

$$\mathcal{R}(\mathcal{E}(|\psi\rangle)) = \mathcal{R}(\sum_j E_j|\psi\rangle\langle\psi|E_j^\dagger) \tag{7.24}$$

$$= \sum_{kj} U_k^\dagger P_k k E_j|\psi\rangle\langle\psi|E_j^\dagger P_k^\dagger U_k \tag{7.25}$$

$$= \sum_{kj} d_k \delta_{jk}|\psi\rangle\langle\psi| \tag{7.26}$$

$$= |\psi\rangle\langle\psi|\sum_k d_k \tag{7.27}$$

which is proportional to the original state, as desired.

7.5. *Fault-tolerance and scalability*

Classical error correction is useful not only for correction data transmission errors, but also, for errors which occur *during computation*, due to faulty gates. Von Neumann recognized this potential in the mid 1950's, when building computers from failure-prone vacuum tubes; his work led to the following theorem:
Theorem: A circuit containing N error-free gates can be simulated with probability of error less than ϵ, using $O(\text{poly}(\log 1/\epsilon)N)$ faulty gates, which fail with probability p, so long as $p < p_{th}$.

p_{th} is some threshold probability that is constant with respect to the circuit size N, determined by properties of the code. The three ingredients behind this theorem are (1) to compute on encode states, (2) to correct errors periodically, and (3) to control error propagation.

Fascinatingly, it is now known that such fault tolerant constructions are also possible with quantum circuits. Indeed, one of the most interesting facts about quantum codes is that certain operations can be performed on encoded states without leaving the code space. The Steane 7-qubit code, with codewords

$$
\begin{aligned}
|0_L\rangle &= \frac{1}{\sqrt{8}}\Big[|0000000\rangle + |1010101\rangle + |0110011\rangle + |1100110\rangle \\
&\quad + |0001111\rangle + |1011010\rangle + |0111100\rangle + |1101001\rangle\Big], \quad (7.28)
\end{aligned}
$$

and $|1_L\rangle = XXXXXXX|0_L\rangle$ has the useful property that X, Y, H, and S gates can all be done to the encoded logical qubit simply by performing the same gate to bitwise, to the seven physical qubits. Similarly, CNOT can be performed bitwise between two encoded qubits, by performing seven CNOT gates between pairs of qubits in the codewords.

However, it turns out that X, S, CNOT, and the gates they generate by combination, are not universal for quantum computation; this set is known as the Clifford group operations, and by the Gottesman-Knill theorem, if a quantum circuit is composed from them, and qubits are initialized to, and measured in the computational basis, then the circuit can be efficiently simulated by a classical computer!

A universal set of gates is possible to construct, but this requires some special circuits wich involve unencoded states. That can be cone, for example, to implement the T gate, obtaining a univesal set from which any quantum gate can be efficiently approximated. Unfortunately, this construction drives down the value of p_{th}, which is currently estimated, in the most optimistic scenario, to lie around 10^{-3} or 10^{-4}.

The achievability of robust, fault-tolerant behavior in quantum information systems is a key goal for both experimental and theoretical work in this field. Natural mechanisms have been found for classical systems to obtain robust behavior; for quantum systems the challenge is to experimentally determine p_{th} for a given system, and engineer to reach below this. Alternatively, clever new strategies may be found to raise p_{th}. The metrology, comminucation, and communication ideas of quantum information all depend on this, and clearly, the potential new quantum resources are well worth reaching for the threshold.

References

[1] M. A. Nielsen and I. L. Chuang. *Quantum computation and quantum information.* Cambridge University Press, Cambridge, 2000.

[2] T. M. Cover and J. A. Thomas. *Elements of Information Theory.* John Wiley and Sons, New York, 1991.

Course 2

MESOSCOPIC STATE SUPERPOSITIONS AND DECOHERENCE IN QUANTUM OPTICS

S. Haroche

Laboratoire Kastler Brossel, Département de Physique
Ecole Normale Supérieure, 24 rue Lhomond F-75005 Paris
and Collège de France, 11 place M. Berthelot, F-75005 Paris

D. Estève, J.-M. Raimond and J. Dalibard, eds.
Les Houches, Session LXXIX, 2003
Quantum Entanglement and Information Processing
Intrication quantique et traitement de l'information
© 2004 Elsevier B.V. All rights reserved

Contents

The study of mesoscopic superpositions of states, the so–called Schrödinger cats, has become a very active field of theoretical and experimental physics. Mesoscopic superpositions play an important role in quantum information science, where systems of qubits made of atoms or photons in entangled configurations of increasing complexity are manipulated [1, 2]. These superpositions are notoriously fragile, loosing rapidly their quantum coherence. The coupling of mesoscopic systems to their environment very efficiently blurs the interference effects between their states. This phenomenon, called decoherence, has been extensively studied in various contexts [3–7]. Monitoring and controlling decoherence, as well as correcting for its effects, are among the essential goals of quantum information physics [1].

At a more fundamental level, preparing mesoscopic state superpositions and studying their decoherence is a way to test the measurement process in quantum physics [8]. In an usual quantum measurement, the microsystem is coupled to a macroscopic meter. In the absence of decoherence, this coupling would result in an entanglement between these two systems and in the generation of Schrödinger-cat-like states of the meter. These superpositions are never observed, though, because the system very quickly evolves under the effect of environment induced decoherence into a statistical mixture of states. The coupling to the environment has the effect of leaving unaltered some states, the "pointer states" [3, 9] of the meter which are correlated to the eigenstates of the measured microscopic system, while the coherences between these states is rapidly destroyed. Eluciding the mechanisms by which the robust pointer states are selected out of the huge ensemble of fragile mesoscopic superpositions is an essential aspect of decoherence theory [7, 10].

The study of mesoscopic states and their decoherence has brought closer theorists and experimentalists from disparate domains of physics. Atomic and solid state physicists, in particular, have developed a common interest in quantum information science. The Les Houches Summer School in which this course was given has been a manifestation of this common interest, with experts and students coming from atomic and condensed matter physics discussing together and benefitting from each other expertise.

These Lecture Notes provide an introduction to mesoscopic state studies in quantum optics, a field of physics which has become a testing ground for the manipulation of systems of increasing complexity. On the theoretical side, the evolution of the atomic and photonic systems of quantum optics is often entirely calculable, making the physics understandable in terms of very simple models. On the experimental side, the sophistication of modern laser technology allows for a very delicate control of the systems under study, giving in particular the possibility to manipulate their particles one at a time, with unsurpassed precision. Schrödinger cat states of several kinds have been proposed in quantum optics [11–15], and some versions of these states realized and studied in the laboratory [16–19]. These studies have illustrated various aspects of decoherence theory and opened a 'bottom up' approach to the mesoscopic frontier, complementary to the 'top down' approach of condensed matter physics.

The goal of this Course was threefold. I intended first to present the main ideas and methods of quantum optics to an audience which was coming, in part, from a different physics community. I wanted also to show, on simple examples, how mesoscopic states could be prepared and studied using the conceptually simple methods of quantum optics. Finally, I wished to analyze simple models of decoherence and describe textbook experiments which illustrate clearly the fundamental connexion between environment induced decoherence, entanglement and complementarity. These Notes start by an overview of quantum optics (section 1), followed by a description of beam splitters and particle interference effects (section 2). Section 3 is devoted to the description of Schrödinger cats in Cavity Quantum Electrodynamics experiments. In section 4 proposals to generate and study mesoscopic state superpositions of atoms in Bose Einstein condensates are described. Finally, these Notes conclude with a brief review of other kinds of quantum optics and atomic physics experiments dealing with mesoscopic state superpositions.

1. An overview of quantum optics

This first section presents an overview of the quantum optics formalism [20–25]. We start by a reminder about the analogy between the electromagnetic field and a collection of quantized harmonic oscillators characterized by their annihilation and creation operators. We introduce the main observables of the field (energy, photon number, field quadratures). We describe some important states of the field associated to a single mode, giving a special attention to the quasi-classical coherent states and to their superpositions (Schrödinger cat states). We define then two convenient field representations in phase space, the Q and W functions, and summarize their main properties. We finally recall how to describe the coupling of the field with atoms and we show how an atomic source can generate single photons, while a classical current creates coherent states. We end the section by describing a simple model of photon counter.

1.1. Field as a sum of harmonic oscillators

We start by recalling briefly the general formalism of field quantization in Coulomb gauge, which is described in more details in [26]. The vector potential $\mathbf{A}(\mathbf{r}, t)$ of a classical electromagnetic field, or equivalently its electric field $\mathbf{E}(\mathbf{r}, t)$ or magnetic field $\mathbf{B}(\mathbf{r}, t)$, can be expanded along plane waves of polarization $\boldsymbol{\epsilon}_j$ and wave vector $\mathbf{k}_j$ according to :

$$\mathbf{A}(\mathbf{r}, t) = \sum_j A_j \left[a_j \boldsymbol{\epsilon}_j e^{i\mathbf{k}_j.\mathbf{r}} + a_j^* \boldsymbol{\epsilon}_j^* e^{-i\mathbf{k}_j.\mathbf{r}} \right] , \quad (1.1)$$

$$\mathbf{E}(\mathbf{r}, t) = -\frac{d\mathbf{A}(\mathbf{r}, t)}{dt} = i \sum_j E_j \left[a_j \boldsymbol{\epsilon}_j e^{i\mathbf{k}_j.\mathbf{r}} - a_j^* \boldsymbol{\epsilon}_j^* e^{-i\mathbf{k}_j.\mathbf{r}} \right]$$

$$(E_j = \omega_j A_j) \quad (1.2)$$

$$\mathbf{B}(\mathbf{r}, t) = \nabla \times \mathbf{A}(\mathbf{r}, t)$$

$$= i \sum_j \left(\frac{E_j}{c} \right) \left[a_j \boldsymbol{\kappa}_j \times \boldsymbol{\epsilon}_j e^{i\mathbf{k}_j.\mathbf{r}} - a_j^* \boldsymbol{\kappa}_j \times \boldsymbol{\epsilon}_j^* e^{-i\mathbf{k}_j.\mathbf{r}} \right] . \quad (1.3)$$

In these expressions, the a_j coefficients are C-number field coordinates and $\boldsymbol{\kappa}_j = \mathbf{k}_j / |\mathbf{k}_j|$ is the unit vector defining the wave vector direction. There are two modes corresponding to two orthogonal polarizations $\boldsymbol{\epsilon}_j$ normal to each wave vector $\mathbf{k}_j$. The time evolution of the field coordinates simply writes:

$$a_j(t) = a_j(0) e^{-i\omega_j t} , \quad (1.4)$$

where $\omega_j = c|\mathbf{k}_j|$ is the angular frequency of mode j. The A_j's and E_j's are adjustable normalization constants. The continuous sum over the field modes is made discrete by introducing a cubic 'quantization box' of linear dimension L, whose infinite limit is taken at the end of calculations. The wave vector coordinates $k_{j_x}, k_{j_y}, k_{j_z}$ are integer multiples of $2\pi/L$. The sum over j in Eqs. (1.1-1.3) is thus equivalent to a continuous three dimension integral.

The field hamiltonian H is obtained by evaluating the sum over space of the energy density:

$$H = \frac{\varepsilon_0}{2} \int (\mathbf{E}^2 + c^2 \mathbf{B}^2) d^3\mathbf{r} = \varepsilon_0 L^3 \sum_j E_j^2 \left(a_j a_j^* + a_j^* a_j \right) . \quad (1.5)$$

This equation can be written as:

$$H = \sum_j \frac{\hbar \omega_j}{2} \left(a_j a_j^* + a_j^* a_j \right) , \quad (1.6)$$

provided the normalization constant is taken as $E_j = \sqrt{\hbar\omega_j/2\epsilon_0 L^3}$, where $\hbar$ is Planck's constant and ϵ_0 the vacuum permittivity. As shown below, this normalization choice makes obvious the analogy between each field mode and a mechanical harmonic oscillator. This expression of H can be finally rewritten as:

$$H = \sum_j \hbar\omega_j \left(X_j^2 + P_j^2\right) , \tag{1.7}$$

where we have introduced the real and imaginary parts of the field coordinates as:

$$a_j = \left(\sqrt{m\omega_j/2\hbar}\right) x_j + (i/\sqrt{2m\hbar\omega_j}) p_j = X_j + iP_j . \tag{1.8}$$

We recognize here the hamiltonian of a sum of independent harmonic oscillators whose normalized coordinates in phase space are X_j and P_j.

The standard quantization procedure is then straightforward [26]. As for a mechanical oscillator, the classical field coordinates are replaced by non-commuting operators (marked for the moment with 'hats'), obeying the relations:

$$\left[\hat{x}_j, \hat{p}_k\right] = i\hbar\delta_{jk} \tag{1.9}$$

$$\left[\hat{X}_j, \hat{P}_k\right] = \frac{i}{2}\delta_{jk}; \quad \left[\hat{a}_j, \hat{a}_k^\dagger\right] = \delta_{jk} \tag{1.10}$$

These 'hats' will be omitted in the following when there will be no confusion. In the Heisenberg picture, the field coordinates become time-dependent operators $\hat{a}(t)$, one for each plane wave mode of the field:

$$\hat{a}\,(t) = \hat{a}\,(0)\, e^{-i\omega t} \tag{1.11}$$

The electric field in each mode $(\mathbf{k}, \boldsymbol{\epsilon})$ becomes likewise a time dependent operator, linear superpositions of the field operator a and its hermitian adjunct $a^\dagger$ (we omit the indices in the expression of the field operators, when there is no possible confusion):

$$\mathbf{E}_{\mathbf{k},\boldsymbol{\epsilon}}\,(\mathbf{r}, t) = i\sqrt{\frac{\hbar\omega}{2\varepsilon_0 L^3}} \left(\boldsymbol{\epsilon} a(t) e^{i\mathbf{k}.\mathbf{r}} - \boldsymbol{\epsilon}^* a^\dagger(t) e^{-i\mathbf{k}.\mathbf{r}}\right) , \tag{1.12}$$

while the free field Hamiltonian in each mode takes the familiar form:

$$H = \hbar\omega \left(a^\dagger a + \frac{1}{2}\right) . \tag{1.13}$$

In the Schrödinger picture these operators are given by similar expressions in which a and $a^\dagger$ are time independent. The field's expression [Eq.(1.12)] contains

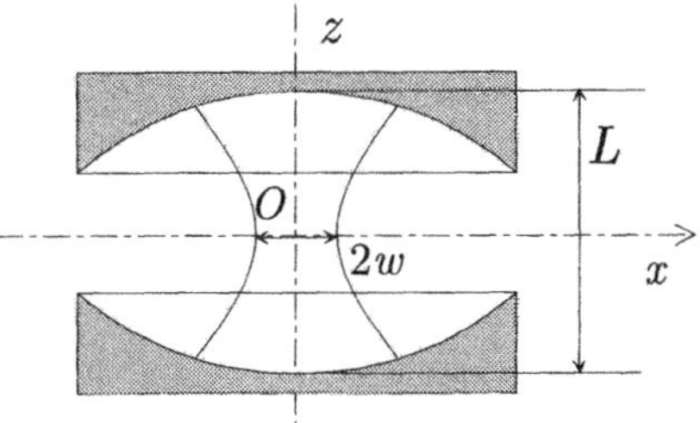

Fig. 1. Scheme of a microwave Fabry-Perot cavity. The two spherical mirrors separated by L sustain a Gaussian mode with waist w. The cavity axis normal to the mirrors is Oz. The sinusoidal standing wave pattern (with field nodes and antinodes along Oz) is not represented. Atoms propagate along Ox axis.

explicitely the expression of the vacuum field fluctuations in the quantization box of size L^3: $E_0 = \sqrt{\hbar\omega/2\varepsilon_0 L^3}$. This quantity also represents the field per photon in the L^3 volume. This field goes to 0 when L is taken to infinity, in usual situations where the quantization box is merely a computation convenience.

Special mention must be made of Cavity Quantum Electrodynamics (CQED) [27–30] where the quantization volume V remains actually finite, with a well defined meaning: it is the physical volume of the real cavity in which the field is stored. Instead of expanding the field in plane waves, it is then more convenient to develop it along the standing wave modes of the cavity. In one of these modes, the field operator writes (Schrödinger picture):

$$\mathbf{E}(\mathbf{r}) = iE_0\left(\boldsymbol{\epsilon} a f(\mathbf{r}) - \boldsymbol{\epsilon}^* a^\dagger f^*(\mathbf{r})\right) \tag{1.14}$$

where $f(\mathbf{r})$ is a scalar function which describes the spatial variation of the field in the mode (we consider here for simplicity that the mode polarization $\boldsymbol{\epsilon}$ is constant). The effective mode volume is then:

$$V = \int |f(\mathbf{r})|^2 \, d^3\mathbf{r}\,, \tag{1.15}$$

and the field per photon:

$$E_0 = \sqrt{\frac{\hbar\omega}{2\varepsilon_0 V}} \tag{1.16}$$

As a specific example, consider the case of a Fabry-Perot cavity made of two spherical mirrors facing each other (Figure 1). This is the geometry of the microwave CQED experiments described later in this chapter [30]. The mode is then a standing wave with a Gaussian transverse profile and a sinusoidal field variation in the longitudinal direction normal to the mirrors, separated by the distance L. The waist w characterizes the width of the Gaussian. The mode volume is $V = \pi L w^2/4$. For the specific parameters of the experiment ($L = 2.7$ cm; $w = 6$ mm) we have $V = 0.6\ \text{cm}^3$.

The field quadrature operators in each mode correspond to the usual oscillator position and momentum coordinates. In Schrödinger picture they are expressed as:

$$X = \frac{a + a^\dagger}{2} \quad ; \quad P = \frac{a - a^\dagger}{2i} = \frac{e^{-i\pi/2}a + e^{i\pi/2}a^\dagger}{2} . \tag{1.17}$$

More generally, we define phase quadratures as linear combinations of a and $a^\dagger$:

$$X_\varphi = \frac{e^{-i\varphi}a + e^{i\varphi}a^\dagger}{2} \; ; X_{\varphi+\pi/2} = \frac{e^{-i\varphi}a - e^{i\varphi}a^\dagger}{2i} . \tag{1.18}$$

They satisfy the commutation rules:

$$\left[X_\varphi, X_{\varphi+\pi/2}\right] = \frac{i}{2} , \tag{1.19}$$

which correspond to the uncertainty relations:

$$\Delta X_\varphi \Delta X_{\varphi+\pi/2} \geq \frac{1}{4} , \tag{1.20}$$

where ΔX_φ and $\Delta X_{\varphi+\pi/2}$ are conjugate phase quadrature fluctuations.

We recall below how these quadrature operators are measured in quantum optics. The eigenstate of the quadrature operator X_φ corresponding to the real and continuous eigenvalue x is defined by the eigenvalue equation:

$$X_\varphi \left|x\right\rangle_\varphi = x \left|x\right\rangle_\varphi . \tag{1.21}$$

The non-normalizable $|x\rangle_\varphi$ states obey the orthogonality and closure relationships:

$$_\varphi\!\left\langle x \mid x'\right\rangle_\varphi = \delta(x - x'); \quad \int |x\rangle_{\varphi\,\varphi} \langle x|\, dx = \mathbb{1} , \tag{1.22}$$

and the transformation from the $|x\rangle_\varphi$ basis to the conjugate basis $|x\rangle_{\varphi+\pi/2}$ is a simple Fourier transform:

$$|x\rangle_{\varphi+\pi/2} = \int dy\, |y\rangle_{\varphi\varphi} \langle y \mid x\rangle_{\varphi+\pi/2} = \frac{1}{\sqrt{\pi}} \int dy\, e^{2i\,x\,y}\, |y\rangle_\varphi \tag{1.23}$$

Note the unusual factor 2 in the exponent of the Fourier transformation, due to the normalization we have chosen for the conjugate variables [factors $1/2$ in Eqs.(1.17)and (1.18)].

The phase quadrature eigenstates are the natural generalization for fields of the usual position and momentum basis states of a mechanical oscillator. A more fundamental state basis widely used in quantum optics is formed by the energy eigenstates. The operator $a^\dagger a$ appearing in Eq (1.13) has a non-degenerate spectrum made of all non-negative integers. The corresponding eigenstates are the Fock states, or photon number states noted $|n\rangle$. The ground state $|0\rangle$ of the field in each mode, called the vacuum state, is the eigenstate of $a^\dagger a$ with eigenvalue 0. It obeys the equation:

$$a|0\rangle = 0 \,. \tag{1.24}$$

The expectation value $\langle 0|\, X_\varphi \,|0\rangle$ of any phase quadrature in vacuum is zero while the quadrature fluctuations are isotropic and correspond to the minimal value compatible with Heisenberg uncertainty relations:

$$\Delta X_\varphi^{(0)} = \sqrt{\langle 0|\, X_\varphi^2 \,|0\rangle} = 1/2 \,. \tag{1.25}$$

The probability distribution $P^{(0)}(x)$ of the field quadrature in vacuum is a Gaussian, like the distribution of the positions of a ground state mechanical oscillator:

$$P^{(0)}(x) = \left|_\varphi \langle x \mid 0\rangle\right|^2 = \left(\frac{2}{\pi}\right)^{1/2} e^{-2x^2} \,. \tag{1.26}$$

In summary, the vacuum field in each mode has isotropic Gaussian fluctuations around zero field.

The n-photon Fock state in a field mode obeys the eigenvalue equation:

$$a^\dagger a|n\rangle = n|n\rangle \,. \tag{1.27}$$

The action of the field operators a and $a^\dagger$ on this state are:

$$a\,|n\rangle = \sqrt{n}\,|n-1\rangle \quad ; \quad a^\dagger\,|n\rangle = \sqrt{n+1}\,|n+1\rangle \tag{1.28}$$

hence the name of photon annihilation and creation operators given to a and $a^\dagger$.

The n-photon state is generated by repeated operation of the photon creation operator on the vacuum, according to:

$$|n\rangle \;=\; \frac{a^{\dagger n}}{\sqrt{n!}}\,|0\rangle \,. \tag{1.29}$$

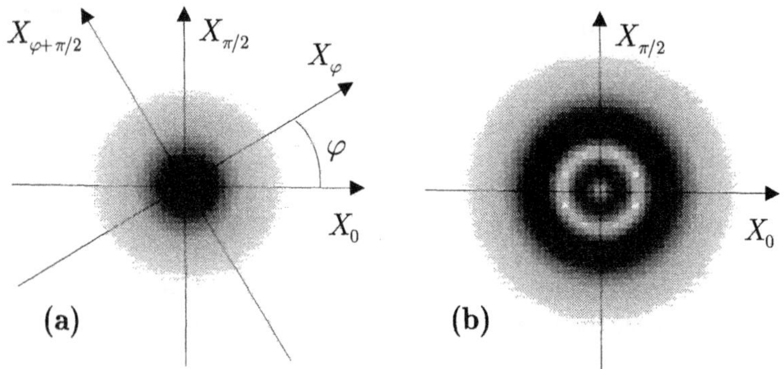

Fig. 2. The phase space of a field mode. (a) Graphical illustration of the X_0, $X_{\pi/2}$, X_φ and $X_{\varphi+\pi/2}$ quadratures together with a representation of the vacuum state wavefunction. (b) Graphical representation of a 3-photon Fock state.

Fock states have isotropic non-minimal fluctuations of their field quadratures:

$$\langle n| X_\varphi |n\rangle = 0; \quad \Delta X_\varphi^{(n)} = \sqrt{\langle n| X_\varphi^2 |n\rangle} = \frac{1}{2}\sqrt{2n+1}\,. \tag{1.30}$$

The probability distribution $P^{(n)}(x)$ of a quadrature in state $|n\rangle$ can be deduced from the analogous expression giving the square of the wave function of n quanta in a harmonic oscillator:

$$P^{(n)}(x) = \left|_\varphi \langle x \mid n\rangle\right|^2 = \left(\frac{2}{\pi}\right)^{1/2} \frac{1}{2^n n!} e^{-2x^2} \left(H_n(x\sqrt{2})\right)^2 \tag{1.31}$$

where $H_n(x)$ is the Hermite polynomial of rank n.

A qualitative picture of Fock states is shown in Fig. 2. The phase space of a single mode is represented with the two X_0 and $X_{\pi/2}$ quadratures as reference axes. The quadrature X_φ is associated with a direction making the angle φ with the X_0 direction. An ensemble of X_φ measurements in the vacuum field yields a Gaussian distributions of points along this direction. Measurements along all possible quadrature directions gives an isotropic Gaussian cloud centered at phase space origin, as shown in Fig. 2(a). For a Fock state with $n > 0$ the quadrature distribution along any direction exhibits maxima separated by dark fringes corresponding to the zero of the Hermite polynomials. The $n = 3$ state is represented in Fig. 2(b), as an isotropic cloud of points distributed over concentric rings centered at phase space origin.

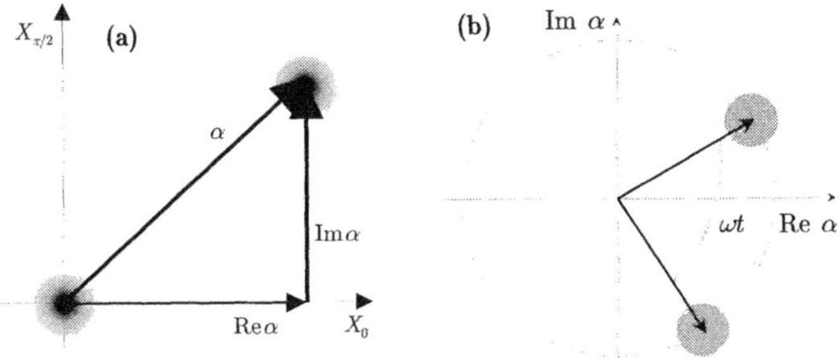

Fig. 3. Coherent state. (a) Pictorial representation of the action of the displacement operator on the vacuum state. The displacement by a complex amplitude α amounts to a displacement by $\mathrm{Re}\,\alpha$ along the X_0 quadrature axis, followed by a displacement by $\mathrm{Im}\,\alpha$ along the $X_{\pi/2}$ quadrature. (b) Time evolution of a coherent state.

1.2. Coherent states

Fock states contain no phase information. Quadrature field eigenstates have a well defined phase, but they are non-physical since they are non-normalizable and have an infinite mean energy. Their amplitude along one phase direction is perfectly determined, while it is totally random in the conjugate orthogonal direction (infinite squeezing). To describe situations where the phase of the field is relevant, it is thus often more convenient to expand the field neither on Fock nor on quadrature field eigenstates, but on the basis of so-called coherent states [31, 32], which are more physical and experimentally much more easily accessible. We recall briefly in this section the definition and the main properties of these states.

A coherent state of a single field mode is defined as resulting from the translation of the vacuum field in phase space. This translation is represented, in its most general form by the unitary operator:

$$D(\alpha) = e^{\alpha a^{\dagger} - \alpha^{*} a} \tag{1.32}$$

where α is an arbitrary C-number whose real and imaginary parts are the projections along the X_0 and $X_{\pi/2}$ directions respectively of the two–dimensional translation vector. Applying this translation amounts to displacing the vacuum Gaussian 'packet' in a given direction, without changing its shape. The translated vacuum state is the coherent state simply noted as $|\alpha\rangle$:

$$|\alpha\rangle = D(\alpha)|0\rangle \ . \tag{1.33}$$

The translated ‘packet’, whose evolution is determined in the Schrödinger picture by the free field Hamiltonian [Eq. (1.13)], subsequently rotates at frequency ω in phase space, without deformation, according to the equation:

$$|\Psi(t)\rangle = e^{-i\omega t/2}|\alpha e^{-i\omega t}\rangle \,. \tag{1.34}$$

This corresponds to the best possible approximation of a classical free oscillator motion, with minimal wave packets uncertainties for any two conjugate quadratures. Fig. 3(a,b) shows how the vacuum state is transformed by translation into a coherent state of well defined amplitude and phase and how this state freely evolves in phase space. For simplicity, the Gaussian clouds of points have been replaced in Fig. 3(b) by uncertainty circles of radius unity.

In order to make the translation operation more explicit, it is convenient to split the exponential in Eq.(1.34) in two, separating the contributions of the real and imaginary parts of α. For this, we make use of the Glauber relation [24] $e^{A+B} = e^A e^B e^{-[A,B]/2}$ (valid if $[A, B]$ commutes with A and B) and we get:

$$D(\alpha) = e^{-i\alpha_1\alpha_2} \exp(2i\alpha_2 X_0) \quad \exp(-2i\alpha_1 X_{\pi/2}) \tag{1.35}$$

Using a mechanical oscillator analogy, the displacement $D(\alpha)$ can thus be viewed as a translation along space by an amount $\alpha_1 = Re(\alpha)$, followed by a ‘momentum kick’ of magnitude $\alpha_2 = \mathrm{Im}\,(\alpha)$. This kick can be produced by the action of a delta-like impulsive force on the oscillator. The operation is completed by a global phase shift of the system’s state. Note that the order of these two translations along X_0 and $X_{\pi/2}$ can be exchanged, provided the overall phase shift is replaced by its opposite (the two translations along the conjuguate directions in phase space do not commute).

Let us compute the probability amplitude for finding the value x when measuring the quadrature operator X_0 on a field in state $|\alpha\rangle$. A straightforward calculation using Eqs.(1.33) and (1.35) yields:

$$\langle x \mid \alpha\rangle = \left(\frac{2}{\pi}\right)^{1/4} \exp(-i\alpha_1\alpha_2)\exp(2i\alpha_2 x)\exp[-(x-\alpha_1)^2]\,. \tag{1.36}$$

We recognize in this expression a translation by the amount α_1 of the ground state wave packet, accompanied by a phase modulation at frequency α_2 which describes the momentum kick of the state. The probability for finding the value x for the quadrature is thus:

$$P(x) = \left(\frac{2}{\pi}\right)^{1/2} \exp\left[-2\,(x-\alpha_1)^2\right] \tag{1.37}$$

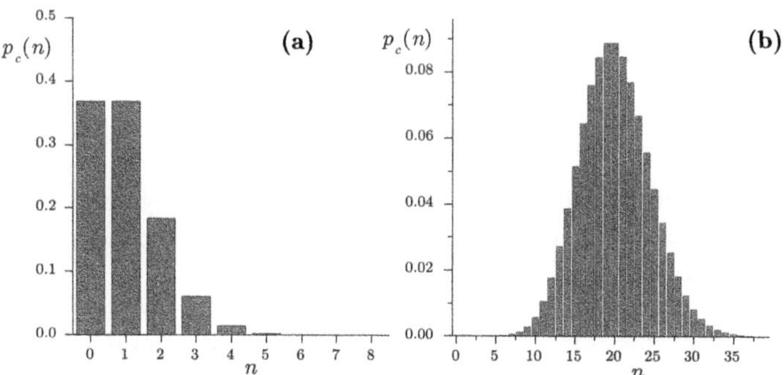

Fig. 4. Photon number statistical distributions. (a) Coherent field with $\overline{n} = 1$ photons on the average. (b) Coherent field with $\overline{n} = 20$.

If the coherent state is left to evolve freely, this probability becomes at time t:

$$P(x,t) = \left(\frac{2}{\pi}\right)^{1/2} \exp\left[-2\left(x - \alpha_1 \cos\omega t - \alpha_2 \sin\omega t\right)^2\right] . \tag{1.38}$$

We will see below how the translation in phase space can be physically implemented when the oscillator is not a material particle, but a field in a cavity.

An alternative and useful expression of the displacement operator is obtained by using again the Glauber relation, separating this time the a and $a^\dagger$ terms:

$$D(\alpha) = \exp\left[\alpha a^\dagger - \alpha^* a\right] = \exp\left(-\frac{|\alpha|^2}{2}\right) \exp(\alpha a^\dagger) \exp(-\alpha^* a) . \tag{1.39}$$

This form corresponds to the 'normal ordering' in quantum optics. If we expand the exponential of operators along increasing powers, all the a^n terms are on the right and the $a^{\dagger n}$ terms on the left. The normal ordering of $D(\alpha)$ makes it straightforward to expand a coherent state along the Fock basis. The action of the $\exp(-\alpha^* a)$ operator placed on the right leaves the vacuum unchanged, since only the zero order term in the expansion along powers of a yields a non zero result. Combining Eqs.(1.33)and (1.39), one finds easily:

$$|\alpha\rangle = \sum_n C_n(\alpha)\,|n\rangle \quad \text{with} \quad C_n(\alpha) = \exp\left(-\frac{|\alpha|^2}{2}\right) \frac{\alpha^n}{\sqrt{n!}} . \tag{1.40}$$

The distribution of photon numbers in a coherent state obeys a Poisson statistics. Fig. 4 shows this distribution for $\alpha = 1$ and $\alpha = \sqrt{20}$. The average photon number $\overline{n}$ and photon number variances Δn are:

$$\overline{n} = |\alpha|^2 \quad ; \quad \frac{\Delta n}{\overline{n}} = \frac{1}{|\alpha|} = \frac{1}{\sqrt{\overline{n}}} . \tag{1.41}$$

The relative fluctuation of the photon number is thus inversely proportional to the square root of the average photon number. For large fields, this fluctuation becomes negligible (classical limit).

It is important to notice that coherent states are eigenstates of the photon annihilation operator. This essential property is easily derived from Eqs.(1.28) and (1.40):

$$a\,|\alpha\rangle = \alpha\,|\alpha\rangle \quad \text{and} \quad \langle\alpha|\,a^\dagger = \langle\alpha|\,\alpha^* \tag{1.42}$$

It is also useful to recall the expression of the scalar product of two coherent states:

$$\langle\alpha \mid \beta\rangle = e^{-|\alpha|^2/2\,-|\beta|^2/2\,+\alpha^*\beta} \quad ; \quad |\langle\alpha \mid \beta\rangle|^2 = e^{-|\alpha-\beta|^2} , \tag{1.43}$$

which shows that the overlap of two such states decreases exponentially with their 'distance' in phase space. Although they are never strictly orthogonal, they become practically so when the distance of their centers is much larger than 1, the radius of the coherent state uncertainty circle. The coherent states of a field mode constitute a complete set of states in the Hilbert space of this mode. This is expressed by a closure relationship, easily derived by using the Fock state expansion of coherent states and the completeness of these Fock states:

$$\frac{1}{\pi}\int d\alpha_1 d\alpha_2\,|\alpha\rangle\,\langle\alpha| = \mathbb{1} . \tag{1.44}$$

Note however that the coherent state basis, being made of non-orthogonal states, is over-complete. This means in particular that, although any field state can be expanded on it, this expansion is not unique.

1.3. Schrödinger cats as superpositions of coherent states

Among all the possible field states, we will give a special attention to the superpositions of two quasi-orthogonal coherent states, represented in space phase by two non-overlapping circles. These states are prototypes of Schrödinger cats [14, 15]. We will see later how they can be prepared and used to study the phenomenon of decoherence. We give here only some of their remarkable properties. As a simple example of such a cat state, let us consider a linear superposition with equal

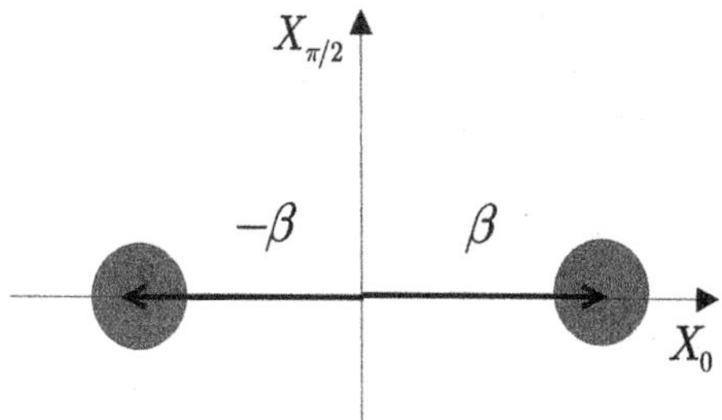

Fig. 5. Pictorial representation of a π-phase cat in phase space.

weights of two coherent states with opposite phases, represented in phase space by two circles whose centers are on the X_0 axis, symmetrical with respect to the origin (see Fig 5). This superposition, called a π-phase cat in the following, writes:

$$\left|\Psi_{cat}^{even}\right\rangle = \frac{|\beta\rangle + |-\beta\rangle}{\sqrt{2\left(1+e^{-2|\beta|^2}\right)}} \approx (1/\sqrt{2})\left(|\beta\rangle + |-\beta\rangle\right) , \tag{1.45}$$

where β is the amplitude of the field (real in this case). The superscript 'even' in the state expression will be explained below. The denominator in the first right hand side term insures the normalization of the state, taking into account the overlap of the two $|\beta\rangle$ and $|-\beta\rangle$ coherent states. If $|\beta| \gg 1$, this overlap can be neglected and the cat state is expressed by the simpler form given by the second right-hand-side term in Eq.(1.45). The coherence between the two states is an essential feature which distinguishes it from a mere statistical mixture. This is made clear by expressing the field density operator:

$$\rho_{cat} \approx \frac{1}{2}\left(|\beta\rangle\langle\beta| + |-\beta\rangle\langle-\beta| + |\beta\rangle\langle-\beta| + |-\beta\rangle\langle\beta|\right) . \tag{1.46}$$

The 'cat state' coherence is described by the off-diagonal part of this density operator (last two terms in the right hand side of Eq.(1.46). This coherence is displayed by analyzing the field quadrature distribution in this state, with a proper choice of phase. Suppose first that we measure the quadrature X_0 along the direction of the cat alignment. It is straightforward that the probability distribution for this quadrature is merely the sum of two Gaussians, centered at $\pm\beta$:

$$P_0^{(cat)}(x) \approx \frac{1}{\sqrt{2\pi}}\left(e^{-2(x-\beta)^2} + e^{-2(x+\beta)^2}\right) . \tag{1.47}$$

The probability amplitudes for measuring a value x in state $|+\beta\rangle$ and $|-\beta\rangle$ do not appreciably overlap and thus cannot interfere: the resulting distribution is simply the sum of the distributions corresponding to the two state components and the state coherence is not apparent. If we measure instead the quadrature $X_{\pi/2}$ along the direction orthogonal to the cat alignment, the probability distribution writes:

$$P_{\pi/2}^{(cat)}(x) \approx \frac{1}{2}\left|{}_{\pi/2}\langle x \mid \beta\rangle + {}_{\pi/2}\langle x \mid -\beta\rangle\right|^2 , \tag{1.48}$$

where the index $\pi/2$ in the ${}_{\pi/2}\langle x|$ bra indicates an eigenstate of $X_{\pi/2}$, related to those of X_0 by a Fourier transform [see Eq.(1.23)]. The probability $P_{\pi/2}(x)$ is the square of the sum of two amplitudes which are both non zero. These amplitudes are easy to compute. The scalar product of $|x\rangle_{\pi/2}$ with $|\beta\rangle$ is equal to the product of $|x\rangle_0$ with $|-i\beta\rangle$, as a mere rotation in phase space indicates. Using Eq.(1.36), we immediately get:

$$\begin{aligned} {}_{\pi/2}\langle x \mid \beta\rangle &= {}_0\left\langle x \,\middle|\, \beta e^{-i\pi/2}\right\rangle = {}_0\langle x \mid -i\beta\rangle \\ &= \left(\frac{2}{\pi}\right)^{1/4} \exp(-2i\beta x)\exp(-x^2) \end{aligned} \tag{1.49}$$

and:

$$P_{\pi/2}^{(cat)} \approx \left(\frac{2}{\pi}\right)^{1/2} e^{-2x^2}(1+\cos 4\beta x)\,. \tag{1.50}$$

The probability distribution is a Gaussian centered at $x = 0$, modulated by an interference term with fringes having a period $1/4\beta$ inversely proportional to the 'cat size' β. This interference term is a conspicuous signature of the coherence of the state superposition. The distribution of any phase quadrature X_φ can be obtained in the same way. The interference term exists only when φ is close to $\pi/2$. A graphical representation is very convenient to understand why it is so (Fig. 6). For a coherent state, a field quadrature takes non zero values in an interval corresponding to the projection of the state uncertainty circle on the direction of the quadrature. For a Schrödinger cat state, there are two such intervals, corresponding to the two state components. If $\beta \gg 1$ and $\varphi = 0$ [Fig. 6(a)], the two intervals are non-overlapping and there is no interference. For a φ value between 0 and $\pi/2$ [Fig. 6(b)], the two intervals are closer than for $\varphi = 0$, resulting in two still non-overlapping Gaussians without interference. It is only when φ gets very close to $\pi/2$ that the two projected intervals overlap along the direction of the quadrature, leading to a large interference term [Fig. 6(c)].

Instead of rotating the quadrature for a given 'frozen' cat state, let us choose a given quadrature (e.g. X_0) and consider the evolution of its probability distribution for a freely evolving cat state. This evolution amounts to a rotation in

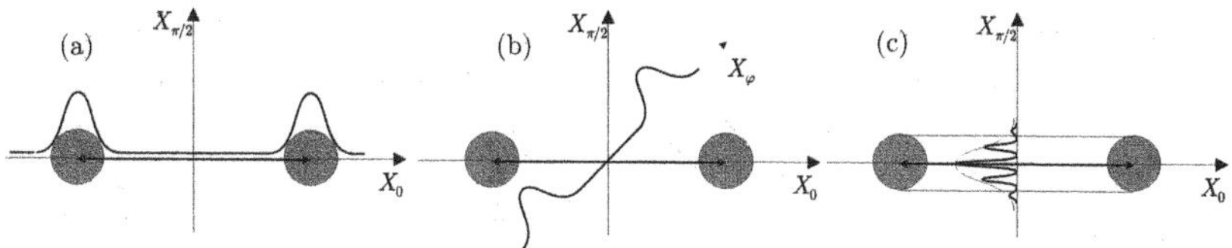

Fig. 6. Quadratures of a π phase Schrödinger cat. (a) The X_0 quadrature exhibits well separated Gaussian peaks corresponding to the cat's components. (b) For $X_{\pi/4}$,, the Gaussian peaks distance is reduced. (c) For $X_{\pi/2}$ the two peaks merge and fringes show up.

the same direction of the two coherent state components which stay opposite to each other in phase space. Their projections along the (now fixed) direction of the field quadrature oscillate with opposite phases, corresponding to two wave packets periodically colliding at $x = 0$. When the two packets merge, fringes with 100 % contrast appear under the Gaussian envelope. This periodic fringe pattern is a signature of the cat state coherence. It would not exist if the field were described by an incoherent superposition of coherent states [first two terms in Eq.(1.46)]. The fringes are getting narrower when the amplitude of the field is increased. We will see later that these fringes are very fragile and efficiently washed out by decoherence.

Another aspect of the coherence of cat states made with fields of opposite phases is revealed by considering their photon number distribution. The cat state given by Eq. (1.45) develops only along even number states, since the probability for finding n photons in it is proportional to $1 + (-1)^n$. For this reason, we have added the superscript 'even' in its name. Similarly the Schrödinger cat state with opposite amplitudes: $|\psi^{odd}\rangle = (1/\sqrt{2})[|\beta\rangle - |-\beta\rangle]$ develops only along the odd photon number states since the probability for finding n photons in it is proportional to $1 - (-1)^n$. We will call it an "odd phase cat". The periodicity of the photon number is related to the coherence of the state, since a statistical mixture of $|\beta\rangle$ and $|-\beta\rangle$ contains obviously all photon numbers. We can thus say that the photon number distribution, with its "dark fringes" corresponding to zero probability for odd or even photon numbers, is a signature of the even and odd cats coherence, as is the existence of dark fringes in their $X_{\pi/2}$ quadrature.

It is finally convenient for this discussion to introduce the photon number parity operator $\mathcal{P}$ [24] which admits as eigenstates all the superpositions of even photon numbers with the eigenvalue $+1$ and all the superpositions of odd photon number states with the eigenvalue -1. An obvious expression for this operator is:

$$\mathcal{P} = e^{i\pi a^\dagger a} \, . \tag{1.51}$$

The odd and even phase cats $|\beta\rangle \pm |-\beta\rangle$ are eigenstates of $\mathcal{P}$ with the $+1$ and -1 eigenvalues. Note also, from Eq.(1.42), that the action of the annihilation operator on an even (odd) phase cat results in the switching of the cat parity according to:

$$a[|\beta\rangle \pm |-\beta\rangle] = \beta[|\beta\rangle \mp |-\beta\rangle] \,. \tag{1.52}$$

Let us note finally the action of the parity operator on a quadrature eigenstate:

$$\mathcal{P}|x\rangle_\varphi = |-x\rangle_\varphi \,, \tag{1.53}$$

This is a direct consequence of the parity of the Hermite polynomials, proportional to the scalar product $\langle n|x\rangle$, which develop on even (odd) powers of x if n is even (odd). Eq.(1.53) can be taken as an alternative definition of $\mathcal{P}$, as the operator which reflects any quadrature eigenstate into the state with an opposite eigenvalue.

1.4. Field representations: the Q and W functions

We have extensively used pictorial phase-space representations in the description of various field states. They have remained however up to this point rather qualitative, fuzzy distributions of points or uncertainty circles. In this section, we show that the formalism of quantum optics can make these phase space pictures fully quantitative by associating to each state of the field two functions taking real values in phase space. These functions, called Q and W, are usually represented by three-dimensional plots in which the horizontal plane is the phase space and the vertical axis represents the value of the function. They give a vivid description of any single mode field state, whether it is pure or a statistical mixture. They can be deduced form one another and the knowledge of one of them is sufficient to reconstruct the field density operator. In other words, the Q or the W function contains all the quantum information there is about the field state. A full discussion of the phase space representation of fields in quantum optics can be found in [22]. We recall here only the definition of these functions and some of their important properties.

1.4.1. The Q function

Given a field state described by its density operator ρ, its Q function at the point in phase space associated to the C-number α is defined as:

$$Q(\alpha) = \frac{1}{\pi}\langle\alpha|\,\rho\,|\alpha\rangle = \frac{1}{\pi}Tr\,[\rho\,|\alpha\rangle\,\langle\alpha|] \,. \tag{1.54}$$

The Q function thus represents, for a pure case, the square of the overlap of the state with a coherent one. It is a real and positive quantity. Its normalization is

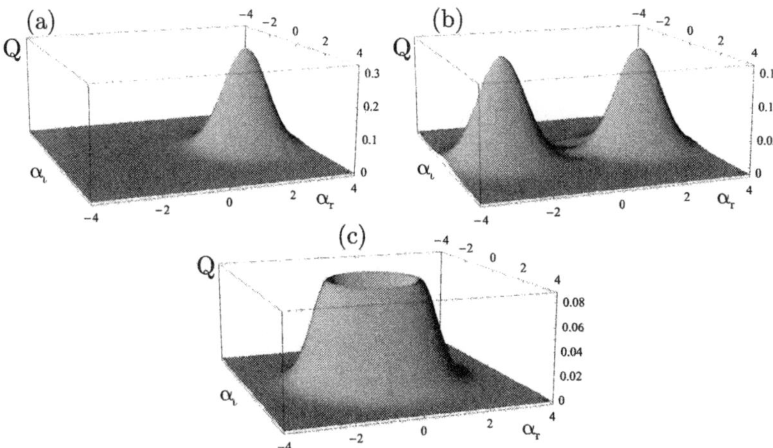

Fig. 7. Plots of Q functions. (a) 5 photons coherent state (real amplitude). (b) Schrödinger cat state, superposition of two 5 photon coherent states with opposite phases. Note the weak interference pattern near the origin. (c) Two-photon Fock state.

chosen to insure that the integral of Q over the whole phase space is equal to 1. An alternative definition of Q is given by using Eq.(1.33):

$$Q(\alpha) = \frac{1}{\pi}\text{Tr}[|0\rangle\langle 0|D(-\alpha)\rho D(\alpha)] \tag{1.55}$$

The Q function for the coordinate α is the expectation value, in the state of the field translated by $-\alpha$, of the projector on the vacuum. We will see in section 3 how this definition can be directly exploited for an experimental determination of Q in CQED.

It results directly from Eqs.(1.54) and (1.43) that the Q function of a coherent state is a Gaussian centered at the value corresponding to the complex amplitude of the state [see Fig. 7(a)]. The Q function of a Schrödinger cat of the form [Eq.(1.45)] is essentially the superposition of two Gaussians, centered at $\pm\beta$ [Fig. 7(b)]. There is a small additional interference term taking non zero values between these two Gaussians, but it is of the order of the scalar product of the two cat components, vanishingly small as soon as they are separated. The Q function is thus not appropriate to describe, in practice, the coherence of a cat state. We have finally shown in Fig. 7(c) the Q function of a $n = 2$ Fock state.

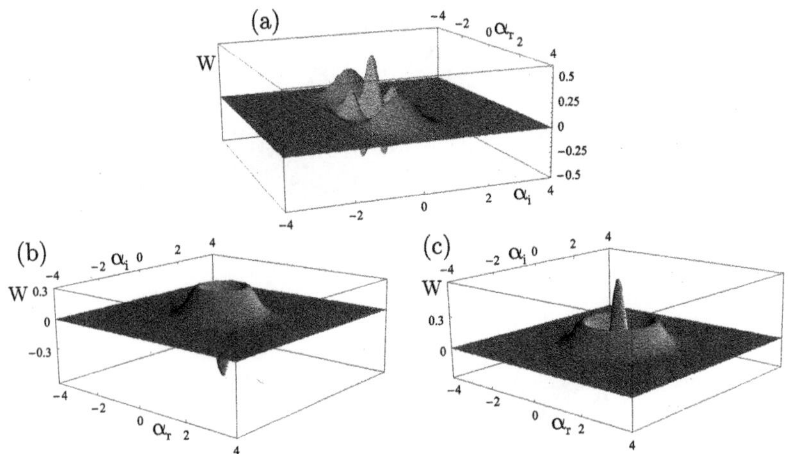

Fig. 8. Plot of W functions. (a) Schrödinger cat, superposition of two 5-photon coherent states with opposite phases. (b) One-photon Fock state. (c) Two-photon Fock state.

1.4.2. The W function

Given a field state described by its density operator ρ, its W function at the point in phase space whose C-number coordinate is $\alpha = x + ip$ is defined as:

$$W(x, p) = \frac{1}{\pi} \int dx' e^{-2ix'p} \langle x + \frac{x'}{2} | \rho | x - \frac{x'}{2} \rangle . \tag{1.56}$$

It appears as the Fourier transform of a function built on off-diagonal matrix elements of the field density operator in a quadrature representation. This expression was derived first by Wigner in 1932 [33] (hence the name W or Wigner function) in order to build a phase space distribution for a quantum particle resembling as closely as possible the probability distribution of classical statistical physics. The W function is, as Q, a real function, whose integral in phase space is equal to 1. Contrary to Q, however, it can take negative values in some domains of phase space. These negative values are, as shown below, a signature of non-classical behavior for the corresponding states.

From the definition [Eq.(1.56)] and the expressions of the coherent states in a quadrature basis [Eq.(1.36)], the W function of coherent states are easy to compute. They are, like Q, Gaussian functions centered at the value corresponding to the complex amplitude of the state, but their width is $\sqrt{2}$ times smaller. The W function of a π-phase cat state is, as its Q function, a superposition of two Gaussian peaks with, in addition, a large interference pattern between these peaks presenting oscillations, with an alternance of positive and negative ridges [Fig.

8(a)]. This pattern is a signature of the state coherence, lacking in the W function of a statistical mixture of coherent states. The W function is thus much better adapted than Q for the study of a mesoscopic state coherence. Fig. 8(b) and (c) present the W function of the $n = 1$ and $n = 2$ Fock states which also exhibit negative parts as typical non-classical features.

By inverse Fourier transform of Eq.(1.56), the matrix elements of the field density operator are expressed as:

$$\langle x + \frac{x'}{2}|\rho|x - \frac{x'}{2}\rangle = \int dp e^{2ix'p} W(x, p) , \tag{1.57}$$

which shows explicitely that the field density operator is fully determined by the knowledge of W. In particular the diagonal elements, representing the probability density of the quadrature X_0, are given by:

$$\langle x|\rho|x\rangle = \int dp W(x, p) . \tag{1.58}$$

This equation expresses an important property that W shares with a classical probability distribution for a particle in phase space. The probability that the X_0 quadrature takes a given value x is obtained by integrating the W function for this x value, along all the possible values of the orthogonal quadrature. In the same way, the probability for finding a classical particle at one point is given by integrating its phase space probability density over its momentum and vice versa. We have chosen here to use the X_0 and $X_{\pi/2}$ axes as reference frame in phase space. Any other set of orthogonal quadratures is of course equally possible. Transforming from one to the other amounts to a rotation in phase space, with a simple linear coordinate transformation for the W function. The integration property is obviously independent upon the coordinate frame. In other words, the density probability for any quadrature is given by summing W over the orthogonal one. We have seen above that some states, such as the Schrödinger cats or the Fock states have for some quadratures values occuring with 0 probability ('dark fringes'). The existence of these destructive interferences is an indication of non-classicality. This means that the integral of the W function along the orthogonal quadrature vanishes, which is possible only if W presents alternances of positive and negative values. We thus understand that negative values of W are related to non-classicality. We will show in section 2 that the integral property of W expressed by Eq.(1.58) is at the basis of a very general method to determine the W function of a light field, involving the measurement of the fluctuations of all its quadratures (quantum tomography).

We conclude this section by giving an alternative expression of W which will also be useful in the following. To obtain it, we let the reader demonstrate the simple translation relations:

$$|x - \frac{x'}{2}\rangle = e^{-i(x-x')p} D(x + ip)| - \frac{x'}{2}\rangle \,, \tag{1.59}$$

and

$$\langle x + \frac{x'}{2}| = \langle \frac{x'}{2}|D(-x - ip)e^{i(x+x')p} \,, \tag{1.60}$$

which follow directly from the definition of the displacement operators. Replacing $|x \pm x'/2\rangle$ in Eq.(1.57) by the expressions given by Eqs.(1.59) and (1.60), and noting that $\mathcal{P}|x'/2\rangle = |-x'/2\rangle$ [see Eq. (1.53)], we immediately get:

$$W(x, p) = \frac{2}{\pi} Tr[D(-\alpha)\rho D(\alpha)\mathcal{P}] \,, \tag{1.61}$$

which shows that the W function for the coordinate α is the expectation value in the state translated by $-\alpha$ of the field parity operator. The derivation presented here is due to L. Davidovich (private communication). The expression (1.61) is usually derived in a more complex way [24]. We will see in section 3 how this definition can be directly exploited for an experimental determination of W in CQED.

1.5. Coupling between field and charges: modeling sources and photon detectors

We have so far considered free fields, without describing the material systems they are coupled to. The field interacts in fact with charged particles (atomic electrons, currents ...). These interactions are essential to describe the field emission (sources) and its measurement (detectors). The Hamiltonian describing the evolution of the field +charges system writes in Coulomb gauge, at the non-relativistic limit [26, 34]:

$$\begin{aligned} H_{\text{charges + field}} &= \sum_i \frac{1}{2m_i} (\mathbf{p}_i - q_i \mathbf{A}(\mathbf{r}_i))^2 + U(\mathbf{r}_1, \mathbf{r}_2, ..., \mathbf{r}_i, ...) \\ &\quad + \sum_j \hbar\omega_j \left(a_j^+ a_j + \frac{1}{2} \right) . \end{aligned} \tag{1.62}$$

The $q_i, m_i, \mathbf{r}_i, \mathbf{p}_i$ are respectively the charge, mass, position and momentum of particle i (electrons and nuclei inside atoms or molecules). $U(\mathbf{r}_1, \mathbf{r}_2, ..., \mathbf{r}_i, ...)$ is the Coulomb potential of the charges, which depends only on their positions. The classical version of this Hamiltonian yields the dynamical equation for the charges driven by the Lorentz force produced by the field, and the Maxwell equations describing the evolution of the fields produced by the distribution of charges and currents. This coupled evolution leads in general to a complicated dynamics. We recall here only some simple situations in which either single photon Fock states or quasi-classical coherent states of the field are generated.

1.5.1. Single atom coupled to field: Rabi oscillation, spontaneous emission and preparation of Fock states

Consider first a one-electron atom coupled to a field mode in a cavity [27–30]. The atom has two relevant energy levels e and g separated by a transition resonant or nearly resonant with the mode, so that one photon processes are dominant (we neglect the $\mathbf{A}^2$ term creating or annihilating 0 or 2 photons). Eq.(1.62) then reduces to:

$$\begin{aligned} H_{\text{charges + field}} &= H_{\text{at}} + H_{\text{field}} + H_{\text{int}} \\ H_{\text{at}} &= \frac{p^2}{2m} + U(\mathbf{r}) \quad ; \quad H_{\text{field}} = \hbar\omega\left(a^+a + \frac{1}{2}\right) \\ H_{\text{int}} &= -\frac{q}{m}\frac{E}{\omega}[(\mathbf{p}.\boldsymbol{\epsilon})af(\mathbf{r}) + (\mathbf{p}.\boldsymbol{\epsilon}^*)a^\dagger f^*(\mathbf{r})] \,. \end{aligned} \tag{1.63}$$

This Hamiltonian consists of three terms, the free hydrogen-like atom hamiltonian H_{at}, the free field hamiltonian H_{field} and the coupling term H_{int}. Projecting explicitely on the two atomic levels e and g and making the rotating wave approximation (RWA) which amounts to neglecting far off-resonant couplings [34], we get:

$$H_{\text{int}} \approx -\frac{q}{m}\frac{E}{\omega}[f(\mathbf{r})\ (\mathbf{p}.\boldsymbol{\epsilon})_{eg}\ |e\rangle\ \langle g|\, a + f^*(\mathbf{r})(\mathbf{p}.\boldsymbol{\epsilon}^*)_{ge}\ |g\rangle\ \langle e|\ a^\dagger] \,. \tag{1.64}$$

The a term in the right hand side describes photon absorption processes accompanied by the g to e atom jump, while the $a^\dagger$ term accounts for photon emission combined to e to g atom transitions. The RWA approximation amounts to disregarding two other off-resonant processes in which the atom and the field get excited or de excited together. To simplify this expression further, we can combine the atomic matrix elements and the field parameters into a single constant describing the strength of the atom field coupling, which is called the vacuum Rabi frequency $\Omega(\mathbf{r})$:

$$\begin{aligned} H_{\text{int}} &\approx \frac{\hbar\Omega(\mathbf{r})}{2}[|e\rangle\ \langle g|\, a + |g\rangle\ \langle e|\, a^\dagger] \,; \\ \Omega(\mathbf{r}) &= -\frac{q}{m} f(\mathbf{r})(\mathbf{p}.\boldsymbol{\epsilon})_{eg}\sqrt{\frac{2}{\hbar\omega\varepsilon_0 V}} \,. \end{aligned} \tag{1.65}$$

The Rabi frequency is a function of the atom's position since the field strength seen by the atom depends on its location inside the mode via the $f(\mathbf{r})$ function. We will assume, without loss of generality, that $f(\mathbf{r})$, $(\mathbf{p}.\boldsymbol{\epsilon})_{eg}$ and thus $\Omega(\mathbf{r})$ are real. If the atom is fixed in space, we can disregard the $\mathbf{r}$ parameter and consider the Rabi frequency as a constant Ω. To describe an experiment in which an atom is moving across the field mode, we can also, in general, replace the spatially dependent $\Omega(\mathbf{r})$ quantity by an effective spatial average and drop the explicit $\mathbf{r}$ dependence. The system's evolution ruled by the Hamiltonian given by Eq.(1.65) corresponds to the Rabi oscillation in which the atom and the field reversibly exchange their energy. Assume that the atom is prepared at time $t = 0$ in level e with no photon in the field. The system evolves at time t into a linear superposition of the two states $|e, 0\rangle$ and $|g, 1\rangle$ corresponding respectively to the atom in e without photon or to the atom in g with 1 photon present:

$$|\Psi(0)\rangle = |e\rangle\,|0\rangle \to |\Psi(t)\rangle = \cos(\Omega t/2)\,|e\rangle\,|0\rangle - i\sin(\Omega t/2)\,|g\rangle\,|1\rangle\ . \tag{1.66}$$

The two probability amplitudes oscillate at the Rabi frequency Ω. In general, the atom-field state is non-separable, a manifestation of entanglement. For $\Omega t = \pi$, however, there is no entanglement and a one photon Fock state is deposited in the mode.

The reversibility of the Rabi oscillation is a special feature of the single field mode situation which requires in practice that the atom must be placed into a high-Q cavity sustaining the mode, with negligible field losses. In the more general situation where the atom interacts with the field in free space, its coupling to a continuum of field modes must be considered. Focusing again on a two-level atom in the RWA approximation, we get the following atom-field Hamiltonian:

$$H'_{\text{int}} = -\frac{q}{m}\sum_j \frac{E_j}{\omega_j}(\mathbf{p}.\boldsymbol{\epsilon}_j)_{eg}\, e^{i\mathbf{k}_j.\mathbf{r}}\,|e\rangle\,\langle g|\,a_j + h.c.\,. \tag{1.67}$$

The evolution of the atom-field system ruled by this Hamiltonian has been first described by Wigner and Weisskopf [35] and analyzed since in many papers and text books. When the atom is initially excited and the field in vacuum, the situation corresponds to the well-known spontaneous emission phenomenon, which, in free space, is irreversible. Solving for the system's evolution in the interaction representation, and neglecting Lamb-shift terms irrelevant to this discussion, we get:

$$|\Psi(0)\rangle = |e, 0\rangle \to |\Psi(t)\rangle = e^{-\Gamma t/2}\,|e, 0\rangle +$$
$$\left(\frac{1}{i\hbar}\right)\sum_j \langle g, 1_j|\,H'_{\text{int}}\,|e, 0\rangle\,\frac{1 - e^{-\Gamma t/2}e^{-i(\omega_{eg}-\omega_j)t}}{i\left(\omega_{eg} - \omega_j - i\Gamma/2\right)}\,|g, 1_j\rangle\ . \tag{1.68}$$

The final system's state is made of two parts, an initial state which is decaying exponentially in time and a final state consisting of an atom in the lower state with a photon distributed in a continuum of modes. The spontaneous decay is described by the natural rate Γ, simply expressed in terms of the matrix element $(\mathbf{p}.\boldsymbol{\epsilon})_{eg}$ and the atomic frequency ω_{eg}. After a time $t \gg 1/\Gamma$, the field ends up in a superposition of one-photon states belonging to each possible final mode $|1_i\rangle$, (Lorentzian distribution in frequency, with a width Γ). The corresponding one-photon wave packet writes [we leave the reader compute the exact expression of the c_i's from Eqs.(1.68) and (1.67)]:

$$|\Psi(t \gg 1/\Gamma)\rangle = \sum_j c_j \, a_j^+ \, |0\rangle \ . \tag{1.69}$$

This represents a one-photon packet whose wave vector angular distribution and polarization are determined by the matrix elements $(\mathbf{p}.\boldsymbol{\epsilon})_{eg}$ which appear in the expression of the c_j's.

This discussion can be generalized to more complex situations, involving a multilevel atom. For example, an atom with three levels e, g and i defining two cascading transitions, will emit, starting from e in vacuum, a pair of correlated photons whose direction of emission and polarizations depend on the matrix element of the electron momentum operator between the relevant atomic states. Such cascading transitions are useful for the preparation of entangled photons [36].

1.5.2. Classical current coupled to field: generation of coherent states

Single atom emitters, as just recalled, spontaneously generate Fock states of the field. Coherent states, on the other hand, can be produced by coupling the field to a monochromatic classical charge distribution [26]. If the charges motion is imposed (independent of field) and classical (negligible fluctuations of $\mathbf{r}_i$ and $\mathbf{p}_i$), the field Hamiltonian reduces to:

$$H_{\text{field}}^{\text{classical source}} = \sum_j \hbar\omega_j \left(a_j^\dagger a_j + \frac{1}{2} \right) - \int \mathbf{A}(\mathbf{r}).\mathbf{J}(\mathbf{r}, t) d^3\mathbf{r} \ . \tag{1.70}$$

in which $\mathbf{J}(\mathbf{r}, t)$ is the classical current density which can be expressed as a distribution over the particles velocities:

$$\mathbf{J}(\mathbf{r}, t) = \sum_i q_i \mathbf{v}_i(t) \delta\left(\mathbf{r} - \mathbf{r}_i(t)\right) \quad ; \mathbf{v}_i = \frac{d\mathbf{r}_i}{dt} \ . \tag{1.71}$$

This classical current Hamiltonian can be derived from the expression of the classical field Lagrangian [26]. It can also be deduced from the fully quantum hamiltonian $H_{\text{charges + field}}$ given by Eq.(1.62), letting the m_i go to infinity while keeping $\mathbf{p}_i/m_i = \mathbf{v}_i$ finite. The charges motion is then independent of $\mathbf{A}$. We must however retain the term linear in $\mathbf{A}$ in the Hamiltonian to describe the action of the imposed currents on the field. The interaction term becomes:

$$H_{\text{int}}^{\text{classical}} = -\sum_i q_i \mathbf{v}_i(t).\mathbf{A}(\mathbf{r}_i(t)) \ , \tag{1.72}$$

which is identical to the current-field interaction term in Eq.(1.70). Consider the case of a monochromatic current defined as:

$$\mathbf{J}(\mathbf{r}, t) = \mathbf{J}_0(\mathbf{r})e^{-i\omega t} \ . \tag{1.73}$$

and assume that this current is coupled to a resonant field mode in a cavity. Making again the RWA approximation, we can write the interaction as:

$$H_{\text{int}}^{\text{classical}} = -\int \mathbf{J}(\mathbf{r}, t).\mathbf{A}(\mathbf{r})d^3\mathbf{r} \approx -(s_0\, e^{i\omega t}a + s_0^* e^{-i\omega t}a^\dagger) \ , \tag{1.74}$$

where

$$s_0 = \sqrt{\frac{\hbar}{8\varepsilon_0\omega V}} \int f(\mathbf{r})\, (\mathbf{J}_0(\mathbf{r}).\boldsymbol{\epsilon})\, d^3\mathbf{r} \tag{1.75}$$

describes the overlap between the mode and the distribution of the source currents. Going in the interaction picture, the atom field coupling becomes:

$$\tilde{H}_{\text{int}}^{\text{classical}} = e^{iH_{\text{field}}t/\hbar} H_{\text{int}}^{\text{classical}} e^{-iH_{\text{field}}t/\hbar} = -\left(s_0 a + s_0^* a^\dagger\right) \ , \tag{1.76}$$

and the evolution of the field, supposed to be in vacuum at time $t = 0$, can be expressed as:

$$\begin{aligned} \left|\tilde{\Psi}(t)\right\rangle &= e^{-i\tilde{H}_{\text{int}}t/\hbar}\,|0\rangle = e^{i(s_0ta + s_0^*ta^\dagger)/\hbar}\,|0\rangle \\ &= D(is_0t/\hbar)\,|0\rangle = |\alpha = is_0t/\hbar\rangle \ . \end{aligned} \tag{1.77}$$

The unitary atom evolution operator is then identical to a field displacement in phase space. The classical current source generates a coherent state in the mode, whose amplitude increases linearly with time (mean photon number scaling as t^2). We further note that the complex amplitude of this coherent field is given by an expression which does not contain Planck's constant:

$$|is_0t/\hbar|\sqrt{\frac{\hbar\omega}{2\varepsilon_0 V}} = \frac{t}{4\varepsilon_0 V}\int d^3\mathbf{r}\, f(\mathbf{r})\, (\mathbf{J}_0(\mathbf{r}).\boldsymbol{\epsilon}) \ . \tag{1.78}$$

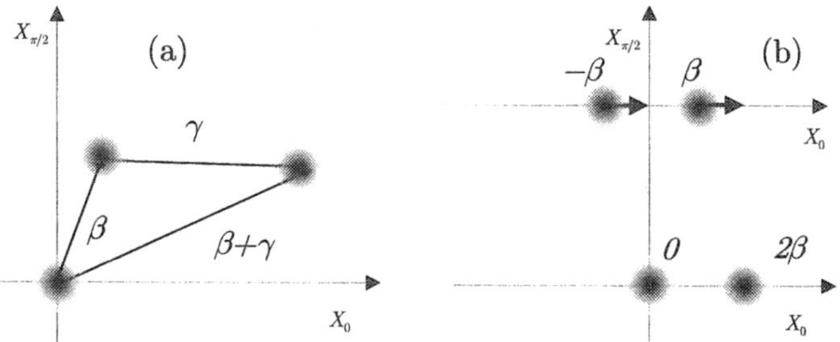

Fig. 9. Translation of field states in phase space. (a) a coherent field $|\beta\rangle$ translated by a displacement operator with the amplitude γ becomes $|\beta+\gamma\rangle$ (up to a global phase). (b) A π-phase cat $|\beta\rangle+|-\beta\rangle$ displaced by β yields an amplitude cat $|2\beta\rangle+|0\rangle$ (state normalization omitted). For the sake of clarity, the initial and final states are represented with different axis sets.

This is thus the same amplitude as the one given by a fully classical description. The field emitted by a classical imposed source is a coherent state whose amplitude is given by the classical solution of Maxwell equations. An arbitrary translation $D(\gamma)$ of the field in phase space is realized by coupling it during a given time t to a classical source with appropriate amplitude and phase ($\gamma = is_0 t/\hbar$). The same analysis applies if there is already an initial field in the cavity. Coupling for a given time the cavity field to a resonant source of classical current thus provides a simple way of displacing a coherent state in phase space. Assuming that the field is initially in state $|\beta\rangle$, we get:

$$\begin{aligned} \left|\tilde{\Psi}(t)\right\rangle &= D(\gamma = is_0 t/\hbar)\,|\beta\rangle \\ &= \exp(\gamma a^\dagger - \gamma^* a)\;\exp(\beta a^\dagger - \beta^* a)\;|0\rangle \\ &= e^{i(\beta_1\gamma_2 - \beta_2\gamma_1)}\,|\beta+\gamma\rangle\;. \end{aligned} \tag{1.79}$$

We realize in this way the translation defined by the vector represented by the C-number γ, combined with a global phase shift of the field state. This translation is schematically represented by the diagram of Fig. 9(a). A π phase cat can be translated in the same way, according to:

$$\left|\tilde{\Psi}(t)\right\rangle = \left(\frac{1}{\sqrt{2}}\right)\left[e^{i\beta\gamma_2}\,|\beta+\gamma\rangle + e^{-i\beta\gamma_2}\,|-\beta+\gamma\rangle\right] \tag{1.80}$$

and in the special case where $\beta = \gamma$ (real) we get:

$$\left|\tilde{\Psi}(t)\right\rangle = \left(\frac{1}{\sqrt{2}}\right)\left[|2\beta\rangle + |0\rangle\right] \tag{1.81}$$

In this case the initial cat, superposition of two coherent fields with opposite phases becomes an 'amplitude cat', superposition of the vacuum with a coherent field [see Fig 9(b)].

1.6. Modeling a photon counting process

To conclude this overview, we finally turn to the description of field detection. We summarize here the formalism introduced by R.Glauber in the first Les Houches school devoted to quantum optics in 1964 [37]. As an elementary photon counter, we consider an atom with a ground state g and a continuum of excited states e_i (corresponding to electronic states above the ionization threshold). We focus now on the photon absorption process starting with the atom in level g. The field with a given polarization $\boldsymbol{\epsilon}$ is in an initial state involving a set of field modes labeled by the index j. The absorption is governed by the atom-field coupling written in the RWA approximation as:

$$H_{\rm int}^{\rm Det} \approx -\frac{q}{m}\mathbf{p}.\mathbf{A}(\mathbf{r}) \approx -\frac{q}{m}\sum_i (\mathbf{p}.\boldsymbol{\epsilon})_{ig}\,|e_i\rangle\,\langle g|\sum_j \frac{E_j}{\omega_j} a_j e^{i\mathbf{k}_j.\mathbf{r}} + \text{h.c.}\,. \quad (1.82)$$

We assume that the excited atomic continuum extends over a relatively narrow absorption band whose center defines an average transition frequency ω_0. The field modes relevant for the absorption are the ones whose frequency ω_j are close to ω_0, so that we can rewrite Eq.(1.82) as:

$$H_{\rm int}^{\rm Det} \approx -\sum_i \frac{q}{im\omega_0}(\mathbf{p}.\boldsymbol{\epsilon})_{ig}\,|e_i\rangle\,\langle g|\sum_j iE_j a_j e^{i\mathbf{k}_j.\mathbf{r}} + \text{h.c.}\,. \quad (1.83)$$

We recognize in the sums over the j modes in this equations the so-called positive and negative frequency parts of the electric field operator, which respectively annihilate and create a photon:

$$E^+(\mathbf{r}) = i\sum_j E_j a_j e^{i\mathbf{k}_j.\mathbf{r}} \quad ; \quad E^-(\mathbf{r}) = -i\sum_j E_j a_j^\dagger e^{-i\mathbf{k}_j.\mathbf{r}}\,. \quad (1.84)$$

Regrouping atomic parameters in a single coupling constant κ_{ig}, we express the interaction Hamiltonian in terms of these positive and negative electric fields as:

$$H_{\rm int}^{\rm Det} = -\sum_i \left(\kappa_{ig}\,|e_i\rangle\,\langle g|\,E^+(\mathbf{r}) - \kappa_{ig}^*\,|g\rangle\,\langle e_i|\,E^-(\mathbf{r})\right)\,. \quad (1.85)$$

A last convenient transformation of this expression is obtained by going into the representation picture, changing $H_{\rm int}^{\rm Det}$ into

$$\tilde{H}_{\rm int}^{\rm Det} = e^{i(H_{\rm at}+H_{\rm field})t/\hbar)}\,H_{\rm int}^{\rm Det} e^{i(H_{\rm at}+H_{\rm field})t/\hbar}\,. \quad (1.86)$$

The atom-field coupling then involves the electric field operators in the Heisenberg point of view, as time-dependent operators:

$$\tilde{H}_{\rm int}^{\rm Det} = -\sum_i \left(\kappa_{ig} e^{i\omega_{ig}t} \left|e_i\right\rangle \left\langle g\right| E^+(\mathbf{r},t) - \kappa_{ig}^* e^{-i\omega_{ig}t} \left|g\right\rangle \left\langle e_i\right| E^-(\mathbf{r},t)\right) . \quad (1.87)$$

Assuming that at $t = 0$ the system is in state $|\tilde{\Psi}(0)\rangle = |g\rangle|\tilde{\Psi}_{\rm field}\rangle$ we expect at first order of perturbation theory that the state at time t is:

$$\begin{aligned}\left|\tilde{\Psi}(t)\right\rangle &\approx |g\rangle\left|\tilde{\Psi}_{\rm field}\right\rangle + \frac{1}{i\hbar}\int_0^t dt'\,\tilde{H}_{\rm int}^{\rm Det}(t')\,|g\rangle\left|\tilde{\Psi}_{\rm field}\right\rangle \\ &= |g\rangle\left|\tilde{\Psi}_{\rm field}\right\rangle - \frac{1}{i\hbar}\sum_i \kappa_{ig}\,|e_i\rangle \int_0^t dt' e^{i\omega_{ig}t'} E^+(\mathbf{r},t')\left|\tilde{\Psi}_{\rm field}\right\rangle . \quad (1.88)\end{aligned}$$

We finally get the total photo-ionization probability at time t by summing the contributions of all final states:

$$P_e(t) = \sum_i \left\langle \tilde{\Psi}(t) \,\middle|\, e_i \right\rangle \left\langle e_i \,\middle|\, \tilde{\Psi}(t) \right\rangle , \quad (1.89)$$

which directly yields the total atomic excitation probability:

$$P_e(t) = \frac{1}{\hbar^2}\sum_i \left|\kappa_{ig}\right|^2 \int_0^t\int_0^t dt'dt'' e^{i\omega_{ig}(t'-t'')} \times \left\langle \tilde{\Psi}_{\rm field}\right| E^-(\mathbf{r},t'')E^+(\mathbf{r},t')\left|\tilde{\Psi}_{\rm field}\right\rangle . \quad (1.90)$$

Summing over the final atom states i is equivalent to an integration over a continuum. We assume the variation of the coupling $|\kappa_{ig}|^2$ slow enough to replace it by a constant $|\kappa_0|^2$ over the continuum width. Integration of the exponential over ω_{ig} introduces the function $2\pi\delta(t' - t'')$. Hence:

$$P_e(t) = \frac{2\pi}{\hbar^2}|\kappa_0|^2 \int_0^t dt' \left\langle \tilde{\Psi}_{\rm field}\right| E^-(\mathbf{r},t')E^+(\mathbf{r},t')\left|\tilde{\Psi}_{\rm field}\right\rangle \quad (1.91)$$

The derivative $dP_e(t)/dt$ represents the probability to detect a photo-ionization process per time unit at time t, with an atom at $\mathbf{r}$. It is the single photon counting rate $w_1(\mathbf{r},t)$:

$$w_1(\mathbf{r},t) = \frac{dP_e(t)}{dt} = \frac{2\pi}{\hbar^2}|\kappa_0|^2 \left\langle \tilde{\Psi}_{\rm field}\right| E^-(\mathbf{r},t)E^+(\mathbf{r},t)\left|\tilde{\Psi}_{\rm field}\right\rangle . \quad (1.92)$$

which is proportional to the mean value in the field state of the Hermitian operator $E^-(\mathbf{r},t)E^+(\mathbf{r},t)$, product of the negative and positive frequency parts of the electric field (the positive frequency part is at right). If the field is defined by its density operator ρ_{field} , the counting rate is:

$$w_1(\mathbf{r},t) = \frac{2\pi}{\hbar^2}\,|\kappa_0|^2\,\mathrm{Tr}\left[\rho_{\text{field}}E^-(\mathbf{r},t)E^+(\mathbf{r},t)\right]\,. \tag{1.93}$$

We similarly define the double counting rate describing the probability to detect a photo-ionization event at $\mathbf{r}$, in a unit time interval around t and a photo-ionization event at $\mathbf{r}'$, in a unit time interval around t':

$$w_2(\mathbf{r},t;\mathbf{r}',t') \propto \mathrm{Tr}\left[\rho_{\text{field}}E^-(\mathbf{r},t)E^-(\mathbf{r}',t')E^+(\mathbf{r}',t')E^+(\mathbf{r},t)\right]\,. \tag{1.94}$$

The basic formulas (1.93) and (1.94) are essential to describe field measurements in a wide variety of quantum optics experiments.

2. Beam splitters and interferences in quantum optics

We analyze in this section interference and field amplitude measurement experiments in quantum optics. This leads us to consider multi-mode fields. An essential tool to couple two modes is the linear beam splitter. We present a simple model of this device and describe how a combination of two of them makes it possible to realize text book illustrations of quantum interferences involving one or two photons. We show also that a linear beam splitter is useful to realize a homodyne detection of the field quadratures. It also provides a model of the coupling of a field with its environment, leading to a simple description of decoherence phenomena. We then consider non-linear generalizations of the beam splitter, leading directly to the realization of multi-particle interferences involving Fock states. We analyze the special features of these interferences, their extreme sensitivity to decoherence and we finally describe an ion trap experiment simulating them.

2.1. Linear beam splitters

We consider now two propagating waves (a) and (b) coupled by a beam-splitter (partly reflecting, partly transmitting plate). The two waves propagate at right angle and the beam splitter lies at the beam intersection, at 45 degrees of both beams (Fig. 10). We assume that the modes are geometrically matched and have the same polarization (orthogonal to the incidence plane). The spatial structure of mode (b) reflected by the beam-splitter exactly coincides with the transmitted (a) mode. Classically, the two field mode amplitudes are mixed by the beam

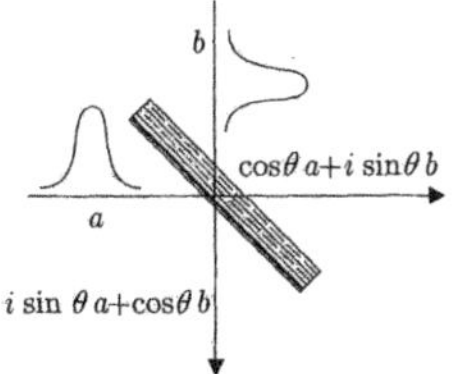

Fig. 10. Two freely propagating modes coupled by a semi-transparent beam splitter.

splitter. An impinging field in mode (a) of amplitude E_0 results in a transmitted amplitude $t_r E_0$ in mode (a) and a reflected amplitude $r_e E_0$ in mode (b), where t_r and r_e are the complex beam splitter transmission and reflexion amplitude coefficients. Fresnel reflexions laws furthermore entail that t_r/r_e is pure imaginary ($\pi/2$ phase shift between transmitted and reflected fields). We restrict ourselves to the ideal loss-less situation with $|t_r|^2 + |r_e|^2 = 1$. This makes it convenient to introduce a beam splitter mixing angle θ and to define $t_r = \cos\theta$ and $r_e = i\sin\theta$, with the i factor accounting for the Fresnel phase shift. Calling a and b the field operators of modes (a) and (b) (which commute together), it is natural to assume that the classical beam splitter action translates into a quantum transformation which affects the two modes according to the following unitary transformation:

$$U^\dagger(\theta)aU(\theta) = \cos\theta a + i\ \sin\theta b\ ;\ U^\dagger(\theta)bU(\theta) = i\sin\theta a + \cos\theta b\ , \quad (2.1)$$

in which U can be viewed as the evolution operator $e^{-iHt/\hbar}$ associated to the Hamiltonian:

$$H = -\hbar g_b(ab^\dagger + a^\dagger b)\ , \quad (2.2)$$

acting on the field during time $t = \theta/g_b$. The connexion between Eqs.(2.1) and (2.2) can easily be made by using the well known Baker-Hausdorff relation [24]:

$$\begin{aligned}\exp(iG\theta)a\exp(-iG\theta) &= a + i\theta\,[G,a] + \frac{i^2\theta^2}{2!}\,[G,[G,a]] + \ldots \\ &+ \frac{i^n\theta^n}{n!}\,[G,[G,\ldots[G,a]]] + \ldots\ , \end{aligned} \quad (2.3)$$

in which we define $G = -(ab^\dagger + a^\dagger b)$. The nested commutators in the right hand side of Eq.(2.3) are readily estimated ($[G,a] = b$; $[G,[G,a]] = a$...) so that all the odd-order terms in the development are proportional to b, while the even order terms are proportional to a. The coefficients of a and b coincide with the series expansion of the sine and cosine functions and we retrieve Eq.(2.1) The beam splitter thus appears as a linear device implementing between the two beams a coupling described by the Hamiltonian (2.2). This coupling merely describes a symmetrical exchange of photons between the two modes.

In a real experiment, the fields have a slowly varying envelope and pass during a finite time τ over the plate. The process can be seen as a collision between two wave packets (one of which may be the vacuum), mixed by the beam-splitter. To describe this collision, we should in all rigor consider that each packet is a superposition of modes with wave vector $\mathbf{k}$ (corresponding to field operators $a_{\mathbf{k}}$ distributed in frequency over an interval $c\Delta k = 1/\tau$). This point of view is mathematically cumbersome. An equivalent but much simpler model describes the fields as stationary plane waves whose coupling to the beam splitter is 'switched on' adiabatically during the time interval τ around 0. This amounts to considering a coupling of the form:

$$H = -\hbar g_{bs}(t)(ab^\dagger + a^\dagger b) \qquad \text{with} \quad \int dt \; g_{bs}(t) = \theta \, , \tag{2.4}$$

where $g_{bs}(t)$ is a function of time of width τ whose precise shape does not matter. In order to describe the field evolution produced by the beam-splitter, we can adopt two equivalent view points: the Heisenberg picture considers that the field remains in the time-independent state $|\Psi_{\text{field}}\rangle$, while the operators evolve from their initial to their final form within time τ [transformation expressed by Eq.(2.1)]. The Schrödinger picture considers conversely that the operators are time-independent and that the system's states evolve under the effect of U. We will adopt, depending upon the problems we have to solve, one or the other of these two equivalent perspectives.

2.2. *Schrödinger picture: effect of the beam splitter on some field states*

We start by describing the effect of the beam splitter on various initial field states, in the Schrödinger picture. We consider first situations where (a) is initially excited while (b) is in vacuum. The simplest case corresponds to a single photon impinging in (a). The initial state is then written as $|1, 0\rangle$ [in the following the first label in the kets refers to mode (a), the second to (b)]. With a sequence of obvious equalities, we get:

$$\begin{aligned} |1_a, 0_b\rangle \to |\Psi\rangle &= U(\theta)\,|1_a, 0_b\rangle = U(\theta)a^\dagger\,|0,0\rangle = U(\theta)a^\dagger U^\dagger(\theta)\,|0,0\rangle \\ &= (\cos\theta\; a^\dagger + i\sin\theta b^\dagger)\,|0,0\rangle = \cos\theta\,|1,0\rangle + i\sin\theta\;|0,1\rangle \end{aligned} \tag{2.5}$$

The photon is then in general coherently split between the two modes. The final energy distribution corresponds to the classical partition. Most importantly, the final state generally exhibits entanglement between the two modes. If now n

photons impinge in (a), a similar calculation yields:

$$
\begin{aligned}
|n_a, 0_b\rangle \rightarrow |\Psi\rangle &= U(\theta)\,|n_a, 0_b\rangle = U(\theta)\frac{(a^+)^n}{\sqrt{n!}}\,|0,0\rangle \\
&= \frac{1}{\sqrt{n!}}U(\theta)(a^+)^n U^+(\theta)\,|0,0\rangle \\
&= \frac{1}{\sqrt{n!}}(\cos\theta\ a^+ + i\sin\theta\ b^+)^n\,|0,0\rangle \\
&= \sum_{p=0}^{n}\binom{n}{p}^{1/2}(\cos\theta)^{n-p}(i\sin\theta)^p\,|n-p, p\rangle \quad , (2.6)
\end{aligned}
$$

where $\binom{n}{p} = n!/[p!(n-p)!]$ is the binomial coefficient. The final quantum state is a coherent superposition of terms corresponding to all possible partitions of the n incoming photons between the two modes. This state is in general massively entangled. The binomial distribution is well known in statistical physics. The photon distribution between (a) and (b) is the same as the distribution of n molecules at thermal equilibrium between two compartments whose volumes are in the ratio $\tan\theta$. The average photon number in the two beams are $n_a = n\sin^2\theta$ and $n_b = n\cos^2\theta$. The fluctuations of these numbers correspond to the usual partition noise: $\Delta n_a = \Delta n_b = \sqrt{n}\sin\theta\cos\theta$. For a symmetrical beam splitter ($\theta = \pi/4$), half of the photon on average find their way in each beam with $\Delta n_a = \Delta n_b = \sqrt{n}/2$.

Suppose now that the initial field in (a) is in a coherent state $|\alpha\rangle$. The beam splitter then transforms it according to:

$$
\begin{aligned}
|\alpha_a, 0_b\rangle \quad &\rightarrow |\Psi\rangle = U(\theta)\,|\alpha_a, 0_b\rangle = U(\theta)e^{\alpha a^\dagger - \alpha^* a}U^\dagger(\theta)\,|0,0\rangle \\
&= \exp(\alpha\left[\cos\theta a^\dagger + i\sin\theta b^\dagger\right] - \alpha^*\left[\cos\theta a - i\sin\theta b\right])\,|0,0\rangle \\
&= |\alpha\cos\theta\rangle_a\,|i\alpha\sin\theta\rangle_b\ . \quad (2.7)
\end{aligned}
$$

This series of equalities is easily demonstrated with the help of Eq.(1.33) and (2.1) and making use of the operator identity $Uf(A)U^\dagger = f(UAU^\dagger)$, valid whenever f can be expanded in power series. The output state thus appears as a separable tensor product of coherent states in the two modes, with amplitudes having the same values as in the corresponding classical situation. The beam splitter coherent state dynamics, entirely classical even for very low average photon numbers, is thus very different from the one of Fock states. The non-entanglement of coherent states involved in linear coupling operations is a very basic feature of these states, with far reaching consequences as discussed below in the context of decoherence.

Note that, as far as the photon number distribution is concerned, a linear beam splitter acts as if it were dispatching the quanta independently in the two modes, according to the classical energy partition probability. As a result, the photon distributions are very similar for an initial Fock state or a coherent state. Note however that for a symmetrical partition, the photon number fluctuation is equal to $\sqrt{\overline{n}/2}$ for an initial coherent state, i.e. $\sqrt{2}$ times larger than for an initial Fock state with the same number of photons. This is because the fluctuation of the photon number of a beam-split coherent state cumulates the initial Poisson noise of this state with the beam-splitter partition noise. Note also that this independent photon partition does not realize a separation of the Schrödinger cat type between the two modes, for which we expect the photons to be in a superposition of a state where they are all channeled in (a) with a state in which they are all channeled in (b). Such a situation corresponds to a bi-modal photon distribution in each mode, peaked at $n_a = 0, n_b = n$ and $n_a = n, n_b = 0$, very different from the partition realized by the beam splitter with its single peak at $n_a = n_b = n/2$. We will see below that the Schrödinger cat situation generally requires a different kind of non-linear beam splitting device.

A linear beam splitter can also be used to mix two excited modes. As a simple example, suppose that (a) and (b) initially contain a single photon. The output state is then:

$$\begin{aligned}|1_a, 1_b\rangle \quad &\rightarrow |\Psi\rangle = U(\theta)\,|1_a, 1_b\rangle = U(\theta)a^\dagger b^\dagger U^\dagger(\theta)\,|0,0\rangle \\ &= (\cos\theta\, a^\dagger + i\sin\theta b^\dagger)(i\sin\theta a^\dagger + \cos\theta b^\dagger)\,|0,0\rangle \\ &= \frac{i\sin 2\theta}{\sqrt{2}}(|2,0\rangle + |0,2\rangle) + \cos 2\theta\,|1,1\rangle\ , \end{aligned} \tag{2.8}$$

which is generally an entangled state. In the special case of a symmetrical beam-splitter, we get:

$$|1,1\rangle \rightarrow \frac{1}{\sqrt{2}}(|2,0\rangle + |0,2\rangle) \quad , \tag{2.9}$$

which is a superposition of two photons in (a) with two photons in (b). This looks like a kind of 'Schrödinger kitten' built on Fock states, but the situation cannot be generalized to larger photon numbers and the fact that non-linear beam splitting elements are required to prepare large Schrödinger cats remains true. The fact that the two photons are, in the symmetrical case, bunched in the same mode of the beam splitter can be seen as a consequence of the bosonic nature of photons. It is also (and not independently) a quantum interference effect. The probability amplitude for getting one photon in each mode cancels, because of the exact destructive interference between two indistinguishable quantum paths. In one of them, both photons are transmitted with a probability amplitude $t_r \times t_r =$

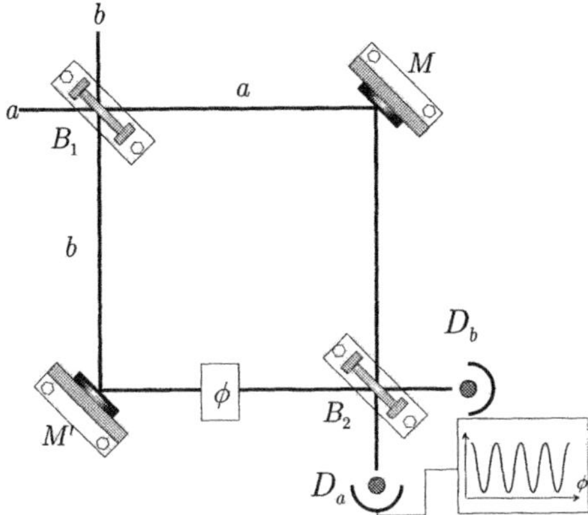

Fig. 11. Scheme of a Mach Zehnder interferometer

1/2 whereas in the other they are both reflected with a probability amplitude $ir_e \times ir_e = -1/2$. This process and the resulting state are non-classical. The lack of photon coincidences in the two final output modes when the beam splitter is exactly balanced has been observed first by Ou and Mandel in a pioneering two-photon interference experiment [38].

This photon bunching effect occurs because the polarizations of the two photons are the same. Photons with disparate polarizations, on the other hand, behave differently. Photons with orthogonal polarization states, antisymmetric versus photon exchange for instance, emerge in an antisymmetric superposition of modes, with exactly one photon in each. This is required to preserve the overall state symmetry with respect to photon exchange. Photon entanglement and Bell state analysis of photon pairs are based on this feature, which we will not discuss further here.

2.3. The Mach-Zehnder interferometer

Beam splitters are key elements in optical interferometers. We describe here the Mach Zehnder device for its simplicity and high symmetry. It is made of two identical 50 % beam splitters B_1 and B_2 with two folding mirrors M and M' (see Fig. 11). The photons cross the apparatus along two paths (a) and (b). A retarding plate shifts the phase in path (b) by a tunable angle φ. The field can be detected in both outputs with photon counters D_a and D_b. Consider first an experiment performed with a single photon impinging in mode (a), the initial state of the field being $|1, 0\rangle$. The successive operations on this state of B_1, the phase shifter and B_2 lead to the following transformations:

$$
\begin{aligned}
|1,0\rangle \quad & \rightarrow \frac{1}{\sqrt{2}}(|1,0\rangle + i\,|0,1\rangle) \rightarrow \frac{1}{\sqrt{2}}(|1,0\rangle + ie^{i\varphi}\,|0,1\rangle) \\
& \rightarrow \quad \frac{1}{2}(|1,0\rangle + i\,|0,1\rangle) + i\frac{e^{i\varphi}}{2}(i\,|1,0\rangle + |0,1\rangle) \\
& = \frac{1}{2}(1-e^{i\varphi})\,|1,0\rangle + \frac{1}{2}(1+e^{i\varphi})\,|0,1\rangle \;. \qquad (2.10)
\end{aligned}
$$

We then get simply the photon counting rate S_a in D_a and S_b in D_b as:

$$
\begin{aligned}
S_a \quad & \propto \langle\Psi|\, a^\dagger a\,|\Psi\rangle = \langle\Psi|\, a^\dagger\,|0,0\rangle\,\langle 0,0|\, a\,|\Psi\rangle \\
& = |\langle 00|\, a\,|\Psi\rangle|^2 = \frac{1}{2}(1-\cos\varphi) \\
S_b \quad & \propto \frac{1}{2}(1+\cos\varphi)\;. \qquad (2.11)
\end{aligned}
$$

Fringes with 100 % contrast are observed in both detectors, with a π-phase shift between the two. The interference occurs in the probability for detecting each photon crossing the apparatus, which is in agreement with the famous statement by Dirac that in this kind of interferometer at least, 'photons interfere with themselves'. Note however that this statement is challenged by other quantum optics experiments, such as the Ou and Mandel one quoted above. We will come back to this point later. Even if the process described by Eq.(2.10) is intrinsically a single photon one, it has to be recorded by accumulating statistics over many photon events, repeating the same experiment for different values of the phase shift φ. Observation of such single photon interference signals have been made [39]. They are a vivid illustration of wave-particle complementarity.

Let us now consider the situation where the initial field is in an n photon Fock state. A similar analysis yields the field final state:

$$
|\Psi\rangle = \frac{1}{\sqrt{n!}}\left[\frac{1}{2}\left(1-e^{i\varphi}\right)a^\dagger + \frac{1}{2}\left(1+e^{i\varphi}\right)b^\dagger\right]^n |0,0\rangle \;, \qquad (2.12)
$$

which shows that we are in fact repeating n time the single photon experiment. The interferometer dispatches photons in both arms according to a binomial law with probability $p = (1-\cos\varphi)/2$ in (a) and $q = 1-p$ in (b). The counting rate is proportional to the mean number of photons, i.e. $n(1-\cos\varphi)/2$ in (a) and $n(1+\cos\varphi)/2$ in (b). The signal has the same shape as in the one photon case. To maximize the signal, it is convenient to detect the difference of the two output channels $S_b - S_a = n\cos\varphi$.

Let us also consider the case of an initial coherent state $|\alpha\rangle$ impinging in mode (a). We leave the reader to show that the final field state is then $|(\alpha/2)(1-$

$e^{i\varphi}), (\alpha/2)(1+e^{i\varphi})\rangle$, corresponding to an unentangled output in which the field in each arm is coherent, with an amplitude corresponding to the classical one. The average photon counting signal is the same as in the one or n-photon cases.

Interferometers are usually designed to measure with precision optical phase changes. It is thus relevant to discuss the fundamental limits set by quantum laws to the sensitivity η of the device. It is defined as the inverse of the smallest phase change $\delta\varphi_{min}$ which can be detected. It is equal to the slope of the fringes $d(S_b - S_a)/d\varphi = n\sin\varphi$, divided by the smallest observable variation of the signal $d(S_b - S_a)_{min} = \Delta(n_b - n_a)$. For a Fock state input, the fluctuation of the photon numbers in the output channels is simply given by the binomial partition law:

$$d(S_b - S_a)/d\varphi = n\sin\varphi\ ;\ \eta = (\delta\varphi)_{\rm min}^{-1} = \frac{d(S_b - S_a)/d\varphi}{\delta(S_b - S_a)_{\rm min}} = \sqrt{n}\ . \qquad (2.13)$$

The sensitivity of the interferometer is thus equal to the square root of the photon number crossing the device during the measurement time. In this case, the sensitivity is independent upon the phase setting. As the phase evolves, the fringe slope and the partition noise vary in the same way and their ratio remains constant. In particular, at dark fringe, there is no signal variation to first order, but there is no noise either.

For a coherent state input and an output detecting the difference of the (a) and (b) counting rates, a similar argument yields $\eta = \sqrt{n}\sin\varphi$. The sensitivity now depends on the phase setting and, for optimum sensitivity, one must set the interferometer at a phase corresponding to maximum fringe slope ($\varphi = \pi/2$). The sensitivity is then again equal to $\sqrt{n}$. This value is the so-called 'standard quantum limit' or 'shot noise limit'. For a coherent state, this limit depends on the combined effects of the intrinsic photon noise of the initial field and the partition noise added by the interferometer. It is interesting to note that the same limit is obtained for a Fock state input, which has no intrinsic photon noise. We have considered here, for simplicity and symmetry, a signal based on the difference of the two output ports. In some devices, only one output detector is used. In this case, it is easy to show that a coherent state input yields a maximum sensitivity, again equal to $\sqrt{n}$, for a phase setting corresponding to a dark fringe. We should also note that these results can be substantially modified if mode (b), instead of being initially empty, contains a non-classical squeezed state of light [40]. The sensitivity can then be increased beyond the standard quantum limit discussed above. We will not describe it any further here. Another way to beat the standard quantum limit is to use multiparticle interferometers, as we will discuss later on.

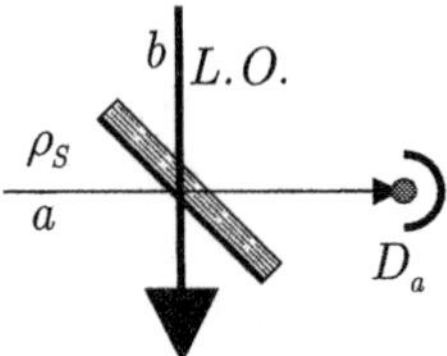

Fig. 12. Scheme of the homodyne detection of an unknown field ρ_S.

2.4. *Homodyne detection of field quadratures*

The experiments described so far are not sensitive to the phase of the field. Basically the same information is extracted by a Mach-Zehnder from a coherent field and from a Fock state, which has no defined phase. To get an information about the phase, one must resort to another kind of interference method which mixes the field, not with itself, but rather with another reference beam, called the local oscillator (L.O.). If this reference has the same frequency as the field to be measured, the method is referred to as 'homodyne' detection. The mixing is performed again with a beam-splitter according to the scheme shown in Fig.12 . We now use a highly transmitting plate ($T = t_r^2 = \cos^2\theta \approx 1-\theta^2$). The 'signal' field to be measured is described by its density operator ρ_s. The L.O. reference field, of same frequency, is a coherent state $|\beta\rangle = |\beta_0 e^{i\varphi}\rangle$. Only a small fraction of the intense L.O. signal is reflected into the output mode carrying the signal beam. The beating between the reflected L.O. beam and the transmitted signal is detected by measuring the photon counting rate w_a in the transmitted signal mode. For the sake of simplicity, we restrict ourselves to experiments in which only mode (a) is detected, the photons leaking into mode (b) being discarded. Taking into account Eqs.(1.93), (2.1) and (1.42), we obtain:

$$\begin{aligned} w_a \quad & \propto \quad \mathrm{Tr}\left[\rho_s\,|\beta\rangle\,\langle\beta|\,(\cos\theta\; a^\dagger - i\,\sin\theta b^\dagger)(\cos\theta a + i\,\sin\theta b)\right] \\ & = \sin^2\theta\,\langle\beta|\,b^\dagger b\,|\beta\rangle + \cos^2\theta\;\mathrm{Tr}(\rho_s a^\dagger a) \\ & \quad + i\,\sin\theta\cos\theta\;\mathrm{Tr}\left[\rho\,|\beta\rangle\,\langle\beta|\,(a^\dagger b - a b^\dagger)\right]\,. \end{aligned} \tag{2.14}$$

The first term in the right hand side of this equation is a constant background representing the intensity of the reflected L.O. beam, equal to $(1-T)\beta_0^2$. The second term corresponds to the intensity of the signal transmitted by the beam-splitter, equal to $T\langle a^\dagger a\rangle_s$, where the $\langle\ \rangle_s$ bracket denotes an average in the initial signal field state. The last 'beating' term, which is the dominant one if β_0 is large enough, contains the information about the signal field phase. It can be written as:

$$\begin{aligned} w_{beat} &= -2\sqrt{T(1-T)}\beta_0 \frac{\left\langle a^\dagger e^{i\varphi} - a e^{-i\varphi}\right\rangle_s}{2i} \\ &= 2\sqrt{T(1-T)}\beta_0 \left\langle X_{\varphi+\pi/2}\right\rangle_s . \end{aligned} \tag{2.15}$$

We measure in this way the field quadrature which is $\pi/2$ out of phase with the L.O. By sweeping the phase of the local oscillator, we can thus measure the expectation value of any field quadrature.

In fact, the method has the potential to extract even more information. By measuring the fluctuations of the photo-detection signal, we can also obtain the probability distribution of any signal field quadrature. Repeating the measurement for a large set of quadrature phases, we get an information which can be exploited to reconstruct the field Wigner function, and hence its density operator. We have seen in section 1.4 that the probability density of a given quadrature is the integral of the W function along a line in phase space orthogonal to the direction of this quadrature. Measuring the quadrature fluctuations for all possible phases thus amounts to determining the integrals of W along all possible directions in the phase plane. A procedure known as the Radon transform can then be used to find out W from the knowledge of these integrals [41]. This Radon transform is employed in a different context in medical tomography. Here, the integrated optical density of an inhomogeneous medium irradiated by X rays is measured along different directions in a plane and the Radon inversion is used to reconstruct the density of the absorbing medium in this plane. In this way, pictures of the inner parts of the body are made. Quantum tomography is quite analogous to medical tomography. The field to be measured is mixed via a beam splitter with L.O. references of various phases and the fluctuations of the homodyne beat signal are detected. Then a Radon inversion procedure reconstructs W and hence the field density operator. The experiment has been performed for squeezed quadrature fields [42] and Fock states [43, 44]. It remains to be done for Schrödinger cat states. We will not describe in any more details quantum tomography and rather focus on some interesting properties of the quadrature fluctuations in a Schrödinger cat state.

We have seen that the quadrature of the field orthogonal to the cat state alignment presents a large interference term. The homodyne current reveals it as a modulation around its average, with an alternance of bright and dark 'fringes'. In other words, a quantum interference signal is encoded in the fluctuations of the homodyne current, which itself results from a classical interference between the signal and the L.O. It is instructive to compute explicitely the fluctuations of this homodyne signal. It is indeed exceedingly fragile and, in a single detector scheme, very sensitive to the losses of the beam-splitter used for the homodyning process. This device must let a small non-zero fraction of the signal leak into mode (b). This loss into the environment is responsible for a strong reduction of the quantum interference signal, signature of the coherence of the Schrödinger cat state. We thus get a glimpse at decoherence in a specific situation.

Let us assume that the signal field injected in mode (a) is prepared in an even phase-cat state $|\alpha\rangle + |-\alpha\rangle$ (α real). Since the quantum interference signal is encoded in the field quadrature $\pi/2$ out of phase with the cat components (see section 1), the L.O. state amplitude β must have the same phase as α and is also real. Adopting the Schrödinger picture, the evolution of the field in the homodyning apparatus writes:

$$\begin{aligned} |\Psi\rangle_0 &= \frac{1}{\sqrt{2}}(|\alpha\rangle + |-\alpha\rangle)\,|\beta\rangle \rightarrow \\ |\Psi\rangle_f &= \frac{1}{\sqrt{2}}\left[|\alpha\ \cos\theta + i\beta\sin\theta\rangle\,|i\alpha\sin\theta + \beta\cos\theta\rangle\right] \\ &+ \frac{1}{\sqrt{2}}\left[|-\alpha\ \cos\theta + i\beta\sin\theta\rangle\,|-i\alpha\sin\theta + \beta\cos\theta\rangle\right] \end{aligned} \tag{2.16}$$

the global final state $|\Psi_f\rangle$ is the sum of two terms, corresponding to the two components of the initial cat state. Each of these terms is the tensor product of two states, the first referring to the measured mode (a), the second to the undetected 'environment' mode (b). In order to analyze the fluctuations of the photo-current in mode (a), one must perform experiments with a large number of 'cat' realizations and accumulate data. The fluctuation in the number of photons detected by D_a reflects the fluctuations of the field quadrature of interest. We must thus compute the probability $P(n)$ to detect n photons in (a), without looking at (b). This is obtained by tracing over mode (b), obtaining the reduced density operator of the field in mode (a) and evaluating the probability to find n photons in this field. This standard procedure leads in a straightforward way to:

$$\begin{aligned} P_a(n) &= \frac{1}{2}\left(|\langle n \mid \alpha\cos\theta + i\beta\sin\theta\rangle|^2 + |\langle n \mid -\alpha\cos\theta\ + i\beta\sin\theta\rangle|^2\right) + \\ &+\mathrm{Re}\big(\langle -\alpha\cos\theta + i\beta\ \sin\theta \mid n\rangle\,\langle n \mid \alpha\cos\theta + i\beta\ \sin\theta\rangle \times \\ &\times \langle -i\alpha\sin\theta + \beta\cos\theta \mid i\alpha\sin\theta + \beta\cos\theta\rangle\big) \end{aligned} \tag{2.17}$$

The last term in this equation represents an n-dependent interference term in the photon number distribution. This terms contains an n-independent factor [last line in Eq.(2.17)], equal to the overlap of the two final states of the field in the undetected mode (b). As soon as these states become distinguishable, i.e. quasi orthogonal, the interference term vanishes. This is a typical manifestation of complementarity. The fact that the environment contains an information about the state of the field in the interferometer destroys the interference of the 'cat', i.e. its quantum coherence. The distance in phase space of the two field components in (b), equal to $2\alpha\sin\theta$, is of the order of unity as soon as one photon on average has leaked into mode (b). This is a general feature of decoherence: the mesoscopic superposition exhibits coherence as long as not a single quantum is lost in the environment. This condition is of course a drastic one, very difficult to fulfill when the mean number of quanta is large. We will encounter this condition again and again in the following.

The calculation of the photon number fluctuation in (a) is somewhat cumbersome, but straightforward. The interference amplitude term is easily determined from the general expression of coherent states scalar products [Eq. (1.43)]:

$$\begin{aligned}&\langle -i\alpha\sin\theta\ +\beta\ \cos\theta \mid i\alpha\sin\theta\ +\beta\ \cos\theta\rangle\ = \\ &\quad\exp(-2\alpha^2\sin^2\theta)\ \exp(2i\alpha\beta\sin\theta\cos\theta)\end{aligned} \tag{2.18}$$

To complete the calculation, we assume that $\beta\sin\theta/\alpha\cos\theta \gg 1$ (large L.O. field), which implies that the average number of L.O. photons reflected in mode (a) is large. We can then replace the Poisson distribution of the L.O. photon number by a Gaussian approximation and we obtain:

$$\begin{aligned}&|\langle n \mid \alpha\cos\theta + i\beta\sin\theta\rangle|^2 = |\langle n \mid -\alpha\cos\theta + i\beta\sin\theta\rangle|^2 \\ &\approx |C_n(\beta\sin\theta)|^2 = e^{-\beta^2\sin^2\theta}\frac{(\beta\sin\theta)^{2n}}{n!} \\ &\propto \exp(-\frac{(n-\beta^2\sin^2\theta)^2}{2\beta^2\sin^2\theta})\,,\end{aligned} \tag{2.19}$$

$$\begin{aligned}&\langle -\alpha\ \cos\theta\ +i\beta\ \sin\theta \mid n\rangle\,\langle n \mid \alpha\ \cos\theta\ +i\beta\ \sin\theta\rangle \\ &\approx \quad |C_n(\beta\sin\theta)|^2\left(1-\frac{i\alpha\cos\theta}{\beta\sin\theta}\right)^{2n} \\ &\approx |C_n(\beta\sin\theta)|^2\ \exp(-2in\frac{\alpha\cos\theta}{\beta\sin\theta})\,.\end{aligned} \tag{2.20}$$

Taking then Eqs.(2.17), (2.19) and (2.20) into account, we find:

$$\begin{aligned}P_a(n) \quad &\propto \exp(-\frac{(n-\beta^2\sin^2\theta)^2}{2\beta^2\sin^2\theta}) \\ &\left[1+\cos 2\left(n\frac{\alpha\cos\theta}{\beta\sin\theta}-\alpha\beta\sin\theta\cos\theta\right)e^{-2\alpha^2\sin^2\theta}\right]\,.\end{aligned} \tag{2.21}$$

We finally remark that the fluctuating photon number n in mode (a) is directly related to the fluctuating quadrature of the field x by the following relation, deduced from Eq.(2.15):

$$n = \beta^2 \sin^2\theta + 2\beta \sin\theta \cos\theta \; x = \beta^2(1-T) + 2\beta\sqrt{T(1-T)}\; x \; . \qquad (2.22)$$

Substituting (2.22) into (2.21), we get the distribution of the quadrature signal as:

$$P(x) \propto e^{-2T} x^2 \left[1 + \cos(4\alpha T x) \; \exp\left[-2\alpha^2(1-T)\right]\right] . \qquad (2.23)$$

For $T = 1$, this expression is identical to Eq.(1.50). The fringe contrast vanishes however as soon as $\alpha\theta > 1$. This is a clear indication of the great fragility of the cat state coherence. It also shows that, when dealing with mesoscopic superpositions, we have to consider explicitely the actual way they are observed and the details of the experimental apparatus. In the version of the experiment we have described, the single detector homodyning scheme discards the information in mode (b), which entails decoherence. This could in principle be avoided by detecting the photons in (b) too and by combining the detections in both arms (balanced homodyne detection). It remains however that any photon loss in the apparatus able to 'give away' the state of the field must be avoided, as we will show below in a more general way.

2.5. *Beam splitters as couplers to environment: relaxation of a coherent state and of a Schrödinger cat*

As we have recalled above, the linear beam splitter which couples two modes of the field via an equation of the form (2.1) is an useful model to study various fundamental effects in quantum optics. We have seen in particular that this linear device leaves coherent states immune to entanglement, preserving the classical character of these states. We show in this subsection that this immunity to entanglement survives if the coherent state is coupled not to one single mode, but to a continuum of modes initially in vacuum. This situation describes in quite general terms field damping, the mode continuum being a model for a reservoir. For example a field stored in a cavity made of mirrors facing each other is coupled to reservoir modes by the scattering of light on mirrors imperfections and this scattering process can be modeled as a linear coupling between the cavity and the reservoir modes quite similar to the coupling performed by an ensemble of beam-splitters. We will understand, with this simple model, that coherent states appear as the natural 'pointer states' in quantum optics. We will then consider the coupling to the same environment of a Schrödinger cat state and show that such superpositions evolve very quickly towards an incoherent mixture of their 'pointer state' components. This will generalize the results of section 2.4. where we had considered not the complete Schrödinger cat density operator, but rather the evolution of one specific observable, namely one of the field quadrature fluctuations.

2.5.1. *Relaxation of a coherent state*

We first consider a field mode (annihilation operator a, frequency ω) coupled to a large set of other modes at frequencies ω_i described by their annihilation operators b_i. The environment modes span a wide frequency range. However the relaxation of (a) is mainly due to those modes with a frequency very close to ω (as required by energy conservation). For a qualitative approach, we thus assume a resonant coupling of (a) with each mode (b_i) through a linear hamiltonian describing a beam-splitter-like photon exchange:

$$H_i = -\hbar g_i (a b_i^\dagger + a^\dagger b_i) \ , \tag{2.24}$$

where the g_i are coupling constants depending smoothly on i. The mode (a) contains initially a coherent state $|\alpha\rangle$. Assuming that the reservoir is at zero temperature, all modes (b_i) are initially in vacuum. The action of the coupling Hamiltonian $\sum_i H_i$ during a short time interval $\delta\tau$ (much shorter than the characteristic relaxation time, but much longer than the field period) is thus equivalent to the coupling of mode (a) to modes (b_i) by a set of beam-splitters having each an amplitude transmission $\cos\theta_i \approx 1 - (g_i\delta\tau)^2/2$ very close to 1. Since mode (a) is not appreciably depleted during time $\delta\tau$, we can sum up independently the actions of these beam splitters acting 'in parallel'. Coupling to mode (b_i) alone transforms the initial state $|\alpha\rangle|0\rangle_i$ into $|\alpha\cos\theta_i\rangle|i\alpha\sin\theta_i\rangle_i$ [see Eq(2.7)]. Expanding the transmission and reflexion coefficients in powers of $g_i\delta\tau \ll 1$ and summing up all relevant modes, we get the global state of the field at time $\delta\tau$:

$$|\psi(\delta\tau)\rangle \approx \left|\alpha\left(1 - \sum_i \frac{g_i^2\delta\tau^2}{2}\right)\right\rangle_a \prod_i |i\alpha\, g_i\delta\tau\rangle_i \ . \tag{2.25}$$

Mode (a) still contains a coherent state whose amplitude is slightly reduced, but which remains unentangled with the reservoir [the situation is much more complex, as discussed above, if (a) is initially in a Fock state]. The amplitude reduction corresponds to an energy transfer into the environment modes. In order to estimate the amplitude loss, we must count the number of environment modes participating in the process. During the short time interval $\delta\tau$, all modes in a frequency interval of the order of $1/\delta\tau$ around ω are appreciably coupled to (a), as can be guessed from a simple energy-time uncertainty relation argument. The sum over i in Eq.(2.25) thus contains a number of terms of the order of $1/\delta\tau$.

Each of these terms being proportional to $\delta\tau^2$, the total amplitude decrease is, during this short time interval, quasi-linear in time. We can write:

$$|\psi(\delta\tau)\rangle \approx \left|\alpha\,(1-\frac{\kappa}{2}\delta\tau)\right\rangle_a \prod_i |i\alpha\, g_i\delta\tau\rangle_i\ , \tag{2.26}$$

where κ is a constant depending upon the mode density in the environment and the distribution of the coupling constants g_i. Note that the argument developed here is quite analogous to the one used to derive qualitatively the Fermi Golden Rule in standard perturbation theory in the presence of a continuum of final states.

For describing the system's evolution during the next time interval $\delta\tau$, the initial state of the environment is a priori slightly different. Since the leaking amplitude is however diluted among a large number of modes, the process where some amplitude would leak back from the environment into (a) is very unlikely. It is thus safe, for the computation of the (a) mode evolution to assume that all modes (b_i) remain practically empty all the time (so called 'Born approximation' in relaxation theory). We can thus consider independently the amplitude reductions resulting from successive time intervals. At an arbitrary time t, the state of (a) is thus still coherent, with the amplitude:

$$\alpha(t) \approx \alpha\left(1-\frac{\kappa\delta\tau}{2}\right)^{t/\delta\tau} \approx \alpha\, e^{-\kappa t/2}\,. \tag{2.27}$$

At any time, the mode (a) still contains a coherent state, unentangled with the mode reservoir. The coherent states, impervious to entanglement when they are coupled to a single beam splitter, keep this remarkable property when a large set of beam-splitters couples them to a big environment. They keep their form of coherent state throughout the evolution, merely loosing their amplitude as their energy leaks into the reservoir. They are the 'pointer states' of the field decoherence process.

The amplitude of the field is exponentially damped with the rate $\kappa/2$ and the field energy decays with the time constant $T_{cav} = 1/\kappa$ which can be identified with the experimental cavity damping time. In a pictorial representation, the disk representing the coherent state follows a logarithmic spiral in phase space, reaching the origin after an infinite time.

Note that the environment modes also contain at any time coherent states resulting from the accumulation of tiny coherent amplitudes along the successive time intervals. The global mode-environment state can be written as:

$$|\alpha e^{-\kappa t/2}\rangle \prod_i |\beta_i\rangle_i\ , \tag{2.28}$$

where the partial amplitudes β_i are such that:

$$\sum_i |\beta_i|^2 = \overline{n}(1 - e^{-t/T_{cav}}) , \tag{2.29}$$

a relation resulting simply from the total energy conservation. Note that a more complete derivation of the relaxation of a coherent state can be found in [24,26].

2.5.2. *Relaxation of a Schrödinger cat state*

Field state which are not coherent can always be considered as superpositions of coherent states [we can use the closure relationship (1.44) to expand them]. They generally get entangled to the environment. As a result, the field evolves from a pure state into a statistical mixture whose density matrix is obtained by tracing over the unobserved environment modes. In the process, quantum coherences are washed out. This is the decoherence phenomenon, which we have already glimpsed at by looking at a quadrature fluctuation of a Schrödinger cat state in section 2.4. We will describe now completely the decoherence of a Schrödinger cat state, by following the full evolution of its field density operator. The results obtained in subsection 2.5.1 about the relaxation of a coherent state will be very useful for this computation. We assume that we start at time $t = 0$ with a field in mode (a), prepared in the superposition $|\Psi_{cat}\rangle = [|\alpha e^{i\phi}\rangle + |\alpha e^{-i\phi}\rangle]/\sqrt{2}$. Following the same reasoning as above, we can write after a short time $\delta\tau$ the combined state of the (a) mode and the environment as:

$$\begin{aligned} |\Psi(\delta\tau)\rangle_{(a)+(E)} &= \frac{1}{\sqrt{2}}\Big[\Big|\alpha e^{i\phi}\left(1-\frac{\kappa\delta\tau}{2}\right)\Big\rangle\prod_i |ig_i\alpha e^{i\phi}\delta\tau\rangle_i + \\ &\Big|\alpha e^{-i\phi}\left(1-\frac{\kappa\delta\tau}{2}\right)\Big\rangle\prod_i |ig_i\alpha e^{-i\phi}\delta\tau\rangle_i\Big] . \end{aligned} \tag{2.30}$$

The two products of environment states appearing in this equation will be called $|\mathcal{E}_{\pm\phi}\rangle$. These two states are correlated to the two components of the cat state in mode (a). As soon as they are orthogonal, the field in mode (a) is maximally entangled with the environment and decoherence has occurred. It is thus important to compute the overlap of the environment states $\langle\mathcal{E}_{-\phi}|\mathcal{E}_{+\phi}\rangle$. It appears as a product over i of partial overlap integrals given by:

$$\langle ig_i\alpha e^{i\phi}\delta\tau|ig_i\alpha e^{-i\phi}\delta\tau\rangle = \exp(-2\alpha^2 g_i^2\delta\tau^2\sin^2\phi)e^{-i\alpha^2 g_i^2\delta\tau^2\sin 2\phi} , \tag{2.31}$$

and the global state overlap in the environment thus writes:

$$\langle\mathcal{E}_{+\phi}|\mathcal{E}_{-\phi}\rangle = \exp\left[-2\alpha^2\sin^2\phi\left(\sum_i g_i^2\right)\delta\tau^2\right]e^{-i\alpha^2\delta\tau^2\sin 2\phi\sum_i g_i^2}$$

$$= \exp(-2\alpha^2 \sin^2 \phi\kappa\delta\tau)e^{-i\alpha^2 \sin 2\phi\kappa\delta\tau} \, . \tag{2.32}$$

Decoherence is completed when this global overlap vanishes. This appears clearly by computing the mode (a) density matrix at time $\delta\tau$, obtained by tracing over the environment:

$$\begin{aligned}\rho_{cat}(\delta\tau) \quad &= \mathrm{Tr}_E \left[|\Psi(\delta\tau)\rangle_{(a)+(E)(a)+(E)}\langle\Psi(\delta\tau)| \right] \\ &= \frac{1}{2}|\alpha(\delta\tau)e^{i\phi}\rangle\langle\alpha(\delta\tau)e^{i\phi}| + \frac{1}{2}|\alpha(\delta\tau)e^{-i\phi}\rangle\langle\alpha(\delta\tau)e^{-i\phi}| + \\ \frac{1}{2}&\langle\mathcal{E}_{+\phi}|\mathcal{E}_{-\phi}\rangle|\alpha(\delta\tau)e^{i\phi}\rangle\langle\alpha(\delta\tau)e^{-i\phi}| + \text{h.c.} \, .\end{aligned} \tag{2.33}$$

The coherence of the field state, given by the off diagonal contributions in the above expression, is indeed proportional to the overlap integral $\langle\mathcal{E}_{-\phi}|\mathcal{E}_{+\phi}\rangle$ and vanishes with it. This is again a manifestation of complementarity, the cat state loosing its coherence as soon as an information about the phase of the field unambiguously leaks in the environment. According to Eq.(2.32), the cat coherence disappears in a characteristic time T_D given by:

$$T_D = \frac{T_{cav}}{2\alpha^2 \sin^2 \phi} = \frac{T_{cav}}{2\overline{n} \sin^2 \phi} \, . \tag{2.34}$$

The denominator in this expression is directly related to the square of the 'distance' in phase space of the two cat components $D^2 = 4\overline{n} sin^2\phi$ so that we can rewrite Eq.(2.34) as:

$$T_D = 2\frac{T_{cav}}{D^2} \, . \tag{2.35}$$

When there are many photons in the field, the cat state decoherence time, inversely proportional to the square of the distance between its components, is much shorter than the damping time of the field energy T_{cav}. We find again, in a quantitative way, the results obtained by the simple models considered above. Decoherence is due to leakage of information (here about the field's phase) in environment. There is a small amount of information in each mode (i), but there are many of them and decoherence is in general very fast. The cat state evolution is especially simple when the average number of photons is large, because decoherence occurs in a very short time during which the decay of the coherent component amplitudes is negligible. Once decoherence has occurred, the subsequent evolution of the system under the effect of relaxation is trivial. The density operator has become an incoherent sum of two coherent state components relaxing independently towards vacuum with the time constant T_{cav}.The dynamics of field relaxation becomes analytically more complicated when the initial average photon number is of the order of unity, because the evolution cannot be separated into an initial decoherence stage followed by a subsequent relaxation. The general model introduced in this subsection can however be used to obtain explicit expressions of the field density matrix evolution over an arbitrary long time [45,46]. We will not discuss this problem any further here.

2.6. *Non-linear beam splitters*

We have noticed in section 2.2 that linear beam splitters channel photons one by one in their two output modes and are thus unable to produce large mesoscopic superpositions of photons. We consider now a non-linear variant of the beam splitter which can, at least in theory, realize a multiparticle channeling, sending together all the photons in one arm or the other. We first consider a simple theoretical model. We then show how this model can be implemented in a real experiment.

2.6.1. *A simple model of non-linear beam splitter*

We consider now a slab of non-linear material whose action on a two-mode field is described by the following non-linear coupling:

$$H_n = -\hbar g_{nl}\left[a\left(b^\dagger\right)^n + a^\dagger\,(b)^n\right]\,. \tag{2.36}$$

This Hamiltonian realizes the reversible conversion of one photon of mode (a) into n photons of mode (b), according to the simple relations:

$$\begin{aligned}\left[a\left(b^\dagger\right)^n + a^\dagger\,(b)^n\right]|1_a,0_b\rangle &= \sqrt{n!}\,|0_a,n_b\rangle\\ \left[a\left(b^\dagger\right)^n + a^+\,(b)^n\right]|0_a,n_b\rangle &= \sqrt{n!}\,|1_a,0_b\rangle\end{aligned} \tag{2.37}$$

To conserve energy, the process must down-convert the photons, producing in mode (b) quanta with an energy n time smaller than in mode (a). As in section 2.1, we assume that the evolution of a field crossing this device is unitary, with an adiabatic switching on and off of the coupling g_{nl} describing in a simple way the passage of a light pulse on the plate. Making again use of the Baker-Hausdorff formula, we can easily show that the $|1,0\rangle$ and $|0,n\rangle$ states [where the first and second label refer respectively to the photon numbers in modes (a) and (b)] are related by a unitary transformation, generalizing to the non-linear case the usual beam splitter equations:

$$\begin{aligned}\exp[ig_{nl}t(a\left(b^\dagger\right)^n + a^\dagger\,(b)^n)]\,|1,0\rangle &= \cos\theta\,|1,0\rangle + i\sin\theta\,|0,n\rangle\\ \exp[ig_{nl}t(a\left(b^\dagger\right)^n + a^\dagger\,(b)^n)]\,|0,n\rangle &= i\sin\theta\,|1,0\rangle + \cos\theta\,|0,n\rangle\ .\end{aligned} \tag{2.38}$$

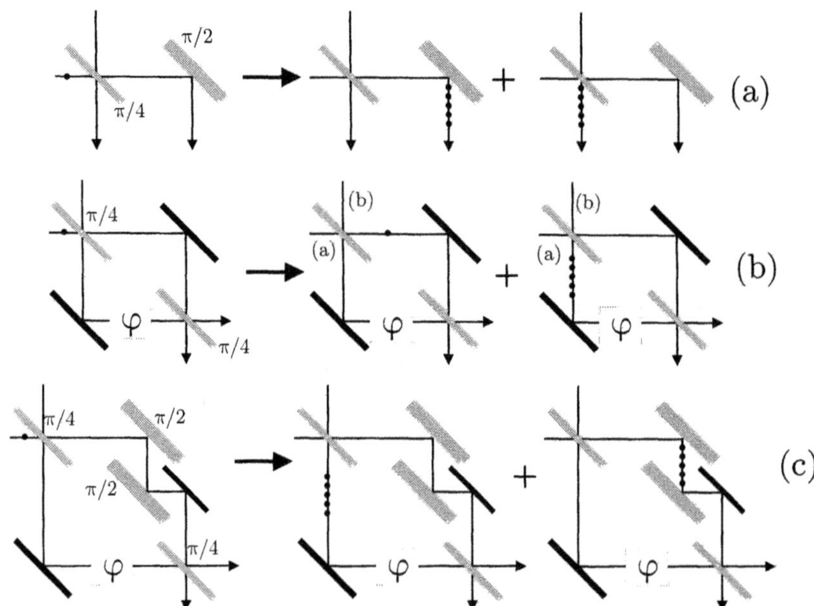

Fig. 13. Non-linear beam splitters. (a) a $(\pi/4)$ and a $(\pi/2)$ non linear beam splitter prepare a $|0, n\rangle + |n, 0\rangle$ mesoscopic superposition. (b) a Mach-Zehnder apparatus with two $\pi/4$ non linear beam splitters realizes an interference between a one-photon and an n-photon arm. (c) a variant with two additional $\pi/2$ non-linear beam splitters realizes an interference between two paths involving each n photons

The coupling angle is now defined as $\theta = g_{nl}t/\sqrt{n!}$. Note that these equations are valid only if we restrict ourselves to initial states of the form $|1, 0\rangle$. For $\theta = \pi/4$ the plate prepares with equal weights a superposition of one photon in (a) and n photons in (b) according to:

$$|1, 0\rangle \to \frac{1}{\sqrt{2}} \left(|1, 0\rangle + i\,|0, n\rangle\right) . \tag{2.39}$$

whereas a plate with $\theta = \pi/2$ converts with 100% efficiency one (a) photon into n photons in (b):

$$|1, 0\rangle \to i\,|0, n\rangle . \tag{2.40}$$

With such devices, it is possible, at least in principle, to prepare mesoscopic superpositions of field states involving n photons all channeled in one mode or the other. Fig. 13(a) shows a simple realization using a $\theta = \pi/4$ non linear beam splitter followed by a $\theta = \pi/2$ one. The final state is in this case of the form $|0, n\rangle + |n, 0\rangle$. It corresponds to a partition of the photons between the two modes very different from the one given by a linear beam splitter, which realizes a near equipartition as discussed in section 2.1. Instead of superposing field states belonging to the same mode, as in the phase cats described above, we are here superposing states distributing photons between two modes. The field state generated by this combination of non linear beam splitters has clearly a Schrödinger cat-like character. This cat state can be called 'non-local', since it exhibits an entanglement between two modes propagating in different regions of space. Each of the components can be considered as mesoscopic when $n \gg 1$, and the two components are orthogonal to each other, corresponding to two classically distinguishable states. We will see in section 3 that similar 'breeds' of non-local cat states can be generated in CQED by exploiting another kind of non-linear matter-field interaction.

Combining non-linear beam splitters, we can also design new kinds of multiparticle Mach-Zehnder interferometers in which many particles can follow together two different paths. These devices are intrinsically much more sensitive to phase shifts than single photon interferometers and can potentially be useful for applications. They are also, as we will see, much more sensitive to decoherence than single particle devices. We have sketched in Fig. 13(b) the principle of such an interferometer (for another scheme see [47]). As in the usual Mach-Zehnder, two $\theta = \pi/4$ non linear plates are separating and recombining the two modes, and mirrors are used to fold them. The final detection is made either in mode (a) (single photon channel) or in mode (b) (n-down converted photons channel). A phase shifter producing a phase shift φ per photon is introduced in the n-photon channel between the two beam splitters. With this device, an interference is produced between two paths, one corresponding to a single photon, the other to n-photons.

A variant can be easily designed in which two $\theta = \pi/2$ plates are introduced in the (a) mode between the two $\theta = \pi/4$ plates in order to reversibly produce and annihilate a n-photon state in this mode. In this way, the apparatus is a genuine n-particle interferometer in which n particles are simultaneously all present in one arm and in the other [see Fig. 13(c)]. The addition of the two $\theta = \pi/2$ plates is however not essential, since it does not change the principle of the interferometer operation. We will disregard it in the following and analyze only the system described by Fig. 13(b). The important point is that we have no way of knowing whether the n-photon exiting in mode (b) have been produced in the first non-linear beam splitter or in the second one. This ambiguity leads to an n-particle interference. A simple analysis similar to the one carried above in the linear Mach-Zehnder case shows that the final state of the field now writes:

$$|\Psi_f\rangle = \frac{1}{2}(1 - e^{in\varphi})\,|1, 0\rangle + \frac{i}{2}(1 + e^{in\varphi})\,|0, n\rangle \ , \tag{2.41}$$

and the mean photon number detected in (a) and in (b) writes:

$$S_a = \frac{1 - \cos n\varphi}{2} \; ; \qquad S_b = n\frac{1 + \cos n\varphi}{2} \; . \tag{2.42}$$

Both signals exhibit now an interference pattern with a 100% fringe contrast, but with an interfrange separation n-time smaller than in the corresponding linear device. It is important to note that for such a multiparticle interferometer, Dirac's statement about 'each photon interfering with itself' does not apply. Clearly, the fringes result here from a quantum interference process involving globally the n particles which can follow two paths associated to orthogonal states. This kind of interference was unknown in Dirac's time and only linear processes were considered. Even then, though, Dirac's statement was in trouble. As we have seen above in section 2.2, a linear beam splitter on which single photon impinge in the two input ports can give rise to interference phenomena which cannot be explained in term of single photon effects [38].

The multiphoton interferometer yielding narrower fringes, it is natural to wonder whether it has a larger sensitivity to phase changes than its linear counterpart. To find it out, we have to compute the derivative of the fringe signal with respect to φ and its fluctuation.We restrict our calculation to mode (b) [the (a) output yielding the same result]. The photon number fluctuation is given by:

$$\begin{aligned} \Delta S_b &= \left(\langle \Psi_f | \, (b^\dagger b)^2 \, | \Psi_f \rangle - \langle \Psi_f | \, b^\dagger b \, | \Psi_f \rangle^2 \right)^{1/2} \\ &= \left(\frac{n^2}{2} (1 + \cos n\varphi) - \frac{n^2}{4} (1 + \cos n\varphi)^2 \right)^{1/2} = \frac{n}{2} \sin n\varphi \; . \end{aligned} \tag{2.43}$$

It is $\sqrt{n}$ time larger than for a linear device with the same number of photons.The slope of the signal, proportional to n^2 is however n time larger. As a result, the sensitivity η is $\sqrt{n}$ time larger than for an usual interferometer, becoming n instead of $\sqrt{n}$. The exact computation:

$$\eta = (\delta\varphi_{\min})^{-1} = \frac{(\delta S_b / \delta\varphi)}{\Delta S_b} = \frac{(n^2/2) \sin n\varphi}{(n/2) \sin n\varphi} = n \; , \tag{2.44}$$

yields, as for the linear device with a Fock state input, a sensitivity independent upon the phase setting.

The increased sensitivity of multiparticle interferometers makes them very attractive for high resolution measurements in spectroscopy or metrology. This advantage has however to be balanced with the difficulty in designing them and with the fact that these devices are intrinsically much more sensitive to decoherence. This point is illustrated by the simple decoherence model sketched in Fig. 14. A loss mechanism is introduced in mode (b) by inserting on the beam path a

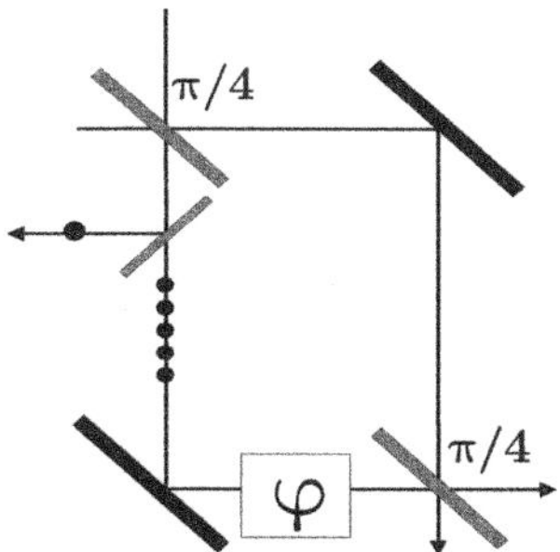

Fig. 14. Decoherence in a non-linear interferometer. Losses are modeled by a high transmission linear beam-splitter inserted in the n photon arm of the interferometer.

linear beam splitter with a small reflection coefficient ϵ. This couples the mode containing n photons with an initially empty environment mode (e) (whose annihilation operator will be called c_e). The probability to loose one of the n photons in this environment is $\sim n\epsilon^2$. We will show that as soon as this probability becomes of the order of 1, the interferometer fringe contrast collapses. In other words, as already noticed for a Schrödinger cat coupled to a reservoir, the coherence of the n-photon state superposition is lost as soon as one particle, potentially able to give an information about the path followed by the system, is leaking in the environment.

Let us follow, in the Schrödinger point of view, the state of the field as it progresses across the interferometer. We now have to keep track of the three (a), (b) and (e) modes. The three successive labels in each ket refer to the photon numbers in these modes. The first non-linear beam splitter corresponds to the transformation:

$$|1_a, 0_b; 0_e\rangle \rightarrow \left(\frac{1}{\sqrt{2}}\right) \left(|1_a, 0_b; 0_e\rangle + i\, |0_a, n_b; 0_e\rangle\right) . \tag{2.45}$$

The coupling to the environment then changes the global field state into:

$$\begin{aligned} &\rightarrow \left(\frac{1}{\sqrt{2}}\right) \Big(|1_a, 0_b; 0_e\rangle + i \cos^n \varepsilon\, |0_a, n_b; 0_e\rangle + \\ &+ \sum_{p=1}^{n} c_p\, |0_a, (n-p)_b; p_e\rangle\Big) , \end{aligned} \tag{2.46}$$

where the c_p are coefficients given by the binomial law which we do not need to write explicitely. We then take into account the effect of the phase shifter in mode (b):

$$\rightarrow \quad \left(\frac{1}{\sqrt{2}}\right) \left(|1_a, 0_b; 0_e\rangle + i \cos^n \varepsilon e^{in\varphi} |0_a, n_b; 0_e\rangle + \right.$$

$$\left. \sum_{p=1}^{n} c_p e^{i(n-p)\varphi} |0_a, (n-p)_b; p_e\rangle \right) , \tag{2.47}$$

and, finally, the action of the second non-linear beam splitter mixing again the (a) and (b) modes:

$$\rightarrow \left| \Psi_f^{(S+E)} \right\rangle = \left(\frac{1}{2}\right) [(|1_a, 0_b; 0_e\rangle + i\, |0_a, n_b; 0_e\rangle) +$$

$$i \cos^n \varepsilon e^{in\varphi} (i\, |1_a, 0_b; 0_e\rangle + |0_a, n_b; 0_e\rangle)] \qquad +$$

$$\left(\frac{1}{\sqrt{2}}\right) \sum_{p=1}^{n} c_p e^{i(n-p)\varphi} |0_a, (n-p)_b; p_e\rangle \, . \tag{2.48}$$

To obtain this final expression, which is an exact one within the assumptions of our model, we have to remark that the non-linear process coupling the (a) and (b) modes has no action on a field state in which the (b) mode contains less than n photons. The expression obtained for the field can be used to compute the photon counting rate in mode (b) and we get:

$$\begin{aligned} S_b &= \left\langle \Psi_f^{(S+E)} \right| b^\dagger b \left| \Psi_f^{(S+E)} \right\rangle_f \\ &= \frac{n}{4} \left| 1 + \cos^n \varepsilon e^{in\varphi} \right|^2 + \frac{1}{2} \sum_{p=1}^{n} |c_p|^2 (n-p) \\ &= \frac{n}{4} [1 + \cos^{2n} \varepsilon + 2 \cos^n \varepsilon \cos(n\varphi)] + \frac{n}{2} (\cos^2 \varepsilon - \cos^{2n} \varepsilon) \, . \end{aligned} \tag{2.49}$$

The sum over p in the second line of Eq.(2.49) is easily performed by using a sum rules satisfied by the c_p binomial coefficients. We leave the reader figure out how one goes from the second to the third line in this equation. The φ dependent term in the final formula given by Eq.(2.49) contains the interference signal. Its amplitude is proportional to $cos^n \epsilon$ which, for small ϵ is equivalent to $\exp(-n\epsilon^2/2)$. The fringe contrast thus decreases exponentially with the size n of the multiparticle system and becomes negligibly small when $n\epsilon^2 > 1$, i.e. when more than one photon on average has been lost in the environment.

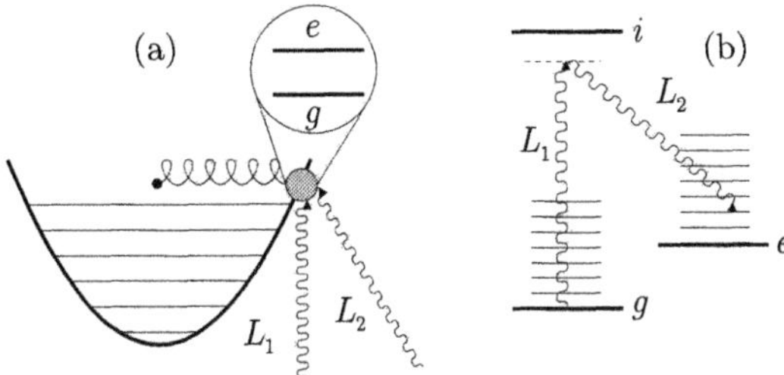

Fig. 15. Principle of an ion trap non-linear beam splitter experiment. (a) Ion in an harmonic potential, showing the quantized motional levels and the ion's internal structure. Lasers L_1 and L_2 induce Raman transitions between the internal and vibration states. (b) Energy levels diagram.

2.6.2. *Simulation of multiparticle interferometry in an ion trap experiment*

Realizing non-linear beam splitters of the kind just described would be very difficult. It would require in particular a medium exhibiting strong non-linearity induced by a single photon, a situation which is encountered so far only in CQED, albeit in a quite different context. We will come back to CQED experiments in section 3. We conclude this section by describing an experiment in which the multiphoton interferometry is simulated in a system where photons are replaced by phonons [48]. The physics of a non-linear Mach-Zehnder apparatus can indeed be mimicked with a single ion in a trap [49], whose evolution is described by equations quite similar to the one we have encountered above in section 2.6.1. The ion is excited by two laser beams inducing a Raman transition between two of its internal energy levels. At the same time, the momentum exchange between the laser beams and the ion results in an excitation of vibration quanta of the ion oscillation in the trap. The coupling of the ion with the laser realizes the dynamics of a non-linear beam splitter of the kind described above, in which the excitation of a single photon in mode (a) is replaced by the internal excitation of the ion and the n photons in mode (b) become n phonons of the ion vibration.

The system we are considering is sketched in Fig. 15(a), with its energy levels shown in Fig. 15(b). We call e and g the two relevant energy levels of the ion (two hyperfine ground states separated by the energy $\hbar\omega_{eg}$), n the vibration quanta of its external motion in a trap potential, which we assimilate to a one-dimension harmonic oscillator vibrating along Ox (angular frequency ω_p). Two laser beams (frequencies ω_1 and ω_2, wave vectors projections along Ox: k_1 and k_2) induce a Raman process involving a virtual transition to an excited level i of the ion. The process is tuned to be resonant with a transition which simultaneously changes the internal state of the ion and its vibrational excitation, the laser frequencies obeying the condition:

$$\omega_1 - \omega_2 = \omega_{eg} + (n' - n)\omega_p \ . \tag{2.50}$$

The coupling between the ion and the laser beams, considered here as classical fields (with complex amplitudes α_1, α_2 and polarizations ϵ_1, ϵ_2), is described by an effective hamiltonian:

$$\begin{aligned} H_{\text{Raman}}(t) &= -\frac{q^2}{m^2}\frac{\hbar}{2\varepsilon_0 V\sqrt{\omega_1\omega_2}}\alpha_1\alpha_2^*\left[\frac{\langle e|\,(\mathbf{p}.\epsilon_2^*)\,|i\rangle\,\langle i|\,\mathbf{p}.\epsilon_1)\,|g\rangle}{\hbar(\omega_{ig}-\omega_1)}\right]\times \\ &\quad \times |e\rangle\,\langle g|\,e^{-i(\omega_1-\omega_2)t}e^{i(k_1-k_2)x} + h.c.\,, \end{aligned} \tag{2.51}$$

where ω_{ig} is the frequency of the transition between levels i and g. The diadic operators $|e\rangle\langle g|$ $(|g\rangle\langle e|)$in this equation are atomic excitation (deexcitation) operators describing jumps of the ion from e to g or back, quite analogous to the a and $a^\dagger$ operators describing the jumps of the (a) mode of the field from the 0 to 1 photon state and back in the non-linear-beam splitter model. The $\exp i(k_1 - k_2)x$ term in Eq.(2.51) describes the phase dependence of the fields seen by the ion at position x. This position is an operator, superposition of the b and $b^\dagger$ ion vibration quanta annihilation and creation operators. Expanding the exponentials in powers introduces operators of the form b^n and $b^{\dagger n}$, quite similar to the expressions appearing in Eq.(2.36). More precisely, we start by writing the exponential in Eq.(2.51) as:

$$e^{i(k_1-k_2)x} = \exp i\sqrt{\frac{\hbar}{2M\omega_p}}\Delta k\ (b + b^\dagger) = \exp[i\eta_L(b+b^\dagger)] \quad , \tag{2.52}$$

where $\Delta k = k_1 - k_2$, M is the mass of the ion and where we have introduced the dimensionless Lamb-Dicke parameter:

$$\eta_L = \left(\frac{\hbar\Delta k^2}{2M\omega_p}\right)^{1/2} = \left(\frac{E_{\text{recoil}}}{\hbar\omega_p}\right)^{1/2} \tag{2.53}$$

This quantity appears naturally when discussing the momentum exchange between an atom and light, especially in ion trap experiments (see [49] and the lecture notes by R. Blatt and D. Wineland in this volume). It can be defined as the square root of the ratio of the Raman-induced ion recoil energy by the vibration quantum. We will assume here that the Lamb-Dicke parameter is very small compared to unity, which allows us to expand in series the right hand side of Eq.(2.52) and retain the lowest non-vanishing term. We thus get:

$$H_{\text{Raman}}(t) = \frac{\hbar\Omega_R}{2}e^{-\eta_L^2/2}e^{-i(\omega_1-\omega_2)t}\,|e\rangle\,\langle g|\sum_{p,q}\frac{(i\eta_L)^{p+q}(b^\dagger)^p b^q}{p!q!}$$

$$+h.c.. \tag{2.54}$$

In order to simplify this expression we have included the laser amplitudes, atom laser detunings and atomic matrix elements appearing explicitely in Eq.(2.52) into a single parameter Ω_R which we will call the Raman Rabi frequency. When the resonance condition (2.50) is fulfilled for an integer value $\Delta n = n' - n$, the leading contributions in H_{Raman} are such that $p - q = \Delta n$. We can thus have either $p = \Delta n, q = 0$ or $p = \Delta n + 1, q = 1$ etc. The first term is dominant (lowest order in η_L). Thus, for this resonance and in interaction picture:

$$\tilde{H}_{\text{Raman}} \approx \frac{\hbar\Omega_R}{2}\,(i\eta_L)^{\Delta n}\,|e\rangle\,\langle g|\left(b^\dagger\right)^{\Delta n} + h.c.. \tag{2.55}$$

This hamiltonian looks like (2.36) in which one replaces the annihilation operator of one photon in mode (a) by the operator changing $|g\rangle$ into $|e\rangle$. These states thus play respectively the roles of the states $|1\rangle_a$ and $|0\rangle_a$ in the multi-particle Mach-Zehnder model [the analogy between a and $a^\dagger$ on one hand and the atomic operators $|e\rangle\langle g|$ and $|g\rangle\langle e|$ on the other hand holds only if there is no more than one photon in mode (a)].The phonon operators b and $b^\dagger$ play on the other hand the role of the photon operators in mode (b). In other words, the ion undergoing the Raman process 'emulates' the n-photon interferometer. By applying Raman pulses of convenient durations, one can simulate the successive non-linear beam splitter operations.

This experiment has been realized by Liebfried *et al*, using a Be ion [48]. We summarize here its main steps. First, the ion is prepared in level g and in its vibration ground state (in a state equivalent to $|1_a, 0_b\rangle$). One then applies a Raman pulse simulating an n-photon $\pi/4$ non-linear beam splitter. This is achieved by choosing the laser resonance condition to realize the situations $\Delta n =$ 1, 2 or 3 and by adjusting the pulse duration. An ion state equivalent to $(|1,0\rangle + i|0,n\rangle)/\sqrt{2}$ is thus prepared. The n-phonon state (equivalent to $|0,n\rangle$) is then phase-shifted by changing by $\delta\omega_p$ the ion vibration frequency during a time t. This is performed by applying a bias electric field to the trap electrodes. A phase shift $\varphi = \delta\omega_p t$ per phonon is achieved in this way.

Finally, a second Raman pulse simulating a second $\pi/4$ n-photon non-linear beam splitter is applied and the ion in state g is detected by recording its fluorescence. A detection laser is switched on, bringing the ion from g into an excited state j from which a photon is emitted in transition back to g. This photon is detected. The cycle of excitation-detection is repeated a large number of times on the closed $g \to j$ transition, resulting in a large number of photons being collected when the ion is in level g only. This detection process is selective (e is not seen). It simulates the detection in mode (a) of the n-photon Mach-Zehnder. The experiment is repeated as φ is varied and fringes versus φ are obtained. Fig.

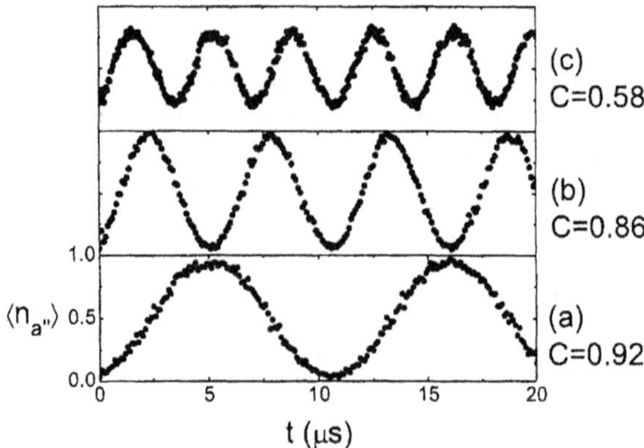

Fig. 16. Trapped ion implementation of a non-linear interferometer. Fringes obtained when varying the interferometer phase φ for $n = 1, 2, 3$ (from bottom to top). The fringes contrast values are indicated on the right. From [48].

16 shows the experimental recordings for $n = 1, 2$ and 3. The narrowing of the fringe spacing, proportional to $1/n$ is clearly observed, as well as reduction of the fringe contrast, an illustration of the strong sensitivity to decoherence of multi-particle interference effects.

3. Schrödinger cats in cavity QED

We have seen in the previous section that a basic ingredient for the generation of a multiparticle mesoscopic state superposition is the existence of a strong non-linearity in the system. Another example of mesoscopic state superposition induced by non-linearity is provided by cavity quantum electrodynamics (CQED) [29]. Here, the strong non-linear interaction of a single atom with a cavity field made of tens of photons results in the generation of cat states, entangled with the atom [30]. These cat states have been observed in experiments and their decoherence has been studied [16, 19]. We describe in this section the theory of these experiments and discuss their relationship with the well known collapse and revival effect of the Rabi oscillation [50]. The graininess of the photon field plays an essential role in the generation of the field mesoscopic superpositions, all these effects vanishing at the classical limit where the photon number goes to infinity (the atom-field interaction remaining finite). We will see also that the concepts of quantum interference and complementarity play a very important role to analyze the physics of these Schrödinger cat systems.

3.1. A reminder on microwave cavity QED and the Jaynes-Cummings model

In CQED, single atoms interact with one mode of the field in a high Q resonator. Dissipation is kept to a minimum, so that the coherent features of the atom-field interaction manifest themselves most conspicuously. In microwave CQED with superconducting cavities [30], the photon storage time is by far the longest relevant time in the experiment, so that many atoms, crossing successively the cavity, interact with the same field, imprinting in and extracting information from it. A detailed description of CQED techniques is beyond the scope of this course. Our goal is only to show how these techniques make it possible to generate mesoscopic superpositions of field states and study their evolution. In this introductory sub-section, we briefly recall the formalism and the notations which will be useful for the description of our system and the analysis of the experiments.

Basically, microwave CQED implements as closely as possible the Jaynes-Cummings (J-C) model of quantum optics [51], consisting in a two-level atom interacting with a single field mode. The atoms cross the cavity with a well defined velocity, which allows us to control the atom-field interaction time. They are prepared in a state belonging to a subspace made of two levels e and g. In practice, the experiments are performed with a Rubidium atomic beam and the levels of interest are two circular Rydberg states of high principal quantum number (equal to 51 and 50 respectively). The transition between these levels, at frequency $\omega_0/2\pi = 51$ GHz, is resonant or nearly resonant with a mode of the cavity (frequency ω), made of two superconducting mirrors facing each other. We call a and $a^\dagger$ the annihilation and creation operators of one photon in the mode. We will not describe here how the atomic states are prepared and detected and how the cavity is made and tuned [see [30] and lecture notes by M. Brune in this Volume].

If we neglect the atom and field relaxation processes, the 'atom + cavity' field evolution is ruled by the interaction Hamiltonian given by Eq.(1.65). Adding the field and the atom's energy terms, we can write the total atom-field Hamiltonian as:

$$H_{JC} = \hbar\omega(a^\dagger a+1/2)+\frac{\hbar\omega_0}{2}(|e\rangle\langle e|-|g\rangle\langle g|)+\frac{\hbar\Omega}{2}(a^\dagger|g\rangle\langle e|+a|e\rangle\langle g|)\,, \quad (3.1)$$

where Ω is the vacuum Rabi frequency, which we have already introduced in section 1.5. This coupling constant depends on the position of the atom. We assume it to be a constant, for sake of simplicity (see remark about this in section 1.5.1 above).

The eigenstates of the 'field + atom' Hamiltonian, when the interaction is neglected, are the tensor products $|e, n\rangle$, $|g, n\rangle$ describing the atom in e or g with a defined photon number n in cavity. The interaction H_{int} couples these states

2 by 2, inside uncoupled two-level manifolds S_n (spanned by the kets $|e,n\rangle$, $|g,n+1\rangle$). The eigenstates and eigenergies of the J-C Hamiltonian are thus given by the exact diagonalization of two-level systems. Using a standard procedure, it is convenient to introduce the Pauli matrices notation and write the projection of H_{JC} inside the S_n manifold (calling P_n the projector on S_n) as:

$$\begin{aligned}\frac{1}{\hbar}P_n H_{JC} P_n &= (n+1)\omega + \frac{1}{2}\begin{pmatrix} \omega_0-\omega & \Omega\sqrt{n+1} \\ \Omega\sqrt{n+1} & -(\omega_0-\omega)\end{pmatrix} \\ &= (n+1)\omega + \frac{1}{2}\left[(\omega_0-\omega)\,\sigma_z + \Omega\sqrt{n+1}\sigma_x\right]. \end{aligned} \tag{3.2}$$

It is then handy to define an n-dependent coupling angle θ_n by the relation:

$$\theta_n = \arctan\left(\frac{\Omega\sqrt{n+1}}{\omega_0-\omega}\right) \quad (0 \le \theta_n < \pi)\,. \tag{3.3}$$

With this notation, simple expressions for the atom-field eigenstates are obtained:

$$\begin{aligned}|+,n\rangle &= \cos\frac{\theta_n}{2}\,|e,n\rangle + \sin\frac{\theta_n}{2}\,|g,n+1\rangle \quad ; \\ |-,n\rangle &= \sin\frac{\theta_n}{2}\,|e,n\rangle - \cos\frac{\theta_n}{2}\,|g,n+1\rangle \,, \end{aligned} \tag{3.4}$$

the corresponding energies being:

$$\frac{1}{\hbar}E_{\pm,n} = (n+1)\omega \pm \frac{1}{2}\sqrt{(\omega_0-\omega)^2 + \Omega^2(n+1)}\,. \tag{3.5}$$

Exact resonance ($\omega_0 = \omega$) corresponds to $\theta_n = \pi/2$. Eqs.(3.4) and (3.5) then become:

$$|+,n\rangle = \frac{1}{\sqrt{2}}(|e,n\rangle + |g,n+1\rangle) \quad ; \; |-,n\rangle = \frac{1}{\sqrt{2}}(|e,n\rangle - |g,n+1\rangle)\,, \tag{3.6}$$

$$\frac{1}{\hbar}E_{\pm,n} = (n+1)\omega \pm \frac{1}{2}\Omega\sqrt{(n+1)}\,. \tag{3.7}$$

The eigenstates of the atom-field system ('dressed states' [34, 52]) are very convenient to compute the atom field evolution. We have already seen in section 1 that an atom, initially in e and resonantly coupled to the vacuum field $|0\rangle$ undergoes a reversible Rabi oscillation. A similar effect occurs if the field is initially in an n Fock state. It is then convenient to expand the initial $|e,n\rangle$ atom-field state on the dressed levels, by inverting Eq.(3.6). We then get in a straightforward way:

$$|\Psi(0)\rangle = |e, n\rangle = \frac{1}{\sqrt{2}}(|+, n\rangle + |-, n\rangle) \rightarrow$$
$$|\Psi(t)\rangle = \frac{1}{\sqrt{2}} e^{-i(n+1)\omega t} (e^{-i\Omega\sqrt{n+1}t/2} |+, n\rangle + e^{i\Omega\sqrt{n+1}t/2} |-, n\rangle) \,. \quad (3.8)$$

and the probability for finding the atom at time t in level e:

$$P_e(t) = |\langle e, n \mid \Psi(t)\rangle|^2 = \cos^2 \frac{\Omega\sqrt{n+1}}{2} t \,. \quad (3.9)$$

The Rabi oscillation induced by an n-photon field occurs at a frequency $\sqrt{n+1}$ time larger than in vacuum. A Fock state being difficult to prepare, the phenomenon is usually observed with a coherent field, superposition of Fock states. We must then weight the different oscillation terms, whose frequency is n-dependent, by the probability to have n photons in the field, given by a Poisson law. We find:

$$P_e(t) = \sum_n |C_n|^2 \, |\langle e, n \mid \Psi(t)\rangle|^2 = e^{-\bar{n}} \sum_n \frac{\bar{n}^n}{n!} \frac{1 + \cos(\Omega\sqrt{n+1}t)}{2} \,. \quad (3.10)$$

The classical limit of this expression is obtained by taking $\bar{n} \rightarrow \infty$ and $\Omega \rightarrow 0$ with $\Omega\sqrt{\bar{n}} = \Omega_{cl}$ remaining finite. We then get a steady Rabi oscillation at frequency Ω_{cl} (classical limit). As soon as the photon number is finite however the distribution of n values leads to a spread of Rabi frequencies and to a collapse of the oscillation. We qualitatively estimate the collapse time T_{collapse} by expressing that the phase variation of the oscillation over the width $\sqrt{\bar{n}}$ of the Poisson law is of the order of 2π:

$$\Omega \left. \frac{d(\sqrt{n+1})}{dn} \right|_{n=\bar{n}} \times \sqrt{\bar{n}}\, T_{\text{collapse}} = \pi \rightarrow T_{collapse} = \frac{2\pi}{\Omega} \,. \quad (3.11)$$

The Rabi oscillation in a coherent field with a finite photon number thus collapses after a time T_{collapse} whose order of magnitude is the Rabi oscillation period in vacuum. This period is typically 20 μs in our CQED experiments. This is however only part of the story and the atom-field evolution involves much more physics. The above calculation does not tell us what happens to the system at longer times. Furthermore, we have so far focused on the atom's state and we have disregarded the field evolution and its correlations to the atom. We are now turning our attention to these important features.

3.2. Rabi oscillation in a mesoscopic field: collapse and revival revisited

The exact evolution equation for the 'atom + coherent field' system is obtained by adding the contributions of the various n states in the expansion of the coherent state. Starting at time $t = 0$ from the $|e, \alpha\rangle$ state representing an initially excited atom and a coherent field of complex amplitude α, we get at time t:

$$
\begin{aligned}
|\Psi(0)\rangle &= |e,\alpha\rangle = \frac{1}{\sqrt{2}}\sum_n C_n(|+,n\rangle + |-,n\rangle) \rightarrow \\
|\Psi(t)\rangle &= \frac{1}{\sqrt{2}}\sum_n C_n \times \\
&\qquad \times e^{-i(n+1)\omega t}\left(e^{-i\Omega\sqrt{n+1}t/2}|+,n\rangle + e^{i\Omega\sqrt{n+1}t/2}|-,n\rangle\right) \\
&= \frac{1}{2}\sum_n C_n e^{-i(n+1)\omega t}\Big\{e^{-i\Omega\sqrt{n+1}t/2}(|e,n\rangle + |g,n+1\rangle) + \\
&\qquad + e^{i\Omega\sqrt{n+1}t/2}(|e,n\rangle - |g,n+1\rangle)\Big\}\,.
\end{aligned} \tag{3.12}
$$

where the C_n are Poisson coefficients given by Eq.(1.40). If we replace in this equation $|g, n+1\rangle$ by $|g, n\rangle$ (disregarding the difference between $\sqrt{n+1}$ and $\sqrt{n}$) and if we neglect the variation of the Rabi frequency $\Omega\sqrt{n+1}$ over the width of the photon number distribution, we obtain the classical limit in which the atom and field states are factorized:

$$
|\Psi(t)\rangle \approx \left(e^{-i\Omega_{cl}t/2}\frac{|e\rangle + |g\rangle}{\sqrt{2}} + e^{i\Omega_{cl}t/2}\frac{|e\rangle - |g\rangle}{\sqrt{2}}\right)|\alpha\rangle\,. \tag{3.13}
$$

At this limit, the field is unaffected by the coupling. The atom on the other hand is in a linear superposition of the two 'dipole states' $(|e\rangle \pm |g\rangle)/\sqrt{2}$. Each term of this superposition has a probability amplitude which evolves at frequency $\pm\Omega_{cl}$ and the Rabi oscillation thus appears as a quantum interference effect between these two amplitudes. This interference can be observed because the two 'paths' (atom in $|e\rangle + |g\rangle$ or in $|e\rangle - |g\rangle$) are fully indistinguishable.

When the coherent field is mesoscopic, i.e. contains a large but finite average number of photons $\overline{n}$, the effects neglected in the classical limit must be taken into account. The atom and the field are then no longer factorized. Their entanglement evolves as a function of time, appearing and disappearing quasi-periodically. Since the evolution equation has an exact solution, we could of course compute it and analyze the atom-field state in its full complexity. This approach is not very transparent however and we prefer to present here an approximate solution which has the merit of displaying the physics of the problem in a simple way. We follow closely the calculation derived in [53–55]. We notice first that the difference between $\sqrt{n+1}$ and $\sqrt{n}$ has its most important effect in the rapidly evolving phase factors, but can be neglected in the expression of the slowly varying C_n coefficients. To evaluate correctly the phase factors, we make the following replacement,which takes into account the graininess of the photon number:

$$\sqrt{n+1} \approx \sqrt{n} + \frac{1}{2\sqrt{n}} \approx \sqrt{n} + \frac{1}{2\sqrt{\bar{n}}} \, . \tag{3.14}$$

We then get the following expression for the atom-field state:

$$\begin{aligned} |\Psi(t)\rangle \quad & \approx \frac{1}{2}\sum_n C_n e^{-i(n+1)\omega t} e^{-i\Omega\sqrt{n}t/2} \, |n\rangle \, (e^{-i\frac{\Omega t}{4\sqrt{\bar{n}}}} \, |e\rangle + e^{i\omega t} \, |g\rangle) \\ & + \frac{1}{2}\sum_n C_n e^{-i(n+1)\omega t} e^{i\Omega\sqrt{n}t/2} \, |n\rangle \, (e^{i\frac{\Omega t}{4\sqrt{\bar{n}}}} \, |e\rangle - e^{i\omega t} \, |g\rangle) \, . \end{aligned} \tag{3.15}$$

In this formula, the Rabi frequency $\Omega\sqrt{n}$ is still a function of n. We approximate it by its expansion in powers of $n - \bar{n}$, limiting the development to its second order:

$$\sqrt{n} \approx \frac{\sqrt{\bar{n}}}{2} + \frac{n}{2\sqrt{\bar{n}}} - \frac{1}{8\bar{n}^{3/2}}(n - \bar{n})^2 \ldots \tag{3.16}$$

We then get after a straightforward calculation:

$$\begin{aligned} |\Psi(t)\rangle \approx & \frac{1}{2} e^{-i\Omega\sqrt{\bar{n}}t/4} \sum_n C_n e^{-i(n+1)\omega t} e^{-i\frac{\Omega n t}{4\sqrt{\bar{n}}}} e^{i\frac{\Omega(n-\bar{n})^2 t}{16\bar{n}^{3/2}}} \, |n\rangle \\ & \times (e^{-i\frac{\Omega t}{4\sqrt{\bar{n}}}} \, |e\rangle + e^{i\omega t} \, |g\rangle) + \\ + & \frac{1}{2} e^{i\Omega\sqrt{\bar{n}}t/4} \sum_n C_n e^{-i(n+1)\omega t} e^{i\frac{\Omega n t}{4\sqrt{\bar{n}}}} e^{-i\frac{\Omega(n-\bar{n})^2 t}{16\bar{n}^{3/2}}} \, |n\rangle \\ & \times (e^{i\frac{\Omega t}{4\sqrt{\bar{n}}}} \, |e\rangle - e^{i\omega t} \, |g\rangle) \end{aligned} \tag{3.17}$$

Each of the two terms in the above expression is a product of an atomic by a field state. We recognize in the field states (sums over n) the expressions of coherent-like states whose phase is spread by a second order phase diffusion term [described by the $\exp(\pm\Omega(n - \overline{n})^2 t/16\overline{n}^{3/2})$ terms]. It is easy to show that this phase spreading occurs in a characteristic time T_{spread} of the order of $\pi\overline{n}/\Omega$. Restricting ourselves to time much shorter than this limit, we will first neglect this phase spreading. A simple manipulation then yields:

$$
\begin{aligned}
|\Psi(t)\rangle \approx & \frac{1}{2}e^{-i\Omega\sqrt{\bar{n}}t/4}e^{-i\omega t/2}\left|\alpha e^{-i(\omega+\Omega/4\sqrt{\bar{n}})t}\right\rangle \times \\
& \times(e^{-i\Omega t/4\sqrt{\bar{n}}}e^{-i\omega t/2}\,|e\rangle + e^{i\omega t/2}\,|g\rangle) + \\
& +\frac{1}{2}e^{i\Omega\sqrt{\bar{n}}t/4}e^{-i\omega t/2}\left|\alpha e^{-i(\omega-\Omega/4\sqrt{\bar{n}})t}\right\rangle \times \\
& \times(e^{i\Omega t/4\sqrt{\bar{n}}}e^{-i\omega t/2}\,|e\rangle - e^{i\omega t/2}\,|g\rangle)\,. \qquad (3.18)
\end{aligned}
$$

We finally adopt the interaction picture, in order to get rid of the obvious free atom and field evolution terms. We write the atom-field state in a compact form which displays immediately the entanglement features of the system:

$$
\left|\tilde{\Psi}(t)\right\rangle \approx \frac{1}{\sqrt{2}}\left[\left|\Psi_{at}^{+}(t)\right\rangle\otimes\left|\Psi_{f}^{+}(t)\right\rangle + \left|\Psi_{at}^{-}(t)\right\rangle\otimes\left|\Psi_{f}^{-}(t)\right\rangle\right]. \qquad (3.19)
$$

The global atom-field state is thus expressed in terms of four normalized atom and field states given by the relations:

$$
\left|\Psi_{at}^{\pm}(t)\right\rangle = \frac{1}{\sqrt{2}}e^{\mp i\Omega\sqrt{\bar{n}}t/2}\left[(e^{\mp i\Omega t/4\sqrt{\bar{n}}}\,|e\rangle \pm |g\rangle)\right], \qquad (3.20)
$$

$$
\left|\Psi_{f}^{\pm}(t)\right\rangle = e^{\pm i\Omega\sqrt{\bar{n}}t/4}\left|\alpha e^{\mp i\Omega t/4\sqrt{\bar{n}}}\right\rangle. \qquad (3.21)
$$

Eq.(3.19) describes a bipartite state whose degree of entanglement varies with time. The two atomic states $|\Psi_{at}^{\pm}\rangle$ are ‘dipole states’ superpositions, characterized by a phase difference between e and g evolving at frequency $\pm\Omega/\sqrt{\bar{n}}$. The two field states $|\Psi_{f}^{\pm}\rangle$ are coherent states whose complex amplitude phases also evolve at the same frequencies, in opposite directions in phase space. In the global wave function [Eq.(3.19)], the atom and field components in each term of the superposition stay locked in phase. The system's evolution is thus described by two very different frequencies. The global phases of the atom and field states rotates ‘fast’ at the Rabi frequency $\pm\Omega\sqrt{\bar{n}}$, while the ‘internal’ phases of the atom dipole and field coherent state rotate slowly at the $\bar{n}$ time smaller frequency $\pm\Omega/\sqrt{\bar{n}}$. At the classical limit ($\bar{n}\to\infty$), the former frequency remains finite while the latter goes to zero. We then retrieve the result given by Eq.(3.13). Let us recall that Eq.(3.19) is approximate and can be used only for $t \ll \pi\bar{n}/\Omega$. It is also valid for $\bar{n}$ not too small (mesoscopic regime). More exact expressions of the entangled atom-field system, valid for longer times, are given by Eqs. (3.12), (3.15) and (3.17).

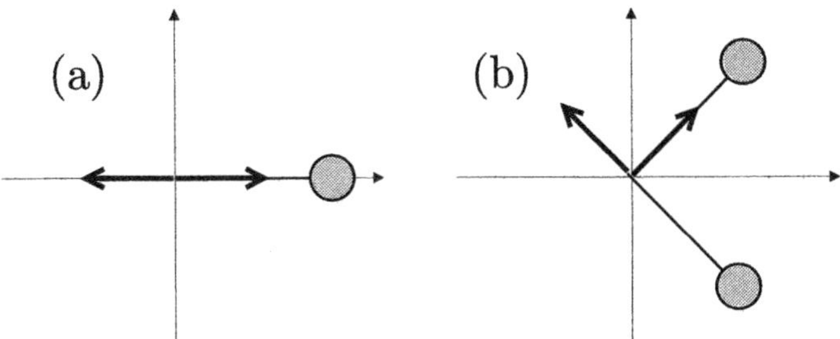

Fig. 17. Pictorial representation of the atomic and field state evolution in the phase plane. (a) initial situation. (b) at a later time, the atom and the field get entangled.

It is instructive to view the system's evolution in a pictorial way (Fig. 17). To exhibit clearly the correlated phase evolutions of the atom and field components, we describe the two level atom as a pseudo-spin evolving in the equatorial plane of a Bloch sphere and we make this plane coincide with the phase space plane of the field quadratures. The atomic states are represented as pseudo-spin vectors and the field coherent states are pictured as circles. At time $t = 0$, the initial e state appears as a superposition of the $|e\rangle + |g\rangle$ and $|e\rangle - |g\rangle$ states represented by two opposite Bloch vectors while the field state is a circle whose center lies on the X_0 quadrature axis [Fig. 17(a)]. As time evolves, the two dipole Bloch vectors start to rotate in opposite directions, while the field splits into two components which stay respectively in phase and π out of phase with the corresponding atomic dipole vectors [Fig. 17(b)]. This simple picture shows that the two subsystems get entangled as soon as the uncertainty circles associated to the field components cease to overlap. It also shows that the field evolves into a Schrödinger cat-like state, superposition of two coherent fields with opposite phases. The entanglement and the generation of the cat states occur within a time of the order of the slow phase drift period, $\sqrt{\bar{n}}/\Omega$. This time goes to infinity at the classical limit. The atom-field entanglement and the generation of Schrödinger cats of the field are thus quantum effects, directly linked to the grainines of the field state in the mesoscopic regime.

The overlap integral of the two field components is readily computed, using Eq.(1.43):

$$\begin{aligned}\left\langle \Psi_f^-(t) \,\middle|\, \Psi_f^+(t)\right\rangle &= e^{i\Omega\sqrt{\bar{n}}t/2}\left\langle \alpha e^{i\Omega t/4\sqrt{\bar{n}}} \,\middle|\, \alpha e^{-i\Omega t/4\sqrt{\bar{n}}}\right\rangle \\ &= e^{i\Omega\sqrt{\bar{n}}t/2} e^{-\bar{n}[1-\exp(-i\Omega t/2\sqrt{\bar{n}})]} \,. \end{aligned} \tag{3.22}$$

At short times such that $\Omega t < 1$, we can expand the exponent in Eq.(3.22) as:

$$-\bar{n}[1-\exp(-i\Omega t/2\sqrt{\bar{n}})] = -\frac{i\Omega t\sqrt{\bar{n}}}{2} - \frac{\Omega^2 t^2}{8} + \ldots , \tag{3.23}$$

and approximate the field components scalar product as:

$$\left\langle \Psi_f^-(t) \,\middle|\, \Psi_f^+(t) \right\rangle \approx e^{-\Omega^2 t^2/8} . \tag{3.24}$$

The overlap between the two field states decays as a Gaussian function of time, over a time scale of the order of the vacuum Rabi period $1/\Omega$. This is indeed the time it takes for the dephasing between the two components to reach a value $1/\sqrt{\bar{n}}$, of the order of the phase uncertainty of the coherent field. The decay of the field components overlap is directly related to the collapse of the atomic Rabi oscillation. The field scalar product is indeed involved in the expression of the atomic reduced density operator which writes, at short times:

$$\begin{aligned}\tilde{\rho}_{at}(t) \quad &= \mathrm{Tr}_{\mathrm{field}}\left(\left|\tilde{\Psi}(t)\right\rangle\left\langle\tilde{\Psi}(t)\right|\right) \approx \\ &\frac{1}{2}\left|\Psi_{at}^+(t)\right\rangle\left\langle\Psi_{at}^+(t)\right| + \frac{1}{2}\left|\Psi_{at}^-(t)\right\rangle\left\langle\Psi_{at}^-(t)\right| + \\ &\frac{1}{2}\left(\left|\Psi_{at}^+(t)\right\rangle\left\langle\Psi_{at}^-(t)\right| + \left|\Psi_{at}^-(t)\right\rangle\left\langle\Psi_{at}^+(t)\right|\right)e^{-\Omega^2 t^2/8} .\end{aligned} \tag{3.25}$$

This operator, expressed in the basis of the atomic dipole states, thus loses its off-diagonal elements as soon as the field components are separated. The atomic coherence is directly related to the contrast of the Rabi oscillation signal which can be written as:

$$\begin{aligned}P_e(t) \quad &\approx \langle e|\,\tilde{\rho}_{at}(t)\,|e\rangle \approx \frac{1}{2}\left|\langle e \mid \Psi_{at}^+(t)\rangle\right|^2 + \frac{1}{2}\left|\langle e \mid \Psi_{at}^-(t)\rangle\right|^2 + \\ &+\mathrm{Re}\,\left\langle e \mid \Psi_{at}^+(t)\right\rangle\left\langle\Psi_{at}^-(t) \mid e\right\rangle e^{-\Omega^2 t^2/8}\end{aligned} \tag{3.26}$$

The collapse of the Rabi oscillation appears as a complementarity effect. The atomic interference is washed out when the field components (to which these dipole states are locked) carry an information about these states. This happens as soon as these field components become distinguishable, i.e. quasi orthogonal. Combining Eqs.(3.26) and (3.20), we get an analytical expression for the Rabi oscillation signal which confirms in quantitative terms the qualitative description given above :

$$P_e(t) \approx \frac{1}{2}[1 + e^{-\Omega^2 t^2/8}\cos(\Omega\sqrt{\bar{n}}t)] \tag{3.27}$$

When the Rabi oscillation has collapsed, the atom-field system presents entanglement. Its state is indeed a superposition with equal weights involving nearly orthogonal field states. A Schrödinger cat state of the field is created, generally entangled to the atom. A field state, de-correlated from the atom can be obtained by measuring the atom's state. Detecting the atom in e or g projects the field in a superposition of the two quasi orthogonal components. Defining $\Phi = \Omega t/4\sqrt{\overline{n}}$ and using Eq.(3.19), we get the expression of the field after detection of the atom in e:

$$\left|\tilde{\Psi}_{\text{field}}(t)\right\rangle_{\text{atom in }e} = \frac{1}{\sqrt{2}}\left[e^{-i(\overline{n}+1)\Phi}\left|\alpha\, e^{-i\Phi}\right\rangle + e^{i(\overline{n}+1)\Phi}\left|\alpha\, e^{i\Phi}\right\rangle\right] . \tag{3.28}$$

The state of the 'cat' is then conditioned to the measurement of the atom (a different cat state is prepared if atom is detected g). In fact, measuring the atom is not always required to prepare a field cat state, separate from the atom. When the phase of the field components has rotated by $\pi/2$, a π–phase cat with two components having opposite phases is produced, which is at this time not entangled with the atom:

$$\begin{aligned}\left|\tilde{\Psi}(t = 2\pi\sqrt{\overline{n}}/\Omega)\right\rangle &\approx \frac{1}{2}\left[e^{-i\pi\overline{n}/2}\left|\alpha\, e^{-i\pi/2}\right\rangle - e^{i\pi\overline{n}/2}\left|\alpha e^{+i\pi/2}\right\rangle\right] \\ &\otimes\left[e^{-i\pi/2}\left|e\right\rangle + \left|g\right\rangle\right]\end{aligned} \tag{3.29}$$

The disentanglement occurs at this specific time because the two atomic dipole states, initially π-out of phase, have evolved into the same e, g superposition, thus resulting in a factorization of the atom-field wave function.

Let us now consider still longer atom-field interaction times. After a time $T_{\text{revival}} = 4\pi/\Omega\sqrt{\overline{n}}$, such that the fields components have rotated by π, the two field components are again overlapping, with a common phase opposite to the initial one. The atomic dipole states are then again orthogonal, their initial phases being exchanged. The coherence of the reduced density operator is then restored and the Rabi oscillation revives, hence the name given to this specific time. This revival effect, predicted long ago, appears here, as does the collapse phenomenon, as a manifestation of complementarity. At the revival time, the field does not contain any information about the dipole state in which the atom is and the interference between the corresponding atomic probability amplitudes reappears.

The above analysis is only qualitative, since it relies on an approximate expression of the atom-field state. It is also possible to compute exactly the Rabi oscillation signal given by Eq.(3.10) for a finite average photon number. Fig. 18 shows the result of this exact calculation, for $\overline{n} = 15$ photons. The collapse and revival are clearly visible. According to the simplified model, the Rabi oscillations should reappear with a 100% contrast. The numerical calculation shows that

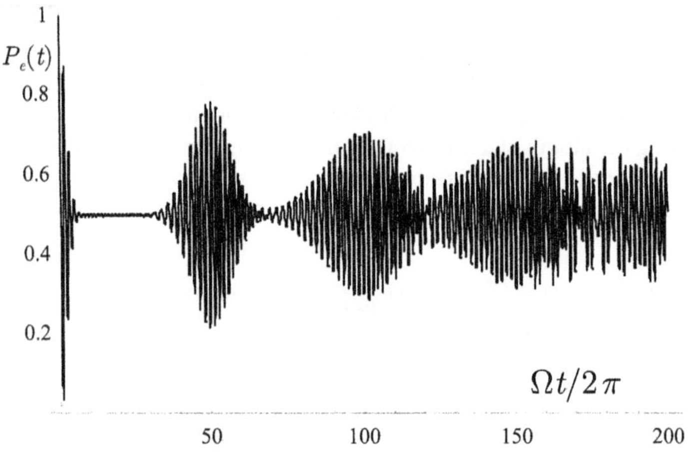

Fig. 18. Rabi oscillation collapse and revivals: computed probability $P_e(t)$ for finding the atom in state $|e\rangle$ versus the interaction time t in units of $2\pi/\Omega$. The cavity contains initially an $\overline{n} = 15$ photons coherent field.

the revival has rather a contrast of about 50%. The simple model also predicts a succession of identical revivals, as the two field components merge together periodically. The exact numerical calculation shows instead that the successive revivals have a decreasing amplitude and an increasing time duration, resulting in a smearing out of the collapse and revival features after two or three revivals. The differences between the exact results and the simple model predictions are due to the drastic approximation made by neglecting terms of second order in Eq.(3.17). Even if the complete phase diffusion of the field takes a long time of the order of $\overline{n}/\Omega$ the phase spreading after a time $\sqrt{\overline{n}}/\Omega$ is of the order of $\pi/\sqrt{\overline{n}}$ which is precisely of the order of the fluctuation of the initial coherent field phase. In other words, the phase of the field spreads by a factor of the order of 2 between time 0 and $4\pi\sqrt{\overline{n}}/\Omega$. Not surprisingly, a calculation neglecting this spreading effect is only qualitative.

Rabi oscillation collapses and revivals have been observed in experiments performed with very small coherent field (photon numbers of the order unity) [56]. Similar collapse and revival phenomena have been observed with Rydberg atoms in thermal or micromaser fields [57]. They have also been seen in studies in which two internal levels of a trapped ions are coupled via a Jaynes-Cummings-type Hamiltonian to a few vibration quanta of the trap potential [58]. One should note that the present analysis, with its specific approximations valid for mesoscopic fields containing tens of photons, does not describe quantitatively these

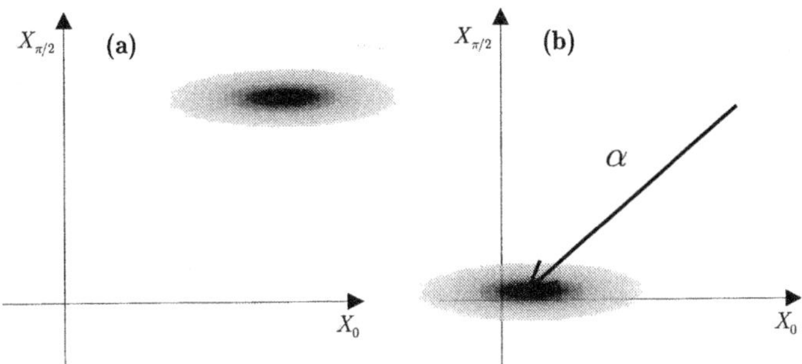

Fig. 19. Principle of homodyne detection. (a) original field. (b) translated field.

experiments. Mesoscopic collapses and revivals have not yet been observed, because the expected revival time, proportional to $\sqrt{\bar{n}}$, is too long. We mention below a related experiment in which Rabi oscillation revivals are induced at an earlier time.

3.3. Observing the Schrödinger cat by a homodyne Q function measurement

The generation of a Schrödinger cat in the Rabi oscillation experiment manifests itself by a phase splitting effect. In order to observe the phase distribution of the field resonantly interacting with the atom, we need to implement a homodyne technique adapted to microwave CQED [19]. In an usual homodyne detection, one would employ a beam splitter to mix the signal with a reference field and a quadratic photo-detector to measure the resulting beating (see section 2). This cannot be done in microwave CQED because the field stored in the cavity cannot be perturbed by a beam splitter or an usual detector. We have developed a variant of homodyning in which a reference coherent field is added to the field to be measured by coupling the cavity to a classical source, the resulting field being detected by an absorbing 'probe' atom crossing subsequently the cavity. This method amounts in effect to a direct measurement of the Q function of the field in the cavity.

The principle of the method is sketched in Fig. 19. We assume that the cavity stores a field described by an operator density ρ_f, qualitatively represented by a shaded area in phase space [Fig. 19(a)]. In order to measure this field, we couple the cavity to a classical pulse of microwave of convenient phase and duration. This amounts to a translation in phase space (see section 1), the initial field being changed into the one described by the density operator $D(\alpha)\rho_f D^{-1}(\alpha)$ [Fig. 19(b)]. In order to measure this field, we then send across the cavity a probe atom, prepared in the lower level g and, by repeating the experiment many times, we measure the probability that the probe atom remains in g after the cavity crossing. We assume furthermore that the probe atoms have a totally random distribution of interaction times with the field, so that it is legitimate to average over time their transition probability. It is then easy to show that the average probability P_g for finding the probe atoms exiting the cavity in g is directly related to the probability that the field in the cavity contains 0 photons. Calling p_n the photon number distribution of this field, we have indeed $P_g = p_0 + \sum_{n>0} p_n \overline{\cos^2(\Omega\sqrt{n}t/2)}$, with the bar over the $\cos^2$ term meaning an average over time. Assuming that this average yields a $1/2$ value and taking into account the normalization of the p_n's ($\sum_{n>1} p_n = 1 - p_0$), we thus get $P_g = (1 + p_0)/2$ and $p_0 = 2P_g - 1$. We finally note that p_0 is simply the diagonal matrix element in vacuum of the translated field density operator, which we can write as:

$$p_0 = 2P_g - 1 = \langle 0|D(-\alpha)\rho_f D(\alpha)|0\rangle\ , \tag{3.30}$$

and we conclude that measuring P_g directly yields the initial field Q function (see Section 1) at the point in phase space represented by α [compare Eq.(3.30) with (1.55)]. By sweeping this parameter and resuming the same procedure, we sample the $Q(\alpha)$ function of the initial field in the cavity. We have just described an ideal situation. Realizing a velocity distribution of probe atoms corresponding to a completely random atom-cavity interaction time is not easy. In practice, we are using probe atoms with a broad Maxwellian velocity distribution. The interaction time, inversely proportional to the velocity, does not have a flat distribution. This results in a sum over n of the p_n's in the expression of P_g which does not exactly reduces to $1 - p_0$. The probe atom signal remains however a fair approximation of $Q(\alpha)$ and yields information about the phase distribution of the field.

We have implemented this method for investigating the 'cat states' produced by the interaction of a single resonant atom with the cavity field [19]. The procedure involves the following successive steps. We first inject a coherent field $|\beta\rangle$ in the cavity, with well defined phase and average photon number (determined by an independent calibration experiment based on the measurement of the light shift induced by this field on an auxiliary atom). Immediately after this field has been injected, we send a resonant atom in level e which undergoes a Rabi oscillation in the field. The interaction time of this atom with the field is adjusted to one of two preset values (32 and 52 μs) by selecting the atom's velocity. The resulting field is then measured by the method outlined above. A reference coherent field with amplitude α is injected and a probe atom in g is subsequently sent across the cavity, this atom being finally detected by a state selective detector. Since we know the amplitude of the field to be measured (equal to the initial amplitude of the coherent field β, diminished by the natural cavity decay during the experimental sequence), the reference field is adjusted to the same amplitude and we vary only its phase ϕ. We finally extract from the data the probability P_g and plot it versus ϕ thus exhibiting the phase splitting effect produced by the Rabi oscillation. More experimental details can be found in [19].

Fig. 20(a) shows the $P_g(\phi)$ function for the two selected interaction times. The signals are plotted versus ϕ for different mean photon numbers $\overline{n}$ in the range 15 to 36. The splitting by a single atom of the field into two symmetrical components with different phases is clearly visible. For a given interaction time, the splitting decreases with field amplitude and for a given $\overline{n}$ the separation increases with time. Fig. 20(b) summarizes the results by plotting the phases of the two components as a function of the dimensionless parameter $\Phi = \Omega t/4\sqrt{\overline{n}}$. The dotted lines correspond to the phases predicted by the simple model discussed above. The solid line results from a numerical simulation solving the exact field equation of motion and taking into account cavity damping. The agreement between the experimental points and the solid line is very good. The maximum phase splitting observed, for $\overline{n} = 15$ photons and an interaction time of 52 μs is 90 degrees.

We have also checked the correlation between the atomic state and the field phase by selectively preparing $|\Psi_{at}^{+}\rangle$ or $|\Psi_{at}^{-}\rangle$ at the beginning of the interaction [19], within a time short enough so that the slow drift of the atom and field phases can be neglected. To prepare $|\Psi_{at}^{+}\rangle$, the atom, initially in g, first performs a $\pi/2$ Rabi pulse according to the transformation:

$$|g\rangle \rightarrow \frac{1}{\sqrt{2}}[e^{-i\pi/4}|\Psi_{at}^{+}(0)\rangle - e^{i\pi/4}|\Psi_{at}^{-}(0)\rangle] . \tag{3.31}$$

The atom is then detuned by Stark effect with respect to the cavity, during a time much shorter than the Rabi period. A pulse of electric field is applied between the cavity mirrors, whose effect is to shift by $\pi/2$ the relative phase of the e and g states [30]. The sequence of Rabi and Stark pulses transforms the initial g state, superposition of the interfering $|\Psi_{at}^{+}(0)\rangle$ and $|\Psi_{at}^{-}(0)\rangle$ states into $|\Psi_{at}^{+}(0)\rangle$ alone. The system ends up in the slowly evolving quasi-stationary state described by the first term in the right hand side of Eq.(3.19). The atom and the field subsequently drift in phase in only one direction. We observe that the Rabi oscillation is frozen from then on. A homodyne measurement of the field phase after the atom exits from C reveals, as expected, only a single phase shifted component (open circles in Fig. 21). Similarly, one prepares $|\Psi_{at}^{-}(0)\rangle$ by applying the same Rabi and Stark switching pulse sequence starting from level e. This state couples to the other component of the field as revealed by the subsequent homodyne detection (solid squares in Fig. 21). This experiment clearly demonstrates correlations between the atomic and field state phases.

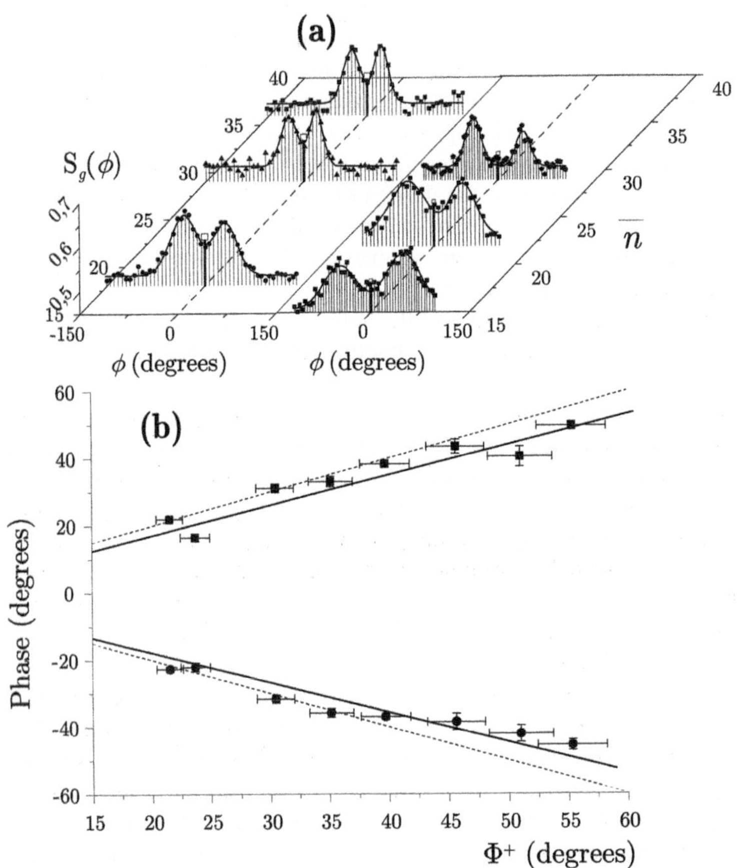

Fig. 20. (a) Field phase distribution for $\bar{n}$ values in the range 15 to 36. The interaction time is 32 μs (left) or 53 μs (right). The points are experimental and the curves are fits on a sum of Gaussians. (b) Phases of the two field components versus $\Phi = \Omega t/4\sqrt{\bar{n}}$. Dotted and solid lines are theoretical (see text). Adapted from [19].

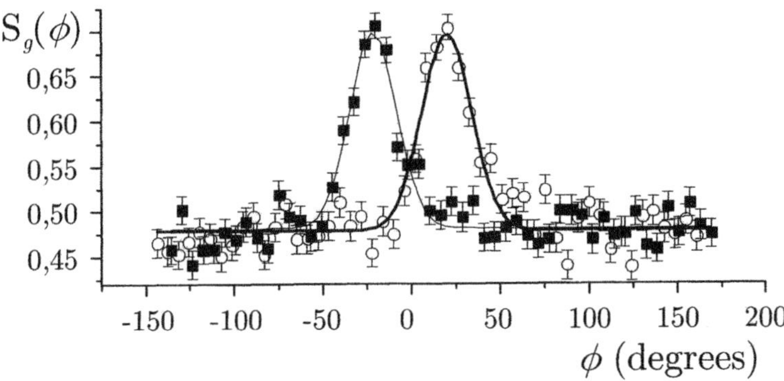

Fig. 21. Field phase distributions following preparation of atomic states $|\phi_a^{\pm}(0)\rangle$ by combination of Rabi and Stark pulses ($\overline{n} = 27$; interaction time 32 μs). Open circles: preparation of $|\phi_a^{+}(0)\rangle$. Solid squares: preparation of $|\phi_a^{-}(0)\rangle$. Solid lines are Gaussian fits. From [19].

We now turn to the analysis of the mesoscopic state coherence. Let us first stress that the experimental parameters are consistent with a survival of this coherence. Fig. 22 shows, for $\overline{n} = 36$ and an interaction time of 32 μs, the expected Wigner function $W(\beta_x + i\beta_y)$ of the field in the cavity. It results from the explicit numerical simulation including the actual experimental parameters. The field state is computed conditioned to an atom found in g, 48 μs after its crossing the cavity axis. We have chosen to compute the W function, and not the Q one, because the former is expected to contain a conspicuous interference term describing the coherence of the cat state whereas the later is largely insensitive to the coherence (see Section 1). We see indeed that this W function clearly exhibits the two separate field components and the interferences which are a clear signature of a mesoscopic quantum coherence. The square of the distance in phase space between the two components, $D^2 = 4\overline{n}\sin^2(\Omega t/4\sqrt{\overline{n}})$ is a measure of the mesoscopic character of this superposition. In the range of $\overline{n}$ values we have explored, D^2 is nearly constant versus $\overline{n}$, equal to 20 for the short interaction time (32 μs) and to 40 for the long one (52 μs). The theoretical decoherence time of the Schrödinger cat state superposition is $2T_{cav}/D^2$ [Eq.(2.35)]. With our cavity decay time $T_{cav} = 1$ ms, we thus expect a decoherence time of 85 μs for the cat prepared within 32 μs, which indicates that the field retains its macroscopic coherence over the duration of the experimental sequence, as confirmed by its Wigner function.

We should note however that the above discussion is based on theoretical arguments. The experimental determination of the field Wigner function in this experiment remains to be done. In the mean time, we have performed another

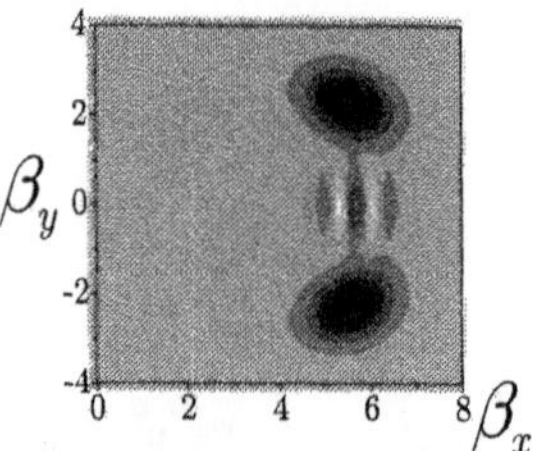

Fig. 22. Computed cavity field Wigner function $W(\beta_x + i\beta_y)$ for $\overline{n} = 36$ and a 32 μs interaction time. Adapted from [19].

simpler coherence test, based on the observation of the Rabi oscillation revival phenomenon. We have already mentioned that we cannot wait long enough for the revival to occur spontaneously. We can however induce the phenomenon at an earlier time and the observation of the reappearing Rabi oscillation is a clear indication of the cat state coherence survival over that time. After collapse of the Rabi oscillation, we apply to the atom, at time T, a Stark pulse switching the signs of the quantum amplitudes associated to e and g. This procedure implements a scheme for decoherence tests in CQED proposed in [59]. According to Eq.(3.19), the Stark pulse suddenly exchanges the atomic states correlated to the $|\Psi_f^+(t)\rangle$ and $|\Psi_f^-(t)\rangle$ field components. The atom-field coupling resumes afterwards, reversing the sign of the rotation of the two field components. At time $2T$, the two fields are back in phase and the Rabi oscillation revives, revealing the coherent nature of the atom-cavity state. This induced revival, reminiscent of a spin echo, is under investigation at the present time and will be published after these notes are completed. We describe in the next section another experiment revealing a 'cat' state coherence, performed in a related dispersive CQED experiment.

3.4. *Dispersive cats in cavity QED*

The coupling of a resonant two-level atom with a field mode in a mesoscopic coherent state singles out two atomic 'dipole states' $|\Phi_{at}^{\pm}\rangle$, equal weight superpositions of e and g, whose phase evolves slowly during the interaction (at the classical limit, this phase is stationary). These atomic dipole states can be seen as eigenstates of the interaction, corresponding to two different atomic indices seen by the field interacting with the atom in the cavity. The atom, initially in level e, is prepared in a superposition of these two eigenstates of the refractive index and the field thus evolves into a superposition of two components with different phases. This interpretation can be extended to the case when the atom and the field are off-resonant (dispersive coupling) with a detuning δ between the atom and the field mode. In this case, the interpretation in term of refractive index is even more appropriate since an off-resonant atom does not exchange any energy with the field and thus really behaves as a piece of transparent dielectric medium as far as its coupling to the field is concerned. We now show that Schrödinger cat states of the field can be generated by exploiting this dispersive interaction and we recall briefly the experiments we have performed with these states.

When the atom-field coupling is not resonant, an exact diagonalization of the Hamiltonian is still possible [Eqs.(3.4) and (3.5)]. It is however more transparent physically to do a perturbative treatment [27, 28]. We assume $\delta = \omega_0 - \omega > 0$, with mixing angles $\theta_n \ll 1$ (the case $\delta < 0$ and $\theta_n \approx \pi$ is treated similarly, with minor obvious changes in the equations). We develop the eigenstates and eigenenergies in power of $\Omega\sqrt{n+1}/\delta$. Eqs.(3.4) and (3.5) yield (to first order for states and to second order for energies):

$$
\begin{aligned}
|+,n\rangle &\approx |e,n\rangle + \frac{\Omega\sqrt{n+1}}{2\delta}\,|g,n+1\rangle \\
|-,n\rangle &\approx -\,|g,n+1\rangle + \frac{\Omega\sqrt{n+1}}{2\delta}\,|e,n\rangle \\
\frac{1}{\hbar}E_{+,n} &= (n+\frac{1}{2})\omega + \frac{\omega_0}{2} + \frac{\Omega^2(n+1)}{4\delta} \\
\frac{1}{\hbar}E_{-,n} &= (n+\frac{3}{2})\omega - \frac{\omega_0}{2} - \frac{\Omega^2(n+1)}{4\delta}\;.
\end{aligned}
\qquad (3.32)
$$

The energy development is valid if the parameter $\Omega^2(n+1)/\delta^2$ is much smaller than 1. We will assume the condition:

$$
\frac{\Omega\sqrt{n+1}}{\delta} \leq \frac{1}{3}\;. \qquad (3.33)
$$

We see on Eq.(3.32) that the eigenstates of the atom-field system $|+,n\rangle$ and $|-,n\rangle$ ('dressed states') are very close to the uncoupled states $|e,n\rangle$ and $|g,n+1\rangle$ which are slightly 'contaminated' (to first order for the states and to second order for the energies) by the atom-field coupling. The energies of these levels are shifted to second order by an amount linear in n (light shift effect).

Assume now that we couple an atom in level e with a coherent field $|\alpha\rangle$ and let the two systems interact for a time t. Expanding the coherent state on a Fock state basis and taking into account that $|e,n\rangle$ is very close to the $|+,n\rangle$ dressed state, we get:

$$
\left|\Psi_{e,\alpha}(0)\right\rangle \quad = |e\rangle\,|\alpha\rangle = \sum_n C_n\,|e,n\rangle \Rightarrow
$$

$$\left|\Psi_{e,\alpha}(t)\right\rangle \approx \sum_n C_n e^{-i(n+1/2)\omega t} e^{-i\omega_0 t/2} e^{-i\Omega^2(n+1)t/4\delta} |e,n\rangle\,, \tag{3.34}$$

which, in interaction picture yields:

$$\left|\tilde{\Psi}_{e,\alpha}(t)\right\rangle \approx \sum_n C_n\, e^{-i\Omega^2(n+1)t/4\delta} |e,n\rangle \quad = e^{-i\Omega^2 t/4\delta} |e\rangle \otimes \left|\alpha e^{-i\Omega^2 t/4\delta}\right\rangle . \tag{3.35}$$

Similarly, for an atom initially in level g we obtain:

$$\left|\tilde{\Psi}_{g,\alpha}(t)\right\rangle \approx \sum_n C_n e^{+i\Omega^2 nt/4\delta} |g,n\rangle \quad = |g\rangle \otimes \left|\alpha e^{+i\Omega^2 t/4\delta}\right\rangle \tag{3.36}$$

The cavity field remains in a coherent state, phase shifted by an angle $\pm\chi = \pm\Omega^2 t/4\delta$ depending upon whether the atom is in level e or g. This effect is interpreted by attributing to the atom an index N_i which, at low field intensity, writes:

$$N_i = 1 \pm \frac{\Omega^2}{4\delta\omega_c}\,, \tag{3.37}$$

with the $+$ and $-$ signs corresponding respectively to an atom in e or g. With the parameters of our microwave CQED experiment ($\Omega = 2\pi \times 50$ kHz, $\omega = 2\pi \times 51$ GHz and $\delta = 3\Omega$) we find $|N_i - 1| = 10^{-7}$, which is a huge value for a single atom effect. Note that this index is linear for extremely low fields only and saturates for average photon numbers of the order of $(\delta/\Omega)^2$. Note also, in addition to the classical phase shift of the coherent state, the global quantum phase shift of the system's state when the atom is in level e. This is a cavity Lamb shift effect (dephasing associated to an excited state in presence of cavity vacuum).

In order to observe directly the phase shifts induced on a coherent field by a non-resonant atom, we have performed the experiment in the same way as for the resonant case, by employing the homodyne method adapted to CQED (results not yet published). The signals are very similar to the ones shown in Fig. 20. The only difference lies in the preparation of the atomic initial states, which must be eigenstates of the refractive index. These are linear combinations of e and g in the resonant case whereas they are simply the e and g states in the non-resonant situation. It is thus easier to prepare these states in the non-resonant case and we observe the two field components, with a positive or a negative phase depending on whether the atom is in e or in g.

By preparing the atom in a superposition of e and g one can also entangle the atom and field systems and prepare superposition states of the field of the Schrödinger cat type. Here again, the experiment is very similar to the one performed in the resonant case. Whereas in the resonant situation the initial atomic

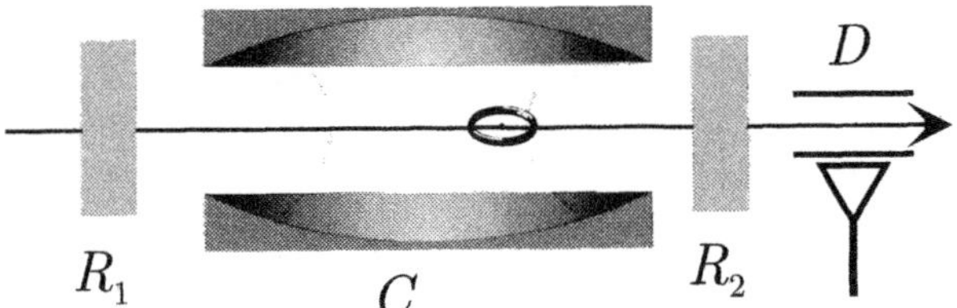

Fig. 23. Scheme of the dispersive Schrödinger cat experimental set-up

state e was already in a superposition of the eigen-atomic dipoles states, we must now achieve this preparation by an explicit manipulation, preparing first the atom in e, then submitting it to a $\pi/2$ pulse of classical radiation R_1, using an auxiliary low-Q field mode acting on the atom before it enters the cavity. With a proper choice of the phase of the R_1 pulse, the atom-field system can evolve according to:

$$|e\rangle\,|\alpha\rangle \to \frac{1}{\sqrt{2}}(|e\rangle + e^{-i\chi}\,|g\rangle)\,|\alpha\rangle \to \frac{e^{-i\chi}}{\sqrt{2}}[|e\rangle\left|\alpha e^{-i\chi}\right\rangle + |g\rangle\left|\alpha e^{i\chi}\right\rangle] \,, \quad (3.38)$$

where the first arrow represents the effect of the classical pulse and the second the one of the atom-cavity field coupling. The phase of the classical pulse R_1 can be adjusted to induce a quantum phase difference between e and g precisely equal to $-\chi$, which compensates for the Lamb shift of the e state and insures that the two components of the field correlated to e and g have quantum amplitudes with the same phase. This condition is not essential, but it simplifies the expressions of the Schrödinger cat states, by making them genuine even and odd parity states when $\chi = \pi/2$ (see below). In the experiments we have performed so far, the classical dephasing 2χ between the two field components was typically of the order of 50 to 100 degrees for fields containing on average a few photons. Note that the non-resonant phase splitting is always smaller than the one observed in resonant case for the same number of photons and interaction time [the angle χ computed under the condition (3.33) is smaller than $\Omega t/4\sqrt{\overline{n}}$].

In the entangled state given by Eq.(3.38), the phase of the field acts as a meter pointing to the atomic energy. Detecting the atom in e or in g would result in a collapse of the state into one of the two components and the ambiguity of the cat state would be lost. In order to maintain the quantum ambiguity, we must mix again the two states e and g after the atom exits the cavity and perform the atomic detection only after. This second mixing occurs in a classical microwave zone R_2 identical to R_1. By a proper choice of phase, this second pulse can induce the transitions $|e\rangle \to [|e\rangle + |g\rangle]/\sqrt{2}$; $|g\rangle \to [|g\rangle - |e\rangle]/\sqrt{2}$. The set-up used for this experiment is sketched in Fig. 23. It shows the cavity mode sandwiched between the two zones R_1 and R_2, a set up reminiscent of a Ramsey separated fields interferometer (more on this below). The combination of the R_2 pulse followed by an energy selective field ionization Rydberg atom detector is equivalent to a coherent device detecting the two orthogonal superpositions $|g\rangle \pm |e\rangle$. Detecting the atom in one of these states, when the system has been prepared in the state given by Eq.(3.38), projects the field into one of the two Schrödinger cat states $|\alpha e^{-i\chi}\rangle \pm |\alpha e^{i\chi}\rangle$. Here again, there is a complete analogy with the similar experiments in the resonant case.

The dispersive cat experiment [16] has been performed before the resonant ones [19]. At the time the dispersive experiment was made, the direct homodyne method adapted to CQED was not yet designed and the phase splitting of the cat was observed indirectly, via an effect quite similar to the collapse of the Rabi oscillation. We have detected the Ramsey fringes of the atom subjected to the two R_1 and R_2 pulses sandwiching the cavity [16]. By sweeping the frequency of these pulses around the atomic resonance, we have observed a modulation of the probability to detect the atom in e or in g. These so-called Ramsey fringes are a manifestation of a quantum interference. As long as we have no way of knowing whether the atom (initially prepared in e and finally detected in g) has undergone the transition in R_1 or in R_2, we must associate a probability amplitude to each of these possibilities and the transition probability contains an interference term between them, responsible for the fringe signal. There is a strong analogy between a Ramsey interferometer and the Mach-Zehnder studied in Section 2, the beam splitters being replaced by the mixing pulses. To underscore this analogy, Fig. 24(a) represents a diagram showing the two paths followed in Hilbert space by the two-level atom as it crosses the apparatus. The upper horizontal line represents level e and the lower level g. The topological analogy with a spatial interferometer is striking (compare with Fig. 11).

Let us focus now on the effect of the coherent field in the cavity placed between the Ramsey zones. This field acts, as seen above, as a meter pointing towards the atomic energy. As soon as the two components of the field cease to overlap, this meter could tell us unambiguously whether the atom crosses the interferometer in e or in g and the Ramsey interference should vanish. This is again complementarity in action, an effect fully similar to the Rabi oscillation collapse analyzed above. The Ramsey interference between the states e and g merely replaces the Rabi oscillation corresponding to the interference between the dipole states $|e\rangle + |g\rangle$ and $|e\rangle - |g\rangle$. To observe this effect, we have recorded the Ramsey fringes for a coherent field containing on average $\overline{n} = 9$ photons and for various detunings δ. Since the cat states dephasing is inversely proportional to δ, the overlap between the two field components increases as δ is decreased [from top to bottom in Fig. 24(b)]. The Ramsey fringe contrast accordingly decreases. The fringe amplitude directly measures the scalar product of the two

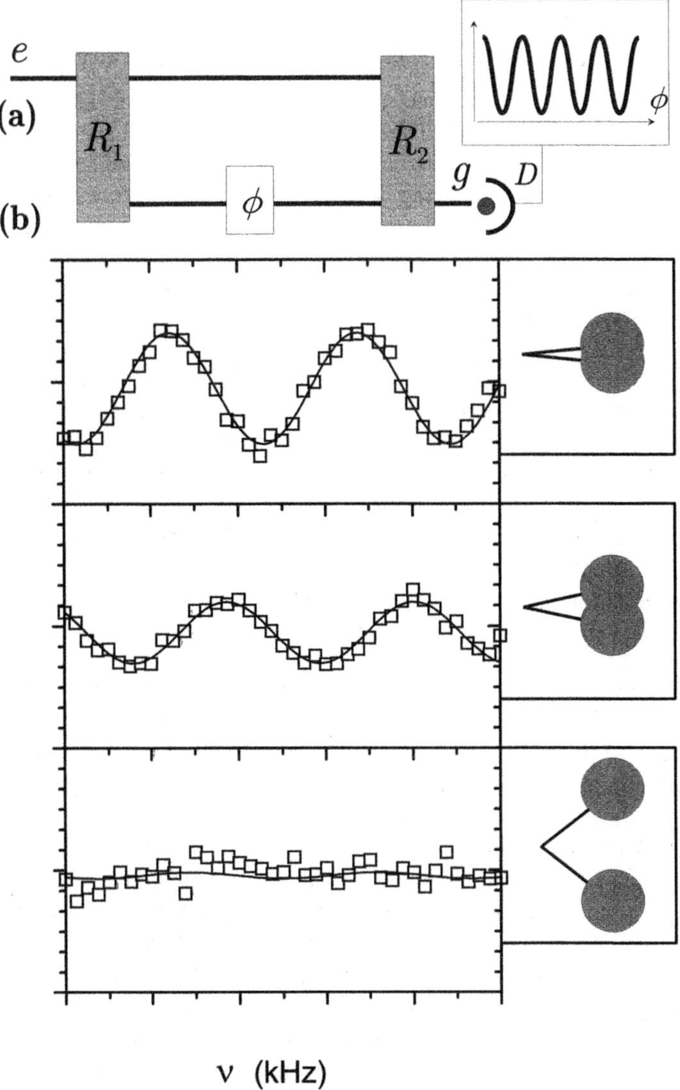

Fig. 24. (a) Scheme of a Ramsey interference process showing the two paths followed by the atom in level e or g. (b) Collapse of Ramsey fringes when the cavity C between R_1 and R_2 initially contains a coherent field with an amplitude $|\alpha| = \sqrt{9.5} = 3.1$, The atom-cavity detuning is $\delta/2\pi = 712$, 347 and 104 kHz respectively, from top to bottom. Points are experimental and curves are sinusoidal fits. Insets show the phase space representation of the final field components in C. Adapted from [16].

field component states.

3.5. *Testing the coherence of π-phase cats: a parity measurement*

Let us now focus on special kind of cats, namely the even or odd superposition of fields with opposite phases. These so called π-phase cats have been studied in Section 1. They can be generated in our Ramsey CQED set up, according to the method outlined above, by letting the dispersive coupling between the atom and a coherent field act for a time such that $\chi = \pi/2$. Whether an even or an odd phase cat is prepared depends upon the random state in which the atom is finally detected after the second Ramsey zone ('even cat' if the atom is found in g, 'odd cat' if it is found in e)

It is instructive to analyze this cat state preparation as a quantum measurement of the field photon number parity $\mathcal{P}$. The procedure by which the cat is prepared, with its succession of two Ramsey pulses applied before and after the atom interacts with the coherent field in the cavity can in fact be described as an ideal measurement of the photon parity. The field in the cavity does affects the phase of the atomic coherence in the Ramsey interferometer. This effect is the counterpart, on the atom, of the phase change induced by it on the field. The phase shift of the atomic coherence can be computed from Eq.(3.32) by estimating the phase accumulation between the dressed levels $|+, n\rangle$ (close to $|e, n\rangle$) and $|-, n-1\rangle$ (close to $|g, n\rangle$) during the time the atom crosses C. It is easy to show that the condition for creating a π-phase cat state is that the atomic coherence accumulates a phase shift of π when the atom interacts with the field of a single photon in the cavity. The same phase shift is obviously accumulated (modulo 2π) for any odd photon number. Conversely, if the cavity contains the vacuum, or any even photon number, the phase shift of the atomic coherence accumulates no phase shift (modulo 2π). In other words, the phase between the two arms of the Ramsey interferometer has a well defined value for all even photon numbers, and is shifted by π for all odd photon numbers, provided the condition for creating a π-phase cat is satisfied. Suppose now that the interferometer is set at a 'bright fringe' for finding the atom in level g if the photon number is even. It it is then also at a bright fringe for finding the atom in e if the photon number is odd. In other words, using this setting of the interferometer makes the atomic state act as a 'pointer' measuring the parity of the photon number. We can remark now that the initial coherent field in the cavity can always be expressed as a linear superposition of even and odd cat states according to:

$$|\beta\rangle = \frac{1}{2}[|\beta\rangle + |-\beta\rangle] + \frac{1}{2}[|\beta\rangle - |-\beta\rangle]\,. \tag{3.39}$$

This analysis leads us to interpret the cat generation, when the atom has interacted with the field and has been detected, as a collapse resulting from an information acquisition about the field. The atom, combined with the Ramsey interferometer, is a measuring device for the photon parity. Since the state to be measured is an equal weight superposition of two eigenstates with different parity eigenvalues, the outcome of the measurement is random, and after it the field is found in the corresponding eigenstate, i.e. in the even cat sate $[|\beta\rangle + |-\beta\rangle]/\sqrt{2}$ if the atom is found in g, in the odd cat state $[|\beta\rangle - |-\beta\rangle/\sqrt{2}$ if the atom is found in e. This description of the experiment is fully equivalent to the one given above, in terms of phase shifts of opposite signs experienced by the two field components interacting with the atom.

These parity considerations shed an interesting light on some already discussed properties of the π-phase cats. We have shown in Section 2 that the coherence of the cat state is destroyed as soon as a single photon is lost in the environment. According to Eq.(1.52) a photon annihilation process switches the parity of a π phase cat state. When the parity of the superposition is changed, the sign of the interference term in the cat quadrature signal, signature of the state coherence, is also changed. In fact, we do not know for sure when a photon is lost if we do not look at the environment. We can only know the probability for a photon to be lost. As soon as this probability is 50%, we can say that the system is in a statistical mixture of even and odd cat states. The interference term is then canceled. We retrieve in this way the conclusions reached earlier about the fragility of this state.

If we could observe the environment and detect the photons lost into it one by one, we could 'see' the cat state undergo quantum jumps between even and odd states, remaining in a coherent superposition whose phase would evolve under a stochastic process. This approach is the one adopted in the Monte Carlo calculations in quantum optics [60], in which single stochastic trajectories of a quantum system are followed and the density matrix reconstructed by summing over these trajectories. To say that the coherence of a cat state is lost is merely a statement about the fact that we are renouncing to follow the evolution of the environment.

In fact, observing the quantum jumps of the field under the effect of decoherence does not even require a direct detection of the environment. It is enough to measure the photon number parity operator, according to the atomic interferometric method described above. A succession of atoms is sent across the apparatus experiencing the Ramsey pulses and interacting in between with the cavity field. Depending on the cat state parity, these probe atoms will be detected in g or in e. Any parity jump will be revealed by a sudden change of the detected atomic level. Such an experiment, not yet performed, would be a direct way to witness the field quantum jumps and to follow the decoherence process as it happens in real time. The continuous observation of the field's parity could also be used to control the decoherence process, as proposed in [61]. As soon as a change of parity has been observed, one could send across the cavity a single resonant atom whose interaction time would be adjusted so that it would emit one photon in it with near unity probability. This would approximately restore the field parity to the value it had before the parity jump. The experiment would be a very delicate one, requiring two kinds of atom-field interaction, a non-resonant dispersive one for the atoms probing the parity and a resonant one for the 'correcting' atoms. Such a procedure is only one example of the ingenious error correction codes proposed in quantum information to fight the effects of decoherence [1].

The parity analysis clearly explains the correlations between the successive atoms interacting with a π phase cat cavity field. In absence of decoherence, the detection of the first atom, which fixes the cat parity, forces the outcome of the measurement for the second atom (and all the other ones) to be the same. In other words, there is a 100% correlation between the measurement results for two atoms interacting successively with the cavity field. This is a mere statement of the fundamental postulate of quantum physics about measurement. If decoherence has occurred between the two atoms, on the other hand, this correlation is destroyed. Studying the loss of correlation as a function of the delay between the preparing atom and a probe atom thus constitutes a simple way to analyze decoherence. The experiment would consist in repeating many times the sequence, sending a pair of preparing and probe atoms with increasing delays and reconstructing by a statistical analysis the conditional probability for finding the second atom in the same state as the first one. This probability should evolve from 1 to $1/2$ as the delay between the two atoms is increased and the decoherence time could be obtained from this experiment. Such an ideal correlation measurement has not yet been performed on a π-phase cat state. A decoherence experiment along the same line has however been realized on cat state with phase splittings of $\sim$ 100 degrees and $\sim$ 50 degrees between the field components, for average photon numbers between 3 and 5 [16]. The two-atom correlation signal is then not as simple to understand as in the π-phase cat, but the calculation shows that a two atom-correlation signal can again be constructed, which is maximal at zero delay between the two atoms and decreases towards zero as decoherence progresses in the time interval between them. This correlation signal has actually been observed in the experiment, showing a very good agreement with the theoretical predictions. The dependence of the decoherence rate on the separation between the two cat components has also been demonstrated, the process becoming faster as this separation is increased, in good agreements with the results of Section 2.

A last interesting property of the photon number parity operator is worth noticing. As mentioned in section 1, the Wigner function W of a field state is directly obtained by translating this field in phase space and then measuring the average parity $\mathcal{P}$ of the resulting field. This result has lead to an ingenious proposal for the measurement of W in CQED [62]. Given a field in a cavity sandwiched be-

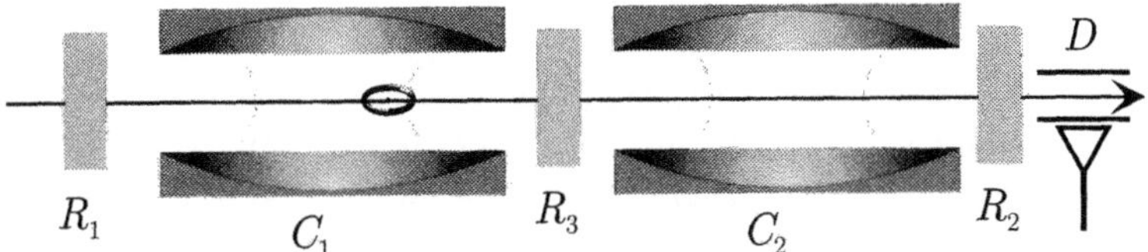

Fig. 25. Scheme of a non-local Scrhödinger cat experiment

tween two Ramsey zones, displace it by a given amount α and send a probe atom across the set-up to measure the field parity according to the method described above. In general the displaced field will not be in a parity eigenstate and the outcome of the measurement will be random. Repeat the experiment many time with the same settings and reconstruct in this way the expectation value of the parity in the displaced field. You obtain, according to Eq.(1.61), the value of $W(\alpha)$. Then change α and repeat the procedure. You sample in this way W at all points in phase space and get the whole W function.

This method is much more direct than the standard quantum optics technique for determining W, which consists in measuring all the quadrature field fluctuations and reconstructing the Wigner function by the inverse Radon transform (see section 2). The parity measurement method has not yet been applied to a cat state, but we have demonstrated it for a one-photon Fock state prepared in the cavity and shown that the Wigner function of this state presents negative values, a clear sign of non-classical behavior [63]. Studying in this way the Wigner function of a π-phase cat and recording the fast decoherence induced decay of its interference term is an important goal of our CQED experimental program.

Let us finally relate this CQED W measurement method to the procedure to determine the Q function described above. In both cases, the field to be measured needs to be translated in phase space by an amount corresponding to the coordinate for which the value of the function is seeked. Once the displacement made, we need to measure either the expectation value of the projection operator in vacuum [for the Q function, see Eq.(1.55], of the photon parity operator [for W, see Eq.(1.61]. These measurements are both performed by statistical methods, accumulating data on probe atoms sent across the cavity. For the Q function, the absorption of resonant probe atoms needs to be detected, whereas for the W function a Ramsey interferometry measurement needs to be performed on non-resonant atoms.

3.6. Non-local cats

Cavity QED methods can be extended to prepare mesoscopic superpositions of field states belonging to spatially distinct cavities [64, 65]. We would thus combine the weirdness of Schrödinger cats with the strangeness of non-locality. A possible experiment, proposed in [62] is sketched in Fig. 25. We now consider two identical cavities C_1 and C_2, successively crossed by an atomic beam. As in the single cavity version, classical $\pi/2$ pulses are applied to the atom in R_1 before C_1 and in R_2 after C_2, prior to detection. A π classical pulse can also be applied to the atoms between C_1 and C_2 in R_3, giving the possibility to exchange the atomic states e and g between the two cavities. The same coherent field β is initially injected in both cavities by coupling them via a T-shaped wave guide to a classical source. Each atom crossing successively the two cavities carries information between them, resulting in an entanglement between field states localized at different positions.

In a first version of the experiment, the R_3 pulse is switched off. The two cavity then experience the same atom-field interaction as in the single cavity case discussed above. The final atom's detection results in the preparation of states of the form:

$$\frac{1}{\sqrt{2}}[|\beta e^{i\chi}, \beta e^{i\chi}\rangle \pm |\beta e^{-i\chi}, \beta e^{-i\chi}\rangle] , \tag{3.40}$$

which are clearly entangled. The coherence of these states and the difference with classical mixtures can be tested, as discussed above, by sending probe atoms across the set-up and performing atom correlation experiments. Alternatively, the R_3 pulse can be activated. In this case, the fields in the two cavity experience phase shifts of opposite signs, resulting in the generation of states of the form:

$$\frac{1}{\sqrt{2}}[|\beta e^{i\chi}, \beta e^{-i\chi}\rangle \pm |\beta e^{-i\chi}, \beta e^{i\chi}\rangle] , \tag{3.41}$$

which are again non-local entangled states. It is also possible, in a final experimental stage, to inject in both cavities a second coherent field of same amplitude as the first, with a phase adjusted to cancel one of the two field components. The resulting field is then:

$$\frac{1}{\sqrt{2}}[|0, 0\rangle \pm | - 2i \sin \chi\beta, -2i \sin \chi\beta\rangle] , \tag{3.42}$$

in the case where R_3 is inactive and:

$$\frac{1}{\sqrt{2}}[|0, -2i \sin \chi\beta\rangle \pm | - 2i \sin \chi\beta, 0\rangle , \tag{3.43}$$

when the switching R_3 is performed. The non-local phase cats are then transformed into 'amplitude cats'. The states described by Eq.(3.43) are particularly interesting. They are coherent superpositions of a state in which the first cavity contains a coherent field while the second is empty with the state representing the reverse situation. In other words, all the photon of the field are found together in one cavity or in the other, with equal probability and a full quantum coherence between the two states. This is the situation we have already encountered in section 2 when describing the $|0, n\rangle + |n, 0\rangle$ state produced by a combination of non-linear beam-splitters. In CQED, it is the non-linear interaction of the field with a single atom which makes this coherent and collective channeling of photons possible. It realizes a situation very different from the ones usually produced by linear beam-splitters in quantum optics. Various tests of the non-local properties of these phase or amplitude non-local cats are possible, including the study of novel kinds of Bell's inequalities [66]. We will not discuss them any further here.

4. Collapse and revivals of matter-waves: proposals for atomic Schrödinger cats

The manipulation of cold atom matter waves in Bose Einstein condensates (BEC) has recently opened the new field of 'atomic quantum optics' in which the coherence properties of matter can be studied and exploited in a way similar to what is done in quantum optics with photons [67]. Whereas non-linearities must be induced in quantum optics by the introduction of an external non-linear medium, they are 'given for free' in BEC atom optics, since the atoms, contrary to photons, naturally interact with each other, introducing spontaneous non-linear Kerr-like effects in the propagation of matter waves. Collapse and revival effects related to the ones of quantum optics are expected in atom optics and Schrödinger cat states of matter could be generated using methods which are direct generalizations of optical processes. In this section, we compare non linear quantum optics with Bose-Einstein condensate matter-wave physics and describe how Schrödinger cat states made of atoms could be produced and studied. Analogies and differences with CQED physics will be stressed. Other discussions about collapse and revivals and Schrödinger cat state generation in BEC can be found in [68–73].

The techniques used to prepare atomic Bose-Einstein condensates, which involve laser and evaporative cooling will not be described here. We will neither discuss the numerous beautiful experiments which have demonstrated over the last nine years the remarkable coherence properties of this new state of matter (see for example the Les Houches lectures by Y. Castin [74]). Our goal is just

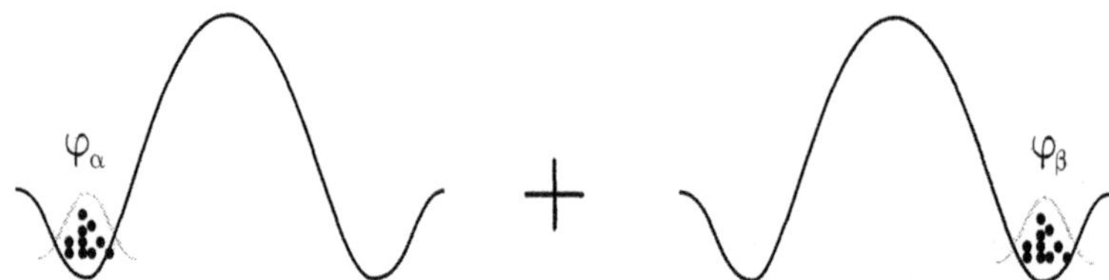

Fig. 26. A non-local cat made of N bosons shared between two potential wells: the particles are in a superposition of a state where they are all in the left well with a state where they are all in the right well.

to describe the principle of matter-wave mesoscopic state superposition experiments in an ideal situation. We will thus assume that a small sample of bosonic atoms (from a few to a few tens) is prepared in the ground state of a potential well. All atoms are supposed to be in the condensed phase (zero temperature limit of BEC) and we wish to exploit the collective properties of this gas of weakly interacting bosons in order to generate and study mesoscopic states. The typical situation we have in mind is sketched in Fig. 26. We would like to prepare the N-identical atoms in a double potential well in a superposition of a state in which they are all in the left well with the similar state in which they are all in the right one. The left and right single particle wave functions (ϕ_α and ϕ_β respectively) are non overlapping and the inter-well barrier is high enough, making tunneling negligible. This is a non-classical situation, reminiscent of the one we have considered in sections 2 and 3 with photons [see Eq.(3.43)]. Although this experiment has not been yet performed, we will show that it can be realistically envisioned, at least for relatively small atom numbers. Before describing the interatomic collisions processes which can be exploited to prepare these states, we will start by recalling the formalism adapted to the description of this system, first in the simplified case of non-interacting bosons.

4.1. *Ideal Bose-Einstein condensate without interactions*

Consider a system of N bosons (without internal structure for simplicity) in a trap. We call $|\phi_\mu\rangle$ (energy ϵ_μ) the basis of single particle energy eigenstates (depending on the trap potential). We adopt the formalism of second quantization. We call a_μ, $a_\mu^\dagger$ the annihilation and creation operators of one boson in state $|\phi_\mu\rangle$. They satisfy the canonical commutation rules of the operators for independent harmonic oscillators. We also call $|0\rangle$ the 'particle vacuum'. The system's state in which n_1 particles are in $|\phi_1\rangle$,... n_μ particles in $|\phi_\mu\rangle$ is:

$$\left|n_1 : \phi_1;\ n_2 : \phi_2; \ldots;\ n_\mu : \phi_\mu; \ldots\right\rangle =$$

$$\frac{1}{\sqrt{n_1!n_2!\ldots n_\mu!\ldots}}(a_1^{\dagger(n_1)}a_2^{\dagger(n_2)}\ldots a_\mu^{\dagger(n_\mu)}\ldots)\,|0\rangle\ , \tag{4.1}$$

while the energy operator of the system is:

$$H=\sum_\mu a_\mu^\dagger a_\mu\ \varepsilon_\mu\ . \tag{4.2}$$

The analogy with a light field whose photons belong to an ensemble of modes is evident. To each one-particle operator

$$V=\sum_{\mu,\nu}|\phi_\mu\rangle\langle\phi_\mu|V|\phi_\nu\rangle\langle\phi_\nu|\ , \tag{4.3}$$

corresponds the second quantization version:

$$\mathcal{V}=\sum_{\mu,\nu}a_\nu a_\mu^\dagger v_{\mu\nu}a_\nu\quad(v_{\mu\nu}=\langle\phi_\mu|\ V\ |\phi_\nu\rangle)\ . \tag{4.4}$$

The operator $\mathcal{V}$ describes, in second quantization, the observable $\sum_{i=1,N}V(i)$, sum of one-particle operators. Let us now describe in this formalism a superposition of states. N particles in state $\lambda_a|\phi_a\rangle+\lambda_b|\phi_b\rangle$ are, in second quantization, described by:

$$|N:\lambda_\alpha\phi_\alpha+\lambda_\alpha\phi_\alpha\rangle=\frac{1}{\sqrt{N!}}(\lambda_\alpha a_\alpha^\dagger+\lambda_\beta a_\beta^\dagger)^N\ |0\rangle\ . \tag{4.5}$$

Note the analogy with N photons in a state superposition of two modes. A particularly interesting superposition situation corresponds to a localized particle. In usual formalism it is described by

$$|\mathbf{r}\rangle=\sum_\mu|\phi_\mu\rangle\langle\phi_\mu|\mathbf{r}\rangle=\sum_\mu\phi_\mu^*(\mathbf{r})|\phi_\mu\rangle\ . \tag{4.6}$$

In second quantization, this state becomes:

$$|1:\ \mathbf{r}\rangle=\sum_\mu\varphi_\mu^*(\mathbf{r})a_\mu^\dagger\ |0\rangle\ , \tag{4.7}$$

which naturally introduces the field operator $\Psi^+(\mathbf{r})$ creating a particle at point $\mathbf{r}$:

$$\Psi^+(\mathbf{r})=\sum_\mu\varphi_\mu^*(\mathbf{r})\ a_\mu^\dagger\ , \tag{4.8}$$

and the adjoint field annihilation operator:

$$\Psi(\mathbf{r}) = \sum_{\mu} \varphi_\mu(\mathbf{r})\, a_\mu \,. \tag{4.9}$$

These operators satisfy the commutation relation of boson fields:

$$\left[\Psi(\mathbf{r}_1), \Psi^+(\mathbf{r}_2)\right] = \delta\left(\mathbf{r}_1 - \mathbf{r}_2\right) , \tag{4.10}$$

and they can be combined to define the second quantization particle density operator:

$$n_D(\mathbf{r}) = \Psi^+(\mathbf{r})\Psi(\mathbf{r}) \,. \tag{4.11}$$

which is the second quantization version of the projector operator $|\mathbf{r}\rangle\langle\mathbf{r}|$. Note the analogy between the annihilation and creation field operators and the positive and negative frequency parts of the electric field operator in quantum optics. The particle density at point $\mathbf{r}$, equal to the mean value in the system's state of the operator $\Psi^+(\mathbf{r})\Psi^-(\mathbf{r})$ is the analog of the photon counting rate, proportional to the expectation value of the product of the negative and positive frequency parts of the field.

We consider now an ideal 'bimodal' condensate at $T = 0$ K evolving in a double potential with two minima separated by a height-adjustable barrier. The ground states of the two wells, when tunneling is negligible, are called $|\phi_\alpha\rangle$ and $|\phi_\beta\rangle$. By lowering the barrier for a given time, one can mix the two states in a controlled way. The situation is very similar to the evolution of a field in a two-mode system under the effect of a linear beam-splitter. We assume that the system is initially prepared in one of the wells, the barrier between the two wells being high enough to make tunneling impossible. The atoms are then all described by the localized wave function ϕ_α. We then suddenly lower the barrier to allow for tunneling. The tunneling Hamiltonian can be written as:

$$H_J = -\hbar J(a_\alpha^\dagger a_\beta + a_\alpha a_\beta^\dagger) , \tag{4.12}$$

where J is a frequency measuring the tunneling rate (which we assume here without loss of generality real). The condensate starts to oscillate between the two wells, its state at time t being:

$$\left|N : \cos(Jt)\phi_\alpha + i\sin(Jt)\phi_\beta\right\rangle = \frac{1}{\sqrt{N!}}(\cos(Jt)a_\alpha^\dagger + i\sin(Jt)a_\beta^\dagger)^N \left|0\right\rangle . \tag{4.13}$$

The evolution is similar to the one produced by a beam-splitter in quantum optics. The case $Jt = \pi/4$ corresponds to a symmetrical beam-splitter. Oscillation between the two wells is also analogous to the Josephson effect in a junction between two superconductors. Any bimodal distribution of the particles between the two wells can be obtained, freezing the system by raising suddenly the barrier at the appropriate time. We should note that the state prepared in this way is a non-entangled product state. The N particles are all in the same one-particle state, a coherent superposition of ϕ_α and ϕ_β. The state (4.13) belongs to a class of bi-modal states taking the general form:

$$\begin{aligned} |N;\theta,\phi\rangle &= \frac{1}{\sqrt{N!}}(\cos\theta a_\alpha^\dagger + \sin\theta e^{i\phi} a_\beta^\dagger)^N |0\rangle \\ &= \sum_p \binom{N}{p}^{1/2} \cos^{N-p}\theta \sin^p\theta e^{ip\phi} |N-p\rangle_\alpha |p\rangle_\beta \ , \end{aligned} \quad (4.14)$$

which we will call 'phase states' for a reason which will become clear below. These states are defined by three parameters, the atom number N, the mixing angle θ, and the phase angle ϕ. In Eq.(4.14), the $\binom{N}{p}$ symbol is the binomial coefficient already introduced in the beam-splitter theory. The phase state (4.14) corresponds to a binomial distribution of particles in the two modes, with a well defined quantum phase ϕ between the two modes. The particle mean numbers and the variance of their difference are given by:

$$N_\alpha = N\cos^2\theta \ ; \ N_\beta = N\sin^2\theta \ ; \ \Delta(N_\alpha - N_\beta) = \sqrt{N}\sin 2\theta \ . \quad (4.15)$$

The fluctuation $\Delta(N_\alpha - N_\beta)$ is maximum, equal to $\sqrt{N}$, for a symmetrical condensate ($\theta = \pi/4$) and zero for a mono-condensate ($\theta = 0$ or $\pi/2$). Let us consider now a symmetrical state. The coherence between Fock states with different particle numbers writes:

$$\begin{aligned} &\langle N-p_1; p_1 \mid N;\pi/4,\phi\rangle \langle N;\pi/4,\phi \mid N-p_2, p_2\rangle \\ &\quad = \frac{1}{2^N}\binom{N}{p_1}^{1/2}\binom{N}{p_2}^{1/2} e^{i(p_1-p_2)\phi} \ . \end{aligned} \quad (4.16)$$

If the phase is unperfectly known [defined by a probability law $P(\phi)$], the system is described by its density operator:

$$\rho = \int_0^{2\pi} d\phi \, |N;\pi/4,\phi\rangle P(\phi) \langle N;\pi/4,\phi| \ , \quad (4.17)$$

and the coherence between states with particle numbers p_1 and p_2 in state $|\phi_\beta\rangle$ becomes:

$$\langle N-p_1, p_1|\rho|N-p_2, p_2\rangle = \frac{1}{2^N}\binom{N}{p_1}^{1/2}\binom{N}{p_2}^{1/2} \times$$

$$\times \int_0^{2\pi} d\phi \; P(\phi) \; e^{i(p_1 - p_2)\phi} \; . \qquad (4.18)$$

This equation exhibits a conjugation relation between the phase and the difference of the number of particles in two Fock states related to each other by a non-zero matrix element of ρ. When $P(\phi)$ is a δ function, ρ has non-diagonal elements on a width $p_1 - p_2$ of order $\sqrt{N}$. The wider $P(\phi)$ is, the narrower the non-diagonal distribution of ρ becomes. At the limit where $P(\phi) = 1/2\pi$ (undetermined phase), ρ is diagonal in $p_1 - p_2$. It is then equivalently described as an incoherent sum of diagonal operators $|N - p, p\rangle\langle N - p, p|$, each representing a projector associated to two condensates with perfectly fixed atom numbers. This complementarity relation between the phase fluctuation $\Delta\phi$ and the width ΔN of the coherent partition of the particle numbers between the two wells is again quite reminiscent of quantum optics. Likewise, by summing projectors on coherent states with different phases, one constructs a field density operator which approaches an incoherent superposition of Fock states when the phase distribution is uniformly distributed.

4.2. *Coherent collisions in BEC: the analogy with the Kerr effect in quantum optics*

4.2.1. *Simple model of elastic binary collisions in a bi-modal condensate*

We have so far neglected the interactions between atoms in the condensate. In fact, the atoms undergo binary elastic collisions which affect the dynamical behavior of the matter waves. Two atoms located at points $\mathbf{r}_1$ and $\mathbf{r}_2$ interact according to a potential $W(\mathbf{r}_1 - \mathbf{r}_2)$ which, at the low energy limit, can be considered as spherically symmetrical ('S–wave scattering'). To a good approximation, the potential can be described as an isotropic contact term:

$$W(\mathbf{r}_1 - \mathbf{r}_2) \approx \frac{4\pi\hbar^2 a_s}{M}\delta(\mathbf{r}_1 - \mathbf{r}_2) \; , \qquad (4.19)$$

where M is the atom mass and a_s the so-called scattering length, a parameter which appears naturally in the quantum description of the collision processes at very low energy [74]. This length can be tuned by applying a magnetic field on the atoms, most efficiently near certain field values corresponding to the so-called Feschbach resonances [75]. The collision probability amplitude depends on the energy level structure of the binary atom system and is strongly affected by the presence of level crossings which are conditioned by the magnetic field. Typically, for collisions between alkali atoms, a_s is of the order of 1 to 10 nm with a sign either positive (repulsive force between atoms) or negative (attractive force). We consider here the repulsive case. The mean value of the interaction energy for a simple condensate of N atoms in state $|\phi_\alpha\rangle$ is:

$$\langle W\rangle = N(N-1)\frac{2\pi\hbar^2 a_s}{M}\int d^3\mathbf{r}\,|\phi_\alpha(\mathbf{r})|^4 \,. \tag{4.20}$$

It appears as the expectation value in the condensate state of the collision Hamiltonian:

$$W = \frac{1}{2}\hbar g_c \hat{N}_\alpha(\hat{N}_\alpha - 1)\,, \tag{4.21}$$

with the coupling g_c defined as:

$$g_c = \frac{4\pi\hbar a_s}{M}\int d^3\mathbf{r}\,|\varphi_\alpha(\mathbf{r})|^4 \,. \tag{4.22}$$

This hamitonian is proportional to the number $N(N-1)/2$ of atom pairs in the condensate, a result which can be understood by a simple book-keeping argument. For a double condensate, we must evaluate separately the collisions between atoms in each mode and the intermode collisions. Assuming to simplify that the interactions are the same in each subsystem and negligible between them (which is obviously the case if the two modes are spatially separated as in Fig. 26), we get:

$$W = W_\alpha + W_\beta = \frac{1}{2}\hbar g_c \hat{N}_\alpha(\hat{N}_\alpha - 1) + \frac{1}{2}\hbar g_c \hat{N}_\beta(\hat{N}_\beta - 1); \tag{4.23}$$

We should stress the extreme simplicity of this model which accounts well for the effects of collisions in the condensate. They are described by a single parameter, the tunable scattering length a_s. The situation is much simpler than in a classical gas where collisions are a source of randomness and irreversibility. In a zero-temperature degenerate condensate, the bosons after a collision process must stay in the same quantum state as before. The collisions constantly reshuffle the particles in the same state, without generating any disorder.

The interactions described by Eqs.(4.21,4.23) are reminiscent of a Kerr Hamiltonian in optics [76]. The non-linear interaction of a light wave with a transparent medium contains terms proportional to the square of the light intensity of the form $\gamma(a^\dagger a)^2$, where γ is a non-linear susceptibility. Its main effect is to make the refractive index of the medium intensity-dependent. These Kerr non linearities give rise to a phase spreading of light beams propagating in the medium over long distances (for example along optical fibers). It could, at least in theory, also lead to phase revival effects and to the generation of Schrödinger cat states of light. This cat generation method has been suggested in the 1980s as a promising way to study mesoscopic field superpositions, but experiments have not been successful, due in part to too strong losses in the non-linear medium. The advent of BEC physics, which naturally introduces a Kerr like coupling of matter waves, has lead to a revival of these early proposals which we will now briefly present.

4.2.2. Collapse and revivals of a condensate bi-modal phase state: the optical analogy

We study now the effect of the atomic collisions on the phase of a bimodal condensate initially prepared in a symmetric phase state. We follow here closely the analysis developed in [44]. We assume that the tunelling between the two modes is negligible (high inter-well barrier). It is convenient to expand the initial phase state along the Fock states. We change the notations and define N_α as $(N/2)+\delta N$ and N_β as $(N/2)-\delta N$. The quantity δN appears as a particle number fluctuation of the order of $\pm\sqrt{N}$. The initial state (time $t=0$) thus writes:

$$\begin{aligned} |\psi(0)\rangle &= |N;\theta=\pi/4,\phi\rangle \\ &= \sum_{N_\alpha=0}^{N} \frac{e^{iN\phi/2}}{2^{N/2}} \sqrt{\frac{N!}{N_\alpha! N_\beta!}}\, e^{-i(N_\alpha-N_\beta)\phi/2} \left|N_\alpha, N_\beta\right\rangle . \end{aligned} \tag{4.24}$$

For large N's, we can develop the factorials (using Stirling formula) and we get the Gaussian approximation:

$$|\psi(0)\rangle \approx \left(\frac{2}{\pi N}\right)^{1/4} \sum_{\delta N} e^{-\delta N^2/N} e^{-i\delta N\phi} \left|\frac{N}{2}+\delta N, \frac{N}{2}-\delta N\right\rangle . \tag{4.25}$$

The sum in this expression, in principle over δN between $-N/2$ and $+N/2$, is in practice restricted to $-\sqrt{N} < \delta N < \sqrt{N}$.

A simple calculation shows that a Fock state with given δN is an eigenstate of the Hamiltonian W, with the eigen energy:

$$\langle W\rangle = \hbar g_c \frac{N}{2}(\frac{N}{2}-1) + \hbar g_c \delta N^2 . \tag{4.26}$$

This collisional energy varies quadratically with δN. The minimum energy state corresponds to the best possible equipartition of the particles between the two modes $\delta N = 0$ if N is even, $\delta N = \pm 1$ if N is odd. Let us now study the effect of these collisions on the evolution of a bimodal phase state of the form (4.25). We apply to each Fock component its collisional dephasing, proportional to time and to its collisional energy (4.26). Up to a global phase factor, we get:

$$|\psi(t)\rangle \approx \left(\frac{2}{\pi N}\right)^{1/4} \sum_{\delta N} e^{-\delta N^2/N} e^{-i\delta N[\phi+(\omega_\alpha-\omega_\beta)t]} \times$$

$$\times e^{-ig_c \delta N^2 t} \left| \frac{N}{2} + \delta N, \frac{N}{2} - \delta N \right\rangle , \tag{4.27}$$

where $\hbar\omega_\alpha$ and $\hbar\omega_\beta$ are the chemical potentials of the two condensate components. The phases in the Fock state expansion present a term linear in δN corresponding to the time evolution of a non-interacting gas and a Kerr-like term quadratic in δN describing the effect of he collisions. The phase coherence of this state is revealed by computing the expectation value of the interference term $\langle a_\beta^\dagger a_\alpha \rangle$ in it:

$$\langle a_\beta^\dagger a_\alpha \rangle = (\frac{N}{2\pi})^{1/2} \sum_{\delta N} e^{-2\delta N^2/N} e^{-i[\phi + (\omega_\alpha - \omega_\beta)t]} e^{-ig_c(2\delta N+1)t} . \tag{4.28}$$

The phase spreading of the last exponential in Eq.(4.28) results in a collapse of this expectation value. The collapse is complete when the phase has fanned out over a full 2π angle over the distribution of δN, whose width is of the order of $2\sqrt{N}$. This occurs within a time T_{collapse}:

$$T_{\text{collapse}} \approx \frac{\pi}{g_c \sqrt{N}} , \tag{4.29}$$

inversely proportional to the collision strength g_c and to the square root of the particle number. This effect is quite analogous to the spreading of the Gaussian wave packet of a free particle in a one dimension propagation problem, the phase ϕ and the fluctuation of the number of bosons δN being replaced by the particle's position and momentum respectively. There is however a big difference between the condensate phase's and the particle position's spreadings. In the latter case the sum over momentum is a continuous integral, whereas, in the condensate situation, the sum over δN is discrete, expressing the graininess of the matter wave. While the free particle wave function spreading is an irreversible process, the phase condensate collapse is reversible. All the phases in Eq.(4.28) are refocused to the same value (modulo 2π) after a time T_{revival}:

$$T_{\text{revival}} = \frac{\pi}{g_c} , \tag{4.30}$$

as well as at all multiples of this time. At these precise moments, the coherence between the two modes of the condensates is fully restored to its initial value. The first revival time is $\sqrt{N}$ times larger than the collapse time, and is independent of the number of particles. It depends only upon the non-linearity g_c introduced by the collisions.

We have limited this discussion, for simplicity, to bi-modal condensates, e.g. prepared in two separated potential wells. It can be generalized to a multiple-well

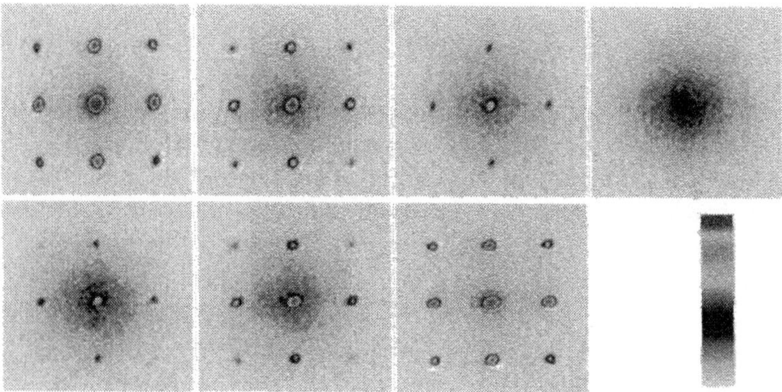

Fig. 27. Signal showing the matter wave collapse and revival in an optical lattice. The successive frames are snapshots of the expanded condensate, corresponding to increasing durations of the intra-well collision process (see text). The time increases from left to right and from top to bottom. At short times, the condensate has a well defined phase, resulting in a Bragg like pattern for the expanded atomic cloud. At the collapse time (last frame at the right of first line) the Bragg pattern is washed out. It then revives at twice the collapse time. Up to five successive collapses and revivals have been observed in this way. Reprinted with permission from [77].

situation, such as a condensate placed in a crystal–like three–dimension lattice. Let us say a few words about this configuration which has been studied extensively theoretically and has recently been investigated in beautiful experiments. They are realized by first preparing a condensate in a magnetic trap, then superposing this trap with an optical lattice made by a combination of three sets of counter propagating light beams creating a three dimensional optical force lattice acting on the atoms. If the wells are shallow enough so that tunneling between them is dominant, the state of the condensate is a big matter wave of the Bloch type, coherent superposition of the ground state wave functions of all the potential wells in the lattice. This state is the multiple well counterpart of the bi-modal states described above. After preparing this initial state, the depth of the lattice is suddenly increased, in a time short compared to the state evolution, and a final lattice configuration is reached in which the tunneling is negligible and the effects of intra-well collisions dominant. These collisions then induce, as in the bimodal case, collapses and revivals of the coherence between any couple of wells in the lattice. Typically, the number of atoms per well is of the order of 2 to 3 and about 10^5 wells are populated in the lattice.

To observe the effect, the optical lattice is suddenly suppressed after having left the collision act during a given time interval. The condensate is then allowed to expand freely for some time, before a picture of the expanded gas is made by absorption of a probe laser beam. When the condensate exhibits inter-well coherence, the expanded gas presents interference terms in its density, reminiscent of a Bragg diffraction pattern. When the inter-well coherence has collapsed, the density interferences vanish and the image of the condensate looks like an incoherent diffraction pattern. The experiment is resumed with various time delays between the sudden raise of the inter-well barriers and the condensate release. A periodic change of the image aspect, going in a reversible way from coherent to incoherent scattering-like patterns is observed (see Fig. 27).

There is a strong analogy between this collision induced effect and the Kerr-induced phase spreading of a coherent field in Quantum Optics. The expectation value of the photonic field operator $\langle a \rangle$ then evolves as $\langle a_\beta^\dagger a_\alpha \rangle$ in the boson situation. In the evolution equation, the non-linear Kerr susceptibility γ merely replaces the collision constant g_c and the number of photons plays the role of the boson number. To show in a quantitative way the evolution of the phase in this problem, we have represented in Fig. 28 adapted from [77] the Q function of a coherent field containing on average 3 photons and subjected to a dephasing Kerr effect. The values of the Q function are indicated by a code of shades. The function is plotted at different times following the field initial preparation, up to the first revival time. The pictures clearly show the fast spreading of the field and its sharp refocusing at T_{revival}. It is remarkable to note that this calculation has been performed for a small average particle number, for which the approximations made in the above model (Stirling formula and Gaussian limit) are not valid. Yet, the basic features of the model remain true.

The Q function evolution displays another remarkable feature. At the time $T_{\text{revival}}/2$, the spreading phase of the field refocuses into two components, separated by π (see Fig. 28). This indicates the transient generation of a π-phase cat state induced by the Kerr non-linearity. The appearance of this state is easy to understand by simple arguments based on the parity properties of the phase cat states discussed above (section 1 and 3). At the time $T_{\text{revival}}/2 = \pi/2\gamma$, the $n = 2p$ even Fock states in the coherent state expansion have undergone a shift $\gamma(2p)^2 \times \pi/2\gamma$, equal to 0 (modulo 2π). The $n = 2p + 1$ odd Fock states, on the other hand, have experienced a phase shift $\gamma(2p+1)^2 \times \pi/2\gamma$, equal to $\pi/2$ (modulo 2π). To make explicit the effect of this parity dependent phase shift, it is convenient to expand the initial coherent state as a sum of even and odd cat components, according to Eq.(3.39), and to multiply the even and odd parts by the corresponding phase factor, equal to 1 for the even part, to $e^{i\pi/2}$ for the odd one. At half the revival time, the field state writes:

$$(1/2)(|\alpha\rangle + |-\alpha\rangle) + (e^{i\pi/2}/2)(|\alpha\rangle - |-\alpha\rangle)$$
$$= \frac{e^{i\pi/4}}{\sqrt{2}}\,|\alpha\rangle + \frac{e^{-i\pi/4}}{\sqrt{2}}\,|-\alpha\rangle\ . \tag{4.31}$$

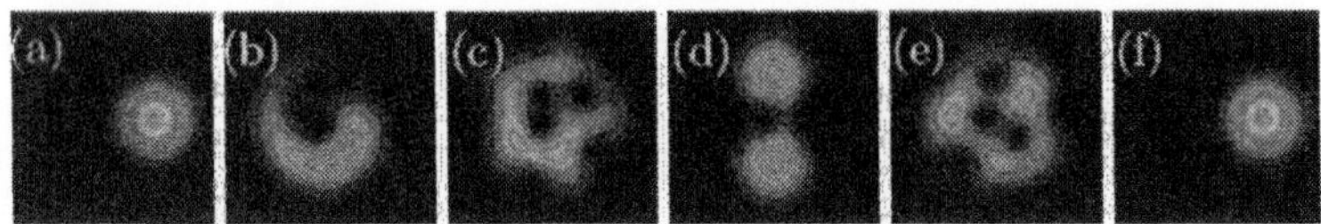

Fig. 28. Evolution of the Q function of an initially coherent field with $\alpha = \sqrt{3}$ in a Kerr non-linear medium. Time for frames (a) to (f) is 0, 0.1, 0.4, 0.5, 0.9 and 1 in units of π/γ. Note the formation of a π phase cat at time $0.5\pi/\gamma$ [frame (d)]. Adapted from [77].

It is a cat state, coherent superposition of two coherent components with opposite phases (although it is not a parity state, since the complex amplitudes of the superposition are not in phase or π out of phase with respect to each other). This process of preparation of cat state has been first described in a paper by Yurke and Stoler [11]. We will show in the next subsection that there is a way to translate this proposal in the condensed boson situation and to use the Kerr-like non–linearity produced by the elastic collisions in the condensate to generate Schrödinger cat states of matter waves.

Let us note, to conclude this subsection, that the collapse and revival phenomenon in photonic or atomic waves presents strong similarities, and also notable differences, with the phenomena bearing the same name in CQED and described in Section 3. Note first that it is not the phase of the field which collapses in the CQED situation, but rather the phase of the Rabi oscillation observed on an atomic signal. Note also that the collapse time of the Rabi oscillation is independent of the photon number for a given vacuum Rabi coupling, whereas it is inversely proportional to the square root of this number in the Kerr collapse. In both cases, the revival occurs after a delay equal to the collapse time multiplied by $\sqrt{n}$. In both the CQED and the Kerr non–linear optics case, Schrödinger cat states with phase opposite components are produced at half the revival time, but the mechanism of their generation are quite different.In the CQED case, cat states are produced via atom-field entanglement as soon as the collapse has occurred and the dephasing between the cat components evolves continuously, going through a maximum at half the revival time. In the Kerr case, there is only one quantum system (the field) and hence no entanglement. The cat state appears only at specific times, and emerges from a continuous evolution of a phase evolving state. Finally, in the CQED case, only the first revival is clearly marked, while the subsequent ones are smeared out and merged into each other. In a Kerr medium with a pure second order term, the revivals are periodical.

4.3. *Proposal to prepare a Schrödinger cat state in a bi-modal condensate*

The optical Kerr effect analogy suggests that a Schrödinger cat of matter-wave should appear at half the phase revival time $T_{\text{revival}}/2 = \pi/2g$. At this specific time, the collisional phase appearing in Eq.(4.27) takes the values:

$$g_c(\delta N)^2 T_{\text{revival}}/2\hbar = \frac{\pi}{2}(\delta N)^2 = \{ \begin{array}{c} 0 \text{ for } \delta N \text{ even} \\ \pi/2 \text{ for } \delta N \text{ odd} \end{array} \quad (\text{modulo } 2\pi)\ . \tag{4.32}$$

If we express the initial phase state as a sum of a component whose δN's are even and a component whose δN's are odd, the evolution produced by the collisions up to time $T_{\text{revival}}/2$ leaves the amplitude of the first component unaltered and phase shifts by $\pi/2$ the amplitude of the second. The separation between even and odd terms in δN is easy to perform analytically, at least when N is even, a situation which we assume fulfilled from now on for simplicity. Inspired by the trick used to write a coherent state as a sum of even and odd π out of phase cats, let us write in the same vein the phase bi-modal state as a superposition of two parity states of the form:

$$\begin{aligned} \left|\psi^{even}\right\rangle &= \frac{1}{2}\frac{1}{2^{N/2}}\left[(a_\alpha^+ + e^{-i\bar{\phi}}a_\beta^+)^N\,|0\rangle + (a_\alpha^+ - e^{-i\bar{\phi}}a_\beta^+)^N\,|0\rangle\right], \\ \left|\psi^{odd}\right\rangle &= \frac{1}{2}\frac{1}{2^{N/2}}\left[(a_\alpha^+ + e^{-i\bar{\phi}}a_\beta^+)^N\,|0\rangle - (a_\alpha^+ - e^{-i\bar{\phi}}a_\beta^+)^N\,|0\rangle\right]. \end{aligned} \tag{4.33}$$

If $N/2$ is an even integer, the first one obviously contains only even powers of δN and the second only odd powers. The δN parity of these states is exchanged if $N/2$ is odd. We have included in the definition of these states the free evolution phase factor (linear in δN) up to the time $T_{\text{revival}}/2$ (phase $\bar{\phi}$ in these expressions). We consider for sake of definiteness the case $N/2$ even and leave the reader to adapt the solution to the $N/2$ odd case. Realizing that the initial state is the sum of the 'even' and 'odd' states defined above and taking into account the evolution of their phase under the effect of collisions up to the time $T_{\text{revival}}/2$, we find an expression which reminds us of Eq.(4.31):

$$\begin{aligned} |\psi(t = \pi/2g)\rangle &= \frac{1}{2^{N/2}}\frac{e^{-i\pi/4}}{\sqrt{2}}(a_\alpha^\dagger + e^{-i\bar{\phi}}a_\beta^\dagger)^N\,|0\rangle + \\ &+ \frac{1}{2^{N/2}}\frac{e^{i\pi/4}}{\sqrt{2}}(a_\alpha^\dagger - e^{-i\bar{\phi}}a_\beta^\dagger)^N\,|0\rangle\ . \end{aligned} \tag{4.34}$$

The system's state appears now as a superposition of N particles all in the state $\frac{1}{\sqrt{2}}[|\phi_\alpha\rangle + e^{-i\bar{\phi}}|\phi_\beta\rangle]$ with N particles all in the state $\frac{1}{\sqrt{2}}[|\phi_\alpha\rangle - e^{-i\bar{\phi}}|\phi_\beta\rangle]$, obviously a Schrödinger cat situation.

The recipe to prepare such a state could, in principle, follow the stages schematized in Fig. 29, from top to bottom. We realize first a double potential well and start with a condensate with an even number of N bosons prepared in the left well (α), separated from the right well (β) by a high barrier [Fig. 29(a)]. The barrier

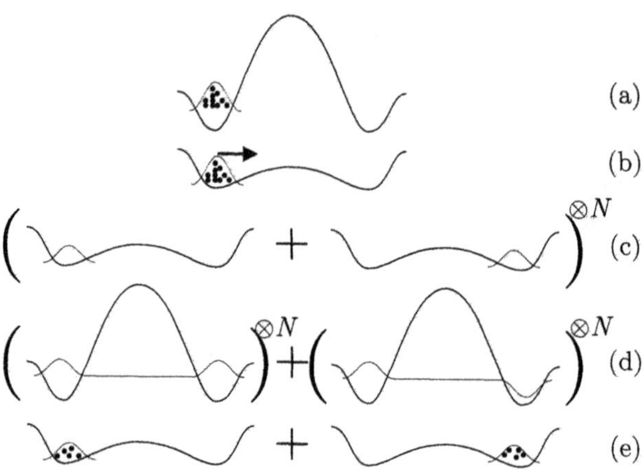

Fig. 29. Creation of a matter Schrödinger cat: the successive stages of the non-local cat preparation are sketched from top to bottom (see text).

is then suddenly lowered at a height such that $J \gg g_c$ [Fig. 29(b)] and we wait a time such that $Jt = \pi/4$: the tunnel effect creates a phase state [Fig. 29(c)]. Note that this is not yet a cat, since all the particles occupy the same quantum state. This is a mere tensor product of N particles in the same wave function, which is expressed by the N exponent at the right of the bracket around the sketch representing this state . The effect of collisions is negligible during this first fast stage. We then suddenly separate the two wells by raising the barrier. The tunnel effect disappears. The collisions become more efficient in each well, due to the extra confinement produced by deepening the wells. Enough time is left for the collisions to achieve the appropriate dephasing. At time $T_{\text{revival}}/2$, a cat state is prepared [Fig. 29(d)]: we have now N atoms, all occupying either one of two orthogonal wave functions. The one particle wave functions are delocalized in both wells, but they are still spatially overlapping. It is possible at this stage to spatially separate the two components of the cat state. To achieve this, the tunnel effect is reestablished by a sudden lowering of the barrier and the system is left to evolve for a quarter period of the coherent oscillation between the two wells (we assume that the collisions have a negligible effect during this time): the N atoms localize at left and at right, in a mesoscopic state superposition [Fig. 29(e)]. The inter-well barrier can then be finally raised, suppressing the tunneling and freezing the non-local cat state. One way to demonstrate the coherence of the cat state generated in this way would be to leave the system alone after the time $T_{\text{revival}}/2$ without raising the inter-well barrier, until the two parts of the cat recombine at time T_{revival}. The recurrence of the bimodal state coherence at this time could be checked by letting the condensate expand and studying the interference term in the atomic density of the expanded cloud, using a method similar to [77].

We have presented here a very general principle, without discussing the experimental limitations and the causes of decoherence of the system. The loss of a single atom from the condensate during this elaborate succession of operations would be lethal for the cat because the lost atom (or molecule formed by three body recombination) would be in a quantum state entangled with the superposition and its detection would collapse it. As in the CQED photonic case, the method seems to be restricted in practice to the preparation and detection of mesoscopic systems made of a few to a few tens of particles. Their study would nevertheless be of great interest to investigate, on this new system, the quantum-classical boundary.

5. Conclusion: a brief comparison with other mesoscopic state superpositions in quantum optics

We have described various kinds of Schrödinger cat states in quantum or atomic optics. These states are characterized by their extreme fragility and sensitivity to decoherence, which occurs at a rate essentially proportional to the number of particles in the system. This puts severe limits to the size of these cat states. They have to be built within a finite time, to let the processes responsible for the preparation of the superposition to take place. This preparation time must be shorter than the decoherence time of the final cat state and this sets, in practice, an absolute limit to the number of particles in the system. In CQED, the maximum number of photons involved in cat states could not exceed a few hundred, even if the technology of cavities were considerably improved. (A discussion of the maximum size of these CQED cats can be found in [78]). In BEC physics, the atomic cats envisioned in section 4 as well as in other studies [69, 71–73] seem also to be at most of a few tens to a few hundred atoms.

The environment-induced decoherence process we have studied here is of a fundamental quantum nature. In an ideal situation, the system and the environment are both initially, in a pure state. They get entangled together and the coherence of the system is lost when we trace over the environment, expressing in this way our inability to keep track of its complex state. If it were possible to detect all the particles in the environment, the coherence lost in it could be retrieved and cat state interferences restored, in correlation with the recorded state of the environment. This kind of procedure, called quantum erasure, has been demonstrated in some simple cases, but it becomes unpractical for large environments. Other, more realistic methods to preserve the cat state coherence consist in watching the system itself and using this information to correct for decoherence [61]. This kind of error correction procedure, essential for quantum computing, remains largely to be investigated experimentally. Note also that we must distinguish the decoherence produced by the quantum entanglement of the system with its environment from the more mundane relaxation processes, often called decoherence too, which are induced by a classical noise in the apparatus (e.g. stray fields). This latter form of decoherence is in principle much easier to correct for than the former, since it is of a classical deterministic origin.

The methods we have described to generate these cat states exploit non-linear processes a the single particle level which must be precisely controlled and tuned (interaction of a field with a single atom in CQED or coherent cold atom collisions in BEC). The observation of the cat coherence involves the detection of multiparticle interference effects. Moreover, the experiments generally require low temperature technology (cryogenic atomic beams in the microwave CQED experiments or ultra cold samples of bosons in the BEC proposals). All these experiments are thus extremely difficult, making the exploration in this way of the quantum-classical boundary a very challenging task.

Other experiments investigating entanglement in large atomic samples should be mentioned. They involve two atomic samples at room temperature which get correlated by a light beam propagating through them [79]. The detection of this beam projects the two ensembles into an entangled state, in a process which bears some similarity with the procedure described in section 4.6 where a single atom crossing two cavities was detected, resulting in the collapse of the two cavity field into an entangled state (see P. Zoller Lectures in this Volume). The atomic states in these experiments are conveniently described as large angular momentum states on a sphere (collective Bloch vectors). The length of these vectors is very large since the number of atoms is macroscopic, but the distance on the Bloch sphere of the states involved in the entanglement is relatively small, of the same order as the distance between the cat state components in the experiments discussed in these Notes.

Non-local superpositions of states involving 'big' systems at the atomic scale have also been observed in another kind of experiment which appears, at least in its principle, simpler than the ones described in these lectures. Big C_{60} molecules and even biomolecules have been made to interfere in a Young apparatus [80], demonstrating that these objects can exist in state superpositions corresponding to hundreds of nanometers of spatial separation. Since these molecules contain several tens of atoms and that larger molecules can probably be made to interfere in the same way, the question which arises is whether such systems are more 'macroscopic' that the cat states described above. In terms of number of elementary particles involved or absolute mass, the answer is yes. But the kind of interferences observed in both cases are very different. In the bosonic or photonic cat states, the particles channeled in the two arms of the interferometer are very weakly interacting (atoms) or completely independent of each other (photons). To split them collectively in different paths is a delicate multiparticle process, sensitive to all kinds of possible perturbations. In the C_{60} experiments, on the other hand, the atoms are strongly bound together and cannot be split apart by the beam splitter. The molecule essentially behaves as an unbreakable object and, in this sense, must be considered as a single interfering particle, even if its mass is large. The 'interference of a big molecule with itself' involves only one relevant parameter of the molecule, the position of its center of mass. This variable is coupled to an 'internal' environment (the rotation and vibration states of the molecule), itself interacting with an external environment, the radiation field. The entanglement between the center of mass and these two environments is usually weak. The main cause of decoherence is the radiation of the molecule while it travels across the interferometer, which can result in photons carrying away an information about its path. The photons should have a short enough wavelength, which means that the temperature of the molecule should be very high for this process to be effective. The real difficulty of these experiments is not decoherence, but the shrinking of the molecule de Broglie wavelength which makes interference fringes more and more difficult to observe when the mass is increased.

As shown by this brief discussion, there is a wide variety of effects which can be investigated with the techniques of quantum optics and atomic physics at the boundary between the quantum and the classical worlds. To define the limit between these words in terms of number of particles or mass is not simple and the relevant parameters to consider depend upon the kind of experiment which is performed. The realm of mesoscopic studies becomes even much larger with the possibilities opened by new condensed matter nano-technologies involving superconductors [81] or quantum dots. It would be very interesting to compare, in the context of quantum information science, the kind of cat states realized in quantum optics to those which are made or discussed in solid state physics. This task would have to wait another Les Houches School.

References

[1] M.A. Nielsen and I.L. Chuang. *Quantum computation and quantum information*. Cambridge University Press, Cambridge, 2000.

[2] D. Bouwmeester, A. Ekert, and A. Zeilinger. *The physics of quantum information*. Springer, Berlin, 2000.

[3] W. H. Zurek. Pointer basis of quantum apparatus: Into what mixture does the wave packet collapse? *Phys. Rev. D*, 24(6):1516, 1981.

[4] A.O. Caldeira and A.J. Leggett. Quantum tunneling in a dissipative system. *Ann. Phys (N.Y.)*, 149:374, 1983.

[5] A.O. Caldeira and A.J. Leggett. Influence of damping on quantum interference: An exactly soluble model. *Phys. Rev. A*, 31:1059, 1985.

[6] E. Joos and H. D. Zeh. The emergence of classical properties through interaction with the environment. *Z. Phys. B*, B59:223, 1985.

[7] W. H. Zurek. Decoherence, einselection, and the quantum origins of the classical. *Rev. Mod. Phys.*, 75:715, 2003.

[8] J. A. Wheeler and W. H. Zurek. *Theory of measurement.* Princeton University Press, Princeton, 1983.

[9] W. H. Zurek. Decoherence and the transition from quantum to classical. *Phys. Today*, 44(10):36, octobre 1991.

[10] J.P. Paz and W.H. Zurek. Quantum limit of decoherence: environment-induced superselection of energy eigenstates. *Phys. Rev. Lett.*, 82:5181, 1999.

[11] B. Yurke and D. Stoler. Generating quantum mechanical superpositions of macroscopically distinguishable states via amplitude dispersion. *Phys. Rev. Lett.*, 57:13, 1986.

[12] B. Yurke, W. Schleich, and D.F. Walls. Quantum superpositions generated by quantum nondemolition measurements. *Phys. Rev*, A42:1703, 1990.

[13] C.M. Savage, S.L. Braunstein, and D.F. Walls. Macroscopic quantum superpositions by means of single-atom dispersion. *Opt. Lett.*, 15:628, 1990.

[14] M. Brune, S. Haroche, J.-M. Raimond, L. Davidovich, and N. Zagury. Manipulation of photons in a cavity by dispersive atom-field coupling: Quantum non demolition measurements and generation of Schrödinger cat states. *Phys. Rev. A*, 45:5193, 1992.

[15] V. Buzek, H. Moya-Cessa, P.L. Knight, and S.D.L. Phoenix. Schrödinger-cat states in the resonant jaynes-cummings model: Collapse and revival of oscillations of the photon-number distribution. *Phys. Rev. A*, 45:8190, 1992.

[16] M. Brune, E. Hagley, J. Dreyer, X. Maître, A. Maali, C. Wunderlich, J.-M. Raimond, and S. Haroche. Observing the progressive decoherence of the meter in a quantum measurement. *Phys. Rev. Lett.*, 77:4887, 1996.

[17] C. Monroe, D. M. Meekhof, B. E. King, and D. J. Wineland. A "Schrödinger cat" superposition state of an atom. *Science*, 272:1131, 1996.

[18] Q.J. Myatt, B.E. King, Q.A. Turchette, C.A. Sackett, D. Kielpinski, W.M. Itano, C. Monroe, and D.J. Wineland. Decoherence of quantum superpositions through coupling to engineered reservoirs. *Nature (London)*, 403:269, 2000.

[19] A. Auffeves, P. Maioli, T. Meunier, S. Gleyzes, G. Nogues, M. Brune, J.M. Raimond, and S. Haroche. Entanglement of a mesoscopic field with an atom induced by photon graininess in a cavity. *Phys. Rev. Lett.*, 91:230405, 2003.

[20] R. Loudon. *The Quantum Theory of Light.* Oxford University Press, Oxford, 1983.

[21] W. Vogel, D.G. Welsch, and S. Wallentowitz. *Quantum optics: an introduction.* Wiley, Berlin, second edition, 2001.

[22] W.P. Schleich. *Quantum optics in phase space.* Wiley, Berlin, 2001.

[23] D.F. Walls and G.J. Milburn. *Quantum Optics.* Springer Verlag, New York, 1995.

[24] S.M. Barnett and P.M. Radmore. *Methods in theoretical quantum optics.* Oxford University Press, Oxford, 1997.

[25] M.O. Scully and M.S. Zubairy. *Quantum Optics.* Cambridge University Press, Cambridge, 1997.

[26] C. Cohen-Tannoudji, J. Dupont-Roc, and G. Grynberg. *An Introduction to Quantum Electrodynamics.* Wiley, New York, 1992.

[27] S. Haroche. Cavity quantum electrodynamics. In J. Dalibard, J.-M. Raimond, and J. Zinn-Justin, editors, *Fundamental Systems in Quantum Optics, Les Houches Summer School, Session LIII*, page 767. North Holland, Amsterdam, 1992.

[28] S. Haroche and J.-M. Raimond. Manipulation of non-classical field states in a cavity by atom interferometry. In P. Berman, editor, *Advances in Atomic and Molecular Physics, supplement 2*, page 123. Academic Press, New York, 1994.

[29] P.R. Berman. *Cavity quantum electrodynamics, Advances in atomic, molecular and optical physics, supplement 2*. Academic Press, Boston, 1994.

[30] J.-M. Raimond, M. Brune, and S. Haroche. Manipulating quantum entanglement with atoms and photons in a cavity. *Rev. Mod. Phys.*, 73:565, 2001.

[31] R. J. Glauber. Coherent and incoherent states of the radiation field. *Phys. Rev.*, 131(6):2766, 1963.

[32] U. M. Titulaer and R. J. Glauber. *Phys. Rev.*, 145:1041, 1966.

[33] E. P. Wigner. On the quantum correction for thermodynamic equilibrium. *Phys. Rev.*, 40:749, 1932.

[34] C. Cohen-Tannoudji, J. Dupont-Roc, and G. Grynberg. *Photons and Atoms*. Wiley, New York, 1992.

[35] NV. Weisskopf and E. Wigner. *Z. Phys*, 63:54, 1930.

[36] A. Aspect, J. Dalibard, and G. Roger. Experimental test of Bell's inequalities using time-varying analysers. *Phys. Rev. Lett.*, 49(25):1804, 1982.

[37] R.J. Glauber. Optical coherence and photon statistics. In C. de Witt, A. Blandin, and C. Cohen-Tannoudji, editors, *Quantum Optics and Electronics, Les Houches Summer School*. Gordon and Breach, London, 1965.

[38] Z.Y. Ou and L. Mandel. Further evidence of nonclassical behavior in optical interference. *Phys. Rev. Lett.*, 62:2941, 1989.

[39] P. Grangier, G. Roger, and A. Aspect. Experimental evidence for a photon anticorrelation effect on a beam splitter: a new light on single-photon interferences. *Europhys. Lett.*, 1:173, 1986.

[40] H.J. Kimble. Quantum flucuations in quantum optics: squeezing and related phenomena. In J. Dalibard, J.-M. Raimond, and J. Zinn-Justin, editors, *Fundamental Systems in Quantum Optics, Les Houches Summer School, Session LIII*, page 545. North Holland, Amsterdam, 1992.

[41] D. T. Smithey, M. Beck, M. G. Raymer, and A. Faridani. Measurement of the Wigner distribution and the density matrix of a light mode using optical homodyne tomography: Application to squeezed states and the vacuum. *Phys. Rev. Lett.*, 70:1244, 1993.

[42] G. Breitenbach, S. Schiller, and J. Mlynek. Measurement of the quantum states of squeezed light. *Nature (London)*, 387:471, 1997.

[43] A.I. Lvovsky, H. Hansen, T. Aichele, O. Benson, J. Mlynek, and S. Schiller. Quantum state reconstruction of the single-photon Fock state. *Phys. Rev. Lett.*, 87:050402, 2001.

[44] A.I. Lvovsky and J. Mlynek. Quantum-optical catalysis: generating nonclassical states of light by means of linear optics. *Phys. Rev. Lett.*, 88:250401, 2002.

[45] J.-M. Raimond, M. Brune, and S. Haroche. Reversible decoherence of a mesoscopic superposition of field states. *Phys. Rev. Lett.*, 79:1964, 1997.

[46] X. Maître, E. Hagley, J. Dreyer, A. Maali, C. Wunderlich, M. Brune, J.-M. Raimond, and S. Haroche. An experimental study of a Schrödinger's cat decoherence with atoms and cavities. *J. Mod. Opt.*, 44:2023, 1997.

[47] J. Jacobson, G. Björk, I. Chuang, and Y. Yamamoto. Photonic de Broglie waves. *Phys. Rev. Lett.*, 74:4835, 1995.

[48] D. Leibfried, B. DeMarco, V. Meyer, M. Rowe, A. Ben-Kish, J. Britton, W. M. Itano, B. Jelenkovic, C. Langer, T. Rosenband, and D. J. Wineland. Trapped-ion quantum simulator: Experimental application to nonlinear interferometers. *Phys. Rev. Lett.*, 89:247901, 2002.

[49] D. Leibfried, R. Blatt, C. Monroe, and D. J. Wineland. Quantum dynamics of single trapped ions. *Rev. Mod. Phys.*, 75:281, 2003.

[50] J.H. Eberly, N.B. Narozhny, and J.J. Sanchez-Mondragon. Periodic spontaneous collapse and revival in a simple quantum model. *Phys. Rev. Lett.*, 44:1323, 1980.

[51] E. T. Jaynes and F. W. Cummings. Comparison of quantum and semiclassical radiation theories with application to the beam maser. *Proc. IEEE*, page 89, 1963.

[52] S. Haroche. Rydberg atoms and radiation in a resonant cavity. In G. Grynberg and R. Stora, editors, *New Trends in Atomic Physics, Les Houches Summer School Session XXXVIII*, page 347. North Holland, Amsterdam, 1984.

[53] J. Gea-Banacloche. Collapse and revival of the state vector in the Jaynes-Cummmings model: an example of state preparation by a quantum apparatus. *Phys. Rev. Lett.*, 65:3385, 1990.

[54] J. Gea-Banacloche. Atom and field evolution in the Jaynes and Cummings model for large initial fields. *Phys. Rev. A*, 44:5913, 1991.

[55] V. Buzek, H. Moya-Cessa, P.L. Knight, and S.J.D. Phoenix. Schrödinger-cat states in the resonant Jaynes-Cummings model: Collapse and revival of oscillations of the photon-number distribution. *Phys. Rev. A*, 45:8190, 1992.

[56] M. Brune, F. Schmidt-Kaler, A. Maali, J. Dreyer, E. Hagley, J.-M. Raimond, and S. Haroche. Quantum Rabi oscillation: a direct test of field quantization in a cavity. *Phys. Rev. Lett.*, 76:1800, 1996.

[57] G. Rempe, H. Walther, and N. Klein. Observation of quantum collapse and revival in a one-atom maser. *Phys. Rev. Lett.*, 58:353, 1987.

[58] D. M. Meekhof, C. Monroe, B. E. King, W. M. Itano, and D. J. Wineland. Generation of nonclassical motional states of a trapped atom. *Phys. Rev. Lett.*, 76:1796, 1996.

[59] G. Morigi, E. Solano, B.-G. Englert, and H. Walther. Measuring irreversible dynamics of a quantum harmonic oscillator. *Phys. Rev. A*, 65:040102, 2002.

[60] J. Dalibard, Y. Castin, and K. Mölmer. Wave-function approach to dissipative processes in quantum optics. *Phys. Rev. Lett.*, 68:580, 1992.

[61] S. Zippilli, D. Vitali, P. Tombesi, and J.M. Raimond. Scheme for decoherence control in microwave cavities. *Phys. Rev. A*, 67:052101, 2003.

[62] L. G. Lutterbach and L. Davidovich. Method for direct measurement of the Wigner function in cavity QED and ion traps. *Phys. Rev. Lett.*, 78:2547, 1997.

[63] P. Bertet, A. Auffeves, P. Maioli, S. Osnaghi, T. Meunier, M. Brune, J.-M. Raimond, and S. Haroche. Direct measurement of the Wigner function of a one-photon Fock state in a cavity. *Phys. Rev. Lett.*, 89:200402, 2002.

[64] L. Davidovich, A. Maali, M. Brune, J.-M. Raimond, and S. Haroche. Quantum switches and non-local microwave fields. *Phys. Rev. Lett*, 71:2360, 1993.

[65] L. Davidovich, M. Brune, J.-M. Raimond, and S. Haroche. Mesoscopic quantum coherences in cavity QED: Preparation and decoherence monitoring schemes. *Phys. Rev. A*, 53:1295, 1996.

[66] K. Banaszek and K. Wodkiewicz. Testing quantum nonlocality in phase space. *Phys. Rev. Lett.*, 82:2009, 1999.

[67] P. Meystre. *Atom optics*. Springer, Berlin, 2001.

[68] E. M. Wright, D. F. Walls, and J. C. Garrison. Collapses and revivals of Bose-Einstein condensates formed in small atomic samples. *Phys. Rev. Lett.*, 77:2158, 1996.

[69] J. I. Cirac, M. Lewenstein, K. Molmer, and P. Zoller. Quantum superposition states of Bose-Einstein condensates. *Phys. Rev. A*, 57:1208, 1998.

[70] A. Sinatra and Y. Castin. Phase dynamics of Bose-Einstein condensates: Losses versus revivals. *Eur. Phys. J. D*, 4:247, 1998.

[71] D. Gordon and C. M. Savage. Creating macroscopic quantum superpositions with Bose-Einstein condensates. *Phys. Rev. A*, 59:4623, 1999.

[72] D. A. R. Dalvit, J. Dziarmaga, and W. H. Zurek. Decoherence in Bose-Einstein condensates: Towards bigger and better Schrödinger cats. *Phys. Rev. A*, 62:013607, 2000.

[73] A. Montina and F. T. Arecchi. Bistability and macroscopic quantum coherence in a Bose-Einstein condensate of ^{7}Li. *Phys. Rev. A*, 66:013605, 2002.

[74] Y. Castin. Bose-Einstein condensates in atomic gases. In R. Kaiser, C. Westbrook, and F. David, editors, *Coherent atomic matter waves, Les Houches summer school series*, page 1. EDP Sciences, Orsay, 2001.

[75] S. Inouye, M.R. Andrews, J. Stenger, H.J. Miesner, D.M. Stanper-Kurn, and W. Ketterle. Observation of Feshbach resonances in a Bose-Einstein condensate. *Nature (London)*, 392:151, 1998.

[76] Y.R. Shen. *The Principles of Non-Linear Optics*. Wiley-interscience, New York, 1984.

[77] M. Greiner, O. Mandel, T. W. Hänsch, and I. Bloch. Collapse and revival of the matter wave field of a Bose-Einstein condensate. *Nature (London)*, 419:51, 2002.

[78] S. Haroche. Breeding non-local Schrödinger cats: a thought experiment to explore the quantum-classical boundary. In J.D. Barrow, P.C.W. Davies, and C.L. Harper, editors, *Science and ultimate reality: quantum theory, cosmology and complexity*, page 1. Cambridge University Press, London, 2004.

[79] B. Julsgaard, A. Kozhekin, and E.S. Polzik. Experimental long-lived entanglement of two macroscopic objects. *Nature (London)*, 413:400, 2001.

[80] L. Hackermüller, S. Uttenthaler, K. Hornberger, E. Reiger, B. Brezger, A. Zeilinger, and M. Arndt. Wave nature of biomolecules and fluorofullerenes. *Phys. Rev. Lett.*, 91:090408, 2003.

[81] C.H. van der Wal, A.C.J. ter Haar, F.K. Wilhelm, R.N. Schouten, C.J.P.M. Harmans, T.P. Orlando, S. Lloyd, and J.E. Moij. Quantum superposition of macroscopic peristent-current states. *Science*, 290:773, 2000.

Course 3

CAVITY QUANTUM ELECTRODYNAMICS

M. Brune

Laboratoire Kastler Brossel,
Ecole Normale Supérieure,
Paris, France

D. Estève, J.-M. Raimond and J. Dalibard, eds.
Les Houches, Session LXXIX, 2003
Quantum Entanglement and Information Processing
Intrication quantique et traitement de l'information
© 2004 Elsevier B.V. All rights reserved

Contents

1. Introduction

One of the most important feature of quantum physics is the concept of entanglement. After interaction, two quantum objects usually behave as a single entity, each of the systems can not any more be described separately. A non-separable entangled state must the be introduced for representing the state of the system as a whole. Such a state presents unbelievable correlation from the point of view of classical logic as pointed out By Einstein Podolsky and Rosen [1] (EPR). Entanglement manifests while performing a measurement on one of the two parts of an EPR pair. It enforces to consider that the other part of the system is instantaneously projected during this measurement independently of the distance separating the two systems. The EPR situation also sits at the heart of quantum measurement theory. While describing quantum mechanically the interaction of a system with a meter, one have to consider at some point a system-meter entangled state whose strangeness was emphasized by the famous Schrödinger cat metaphor [2, 3]. While considering this problem the physics of entangled states provides a new insight in the understanding of the transition between the quantum word of small isolated quantum systems and the classical behavior of macroscopic meters. The concept of decoherence [4, 5], introduced in this context by considering the entanglement of the meter with its environment also relies on the understanding of the behavior of complex entangled states.

Beyond these fundamental problems, entanglement has also be more recently recognized as a powerful tool for manipulating information [6]. The emerging field of quantum information processing opens now the way to the use of entanglement for performing tasks that are impossible to achieve as efficiently with classical logic. Quantum cryptography [7] , whose inviolability relies on quantum physical rules, and teleportation [8] are the most spectacular achievement of this field. New perspectives now rely on advances in the manipulation of isolated particles allowing the preparation of tailored entangled states.

Various techniques are presently used for investigating quantum features related to entanglement in highly controlled systems. The key point is the degree of isolation of the system with respect to the environment. Pioneering experiments where performed with correlated photons. Once entangled, these particle propagate over large distances without interaction with the environment, thereby preserving entanglement until detection. Strongly entangled photon pairs are spontaneously produced by atomic cascades or parametric down-conversion. They have been used to demonstrate the violation of Bell inequalities [9, 10] as well as to implement quantum cryptography [11] and teleportation [12–14]. Triplets of entangled photons have also been generated and used for non-locality tests [15, 16]. This way of getting entangled particles relies however on random irreversible processes. In these experiments, one uses the entangled states that nature gives spontaneously in very specific situations. The method is thus limited to the demonstration of entanglement in relatively simple situations.

Progresses in the manipulation of single isolated massive particles have opened new perspectives by allowing to "synthesize" deterministically complex multiparticle entangled states. The key feature here is the use of strong interactions at the single particle level for generation of entanglement in controlled reversible Hamiltonian processes. Strongly interacting particles however, are also very often strongly coupled to the environment. The difficulty then consists in minimizing this coupling which is responsible for decoherence while preserving strong mutual interactions inside the system. This is presently achieved in two different fields: Ion trapping [17, 18] and microwave Cavity Quantum Electrodynamics (CQED) [19].

This course is devoted to the physics of entanglement in microwave CQED experiments. The heart of this system is a microwave photon trap, made of superconducting mirrors, which stores a few-photon field in a small volume of space for times as long as milliseconds. This field interacts with "circular" Rydberg atoms [20] injected one by one into the cavity. They combine a huge dipole coupling to a single photon with a lifetime (30 ms) three orders of magnitude larger than the cavity crossing time (20 μs). In this system, coupling to the environment is weak enough so that coherent atom-field interaction overwhelms dissipative processes achieving the so called "strong coupling regime". We will focus here on experiments where the strong coupling regime is used to built quantum gates in order to prepare complex multiparticle entangled states. The field of manipulation of *Schrödinger cat states* of the cavity field is investigated in details in the lecture by S. Haroche in this book.

Section 2 of this course is devoted to the description of the strong coupling regime in Rydberg atom CQED [21–24]. The tools of the experiment are briefly presented at the beginning of this section as well as the main characteristics of the strong coupling regime [25–27]. We then present in section 3, how to use the strong atom-cavity to perform various two particles quantum gates. The principle of operation of a quantum phase gate will be discussed. When associating this gate to arbitrary single qubit manipulation, one gets a universal set of gate allowing the step by step preparation of arbitrary multiparticle entangled states. In section 4, we will illustrate this ability by presenting an experimental preparation of a three particles GHZ (Greenberger Horne Zeilinger [15]) entangled state [28].

2. Microwave CQED experiments: The strong coupling regime

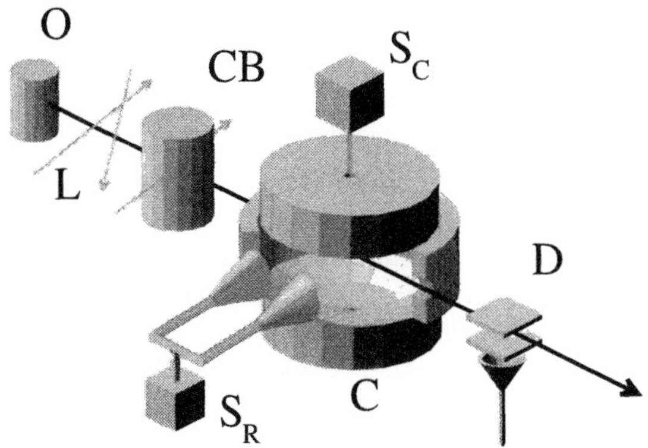

Fig. 1. Experimental set-up

The microwave CQED experiments described in this course all rely on the strong coupling between single two level atoms and a few microwave photons stored in a high Q superconducting cavity. We will recall here the essential properties of the various element of the setup as sketched in fig. 1. A thermal beam of Rubidium atoms originating from oven O is promoted to highly excited circular Rydberg states in the circularization box CB. The excitation scheme is made velocity-selective by a combination of velocity selective optical pumping performed by the lasers L and time of flight techniques [27]. The monokinetic atomic beam then crosses a high Q superconducting cavity mode C tuned close to transition between two circular levels e and g. A small classical field can be injected into C by the classical source S_C. Before and after interaction with C, the atoms can be exposed to microwave pulses generated by the source S_R. These pulses are used for preparation or detection of arbitrary two level superposition states. After leaving C, the atoms are detected one by one in a state selective ionization detector D allowing one to measure whether the atom eventually is in state e or g. Because of the use of superconducting material as well as to the high sensitivity of circular Rydberg atoms to blackbody radiation, all the elements of the set-up, from the circularization to the detection of the atoms, must be cooled down between 1.2 and 0.6 K in a ^{3}He cryostat.

2.1. *The experimental tools and orders of magnitude*

2.1.1. *Circular Rydberg atoms*

They combine a large principal quantum number N with maximal orbital and magnetic quantum numbers $l = |m| = N - 1$. A circular state with principal quantum number N will be referenced as N_c. The wavefunction of the Rydberg electron is a torus whose diameter is $a_0 N^2$. This "large" wavefunction results in a very large dipole coupling between adjacent circular levels. In the experiments described here, the levels e and g are respectively the 51_c and 50_c states. The dipole matrix element between these levels is $d = q a_0 N^2/2 = 1250\ a.u.$. The frequency of this transition is $\nu_{eg} = 51.099$ GHz.

The circular atomic levels are prepared by exciting the valence electron of Rubidium atoms into the 52_c state in a complex process involving 52 photons [20]. The 51_c or 50_c levels are then prepared selectively by a last microwave pulse resonant either on the $52_c \rightarrow 51_c$ one-photon or $52_c \rightarrow 50_c$ two-photon transition at 48.195 GHz and 49.647 GHz, respectively. This process prepares up to 400 circular atoms per preparation pulse. The selectivity of the last microwave transition in a large dc electric field allows the elimination of spurious elliptical levels (all other values of l and m). The purity of the prepared state, measured by a selective spectroscopic method is better than 98%. The stability of circular atoms requires the application of a small electric field providing a physical quantization axis everywhere in the set-up [29]. Under this condition, the atoms prepared in e or g behave as ideal long lived two level atoms while they interact with a nearly resonant cavity mode.

Circular atoms are easily detected by ionization in a relatively small static electric field. As the ionization threshold increases with the binding energy of the levels, one can selectively ionize either e or g in two different detectors. This detection scheme, relies on electron counting. It is extremely sensitive and behave as a meter for the energy of a single atom. It allows measurements on a single realization of the experiment as well as to measure average values of the atomic energy by resuming the same experiment until significant statistics is obtained. The regime of single atom interaction with the cavity is achieved at the expense of low counting rates of typically 0.1 to 0.2 detected atom per preparation pulse (detection efficiency 40%(10)). In this limit, the Poissonian statistic of the number of excited atoms results in a negligible probability to excite two atoms at the same time.

A pulsed velocity selective optical pumping scheme prepares monokinetic Rubidium atoms [27] in the state $5s_{1/2}$ $F = 3$ just after they leave the oven O. This level is the starting point of the circular atoms preparation. The width of the velocity distribution obtained in this way is 10 m/s. It is reduced to 1.5 m/s by time of flight selection between optical pumping and circularization which is also a pulsed process. Due to the control of the atomic velocity and of the time of preparation of Rydberg atoms, one knows the position of the circular atoms inside the setup with a precision of $\pm 1\ mm$. This is used for applying to each atom the proper sequence of controlled interactions with the cavity mode and the auxiliary classical microwave pulses. In particular, the atom-cavity interaction time is adjusted by switching on and off an electric field of about 1 V/cm between the cavity mirrors. The atoms are then tuned in and out of resonance by Stark effect at user controlled programmable times while they cross the cavity.

2.1.2. *The photon box*

The cavity is made of two massive Niobium mirrors in a Fabry-Perot geometry depicted in fig. 1. The two spherical mirrors have a radius of curvature of 40 mm. The distance between the mirrors at center is 27.5 mm. The atoms are nearly resonantly coupled to the TEM_{900} Gaussian mode whose resonance frequency is close to $\nu_{eg} = 51.099\ GHz$. The mode waist, $w_0 = 5.96\ mm$, is close to the wavelength, $\lambda = 6\ mm$. The corresponding mode volume [21] is relatively small ($V \simeq 700\ mm^3$). The microwave electric field amplitude at cavity center $E_0 = \sqrt{h\nu_{eg}/2\epsilon_0 V} = 1.5\ mV/m$ is the essential parameter characterizing the coupling with the atomic dipole. Due to geometrical defects of the mirrors, the degeneracy between the two modes with linear perpendicular polarizations is lifted by about 100kHz. When one atom interacts resonantly with one of these two modes, the coupling with the other one usually plays a negligible role.

A quality factor as high as 3.10^8 corresponding to a photon lifetime of 1 ms is obtained by careful polishing and processing of the mirrors. It is limited by diffusion of photons out of the aperture between the mirrors due to the residual roughness of their surface. These losses do not occur in a closed cavity [30]. However, the closed geometry is not compatible with the electric field needed for stabilizing circular atoms [29]. Diffusion losses are reduced by inserting an aluminum ring nearly closing the opening between the mirrors. The atoms enter the cavity trough 3 mm diameter holes in this ring. Inhomogeneous electric fields in these holes destroy atomic coherence but they do not affect the populations of Rydberg states. An external microwave source is coupled into the cavity mode through small 0.2 mm diameter holes at the center of the mirrors.

2.2. *Resonant atom-field interaction: The vacuum Rabi oscillation*

A detailed description of the atom-cavity interaction can be found in various review papers [21–24] as well as in the lecture by S. Haroche. It relies on the Jaynes-Cummings hamiltonian [31] whose eigenstates are the so called "dressed states" [32] of the atom-field system. The non-degenerate ground state of the system is $|g, 0\rangle$ where 0 stands for the photon number. We are interested here in the dynamics of the two first excited states of the system $|g, 1\rangle$ and $|e, 0\rangle$ which are coupled by an electric dipole transition. This coupling results in a splitting $\hbar\Omega_0 = -2dE_0$ of the first dressed states $|+, 0\rangle$ and $|-, 0\rangle$. An atom initially in state e crossing an empty cavity thus experiences a "vacuum Rabi oscillation": Atom and field exchange periodically one photon at the Rabi frequency $\Omega_0/2\pi = 47\ kHz$.

The corresponding Rabi oscillation signal [26] is presented in fig. 2. It shows the measured average atomic excitation as a function of the atom-field interaction time. The cavity is tuned at resonance with the 51_c to 50_c transition. The mode Q factor is 7.10^7, corresponding to a photon lifetime of 220 μs. Up to four cycles of Rabi oscillations are clearly observed demonstrating the strong coupling regime. The decay of the oscillation signal is due to various imperfections (dark counts, atoms detected in the wrong channel, inhomogeneous stray electric or magnetic fields).

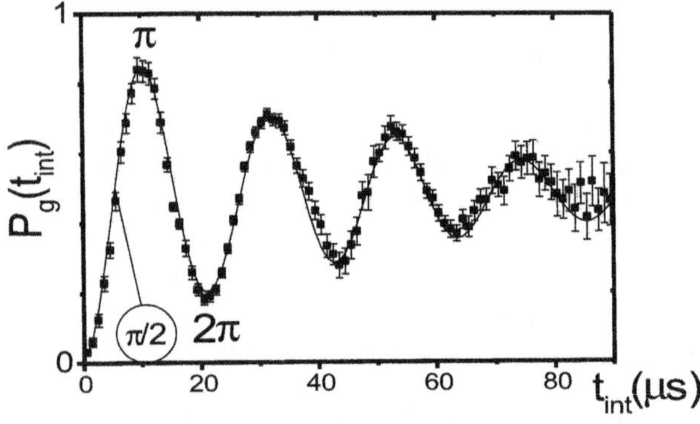

Fig. 2. Rabi oscillation signal: A single atom emits and reabsorbs a single photon. Up to 4 oscillation cycles are observed. Interaction times corresponding to $\pi/2$, π and 2π pulses are marked by labels

If the cavity contains initially n photons, the Rabi oscillation frequency become $(\Omega_0/2\pi)\sqrt{n+1}$ [33]. By observing the Rabi oscillation in a small coherent field [34] stored in C, a discrete spectrum of Rabi frequencies has been observed [22,26]. Note that this spectrum is a direct manifestation of field energy quantization in the cavity mode. This feature has also been used for measuring the photon number distribution of small coherent fields stored in C with up to 1.4 photons on average [26]. Rabi oscillation in small photon number states was also observed in [35].

More recently, the Rabi oscillation in a coherent field has been used to generate phase "Schrödinger cat states" involving fields containing up to 40 photons [36]. The method of preparation and detection of such states is presented in section 2 of the course by S. Haroche in this book.

3. "Quantum logic" operations based on the vacuum Rabi oscillation

The vacuum Rabi oscillation provides important tools for implementing quantum gates performing basic two qubit logic operations. In this section, we briefly present these basic operations. In the next section we will show how they can be combined in order to engineer step by step a three qubit entangled state.

For an atom and a field initially prepared either in $|e, 0\rangle$ or $|g, 1\rangle$, the atom-field wavefunctions after the interaction time t_{int} read respectively :

$$|\psi_e(t_{int})\rangle = cos(\Omega_0 t_{int}/2)|e, 0\rangle + sin(\Omega_0 t_{int}/2)|g, 1\rangle \tag{3.1}$$

$$|\psi_g(t_{int})\rangle = cos(\Omega_0 t_{int}/2)|g, 1\rangle - sin(\Omega_0 t_{int}/2)|e, 0\rangle \tag{3.2}$$

Three basic functions are realized by adjusting the atom-cavity interaction time to specific values corresponding to $\Omega_0 t_{int} = \pi/2, \pi$ or 2π. The π pulse can be used to exchange one excitation between the atom and the cavity mode. In this way, an atom in e can be used to *write* a one-photon field in the cavity. If the atom is prepared in the arbitrary superposition state $c_e|e\rangle + c_g|g\rangle$, it will, after a π pulse, always end up in g and prepare the field in the state: $c_e|1\rangle + c_g|0\rangle$. All the quantum information encoded in the atom by a classical microwave pulse is transferred and stored in the cavity. This writing process being reversible, the stored quantum information can be *read* by another atom prepared in g which is performing an absorbing π pulse. In these processes, the cavity acts as a quantum memory as demonstrated in [27].

In case of an atom prepared in e performing $\pi/2$ pulse in the cavity, the prepared atom-field state is:

$$|\psi_{EPR}\rangle = 1/\sqrt{2}(|e, 0\rangle + |g, 1\rangle) \tag{3.3}$$

It is a maximally entangled state analogous to the singlet state of a pair of spin 1/2. As the atom leaves the cavity after interaction, this state exhibits the non-local quantum correlations which are at the heart of the EPR [1] situation and which characterize vividly the difference between quantum and classical logic through the Bell theorem [9]. Preparation and characterization of $|\psi_{EPR}\rangle$ is presented in [37].

Let us finally consider the 2π Rabi pulse. When the atom is prepared in g the atom-field wavefunction transforms in the following way:

$$\begin{array}{lcl} |g, 1\rangle & \rightarrow & -|g, 1\rangle \\ |g, 0\rangle & \rightarrow & |g, 0\rangle \end{array} \tag{3.4}$$

For a field containing one photon, the 2π pulse leads to a π phase shift of the atom-field state as seen on eq. 3.4. A similar π–phase shift occurs when performing a 2π rotation on a spin 1/2 system [38, 39]. Now if the cavity is initially empty, the system is in the ground state $|g, 0\rangle$. It does not evolve and does not experience any phase shift. In both cases, the field energy (i.e. 0 or 1 photon) is unchanged but the phase of the final state carries information on the photon number. This provides the principle of the QND (quantum non demolition) method of measurement of a 0 or 1 photon field discussed in details in [40].

It also allows one to implement the so cold *Quantum Phase Gate* (QPG) [41]. When combined with arbitrary single qubit operations (i.e. classical microwave pulses applied to single atoms) this two qubits gate is equivalent to the CNOT gate and plays the role of a universal gate for *synthesizing* arbitrary N qubits entangled states.

The QPG transformation simply reads:

$$|a, b\rangle \longrightarrow \exp(i\phi\delta_{a,1}\delta_{b,1})|a, b\rangle \tag{3.5}$$

where $|a\rangle$, $|b\rangle$ stand for the basis states ($|0\rangle$ or $|1\rangle$) of the two qubits and $\delta_{a,1}$, $\delta_{b,1}$ are the usual Kronecker symbols. The QPG leaves the initial state unchanged, except if both qubits are 1, in which case the state is phase–shifted by an angle ϕ. In order to implement the QPG, let us now consider a third atomic level i and let us assume that due to large detunings, this level is not coupled to the high Q cavity mode. To be specific let us consider i as the circular Rydberg state with principal quantum number $N_c = 49$. The transformation corresponding to the 2π Rabi pulse in C is:

$$\begin{array}{lcl} |i, 0\rangle & \longrightarrow & |i, 0\rangle \\ |i, 1\rangle & \longrightarrow & |i, 1\rangle \\ |g, 0\rangle & \longrightarrow & |g, 0\rangle \\ |g, 1\rangle & \longrightarrow & -|g, 1\rangle \end{array} \tag{3.6}$$

When mapping the atomic states i and g on the logical 0 or 1 value of the atomic qubit, it exactly realizes the $\phi = \pi$ QPG. The ability of this gate to generate entangled states can be demonstrated by operating it on a superposition state of either the atomic or field qubit. As an exemple, after preparing the atom-field in the state $1/2(|i\rangle+|g\rangle)(|0\rangle+|1\rangle)$ the operation of the QPG prepares the maximally entangled state:

$$1/2(|i\rangle + |g\rangle)|0\rangle + (|i\rangle - |g\rangle)|1\rangle) \tag{3.7}$$

This equation shows that after interaction with C, the atomic state superposition is phase shifted by π if and only if the cavity contains one photons. Note that the 2π pulse interaction with C leaves the photon number unchanged. Measuring the phase of the atomic superposition state thus amounts to a Quantum Non Demolition (QND) detection of a single photon in C. As shown in [40], this atomic measurement can be implemented using a Ramsey interferometer by applying classical $\pi/2$ pulses to the $g - i$ transition before and after the atom crosses C. This experiment demonstrates that the phase of an atomic superposition state is coherently controlled by the state of a single photon. Symmetrically, we have also demonstrated that the phase of a superposition of the 0 and 1 field states is shifted by π under the operation of the QPG when the atom is prepared in g [41].

4. Step by step synthesis of a three particles entangled state

We present now an experiment where we prepare a set of three entangled qubits consisting of two atoms and a 0 or 1 photon field stored in C [28] by combining elementary quantum gate operations. It is the first example of preparation of a tailored three particle entangled state by a programmed sequence of quantum gates.

4.1. Principle of the preparation of the state

We first recall the sequence of operations used to prepare the three particle entangled state. It was proposed independently in [42] and [43]. The corresponding timing is sketched fig. 3.a. We send across C, initially empty, an atom A_1 initially in e. A $\pi/2$ Rabi pulse prepares the state $|\psi_{EPR}\rangle$ described by eq. 3.3. We then send a second atom A_2. Initially in g, it is prepared, before C, in the state $(|g\rangle + |i\rangle)/\sqrt{2}$ by a Ramsey pulse P_2. This atom interacts with C during its full cavity crossing time (2π Rabi pulse) and performs the QPG operation. Using eq. 3.6, the resulting $A_1 - A_2 - C$ quantum state is :

$$|\Psi_{triplet}\rangle = \frac{1}{2}\left[|e_1\rangle(|i_2\rangle + |g_2\rangle)|0\rangle + |g_1\rangle(|i_2\rangle - |g_2\rangle)|1\rangle\right] \tag{4.1}$$

(the state indices correspond to the atom number). Eq. 4.1 describes a three particle entangled state and can be rewritten as :

$$|\Psi_{triplet}\rangle = \frac{1}{2}\left[|i_2\rangle(|e_1, 0\rangle + |g_1, 1\rangle) + |g_2\rangle(|e_1, 0\rangle - |g_1, 1\rangle)\right] , \tag{4.2}$$

describing an $A_1 - C$ EPR pair whose phase is conditioned to the A_2 state. Since $|\Psi_{triplet}\rangle$ involves two levels for each subsystem, it is equivalent to an entangled state of three spins 1/2. Let us define the states $|+_i\rangle$ ($|-_i\rangle$) (with $i = 1, 2$) as $|+_1\rangle = |e_1\rangle$ ($|-_1\rangle = |g_1\rangle$), $|\pm_2\rangle = (|g_2\rangle \pm |i_2\rangle)/\sqrt{2}$ and $|+_C\rangle = |0\rangle$ ($|-_C\rangle =$

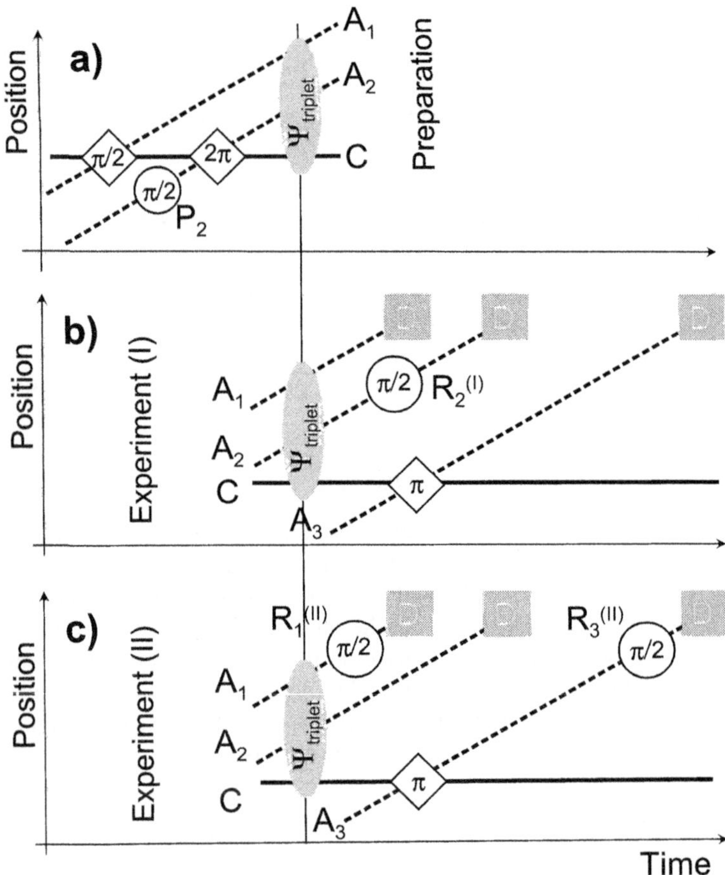

Fig. 3. The entanglement procedure. Qualitative representation of the atoms and cavity space lines during the experiments. The diamonds depict the atom-cavity interactions and the circles the classical pulses produced by S_R. The dark squares are the detection events. a) Preparation of the entangled state $|\Psi_{triplet}\rangle$ sketched by the grey oval. b) Experiment (I): Detection of "longitudinal" correlations. c) Experiment (II): Detection of "transverse" correlations.

$|1\rangle)$. With these notations, $|\Psi_{triplet}\rangle$ takes the form of the Greenberger, Horne and Zeilinger (GHZ) spin triplet [15]:

$$|\Psi_{triplet}\rangle = \frac{1}{\sqrt{2}}\left(|+_1, +_2, +_C\rangle - |-_1, -_2, -_C\rangle\right) , \tag{4.3}$$

Other schemes have been proposed to realize many particle atom-cavity entanglement [44,45].

4.2. Detection of the three-particle entanglement

In order to characterize the state $|\Psi_{triplet}\rangle$, we are able to detect the atomic energy states, but not directly the cavity field. It can, however, be copied onto a third atom A_3 and detected afterwards [27]. The $A_3 - C$ interaction is set so that A_3, initially in g, is not affected if C is empty, but undergoes a π Rabi pulse in a single photon field : $|g, 0\rangle \to |g, 0\rangle$ and $|g, 1\rangle \to -|e, 0\rangle$. Within a phase, A_3 maps the state of C. Thus, by detecting A_1, A_2 and A_3, we measure a set of observable belonging to the three parts of the entangled triplet. If A_3 crosses C before A_1 exits the ring, a three-atom entangled state $|\Psi'_{triplet}\rangle$ would be created between these two events :

$$|\Psi'_{triplet}\rangle = \frac{1}{2}\left[|e_1\rangle(|i_2\rangle + |g_2\rangle)|g_3\rangle - |g_1\rangle(|i_2\rangle - |g_2\rangle)|e_3\rangle\right] \tag{4.4}$$

$$= \frac{1}{2}\left[|i_2\rangle(|e_1, g_3\rangle - |g_1, e_3\rangle) + |g_2\rangle(|e_1, g_3\rangle + |g_1, e_3\rangle)\right] \tag{4.5}$$

Even if A_3 is delayed, its correlations with A_1 and A_2, which reflect those of C, are the same as those described in eq. 4.5. In the following discussion, we thus refer equivalently to C or A_3.

Checking the $A_1 - A_2 - C$ entanglement involves measurements in two different basis. A microwave pulse, after the interaction with C, followed by energy detection in D allows us to probe each atom's pseudo-spin along an arbitrary "quantization axis". In a first experiment (I), whose timing is sketched fig. 3.b, we check "longitudinal" correlations by detecting the "spins" along what we define as the "z axis" (eigenstates $|\pm_i\rangle$ for $i = \{1, 2\}$ and $|+_3\rangle = |e_3\rangle$ and $|-_3\rangle = |g_3\rangle$ for A_3). For A_1 and C (i. e. A_3), this is a direct energy detection. For A_2, a $\pi/2$ analysis pulse $R_2^{(I)}$ on the $i \to g$ transition transforms $|+_2\rangle$ (resp. $|-_2\rangle$) into $|i_2\rangle$ (resp. $|g_2\rangle$). The three atoms should thus be detected in $\{e_1, i_2, g_3\}$ or $\{g_1, g_2, e_3\}$, with equal probabilities. However, these correlations, taken alone, can be explained classically (statistical mixture of $|e_1, i_2, g_3\rangle$ and $|g_1, g_2, e_3\rangle$ states).

A second experiment (II) is required to test the quantum nature of the superposition. We study "transverse correlations" by detecting A_1 and A_2 along the "x axis" (eigenstates $|\pm_{x,i}\rangle = (|+_i\rangle \pm |-_i\rangle)/\sqrt{2}$). A_3 is detected along an axis in the horizontal plane at an angle ϕ from the x direction (eigenstates $|\pm_{\phi,i}\rangle = (|+_i\rangle \pm \exp(+i\phi)|-_i\rangle)/\sqrt{2}$). The timing of experiment (II) is sketched fig. 3.c. Atom A_2 is directly detected in D, since $|\pm_{x,2}\rangle$ coincide with $|g_2\rangle$ and $|i_2\rangle$. A_1 and A_3 undergo, after C, two analysis $\pi/2$ pulses $R_1^{(II)}$ and $R_3^{(II)}$ on the $e \to g$ transition, with a phase difference ϕ. A detection in g amounts to a detection in $|+_x\rangle$ or $|+_\phi\rangle$ for A_1 and A_3 respectively at the exit of C.

For sake of clarity, let us first consider the case of only two atoms (1 and 3) in state:

$$|\Psi'_{EPR}\rangle = \frac{1}{\sqrt{2}}(|e_1, g_3\rangle - |g_1, e_3\rangle) . \tag{4.6}$$

These atoms are analyzed along the x and ϕ directions respectively. When A_1 is detected in $|+_{x,1}\rangle$ (i. e. g_1 in D), A_3 is projected onto $|-_{x,3}\rangle$, since $|\Psi'_{EPR}\rangle$ is the rotation-invariant spin singlet. The detection probability of A_3 in $|+_{\phi,3}\rangle$ (i. e. g_3 in D) thus oscillates versus ϕ between 1 for $\phi = \pm\pi$ and 0 for $\phi = 0, 2\pi$: "Fringes" observed in the joint detection probabilities of the two atoms [37] show that quantum coherence has been transferred between them through the EPR correlations. The phase of the fringes would be shifted by π if the minus sign in eq. 4.6 was changed into a plus. Returning to the three system case, eq. 4.5 shows that similar fringes are expected for the joint detection of A_1 and A_3 corresponding to a given state for A_2. They have the same phase as the EPR fringes described by eq. 4.6 when A_2 is in i_2. They are shifted by π when A_2 is in g_2. This shift results from the action of the $A_2 - C$ phase gate [41] on the $A_1 - C$ EPR pair.

A tight timing is required to have A_1 and A_2 simultaneously inside the ring so that $|\psi_{triplet}\rangle$ is prepared before A_1 losses its coherence in the exit hole of the cavity (it was not the case in the experiments described in section 3.4). A_2 interacts with C for the full atom–cavity interaction time. The π Rabi pulse condition for A_3 is realized with the Stark switching technique. Atom A_1 couples to C 75 μs after the erasing sequence, and should undergo a $\pi/2$ Rabi rotation. It is followed by A_2 after a delay of 25 μs. The separation between A_1 and A_2 is 1.2 cm, twice the cavity waist. Nevertheless, A_1 still interacts with C when A_2 starts its 2π Rabi rotation. Even in this case, an appropriate adjustment of the atom cavity Stark tuning allows to prepare $|\psi_{triplet}\rangle$ with a high fidelity as shown in [28]. Atom A_1 has exited the ring however before A_3 has crossed C, following A_2 after a delay of 75 μs. This timing thus does not permit to prepare $|\Psi'_{triplet}\rangle$ (eq. 4.5). As discussed above, the $A_1 - A_2 - A_3$ correlations nevertheless demonstrate the $A_1 - A_2 - C$ entanglement.

We apply the classical $\pi/2$ microwave pulses when the atom is in an antinode of the standing wave created inside the cavity ring by a classical microwave source S_R. The distance between A_1 and A_2 is such that one is in a node of this wave when the other is in an antinode. In this way, selective pulses may be applied on A_1 and A_2 even if both are simultaneously in the ring. In experiment (I), P_2 and $R_2^{(I)}$ are applied on A_2 on the 54.3 GHz $g \to i$ transition. In experiment (II), $R_1^{(II)}$ and $R_3^{(II)}$ are used to probe the $|\pm_{x,1}\rangle$ and $|\pm_{\phi,3}\rangle$ states. A pulse resonant on the $e \to g$ transition would couple in C through scattering on the mirrors imperfections. A field would then build up in C and spoil quantum correlations. To avoid this, we first apply a π pulse on the $g \to i$ transition transforming the $e - g$ coherence into an $e - i$ one. A $\pi/2$ pulse on the two-photon $e \to i$ transition at 52.7 GHz, which does not feed any field in C, is then used to probe this coherence. States $|+_{x,1}\rangle$ ($|-_{x,1}\rangle$) and $|+_{\phi,3}\rangle$ ($|-_{\phi,3}\rangle$) are mapped by this effective three-photon $\pi/2$ pulse onto i_1 (e_1) and i_3 (e_3) respectively. The results

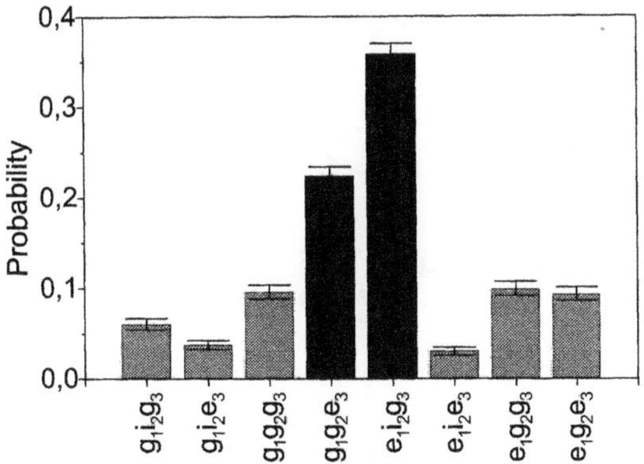

Fig. 4. Longitudinal correlations (experiment I). Histograms of the detection probabilities for the eight relevant detection channels. The two expected channels (g_1, g_2, e_3 and e_1, i_2, g_3, black bars) clearly dominate the others (grey bars), populated by spurious processes. The error bars are statistical.

of experiment (I), fig. 4, are presented as histograms giving the probabilities for detecting the atoms in the eight relevant channels. As expected, the $\{e_1, i_2, g_3\}$ and $\{g_1, g_2, e_3\}$ channels dominate. The total probability of these channels is $P_\parallel = 0.58 \pm 0.02$. The difference between them is due to experimental imperfections. Channel $\{g_1, g_2, e_3\}$ corresponds to one photon stored in the cavity between A_1 and A_3. It is thus sensitive to field relaxation, and leaks into the other $\{g_1\}$ channels. Events with two atoms in the same sample, residual thermal fields, detection errors also contribute to the population of the parasitic channels. Note also that since the experiment involves three levels for each atom, there are altogether 27 detection channels. Fig. 4 presents the channels corresponding to the relevant transitions for each atom: $e \to g$ for A_1 and A_3 ; $g \to i$ for A_2.

The other channels are weakly populated by spurious effects (spontaneous emission outside C, residual thermal photons, influence of the $R_3^{(I)}$ or P_2 pulses on the other atoms, absorption of the cavity field by A_2 due to imperfect 2π Rabi rotation..). The total contribution of these transfer processes is below 15%.

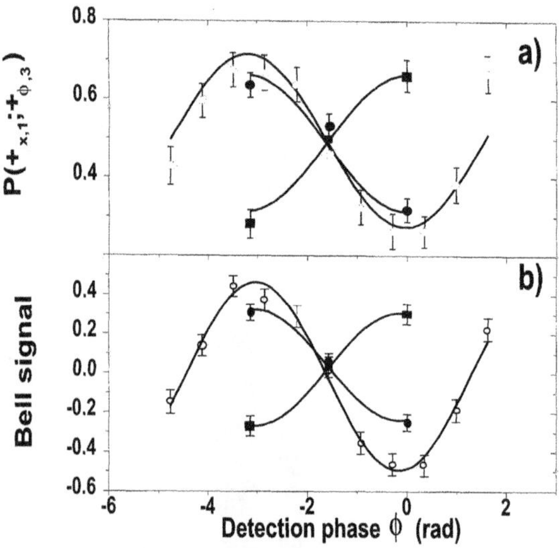

Fig. 5. transverse correlations (experiment II).

For the signals of experiment (II) presented on fig. 5, the relative phase ϕ of $R_3^{(II)}$ and $R_1^{(II)}$ is adjusted by tuning the frequency of the source inducing the $e \rightarrow i$ two-photon transition. Fig. 5.a presents versus ϕ the probability $P(+_{\phi,3}; +_{x,1})$ for detecting A_3 in i (i. e. $|+_{\phi,3}\rangle$) provided A_1 has also been detected in i (i. e. $|+_{x,1}\rangle$). The open circles give the conditional probability when A_2 is not sent. The observed fringes correspond to the two-atom EPR pair situation. The solid circles give the corresponding conditional probability when A_2 is detected in i. Due to very long acquisition times (eight hours for the data in fig. 5), signals have been recorded only for three phase values. The squares correspond to a detection of A_2 in g. The $A_1 - A_3$ correlations are not modified when A_2 is detected in i. When A_2 is detected in g, the $A_1 - A_3$ EPR fringes are shifted by π, as expected. All joint probabilities corresponding to the four possible outcomes for A_1 and A_3 are combined to produce the "Bell signal" [10] which is the expectation value $\langle\sigma_{x,1}\sigma_{\phi,3}\rangle = P_{i_1,i_3} + P_{e_1,e_3} - P_{i_1,e_3} - P_{e_1,i_3}$, where the σ's are Pauli matrices associated to the pseudo-spins and P_{a_1,b_3} is the probability for detecting A_1 in a and A_3 in b ($\{a, b\} = \{i, e\}$). We plot fig. 5.b the Bell signal versus ϕ. The open circles correspond again to no A_2 atom sent, the solid circles and squares to A_2 detected in i or g, respectively. The π phase shift of the $A_1 - A_3$ EPR correlations, conditioned to the A_2 state, is conspicuous. The fringes visibility is $2V_\perp = 0.28 \pm 0.04$.

Due to experimental imperfections, the first stage fig. 3.a of our experiment does not prepare the pure state $|\Psi_{triplet}\rangle$, but rather a mixed state described by a density matrix ρ. The set-up efficiency is thus characterized by a fidelity $F = \langle\Psi_{triplet}|\rho|\Psi_{triplet}\rangle$. If the detection stages (fig. 3.b and 3.c) were perfect, F would be equal to the sum $P_\parallel/2 + V_\perp$ [18]. The value of this quantity, 0.43, is however affected by known detection errors and F is actually larger. Trivial imperfections can occur at three different stages: The mapping of the cavity state onto A_3, the classical microwave pulses $R_2^{(I)}$, $R_1^{(II)}$ and $R_3^{(II)}$, and the energy state-selective atom counting. We have determined these errors independently by additional single atom experiments. Taking them into account, we determine a fidelity $F = 0.54 \pm 0.03$. The three kinds of errors listed above account respectively for corrections of 0.03, 0.05 and 0.03 to the raw 0.43 value. The fact that F is larger than 0.5 ensures that genuine three particle entanglement is prepared here [18].

The combined results of experiments (I) and (II) demonstrate the step by step engineered entanglement of three qubits, manipulated and addressed individually. By adjusting the various pulses, the experiment could be programmed to prepare other tailored multiparticle state. In particular, the generalization of our experiment for preparing multiparticle generalizations of the GHZ triplet [46] are straightforward. These states are generated by a simple iteration of the present scheme [43, 44]. After having prepared the $A_1 - C$ pair in the state described by eq. 4.1, one sends a stream of atoms $A_2 - A_3 - \cdots - A_n$ all prepared in $(|i\rangle + |g\rangle)/\sqrt{2}$ and undergoing, if in g, a 2π Rabi rotation in a single photon field. Since this rotation does not change the photon number, the 0 photon (resp. 1 photon) part of the $A_1 - C$ system gets correlated to an $A_2 - A_3 - \cdots - A_n$ state with all $n-1$ atoms in $(|i\rangle + |g\rangle)/\sqrt{2}$ (resp $(|i\rangle - |g\rangle)/\sqrt{2}$), preparing the entangled state :

$$|\Psi\rangle = \frac{1}{\sqrt{2}}\left(|+_1, +_2, \cdots, +_C\rangle - |-_1, -_2, \cdots, -_C\rangle\right) . \tag{4.7}$$

This state presents non–local $n+1$ particles correlations which could be investigated by the techniques presented here.

Similar controlled and reversible manipulations of many particle entanglement can be performed with other systems. Complex spin manipulations have been demonstrated with nuclear magnetic resonance [47]. These experiments involve however macroscopic samples near thermal equilibrium without clear-cut entanglement [48]. Reversible entanglement with massive particles has also been realized with trapped ions [49]. The generation of an EPR pair [50] and, recently, of four ion entanglement [18] have been reported. In these experiments, strong coupling requires the ions to be only a few micrometers apart and the difficulty is to address them individually. The entangled multi-particle state is prepared in a collective process, involving all qubits at once. Individual addressing of ions is possible in larger ions traps as demonstrated with calcium [51] but controlled quantum logic operations have not yet been demonstrated in this context. In contrast to ion traps, our CQED experiment manipulates particles at centimeter-scale distances, ideal conditions for separate qubit control.

5. Direct atom-atom entanglement: cavity-assisted collision

The atom-atom entangling procedures outlined above rely on the exchange of a photon between the atom and the cavity. The quantum information is transiently stored as a superposition of the zero and one photon states. These schemes are thus sensitive to cavity losses, the main cause of decoherence in our experiments (the atomic lifetime being much longer than the cavity damping time).

It is possible to circumvent this problem by entangling two atoms directly, in a collision process assisted by the non-resonant cavity modes [52]. The first atom (A_1) is initially in e and the second (A_2) in g. The atoms have now different velocities, so that the second catches up the first at cavity center, before exiting first from C. The two cavity modes M_a and M_b are now detuned from the $e \rightarrow g$ transition frequency by amounts Δ and $\Delta + \delta$, greater than Ω. Due to energy conservation, real photon emission cannot occur in this case. Atom A_1 can, however, virtually emit a photon immediately reabsorbed by A_2. This leads to a Rabi oscillation between states $|e, g\rangle$ and $|g, e\rangle$ and thus to atom/atom entanglement generation for most interaction times.

The situation is reminiscent of a resonant van der Waals collision in free space, which can also produce atom-atom entanglement for small enough impact parameters [53]. In the present case, the detuned cavity modes considerably enhance the atom-atom interaction. Note that, in this peculiar "collision" process, the actual distance between the atoms is irrelevant, provided they both interact with the modes.

The quantum amplitudes associated to states $|e, g\rangle$ and $|g, e\rangle$ are periodic functions of the collision duration (which depends on the atomic velocities). The oscillation frequency associated to this second order collision process is $(\Omega^2/4)[1/\Delta + 1/(\Delta + \delta)]$. By repeating the experiment, we reconstruct the probabilities P_{eg} and P_{ge} for finding finally the atom pair in states $|e, g\rangle$ and $|g, e\rangle$. We plot these probabilities versus the dimensionless parameter $\eta = \omega[1/\Delta +$

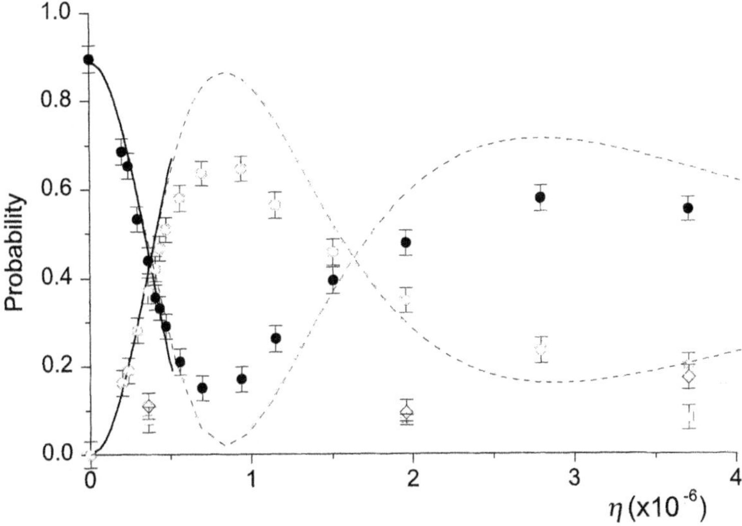

Fig. 6. Cavity assisted collision. Joint detection probabilities P_{eg} and P_{ge} versus the parameter η. Points are experimental. Solid lines for small η values correspond to a simple analytical model based on second order perturbation theory. The dashed lines (large η) present the results of a numerical integration of the system evolution (adapted from [52])

$1/(\Delta + \delta)]$ (see Fig. 6). The oscillations of P_{eg} and P_{ge} as a function of η are well accounted for by theoretical models (solid and dashed lines in Fig. 6).

We have realized the situation of maximum entanglement by adjusting η to the value corresponding to $P_{eg} = P_{ge} = 0.5$. As for the sequential EPR pair generation scheme presented above, we have checked the coherent nature of the pair by performing measurements of observables whose eigenstates are superpositions of energy states.

Since this entanglement procedure implies only a virtual photon exchange with the detuned cavity mode, it is, in first order, insensitive to the cavity damping time or to a stray thermal field in the cavity modes. It thus opens interesting perspectives for demonstrating elementary steps of quantum logic with moderate Q cavities at finite temperature.

We have shown theoretically that the two-qubit Grover search algorithm [54] could be realistically implemented in our set-up with two cavity-assisted collisions between two atoms, performed during the common interaction of the atoms with the cavity mode [55].

6. Conclusion and perspectives

The circular Rydberg atoms already made it possible to operate interesting quantum entanglement processing sequences. The extension to much more complex algorithms requires some improvements of the present set-up.

The fidelity is limited by the imperfections of the elementary gates and by the cavity losses. A better control of the stray fields in the set-up, which seem to be a major cause of imperfections, could improve noticeably this fidelity. The cavity losses could be reduced with a new mirror technology. Encouraging tests indicate that longer cavity damping times are realistically within reach. Moreover, the cavity-assisted collision process makes it possible, in principle, to realize high-fidelity quantum gates with a moderate quality factor.

The main limitation to the scalability thus appears to be the atomic preparation scheme. As discussed above, we operate with single atom samples at the expense of data taking times growing exponentially with the number of qubits. A recent improvement of the field ionization detector efficiency will allow us to run sequences with four or five atomic samples within realistic times.

Further extensions require a deterministic preparation of single Rydberg atom samples. The "dipole blockade" mechanism is a very promising tool [56]. In a dense sample of ground state atoms, the frequency of the transition between a one- and a two-Rydberg sample is displaced by a great amount from the transition producing the first Rydberg, due to the very strong dipole-dipole interaction between these atoms. The laser excitation thus produces a single Rydberg state, with a high probability. This low angular momentum state can be efficiently transferred later to the circular state. We are now developing an experiment based on "atom chips" [57] techniques to explore the feasibility of a deterministic Rydberg atom preparation.

This "atom pistol" would make complex sequences accessible. The maximum number of operations foreseeable is of the order of the atomic lifetime (30 ms) divided by the gate time (10 to 30 μs). This sets a fundamental limit of a few thousand quantum operations. This is far from what is required for a massive quantum computation with error correction. This number of operations is nevertheless competitive with other techniques and large enough to test interesting quantum algorithms and error correction procedures.

Let us note also that these experiments are well-suited to the exploration of other basic quantum mechanisms essential for quantum information processing. In particular, mesoscopic coherent fields stored in the cavity provide an unprecedented tool for an in-depth study of the decoherence mechanisms [36, 58]. We are envisioning an experiment with two superconducting cavity. This opens the way to the generation of non-local mesoscopic states (EPR pairs made of mesoscopic cavity fields) and allows new tests of our understanding of the decoherence process.

References

[1] A. Einstein, B. Podolosky and N. Rosen, Phys. Rev. **47**, 777 (1935).

[2] E. Schrödinger, Naturwissenschaften, **23**, 807, 823, 844 (1935); reprinted in english in [8].

[3] J.A. Wheeler and W.H. Zurek, *Quantum Theory of Measurement*, Princeton Univ. Press (1983).

[4] W.H. Zurek, Physics Today, **44**, 10 p. 36 (1991).

[5] W.H. Zurek, Phys. Rev. D **24**, 1516 (1981) and **26**, 1862 (1982); A.O. Caldeira and A.J. Leggett, Physica A **121**, 587 (1983); E. Joos and H.D. Zeh, Z. Phys. B **59**, 223 (1985); R. Omnès, *The Interpretation of Quantum Mechanics*, Princeton University Press, (1994).

[6] C. H. Bennett, D. P. DiVincenzo, Nature, **404**, 247 (2000).

[7] C. H. Bennett, G. Brassard, A. Ekert, Scientific American, October 1992, p.50.

[8] C. H. Bennett, G. Brassard, C. Crepeau, R. Jozsa, A. Peres, W.K. Wootters, Phys. Rev. Lett. **70**, 1895 (1993).

[9] J.S. Bell, Physics **1**, 195 (1964); J. F. Clauser, M. A. Horne, A. Shimony, R. A. Holt, Phys. Rev. Lett. **23**, 880 (1969).

[10] A. Zeilinger, Rev. Mod. Phys. **71**, S288 (1998).

[11] J. G. Rarity, P. C. M. Owens, P. R. Tapster, J. Mod. Opt. **41**, 2435 (1994).

[12] D. Bouwmeester, Pan Jian-Wei, K. Mattle, M. Eibl, H. Weinfurter, A. Zeilinger, Nature, **390**, 575 (1997).

[13] D. Boschi, S. Branca, F. De Martini, L. Hardy, S. Popescu, Phys. Rev. Lett. **80**, 1121 (1998).

[14] A. Furusawa, J.L. Sorensen, S.L. Braunstein, C.A. Fuchs, H.J. Kimble, E.S. Polzik, Science **282**, 706 (1998).

[15] D. M. Greenberger, M. A. Horne, A. Zeilinger, Am. J. Phys. **58**, 1131 (1990).

[16] J. W. Pan, D. Bouwmeester, M. Daniell, H. Weinfurter, A. Zeilinger, Nature **403**, 515 (2000).

[17] D.M. Meekhof, C. Monroe, B.E. King, W.M. Itano and D.J. Wineland, Phys. Rev. Let., **76**, 1796 (1996).

[18] C.A. Sackett, D. Kielpinski, B.E. King, C. Langer, V. Meyer, C.J. Myatt, M. Rowe, Q.A. Turchette, W.M. Itano, D.J. Wineland, C.C. Monroe, Nature **404**, 256 (2000).

[19] S. Haroche and J.M. Raimond, Cavity Quantum Electrodynamics. *Scientific American* **268**, 54 (1993).

[20] R.G. Hulet and D. Kleppner, Phys. Rev. Lett. **51**, 1430 (1983). P. Nussenzveig, F. Bernardot, M. Brune, J. Hare, J.M. Raimond, S Haroche and W. Gawlik, Phys. Rev. A **48**, 3991 (1993).

[21] S. Haroche, in *Fundamental systems in quantum optics, les Houches Summer School Session LIII*, J. Dalibard, J.M. Raimond, and J. Zinn-Justin, eds. (North Holland, Amsterdam,1992), p. 767. S. Haroche, in *New Trends in Atomic Physics, les Houches Summer School Session XXXVIII*, G. Grynberg and R. Stora, eds. (North Holland, Amsterdam, 1984), p. 347.

[22] J.M. Raimond and S. Haroche, in *Quantum fluctuations, les Houches Summer School Session LXIII*, S. Reynaud E. Giaccobino and J. Zinn-Justin, eds. (North Holland, Amsterdam,1197), p. 309.

[23] S. Haroche and J.M. Raimond in *Avances in Atomic and Molecular Physics, supplement 2*, D. Bates and B. Bederson ed. (Academic Press, New York, 1985) p.347. S. Haroche and J.M. Raimond in *Avances in Atomic and molecular physics, supplement 2*, P. Berman ed. (Academic Press, New York, 1994) p.123.

[24] G. Raithel, C. Wagner, H. Walther, L.M. Narducci and M.O. Scully, Adv. At. Mol. Phys. (Supplement 2), 57 (1994).

[25] F. Bernardot, P. Nussenzveig, M. Brune, J.M. Raimond and S. Haroche, Euro. Phys. Lett. **17**, 33 (1992).

[26] M. Brune, F. Schmidt-Kaler, A. Maali, J. Dreyer, E. Hagley, J.M. Raimond and S. Haroche, *Phys. Rev. Lett.* **76**, 1800 (1996).

[27] X. Maître, E. Hagley, G. Nogues, C. Wunderlich, P. Goy, M. Brune, J.M. Raimond and S. Haroche, *Phys. Rev. Lett.* **79**, 769 (1997).

[28] A. Rauschenbeutel, G. Nogues, S. Osnaghi, P. Bertet, M. Brune, J.M. Raimond and S. Haroche, Science, **288**, 2024 (2000).

[29] M. Gross and J. Liang, Phys. Rev. Lett. **57**, 3160 (1986).

[30] G. Rempe, H. Walther and N. Klein, Phys. Rev. Lett,. **58**, 353 (1987).

[31] E.T. Jaynes and F.W. Cummings, Proc. IEEE, **51**, 89 (1963).

[32] C. Cohen-Tannoudji, J. Dupont-Roc and G. Grynberg, Photons et atomes, Introduction à l'électrodynamique quantique (Interéditions et Editions du CNRS 1987). English translation: Photons and Atoms, Introduction to Quantum Electrodynamics (Wiley, New York 1989).

[33] J.H. Eberly, N.B. Narozhny and J.J. Sanchez-Mondragon, Phys. Rev. Lett. **44**, 1323 (1980).

[34] R. Glauber, Phys. Rev. **131** 2766 (1963).

[35] B.T.H. Varcoe, S. Brattke and H. Walther, Nature, 403, 743-746 (2000).

[36] A. Auffeves, P. Maioli, T. Meunier, S. Gleyzes, G. Nogues, M. Brune, J. M. Raimond, and S. Haroche Phys. Rev. Lett. **91**, 230405 (2003)

[37] E. Hagley, X. Maître, G. Nogues, C. Wunderlich, M. Brune, J.M. Raimond and S. Haroche, Phys. Rev. Lett. **79**, 1 (1997).

[38] H. Rauch, A. Zeilinger, G. Badurek and A. Wilfing, Phys. Lett., **54 A**, 425 (1975).

[39] S.A. Werner, R. Colella, A.W. Overhauser and C.F. Eagen, Phys. Rev. Lett., **35**, 1053 (1975).

[40] G. Nogues, A. Rauschenbeutel, S. Osnaghi, M. Brune, J.M. Raimond and S. Haroche, Nature **400**, 239 (1999).

[41] A. Rauschenbeutel, G. Nogues, S. Osnaghi, P. Bertet, M. Brune, J.M. Raimond and S. Haroche, Phys. Rev. Lett. **83**, 5166 (1999).

[42] S. Haroche *et al.*, in Laser spectroscopy 14, R. Blatt, J. Eschner, D. Leibfried, F. Schmidt-Kaler eds. (World Scientific, New York, 1999) p. 140.

[43] S.B. Zheng, J. of Opt. B **1**, 534 (1999).

[44] S. Haroche, in Fundamental problems in quantum theory, D. Greenberger, A. Zeilinger, eds., Ann. N.Y. Acad. Sci. **755**, 73 (1995).

[45] B. T. H. Varcoe, S. Brattke, B.-G. Englert, H. Walther, in Laser spectroscopy 14, R. Blatt, J. Eschner, D. Leibfried, F. Schmidt-Kaler eds. (World Scientific, New York, 1999) p. 130.

[46] N. D. Mermin, Phys. Rev. Lett. **65**, 1838 (1990).

[47] N. A. Gershenfeld and I. L. Chuang, Science, **275**, 350, (1997); D. G. Cory, A. F. Fahmy and T. F. Havel, Proc. Natl. Acad. Sci. USA **94**, 1634 (1997); J. A. Jones, M. Mosca and R. H. Hansen, Nature **393**, 344 (1998); D. G. Cory et al, Phys. Rev. Lett. **81**, 2152 (1998).

[48] S.L. Braunstein, C.M. Caves, R. Jozsa, N. Linden, S. Popescu and R. Schack, Phys. Rev. Lett. **83**, 1054 (1999).

[49] C. Monroe, D.M. Meekhof, B.E. King, W.M. Itano and D.J. Wineland, Phys. Rev. Lett. **75**, 4714 (1995).

[50] Q. A. Turchette, C.S. Wood, B.E. King, C.J. Myatt, D. Leibfried, W.M. Itano, C. Monroe and D.J. Wineland, Phys. Rev. Lett. **81**, 3631 (1998).

[51] H. C. Nägerl, D. Leibfried, H. Rohde, G. Thalhammer, J. Eschner, F. Schmidt-Kaler, and R. Blatt, Phys. Rev. A **60**, 145 (1999).

[52] S. Osnaghi, P. Bertet, A. Auffeves, P. Maioli, M. Brune, J.M. Raimond, and S. Haroche, Phys. Rev. Lett. **87**, 037902 (2001).

[53] D. Jacksh, I. Cirac, P. Zoller, S. Rolston, R. Côté, M. Lukin, Phys. Rev. Lett. **85**, 2208 (2000).

[54] L. K. Grover, Phys. Rev. Lett. **79**, 325 (1997).

[55] F. Yamaguchi, P. Milman, M. Brune, J.-M. Raimond and S. Haroche, Phys. Rev. A **66**, 010302 (2002).

[56] M. Lukin, M. Fleischhauer, R. Côté, L. Duan, D. Jacksch, I. Cirac and P. Zoller, Phys. Rev. Lett. **87**, 037901 (2001).

[57] J. Reichel, W. Hänsel and T. W. Hänsch, Phys. Rev. Lett., **83**, 3398 (1999).

[58] M. Brune, E. Hagley, J. Dreyer, X. Maître, A. Maali, C. Wunderlich, J.M. Raimond and S. Haroche, Phys. Rev. Lett. **77**, 4887 (1996).

[59] We acknowledge support from the European Community and JST (ICORP "Quantum Entanglement" project).

Course 4

QUANTUM OPTICAL IMPLEMENTATION OF QUANTUM INFORMATION PROCESSING

P. Zoller, J. I. Cirac, Luming Duan and J. J. García-Ripoll

[1] *Institute for Theoretical Physics, University of Innsbruck, and Institute of the Austrian Academy of Sciences for Quantum Optics and Quantum Information, A-6020 Innsbruck, Austria*
[2] *Max-Planck-Institut für Quantenoptik, Hans-Kopfermann-Str. 1, Garching, D-85748, Germany*
[3] *Department of Physics and FOCUS center, University of Michigan, Ann Arbor, MI 48109*

D. Estève, J.-M. Raimond and J. Dalibard, eds.
Les Houches, Session LXXIX, 2003
Quantum Entanglement and Information Processing
Intrication quantique et traitement de l'information
© *2004 Elsevier B.V. All rights reserved*

Contents

1. Introduction

Quantum optical systems are one of the very few examples of quantum systems, where complete control on the single quantum level can be realized in the laboratory, while at the same time avoiding unwanted interactions with the environment causing decoherence. These achievements are illustrated by storage and laser cooling of single trapped ions and atoms, and the manipulation of single photons in Cavity QED, opening the field of engineering interesting and useful quantum states. In the mean time the frontier has moved towards building larger composite systems of a few atoms and photons, while still allowing complete quantum control of the individual particles. The new physics to be studied in these systems is based on entangled states, both from a fundamental point of testing quantum mechanics for larger and larger systems, but also in the light of possible new applications like quantum information processing or precision measurements [1,2].

Guided by theoretical proposals as reviewed in [3], we have seen extraordinary progress in experimental AMO physics during the last few years in implementing quantum information processing. Highlights are the recent accomplishments with ion traps [4], cold atoms in optical lattices [5], cavity QED (CQED) [6] and atomic ensembles [7]. Below we summarize some of the theoretical aspects of implementing quantum information processing with quantum optical systems. In particular, in Sec. 2 we discuss quantum computing with trapped ions. Sec. 3 demonstrates cold coherent collisions as a mean to entangle atoms in an optical lattice. Finally, Sec. 4 reviews atomic ensembles.

2. Trapped ions

Trapped ions is one of the most promising systems to implement quantum computation [3, 8, 9]. In this section we describe the theory of quantum information processing with a system of trapped ions. On the experimental side remarkable progress has been reported during the last two years in realizing some of the these ideas in the laboratory [10–12], as explained in the lecture notes by R. Blatt and D. Wineland.

Ion trap quantum computing, as first proposed in Ref. [8], stores qubits in longlived internal states of single trapped ion. Single qubit gates are performed by coupling the qubit states to laser light for an appropriate period of time. In general, this requires that single ions can be addressed by laser light. Two qubit gates can be achieved by entangling ions via collective phonon modes. Depending on the specific protocol this requires the initialization of the phonon bus in a pure initial state, e.g. laser cooling to the motional ground state in ion traps. However, recently specific protocols for "hot gates" have been developed which loosen these requirements (see [3, 13] and references cited). The unitary operations, which can be decomposed in a series of single and two-qubit operations on the qubits, can either be performed *dynamically*, i.e. based on the time evolution generated by a specific Hamiltonian, or *geometrically* as in holonomic quantum computing [14]. Finally, read out of the atomic qubit is accomplished using the method of quantum jumps [15].

An essential feature of ion trap quantum computers is the scalability to a large number of qubits. This is achieved by moving ions from a storage area, either to address the ions individually to perform the single qubit rotation, or by bringing pairs of ions together to perform a two-qubit gate. Moving ions does not affect the qubit stored in the internal electronic or hyperfine states, and heating of the ion motion can cooled in a nondestructive way by sympathetic cooling [16–18].

In our discussion below we will start with a brief outline of manipulation of trapped ions by laser light. We then proceed to illustrate ion trap quantum computing with two specific examples. We will first discuss in some detail the basic physical ideas and requirements of the original ion trap proposal [8]. Our emphasis is on the two qubit gate, and in direct relation to experimental work described by R. Blatt and D. Wineland. As a second example, we discuss the most recent proposal for a fast and robust 2-qubit gates for scalable ion trap quantum computing, based on laser coherent control techniques [13]. This 2-qubit gate can be orders of magnitude faster than the time scale given by the trap period, thus overcoming previous speed limits of ion trap quantum computing, while at the same time relaxing the experimental constraints of individual laser addressing of the ions and cooling to low temperatures.

2.1. *Modelling a single trapped ion*

In this section we give a theoretical description of quantum state engineering in a system of trapped and laser cooled ions. The development of the theory begins with the description of Hamiltonians, state preparation, laser cooling and state measurements for single ions, and then followed by a generalization to the case of many ions. This serves as the basis of our discussion of quantum computer models.

We describe a single trapped ion driven by laser light as a two-level atom $|g\rangle, |e\rangle$ moving in a 1D harmonic confining potential [3] with Hamiltonian ($\hbar =$

1)

$$H = \nu a^\dagger a - \frac{1}{2}\Delta\sigma_z + \frac{1}{2}\Omega\{\sigma_+ e^{i\eta(a+a^\dagger)} + \text{h.c.}\}. \tag{2.1}$$

Here the first term is the harmonic oscillator Hamiltonian for the center-of-mass motion of the ion with trap frequency ν. We have denoted by a and $a^\dagger$ the lowering and raising operators, respectively, which can be expressed in terms of the position and momentum operators as $\hat{x} = \sqrt{1/2M\nu}(a + a^\dagger)$ and $\hat{p} = i\sqrt{M\nu/2}(a^\dagger - a)$ with M the ion mass. The second and third term in 2.1 describe the driven two-level system in a rotating frame using standard spin-$\frac{1}{2}$ notation, $\sigma_+ = (\sigma_-)^\dagger = |e\rangle\langle g|$ and $\sigma_z = |e\rangle\langle e| - |g\rangle\langle g|$. This internal atomic Hamiltonian is written in a frame rotating with the optical frequency. We denote by $\Delta = \omega_L - \omega_{eg}$ the detuning of the laser with ω_L the laser frequency and by ω_{eg} the atomic transition frequency, and Ω is the Rabi frequency for the transition $|g\rangle \to |e\rangle$. In writing 2.1 we have assumed that the atom is driven by a running laser wave with wave vector $k_L = 2\pi/\lambda_L$ along the oscillator axis. Transitions from $|g\rangle$ to $|e\rangle$ are associated with a momentum kick to the atom by absorption of a laser photon, as described by $\exp(ik_L\hat{x}) \equiv \exp(i\eta(a + a^\dagger))$, which couples the motion of the ion (phonons) to the internal laser driven dynamics.

In Eq. (2.1) we have defined a Lamb-Dicke parameter $\eta = 2\pi a_0/\lambda_L$ with $a_0 0\sqrt{1/2M\nu}$ the ground state size of the oscillator and λ_L the laser wave length. In the Lamb-Dicke limit $\eta \ll 1$ we can expand the atom laser interaction: $H_{AL} = \frac{1}{2}\Omega\{\sigma_-[1 + i\eta(a+a^\dagger) + \mathcal{O}(\eta^2)]$+h.c.$\}$. The resulting Hamiltonian can be further simplified if the laser field is sufficiently weak so that only pairs of bare atom + trap levels are coupled resonantly. We denote by $|g\rangle|n\rangle$ and $|e\rangle|n\rangle$ the eigenstates of the bare Hamiltonian $H_0 = \nu a^\dagger a - \frac{1}{2}\Delta\sigma_z$, where the internal two-level system is in the ground (excited) state and n is the phonon excitation number of the harmonic oscillator. When tuning the laser to atomic resonance $\Delta = \omega_L - \omega_{eg} \approx 0$, i.e. $|\omega_L - \omega_{eg}| \ll \nu$, the transitions changing the harmonic oscillator quantum number n are off-resonance and can be neglected. In this case the Hamiltonian (2.1) can be approximated by

$$H_0 = \nu a^\dagger a - \frac{1}{2}\Delta\sigma_z + \frac{1}{2}\Omega(\sigma_+ + \text{h.c.}) \quad (\Delta \approx 0), \tag{2.2}$$

On the other hand, for laser frequencies close to the lower (red) motional sideband resonance $\Delta \approx -\nu$, i.e. $|\omega_L - (\omega_{eg} - \nu)| \ll \nu$, only transitions decreasing the quantum number n by one are important, and H can be approximated by a Hamiltonian of the Jaynes-Cummings type:

$$H_{\text{JC}} = \nu a^\dagger a - \frac{1}{2}\Delta\sigma_z + \frac{1}{2}\Omega(i\eta\sigma_+ a + \text{h.c.}) \quad (\Delta \approx -\nu). \tag{2.3}$$

Similarly, for tuning to the upper (blue) sideband $\Delta \approx +\nu$,i.e. $\omega_L - (\omega_{eg} + \nu)| \ll \nu$, only transitions increasing the quantum number n by one contribute, so that H can be approximated by the *anti*-Jaynes-Cummings Hamiltonian

$$H_{\mathrm{AJC}} = \nu a^\dagger a - \frac{1}{2}\Delta\sigma_z + \frac{1}{2}\Omega(i\eta\sigma_+ a^\dagger + \mathrm{h.c.}) \quad (\Delta \approx +\nu). \tag{2.4}$$

(see Fig. 1) For the above approximations to be valid we require that the effective Rabi frequencies to the non-resonant states have to be much smaller than the trap frequency, i.e. we must spectroscopically resolve the motional sidebands.

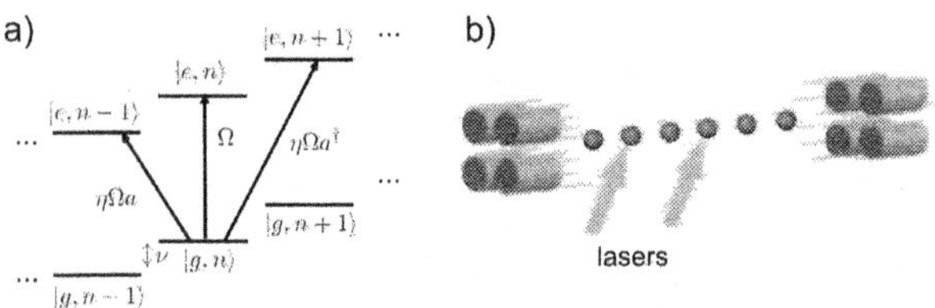

Fig. 1. a) Coupling to the atom + trap levels according to the Hamiltonians (2.2), (2.3 and (2.4, respectively, in lowest order Lamb-Dicke expansion. b) ion trap quantum computer 1995 (schematic)

The eigenstates of the Hamiltonians H_0, $H_{\mathrm{JC}\pm}$ and $H_{\mathrm{AJC}\pm}$ are the dressed states. These states are familiar from cavity QED, and are obtained by diagonalizing the $2x2$ matrices of nearly degenerate states. Applying a laser pulse on resonance, $\Delta = 0$, will according to (2.2) induce Rabi flopping between the states $|g\rangle|n\rangle$ and $|e\rangle|n\rangle$, while a laser tuned for example to the lower motional sideband $\Delta = -\nu$ will lead to Rabi oscillations coupling $|g\rangle|n\rangle$ and $|e\rangle|n-1\rangle$. The above Hamiltonians are basic building blocks to engineer general quantum states of motion. As an example, a laser pulse applied on the carrier frequency ($\Delta = 0$) to a state $(\alpha|g\rangle + \beta|e\rangle)|0\rangle$ will induce a general Rabi rotation without affecting the phonon state, i.e. perform a single qubit rotation. On the other hand, a π-pulse with duration $T = \pi/\eta\Omega$ on the red sideband will swap an initial superposition of qubits to a corresponding superposition of phonon states, $(\alpha|g\rangle+\beta|e\rangle)|0\rangle \to |g\rangle(\alpha|0\rangle+\beta|1\rangle)$. These processes will be the building blocks for the quantum gate discussed below.

We note that the interaction time for the above processes must always be much longer than the trap period $1/\nu$. On the other hand, when we apply a short laser pulse to the ion much less than the trap period $1/\nu$, i.e. we do not spectroscopically resolve the sidebands, and we can ignore the trap motion during the time duration of the pulse. A π-pulse to the two level atom is thus accompanied by a momentum kick to the motional state $|g\rangle|\mathrm{motion}\rangle \to |e\rangle e^{ik_L\hat{x}}|\mathrm{motion}\rangle$, $|e\rangle|\mathrm{motion}\rangle \to |e\rangle e^{-ik_L\hat{x}}|\mathrm{motion}\rangle$. In particular, if we choose a coherent state

$|\alpha\rangle_{\rm coh}$ to represent the motion, we will shift the coherent states $|g\rangle|\alpha\rangle_{\rm coh} \to |e\rangle|\alpha + i\eta\rangle_{\rm coh}$, $|e\rangle|\alpha\rangle_{\rm coh} \to |g\rangle|\alpha - i\eta\rangle_{\rm coh}$. Furthermore, if we apply a short π-pulse in the direction $+k_L$ followed by a pulse from the opposite direction $-k_L$, we achieve a transformation

$$\begin{aligned} |g\rangle|\alpha\rangle_{\rm coh} &\to |g\rangle|\alpha + 2i\eta\rangle_{\rm coh} \\ |e\rangle|\alpha\rangle_{\rm coh} &\to |e\rangle|\alpha - 2i\eta\rangle_{\rm coh} \end{aligned}$$

This process will be the basic element of the high speed 2 qubit gate at the end of this section.

2.2. *Ion trap quantum computer '95*

We describe in some detail the 2-qubit gate in the original ion trap proposal, as illustrated in Fig. 2.2. [8]. In the ion trap quantum computer'95 qubits are represented by the long-lived internal states of the ions, with $|g\rangle_j \equiv |0\rangle_j$ the ground state, and $|e_0\rangle_j \equiv |1\rangle_j$ a (metastable) excited state ($j = 1, \ldots, N$). In addition, we assume that there is a second metastable excited state $|e_1\rangle$ which plays below the role of an auxiliary state. In this system separate manipulation of each individual qubit is accomplished by addressing the ions with different laser beams and inducing a Rabi rotation. The heart of the proposal is the implementation of a two-qubit gate between two (or more) arbitrary ions in the trap by exciting the collective quantized motion of the ions with lasers, i.e. the collective phonon mode plays the role of a quantum data bus. For this we assume that the collective phonon modes have been cooled initially to the ground state.

Single qubit rotations can be performed tuning a laser on resonance with the internal transition ($\Delta_j = 0$) with polarization $q = 0$, $|g\rangle_j \to |e_0\rangle_j$. In an interaction picture the corresponding Hamiltonian is

$$\hat{H}_j = \frac{1}{2}\Omega\left[|e_0\rangle_j\langle g|e^{-i\phi} + |g\rangle_j\langle e_0|e^{i\phi}\right]. \tag{2.5}$$

For an interaction time $t = k\pi/\Omega$ (i.e., using a $k\pi$ pulse), this process is described by the following unitary evolution operator

$$\hat{V}_j^k(\phi) = \exp\left[-ik\frac{\pi}{2}(|e_0\rangle_j\langle g|e^{-i\phi} + h.c.)\right], \tag{2.6}$$

so that we achieve a Rabi rotation

$$\begin{aligned} |g\rangle_j &\longrightarrow |g\rangle_j \cos(k\pi/2) - |e_0\rangle_j i e^{i\phi} \sin(k\pi/2), \\ |e_0\rangle_j &\longrightarrow |e_0\rangle_j \cos(k\pi/2) - |g\rangle_j i e^{-i\phi} \sin(k\pi/2). \end{aligned}$$

When we work with N ions, the ion chain supports N longitudinal modes, of which the center of mass mode, $\nu_1 = \nu$, is energetically separated from the rest, $\nu_k >= \sqrt{3}\nu$ $(k > 1)$. If the laser addressing the j-th ion is tuned to the lower motional sideband of, for example, the center-of-mass mode, we have in the interaction picture the Hamiltonian

$$H_{j,q} = \frac{\eta}{\sqrt{N}}\frac{\Omega}{2}\left[|e_q\rangle_j\langle g|ae^{-i\phi} + |g\rangle_j\langle e_q|a^\dagger e^{i\phi}\right]. \tag{2.7}$$

Here $a^\dagger$ and a are the creation and annihilation operator of the center-of-mass phonons, respectively, Ω is the Rabi frequency, ϕ the laser phase, and η is the Lamb-Dicke parameter. The subscript $q = 0, 1$ refers to the transition excited by the laser, which depends on the laser polarization.

If this laser beam is on for the time $t = k\pi/(\Omega\eta/\sqrt{N})$ (i.e., using a $k\pi$ pulse), the evolution of the system will be described by the unitary operator:

$$\hat{U}_j^{k,q}(\phi) = \exp\left[-ik\frac{\pi}{2}(|e_q\rangle_j\langle g|ae^{-i\phi} + h.c.)\right]. \tag{2.8}$$

It is easy to prove that this transformation keeps the state $|g\rangle_j|0\rangle$ unaltered, whereas

$$|g\rangle_j|1\rangle \longrightarrow |g\rangle_j|1\rangle\cos(k\pi/2) - |e_q\rangle_j|0\rangle ie^{i\phi}\sin(k\pi/2),$$
$$|e\rangle_j|0\rangle \longrightarrow |e_q\rangle_j|0\rangle\cos(k\pi/2) - |g\rangle_j|1\rangle ie^{-i\phi}\sin(k\pi/2),$$

where $|0\rangle$ ($|1\rangle$) denotes a state of the CM mode with no (one) phonon.

Let us now show how a two-bit gate can be performed using this interaction. We consider the following three–step process (see Fig. 2):

(i) A π laser pulse with polarization $q = 0$ and $\phi = 0$ excites the m-th ion. The evolution corresponding to this step is given by $\hat{U}_m^{1,0} \equiv \hat{U}_m^{1,0}(0)$ (Fig. 2a).

(ii) The laser directed on the n–th ion is then turned on for a time of a 2π-pulse with polarization $q = 1$ and $\phi = 0$. The corresponding evolution operator $\hat{U}_n^{2,1}$ changes the sign of the state $|g\rangle_n|1\rangle$ (without affecting the others) via a rotation through the auxiliary state $|e_1\rangle_n|0\rangle$ (Fig. 2b).

(iii) Same as (i).

Thus, the unitary operation for the whole process is $\hat{U}_{m,n} \equiv \hat{U}_m^{1,0}\hat{U}_n^{2,1}\hat{U}_m^{1,0}$ which is represented diagrammatically as follows:

$$\begin{array}{ccccccc}
 & \hat{U}_m^{1,0} & & \hat{U}_n^{2,1} & & \hat{U}_m^{1,0} & \\
|g\rangle_m|g\rangle_n|0\rangle & \longrightarrow & |g\rangle_m|g\rangle_n|0\rangle & \longrightarrow & |g\rangle_m|g\rangle_n|0\rangle & \longrightarrow & |g\rangle_m|g\rangle_n|0\rangle, \\
|g\rangle_m|e_0\rangle_n|0\rangle & \longrightarrow & |g\rangle_m|e_0\rangle_n|0\rangle & \longrightarrow & |g\rangle_m|e_0\rangle_n|0\rangle & \longrightarrow & |g\rangle_m|e_0\rangle_n|0\rangle, \\
|r_0\rangle_m|g\rangle_n|0\rangle & \longrightarrow & -i|g\rangle_m|g\rangle_n|1\rangle & \longrightarrow & i|g\rangle_m|g\rangle_n|1\rangle & \longrightarrow & |e_0\rangle_m|g\rangle_n|0\rangle, \\
|e_0\rangle_m|e_0\rangle_n|0\rangle & \longrightarrow & -i|g\rangle_m|e_0\rangle_n|1\rangle & \longrightarrow & -i|g\rangle_m|r_0\rangle_n|1\rangle & \longrightarrow & -|e_0\rangle_m|e_0\rangle_n|0\rangle.
\end{array} \tag{2.9}$$

The effect of this interaction is to change the sign of the state only when both ions are initially excited. Note that the state of the CM mode is restored to the vacuum state $|0\rangle$ after the process. Equation (2.9) is phase gate $|\epsilon_1\rangle|\epsilon_2\rangle \rightarrow (-1)^{\epsilon_1\epsilon_2}|\epsilon_1\rangle|\epsilon_2\rangle$ ($\epsilon_{1,2} = 0, 1$) which together with single qubit rotations becomes equivalent to a controlled-NOT.

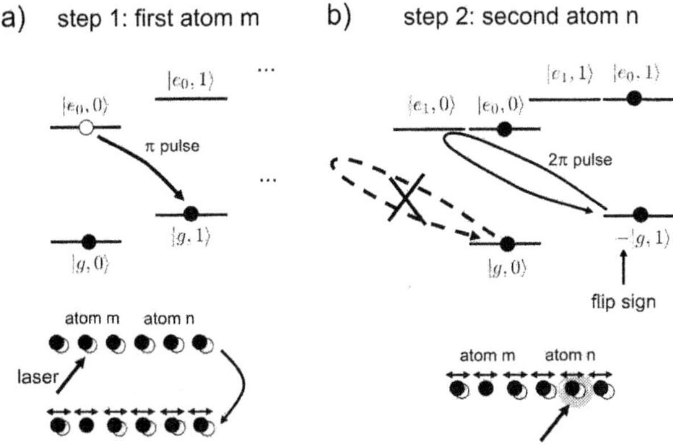

Fig. 2. The two-qubit quantum gate with trapped ions [8]. a) First step according to(2.9): the qubit of the first atom is swapped to the photonic data bus with a π-pulse on the lower motional sideband, b) Second step: the state $|g, 1\rangle$ acquires a minus sign due to a 2π-rotation via the auxiliary atomic level $|r_1\rangle$ on the lower motional sideband.

2.3. *Fast and robust 2-qubit gates for scalable ion trap quantum computing*

Scalability of ion trap quantum computing is based on storing a set of ions in a memory area, and moving ions independently to a processing unit: in particular one must bring together *pairs of ions* to perform a two-qubit gate [16–18]. Basic steps towards this goal have already been demonstrated experimentally [12]. An important question to be addressed is to identify the current limitations of the two–qubit gates with trapped ions (given the fact that one–qubit gates are significantly simpler with those systems). The ideal scheme should [19]:

(i) be independent of temperature, so that one does not need to cool the ions to their ground state after they are moved to or from their storage area);
(ii) require no addressability (to allow the ions to be as close as possible during the gate to increase their interaction strength), and
(iii) be fast, in order to minimize the effects of decoherence during the gate, and to speed up the computation.

This last property has been identified as a key limitation [4]: in essentially all schemes suggested so far [8, 20–23] one has to resolve spectroscopically the motional sidebands of the ions with the exciting laser, which limits the laser intensity and therefore the gate time. The coherent control-gate gate between pairs of ions [13] analyzed below overcomes this problem by not using spectral methods to couple the ion motion to the internal states but rather mechanical effects.

As our model we consider two ions in a one–dimensional harmonic trap, interacting with a laser beam on resonance. The Hamiltonian describing this situation can be written as $H = H_0 + H_1$, where $H_0 = \nu_c a^\dagger a + \nu_r b^\dagger b$ describes the motion in the trap and

$$H_1 = \frac{1}{2}\Omega(t)\left[\sigma_1^+ e^{i\eta_c(a^\dagger+a)+\frac{1}{2}\eta_r(b^\dagger+b)} + \sigma_2^+ e^{i\eta_c(a^\dagger+a)-\frac{1}{2}\eta_r(b^\dagger+b)}\right] + \text{h.c.} \tag{2.10}$$

Here, a and b are the annihilation operators center-of-mass and stretching mode, respectively, and $\nu_c = \nu$ and $\nu_r = \sqrt{3}\nu_c$ the corresponding frequencies. We denote by $\eta_c = \eta/\sqrt{2}$ and $\eta_r = \eta\sqrt[4]{4/3}$ are to associated Lamb–Dicke parameters. Note that the Rabi frequency Ω is the same for both ions, i.e. we have not assumed individual addressing.

In the following we will consider two different kind of processes:

(i) Free evolution, in which the laser is switched off ($\Omega = 0$) for a certain time;
(ii) Sequences of pairs of very fast laser pulses, each of them coming from opposite sides, with duration δt long enough to form a π-pulse ($\Omega\delta t = \pi$), but very short compared to the period of the trap ($\nu\delta t \ll 1$).

Processes (i) and (ii) will be alternated: at time t_1 a sequence of z_1 pulses is applied, followed by free evolution until at time t_2 another sequence of z_2 pulses is applied followed by free evolution and so on. The numbers z_k are integers, whose sign indicates the direction of the laser pulses. We can visualize the motion of the ions as a trajectory in phase space. This is illustrated in Fig. 3 for the center-of-mass state of a single ion (X_c, P_c), where $(X_c + iP_c)/\sqrt{2} = \langle a\rangle$. The time evolution consists of a sequence of kicks (vertical displacements), which are interspersed with free harmonic oscillator evolution (motion along the arcs). The question is now whether we can find a pulse sequence, such that the final phase space point (solid line) is *restored* to the one corresponding to a *free* harmonic evolution (dashed circle). In an appendix at the end of this section we show that this can be achieved if the pulse sequence satisfies a *commensurability condition*

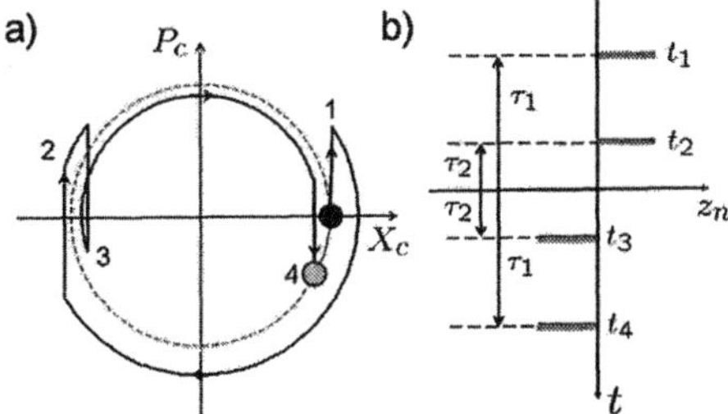

Fig. 3. a) Trajectory in phase space of the center-of-mass state of the ion (X_c, P_c) (where $(X_c + iP_c)/\sqrt{2} = \langle a \rangle$) during the 2-qubit gate (solid line), connecting the initial state (black filled circle) to the final state (grey filled circle) at the gate time T. The time evolution consists of a sequence of kicks (vertical displacements), which are interspersed with free harmonic oscillator evolution (motion along the arcs). A pulse sequence satisfying the commensurability condition (2.11) guarantees that the final phase space point is restored to the one corresponding to a free harmonic evolution (dashed circle). The particular pulse sequence plotted corresponds to a four pulse sequence given in the text (Protocol I). Figure b) shows how the laser pulses (bars) distribute in time for this scheme.

for the center-of-mass and stretch-mode

$$C_c \equiv \sum_{k=1}^{N} z_k e^{-i\nu t_k} = 0, \quad C_r \equiv \sum_{k=1}^{N} z_k e^{-i\sqrt{3}\nu t_k} = 0\,. \tag{2.11}$$

In this case, the motional state of the ion will not depend on the qubits. Thus the evolution operator is given by (see appendix)

$$\mathcal{U}(\Theta) = e^{i\Theta\sigma_1^z\sigma_2^z} e^{-i\nu_c T a^\dagger a} e^{-i\nu_r T b^\dagger b}, \tag{2.12}$$

where T is the total time required by the gate and

$$\Theta = 4\eta^2 \sum_{m=2}^{N} \sum_{k=1}^{m-1} z_k z_m \left[\frac{\sin[\sqrt{3}\nu \Delta t_{km}]}{\sqrt{3}} - \sin(\nu \Delta t_{km}) \right], \tag{2.13}$$

is a function of the spacing between laser pulses $\Delta t_{km} = t_k - t_m$. Therefore, if (2.11) are fulfilled, and $\Theta = \pi/4$ we will produce a controlled–phase gate (which is equivalent to a controlled–NOT gate up to local operations) which is *completely independent of the initial motional state*, i.e. there are no temperature requirements.

It can be shown [13] that for any value of the time T it is always possible to find a sequence of laser pulses which implements the gate, and therefore the gate operation can be, in principle, arbitrarily fast. We give two simple protocols.

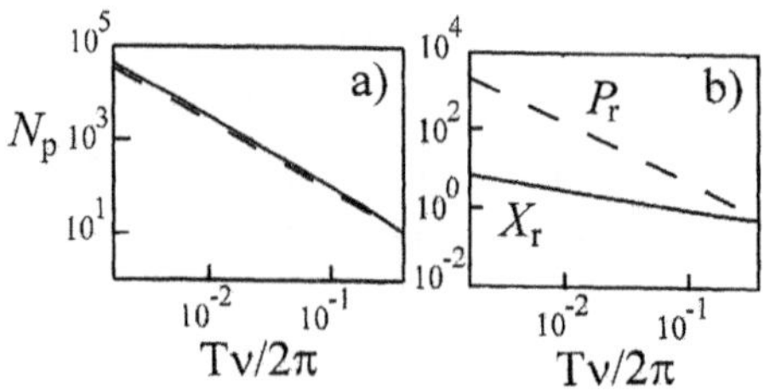

Fig. 4. (a) Log-log plot of the number of pairs of pulses required to produce a phase gate using protocol II, as a function of the duration of the gate, T, for a realistic value [24] of the Lamb–Dicke parameter, $\eta = 0.178$. We plot both the exact result (solid line) and a rough estimate $N_P = 40(\nu T/2\pi)^{-3/2}$ (dashed line) based on perturbative calculations. (b) Maximum relative displacement, X_r (solid), and maximum momentum acquired, P_r (dashed line), for scheme II. These quantities are dimensionless (scaled) versions of the real observables, $X_r = \max[\langle x_r(t)\rangle/a_0]$, and $P_r \max[\langle p_r(t)\rangle a_0/\hbar]$.

Protocol I: This protocol (see Fig. 3) requires the least number of pulses and produces the gate in a fixed time $T \simeq 1.08(2\pi/\nu)$. The sequence of pulses is defined as

$$(z_n/N, t_n) = \{(\gamma, -\tau_1), (1, -\tau_2), (-1, \tau_2), (-\gamma, \tau_1)\}. \qquad (2.14)$$

Here $0 < \gamma = \cos(\theta) < 1.0$ is a real number, which may be introduced by tilting both lasers a small angle θ with respect to the axis of the trap, so that no transverse motion is excited. It is always possible to find a solution to Eq. (2.11) with $\tau_1 \simeq 0.538(4)(2\pi/\nu) > \tau_2 > 0$.

Protocol II: This protocol performs the gate in an arbitrarily short time T. The pulses are now distributed according to

$$(z_n/N, t_n) = \{(-2, -\tau_1), (3, -\tau_2), (-2, -\tau_3), (2, \tau_3), (-3, \tau_2), (2, \tau_1)\}. \qquad (2.15)$$

The whole process takes a time $T = 2\tau_1$ and requires $N_p = \sum |z_n| = 14N$ pairs of pulses. As Fig. 4 shows, the number of pulses increases with decreasing time as $N_p \propto T^{-3/2}$.

Ref. [13] gives a detailed study of the main limitations of the the scheme, and provides quantitative estimates for the gate fidelity. On the list of imperfections is first of all anharmonicities of the restoring forces. The more pulses we apply, the larger the relative displacement of the ions, as Fig. 4(b) shows. When the ions become too close to each other, the increasing intensity of the Coulomb force can lead to a breakdown of the harmonic approximation which is implicit in Eq. (2.10). Imposing an error $E \simeq 10^{-4}$ we estimate the shortest realistic time to be $\nu T \simeq 10^{-3}$ [13]. In addition, laser pulses have a finite duration. However, even for relatively long pulses, we obtain a fidelity which is comparable to the results obtained in current setups [10, 11]. As mentioned before, the scheme is also insensitive to temperature. If the commensurability condition (2.11) is not perfectly satisfied due to, for example, errors in timing of laser pulses, or misalignment of the lasers, then the corresponding contribution to the gate error is still a weak function of temperature.

Finally, we remark that it is not necessary to kick the atoms using pairs of counter-propagating laser beams. The same effect (i.e. a change of sign in η) may also be achieved in current experiments by reverting the internal state of both ions simultaneously. One then only needs a laser beam (aligned with the trap) to kick the atoms, and another laser (orthogonal to the axis of the trap) to produce the NOT gate. The second and more important remark is that it is possible to avoid errors in the laser pulses by using an adiabatic passage scheme (see references cited in [3]) which is insensitive to fluctuations in the laser intensity. In addition, this method also tolerates that the two ions see slightly different laser intensity.

In summary, the new concept of a "coherent control" two-qubit quantum gate allows operations on a time scale up three orders of magnitude faster than the trap frequency, while at the same time requiring no single ion addressing, no Lamb-Dicke assumption, and ground state cooling of the ion, and being robust against imperfections.

Appendix: Derivation of Eqs. (2.11) and (2.12). Here we presents details of the derivation of the commensurability condition (2.11) to achieve the factorization of the motional states according to Eq. (2.12). For a pulse sequence, consisting of kicks interspersed with free harmonic time evolution (Fig. 3), we write $\mathcal{U} = \mathcal{U}_c\mathcal{U}_r$, where $\mathcal{U}_{c,r} = \prod_{k=1}^{N} U_{c,r}(\Delta t_k, z_k)$ has contributions for center-of-mass and relative motion,

$$U_c(t_k, z_k) = e^{-i2z_k\eta_c(a+a^\dagger)(\sigma_1^z+\sigma_2^z)}e^{-i\nu_c\Delta t_k a^\dagger a},$$

$$U_r(t_k, z_k) = e^{-iz_k\eta_r(b+b^\dagger)(\sigma_1^z-\sigma_2^z)}e^{-i\nu_r\Delta t_k b^\dagger b}.$$

The integers z_k indicate the direction of the initial pulse in the sequence of pairs of very fast laser pulses, each of them coming from opposite sites.

In order to fully characterize U, we only have to investigate its action on states of the form $|i\rangle_1|j\rangle_2|\alpha\rangle_c|\beta\rangle_r$, where $i, j = 0, 1$ denote the computational basis, and $|\alpha\rangle$ and $|\beta\rangle$ are coherent states. This task can be easily carried out once we know the action of $\mathcal{U} = \prod_{k=1}^{N} U(\phi_k, p_k)$ on an arbitrary coherent state $|\alpha\rangle$, where

$$U(\phi_k, p_k) = e^{-ip(a+a^\dagger)}e^{-i\phi_k a^\dagger a}.$$

We obtain $\mathcal{U}|\alpha\rangle = e^{i\xi}|\tilde{\alpha}\rangle$, where

$$\tilde{\alpha} = \alpha e^{-i\theta_N} - i\sum_{k=1}^{N} p_k e^{i(\theta_k-\theta_N)},$$

$$\xi = -\sum_{m=2}^{N}\sum_{k=1}^{m-1} p_m p_k \sin(\theta_k - \theta_m) - \Re\left[\alpha \sum_{k=0}^{N} p_k e^{-i\theta_m}\right],$$

with $\theta_k = \sum_{m=1}^{k} \phi_m$.

The crucial point is to realize that if $\sum_{k=1}^{N} p_k e^{i\theta_k} = 0$ the motional state $|\alpha\rangle$ after the evolution is the same as if there was only free evolution (Fig. 1a), and a global phase ξ appears which does not depend on the motional state (Fig. 1a). Translating this result to the operators $\mathcal{U}_c|\alpha\rangle$ and $\mathcal{U}_r|\beta\rangle$, we obtain condition (2.11) for Eq. (2.12) to be valid.

3. Atoms in optical lattices

Bose Einstein condensates (BEC) are a source of a large number of ultracold atoms and, as we will show below, they can also be developed as a tool to provide a *large* number of qubits stored in optical lattices. In a condensate, due to the weak interactions, all atoms occupy the single particle ground state of the trapping potential, corresponding to a product state of the wave function.

This picture is must be revised by inducing a degeneracy in the ground state which is comparable to the number of atoms. For instance, as first proposed in Refs. [25, 26], it is possible to load a BEC in a deep 3D optical lattice forming a perfect Mott insulator phase with one atom per lattice site. The system is no longer a BEC, but an array of a large number of identifiable qubits, that can be entangled in massively parallel operation with spin-dependent lattices [26]. This scenario has recently been realized in the laboratory in a series of remarkable experiments in Munich [27, 28]. Entanglement of atoms in a lattice can also be achieved by dipole-dipole interactions [29, 30]), and the interactions and the speed of the quantum operations may be significantly enhanced using the very strong interactions are obtained between laser excited Rydberg states [31].

3.1. Cold atoms in optical lattices: the Hubbard model

Optical lattices are periodic arrays of microtraps for cold atoms generated by standing wave laser fields. The periodic structure of the lattice gives rise to a series of Bloch bands for the atomic center-of-mass motion. Atoms loaded in an optical lattice from a BEC will only occupy the lowest Bloch band due to the low temperatures. The physics of these atoms can be understood in terms of a Hubbard model with Hamiltonian [25]

$$H = -\sum_{\langle i,j\rangle} J_{ij} b_i^\dagger b_j + \frac{1}{2} U \sum_i b_i^\dagger b_i^\dagger b_i b_i \ . \tag{3.1}$$

Here b_i and $b_i^\dagger$ are bosonic destruction operators for atoms at each lattice site satisfying the bosonic commutation relations $[b_i, b_j^\dagger] = \delta_{ij}$. The tunneling of the atoms between different sites is described by the hopping matrix elements J_{ij}. The parameter U is the onsite interaction of atoms resulting from the collisional interactions. The distinguishing feature of this system is the time dependent control of the parameters J_{ij} (kinetic energy) and U (potential energy) by the intensity of the lattice laser. Increasing the intensity of the laser deepens the lattice potential, and suppresses the hopping while at the same time increasing the atomic density at each lattice site and thus the onsite interaction. For shallow lattices $J_{ij} \gg U$ the kinetic energy is dominant, and the ground state of N atoms will be a superfluid in which all bosonic atoms occupy the lowest momentum state in the Bloch band, $(\sum_i b^\dagger)^N |\mathrm{vac}\rangle$. If $J_{ij} \ll U$, on the other hand, the interactions dominate: for commensurate filling, i.e. when the number of lattice site matches the number of atoms, the ground state becomes a Mott-insulator state $b_1^\dagger \dots b_N^\dagger |\mathrm{vac}\rangle$ (Fock state of atoms). The superfluid-Mott-insulator transition is an example of a so-called quantum phase transition as studied in Ref. [32]. This Mott insulator regime is of particular interest, as it provides a *very large* number of identifiable atoms located in the the array of microtraps provided by the optical lattice, whose internal hyperfine or spin states can serve as qubits [25]. The first experimental realization of the Mott insulator quantum phase transition was recently reported by Bloch and collaborators [27].

3.2. Entanglement via coherent ground state collisions

Entanglement of qubits represented by cold atoms in a Mott-phase can be obtained by combing the collisional interactions (compare the onsite interaction in Eq. 3.1) with a *spin-dependent* optical lattice [26]. Let us assume that qubits $|0\rangle$, $|1\rangle$ are stored in two longlived atomic hyperfine ground states. With an appropriate choice of atomic states and the laser configurations [26] we can generate an optical lattice which is spin-dependent, i.e. atoms in $|0\rangle$ and $|1\rangle$ see a different optical potential. In addition these two optical potentials can change in time, so that both lattices have a tunable separation.

This provides us with a mechanism to *move* atoms *conditional* to the state of the qubit. In particular, we can *collide* two atoms "by hand" , as illustrated in Fig. 5, so that only the component of the wave function with the first atom in $|1\rangle$

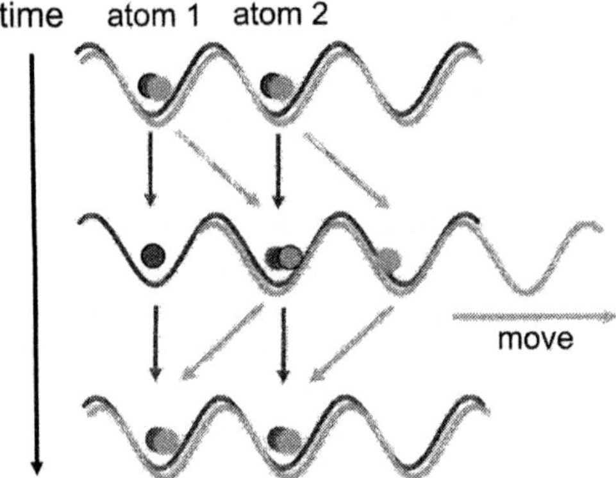

Fig. 5. Controlled collisions of two atoms with internal states $|0\rangle$ and $|1\rangle$ (red and blue circles) in a moveable state-dependent optical lattice (red and blue lattice) to entangle two atoms [26, 28]. This scheme unterlies the quantum simulator on the optical lattice.

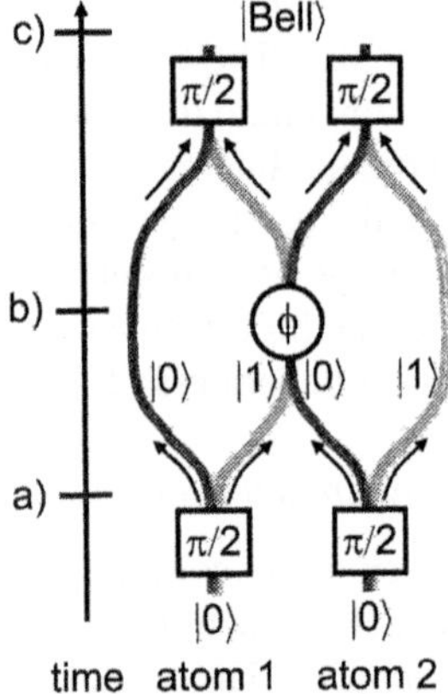

Fig. 6. Ramsey experiment with two atoms colliding in a lattice to generate a Bell state following Ref. [26, 28]. Time evolution is from bottom to top. The two atoms are initially prepared in the product state $|0\rangle|0\rangle$. A $\pi/2$ pulse generates the (unnormalized) superposition state $(|0\rangle + |1\rangle)((|0\rangle + |1\rangle))$ (a). A coherent collision provides a phase shift ϕ conditional to the first atom being in state $|1\rangle$ and the second atom being in state $|0\rangle$, i.e. $|0\rangle|0\rangle + e^{i\phi}|0\rangle|1\rangle + |1\rangle|0\rangle + |1\rangle|1\rangle$ (b). A final $\pi/2$-pulse closes the Ramsey interferometer resulting in the state $(1 - e^{i\phi})|\mathrm{Bell}\rangle + (1 + e^{i\phi}|1\rangle|1\rangle$, which for $\phi = \pi$ is a Bell state.

and the second atom in $|0\rangle$ will pick up a collisional phase ϕ, which entangles the atoms. In fact, this interaction gives rise to a phase gate between adjacent atoms $|1\rangle_i|0\rangle_{i+1} \longrightarrow e^{i\phi}|1\rangle_i|0\rangle_{i+1}$. In Fig. 6 we illustrate a Ramsey type experiment to generate and detect a Bell state via these collisional interactions. Again this idea has been demonstrated recently in a seminal experiment in the Munich group [28]. This conditional quantum logic can also be realized with magnetic, electric microtraps and microoptical dipole traps [33–35].

In more detail, we consider a situation where two atoms with electrons populating the internal states $|0\rangle$ and $|1\rangle$, respectively, are trapped in the ground states $\psi_0^{0,1}$ of two potential wells $V^{0,1}$. Initially, these wells are centered at positions $\bar{x}^0$ and $\bar{x}^1$, sufficiently far apart (distance $d = \bar{x}_1 - \bar{x}_0$) so that the particles do not interact. The positions of the potentials are moved along trajectories $\bar{x}^0(t)$ and $\bar{x}^1(t)$ so that the wavepackets of the atoms overlap for certain time, until finally they are restored to the initial position at the final time. This situation is described by the Hamiltonian

$$H = \sum_{\beta=0,1} \left[\frac{(\hat{p}^\beta)^2}{2m} + V^\beta \left(\hat{x}^\beta - \bar{x}^\beta(t)\right) \right] + u^{01}(\hat{x}^0 - \hat{x}^1). \tag{3.2}$$

Here, $\hat{x}^{0,1}$ and $\hat{p}^{0,1}$ are position and momentum operators, $V^{0,1}\left(\hat{x}^{0,1} - \bar{x}^{0,1}(t)\right)$ describe the displaced trap potentials and u^{01} is the atom–atom interaction term. Ideally, we would like to implement the transformation from before to after the collision,

$$\psi_0^0(x^0 - \bar{x}^0)\psi_0^1(x^b - \bar{x}^1) \to e^{i\phi}\psi_0^0(x^0 - \bar{x}^0)\psi_0^1(x^1 - \bar{x}^1), \tag{3.3}$$

where each atom remains in the ground state of its trapping potential and preserves its internal state. The phase ϕ will contain a contribution from the interaction (collision). The transformation (3.3) can be realized in the *adiabatic limit*, whereby we move the potentials slowly on the scale given by the trap frequency, so that the atoms remain in the ground state. Moving non-interacting atoms will induce kinetic single particle kinetic phases. In the presence of interactions ($u^{\mathrm{ab}} \neq 0$), we define the time–dependent energy shift due to the interaction as

$$\Delta E(t) = \frac{4\pi a_s \hbar^2}{m} \int dx |\psi_0^0\left(x - \bar{x}^0(t)\right)|^2 |\psi_0^1\left(x - \bar{x}^1(t)\right)|^2, \tag{3.4}$$

where a_s is the s–wave scattering length. We assume that $|\Delta E(t)| \ll \hbar\nu$ with ν the trap frequency so that no sloshing motion is excited. In this case, (3.3) still holds with $\phi = \phi^0 + \phi^1 + \phi^{01}$, where in addition to (trivial) single particle kinetic phases ϕ^0 and ϕ^1 arising from moving the potentials, we have a *collisional phase shift*

$$\phi^{01} = \int_{-\infty}^{\infty} dt \Delta E(t)/\hbar. \tag{3.5}$$

If the first atom is in a superposition state of the two qubits, the atomic wave packet would be "split" by moving the state dependent potentials, very much like with a beam splitter in atom interferometry. Thus we can move the potentials of neighboring atoms such that only the $|0\rangle$ component of the first atom "collides" with the state $|0\rangle$ of the second atom

$$\begin{aligned}
|0\rangle_1|0\rangle_2 &\to e^{i2\phi^0}|0\rangle_1|0\rangle_2,\\
|0\rangle_1|1\rangle_2 &\to e^{i(\phi^0+\phi^1+\phi^{01})}|0\rangle_1|1\rangle_2,\\
|1\rangle_1|0\rangle_2 &\to e^{i(\phi^0+\phi^1)}|b\rangle_1|0\rangle_2,\\
|1\rangle_1|1\rangle_2 &\to e^{i2\phi^1}|1\rangle_1|1\rangle_2,
\end{aligned} \tag{3.6}$$

where the motional states remain unchanged in the adiabatic limit, and ϕ^0 and ϕ^1 are single particle kinetic phases. The transformation (3.6) corresponds to a fundamental two–qubit gate. The fidelity of this gate is limited by nonadiabatic effects, decoherence due to spontaneous emission in the optical potentials and collisional loss to other unwanted states, or collisional to unwanted states. According to Ref. [26] the fidelity of this gate operation is remarkably close to one in a large parameter range.

3.3. Application: quantum simulations

Applying the previous method to an optical lattice that has more qubits, we can entangle many atoms with a single lattice movement, i.e. in a highly parallel entanglement operation. While for two atoms we have obtained a Bell state (see Fig. 6), for three atoms this produces a maximally entangled GHZ-state, and for $2D$ lattices this allows the generation of a cluster state, which is the basic resource for universal quantum computing in Briegel *et al.*'s one way quantum computer [36].

The parallelism inherent in the lattice movements makes "atoms in optical lattices" an ideal candidate for a Feynman-type quantum simulator (see the Appendix of this section) for bosonic, fermionic and spin many body systems, allowing simulation of various types and strengths of particle interactions, and $1, 2$ or $3D$ lattice configurations in a regime of many atoms, clearly unaccessible to any classical computer. By a stroboscopic switching of laser pulses and lattice movements combined with collisional interactions one can implement sequences of 1 and 2-qubit operations to simulate the time evolution operator of a many body system [37]. For translationally invariant systems, there is no need to address individual lattice sites, which makes the requirements quite realistic in the light of the present experimental developments. On the other hand, as noted above, Hubbard Hamiltonians with interactions controlled by lasers can also be realized directly with cold bosonic or fermionic atoms in optical lattices. This "analogue" quantum simulation provides a direct way of studying properties of strongly correlated systems in cold atom labs, which in the future may develop into a novel tool of condensed matter physics.

For the near future, we expect that atoms in optical lattices will be used to simulate a variety of other physical systems like, for example, interacting Fermions in 2 Dimensions using different lattice geometries. We also expect an important progress towards loading single (neutral) atoms in different types of potentials (optical, magnetic, etc), and the performance of quantum gates with few of these systems. This would allow to create few atom entangled states which may be used to observe violations of Bell inequalities, or to observe interesting phenomena like teleportation or error correction. As opposed to the trapped ions, at the moment it is hard to predict whether scalable quantum computation will be possible with neutral atoms in optical lattices using the present experimental set–ups. In any case, due to the high parallelism of these systems, we can clearly foresee that they will allow us to obtain a very deep insight in condensed matter physics via quantum simulations.

Appendix: Quantum simulator In brief, the basic concept of the quantum simulator is as follows. Let us consider a quantum system composed of N qubits all initially in state $|0\rangle$. We apply a two–qubit gate (specified by a 4×4 unitary matrix) to the first and second qubit, another one to the second and the third, and so on until we have performed $N-1$ such gates. Now, we measure the last qubit in the basis $|0\rangle, |1\rangle$. Let us denote by p_0 and p_1 the probability of obtaining 0 and 1 in this measurement. Our goal is to determine such probabilities with a prescribed precision (for example, of 1%). A way to determine the probabilities using a classical computer is to simulate the whole process: we take a vector which has 2^N components and multiply it by a $2^N \times 2^N$ matrix every time we simulate the action of a gate. At the end we can calculate the desired probabilities using the standard rules of Quantum Mechanics. However, as soon as N is of the order of 30, we will not be able to store the vector and the matrices in any existing computer. Moreover, the time required to simulate the action of the gates will increase exponentially with the number of qubits. However, with a quantum computer this simulation will required to repeat the same computation of the order of 100 times, and each computation requires only $N-1$ gates. Thus, we see that the quantum computer itself is much more efficient to simulate quantum systems, something that Feynman already pointed out in 1982 [1,38]. Of course, this particular example is artificial, and it is not related to a real problem. However, there exist physical systems which cannot be simulated with classical computers but in which a quantum computer could offer an important insight on some physical phenomena which are not yet understood [38]. For example, one could use a quantum computer to simulate spin systems or Hubbard models, and extract some information about open questions in condensed matter physics. Another possibility is to use an "analogue" quantum computer (as our artificial Hubbard models) to do the job, i.e. to choose a system which is described by the same Hamiltonian which one wants to simulate, but that can be very well controlled and measured.

4. Quantum information processing with atomic ensembles

4.1. Introduction

In the previous section, the quantum computation schemes are based on laser manipulation of single trapped particles. Here, we will show that laser manipulation of macroscopic atomic ensembles can also be exploited for implementation of quantum information processing [39–48]. In particular, we will discuss the uses of this system for continuous variable quantum teleportation and for implementation of quantum repeaters which enable scalable long-distance quantum communication.

The atomic ensemble contains a large number of identical neutral atoms, whose experimental candidates can be either laser-cooled atoms [46, 49, 50], or room-temperature gas [44, 47, 48, 51, 52]. The motivation of using atomic ensembles instead of single-particles for quantum information processing is mainly two-folds: firstly, laser manipulation of atomic ensembles without separate addressing of individual atoms is typically much easier than the laser manipulation of single particles; secondly and more importantly, the use of the atomic ensembles allows for some collective effects resulting from many-atom coherence to enhance the signal-to-noise ratio, which is critical for implementations of some quantum information protocols.

In the next section, we first show the ideas of using atomic ensembles for implementation of scalable long-distance quantum communication. Long-distance quantum communication is necessarily based on the use of photonic channels. However, due to losses and decoherence in the channel, the communication fidelity decreases exponentially with the channel length. To overcome this outstanding problem, one needs to use the concept of quantum repeaters [53], which provide the only known way for robust long-distance quantum communication. The best known method for complete implementation of quantum repeaters with sensible experimental technologies was proposed in Ref. [40]. Significant experimental advances haven been achieved recently towards realization of this comprehensive scheme, and we will briefly review these advances. In the final section, we discuss the use of atomic ensembles for continuous variable quantum information processing. Laser manipulation of atomic ensembles provides an elegant way for realizing continuous variable atomic quantum teleportation [42], and we will review the basic theoretical schemes as well as the following experimental achievements.

4.2. Atomic ensembles for implementation of quantum repeaters

Quantum communication is an essential element required for constructing quantum networks and for secretly transferring messages by means of quantum cryptography. The central problem of quantum communication is to generate nearly perfect entangled states between distant sites. Such states can be used then to implement secure quantum cryptography [54] or to transfer arbitrary quantum messages [55]. The schemes for quantum communication need to be based on the use of the photonic channels. To overcome the inevitable signal attenuation in the channel, the concept of entanglement purification was invented [56]. However, entanglement purification does not fully solve the problem for long-distance quantum communication. Due to the exponential decay of the entanglement in the channel, one needs an exponentially large number of partially entangled states to obtain one highly entangled state, which means that for a sufficiently long distance the task becomes nearly impossible.

The idea of quantum repeaters was proposed to solve the difficulty associated with the exponential fidelity decay [53]. In principle, it allows to make the overall communication fidelity very close to the unity, with the communication time growing only polynomially with the transmission distance. In analogy to fault-tolerant quantum computing [57], the quantum repeater proposal is a concatenated entanglement purification protocol for communication systems. The basic idea is to divide the transmission channel into many segments, with the length of each segment comparable to the channel attenuation length. First, one generates entanglement and purifies it for each segment; the purified entanglement is then extended to a longer length by connecting two adjacent segments through entanglement swapping [55]. After entanglement swapping, the overall entanglement is decreased, and one has to purify it again. One can continue the rounds of the entanglement swapping and purification until a nearly perfect entangled states are created between two distant sites.

To implement the quantum repeater protocol, one needs to generate entanglement between distant quantum bits (qubits), store them for sufficiently long time and perform local collective operations on several of these qubits. The requirement of quantum memory is essential since all purification protocols are probabilistic. When entanglement purification is performed for each segment of the channel, quantum memory can be used to keep the segment state if the purification succeeds and to repeat the purification for the segments only where the previous attempt fails. This is essentially important for polynomial scaling properties of the communication efficiency since with no available memory we have to require that the purifications for all the segments succeeds at the same time; the probability of such event decreases exponentially with the channel length. The requirement of quantum memory implies that we need to store the local qubits in the atomic internal states instead of the photonic states since it is difficult to store photons for a reasonably long time. With atoms as the local information carriers it seems to be very hard to implement quantum repeaters since normally one needs to achieve the strong coupling between atoms and photons with high-finesse cavities for atomic entanglement generation, purification, and swapping [58, 59], which, in spite of the recent significant experimental advances [60–63], remains a very challenging technology.

To overcome this difficulty, a scheme was proposed in Ref. [40] to realize quantum repeaters based on the use of atomic ensembles. The laser manipulation of the atomic ensembles, together with simple linear optics devices and routine single-photon detection, do the whole work for long-distance quantum communication. This scheme combines entanglement generation, connection, and application, with built-in entanglement purification, and as a result, it is inherently resilient to influence of noise and imperfections. Here, we will first explain the basic ideas of this theoretical proposal and then review the recent experimental advances.

4.2.1. Entanglement generation

To realize long-distance quantum communication, first we need to entangle two atomic ensembles within the channel attenuation length. This entanglement generation scheme is based on single-photon interference at photodetectors, which critically uses the fault-tolerance property of the photon detection [64] and the collective enhancement of the signal-to-noise ratio available in a many-atomic ensemble under an appropriate interaction configuration [65].

The system is a sample of atoms prepared in the ground state $|1\rangle$ with the level configuration shown in Fig. 7. This sample is illuminated by a short, off-resonant laser pulse that induces Raman transitions into the state $|2\rangle$ (a hyperfine level in the ground-state manifold with a long coherence time). We are particularly interested in the forward-scattered Stokes light that is co-propagating with the laser. Such scattering events are uniquely correlated with the excitation of the symmetric collective atomic mode S given by $S \equiv \left(1/\sqrt{N_a}\right) \sum_i |g\rangle_i \langle s|$ [65], where the summation is taken over all the atoms. In particular, an emission of the

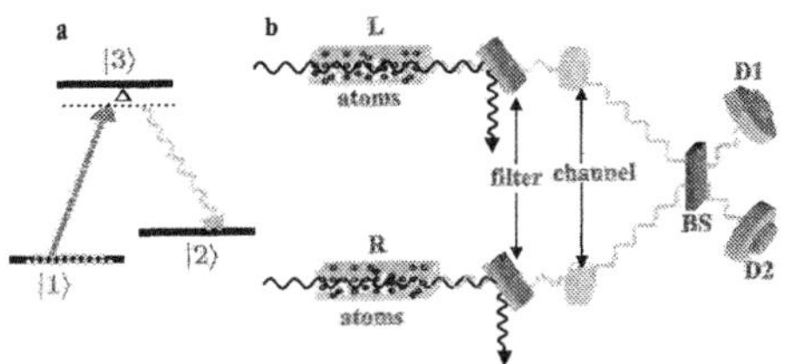

Fig. 7. (a) The relevant level structure of the atoms in the ensemble with $|1\rangle$, the ground state, $|2\rangle$, the metastable state for storing a qubit, and $|3\rangle$, the excited state. The transition $|1\rangle \to |3\rangle$ is coupled by the classical laser with the Rabi frequency Ω, and the forward scattering Stokes light comes from the transition $|3\rangle \to |2\rangle$. For convenience, we assume off-resonant coupling with a large detuning Δ. (b) Schematic setup for generating entanglement between the two atomic ensembles L and R. The two ensembles are pencil shaped and illuminated by the synchronized classical laser pulses. The forward-scattering Stokes pulses are collected after the filters (polarization and frequency selective) and interfered at a 50%-50% beam splitter BS after the transmission channels, with the outputs detected respectively by two single-photon detectors D1 and D2. If there is a click in D1 *or* D2, the process is finished and we successfully generate entanglement between the ensembles L and R. Otherwise, we first apply a repumping pulse to the transition $|2\rangle \to |3\rangle$ on the ensembles L and R to set the state of the ensembles back to the ground state $|0\rangle_a^L \otimes |0\rangle_a^R$, then the same classical laser pulses as the first round are applied to the transition $|1\rangle \to |3\rangle$ and we detect again the forward-scattering Stokes pulses after the beam splitter. This process is repeated until finally we have a click in the D1 *or* D2 detector.

single Stokes photon in a forward direction results in the state of atomic ensemble given by $S^\dagger |0_a\rangle$, where the ensemble ground state $|0_a\rangle \equiv \bigotimes_i |1\rangle_i$.

We assume that the light-atom interaction time is short so that the mean photon number in the forward-scattered Stokes pulse is much smaller than 1. One can assign an effective single-mode bosonic operator a for this Stokes pulse with the corresponding vacuum state denoted by $|0_p\rangle$. The whole state of the atomic collective mode and the forward-scattered Stokes mode can now be written in the following form [65]

$$|\phi\rangle = |0_a\rangle |0_p\rangle + \sqrt{p_c} S^\dagger a^\dagger |0_a\rangle |0_p\rangle + o(p_c), \tag{4.1}$$

where p_c is the small excitation probability.

Now we explain how to use this setup to generate entanglement between two distant ensembles L and R using the configuration shown in Fig. 7. Here, two laser pulses excited both ensembles simultaneously, and the whole system is described by the state $|\phi\rangle_L \otimes |\phi\rangle_R$, where $|\phi\rangle_L$ and $|\phi\rangle_R$ are given by Eq. (4.1) with all the operators and states distinguished by the subscript L or R. The forward scattered Stokes signal from both ensembles is combined at the beam splitter and a photodetector click in either D1 *or* D2 measures the combined radiation from two samples, $a_+^\dagger a_+$ or $a_-^\dagger a_-$ with $a_\pm = \left(a_L \pm e^{i\varphi} a_R\right)/\sqrt{2}$. Here, φ denotes an unknown difference of the phase shifts in the two-side channels. We can also assume that φ has an imaginary part to account for the possible asymmetry of the setup, which will also be corrected automatically in our scheme. But the setup asymmetry can be easily made very small, and for simplicity of expressions we assume that φ is real in the following. Conditional on the detector click, we should apply a_+ or a_- to the whole state $|\phi\rangle_L \otimes |\phi\rangle_R$, and the projected state of the ensembles L and R is nearly maximally entangled with the form (neglecting the high-order terms $o\,(p_c)$)

$$|\Psi_\varphi\rangle_{LR}^{\pm} = \left(S_L^\dagger \pm e^{i\varphi} S_R^\dagger\right)/\sqrt{2}\,|0_a\rangle_L\,|0_a\rangle_R\,. \tag{4.2}$$

The probability for getting a click is given by p_c for each round, so we need repeat the process about $1/p_c$ times for a successful entanglement preparation, and the average preparation time is given by $T_0 \sim t_\Delta/p_c$. The states $|\Psi_r\rangle_{LR}^{+}$ and $|\Psi_r\rangle_{LR}^{-}$ can be easily transformed to each other by a simple local phase shift. Without loss of generality, we assume in the following that we generate the entangled state $|\Psi_r\rangle_{LR}^{+}$.

The presence of noise will modify the projected state of the ensembles to

$$\rho_{LR}\,(c_0,\varphi) = \frac{1}{c_0+1}\left(c_0\,|0_a0_a\rangle_{LR}\,\langle 0_a0_a| + |\Psi_\varphi\rangle_{LR}^{+}\,\langle\Psi_\varphi|\right), \tag{4.3}$$

where the "vacuum" coefficient c_0 is determined by the dark count rates of the photon detectors. It will be seen below that any state in the form of Eq. (4.3) will be purified automatically to a maximally entangled state in the entanglement-based communication schemes. We therefore call this state an effective maximally entangled (EME) state with the vacuum coefficient c_0 determining the purification efficiency.

4.2.2. *Entanglement connection through swapping*

After successful generation of entanglement within the attenuation length, we want to extend the quantum communication distance. This is done through entanglement swapping with the configuration shown in Fig. 8. Suppose that we start with two pairs of the entangled ensembles described by the state $\rho_{LI_1} \otimes \rho_{I_2R}$, where ρ_{LI_1} and ρ_{I_2R} are given by Eq. (4.3). In the ideal case, the setup shown in Fig. 8 measures the quantities corresponding to operators $S_\pm^\dagger S_\pm$ with $S_\pm = \left(S_{I_1} \pm S_{I_2}\right)/\sqrt{2}$. If the measurement is successful (i.e., one of the detectors registers one photon), we will prepare the ensembles L and R into another EME state. The new φ-parameter is given by $\varphi_1 + \varphi_2$, where φ_1 and φ_2 denote the old φ-parameters for the two segment EME states. Even in the presence of realistic noise such as the photon loss, an EME state is still created after a detector

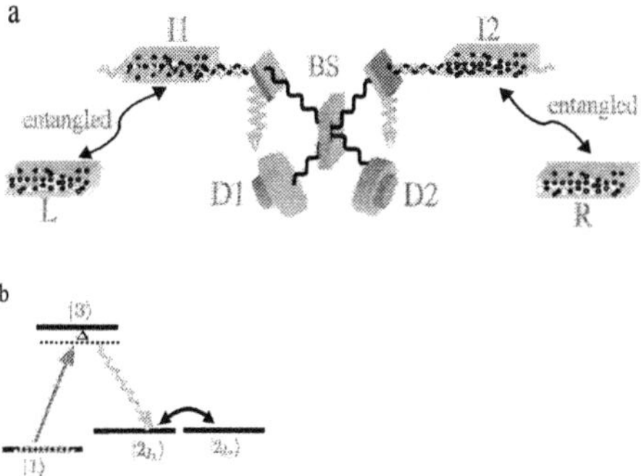

Fig. 8. (a) Illustrative setup for the entanglement swapping. We have two pairs of ensembles L, I_1 and I_2, R distributed at three sites L, I and R. Each of the ensemble-pairs L, I_1 and I_2, R is prepared in an EME state in the form of Eq. (3). The excitations in the collective modes of the ensembles I_1 and I_2 are transferred simultaneously to the optical excitations by the repumping pulses applied to the atomic transition $|2\rangle \to |3\rangle$, and the stimulated optical excitations, after a 50%-50% beam splitter, are detected by the single-photon detectors D1 and D2. If either D1 *or* D2 clicks, the protocol is successful and an EME state in the form of Eq. (3) is established between the ensembles L and R with a doubled communication distance. Otherwise, the process fails, and we need to repeat the previous entanglement generation and swapping until finally we have a click in D1 or D2, that is, until the protocol finally succeeds. (b) The two intermediated ensembles I_1 and I_2 can also be replaced by one ensemble but with two metastable states I_1 and I_2 to store the two different collective modes. The 50%-50% beam splitter operation can be simply realized by a $\pi/2$ pulse on the two metastable states before the collective atomic excitations are transferred to the optical excitations.

click. The noise only influences the success probability to get a click and the new vacuum coefficient in the EME state. The above method for connecting entanglement can be continued to arbitrarily extend the communication distance.

4.2.3. *Entanglement-based communication schemes*

After an EME state has been established between two distant sites, we would like to use it in the communication protocols, such as for quantum teleportation, cryptography, or Bell inequality detection. It is not obvious that the EME state (4.3), which is entangled in the Fock basis, is useful for these tasks since in the Fock basis it is experimentally hard to do certain single-bit operations. In the following we will show how the EME states can be used to realize all these protocols with simple experimental configurations.

Quantum cryptography and the Bell inequality detection are achieved with the setup shown by Fig. 9a. The state of the two pairs of ensembles is expressed as $\rho_{L_1R_1} \otimes \rho_{L_2R_2}$, where $\rho_{L_iR_i}$ $(i = 1, 2)$ denote the same EME state with the vacuum coefficient c_n if we have done n times entanglement connection. The φ-parameters in $\rho_{L_1R_1}$ and $\rho_{L_2R_2}$ are the same provided that the two states are

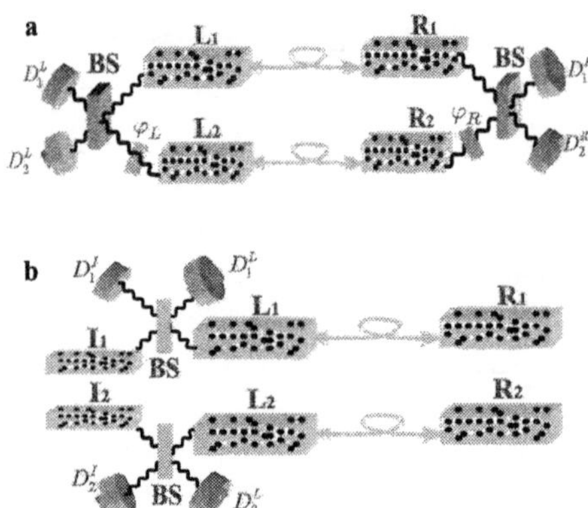

Fig. 9. (a) Schematic setup for the realization of quantum cryptography and Bell inequality detection. Two pairs of ensembles L_1, R_1 and L_2, R_2 have been prepared in the EME states. The collective atomic excitations on each side are transferred to the optical excitations, which, respectively after a relative phase shift φ_L or φ_R and a 50%-50% beam splitter, are detected by the single-photon detectors D_1^L, D_2^L and D_1^R, D_2^R. We look at the four possible coincidences of D_1^R, D_2^R with D_1^L, D_2^L, which are functions of the phase difference $\varphi_L - \varphi_R$. Depending on the choice of φ_L and φ_R, this setup can realize both the quantum cryptography and the Bell inequality detection. (b) Schematic setup for probabilistic quantum teleportation of the atomic "polarization" state. Similarly, two pairs of ensembles L_1, R_1 and L_2, R_2 are prepared in the EME states. We want to teleport an atomic "polarization" state $\left(d_0 S_{I_1}^\dagger + d_1 S_{I_2}^\dagger\right) |0_a 0_a\rangle_{I_1 I_2}$ with unknown coefficients d_0, d_1 from the left to the right side, where $S_{I_1}^\dagger$, $S_{I_2}^\dagger$ denote the collective atomic operators for the two ensembles I_1 and I_2 (or two metastable states in the same ensemble). The collective atomic excitations in the ensembles I_1, L_1 and I_2, L_2 are transferred to the optical excitations, which, after a 50%-50% beam splitter, are detected by the single-photon detectors D_1^I, D_1^L and D_2^I, D_2^L. If there are a click in D_1^I *or* D_1^L and a click in D_2^I *or* D_2^I, the protocol is successful. A π-phase rotation is then performed on the collective mode of the ensemble R_2 conditional on that the two clicks appear in the detectors D_1^I,D_2^L or D_2^I,D_1^L. The collective excitation in the ensembles R_1 and R_2, if appearing, would be found in the same "polarization" state $\left(d_0 S_{R_1}^\dagger + d_1 S_{R_2}^\dagger\right) |0_a 0_a\rangle_{R_1 R_2}$.

established over the same stationary channels. We register only the coincidences of the two-side detectors, so the protocol is successful only if there is a click on each side. Under this condition, the vacuum components in the EME states, together with the state components $S_{L_1}^\dagger S_{L_2}^\dagger |\text{vac}\rangle$ and $S_{R_1}^\dagger S_{R_2}^\dagger |\text{vac}\rangle$, where $|\text{vac}\rangle$ denotes the ensemble state $|0_a 0_a 0_a 0_a\rangle_{L_1 R_1 L_2 R_2}$, have no contributions to the experimental results. So, for the measurement scheme shown by Fig. 7, the ensemble state $\rho_{L_1 R_1} \otimes \rho_{L_2 R_2}$ is effectively equivalent to the following "polarization" maximally entangled (PME) state (the terminology of "polarization" comes from an analogy to the optical case)

$$|\Psi\rangle_{\mathrm{PME}} = \left(S_{L_1}^{\dagger} S_{R_2}^{\dagger} + S_{L_2}^{\dagger} S_{R_1}^{\dagger} \right) / \sqrt{2}\, |\mathrm{vac}\rangle \,. \tag{4.4}$$

The success probability for the projection from $\rho_{L_1 R_1} \otimes \rho_{L_2 R_2}$ to $|\Psi\rangle_{\mathrm{PME}}$ (i.e., the probability to get a click on each side) is given by $p_a = 1/[2\,(c_n + 1)^2]$. One can also check that in Fig. 9, the phase shift ψ_Λ ($\Lambda = L$ or R) together with the corresponding beam splitter operation are equivalent to a single-bit rotation in the basis $\left\{ |0\rangle_\Lambda \equiv S_{\Lambda_1}^{\dagger} \, |0_a 0_a\rangle_{\Lambda_1 \Lambda_2} \,,\; |1\rangle_\Lambda \equiv S_{\Lambda_2}^{\dagger} \, |0_a 0_a\rangle_{\Lambda_1 \Lambda_2} \right\}$ with the rotation angle $\theta = \psi_\Lambda / 2$. Since we have the effective PME state and we can perform the desired single-bit rotations in the corresponding basis, it is clear how to use this facility to realize quantum cryptography, Bell inequality detection, as well as teleportation (see Fig. 9b).

It is remarkable that all the steps of entanglement generation, connection, and applications described above are robust to practical noise. The dominant noise in this system is photon loss, including the contributions from the channel attenuation, the detector and the coupling inefficiencies etc. It the photon is lost, we will never get a click from the detectors, and we simply repeat this failed attempt until we succeed. So this noise only influences the efficiency to register a photon, but has no influence on the final state fidelity if the photon is registered. Furthermore, one can show that the nose influence on the efficiency is actually only moderate in the sense that the required number of attempts for a successful event only increases with the communication distance by a slow polynomial law [40]. So we get high-fidelity quantum communication with a moderate polynomial overhead, which is the essential advantage of the quantum repeater protocol.

4.2.4. *Recent experimental advances*

The physics behind the above scheme for quantum repeaters is based on the definite correlation between the forward-scattered Stokes photon and the long-lived excitation in the collective atomic mode. The correlation comes from the collective enhancement effect due to many-atom coherence (for a single atom, the atomic excitation cannot be correlated with radiation in a certain direction without the use of high-finesse cavities [65]). The entanglement generation, connection, and application schemes described above are all based on this correlation. So the first enabling step for demonstration of this comprehensive quantum repeater scheme is to verify this correlation. Several exciting experiments have been reported on demonstration of this correlation effect [46–48].

The first experiment was reported from Caltech which demonstrate the nonclassical correlation between the emitted photon and the collective atomic excitation. The collective atomic excitation is subsequently transferred to a forward-scattered anti-Stokes photon for measurements (see Sec. 3.2.2), so what one really detects in experiments is the correlation between the pair of Stokes and anti-Stokes photons emitted successively. In the Caltech experiment, the atomic ensemble is a cloud of cold atomic in a magnetic optical trap. To experimentally confirm the correlation between the Stokes and the anti-Stokes photons, one measures the auto-correlations $\tilde{g}_{1,1}$, $\tilde{g}_{2,2}$ of the Stokes and the anti-Stokes fields and the cross correlation $\tilde{g}_{1,2}$ between them. For any classical optical fields (fields with well defined $P-$representations), these correlations should satisfy the Cauthy-Schwarz inequality $\left[\tilde{g}_{1,2}\right]^2 \leq \tilde{g}_{1,1}\tilde{g}_{2,2}$, while for correlations between the non-classical single-photon pairs, this inequality will be violated. In the experiment [46], this inequality was measured to be strongly violated with $[\tilde{g}_{1,2}^2(\delta t) = 5.45 \pm 0.11] \nleq [\tilde{g}_{1,1}\tilde{g}_{2,2} = 2.97 \pm 0.08]$. Here, δt is the time delay between the pair of Stokes and anti-Stokes photons, which is 405 ns in the initial experiment but could be much longer (up to seconds) if one loads the atoms into a far-off-resonant optical trap. Note that δt is basically limited by the spin relaxation time in the ensemble, and for implementation of quantum repeaters it is important to get a long δt to enable storage of quantum information in the ensemble.

Another related experiment was reported from Harvard [47], which uses hot atomic gas instead of the cold atomic ensemble. This experiment also measures the correlation between the Stokes and anti-Stokes fields. The difference is that it is not operated in the single-photon region. Instead, both the Stokes and anti-Stokes fields may have up to thousand of photons. In this limit, there is also some inequality need to be satisfied by the classical fields, and the experiment measures a violation of this inequality by about 4%. The other experiment with room-temperature atomic gas was reported from USTC [48], which operates in the single-photon region as required by the quantum repeater scheme. This experiment uses a similar detection method as the Caltech experiment, and measures a violation of the Cauchy-Schwarz inequality with $\left[g_{1,2}^2(\delta t) = 4.17 \pm 0.09\right] \nleq \left[g_{1,1}g_{2,2} = 3.12 \pm 0.08\right]$, where the time delay δt is observed to be about 2 μs.

4.3. *Atomic ensembles for continuous variable quantum information processing*

In continuous variable quantum information protocols, information is carried by some observables with continuous values. There have been quite a lot of interests in continuous variable information processing, including proposals for continuous variable quantum teleportation, cryptography, computation, error correction, and entanglement purification [2].

Here we will review some recent schemes using atomic ensembles for realization of continuous variable quantum teleportation [41, 42, 44]. Note that atomic quantum teleportation (not realized yet) typically requires strong coupling between the atom and the photon. Collective enhancement in the atomic ensemble

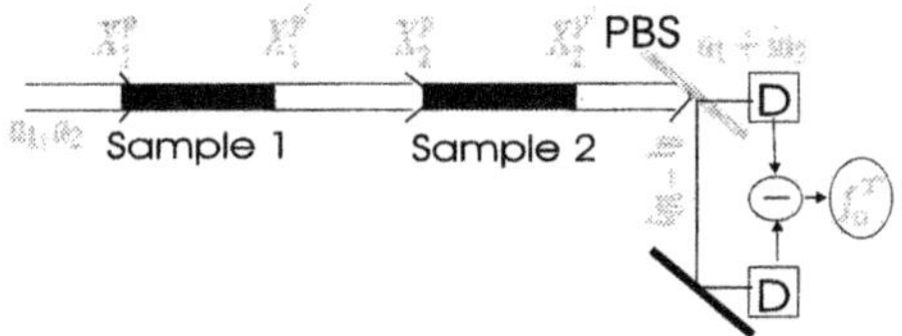

Fig. 10. Schematic setup for Bell measurements. A linearly polarized strong laser pulse (decomposed into two circular polarization modes a_1, a_2) propagates successively through the two atomic samples. The two polarization modes $(a_1 + ia_2)/\sqrt{2}$ and $(a_1 - ia_2)/\sqrt{2}$ are then split by a polarizing beam splitter (PBS), and finally the difference of the two photon currents (integrated over the pulse duration T) is measured.

plays an important role here as it significantly alleviates this stringent requirement. We will briefly explain the idea in Ref. [42] which uses only coherent light to generate continuous variable entanglement between two distant ensembles for atomic quantum teleportation. The scheme in [42] has been followed by the exciting experiment reported in Ref. [44] which demonstrates entanglement between two macroscopic ensembles for the first time.

For an optical field with two circular polarization modes a_1, a_2, one can introduce the Stokes operators by $S_x^p = \frac{1}{2}\left(a_1^\dagger a_2 + a_2^\dagger a_1\right)$, $S_y^p = \frac{1}{2i}\left(a_1^\dagger a_2 - a_2^\dagger a_1\right)$, $S_z^p = \frac{1}{2}\left(a_1^\dagger a_1 - a_2^\dagger a_2\right)$. If the light is linearly polarized along the $\vec{x}$ direction, one can define a pair of canonical operators by $X^p = S_y^p/\sqrt{\langle S_x^p\rangle}$, $P^p = S_z^p/\sqrt{\langle S_x^p\rangle}$ with $\left[X^p, P^p\right] = i$. Similarly, for a polarized atomic ensemble with the collective spin $\overrightarrow{S^a}$ pointing to the $\vec{x}$ direction, one can also define a pair of canonical operators $X^a = S_y^a/\sqrt{\langle S_x^a\rangle}$, $P^a = S_z^a/\sqrt{\langle S_x^a\rangle}$ with $\left[X^a, P^a\right] = i$. When the light passes through the atomic ensemble in an appropriate off-resonant interaction configuration detailed in Ref. [42], the continuous variable operators defined above will transform by the following form

$$\begin{aligned} X^{p\prime} &= X^p - \kappa_c P^a, \\ X^{a\prime} &= X^a - \kappa_c P^p, \\ P^{\beta\prime} &= P^\beta, \ (\beta = a, p), \end{aligned} \tag{4.5}$$

where κ_c is a parameter characterizing the interaction strength whose typical value is around 5.

For quantum teleportation, first one needs to generate entanglement between two distant ensembles 1 and 2. This is done through a nonlocal Bell measurement of the EPR operators $X_1^a - X_2^a$ and $P_1^a + P_2^a$ with the setup depicted by Fig. 10. This setup measures the Stokes operator $X_2^{p\prime}$ of the output light. Using Eq. (3.5), we have $X_2^{p\prime} = X_1^p + \kappa_c \left(P_1^a + P_2^a\right)$, so we get a collective measurement of $P_1^a + P_2^a$ with some inherent vacuum noise X_1^p. The efficiency $1 - \eta$ of this measurement is determined by the parameter κ_c with $\eta = 1/\left(1 + 2\kappa_c^2\right)$. After this round of measurements, we rotate the collective atomic spins around the x axis to get the transformations $X_1^a \to -P_1^a$, $P_1^a \to X_1^a$ and $X_2^a \to P_2^a$, $P_2^a \to -X_2^a$. The rotation of the atomic spin can be easily obtained by applying classical laser pulses. After the rotation, the measured observable of the first round of measurement is changed to $X_1^a - X_2^a$ in the new variables. We then make another round of collective measurement of the new variable $P_1^a + P_2^a$. In this way, both the EPR operators $X_1^a - X_2^a$ and $P_1^a + P_2^a$ are measured, and the final state of the two atomic ensembles is collapsed into a two-mode squeezed state with variance $\delta\left(X_1^a - X_2^a\right)^2 = \delta\left(P_1^a + P_2^a\right)^2 = e^{-2r}$, where the squeezing parameter r is given by

$$r = \frac{1}{2}\ln\left(1 + 2\kappa_c^2\right). \tag{4.6}$$

Thus, using only coherent light, we generate continuous variable entanglement [66] between two nonlocal atomic ensembles. With the interaction parameter $\kappa_c \approx 5$, a high squeezing (and thus a large entanglement) $r \approx 2.0$ is obtainable.

To achieve quantum teleportation, first the ensembles 1 and 2 are prepared in a continuous variable entangled state using the nonlocal Bell measurement described above. Then, a Bell measurement with the same setup as shown by Fig. 10 on the two local ensembles 1 and 3, together with a straightforward displacement of X_3^a, P_3^a on the sample 3, will teleport an unknown collective spin state from the atomic ensemble 3 to 2. The teleported state on the ensemble 2 has the same form as that in the original proposal of continuous variable teleportation using squeezing light [67], with the squeezing parameter r replaced by Eq. (4.6) and with an inherent Bell detection inefficiency $\eta = 1/\left(1 + 2\kappa_c^2\right)$. The quality of teleportation is best described by the fidelity, which, for a pure input state, is defined as the overlap of the teleported state and the input state. For any coherent input state of the sample 3, the teleportation fidelity is given by

$$F = 1/\left(1 + \frac{1}{1 + 2\kappa_c^2} + \frac{1}{2\kappa_c^2}\right). \tag{4.7}$$

Equation (4.7) shows that a high fidelity $F \approx 96\%$ would be possible for the teleportation of the collective atomic spin state with the interaction parameter $\kappa_c \approx 5$.

In the experimental demonstration [44], the atomic ensembles are provided by room-temperature Cesium atomic gas in two separate glass cells with coated wall to increase the spin relaxation time. Each cell is about 3 cm long, containing about 10^{12} atoms. The entanglement is generated through collective Bell measurements by transmitting a coherent light pulse as described above. To confirm and measure the generated entanglement, one needs to transmit another verifying pulse. Trough a homodyne detection of this verifying pulse, one can basically detect the EPR variation $\Delta_{EPR} = \left[\delta\left(X_1^a - X_2^a\right)^2 + \delta\left(P_1^a + P_2^a\right)^2\right]/2$ [66], and $\xi = 1 - \Delta_{EPR}$ serves as a measure of the entanglement, which is zero for separable states and 1 for the maximally entangled state. In this experiment, ξ is measured to be (35 ± 7) %, and this entanglement survives by about 0.5 ms (the relaxation time is measured by changing the time delay between the entangling and the verifying pulses). The demonstrated entanglement will be important for the next-step applications.

5. Conclusions

During the last few years the fields of atomic physics and quantum optics have experienced an enormous progress in controlling and manipulating atoms with lasers. This has immediate implications for quantum information processing, since this progress allows atomic systems to fulfill the basic requirements to implement the basic building blocks of a quantum computer. In this article we have illustrated these statements with two particular systems: trapped ions, neutral atoms in optical lattices and atomic ensembles.

The physics of trapped ions is very well understood. In fact, with the recent experimental results we can foresee no fundamental obstacle to build a scalable quantum computer with trapped ions. Of course, technical development may impose severe restrictions to the time scale in which this is achieved. On the other hand, neutral atoms in optical lattices seem to be ideal candidates to study a variety of physical phenomena by using them to simulate other physical systems. This quantum simulation may turn out to be the first real application of quantum information processing. Atomic ensembles, on the other hand, are ideal to realize quantum communication protocols (e.g. the quantum repeater, and the entanglement of distant atomic ensembles) within setups which are considerably simpler from an experimental point of view than the single atom and ion experiments. There are other quantum optical systems that have experienced a very remarkable progress during the last years, and which may equally important in the context of quantum information. An example is cavity QED, where groups at Caltech, Georgia Tech, Innsbruck, and Munich have trapped single atoms and ions inside cavities, and let them interact with the cavity field, which can be used as single (or entangled) photon(s) generators as well as to build quantum repeaters for quantum communication.

References

[1] M. Nielsen and I. Chuang. *Quantum Computation and Quantum Information.* Cambridge University Press, 2000.

[2] S. Braunstein and A. K. Pati, editors. *Quantum Information with Continuous Variables.* Kluwer Academic Publishers, 2003.

[3] J.I. Cirac, L.M. Duan, D. Jaksch, and P. Zoller. Quantum optical implementation of quantum information processing. In F. De Martini and C. Monroe, editors, *Proceedings of the International School of Physics "Enrico Fermi" Course CXLVIII, Experimental Quantum Computation and Information.* IOS Press, Amsterdam, 2002.

[4] B. G. Levi. *Physics Today*, May 2003.

[5] J. I. Cirac and P. Zoller. *Science*, 301:176, 2003.

[6] J. M. Raimond, M. Brune, and S. Haroche. *Rev. Mod. Phys.*, 73:565, 2001.

[7] M. D. Lukin. 2003.

[8] J. I. Cirac and P. Zoller. *Phys. Rev. Lett.*, 74(20):4091, 1995.

[9] A. Steane. *Appl. Phys. B*, 64:623, 1997.

[10] F. Schmidt-Kaler, H. Häffner, M. Riebe, S. Gulde, G.P.T. Lancaster, T. Deuschle, C. Becher, C.F. Roos, J. Eschner, and R. Blatt. *Nature*, 422:408, 2003.

[11] D. Leibfried, B. DeMarco, V. Meyer, D. Lucas, M. Barrett, J. Britton, W. M. Itano, B. Jelenkovi, C. Langer, T. Rosenband, and D. J. Wineland. *Nature*, 422:412, 2003.

[12] D. Leibfried, B. DeMarco, V. Meyer, M. Rowe, A. Ben-Kish, M. Barrett, J. Britton, J. Hughes, W. M. Itano, B. M. Jelenkovic, C. Langer, D. Lucas, T. Rosenband, and D. J. Wineland. *J. Phys. B*, 36:599, 2003.

[13] J. J. García-Ripoll, P. Zoller, and J. I. Cirac. *Phys. Rev. Lett.*, 91:157901, 2003.

[14] L.-M. Duan, J. I. Cirac, and P. Zoller. *Science*, 292:1695, 2001.

[15] C. W. Gardiner and P. Zoller. *Quantum Noise: A Handbook of Markovian and Non-Markovian Quantum Stochastic Methods with Applications to Quantum Optics.* Springer, 1999.

[16] D. J. Wineland, C. Monroe, W. M. Itano, D. Leibfried, B. E. King, and D. M. Meekhof. *RES J. NIST*, 103:259, 1998.

[17] J. I. Cirac and P. Zoller. *Nature*, 404:579, 2000.

[18] D. Kielpinksi, C. Monroe, and D. J. Wineland. *Nature*, 417:709, 2002.

[19] A. Steane, C. F. Roos, D. Stevens, A. Mundt, D. Leibfried, F. Schmidt-Kaler, and R. Blatt. *Phys. Rev. A*, page 042305, 2000.

[20] A. Sørensen and K. Mølmer. *Phys. Rev. A*, 62:022311, 2000.

[21] A. Sørensen and K. Mølmer. *Phys. Rev. Lett.*, 82(9):1971, 1999.

[22] D. Jonathan, M. B. Plenio, and P. L. Knight. *Phys. Rev. A*, 62:042307, 2000.

[23] G. J. Milburn, S. Schneider, and D. F. V. James. *Fortschritte der Physik*, 48:801, 2000.

[24] B. DeMarco, A. Ben-Kish, D. Leibfried, V. Meyer, M. Rowe, B. M. Jelenkovic, W. M. Itano, J. Britton, C. Langer, T. Rosenband, and D. J. Wineland. *Phys. Rev. Lett.*, 89:267901–1, 2002.

[25] D. Jaksch, C. Bruder, J.I. Cirac, C. Gardiner, and P. Zoller. *Phys. Rev. Lett.*, 81(15):3108, 1998.

[26] D. Jaksch, H.J. Briegel, J.I. Cirac, C.W. Gardiner, and P. Zoller. *Phys. Rev. Lett.*, 82:1975, 1999.

[27] M. Greiner, O. Mandel, T. Esslinger, T.W. Hänsch, and I. Bloch. *Nature*, 415:39, 2002.

[28] O. Mandel, M. Greiner, A. Widera, T. Rom, T.W. Hänsch, and I. Bloch. *Nature*, 425:937, 2003.

[29] G. Brennen, C. Caves, P. Jessen, and I. Deutsch. *Phys. Rev. Lett.*, 82:1060, 1999.

[30] G.K. Brennen, I.H. Deutsch, and C.J. Williams. *Phys. Rev. A*, 65:022313, 2002.

[31] D. Jaksch, J. I. Cirac, P. Zoller, S.L. Rolston, R. Cote, and M. D. Lukin. *Phys. Rev. Lett.*, 85:2208, 2000.

[32] S. Sachdev. *Quantum Phase Transitions*. Cambridge University Press, Cambridge, 1999.

[33] T. Calarco, E.A. Hinds, D. Jaksch, J. Schmiedmayer, J.I. Cirac, and P.Zoller. *Phys. Rev. A*, 61:22304, 2000.

[34] N. Schlosser, G. Reymond, I. Protsenko, and P. Grangier. *Nature*, 411:1024, 2001.

[35] F.B.J. Buchkremer, R. Dumke, M. Volk, T. Muether, G. Birkl, and W. Ertmer. *Laser Physics*, 12:736, 2002.

[36] R. Raussendorf and H. J. Briegel. *Phys. Rev. Lett.*, 86(22):5188, 2001.

[37] E. Jane, G. Vidal, W. Dür, and P. Zoller. *Quantum Information and Computation*, 1:15, 2003.

[38] S. Lloyd. *Science*, 273:1073, 1996.

[39] M. D. Lukin, M. Fleischhuaer, R. Cote, L.-M. Duan, D. Jaksch, J. I. Cirac, and P. Zoller. *Phys. Rev. Lett.*, 87:037901, 2001.

[40] L.-M. Duan, M. D. Lukin, J. I. Cirac, and P. Zoller. *Nature*, 414:413, 2001.

[41] A. Kuzmich and E. S. Polzik. *Phys. Rev. Lett.*, 85(26):5639, 2000.

[42] L.-M. Duan, J. I. Cirac, P. Zoller, and E. S. Polzik. *Phys. Rev. Lett.*, 85(26):5643, 2000.

[43] M. Fleischhauer and M. D. Lukin. *Phys. Rev. Lett.*, 84(22):5094, 2000.

[44] B. Julsgaard, A. Kozhekin, and E. S. Polzik. *Nature*, 413:400, 2001.

[45] L.-M Duan. *Phys. Rev. Lett.*, 88:170402, 2002.

[46] A. Kuzmich, W. P. Bowen, A. D. Boozer, A. Boca, C. W. Chou, L.-M. Duan, and H. J. Kimble. *Nature*, 423:731, 2003.

[47] C. H. van der Wal, M. D. Eisaman, A. Andre, R. L. Walsworth, D. F. Phillips, A. S. Zibrov, and M. D. Lukin. *Science*, 301:196, 2003.

[48] W. Jiang, C. Han, P. Xue, L. M. Duan, and G. C. Guo. *quant-ph/0309175*.

[49] J. Hald, J. L. Sorensen, C. Schori, and E.S. Polzik. *Phys. Rev. Lett.*, 83:1319, 1999.

[50] L. V. Hau, S. E. Harris, Z. Dutton, and C. H. Behroozi. *Nature*, 397:594, 1999.

[51] M. M. Kash, V. A. Sautenkov, A. S. Zibrov, L. Hollberg, G. R. Welch, M. D. Lukin, Y. Rostovtsev, E. S. Fry, and M.O. Scully. *Phys. Rev. Lett.*, 82:5229, 1999.

[52] A. Kuzmich, L. Mandel, and N. P. Bigelow. *Phys. Rev. Lett.*, 85(8):1594, 2000.

[53] H. J. Briegel, W. Dur, J. I. Cirac, and P. Zoller. *Phys. Rev. Lett.*, 81(26):5932, 1998.

[54] A. Ekert. *Phys. Rev. Lett.*, 67(6):661, 1991.

[55] C. H. Bennett, G. Brassard, C. Crepeau, R. Jozsa, A. Peres, and W. K. Woothers. *Phys. Rev. Lett.*, 70(13):1895, 1993.

[56] C. H. Bennett, D. P. Divincenzo, J. A. Smolin, and W. K. Wootters. *Phys. Rev. A*, 54(5):3824, 1996.

[57] J. Preskill. *http://www.theory.caltech.edu/people/preskill/ph229/*, 2001.

[58] J. I. Cirac, P. Zoller, H. J. Kimble, and H. Mabuchi. *Phys. Rev. Lett.*, 78(16):3221, 1997.

[59] S.J. Van Enk, J.I. Cirac, and P. Zoller. *Science*, 279:205, 1998.

[60] J. Ye, D. W. Vernooy, and H.J. Kimble. *Phys. Rev. Lett.*, 83:4987, 1999.

[61] A. Kuhn, M. Hennrich, and G. Rempe. *Phys. Rev. Lett.*, 89:067901, 2002.

[62] J. A. Sauer, K. M. Fortier, M. S. Chang, C. D. Hamley, and M. S. Chapman. *quant-ph/0309052*.

[63] J. McKeever, A. Boca, A. D. Boozer, J. R. Buck, and H. J. Kimble. *Nature*, 425:268, 2003.

[64] C. Cabrillo, J. I. Cirac, P. G-Fernandez, and P. Zoller. *Phys. Rev. A*, 59(2):1025, 1999.

[65] L.-M. Duan, J.I. Cirac, and P. Zoller. *Phys. Rev. A*, 66:023818, 2002.

[66] L.-M. Duan, G. Giedke, J. I. Cirac, and P. Zoller. *Phys. Rev. Lett.*, 84:2722, 2000.

[67] S. L. Braunstein and H. J. Kimble. *Phys. Rev. Lett.*, 80(4):869, 1998.

[61] A. Kuhn, M. Hennrich, and G. Rempe. *Phys. Rev. Lett.*, 89:067901, 2002.

[62] J. A. Sauer, K. M. Fortier, M. S. Chang, C. D. Hamley, and M. S. Chapman. *quant-ph/0309052*.

[63] J. McKeever, A. Boca, A. D. Boozer, J. R. Buck, and H. J. Kimble. *Nature*, 425:268, 2003.

[64] C. Cabrillo, J. I. Cirac, P. G-Fernandez, and P. Zoller. *Phys. Rev. A*, 59(2):1025, 1999.

[65] L.-M. Duan, J.I. Cirac, and P. Zoller. *Phys. Rev. A*, 66:023818, 2002.

[66] L.-M. Duan, G. Giedke, J. I. Cirac, and P. Zoller. *Phys. Rev. Lett.*, 84:2722, 2000.

[67] S. L. Braunstein and H. J. Kimble. *Phys. Rev. Lett.*, 80(4):869, 1998.

Course 5

QUANTUM INFORMATION PROCESSING IN ION TRAPS I

R. Blatt[1,2], H. Häffner[1], C. F. Roos[1], C. Becher[1] and F. Schmidt-Kaler[1]

[1] *Institut für Experimentalphysik, Universität Innsbruck Technikerstrasse 25, A-6020 Innsbruck, Austria*

[2] *Institut für Quantenoptik und Quanteninformation, Österreichische Akademie der Wissenschaften Technikerstrasse 25, A-6020 Innsbruck, Austria*

D. Estève, J.-M. Raimond and J. Dalibard, eds.
Les Houches, Session LXXIX, 2003
Quantum Entanglement and Information Processing
Intrication quantique et traitement de l'information
© 2004 Elsevier B.V. All rights reserved

Contents

Abstract

The scheme of an ion trap quantum computer is described and recent experiments with trapped Ca^+ ions are discussed. Ion trap physics is briefly reviewed, spectroscopic and laser cooling tools are summarized and basic gate operations using composite laser pulses are described. Quantum information processing with ions is exemplified with the operation of the Deutsch–Jozsa algorithm. The realization of the Cirac–Zoller CNOT gate operation is reviewed and several experiments investigating entanglement of ions are summarized.

1. Introduction

Quantum information processing, that is using the laws of quantum mechanics for computational purposes, was proposed and considered first by Feynman and Deutsch [1, 2]. Generalizing the concept of a classical bit of information to the quantum world, the quantum bit (qubit), i.e. the memory unit for quantum information is given by a two–level system and quantum information is considered to be encoded as a superposition of these states $|\psi\rangle$

$$|\psi\rangle = \alpha|0\rangle + \beta|1\rangle, \qquad \text{with} \quad |\alpha|^2 + |\beta|^2 = 1. \tag{1.1}$$

Due to its probabilistic interpretation, at first it appears impossible to predictably handle and process such an elusive quantum state and even more so carry out a whole computation with information provided in this way. Therefore, and since quantum algorithms did not seem to be advantageous for computational purposes, quantum computing was widely considered as a somewhat academic curiosity.

This situation changed drastically in 1994 when P. Shor [3] proposed a quantum algorithm which enables one to factorize large numbers much faster than with any (known) classical computation. Thus, quantum computer technology would, for example, endanger all of today's powerful cryptosystems since they are all based on the fact that factorization of large numbers is a difficult computational problem. Therefore, the discovery of a powerful quantum algorithm for fast factoring and the subsequently found fast algorithm for searching data bases by L. Grover [4] spawned a worldwide search for the development of systems to implement and build a quantum computer.

The requirements for such a quantum processor are nowadays known as the DiVincenzo criteria [5]: Storing and processing quantum information requires (i) scalable physical systems with well-defined qubits, which (ii) can be initialized and have (iii) long lived quantum states in order to ensure long coherence times during the computational process. The necessity to coherently manipulate the stored quantum information requires (iv) a set of universal gate operations between the qubits which must be implemented using controllable interactions of the quantum systems. Finally, determining reliably the outcome of a quantum computation requires (v) an efficient measurement procedure. During the last years, a large variety of physical systems have been proposed and investigated for their use in quantum information processing and are considered in other chapters of this volume.

In the following, quantum information processing is discussed using trapped ions with particular emphasis on a realization using trapped Ca^+ ions where the qubit is encoded in a narrow optical quadrupole transition. A different approach using two hyperfine levels of Be^+ ions is treated in the notes by David Wineland [6] in this volume. For additional work and further experiments on the implementation of ion trap quantum information processing the reader is referred to numerous reviews [7–25].

This chapters is organized as follows: In section2 the general concept of an ion trap quantum computer will be presented. The physics of linear ion traps is briefly described in section 3 including the behavior of ion strings used to realize a quantum register, the spectroscopy of Ca^+–ions as well as laser cooling of ion strings and addressing of individual ions of a string. The basic building blocks for quantum information processing with ions, i.e. the coherent manipulation of the ions' internal and external states, the appearing AC Stark effects and their compensation and basic gate operations are treated in section 4. Elementary quantum information processing is then demonstrated by the operation of the Deutsch–Jozsa algorithm with a single trapped ion in section 5. Scaling an ion trap quantum computer requires the realization of 2-qubit operations working on two ions which is discussed in terms of the Cirac–Zoller CNOT gate operation described in section 6. Using single and two–qubit operations entanglement of ions is achieved and investigated (see section 7).

2. Ion trap quantum computer - the concept

Strings of trapped ions have been proposed in 1995 for quantum computation by Ignacio Cirac and Peter Zoller [26]. With such a system all requirements for a quantum information processor [5] can be met. It is well known from time and frequency standards applications [27] that trapped and lasers cooled ions in Paul traps are especially well isolated from the environment. Thus, they are ideal for carrying quantum information, and qubits can be realized using either narrow optical transitions, long lived hyperfine states or corresponding Zeeman sub-states. Strings of ions can be stored in linear Paul traps [28,29] and thus serve as a quantum register. For initializing qubits the internal states of the ions must be preset in well defined states which is achieved, for example by individually addressing ions in a string with an interacting laser beam. A set of universal quantum gate operations is then given by i) single qubit rotations (which are realized by Rabi oscillations of individual ions) and ii) the equivalent of the XOR operation of classical computers, i.e. the controlled–NOT (CNOT) operation. Applying the CNOT gate the state of a control qubit defines the state of a target qubit, i.e. $|\varepsilon_c, \varepsilon_t\rangle \rightarrow |\varepsilon_c, \varepsilon_c \oplus \varepsilon_t\rangle$ with $\varepsilon_{c,t} = 0, 1$, and $\oplus$ denoting addition modulo 2. This CNOT–gate operation is experimentally the hardest part to implement and one realization was proposed in the seminal paper by Cirac and Zoller (CZ) [26].

Their idea is as follows: In a first step the entire ion string (i.e. the quantum register) is cooled to the ground state of its harmonic motion in the ion trap. Since the mutual Coulomb repulsion spatially separates the ions, any induced motion couples to all ions. Therefore a universal quantum gate can be achieved by addressing an ion (which the algorithms identifies as the controlling qubit) in the string and exciting it with the interacting laser beam in such a way that its internal excited state amplitude is mapped to a single phonon quantum motion of that ion. This phonon, however, is now carried by the entire string. An operation on one of the other ions chosen according to the underlying algorithm as the target qubit now depends on whether or not there is motion in the string.

Any algorithm can be broken down in a series of such one- and two-qubit operations and therefore this set of instructions constitutes a universal quantum gate [30]. Thus, the realization of these quantum gates allows one to build and operate a quantum computer. Moreover, in principle, this concept provides a scalable approach towards quantum computation and has therefore attracted quite some attention.

During the last years several other techniques have been proposed to implement gate operations with trapped ions. Sørensen and Mølmer [31,32], and with a different formulation Milburn [33], proposed a scheme for "hot" quantum gates, i.e. their procedures for gate operations do not require ground state cooling of an ion string. Although successfully applied to trapped Be^+ ions [72], with the trapping parameters currently available, these gate procedures are not easily applicable to Ca^+ ions. Other gates based on ac Stark shifts have been suggested by Jonathan *et al.* [35] and holonomic quantum gates (using geometric phases) have been proposed by Duan *et al.* [34]. A different CNOT–gate operation also based on the ac Stark effect which does not require individual addressing and ground state cooling has been realized with trapped Be^+ ions [79].

3. Physics of ion traps

Trapped ions in ultra high vacuum are considered an ideal tool for quantum computation purposes since they provide atomic particles as qubits which are at rest in free space and thus are well isolated from perturbations as, e.g. by collisions with confining walls or background gas atoms. In order to realize a quantum register, strings of ions in linear rf Paul traps are a natural choice [37]. The linear variant of the Paul trap is based on the quadrupole mass–filter potential [38]. A time varying voltage applied to a set of electrodes provides a quadrupole potential confining the ions in the two directions perpendicular to the z-axis. The motion along the z-axis, however, is not affected. For axial confinement, additional electrodes must be attached. The shape of often used linear ion traps are sketched in Fig. 1. A more detailed treatment of the trap potential for linear ion

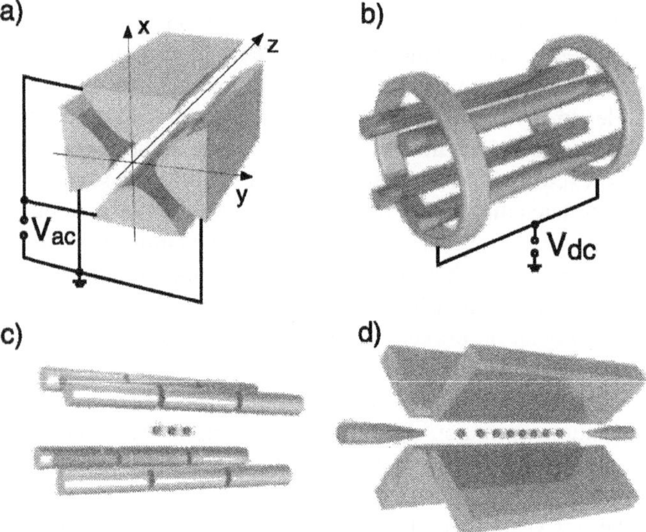

Fig. 1. Linear ion traps: a) Paul mass filter, an ac-voltage V_{ac} is applied between the hyperbolic electrodes, b) linear ion trap with rods replacing the hyperbolic electrodes and additional tips for the axial confinement using a dc-potential V_{dc}, ac-voltages as in a), c) segmented trap (see for details [6], d) knife edges replacing the hyperbolic electrodes and tips for the axial confinement.

traps can be found in the lecture by David Wineland included in this volume and in the literature [39, 40]. For the purposes of quantum information processing here, it suffices to note that to a good approximation the motion of a single ion can be described by harmonic oscillations ω_i along the three different directions $i = x, y, z$ determined by the trap voltages V_{dc} and $V_{ac}\cos(\Omega t)$, where the z-axis is generally referred to as the trap axis, the x, y axes correspond to the transverse coordinates, and Ω is referred to as the trap's driving frequency. Thus we have for the quantized motion in the trap the Hamiltonian

$$H_t = \sum_{i=x,y,z} \hbar\omega_i (a_i^\dagger a_i + \frac{1}{2}) \tag{3.1}$$

where a_i, $a_i^\dagger$ denote the corresponding annihilation and creation operators of the harmonic oscillators along the trap directions x, y, z.

Note that the potential is always created by ac-voltages, i.e. it is explicitly time-dependent and thus the Hamiltonian (3.1) describes the oscillation in a pseudo-potential. In the very center of the trap, i.e. for $x = y = 0$ the harmonic approximation holds exactly, however, any excursion of the ions from the trap axis leads to a residual, though spurious, motion at the rf. frequency Ω of the driving ac-voltage. For precision experiments and for quantum information purposes stray potentials are usually cancelled using additional electrodes and voltages not shown in Fig. 1 [36].

3.1. *Ion strings for quantum computation*

The concept of ion trap quantum computing as proposed by Cirac and Zoller makes use of the Coulomb coupling within a string of ions aligned along the trap z-axis. Therefore, we consider in more detail the confinement and oscillation of an ion string. Here we follow the treatments of A. Steane [7] and D. James [10].

For N ions of mass M and charge e confined in the 3-dimensional harmonic trap potential with (angular) frequencies ω_x, ω_y and ω_z the potential energy is given by [10]

$$U = \frac{M}{2} \sum_{n=1}^{N} (\omega_x^2 x_n^2 + \omega_y^2 y_n^2 + \omega_z^2 z_n^2) + \frac{e^2}{8\pi\epsilon_0} \sum_{\substack{n,m=1 \\ m\neq n}}^{N} \frac{1}{|\mathbf{r}_n - \mathbf{r}_m|}, \tag{3.2}$$

where $\mathbf{r}_n = (x_n, y_n, z_n)$ is the position vector of the nth ion. The first term denotes the trap (pseudo–)potential, while the second describes the mutual Coulomb interactions. In an ideal linear trap with axial frequency ω_z, the two radial frequencies are degenerate if the trap construction is perfectly symmetric:

$$\omega_x = \omega_y = \omega_r \,. \tag{3.3}$$

For the following treatment of equilibrium positions we assume a linear ion string configuration. This is the case only if the radial confinement is sufficiently stronger than the axial confinement and if the number of ions N is not too large, otherwise the ions will assume a zig-zag or even more complicated 3-dimensional geometric configurations. For a linear string we have $x_n^{(0)} = y_n^{(0)} = 0$, where the upper index $^{(0)}$ denotes the ions' equilibrium positions. The axial equilibrium positions $z_n^{(0)}$ are then determined by the following equation:

$$\left[\frac{\partial U}{\partial z_n}\right]_{z_n=z_n^{(0)},\ x_n=y_n=0} = 0\,. \tag{3.4}$$

For $N = 2$ and $N = 3$, this can be solved analytically [7, 10]:

$$\begin{aligned} N=2: &\quad z_1 = -(1/2)^{2/3}\,l,\ z_2 = (1/2)^{2/3}\,l \\ N=3: &\quad z_1 = -(5/4)^{1/3}\,l,\ z_2 = 0,\ z_3 = (5/4)^{1/3}\,l\,, \end{aligned} \tag{3.5}$$

where

$$l = \left(\frac{e^2}{4\pi\epsilon_0 M\omega_z^2}\right)^{1/3} \tag{3.6}$$

represents the natural length scale of the problem. For axial trap frequencies on the order of $\omega_z = 2\pi \cdot 1$ MHz and for $^{40}Ca^+$ ions this amounts to inter-ion-separations of $\Delta z = 5.6\ \mu$m and $\Delta z = 4.8\ \mu$m for $N = 2$ and $N = 3$, respectively. For more than three ions the problem must be solved numerically. The spacing between adjacent ions increases from the center to the edges of the string. The minimum separation Δz_{min} in the crystal occurs between the central ions. Numerical simulations yield [7]

$$\Delta z_{min} \simeq 2.0\, l N^{-0.57}\,. \tag{3.7}$$

This approximation remains accurate for ion numbers of up to $N \simeq 1000$. For an implementation of the Cirac–Zoller quantum computer individual ion addressing is required. This addressing is most difficult for the most closely spaced central ions. For small oscillations around the equilibrium positions ($y_n^{(0)} = 0,\ x_n^{(0)} = 0,\ z_n^{(0)}$) the Lagrangian $L = T - U$ of the system (T being the kinetic energy) takes the form [41]

$$L = \frac{M}{2}\sum_{i=x,y,z}\left(\sum_{n=1}^{N}(\dot{\varrho}_{i|n}^2) - \omega_z^2 \sum_{n,m=1}^{N}(C_{n,m}^i \varrho_{i|n}\varrho_{i|m})\right), \tag{3.8}$$

where $\varrho_{x|n}$, $\varrho_{y|n}$ and $\varrho_{z|n}$ are the displacements of the nth ion from its equilibrium position in x, y and z direction, respectively. The coupling matrices are given by

$$C^z_{n,m} = \begin{cases} 1 + 2\sum\limits_{\substack{p=1 \\ p\neq m}}^{N} \dfrac{l^3}{\left|z_m^{(0)} - z_p^{(0)}\right|^3} & \text{if } n = m\,, \\ \dfrac{-2l^3}{\left|z_m^{(0)} - z_n^{(0)}\right|^3} & \text{if } n \neq m \end{cases} \tag{3.9}$$

and

$$C^x_{n,m} = C^y_{n,m} = \left(\frac{1}{\alpha} + \frac{1}{2}\right)\delta_{n,m} - \frac{1}{2}C^z_{n,m}\,, \tag{3.10}$$

where $\delta_{n,m}$ is the Kronecker delta and $\alpha = (\omega_z/\omega_r)^2$ is a parameter describing the anisotropy of the trap potential. The motions in x, y and z direction are coupled. The radial directions x and y are assumed to be degenerate. As a consequence, the solutions of the problem, obtained by diagonalization of the C^x, C^y and C^z matrices, are two independent sets of normal modes. One set contains normal modes in the axial direction, the other in the radial direction, the latter being two-fold degenerate corresponding to x- and y-directions. There are N axial and $2N$ radial modes.

The principal mode of the *axial modes* is the center-of-mass (COM) mode in which all ions move parallel in the axial direction with the same amplitude. This is the so–called axial COM mode with frequency ω_z, i.e. the axial frequency of a single ion. Higher order axial modes always have higher frequency. The second axial mode is called the breathing mode, where the ions oscillate with amplitudes proportional to their distance from the trap center - in magnitude as well as in sign. Thus for uneven ion numbers the center ion stands still in the breathing mode. The frequency of the breathing mode is $\sqrt{3}\omega_z$, independent of the number of ions in the chain. The motional behavior of higher order modes is described by more complicated eigenvectors. The frequencies of these modes must be calculated numerically and depend slightly on the total ion number. This dependence, however, is so weak that, for example, for all $N \leq 10$ the axial mode frequencies are described with very high accuracy by the list $\{1,\ \sqrt{3},\ \sqrt{29/5},\ 3.051,\ 3.671,\ 4.272,\ 4.864,\ 5.443,\ 6.013,\ 6.576\}$ in units of ω_z [7]. For ions cooled close to the ground state of the trap (cf. section3.3) the spatial extension of the ions' wavepackets is given by the standard deviation of the Gaussian ground state probability distribution of the axial COM mode, which is given by

$$\delta z_{com} = \sqrt{\frac{\hbar}{2NM\omega_z}}\,. \tag{3.11}$$

For two $^{40}Ca^+$ ions and $\omega_z = 2\pi \cdot 1$ MHz one finds $\delta z_{com} = 7.9$ nm which is three orders of magnitude smaller than the separation between the two ions.

The principal *radial mode* is again the center-of-mass mode, all ions oscillating radially in phase and with equal amplitude at the frequency ω_r. This mode is referred to as the radial COM mode. In contrast to the axial modes, however, every higher order mode is at a *lower* frequency [41]. The eigenvectors of the radial modes point radially instead of in the z-direction but are otherwise formally identical to the eigenvectors of the axial modes. As with the axial modes, the frequency of the second, often called rocking, mode can be determined analytically and does not depend on the number of ions in the chain:

$$\omega_{rock} = \sqrt{\omega_r^2 - \omega_z^2}\,. \tag{3.12}$$

The frequencies of higher order modes must be calculated numerically and again depend only very slightly on N.

3.2. *Spectroscopy in ion traps*

An ion trapped in a harmonic potential with frequency ω, interacting with the travelling wave of a single mode laser tuned close to a transition (with frequency ω_a) that forms an effective two-level system, is described by the Hamiltonian [42–45]

$$H = H_0 + H_1 \tag{3.13}$$

$$H_0 = \frac{p^2}{2m_0} + \frac{1}{2}m_0\omega^2 x^2 + \frac{1}{2}\hbar\omega_a\sigma_z \tag{3.14}$$

$$H_1 = \frac{1}{2}\hbar\Omega(\sigma^+ + \sigma^-)\left(e^{i(kx-\omega_l t+\phi)} + e^{-i(kx-\omega_l t+\phi)}\right), \tag{3.15}$$

where k is the wave number, ω_l the frequency and ϕ the phase of the laser radiation. σ_z, σ^+ and σ^- denote the usual Pauli matrices and m_0 is the mass of the ion. Considering for simplicity one–dimensional harmonic motion only, H_0 describes the state of the ion while the laser-ion interaction is contained in H_1, its strength being given by the coupling constant Ω. Here we assume that only a single transition (ω_a being the atomic transition frequency) is close to resonance and that the laser is directed along the x-axis to the ion. The Pauli operators act on the internal atomic states, $|g\rangle$ and $|e\rangle$. Defining the Lamb-Dicke parameter[1],

$$\eta = k\sqrt{\frac{\hbar}{2m_0\omega}}\,, \tag{3.16}$$

[1] If the laser is at an angle β to the oscillation axis, the definition has to be replaced by $\eta = k\cos\beta\sqrt{\hbar/2m\omega}$.

the ion's external degrees of freedom, i.e. its harmonic oscillation, can be expressed in terms of creation and annihilation operators as

$$H_0 = \hbar\omega(a^\dagger a + \frac{1}{2}) + \frac{1}{2}\hbar\omega_a\sigma_z \tag{3.17}$$

$$H_1 = \frac{1}{2}\hbar\Omega\left(e^{i\eta(a+a^\dagger)}\sigma^+ e^{-i(\omega_l t+\phi)} + e^{-i\eta(a+a^\dagger)}\sigma^- e^{i(\omega_l t+\phi)}\right) . \tag{3.18}$$

Here, the rotating wave approximation has been made. In the interaction picture defined by $U = e^{iH_0t/\hbar}$ the Hamiltonian $H_I = U^\dagger HU$ takes the form

$$H_I = \frac{1}{2}\hbar\Omega\left(e^{i\eta(\hat{a}+\hat{a}^\dagger)}\sigma^+ e^{i\phi}e^{-i\Delta t} + e^{-i\eta(\hat{a}+\hat{a}^\dagger)}\sigma^- e^{-i\phi}e^{i\Delta t}\right) , \tag{3.19}$$

with $\hat{a} = ae^{i\omega t}$ and $\Delta = \omega_l - \omega_a$. The laser couples the state $|S, n\rangle$ to all states $|D, n'\rangle$, where n, n' are vibrational quantum numbers. If the laser is tuned close to resonance of a transition $|S, n\rangle \leftrightarrow |D, n + m\rangle$ with fixed m and $n = 0, 1, 2, 3, ...$ (which corresponds to $(\omega_l - \omega_a) \approx m\omega$), coupling to other levels can be neglected, provided the laser intensity is sufficiently low ($\Omega \ll \omega$). In that case, the laser induces a pairwise coupling between the levels $|S, n\rangle$ and $|D, n+m\rangle$. The time evolution of the state $\Psi(t) = \sum_k (c_k(t)|S, k\rangle + d_k(t)|D, k\rangle)$ is governed by the Schrödinger equation $i\hbar\partial_t\Psi = H\Psi$ which is equivalent to the set of coupled equations

$$\dot{c}_n = -i^{(1-|m|)}e^{i\delta t}e^{-i\phi}\left(\Omega_{n+m,n}/2\right) d_{n+m} \tag{3.20}$$

$$\dot{d}_{n+m} = -i^{(1+|m|)}e^{-i\delta t}e^{i\phi}\left(\Omega_{n+m,n}/2\right) c_n . \tag{3.21}$$

$\delta = \Delta - m\omega$ accounts for a detuning of the laser from the transition, the constant

$$\Omega_{n+m,n} := \Omega\|\langle n+m|e^{i\eta(a+a^\dagger)}|n\rangle\| \tag{3.22}$$

is called Rabi frequency. Solutions to these equations show an oscillatory and complete exchange of population between the coupled levels. On resonance, this oscillation takes place with a frequency equal to the Rabi frequency $\Omega_{n+m,n}$. When the laser is detuned from resonance, the population transfer is no longer complete (amplitude $\Omega^2_{n+m,n}/(\delta^2 + \Omega^2_{n+m,n})$), yet it takes place at a higher frequency ($= \sqrt{\delta^2 + \Omega^2_{n+m,n}}$) [44].

Transitions that do not change the number of vibrational quanta ($\leftrightarrow m = 0$) are called *carrier* transitions. A transition is termed *blue sideband* if an absorption process is accompanied by an increase in the motional quantum number ($\leftrightarrow m = +1$) while it is termed *red sideband* if it decreases upon absorption ($\leftrightarrow m = -1$).

For a given Lamb-Dicke parameter η, the coupling strengths $\Omega_{n,n}$, $\Omega_{n-1,n}$ and $\Omega_{n+1,n}$ of carrier, red and blue sideband, respectively, as a function of n can be calculated with eqn. (3.22), using Laguerre polynomials [44]. This calculation considerably simplifies in the so-called Lamb-Dicke regime defined by the condition $\eta^2(2n+1) \ll 1$. In this regime, the atomic wavepacket is confined to a space much smaller than the wavelength of the transition. A first order Taylor expansion in eqn. (3.22) is then a very good approximation:

$$e^{i\eta(a^\dagger+a)} = 1 + i\eta(a^\dagger + a) + \mathcal{O}(\eta^2) \,. \tag{3.23}$$

The coupling strength on the carrier is well approximated by

$$\Omega_{n,n} = (1 - \eta^2 n)\Omega \tag{3.24}$$

On resonance, the interaction Hamiltonian (3.19) reduces to

$$H_I = \frac{1}{2}\hbar\Omega_{n,n}(\sigma^+ e^{i\phi} + \sigma^- e^{-i\phi}) \,. \tag{3.25}$$

Note that in the resonant case the time dependence of H_I vanishes. If the laser is pulsed, then this interaction is available only for a certain pulse duration t and the action of such a carrier pulse is described by the unitary operator

$$R(\theta,\phi) = e^{iH_I t/\hbar} = e^{i\frac{\theta}{2}(e^{i\phi}\sigma^+ + e^{-i\phi}\sigma^-)}, \tag{3.26}$$

where $\theta = \Omega_{n,n}t$. This is the unitary operator representing an elementary single qubit rotation.
The Rabi frequency $\Omega_{n,n'}$ on the sidebands are given by

$$\Omega_{n-1,n} = \eta\sqrt{n}\Omega \tag{3.27}$$

on the red sideband and

$$\Omega_{n+1,n} = \eta\sqrt{n+1}\Omega \tag{3.28}$$

on the blue sideband. On the red sideband ($|S,n\rangle \leftrightarrow |D,n-1\rangle$), the Hamiltonian takes the form

$$H_I = \frac{1}{2}\hbar\eta\Omega(a\sigma^+ e^{i\phi} - a^\dagger\sigma^- e^{-i\phi}) \,, \tag{3.29}$$

which can also be written as

$$H_I = \frac{1}{2}\hbar\Omega_{n-1,n}(a_r\sigma^+ e^{i\phi} - a_r^\dagger\sigma^- e^{-i\phi}) \tag{3.30}$$

with the "normalized" phonon annihilation and creation operators $a_r = a/\sqrt{n}$ and $a_r^\dagger = a^\dagger/\sqrt{n}$, respectively. This definition ensures that the operators $a_r\sigma^+$ and $a_r^\dagger\sigma^-$ do not change the norm of a state vector. Accordingly, the matrix of a resonant laser pulse of duration t on the red sideband is given by

$$R^-(\theta, \phi) = e^{i\frac{\theta}{2}(e^{i\phi}\sigma^+ a_r - e^{-i\phi}\sigma^- a_r^\dagger)}, \tag{3.31}$$

where $\theta = \Omega_{n-1,n}t$.

On the blue sideband ($|S, n\rangle \leftrightarrow |D, n+1\rangle$)

$$H_I = \frac{1}{2}\hbar\eta\Omega(a^\dagger\sigma^+ e^{i\phi} - a\sigma^- e^{-i\phi}) \,. \tag{3.32}$$

By the same argument as for red sideband coupling, this Hamiltonian can be rewritten as

$$H_I = \frac{1}{2}\hbar\Omega_{n+1,n}(a_b^\dagger\sigma^+ e^{i\phi} - a_b\sigma^- e^{-i\phi}) \,, \tag{3.33}$$

where $a_b = a/\sqrt{n+1}$ and $a_b^\dagger = a^\dagger/\sqrt{n+1}$. A resonant laser pulse on the blue sideband is given by

$$R^+(\theta, \phi) = e^{i\frac{\theta}{2}(e^{i\phi}\sigma^+ a_b^\dagger - e^{-i\phi}\sigma^- a_b)}, \tag{3.34}$$

where $\theta = \Omega_{n+1,n}t$.

In summary, we arrive at a level scheme and interactions as depicted in Fig. 2. Spectroscopy of such a trapped ion then reveals the harmonic oscillator structure

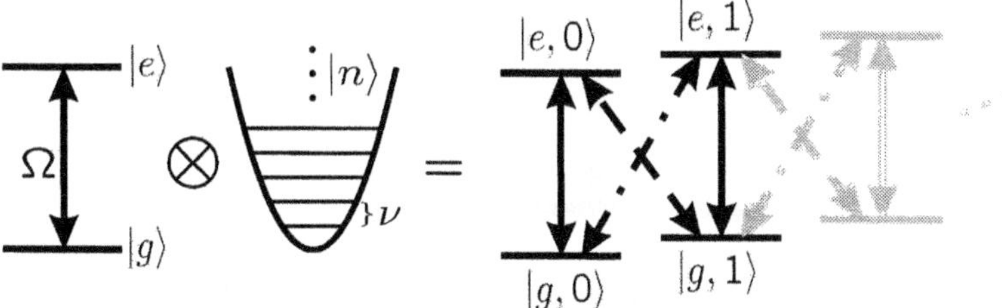

Fig. 2. Level scheme of a harmonically trapped two–level atom: The internal states $|g\rangle$, $|e\rangle$ of the atom are labelled with the respective quantum number $|n\rangle$ of the state of the harmonic oscillator. In the Lamb–Dicke regime only carrier ($|g, n\rangle \leftrightarrow |e, n\rangle$) and first order sideband transitions ($|g, n\rangle \leftrightarrow |e, n \pm 1\rangle$) need to be considered.

as given by the relations above. Note, that precisely the transitions on the first–order sidebands are used for quantum information processing according to the

Cirac–Zoller proposal since they allow one to map internal states to motional states in order to realize the CNOT–gate operation [26].

Internal state detection of a trapped ion is achieved using the electron shelving technique [46]. For this, one of the internal states of the trapped atom is selectively excited to a third short–lived state thereby scattering many photons on that transition if the coupled internal state was occupied. If, on the other hand, the atom's electron resides in the uncoupled state of the qubit (i.e. the electron is shelved in that state) then no photons are scattered and thus the internal state can be detected with an efficiency of nearly 100%. For details, the reader is referred to [40] and the references therein.

In the following we will concentrate on the Innsbruck experiments with Ca^+ ions, experiments with Be^+ are described in the notes by David Wineland [6] and for some other approaches we refer to [16, 17, 25]. Fig. 3 shows the relevant levels of the Ca^+ ion which are populated in the experiment. The qubit is implemented using the narrow quadrupole transition at 729 nm, i.e. $|g\rangle = |S_{1/2}\rangle$ and $|e\rangle = |D_{5/2}\rangle$. For optical cooling and state detection resonance fluorescence on the $S_{1/2}$—$P_{1/2}$ transition is scattered by excitation with 397 nm and 866 nm radiation. The laser at 854 nm is applied to repump the excited state $|e\rangle$, for example after a shelving operation. State measurement is achieved by probing the ground

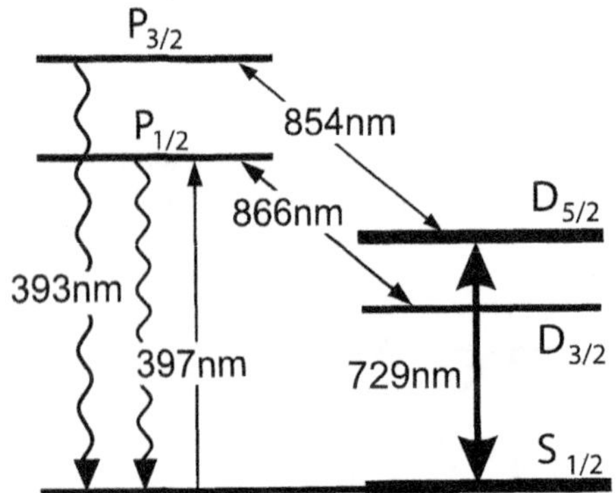

Fig. 3. Level scheme of $^{40}Ca^+$. The qubit is implemented using the narrow quadrupole transition. All states split up into the respective Zeeman sublevels.

state $|g\rangle$ after any manipulation on the quadrupole transition as described above.

3.3. *Laser cooling of ion strings*

An essential ingredient of the Cirac–Zoller scheme is that the initial state of the quantum register is prepared in its motional ground state, i.e. we require at least one of the motional modes (e.g. $n_z = 0$, the axial motion) in the ground state. Other modes need only to be cooled such that the Lamb-Dicke regime is fulfilled.

Laser cooling of trapped and free atoms has been thoroughly described elsewhere, for details we refer the reader to the more recent reviews [40, 47] and the references therein. Usually so-called sideband cooling [48, 49] is used to cool one mode of an ion string to its motional ground state. This is performed by excitation on the first red sideband which causes optical pumping of the vibrational states. As a result the ion(s) reside(s) after the cooling process in the ground state $|g, 0\rangle$ for a single ion or in the state $|ggg\ldots, 0\rangle$ for a string of ions where the quantized ion motion refers to the mode used as the quantum bus of the Cirac–Zoller proposal. This is experimentally achieved using optical pumping schemes involving either Raman transitions [50] or coupled transitions [49, 51]. More elaborate cooling schemes using electromagnetic transparency [52, 53] or sympathetic cooling [54–56] have been investigated and can be applied for quantum information purposes.

3.4. Addressing of individual ions

As outlined above, the realization of the Cirac–Zoller quantum computer starts by preparing the entire quantum register of L ions in the ground state of its harmonic motion, i.e. the initial state vector of the system is prepared as

$$|\Psi\rangle = \sum_{\underline{x}} c_{\underline{x}} |x_{L-1}\ldots x_0\rangle \otimes |0\rangle_{\text{mode}} = \sum_{\underline{x}} c_{\underline{x}} |\underline{x}\rangle \otimes |0\rangle_{\text{mode}} \tag{3.35}$$

where $|\underline{x}\rangle = |x_{L-1}\rangle\cdots|x_0\rangle$, and $|x_k\rangle$ is either $|g\rangle$ or $|e\rangle$ of the k−th ion in the string. Then, the implementation of the CZ CNOT-gate operation requires that individual ions can be addressed in order to rewrite internal information onto the "phonon" mode using sideband transitions. Therefore, the Innsbruck experiments were designed to operate in a regime where the minimum ion distance is on the order of a few μm such that focussing a laser beam at 729 nm is feasible to individually address the single ions [57]. In the current setup, Ca^+ ions are stored with axial trap frequencies of about 1-1.2 MHz and thus the inter-ion distance of two and three ions is on the order of 5 μm. The laser beam at 729 nm is focussed to a waist diameter of approximately 2.5 μm such that with the Gaussian beam neighboring ions are excited with less than 10^{-3} of the central intensity. Beam steering and thus individual addressing is achieved using electrooptic beam deflection for fast switching ($\sim 15\mu$s) between different ion positions [58].

4. Coherent manipulation of quantum information

Quantum information processing requires that individual qubits are coherently manipulated. Together with a 2-qubit controlled-NOT operation this provides a set of universal gate operations. Thus, the ability to perform one and two–qubit rotations is central to quantum information processing. We realize single-qubit rotations by coherent manipulation of the $S_{1/2}(m = -1/2) \leftrightarrow D_{5/2}(m = -1/2)$ transition in Ca^+. Coupling of two qubits requires the precise control of the motional state of a single ion or a string of ions. Both operations can be performed by applying laser pulses at the carrier (i.e. not changing the vibrational quantum number, $\Delta n_z = 0$) or at one of the sidebands of the S–D transition (i.e. laser detuned by $\pm\nu_z$, thus changing the vibrational quantum number by $\Delta n_z = \pm 1$).

All qubit transitions are described as rotations on a corresponding Bloch sphere and they are written as unitary operations $R(\theta, \phi)$ on the carrier (3.26), $R^-(\theta, \phi)$ on the red (3.31) and $R^+(\theta, \phi)$ on the blue (3.34) sidebands. The parameter θ describes the angle of the rotation and depends on the strength and the duration of the applied pulse. ϕ denotes its phase, i.e. the relative phase between the optical field and the atomic polarization and determines the axis about which the Bloch vector rotates. Fig. 4 shows an example of coherent Rabi oscillations for an excitation on the carrier and the blue sideband, respectively. As can be in-

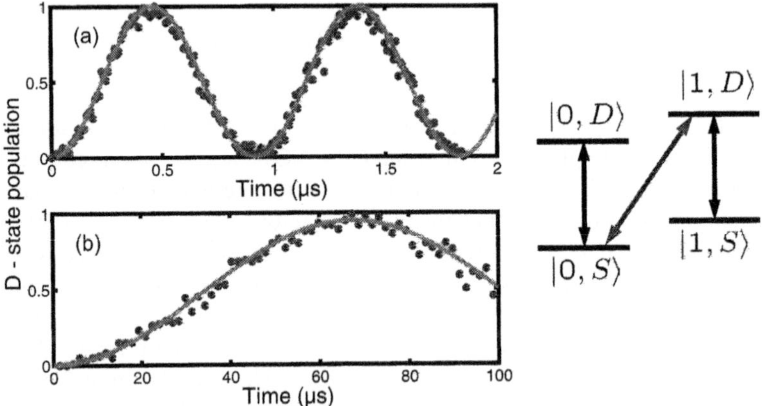

Fig. 4. Rabi oscillations on (a) the carrier and (b) the blue sideband as indicated in the level scheme. Note that the time scale is different for (a) and (b) since the blue sideband coupling is reduced by the Lamb–Dicke factor η (cf. eq. (3.16)).

ferred from the Rabi frequencies pertaining to the population oscillations shown in Fig. 4, typical pulse durations for a π-pulse range from about 1 to several 10 μs for the carrier transition and 50-200 μs on the sideband transition, with the chosen time depending on the desired speed and precision of the operations. Such pulses are the primitives for quantum information processing with trapped ions. By concatenating pulses on the carrier and on the sidebands gate operations and eventually whole quantum algorithms can be implemented. The control of the lasers frequency and phase with an acousto optical modulator is described in [58]. Even the simplest gate operations require several pulses, therefore it is imperative to control the relative optical phases of these pulses in a very precise way or, at least, to keep track of them such that the required pulse sequences lead to the desired operations. This requires the precise consideration of all phases introduced by the light shifts of the exciting laser beams.

4.1. AC Stark shift and its compensation

The laser beam at 729 nm for information processing acts off-resonantly on the various Zeeman components of the S $\leftrightarrow$ D transitions as well as on the other S $\leftrightarrow$ P and D $\leftrightarrow$ P dipole transitions, as can be inferred from the level scheme (cf. Fig. 3). The individual shifts have different signs but compensate each other only to some extent. The combined shifts are typically of the order of several kHz. They have to be carefully controlled since the detuning and the interaction time of the 729 nm laser change depending on the pulse sequence pertaining to gate operations or an entire algorithm.

For a precise measurement of the overall Stark shift we used a Ramsey type interference experiment . Fig. 5 shows the measurement procedure and the result. First, we apply a $\pi/2$–pulse on the carrier, followed by a pulse of variable length at a specific detuning, and eventually the Ramsey–cycle is closed with another $\pi/2$–pulse. Thus, without a phase shift, the population is transferred to the excited state, whereas in the presence of light–shifts the phase of the transition is affected and accordingly this results in the observed Ramsey fringes of the D–state population. The frequency of the observed sinusoidal variation is a direct measure of the light shift. The measured light shifts changes sign for various detunings around the qubit transition [59].

The precise knowledge of magnitude and sign of the light shift allows us to compensate for the unwanted Stark shift created by a gate pulse by applying simultaneously an additional off-resonant pulse. For example, for an excitation on the blue sideband ($\nu_z \simeq 1.7$ MHz) the net Stark shift is negative, thus the simultaneous application of an additional off-resonant pulse which leads to the same positive shift is able to compensate the shift of the gate pulses. In the experiment, we implemented this procedure and showed by measurements as in Fig. 5 that light shifts can indeed be reduced to below 2% of their value without compensation. By means of this compensation technique, arbitrary pulse sequences can be concatenated without the necessity to keep track of the changing phase. Note that this procedure is routinely applied for all quantum information processing experiments discussed below.

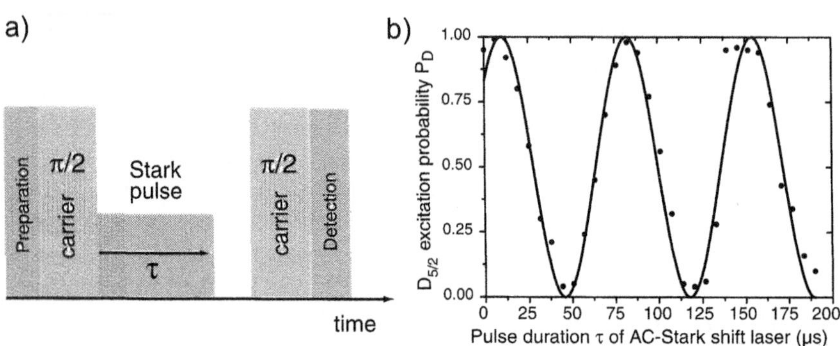

Fig. 5. (a) Ramsey technique to measure light shifts, see text. (b) Observed Ramsey fringes as a function of the applied pulse length. The frequency of the sinusoidally varying signal corresponds to the light shift.

4.2. *Basic gate operations*

For a discussion of the basic gate operations we follow here the more detailed and thorough description of [45]. Although this section considers a single ion coupled to a vibrational mode, the same formalism applies if the ion is part of a string and the mode is a common vibration of the entire crystal if only one ion is optically addressed.

Even a *single* trapped ion can be used as a complete two-qubit quantum computer if we consider the Hilbert space being spanned by its internal levels $|g\rangle$, $|e\rangle$ and a single phonon excitation, i.e. $|n\rangle = |0\rangle, |1\rangle$. A universal set of quantum gates then includes single qubit rotations of the *vibrational* qubit. Such rotations of the vibrational qubit pose certain problems arising from the fact that the mode has many *equidistant* oscillator levels, not just the qubit levels $|n = 0\rangle$ and $|n = 1\rangle$ and the computational space would not be conserved, for example by blue sideband transitions when state $|n = 1\rangle$ is occupied.

Single qubit operations on the *internal* state, on the other hand, can be efficiently performed by carrier pulses. Thus, a natural approach is to swap the internal and vibrational qubit information, then perform the desired operation on the internal qubit and finally swap back. This swap operation would interchange the populations of $|D, 0\rangle$ and $|S, 1\rangle$ and leave the populations of the other two computational basis states ($|S, 0\rangle$ and $|D, 1\rangle$) unaffected. Accordingly, the corresponding matrix in the basis order $\{|S, 0\rangle, |D, 0\rangle, |S, 1\rangle, |D, 1\rangle\}$ reads

$$SWAP = \begin{bmatrix} 1 & 0 & 0 & 0 \\ 0 & 0 & 1 & 0 \\ 0 & 1 & 0 & 0 \\ 0 & 0 & 0 & 1 \end{bmatrix}. \tag{4.1}$$

A red sideband π-pulse $R^-(\pi, \phi)$ would interchange the populations of $|D, 0\rangle$ and $|S, 1\rangle$ as desired (modulo certain phase factors which depend on ϕ). However, such an operation would also (partially) transfer a possible $|D, 1\rangle$ state population (which is supposed to be unaffected) to the $|S, 2\rangle$ state. Again the problem is the conservation of computational space. The task of implementing a $SWAP$ operation, without this problem, can in fact be mastered by the technique of *composite pulses*, which in the framework of NMR have been in use for several years [60,61].

This technique takes advantage of the fact that the Rabi frequency of a sideband transition depends strongly on the motional quantum number, cf. eqn. (3.27). In particular, the $|D, 1\rangle \leftrightarrow |S, 2\rangle$ Rabi frequency is larger by a factor of $\sqrt{2}$ than the Rabi frequency of the $|D, 0\rangle \leftrightarrow |S, 1\rangle$ transition. Using this, it is possible to construct a series of pulses which fulfills the above requirements:

$$R_{swap}(\theta, \phi_0) = R^-(\frac{\pi}{\sqrt{2}}, \phi_0)\; R^-(\frac{2\pi}{\sqrt{2}}, \phi_0 + \phi_{swap})\; R^-(\frac{\pi}{\sqrt{2}}, \phi_0), \tag{4.2}$$

with $\phi_{swap} = \arccos(\cot^2(\frac{\pi}{\sqrt{2}})) \simeq 0.303\pi$. The rotation angle θ here refers to the $|D, 0\rangle \leftrightarrow |S, 1\rangle$ transition. The same light pulse sequence corresponds to the operation

$$R'_{swap}(\theta', \phi_0) = R^-(\pi, \phi_0)\; R^-(2\pi, \phi_0 + \phi_{swap})\; R^-(\pi, \phi_0), \tag{4.3}$$

on $|D, 1\rangle \leftrightarrow |S, 2\rangle$, where the individual rotation angles resulting in θ' are larger by $\sqrt{2}$.

For the swapping pulse sequence, a variation of only the angle ϕ_0 rotates the whole trajectories about the z-axis of the Bloch spheres. This changes nothing in the $|D, 1\rangle \leftrightarrow |S, 2\rangle$ subspace: the overall sequence R'_{swap} always acts like an effective 4π pulse, hence like the identity operation I. In fact, any pulse sequence $R' = R^-(\pi, \phi_1)R^-(2\pi, \phi_2)R^-(\pi, \phi_1)$ with arbitrary ϕ_1 and ϕ_2 meets this requirement. On the $|D, 0\rangle \leftrightarrow |S, 1\rangle$ transition, ϕ_0 plays a similar role as ϕ in a single π-pulse $R^-(\pi, \phi)$. As in that case, no angle ϕ exists which would correspond to a perfect swapping operation. Choosing $\phi_0 = \frac{1}{\sqrt{2}}\pi \simeq 0.707$ in Eq. 4.3 yields for the *SWAP* operation

$$|D, 0\rangle \mapsto -|S, 1\rangle \text{ and } |S, 1\rangle \mapsto |D, 0\rangle\ , \tag{4.4}$$

so that the complete matrix of this operation reads

$$SWAP(\frac{1}{\sqrt{2}}\pi) = \begin{bmatrix} 1 & 0 & 0 & 0 \\ 0 & 0 & 1 & 0 \\ 0 & -1 & 0 & 0 \\ 0 & 0 & 0 & 1 \end{bmatrix}. \tag{4.5}$$

Note that the notation $SWAP(\phi_0)$ for the overall operation as a function of ϕ_0 has been introduced. The unwanted minus sign in $SWAP(\frac{1}{\sqrt{2}}\pi)$ could, in principle, be compensated for by an appropriate series of single qubit rotations added before and after $SWAP(\phi_0)$.

In the complete sequence of an algorithm, the *SWAP* operation will typically appear in pairs. The problem of the unwanted phase factor -1 is solved by swapping the two qubits first with some arbitrary ϕ_0 (for example $SWAP(0)$) and swapping back later with a ϕ_0 different by π ($SWAP(\pi)$). With the help of the composite *SWAP* operation it is possible to perform *all* single qubit rotations within the computational space. Note that the composite SWAP operation can be performed in the very same way by using sideband pulses instead of red sideband pulses with appropriate assignment of logical qubits.

The last missing building block to make our single ion quantum gate toolbox complete is a *universal two-qubit gate* between internal and vibrational qubit. The universal CNOT–gate operation can be realized from a universal controlled-phase gate, i.e. an operation, which changes the sign (or more generally the phase) of the 2nd (target) qubit's wavefunction conditioned on the controlling qubit. Using individual qubit rotations such a phase gate can always be transformed into a CNOT–gate operation [62, 63].

Using composite pulses a phase gate can be implemented which effectively corresponds to a 2π rotation, i.e. $-1 \cdot I$, on the $|S,0\rangle \leftrightarrow |D,1\rangle$ *and* on the $|S,1\rangle \leftrightarrow |D,2\rangle$ transition (despite the difference in Rabi frequency by a factor $\sqrt{2}$). The states $|S,0\rangle$, $|S,1\rangle$ and $|D,1\rangle$ then acquire a -1 sign, while $|D,0\rangle$ remains unaffected if using blue sidebands only. The corresponding matrix reads

$$\Phi' = \begin{bmatrix} 1 & 0 & 0 & 0 \\ 0 & -1 & 0 & 0 \\ 0 & 0 & 1 & 0 \\ 0 & 0 & 0 & 1 \end{bmatrix}. \tag{4.6}$$

Note that a phase factor of -1, which is global in the computational space, has been dropped. A composite pulse sequence which meets these requirements is

$$R_{phase} = R^+(\pi, \frac{\pi}{2})\ R^+(\frac{\pi}{\sqrt{2}}, 0)\ R^+(\pi, \frac{\pi}{2})\ R^+(\frac{\pi}{\sqrt{2}}, 0) \tag{4.7}$$

on the $|S,0\rangle \leftrightarrow |D,1\rangle$ (blue) sideband. On the $|S,1\rangle \leftrightarrow |D,2\rangle$ subspace it reads

$$R'_{phase} = R^+(\sqrt{2}\pi, \frac{\pi}{2})\ R^+(\pi, 0)\ R^+(\sqrt{2}\pi, \frac{\pi}{2})\ R^+(\pi, 0). \tag{4.8}$$

Fig. 6 illustrates the sequences in the Bloch sphere picture. No auxiliary level is required as in the original CZ proposal [26] , which makes the method more universal. The sequence takes a total of 3.4π of sideband rotations.

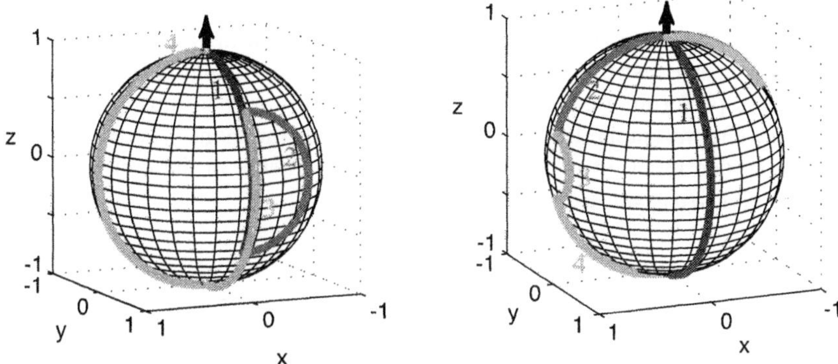

Fig. 6. Bloch sphere trajectories for the composite phase operation Φ'. Left: Bloch sphere of the quasi-two-level-system $|S,0\rangle \leftrightarrow |D,1\rangle$. The initial state is assumed to be $|S,0\rangle$, which is marked by the black arrow. The pulse sequence is R_{phase}. Pulse number 1 of the sequence $(R^+(\frac{\pi}{\sqrt{2}},0))$ rotates the state vector about the x-axis by $\frac{\pi}{\sqrt{2}}$. Pulse number 2 accomplishes a π-rotation about the y-axis. It therefore transforms the state to its mirror image about the x-y-plane. Consequently, laser pulse number 3, which is identical to number 1, rotates the state vector all the way down to the bottom of the sphere. Pulse number 4, just like number 2, represents a π-rotation about the y-axis. The final state thus is identical to the initial one, modulo some phase. It turns out that the acquired phase factor of the overall operation is -1, hence the sequence acts like a single 2π-rotation: $|S,0\rangle \mapsto -|S,0\rangle$ and $|D,1\rangle \mapsto -|D,1\rangle$. Right: The same laser pulse sequence and its action on the $|S,1\rangle \leftrightarrow |D,2\rangle$ subspace. It is easy to understand that the overall sequence brings the state vector back to its starting point (black arrow) modulo some phase. Multiplying the matrices of the individual pulses reveals that the acquired phase is again -1. Thus, R'_{phase} also acts like a single 2π-rotation: $|S,1\rangle \mapsto -|S,1\rangle$ and $|D,2\rangle \mapsto -|D,2\rangle$.

5. Deutsch-Jozsa algorithm

The answer to the question whether a coin is fair (head on one side, tail on the other) or fake (heads or tails on both sides) requires two examinations, i.e. a look on each side. Only one measurement is necessary if the coin is represented by operations of a quantum processor. The associated quantum algorithm is known as the Deutsch–Josza (DJ) algorithm [2, 64].

With the basic gate operations introduced above the composite pulse sequences are applied for an actual computation of the DJ algorithm on a quantum processor based on a single trapped $^{40}Ca^+$ ion driven by laser pulses [65]. Compensating for ac–Stark shifts allows us to achieve the required control over the optical phases of the pulses. With the use of the composite pulse sequence we achieve complete control over the ion's motional and electronic state. The implementation of a quantum algorithm on a single-ion processor serves as a study of the suitability and scalability of these techniques.

The DJ algorithm was experimentally realized first using a bulk nuclear magnetic resonance (NMR) technique [66, 67], encoding quantum bits as nuclear spins and performing ensemble measurements. In order to illustrate the algorithm, we represent the four possible coins by four functions f_i, $(i = 1, 2, 3, 4)$ that map one input bit ($a = 0, 1$ denoting 'which side of the coin') onto one output bit ($f_i(a) = 0, 1$ denoting 'head or tail').

a	$f_1(a)$	$f_2(a)$	$f_3(a)$	$f_4(a)$
0	0	1	0	1
1	0	1	1	0

These functions consist of two constant functions $f_1(a) = 0$, $f_2(a) = 1$, representing the fake coins, and two balanced functions $f_3(a) = \text{ID}\ a$, $f_4(a) = \text{NOT}\ a$, which stand for the fair coins. An unknown function is characterized as constant or balanced by evaluating $f(0) \oplus f(1)$ which yields 0 (or 1) for a constant (or balanced) function ($\oplus$ denotes addition modulo 2). Classically, this evaluation requires two function calls, whereas the DJ quantum algorithm allows one to obtain the desired information with a single evaluation of the unknown f. Fig. 7 shows a circuit diagram which describes the implementation of the Deutsch-Jozsa algorithm using basic quantum operations [62]. The two qubits required for the DJ algorithm are encoded in the electronic state and in the phonon (i.e. vibrational quantum) number of the axial vibration mode of a single trapped ion (cf. Fig. 2). Computational qubit operations are realized by applying laser pulses on the carrier $R(\theta, \phi)$ and on the blue sideband $R^+(\theta, \phi)$ of the electronic quadrupole transition as described above. The rotations $R_y := R(\theta, \pi/2)$

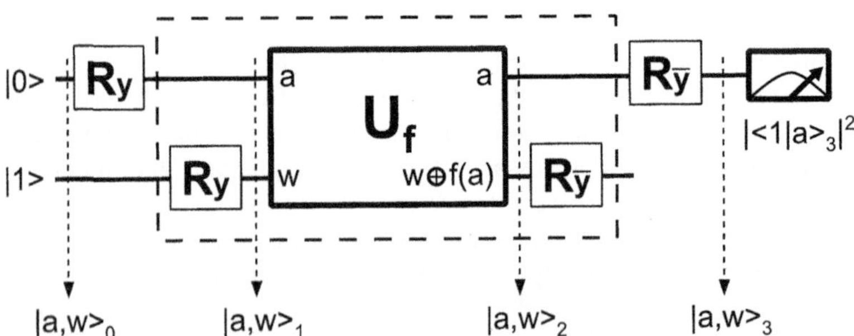

Fig. 7. Quantum circuit for implementing the Deutsch–Jozsa algorithm with basic quantum operations. The upper line shows the input qubit $|a\rangle$ ('which side of the coin' information), the lower line an auxiliary working qubit $|w\rangle$.

Table 1
Functions f_n and their implementation by logical operations. Note that the CNOT–operation corresponds to a gate where the vibrational state being in $|n = 1\rangle$ controls the internal state of the ion, whereas the (zero–controlled) Z–CNOT–operation requires the controlling qubit to be in $|n = 0\rangle$ state to change the target state.

function	Logical operations
f_1	$R_{\overline{y}_w} R_{y_w}$
f_2	$R_{\overline{y}_w}\,\mathsf{SWAP}^{-1}\mathsf{NOT}_a\mathsf{SWAP}\,R_{y_w}$
f_3	$R_{\overline{y}_w}\,\mathsf{CNOT}\,R_{y_w}$
f_4	$R_{\overline{y}_w}\,\mathsf{Z\text{–}CNOT}\,R_{y_w}$

create superpositions $|a\rangle_1 = (|0\rangle + |1\rangle)/\sqrt{2}$ and $|w\rangle_1 = (|0\rangle - |1\rangle)/\sqrt{2}$ from the inputs $|a\rangle_0 = |0\rangle$ and $|w\rangle_0 = |1\rangle$. The box U_{f_n} represents a unitary operation specific to each of the functions f_n, which applies f_n to a and adds the result to w modulo 2. Table 1 lists the logic operations required for transforming $|w\rangle$ into $|w \oplus f_n(a)\rangle$, the corresponding pulse sequences are listed in Table 2. The output of the box is $|a, w\rangle_2 = (|0, w_{in} \oplus f_n(0)\rangle + |1, w_{in} \oplus f_n(1)\rangle)/\sqrt{2}$. Up to an overall sign $|w\rangle$ is left unchanged, but the positive superposition $(|0\rangle + |1\rangle)/\sqrt{2}$ on $|a\rangle$ is transformed into a negative superposition $|a\rangle_2 = (|0\rangle - |1\rangle)/\sqrt{2}$ if f is balanced, otherwise it is unchanged. After the final rotations $R_{\overline{y}}$, a measurement on $|a\rangle$ is performed with result $|a\rangle_3$ = either $|0\rangle$ or $|1\rangle$. Because of the sign change in $|a\rangle_2$ if f is balanced, $|\langle 1|a\rangle_3|^2 = f_n(0) \oplus f_n(1)$, that is, $|a\rangle_3$ yields the desired information whether the function f_n is balanced or constant. The working qubit w resumes its initial value $|w\rangle_3 = |w\rangle_0 = |1\rangle$. For the experiments Ca^+ ions were loaded into a linear Paul trap with an axial frequency $\omega_z = 2\pi \cdot 1.7$MHz [65]. Each experimental cycle starts with Doppler cooling followed by sideband cooling and optical pumping to the $S_{1/2}(m_J = -1/2)$ state. The fidelity of the implemented algorithm is measured by repeating several thousand times the experimental sequence of cooling, initializing both qubits, applying laser pulses for the algorithm and the final measurement. For cases 1, 3 and 4, the fidelity of identifying the function's class with a single measurement exceeds 97%; for case 2, it is above 90%. Note that for the decision whether the function is constant or balanced, only $|\langle 1|a\rangle|^2$ at the end of the algorithm needs to be measured. It was verified that the working qubit $|w\rangle$ is reset to its initial value by reading out the phonon number through a measurement of the Rabi frequency of the blue sideband transition [49, 68]. Due to the experimental errors the measured output of the algorithm slightly deviates from the ideal result. As major sources for this infidelity decoherence of the laser–atom phase was identified, in particular caused

Table 2

Functions f_n and their implementation by laser pulses on the carrier and upper sideband transitions $R(\theta, \phi)$ and $R^+(\theta, \phi)$ according to Eqs. 3.26 and 3.34.

function	Laser pulse sequences
f_1	-
f_2	$R^+(\frac{\pi}{\sqrt{2}}, 0)R^+(\frac{2\pi}{\sqrt{2}}, \phi_{\mathrm{SWAP}})R^+(\frac{\pi}{\sqrt{2}}, 0)$ $R(\frac{\pi}{2}, 0)R(\pi, \frac{\pi}{2})R(\frac{\pi}{2}, \pi)$ $R^+(\frac{\pi}{\sqrt{2}}, \pi)R^+(\frac{2\pi}{\sqrt{2}}, \pi + \phi_{\mathrm{SWAP}})R^+(\frac{\pi}{\sqrt{2}}, \pi)$
f_3	$R^+(\frac{\pi}{\sqrt{2}}, 0)R^+(\pi, \frac{\pi}{2})R^+(\frac{\pi}{\sqrt{2}}, 0)R^+(\pi, \frac{\pi}{2})$
f_4	$R(\pi, 0)R^+(\frac{\pi}{\sqrt{2}}, 0)R^+(\pi, \frac{\pi}{2})R^+(\frac{\pi}{\sqrt{2}}, 0)R^+(\pi, \frac{\pi}{2})R(\pi, 0)$

by ambient magnetic field fluctuations [69]. Moreover, in the implementation of case 2, which requires the most complex pulse sequence, higher laser power of the sideband transitions was used in order to speed up the algorithm and thus reduce the sensitivity to phase decoherence. This in turn caused off-resonant carrier excitation which limited the obtainable fidelity. The ability to trace the evolution of $|\langle 1|a\rangle|^2$ during the quantum algorithm is a major advantage of our state detection technique. For this, the pulse sequence is truncated at a certain time t and $|\langle 1|a(t)\rangle|^2$ is obtained by measuring the probability of finding the ion in the $D_{5/2}$ state. In Fig. 8 this probability is displayed as a function of time for all four cases. The data agree very well with the calculated ideal evolution (solid lines in Fig. 8), demonstrating the high precision of the applied pulse sequence, and especially the control over the optical phases. Note that the solid lines in Fig. 8 show the calculated result without the need for fit parameters.

6. Cirac-Zoller CNOT–gate operation

For the realization of the CZ CNOT–gate operation, two ions are loaded to the linear trap and by means of an intensified CCD camera, the fluorescence is monitored separately for each ion [58]. If no information on a particular qubit is needed, the signal of a photomultiplier tube is used to infer the overall state population. In this case, the exposure time can be reduced to 3.5 ms.

For the two-qubit CNOT operation, Cirac and Zoller proposed to use the common vibration of an ion string to convey the information for a conditional operation (bus-mode) [26]. Accordingly, the gate operation can be achieved with a sequence of three steps after the ion string has been prepared in the ground state $|n_b = 0\rangle$ of the bus-mode. First, the quantum information of the control ion is

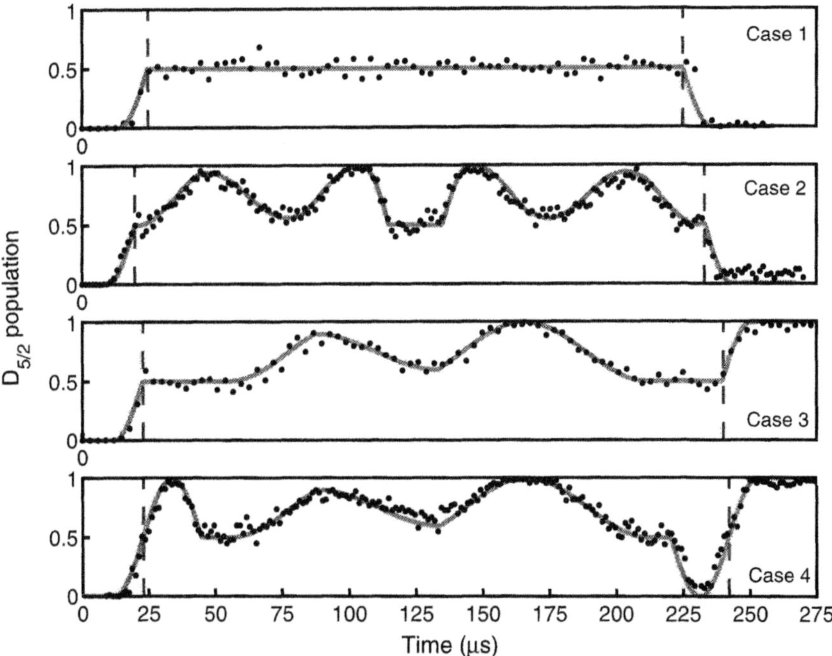

Fig. 8. Deutsch-Josza algorithm: Time evolution of $|\langle 1|a\rangle|^2$. Points are the probabilities, each inferred from 100 measurements, the line shows the ideal evolution. No parameters were adjusted to fit the data. The implementation of the functions $R_{\overline{y}_W} U_{f_n} R_{y_W}$ takes place between the dashed lines. An initial R_a and a final $\overline{R}_y$ rotation on $|a\rangle$, implemented by carrier pulses, complete the algorithm. Taking case 3 as an the example, R_{y_a} lasts from $12\mu s$ to $22\mu s$. Then $R_{\overline{y}_W} U_{f_n} R_{y_W}$ on $|a, w\rangle$ is implemented from $54\mu s$ to $212\mu s$ with the laser tuned to the blue sideband. The laser phase is switched at 87, 133 and 166 μs. The final $R_{\overline{y}_a}$ pulse is applied from 240 to 250 μs.

mapped onto this vibrational mode, the entire string of ions is moving and thus the target ion participates in the common motion. Second, and conditional upon the motional state, the target ion's qubit is inverted [71]. Finally, the state of the bus-mode is mapped back onto the control ion. Note, that this gate operation is not restricted to a two-ion crystal since the vibrational bus mode can be used to interconnect any of the ions in a large crystal, independent of their position.

We realize this gate operation [70] with a sequence of laser pulses. A blue sideband π-pulse, $R^+(\pi, 0)$, on the control ion transfers its quantum state to the bus-mode. Next, we apply the CNOT–gate operation

$$R_{\text{CNOT}} = R(\pi/2, 0)\, R_{phase}\, R(\pi/2, \pi) \tag{6.1}$$

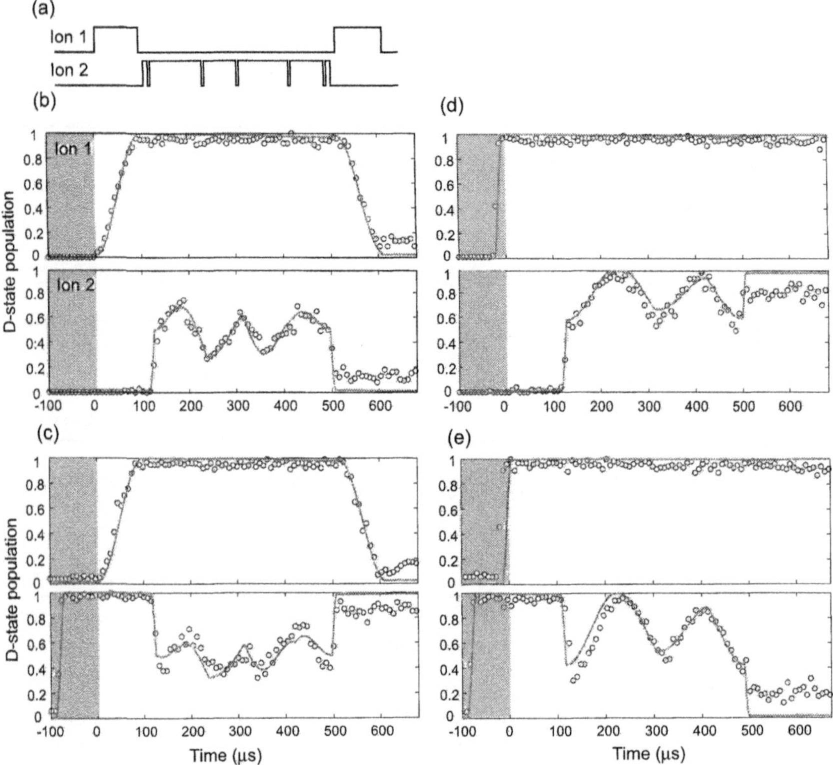

Fig. 9. State evolution of both qubits $|control, target\rangle = |ion\,1, ion\,2\rangle$ under the CNOT operation. (a) CNOT–gate pulse sequence. Evolution of the D-state population after he quantum register is initialized in the states (b) $|S, S\rangle$, (c) $|S, D\rangle$, (d) $|D, S\rangle$, or (e)$|D, D\rangle$ (shaded area, $t \leq 0$). The solid lines indicate the theoretically expected behavior.

to the target ion. Finally, the bus-mode and the control ion are reset to their initial states by another π-pulse $R^+(\pi, \pi)$ on the blue sideband. We apply the gate to all computational basis states and follow their temporal evolution, see Fig. 9. After initialization, the quantum gate pulse sequence (a) is applied: After mapping the first ion's state (control qubit) with a π-pulse of length 95 μs to the bus-mode, the single-ion CNOT sequence (consisting of 6 concatenated pulses) is applied to the second ion (target qubit) for a total time of 380 μs. Finally, the control qubit and bus mode are reset to their initial values with the concluding π-pulse applied to the first ion. To follow the temporal evolution of both qubits during the gate, the pulse sequence (a) is truncated and the $|D\rangle$ state probability

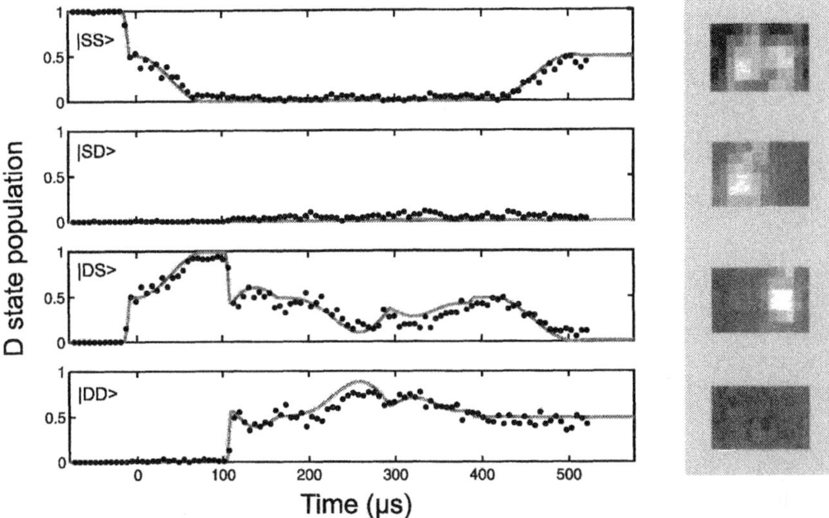

Fig. 10. Left: the controlled-NOT gate operation R_{CNOT} is performed with ions initially prepared in $|S+D, S\rangle$. The data points represent the probability for the ion string to be in the state indicated on the right-hand side by the corresponding CCD image, during the execution of the gate. The measurement procedure is the same as in Fig. 9. Right: CCD images of the fluorescence of the two-ion crystal as measured in different logic basis states: $|SS\rangle$, $|SD\rangle$, $|DS\rangle$, and $|DD\rangle$. The ion distance is 5.3 μm.

is measured as a function of time. Input parameters for the calculated evolution are the independently measured Rabi frequencies on the carrier and sideband transitions and the addressing error. The desired output is reached with a fidelity of 71 to 77%.

If the qubits are initialized in the superposition state $|\text{control, target}\rangle = |S + D, S\rangle$, the CNOT operation generates an entangled state $|S, S\rangle + |D, D\rangle$. The corresponding data are plotted in Fig. 10, left side. At the end of the sequence, near t=500 μs, only the states $|S, S\rangle$ and $|D, D\rangle$ are observed with P_{SS}=0.42(3) and P_{DD}=0.45(3). The phase coherence of both these components is verified by applying additional analysis $\pi/2$-pulses on the carrier transition followed by the projective measurement. From the observed populations prior to the analyzing pulses and the contrast of the oscillation, see Fig. 11, we calculate the fidelity according to the prescription given in Sackett et al. [72], and find a gate fidelity of 0.71(3) [70]. The gate fidelity is well understood in terms of a collection of experimental imperfections [70]. Most important is dephasing due to laser frequency noise and due to ambient magnetic field fluctuations that cause a Zeeman shift of the qubit levels [73]. As quantum computing might be understood as

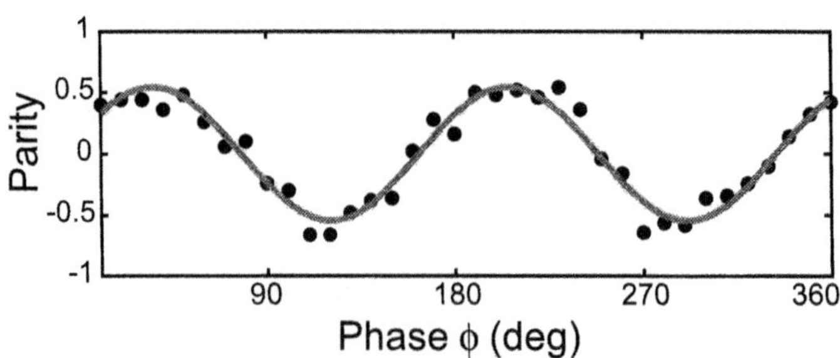

Fig. 11. Analysis of the entangled output state of a CNOT: After the gate operation, we apply $\pi/2$-pulses on the carrier transition, with a phase ϕ, to both ions, and measure the parity $P = P_{SS} + P_{DD} - (P_{SD} + P_{DS})$ as a function of the phase. The quantum nature of the gate operation is proved by observing oscillations with $\cos 2\phi$, whereas a non-entangled state would yield a variation with $\cos\phi$ only. The observed visibility is 0.54(3).

a multi-particle Ramsey interference experiment, a faster execution of the gate operation would help to overcome this type of dephasing errors. However, a different type of error increases with the gate speed: With higher Rabi frequencies, the off-resonant excitation of the nearby and strong carrier transition becomes increasingly important [74] even if the corresponding phase shift is compensated. Additional but minor errors are due to the addressing imperfection, residual thermal excitation of the bus mode and spectator modes as well as laser intensity fluctuations. Higher trap frequencies and the use of hyperfine ground states (in $^{43}Ca^{+}$ ions) coupled by Raman transitions for the qubit will improve the gate fidelity in future experiments.

7. Entanglement and Bell state generation

Quantum state tomography allows the estimation of an unknown quantum state that is available in many identical copies. It has been experimentally demonstrated for a variety of physical systems, among them the quantum state of a light mode [75], the vibrational state of a single ion [76], and the wave packets of atoms of an atomic beam [77]. Multi-particle states have been investigated in nuclear magnetic resonance experiments [78] as well as in experiments involving entangled photon pairs. Here, we describe how this technique is applied for the first time to entangled massive particles on a quantum register for quantum state tomography and studies of coherence [C.F. Ross, G.P.T. Lancaster, M. Riebe, H. Häfner, W. Hänsel, S. Gulde, C. Becher, J. Eschner, F. Schmidt-Kaler, and R. Blatt. Tomography of entangled massive particles. *Phys. Rev. Lett.*, (2004),

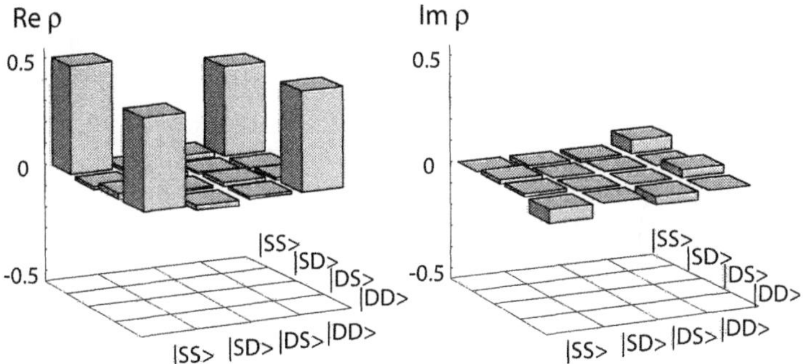

Fig. 12. Real and imaginary part of the density matrix ρ_{Φ_+} that approximates $1/\sqrt{2}(|S, S + D, D\rangle)$. The measured fidelity is F= $\langle\Phi_+|\rho_{\Phi_+}|\Phi_+\rangle = 0.91$.

in press; *quant-ph/0307210*]. In future, this tool may be applied for process tomography, e.g. to study the performance of quantum logic gates [70–72, 79] and algorithms [65].

In the Innsbruck experiment, we first create a Bell state by applying laser pulses to ion 1 and 2 on the blue sideband and on the carrier transition. To produce the Bell state $\Psi_\pm = 1/\sqrt{2}(|S, D\rangle \pm |D, S\rangle)$ we use the pulse sequence $U_{\Psi_\pm} = R_2^+(\pi, \pm\pi/2) \cdot R_2(\pi, \pi/2) \cdot R_1^+(\pi/2, -\pi/2)$ applied to the $|S, S\rangle$ state. Here, the indices 1 (2) refer to pulses applied to ion 1 and 2, respectively. The first pulse $R_1^+(\pi/2, -\pi/2)$ entangles the motional and the internal degrees of freedom. The next two pulses $R_2^+(\pm\pi, \pi/2) \cdot R_2(\pi, \pi/2)$ map the motional degree of freedom onto the internal state of ion 2. Appending another π-pulse on the carrier transition frequency, $R_2(\pi, 0)$, to the sequence $U_{\Psi_\pm}$ produces the state $\Phi_\pm$. This pulse sequence takes less than 200 μs.

The tomographic method consists of individual single-qubit rotations, followed by a projective measurement. For the analysis of the data, we employ a maximum likelihood estimation of the density matrix following the procedure as suggested in [80, 81] and implemented in experiments with pairs of entangled photons [82]. As an example, Fig. 12 shows the reconstructed density matrix ρ of one out of four Bell states. To monitor the evolution of these entangled states in time we introduce a waiting interval before the state tomography. We expect that Bell states of the type $\Psi_\beta = |S, D\rangle + e^{i\beta}|D, S\rangle$ are immune against collective dephasing due to fluctuations of the qubit energy levels or the laser frequency [83]. However, a magnetic field gradient that gives rise to different Zeeman shifts on qubits 1 and 2 leads to a deterministic and linear time evolution of the relative phase $e^{i\beta}$ between the $|S, D\rangle$ and the $|D, S\rangle$ component of the $\Psi_\pm$ states. We plot

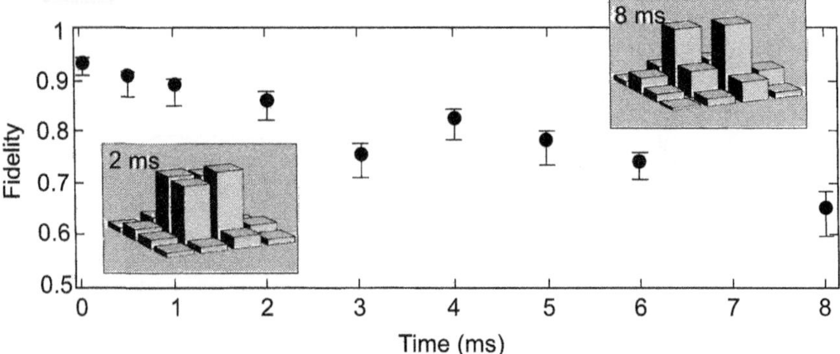

Fig. 13. Decay of the measured fidelity is $F_m = max\{\langle\Psi_\beta|\rho_{\Psi_\beta}|\Psi_\beta\rangle\}$. Inset: Real part of the density matrix after 2 ms and 8 ms.

the maximum overlap $F_m = max\{\langle\Psi_\beta|\rho_{\Psi_\beta}|\Psi_\beta\rangle\}$ for the reconstructed density matrices as a function of time, see Fig. 13, which exhibits a coherence time of a few ms. In some measurements, the lifetime of the $\Psi_\pm$ states exceeds 20 ms. Finally, we specify the entanglement of the four Bell states, using the entanglement of formation [84], and find $E(\Psi_-)=0.79(4)$, $E(\Psi_+)=0.75(5)$, $E(\Phi_+)=0.76(4)$ and $E(\Phi_-)=0.72(5)$.

8. Summary and perspectives

On the road towards a scalable quantum processor [85] with ion traps, single-qubit rotations and a universal two-qubit operations gate have been realized. With trapped Ca^+ ions, we present an experimental setup which allows one to flexibly control a register of two qubits. With the universal set of quantum gates all unitary operations can be implemented. Therefore, arbitrary two-qubit states can be synthesized with high fidelities and analyzed via state tomography. The currently available experiments demonstrate the operation of a small quantum computer and allow one to fathom the basics of experimental quantum information processing.

While the number of qubits and the complexity of these operations is far from what will be eventually needed for competitive factoring, such small-scale quantum computers already offer prospects for several applications. The precision of spectroscopic measurements is usually limited by the available signal-to-noise ratio (SNR) which scales as $\sqrt{N}$ with a sample of N particles. With entangled states, as available in such quantum registers, the SNR scales as N which has immediate impact for example with time and frequency standard application [72]. Moreover, the sensitivity of entangled states with respect to external perturbations may be advantageously exploited to detect and measure decoherence sources. Quantum communication over large distances requires the repetition of the quantum signal using so-called quantum repeaters which, in turn, can be realized even with few qubits [86].

One of the most striking features is that the ion-trap quantum information processor is scalable in principle, i.e. adding more qubits is straightforward and at least up to about 10 qubits this should not pose insurmountable technical difficulties. Larger systems will require special architectures such as ion trap arrays [87], moving ions in structured ion traps [88] or even interconnecting several small ion-trap quantum computers using cavities and photons as a quantum channel [89,90]. While all these techniques require tremendous technical efforts, to the best of current knowledge there are no principal limitations to scaling up an ion-trap quantum computer.

Currently, work is in progress in several laboratories, especially in the group of D. Wineland at NIST (Boulder, Colorado) with Be^+, in the groups of R. Blatt (Innsbruck, Austria) and A. Steane (Oxford, UK) with $^{40}Ca^+$ and $^{43}Ca^+$ and in the group of C. Monroe (Ann Arbor, Michigan) with $^{111}Cd^+$ ions in order to implement so-called logical gates with physical ion-qubits (i.e. individual ions). Encoding quantum information in logical qubits composed of three and eventually five ions will allow one to use error-correction codes for stable quantum information processing [91–96].

The current experiments demonstrate that ion-trap quantum information processors constitute a viable way towards the realization of large-scale quantum computing and they provide ideal means for the engineering of quantum objects and controlling quantum processes at mesoscopic and macroscopic scales.

Acknowledgements

We thank Stephan Gulde, Mark Riebe, Gavin Lancaster, and Wolfgang Hänsel for their continuous support in the laboratory. We gratefully acknowledge support by the European Commission (QUEST, QUBITS and QGATES networks), by the Austrian Science Fund (FWF), and by the Institut für Quanteninformation GmbH. H. H. acknowledges support by the Marie-Curie program of the European Commission.

References

[1] R. P. Feynman. Simulating physics with computers. *Int. J. Theor. Phys.* 21, 467 (1982).

[2] D. Deutsch. Quantum theory,the Church-Turing principle and the universal quantumcomputer. *Proc. R. Soc. Lond. A* 400, 97 (1985).

[3] P. Shor, in Proceedings, *35th Annual Symposium on Foundations of Computer Science*, IEEE Press, Los Alamitos, CA 1994; P. Shor, SIAM J. Comp. 26, 1484 (1997); see also A. Ekert and R. Josza, Rev. Mod. Phys. 68(3), (1996).

[4] L. K. Grover. Quantum mechanics helps in searching for a needle in a haystack. *Phys. Rev. Lett.* 79, 325 (1997).

[5] D. P. DiVincenzo. Dogma and heresy in quantum computing. *Quantum Information and Computation* 1, Special 1-6 (2001).

[6] D. J. Wineland. Quantum information processing in ion traps II. *See lecture notes in this volume.*

[7] A. Steane. The ion trap quantum information processor. *Appl. Phys. B* 64, 623 (1997).

[8] D.J. Wineland. C. Monroe, W.M. Itano, D. Leibfreid, B.E. King, D.M. Meekhof, C. Myatt, and C. Wood. Experimental issues in coherent quantum-state manipulation of trapped atomic ions. *J. Res. Nat. Inst. Stand. Tech.* 1003, 259 (1998).

[9] D.J. Wineland, C. Monroe, W.M. Itano, B.E. King, D. Leibfried, D.M. Meekhof,C. Myatt, and C. Wood. Experimental primer on the trapped ion quantum computer. *Fortschritte der Physik* 46, 363 (1998).

[10] D. F. V. James. Quantum dynamics of cold trapped ions with application to quantum computation. *Appl. Phys. B* 66 181-190 (1998).

[11] H. C. Nägerl, F. Schmidt-Kaler, J. Eschner, R. Blatt, W. Lange, H. Baldauf, H. Walther. Linear Ion Traps for Quantum Computation. In D. Bouwmeester, A. Ekert, A. Zeilinger, editors, *The Physics of Quantum Information.* Springer, Berlin, 2000

[12] J-F. Poyatos, J.I. Cirac, and P. Zoller. Schemes of quantum computation with trapped ions. In S.L. Braunstein, H.-K. Lo, and P. Kok, edirors, Scalable quantum computers, 15, Wiley-VCH, Berlin, 2001.

[13] G.J. Milburn, S. Schneider, and D.F.V. James. Ion trap quantum computing with warm ions. In S.L. Braunstein, H.-K. Lo, and P. Kok, edirors, Scalable quantum computers, 31, Wiley-VCH, Berlin, 2001.

[14] A. Sørensen and K. Mølmer. Ion trap quantum computer with bichromatic light. In S.L. Braunstein, H.-K. Lo, and P. Kok, edirors, Scalable quantum computers, 41, Wiley-VCH, Berlin, 2001.

[15] D.F.V. James. Quantum computation with hot and cold ions: an assessment of proposed schemes. In S.L. Braunstein, H.-K. Lo, and P. Kok, edirors, Scalable quantum computers, 53, Wiley-VCH, Berlin, 2001.

[16] A.M. Steane and D.M. Lucas. Quantum computing with trapped ions, atoms and light. In S.L. Braunstein, H.-K. Lo, and P. Kok, edirors, Scalable quantum computers, 69, Wiley-VCH, Berlin, 2001.

[17] C. Monroe. Quantum information processing with atoms and photons. Nature 416, 238 (2002).

[18] D. J. Wineland. Trapped ions and quantum information processing. In F. De Martini and C. Monroe, editors, *Experimental Quantum Computation and Information, Proc. Int. School of Physics "Enrico Fermi"*, course CXLVIII, 165-196, IOS Press, Amsterdam 2002.

[19] F. Schmidt-Kaler, D-Leibfried, J. Eschner, and R. Blatt. Towards quantum computation with trapped Ca^+ ions. In F. De Martini and C. Monroe, editors, *Experimental Quantum Computation and Information, Proc. Int. School of Physics "Enrico Fermi"*, course CXLVIII, 197-213, IOS Press, Amsterdam 2002.

[20] D.J. Wineland, M. Barrett, J. Britton, J. Chaverini, B. DeMarco, W.M. Itano, B. Jelenković, C. Langer, D. Leibfried, V. Meyer, T. Rosenband, and T. Schätz. Quantum information processing with trapped ions. Phil. Trans. R. Soc. Lond. A 361, 1349 (2003).

[21] S. Gulde, H. Häffner, M. Riebe, G. Lancaster, A. Mundt, A. Kreuter, C. Russo, C. Becher, J. Eschner, F. Schmidt-Kaler, I. L. Chuang, and R. Blatt. Quantum information processing with trapped Ca^+ ions. Phil. Trans. R. Soc. Lond. A 361, 1 (2003).

[22] D.J. Wineland, D. Leibfried, B. DeMarco, V. Meyer, M. Rowe, A. Ben-Kish, M. Barrett, J. Britton, J. Hughes, W.M. Itano, B. M. Jelenković, C. langer, D. Lucas, and T. Rosenband. Quantum information processing and multiplexing with trapped ions. In H. R. Sadeghpour, E.J. Heller, and D.E. Pritchard, editors, *Proc. of the XVIII International Conference on Atomic Physics* p. 263, World Scientific, Singapore, 2003.

[23] S. Gulde, H. Häffner, M. Riebe, G. Lancaster, A. Mundt, A. Kreuter, C. Russo, C. Becher, J. Eschner, F. Schmidt-Kaler, I. L. Chuang, and R. Blatt. Quantum information processing and cavity QED experiments with trapped Ca+ ions. In H. R. Sadeghpour, E.J. Heller, and D.E. Pritchard, editors, *Proc. of the XVIII International Conference on Atomic Physics*, p. 293, World Scientific, Singapore, 2003.

[24] M. Šašura and V. Bužek. Cold trapped ions as quantum information processors. *Journ. of Mod. Opt.* 49, 1593 (2002).

[25] C. Wunderlich and C. Balzer. Quantum measurements and new concepts for experiments with trapped ions. *quant-ph/0305129*, 2003.

[26] J. I. Cirac, P. Zoller. Quantum computations with cold trapped ions. *Phys. Rev. Lett.* 74, 4091 (1995).

[27] R. Blatt, in *Atomic Physics 14*, Proceedings on 14th International Conference on Atomic Physics, Editors: D. J. Wineland, C. E. Wieman, S. J. Smith, p. 219, AIP Press, New York 1995.

[28] I. Waki, S. Kassner, G. Birkl, H. Walther, Phys. Rev. Lett. **68**, 2007 (1992); G. Birkl, S. Kassner, H. Walther, Nature 357, 310 (1992).

[29] M.G. Raizen et al. Ionic crystals in a linear Paul trap. *Phys. Rev. A* 45, 6493 (1992)

[30] D. P. DiVincenzo. Two-bit gates are universal for quantum computation. *Phys. Rev. A* 51, 1015 (1995).

[31] A. Sørensen and K. Mølmer. Quantum Computation with Ions in Thermal Motion. *Phys. Rev. Lett.* 82, 1971 (1999);K. Mølmer and A. Sørensen. Multiparticle Entanglement of Hot Trapped Ions. *Phys. Rev. Lett.* 82, 1835 (1999).

[32] K. Mølmer and A. Sørensen. Entanglement and quantum computation with ions in thermal motion. *Phys. Rev.* A 62, 022311 (2000).

[33] G. J. Milburn. Simulating nonlinear spin models in an ion trap. *quant-ph/9908037*

[34] L. M. Duan, I. Cirac, and P. Zoller. Geometric manipulation of trapped ions for quantum computation. *Science* 292, 1695 (2001).

[35] D. Jonathan, M. B. Plenio, and P. Knight. Fast quantum gates for cold trapped ions. *Phys. Rev. A* 62, 042307 (2000), D. Jonathan and M. B. Plenio. Light-shift-induced quantum gates for ions in thermal motion. *Phys. Rev. Lett.* 87, 127901 (2001).

[36] D. J. Berkeland, J. D. Miller, J. C. Bergquist, W. M. Itano, and D. J. Wineland. Minimization of ion micromotion in a Paul trap. *J. Appl. Phys* 83, 5025 (1998).

[37] see articles in: "The Physics of Quantum Information", ed. by D. Bouwmeester, A. Ekert, and A. Zeilinger, Springer–Verlag, Berlin 2000

[38] J. Prestage, G. J. Dick, and L. Maleki. New ion trap for frequency standard applications. *J. Appl. Phys.* 66, 1013 (1989), G. R. Janik, and L. Maleki. Simple analytic potentials for linear ion traps. *J. Appl. Phys.* 67, 6050 (1990), M. G. Raizen, J. M. Gilligan, J. C. Bergquist, W. M. Itano, and D. J. Wineland. Linear trap for high-accuracy spectroscopy of stored ions. *J. Mod. Optics* 39, 233 (1992).

[39] P. K. Gosh, *Ion traps*, Clarendon, Oxford 1995.

[40] D. Leibfried, R. Blatt, C. Monroe, and D. Wineland. Quantum dynamics of single trapped ions. *Rev. Mod. Phys. 75, 281 (2003).*

[41] D.G. Enzer, M.M. Schauer, J.J. Gomez, M.S. Gulley, M.H. Holzscheiter, P.G. Kwiat, S.K. Lamoreaux, C.G. Peterson, V.D. Sandberg, D. Tupa, A.G. White, R.J. Hughes, and D.F.V. James. Observation of Power-Law Scaling for Phase Transitions in Linear Trapped Ion Crystals. *Phys. Rev. Lett.* 85, 2466, (2000).

[42] C. A. Blockley, D. F. Walls, and H. Risken, Quantum collapse and revivals in a quantized trap. *Europhys. Lett.* 17, 509 (1992).

[43] J. I. Cirac, A. S. Parkins, R. Blatt, and P. Zoller. Non–classcial states of motion in ion traps. *Adv. At. Molec. and Opt. Phys.* 37, 237 (1996).

[44] C. F. Roos. *Controlling the quantum state of trapped ions*. Dissertation, Innsbruck 2000, available from http:\\heart-c704.uibk.ac.at.

[45] S. T. Gulde. *Experimental realization of quantum gates and the Deutsch–Jozsa algorithm with trapped* $^{40}Ca^{+}$ *ions*. Dissertation, Innsbruck 2003, available from http:\\heart-c704.uibk.ac.at.

[46] W. Nagourney, J. Sandberg, and H. Dehmelt. Shelved optical electron amplifier: Observation of quantum jumps. *Phys. Rev. Lett.* 56, 2797 (1986); Th. Sauter, W. Neuhauser, R. Blatt, and P. E. Toschek. Observation of Quantum Jumps. *Phys. Rev. Lett.* 57, 1696 (1986); J. C. Bergquist, R. Hulet, W. M. Itano, and D. J. Wineland. Observation of Quantum Jumps in a Single Atom. *Phys. Rev. Lett.* 57, 1699 (1986).

[47] J. Eschner, G. Morigi, F. Schmidt-Kaler, R. Blatt. Laser cooling of trapped ions. *J. Opt. Soc. Am.* B 20, 1003-1015 (2003).

[48] F. Diedrich, J. C. Bergquist, W. M. Itano, and D. J. Wineland. Laser cooling to the zero-point energy of motion. Phys. Rev. Lett. 62, 403 (1989).

[49] Ch. Roos, Th. Zeiger, H. Rohde, H. C. Nägerl, J. Eschner, D. Leibfried, F. Schmidt-Kaler, and R. Blatt. Quantum State Engineering on an Optical Transition and Decoherence in a Paul Trap. *Phys. Rev. Lett.* 83, 4713 (1999).

[50] C. Monroe, D. M. Meekhof, B. E. King, S. R. Jefferts, W. M. Itano, D. J. wineland, and P. Gould. Resolved-sideband Raman cooling of a bound atom to the 3D zero-point energy. it Phys. Rev. lett. 75, 4011 (1995).

[51] I. Marzoli, J. I. Cirac, R. Blatt, and P. Zoller. Laser cooling of trapped three-level ions: Dsigning two-level systems for sideband cooling. Phys. Rev. A 49, 2771 (1994).

[52] G. Morigi, J. Eschner, and C. H. Keitel. Ground State Laser Cooling Using Electromagnetically Induced Transparency. *Phys. Rev. Lett.* 85, 4458-4461 (2000).

[53] C. F. Roos, D. Leibfried, A. Mundt, F. Schmidt-Kaler, J. Eschner, R. Blatt. Experimental demonstration of ground state laser cooling with electromagnetically induced transparency. *Phys. Rev. Lett.* 85, 5547 (2000),

[54] D. Kielpinski, B. E. King, C. J. Myatt, C. A. Sackett, Q. A. Turchette, W. M. Itano, C. Monroe, and D. J. Wineland. Sympathetic cooling of trapped ions for quantum logic. *Phys. Rev. A* 61, 032310 (2000).

[55] H. Rohde, S. T. Gulde, C. F. Roos, P. A. Barton, D. Leibfried, J. Eschner, F. Schmidt-Kaler, R.Blatt, Sympathetic ground state cooling and coherent manipulation with two-ion-crystals. *J. Opt.* B 3, S34 (2001).

[56] B. B. Blinov, L. Deslauriers, P. Lee, M. J. Madsen, R. Miller, C. Monroe. Sympathetic cooling of trapped Cd^{+} isotopes. *Phys. Rev. A* 65, 040304(R) (2002)

[57] H. C. Nägerl, D. Leibfried, H. Rohde, G. Thalhammer, J. Eschner, F. Schmidt-Kaler, and R. Blatt. Laseraddressing of individual ions in a linear ion trap. *Phys. Rev. A* 60, 145 (1999).

[58] F. Schmidt-Kaler, H. Häffner, S. Gulde, M. Riebe, G.P.T. Lancaster, T. Deuschle, C. Becher, W. Hänsel, J. Eschner, C.F. Roos, and R. Blatt. How t o realize a universal quantum gate with trapped ions. *Appl. Phys. B* 77, 789 (2003).

[59] H. Häffner, S. Gulde, M. Riebe, G. Lancaster, C. Becher, J. Eschner, F. Schmidt-Kaler, and R. Blatt. Precision measurement and compensation of optical Stark shifts for an ion-trap quantum processor. *Phys. Rev. Lett.* 90, 143602 (2003).

[60] M.H. Levitt. Composite pulses. *Prog. NMR Spectroscopy* 18, 61 (1986).

[61] A.M. Childs and I.L. Chuang. Universal quantum computation with two-level trapped ions. *Phys. Rev. A* 63, 012306 (2001).

[62] M.A. Nielsen and I.L. Chuang. *Quantum computation and quantum information.* Cambridge University Press, Cambridge 2000.

[63] I.L. Chuang. *See lecture notes in this volume.*

[64] D. Deutsch and R. Jozsa. Rapid solution of problems by quantum computation. *Proc. R. Soc. London A* 439, 553 (1992).

[65] S. Gulde, M. Riebe, G. P. T. Lancaster, C. Becher, J. Eschner, H. Hf̈fner, F. Schmidt-Kaler, I. L. Chuang, and R. Blatt. Implementation of the Deutsch Josza algorithm on an ion-trap quantum computer. *Nature* 421, 48 (2003).

[66] I.L. Chuang, L.M.K. Vandersypen, X. Zhou, D.W. Leung, and S. Lloyd. Experimental realization of a quantum algorithm. *Nature* 393, 143 (1998).

[67] T.F. Jones and M. Mosca. Implementation of a quantum algorithm to solve Deutsch's problem on a nuclear magnetic resonance quantum computer. *J. Chem. Phys.* 109, 1648 (1998).

[68] D. M. Meekhof, C. Monroe, B. E. King, W. M. Itano and D. J. Wineland. Generation of Nonclassical Motional States of a Trapped Atom. *Phys. Rev. Lett.* 76, 1796 (1996).

[69] F. Schmidt-Kaler, S. Gulde, M. Riebe, T. Deuschle, A. Kreuter, G. Lancaster, C. Becher, J. Eschner, H. Häffner, R. Blatt. The coherence of qubits based on single Ca^+ ions. *J. Phys. B: At. Mol. Opt. Phys.* 36, 623 (2003).

[70] F. Schmidt-Kaler, H. Häffner, M. Riebe, S. Gulde, G. P. T. Lancaster, T. Deuschle, C. Becher, C. F. Roos, J. Eschner, and R. Blatt. Realization of the Cirac-Zoller controlled-NOT gate. *Nature* 422, 408 (2003).

[71] C. Monroe, D. M. Meekhof, B. E. King, W. M. Itano and D. J. Wineland. Demonstration of a fundamental quantum logic gate. *Phys. Rev. Lett.* 75, 4714 (1995).

[72] C.A. Sackett, D. Kielpinski, B. E. King, C. Langer, V. Meyer, C. J. Myatt, M. Rowe, Q. A. Turchette, W. M. Itano, D. J. Wineland, and C. Monroe, Experimental entanglement of four particles. *Nature* 404, 256 (2000).

[73] F. Schmidt-Kaler, H. Häffner, S. Gulde, M. Riebe, G. P. T. Lancaster, T. Deuschle, C. Becher, W. Hänsel, J. Eschner, C. F. Roos, and R. Blatt. How to realize a universal quantum gate with trapped ions. *Appl. Phys. B* 77, 789 (2003).

[74] A. Steane, C. F. Roos, D. Stevens, A. Mundt, D. Leibfried, F. Schmidt-Kaler, and R. Blatt. Speed of ion-trap quantum-information processors. *Phys. Rev. A* 62, 042305 (2000).

[75] D. T. Smithey, M. Beck, M. G. Raymer, A. Faridani. Measurement of the Wigner distribution and the density matrix of a light mode using optical homodyne tomography: Application to squeezed states and the vacuum. *Phys. Rev. Lett.* 70, 1244 (1993).

[76] D. Leibfried, D. M. Meekhof, B. E. King, C. Monroe, W. M. Itano, and D. J. Wineland. Experimental Determination of the Motional Quantum State of a Trapped Atom. *Phys. Rev. Lett.* 77, 4281 (1996).

[77] Ch. Kurtsiefer, T. Pfau, J. Mlynek. Measurement of the Wigner function of an ensemble of helium atoms. *Nature* 386, 150 (1997).

[78] I. L. Chuang, N. Gershenfeld, M. G. Kubinec, D. Leung. Bulk quantum computation with nuclear magnetic resonance: theory and experiment. *Proc. R. Soc. London A* 454, 447 (1998).

[79] D. Leibfried, B. DeMarco, V. Meyer, D. Lucas, M. Barrett, J. Britton, W. M. Itano, B. Jelenković, C. Langer, T. Rosenband and D. J. Wineland. Experimental demonstration of a robust, high-fidelity geometric two ion-qubit phase gate. *Nature* 422, 412 (2003).

[80] Z. Hradil. Quantum-state estimation. *Phys. Rev. A* 55, R1561 (1997).

[81] K. Banaszek, G. M. D'Ariano, M. G. A. Paris, M. F. Sacchi. Maximum-likelihood estimation of the density matrix. *Phys. Rev. A* 61, 010304 (1999).

[82] D. F. V. James, P. G. Kwiat, W. J. Munro, A. G. White. Measurement of qubits. *Phys. Rev.* A 64, 052312 (2001).

[83] D. Kielpinski et al. A Decoherence-Free Quantum Memory Using Trapped Ions. *Science* 91, 1013 (2001).

[84] W. K. Wootters. Entanglement of Formation of an Arbitrary State of Two Qubits. *Phys. Rev. Lett.* 80, 2245 (1998).

[85] Quantum Information Science and Technology Roadmapping Project (ARDA). available from http://qist.lanl.gov/

[86] H.-J. Briegel, W. Dür, J. I. Cirac, and P. Zoller. Quantum Repeaters: The Role of Imperfect Local Operations in Quantum Communication. *Phys. Rev. Lett.* 81, 5932 (1998).

[87] J. I. Cirac and P. Zoller. A scalable quantum computer with ions in an array of microtraps. *Nature* 404, 579 (2000).

[88] D. Kielpinski, C.R. Monroe, and D.J. Wineland, Architecture for a Large-Scale Ion-Trap Quantum Computer, *Nature* 417, 709-711 (2002).

[89] J. I. Cirac, P. Zoller, H. J. Kimble, and H. Mabuchi. Quantum State Transfer and Entanglement Distribution among Distant Nodes in a Quantum Network. *Phys. Rev. Lett.* 78, 3221 (1997).

[90] A. B. Mundt, A. Kreuter, C. Becher, D. Leibfried, J. Eschner, F. Schmidt-Kaler, and R. Blatt. Coupling a Single Atomic Quantum Bit to a High Finesse Optical Cavity. *Phys. Rev. Lett.* 89, 103001 (2002).

[91] Andrew Steane. Simple quantum error correcting codes. *Phys. Rev.* A, 54 , 4741 (1996).

[92] A. M. Steane. Error Correcting Codes in Quantum Theory. *Phys. Rev. Lett.* 77, 793-797 (1996).

[93] R. Laflamme, C. Miquel, J. P. Paz, and W. H. Zurek. Perfect Quantum Error Correcting Code. *Phys. Rev. Lett.* 77, 198-201 (1996).

[94] I. L. Chuang and Y. Yamamoto. Quantum Bit Regeneration. *Phys. Rev. Lett.* 76, 4281-4284 (1996).

[95] H. Mabuchi and P. Zoller. Inversion of Quantum Jumps in Quantum Optical Systems under Continuous Observation. *Phys. Rev. Lett.* 76, 3108-3111 (1996).

[96] S. L. Braunstein and J. A. Smolin. Perfect quantum-error-correction coding in 24 laser pulses. *Phys. Rev.* A 55, 945 (1997).

Course 6

QUANTUM INFORMATION PROCESSING IN ION TRAPS II

D. J. Wineland

National Institute of Standards and Technology, Boulder, CO, 80305-3328

This work is supported by NIST and NSA/ARDA under contract No. MOD-7171.00

D. Estève, J.-M. Raimond and J. Dalibard, eds.
Les Houches, Session LXXIX, 2003
Quantum Entanglement and Information Processing
Intrication quantique et traitement de l'information
© 2004 Elsevier B.V.
Not subject to U.S. copyright

TOMORROW:
starts at

Contents

Abstract

These notes focus on two aspects of ion-trap quantum information processing: (1) linear RF (Paul) ion traps that appear to be scalable for making a processor array, and (2) ion qubits where single- and two-qubit gates are based on driving two-photon stimulated Raman transitions. Reviews on the current status of ion traps for information processing are also cited.

1. Introduction

In 1995, Ignacio Cirac and Peter Zoller described how an ensemble of trapped ions could be used to implement a quantum information processor [1]. Several experimental groups throughout the world have pursued this basic idea, and although a useful device still does not exist, ion-trappers are optimistic that one can eventually be built. In part, this is because it appears that the ion-trap scheme satisfies the main requirements for a quantum computer outlined by DiVincenzo [2]: (1) a scalable system of well-defined qubits, (2) a method to reliably initialize the quantum system, (3) long coherence times, (4) existence of universal gates, and (5) an efficient measurement scheme. Most of these requirements have been demonstrated, and straightforward, albeit technically difficult, paths to solving the remaining problems exist.

Within the limited space of these notes, rather than giving an overview of current capabilities of the ion trap scheme, we focus in some detail on two topics: (1) linear RF (Paul) ion traps that appear to be extendable to making a processor array and (2) ion qubits where single- and two-qubit gates are based on driving two-photon stimulated Raman transitions. For additional and more general accounts of the ion trap scheme the reader is referred to the notes by Rainer Blatt included in this volume and various reviews [3] - [16]. These notes focus on experiments carried out at NIST, but similar work has been and is currently being pursued at Aarhus, Almaden(IBM), Hamburg, Hamilton (McMaster Univ., Ontario), Innsbruck, Los Alamos (LANL), University of Michigan, Garching (MPQ), Oxford, and Teddington (NPL, U.K.).

2. Linear RF (Paul) ion traps

By acting on the net charge of atomic ions, particular arrangements of electromagnetic fields can be used to confine them. For studies of ions at low kinetic energy (less than a few electron volts), two types of traps are typically employed - the Penning trap, which uses a combination of static electric and magnetic fields, and the Paul (RF) trap that confines ions through ponderomotive forces generated by inhomogeneous oscillating electric fields. The operation of these traps is discussed in various reviews (see for example, Refs. [17] - [22]), and in a book by Ghosh [23]. For brevity, we discuss one trap configuration, the linear Paul trap that is derived from the original configuration of Drees and Paul [24]. This trap appears to be particularly useful in the context of quantum information processing because it provides a straightforward way to address and manipulate individual ions; however, other trap configurations would also work. Linear Paul traps with particular application to quantum information processing are discussed by various groups; see for example, Refs. [3] - [16] and [25] - [33].

We assume the trap electrodes are configured as shown in Fig. 1. The purpose of the trap structure is to provide a 3-D harmonic well for the ions. The strength of the well along the axis of the system (the z direction in Fig. 1) is weaker than the strength in the x and y directions, so the configuration of multiple ions in the trap is a linear array. The rods are assumed to be parallel and positioned at the vertices of a square. A (RF) potential $V_0 \cos(\Omega_T t)$ is applied to two diagonally opposite rods; the other rods are segmented into "control" electrode sections that are held at RF ground. A static potential U_o is applied to the outer control electrodes and a potential U_c is applied to the central control electrodes (the RF electrodes have a static potential 0). Near the center of the trapping region this configuration gives rise to a potential that can be approximated by (up to a constant potential)

$$\Phi = \frac{1}{2}\left(1 - \frac{(x^2 - y^2)}{R^2}\right) V_0 \cos \Omega_T t + \frac{(U_x x^2 + U_y y^2 + U_z z^2)}{2R^2}, \tag{2.1}$$

where $U_x = \alpha_{xo}U_o + \alpha_{xc}U_c$, $U_y = \alpha_{yo}U_o + \alpha_{yc}U_c$, and $U_z = \kappa(U_o - U_c)$. From Laplace's equation, $U_x + U_y + U_z = 0$. This tells us that three-dimensional confinement by the static electric potentials is impossible since confinement in one direction ($U_i > 0$) necessarily implies that at least one other direction must be anti-confining ($U_j < 0$).[1] In general, to find the coefficients α_{xo}, α_{xc}, α_{yo}, α_{yc}, and κ, we must solve Laplace's equation numerically for a specific electrode geometry.

[1] For the linear quadrupole mass spectrometer configuration, $U_x = -U_y = U_o = U_c$ and $U_z = 0$ (no confinement along z). In this case, if the electrode surfaces conform to equipotentials (hyperbolic surfaces) of Φ, then R is the distance from the trap axis to the nearest point on each electrode. If the electrode surfaces have different cross sections, R will still be approximately equal to this distance.

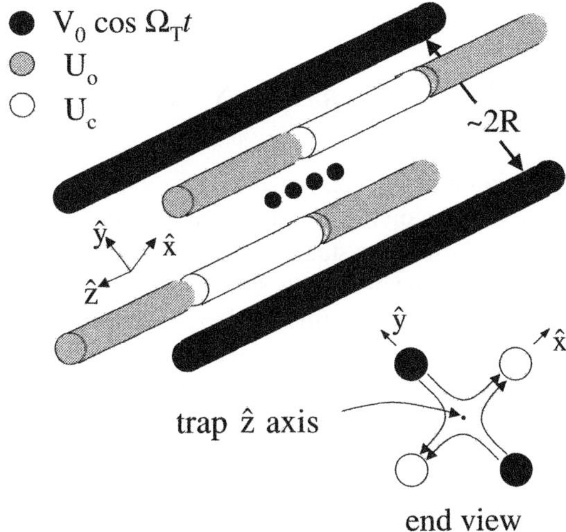

Fig. 1. Electrode configuration for a linear RF (Paul) trap. A common RF potential $V_0 \cos(\Omega_T t)$ is applied to the dark electrodes; the other "control" electrodes are held at RF ground through capacitors (not shown) connected to ground. The lower-right portion of the figure shows the x, y electric fields from the applied RF potential at an instant when the RF potential is positive relative to ground. For certain values of the trap parameters, the RF electric fields create a ponderomotive potential that confines ions to the z axis. A static electric potential well is created along the z axis by applying (for positive ions) a positive potential between the outer (grey) and central (white) control electrodes. Four trapped ions near the center of the trap are depicted as black dots.

From the potential of Eq. (2.1), we can derive the classical equations of motion for a single trapped ion. In the ith direction, $F_i = -\partial(q\Phi)/\partial x_i = m\ddot{x}_i$, where $i \in \{x, y, z\}$, $x_x \equiv x(t)$, $x_y \equiv y(t)$, $x_z \equiv z(t)$, and m and q are the ion's mass and charge. This gives the equations of motion, which can be written in the form

$$\frac{d^2 x_i}{d\xi^2} + \left[a_i - 2q_i \cos 2\xi\right] x_i = 0 \qquad (i \in \{x, y, z\}), \tag{2.2}$$

where

$$a_i \equiv 4qU_i/m\Omega_T^2 R^2, \quad q_x = -q_y \equiv -2qV_0/m\Omega_T^2 R^2, \quad q_z = 0, \tag{2.3}$$

and where time is parameterized according to $\xi \equiv \Omega_T t/2$. In the z direction, Eq. (2.2) reduces to $\ddot{z} + (qU_z/mR^2)z = 0$. Therefore, when $qU_z \propto a_z > 0$, the ions are harmonically bound along the trap axis with frequency $\omega_z = (qU_z/mR^2)^{1/2} = \sqrt{a_z}\Omega_T/2$. In the x and y directions, Eq. (2.2) appears in the standard form of the Mathieu equation [34]. Its solutions can be written

$$x_i(\xi) = Ae^{i\beta_i\xi}\sum_{n=-\infty}^{\infty} C_{2n}e^{i2n\xi} + Be^{-i\beta_i\xi}\sum_{n=-\infty}^{\infty} C_{2n}e^{-i2n\xi} \quad (i \in \{x, y\}).(2.4)$$

Plugging this expression into Eq. (2.2) results in a recursion relation for the coefficients C_{2n} and an expression for β_i in terms of a_i and q_i [34].

When β_i is real the solutions are stable. Several regions in a_i,q_i space give rise to stable solutions [23,34]; one of these is shown in Fig. 2. As a practical matter, we are usually concerned only with this stable region for values $|a_i|, |q_i| \ll 1$. For each point in the stable region, the overall binding potential is a combination of a static potential (the a_i term) and the RF ponderomotive potential (determined by q_i). The line labelled $\beta_i = 1$ corresponds to the so-called Mathieu instability where the component of the ion's oscillation at frequency $\beta_i\Omega_T/2$ (the C_0 term in Eq. (2.4)) is equal to $\Omega_T/2$. This condition defines an unstable point because the ion's motion (at frequency $\Omega_T/2$) can be parametrically excited by the inhomogeneous electric fields oscillating at the trap drive frequency Ω_T. The line labelled $\beta_i = 0$ corresponds to the absence of binding; to the left of this line, the combination of static potential and RF ponderomotive potential is anti-confining. For regions where $a_i < 0$, the static potential is anti-confining; however, this can be compensated for by the RF ponderomotive potential, as in the lower right part of the figure. If we want $\omega_x \simeq \omega_y$, then we must have $a_x \simeq a_y < 0$, since $a_z > 0$. This particular case has been examined in Ref. [35]. Often however, we want ω_x sufficiently different from ω_y to simplify Doppler laser cooling. This is because when $\omega_x = \omega_y$, then the ions' radial motion in a direction perpendicular to the plane defined by the Doppler-cooling laser's $\vec{k}$ vector and the trap's z axis will be heated without bound by photon recoil [36, 37]. By making $\omega_x \neq \omega_y$ and making both the x and y axes be at 45° with respect to the plane defined by the Doppler-cooling laser's $\vec{k}$ vector and the trap's z axis, we achieve optimum Doppler cooling in the x and y directions. Luckily, this is the natural choice for introducing a laser beam into the trap of Fig. 1, that is, along the $\hat{x}\pm\hat{y}$ directions.[2] In any case, making $\omega_x \neq \omega_y$ may mean that a_x and a_y have different signs.

Figure 2 shows only the region of stability for either the x or y direction. However, for 3-D confinement, we must satisfy the stability conditions for the x, y, and z directions simultaneously. This restricts the stability region to a smaller zone than that shown in Fig. 2. To determine the overall stability region we can plot the x and y stability regions on the same graph as shown in Fig. 3 for the

[2] In more realistic situations, such as for the traps described in Ref. [38], the electrode surfaces do not have reflection symmetry about the x and y axes as for the electrode structure in Fig. 1. In this case, the x and y axes will twist as a function of z and one must take care to avoid recoil heating as noted above.

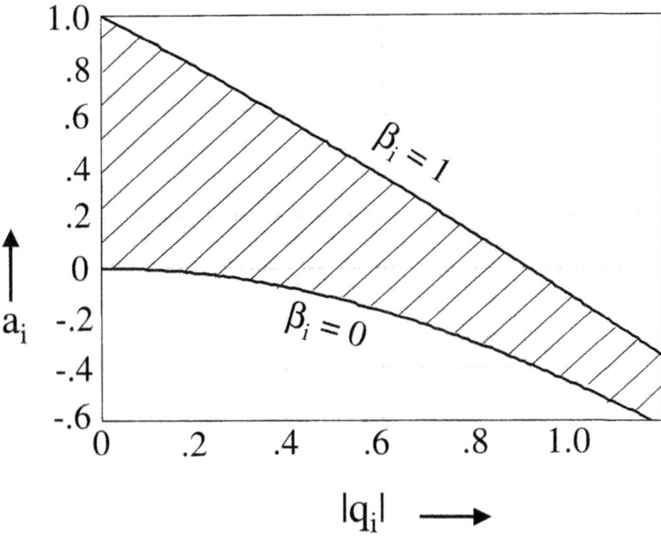

Fig. 2. Region of stable motion (cross-hatched area) in a_i, q_i space ($i \in \{x, y\}$) for a linear Paul trap. The line labelled $\beta_i = 1$ corresponds to a trap oscillation frequency $\omega_i = \Omega_T/2$, the maximum value possible. The line labelled $\beta_i = 0$ corresponds to a trap oscillation frequency $\omega_i = 0$.

particular value $a_z = 0.1$. From Laplace's equation we have $a_y = -a_x - a_z$; we can therefore represent the y stability region as outlined by the dashed curves in Fig. 3. The (cross-hatched) region where the two stability regions overlap gives the overall region of stability.

For $|a_i|, q_i^2 \ll 1$, Eq. (2.4) can be approximated by keeping only the terms with coefficients C_{-1}, C_0, and C_1. We find (up to initial phases)

$$x_i(t) \simeq X_{i0} \cos \omega_i t \left[1 + \frac{q_i}{2} \sin \Omega_T t \right] \quad (i \in \{x, y\}), \tag{2.5}$$

where $\omega_i = \beta_i \Omega_T/2$, $\beta_i \simeq \sqrt{a_i + q_i^2/2}$ ($i \in \{x, y\}$), and X_{i0} is given by initial conditions. We see that the motion is a combination of harmonic motion at the "secular" frequency ω_i with a superimposed "micromotion" component (frequency Ω_T), whose amplitude is proportional to the ion's displacement X_{i0} from the z axis. For the particular case a_i ($\propto U_i$) $= 0$, the ion's secular plus micromotion kinetic energy averaged over one period ($2\pi/\Omega_T$) of the driven motion is constant and equal to $mq_i^2\Omega_T^2 X_{i0}^2/16$. Therefore, the ion's kinetic energy is shared between the driven micromotion and secular motion. When the ion is near the z axis, the kinetic energy is predominantly in the secular motion. At

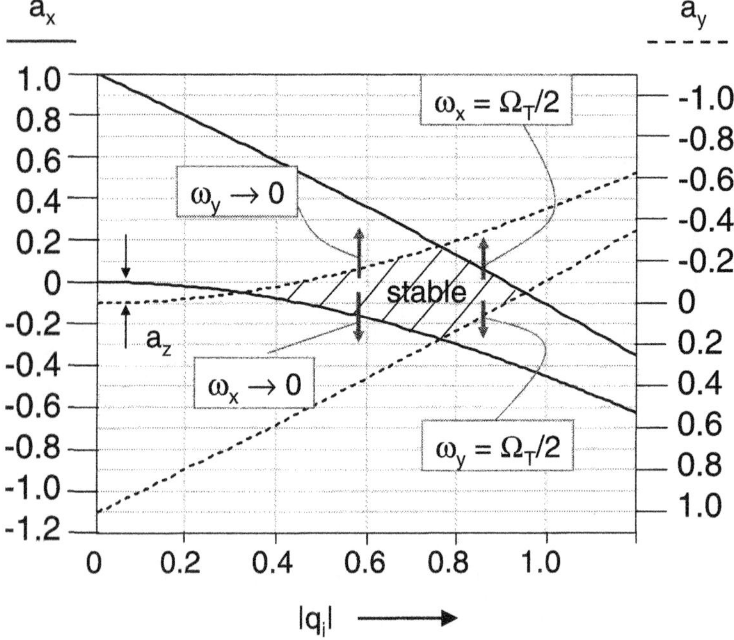

Fig. 3. Region of simultaneous stability (cross-hatched area) for the x and y motion of a trapped ion assuming $a_z = 0.1$ ($\omega_z = 0.158\ \Omega_T$). The vertical scale for the y stability diagram is obtained from Laplace's equation $a_y = -a_z - a_x$. The arrows indicate transitions from stability to instability and the accompanying labels indicate the causes of instability at the boundaries.

the extremes of displacement from the origin ($x_i(t) \simeq X_{i0}$), the kinetic energy is predominantly in the micromotion. As discussed below, we desire to suppress the RF micromotion; one way this can be achieved is by confining a string of ions near the axis of a linear Paul trap.

2.1. *The pseudopotential approximation*

In the limit that $|a_i|, q_i^2 \ll 1$, a "pseudopotential" description of the ion's motion [18, 23] is more intuitive and leads to the same result as in Eq. (2.5). In the pseudopotential approximation we make the (self-consistent) assumption that at a given displacement $\vec{\rho} = x\hat{x} + y\hat{y}$ from the trap axis, the driven micromotion can be derived from the electric field at the mean position of the ion, averaged over one RF period. That is, for the duration ($2\pi/\Omega_T$) of one RF period, we assume that $x_{total} = x + x_\mu(t)$ and $y_{total} = y + y_\mu(t)$, and that x and y (the coordinates of the secular motion) change negligibly. Therefore the we derive the micromotion ($x_\mu(t)$ and $y_\mu(t)$ from electric fields $E_x(x)$ and $E_y(y)$. From the first term of Eq. (2.1), we have $m\ddot{x}_\mu = F_x = -q\partial\Phi/\partial x = -xqV_0\cos\Omega_T t/R^2$, which gives $x_\mu(t) = xqV_0\cos\Omega_T t/m\Omega_T^2 R^2$, equivalent to the second term in Eq. (2.5) (similarly for y_μ). We can derive the pseudopotential U in two different ways. In the first, we note that the micromotion kinetic energy in the x direction, averaged over one period of the micromotion, is given by $\langle KE(x_\mu)\rangle = x^2q^2V_0^2/4m\Omega_T^2R^4$. The x, y pseudopotential is then $U_{x,y} = \langle KE_\mu\rangle = \langle KE(x_\mu)\rangle + \langle KE(y_\mu)\rangle$. Qualitatively, the restoring force originates from the fact that the micromotion kinetic energy increases as $|\vec{\rho}|^2$ and the pseudopotential force $-\vec{\nabla}\langle KE_\mu\rangle$ is towards the trap axis. An alternative way to derive the pseudopotential considers the electric force on an ion with x_{total} and y_{total} given as above. In the x direction, we have $F_x(t) \simeq F_x(\vec{\rho}, t) + (\partial F_x(\vec{\rho}, t)/\partial x)_{\vec{\rho}}\, x_\mu(t) + (\partial F_x(\vec{\rho}, t)/\partial y)_{\vec{\rho}}\, y_\mu(t)$. Averaged over one RF period (and making use of $\vec{\nabla} \times \vec{E} \simeq 0$, which is valid for our case where the trap structures are small compared to the wavelength associated with Ω_T), we find $\langle F_x(t)\rangle = -x(q^2V_0^2/2m\Omega_T^2R^4) = -\nabla_x\langle KE_\mu\rangle$. Therefore, from $\langle F_x(t)\rangle$ and $\langle F_y(t)\rangle$ we find the same pseudopotential U. To find the total potential, we must add the static potential (the second term on the right-hand side of Eq. (2.1)) to the pseudopotential.

Overall then, we can express the ponderomotive pseudopotential as

$$U_{x,y}(\rho) \simeq \frac{q^2V_0^2}{4m\Omega_T^2R^4}(x^2+y^2) + \frac{(U_x x^2 + U_y y^2)}{2R^2}. \tag{2.6}$$

The oscillation frequency of an ion in this pseudopotential is given by ω_i ($= \Omega_T(a_i + q_i^2/2)^{1/2}/2$) as in Eq. (2.5). As discussed below, we will typically want $\omega_{x,y} \gg \omega_z$, in which case $q_{x,y}^2/2 \gg |a_{x,y}|$ (the RF pseudopotential dominates in the x, y directions), so that $\omega_{x,y} \simeq qV_0/(\sqrt{2}m\Omega_T R^2)$. As an example of experimental parameters, typical conditions in the NIST experiments are: $q = 1\ e$, $V_0 = 500\ V$, $\Omega_T/2\pi \simeq 200$ MHz, $m = 9\ u$ (^{9}Be$^+$), and $R = 200\ \mu$m, so that if we can assume the static potential is negligible in the x, y directions, then $\omega_{x,y}/2\pi \simeq$ 6 MHz ($q_i \simeq 0.085$).

As an aside, we note that the principle of optical dipole traps for neutral atoms is entirely analogous to that of the Paul traps for charged particles. For optical dipole force traps, Ω_T is replaced by the optical frequency, and the (spatially inhomogeneous) laser electric field traps the optically active electron(s) to which the atomic core is attached. (In contrast to the electron(s) response, the (charged) atomic core plays a negligible role in the trapping because of its large mass; the pseudopotential strength is proportional to the inverse of the mass, as seen in Eq. (2.6)). The energy of the system is shared between the secular kinetic energy of the atom in the trap and the energy of the electronic excitation. Atoms in optical traps seek regions of minimum $|E(\mathbf{r})|$ for laser tuning to the blue of (higher than) the electron's resonance frequency. This is analogous to the Paul-trap case, where the ion's natural resonant frequency is equal to 0 in the absence of the applied fields (since it is otherwise free), and the applied frequency Ω_T is always blue of this natural resonance frequency. However, an important difference occurs for optical dipole force traps because the electron(s) response is dispersive about its resonance frequency. The electron's motion is out of phase with the electric force when driven above (blue of) its resonant frequency and is in phase when driven below (red of) its resonant frequency. This means that in the optical dipole trap, the atom will seek the maximum of $|E(\mathbf{r})|$ when the laser frequency is less than the relevant atomic transition frequency.

2.2. *Quantized motion and RF micromotion*

For a proper description of quantum information processing using trapped ions, we will need to treat the ions' motion quantum mechanically. For trapping in a static harmonic potential, such as the axial motion in the linear trap described above, the quantization procedure is straightforward and follows the usual procedure for a simple harmonic oscillator. To a good approximation, an ion trapped in a ponderomotive potential can also be described effectively as a simple harmonic oscillator, even though the Hamiltonian is actually time-dependent, so no stationary states exist. For practical purposes, the system can be treated as if the Hamiltonian were that of an ordinary time-independent harmonic oscillator [39]- [45], although the coupling to laser beams is reduced [44–46]. The classical micromotion (the term that includes the factor $\sin \Omega_T t$ in Eq. (2.5)) may be viewed in the quantum picture as causing the ion's wavefunction to oscillate and breathe at the drive frequency Ω_T [44]- [43]. However, this motion is separated spectrally from the secular motion. Since the operations we will consider rely on a resonant or near-resonant interaction at the secular frequencies, we will average over the components of motion at the drive frequency Ω_T. Therefore, to a good approximation, the pseudopotential secular motion behaves as an oscillator in a static potential. The main consequence of the quantum treatment is that transition rates between quantum levels are lowered somewhat [44–46]; however, these changes can be accounted for by experimental calibration. Qualitatively, the effect of the RF micromotion is to smear out the ion's wavefunction, which reduces the interaction with optical radiation. This latter effect can be compensated by increasing the radiation intensity. In any case, for most of the applications discussed here, we will be considering the motion of the ions along the axis of a linear Paul trap where this modification is absent because of the absence of micromotion.

2.3. Static offset potentials

A more serious practical issue concerning micromotion is the effect of static electric fields that displace the mean position of an ion from the trap's axis. These static fields might arise from the Coulomb repulsion between ions if, for example, the ions are forced into a zig-zag pattern around the z axis by an axial potential that is too strong relative to the x, y potential [47]. However, we usually operate with the x, y potential strong enough to confine ions on the axis. A more insidious source of micromotion is displacement from the trap axis caused by stray static electric fields. These fields can shift the x, y potential minimum of the trap to a location away from the z axis where the force from the stray static field is compensated by the ponderomotive force. In this case, a steady RF micromotion ensues even when the ions' secular motion is cold. If the micromotion amplitude is too large, it can significantly alter an ion's absorption spectrum, with important consequences for laser cooling [48–50]. In principle, the extra micromotion should not cause heating if the stray fields are uniform over the entire ion sample since the common-mode RF micromotion does not resonantly couple to the secular motion.[3] However, in practice, RF heating of multiple trapped ions appears to be aggravated if the ions are displaced from the trap axis. This may be related to the RF heating mediated by the nonlinear Coulomb interaction that has been extensively studied at Garching (MPQ) and Almaden (IBM) for two ions in a 3-D Paul trap (for a summary, see Ref. [51]).

Often, the origin of stray static fields is not known in specific experiments. They might be caused by potentials on surfaces outside the immediate trap area. They might also originate from variations in contact potentials (work functions) on trap electrode surfaces caused by different crystal planes, extraneous material adsorbed on the electrode surfaces, or perhaps by electrons residing on insulating layers attached to the electrode surfaces. The adverse effects of insulating layers and surface electrons can be mitigated by the use of heaters in the electrodes [52, 53]; however, it is usually necessary to compensate for these stray fields by applying static correction potentials to the trap electrodes or auxiliary electrodes outside the immediate trap structure.

Different methods can be used to sense micromotion (for a summary see, for example, Ref. [54] and references therein). A common method is to sense it through its effect on scattering of laser light that is tuned near an allowed transition in the ion, usually taken to be the Doppler-cooling transition. It is useful to distinguish two regimes that depend on the ratio γ/Ω_T, where γ is the radiative linewidth of the allowed transition. When $\gamma/\Omega_T \gg 1$, the micromotion fre-

[3] This argument applies to like ions. However, if different kinds of ions are simultaneously trapped, as in the proposals that use sympathetic cooling [4, 33], a displacement from the trap axis gives rise to a differential micromotion between ions, which can cause heating.

quency modulates the ions' absorption profile through the micromotion-induced first-order Doppler shift. By tuning a laser beam to the lower-frequency side of the absorption profile (the natural choice for Doppler cooling), when the maximum Doppler shift is small compared to γ, the fluorescence is modulated at Ω_T, which can be detected. The component of micromotion along the direction of the laser beam can be eliminated by applying additional static correction potentials.[4] When $\gamma/\Omega_T \ll 1$, the absorption spectrum develops discrete RF sidebands on both sides of the carrier; by minimizing these sidebands, the component of micromotion along the direction of the laser beam is minimized. For laser intensities below saturation, the fluorescence for the nth order sideband is proportional to $J_n^2(k_i x_{RF})$ where J_n is the nth-order Bessel function, k_i is the component of the laser beam's wave vector along the direction of micromotion, and x_{RF} is the amplitude of RF micromotion. This method can be difficult to implement because each sideband has multiple values of x_{RF} that null $J_n^2(k_i x_{RF})$, so we must sample more than one sideband to find the condition where $x_{RF} = 0$. Alternatively, for $\gamma/\Omega_T \ll 1$, the time-averaged strength of the carrier fluorescence (proportional to $J_0^2(k_i x_{RF})$) is a good indicator of $x_{RF} = 0$, since the maximum value of J_0^2 $(J_0(0)^2 = 1)$ can be distinguished from the weaker auxiliary maxima of J_0^2. For example, the first auxiliary maximum where $k_i x_{RF} \simeq 3.83$ gives $J_0^2 \simeq 0.16$. For any value of γ/Ω_T, when using laser scattering as a probe, the motion must be sensed with three non-coplanar laser beam directions to ensure that the micromotion is eliminated in all directions. For a linear RF trap the micromotion needs to be sensed in only two non-parallel directions in the x, y plane.

One alternative to using laser scattering to sense micromotion is to measure the mode frequency spectrum of a ion crystal composed of ions of different mass [55]. The technique is based on the fact that when the RF pseudopotential dominates, an ion's restoring potential is inversely proportional to its mass. Therefore, a stray static field transverse to the trap axis will cause displacements from the trap axis that depend on the ion's mass. (The displacements are proportional to the ion mass when the Coulomb interaction between ions can be neglected.) For a linear crystal composed of different mass ions, the stray fields cause the vector joining the ions to be non-parallel to the trap axis. This affects the motional mode spectrum in a way that can be sensed and then used to apply compensating fields that cancel the stray fields. This can be a technical advantage because extra probe laser beams are not required; moreover, the mode spectrum must be measured anyway in the experimental calibration procedure.

Not all consequences of RF micromotion are bad. The $J_0^2(k x_{RF})$ dependence

[4]This argument applies when all electrodes to which RF is applied have the same RF phase. If different phases exist on two or more electrodes, the RF micromotion cannot be completely suppressed by this procedure.

noted above can actually be used as a form of differential selection of ion qubits when the ions are so close together that selection by laser-beam focusing is precluded. This selection method has been used to entangle two ion qubits [46] and might also be used in more general schemes of quantum computation [56].

For the purposes of quantum computing, we will assume for simplicity that the x, y potentials are strong compared to the z potential. Therefore, when a small number of ions are trapped and cooled, each ion seeks the bottom of the trap well, but the mutual Coulomb repulsion between ions results in an equilibrium configuration in the form of a linear array, as indicated in Fig. 1. To give an idea of array size, two like ions in such a trap are spaced by $2^{1/3}s$, and three ions are spaced by $(5/4)^{1/3}s$ where $s \equiv (q^2/(4\pi\epsilon_o m\omega_z{}^2))^{1/3}$. Expressed equivalently, for singly charged ions, the spacing parameter in micrometers is $s(\mu m) = 15.2(M(u)\nu_z^2(\text{MHz}))^{-1/3}$, where the ion's mass is expressed in a.m.u. and the axial z frequency ($\nu_z \equiv \omega_z/2\pi$) in MHz. For ν_z = 5 MHz, two $^9Be^+$ ions are separated by 3.15 μm. For larger numbers of ions, see for example, Refs. [3, 57]. The minimum time to implement most of the ion logic gates that have been devised is on the order of the oscillation period of the ions in the trap. In some of the gates [1] however, we want to be able to individually address ions by laser beam focusing. Since $s \propto \nu_z^{-2/3}$ the requirement of individual addressing by laser beam focusing may therefore limit the gate speeds. (This limitation would be overcome by implementation of the proposals in Refs. [29, 58].)

2.4. *Practical traps and their limitations*

Ion trap structures are usually installed in a high-vacuum enclosure to minimize collisions with background gas. Ions are often loaded by transmitting a beam of neutral atoms of the desired element and an ionizing electron beam through the trap simultaneously. When the atoms are ionized inside the trapping zone, the resulting ions can be trapped. In place of an ionizing electron beam, photoionization has also been used with good success to create Ca^+ [59, 60] and Mg^+ ions [59]; however, for many ions, the required laser wavelengths for photoionization are more difficult to achieve.

To form stable linear arrays of ions, we want $\omega_{x,y} \propto V_0/R^2 \gg \omega_z$. Therefore for high gate rates ($\propto \omega_z$), we want to apply large RF potentials and make the traps as small as possible. An additional reason to make the traps small is to aid in separating individual ions into separate zones [38] in a multiplexed architecture [4, 33]. The need for large RF potentials and small electrode separations can aggravate the problems of RF breakdown, which can occur either through the vacuum or on the surfaces of insulators between electrodes. At the same time, any insulators between electrodes should have small RF loss; otherwise the trap structure may be heated to high temperatures, which can cause outgassing and trap movement from thermal expansion.

For high-fidelity gate operations, we will want to cool the ions' motion to, or very near to, the ground state and maintain this condition during all logic operations. Therefore, all sources of motional heating must be suppressed. Since the center-of-mass modes for a group of ions behave as charged harmonic oscillators, they are very susceptible to heating from fluctuating electric fields. Although the heating from thermal electric fields appears to be manageable [4], the heating that has been observed in experiments at NIST exceeds that of thermal electric fields by more than an order of magnitude [61]. The specific cause of this heating is currently not understood, but it appears that it can be reduced by improving the integrity of electrode surfaces [38]. Clearly, future work must continue to address this problem; however, heating from external fields can be mitigated with sympathetic cooling [4, 33].

We will also want to maintain a vacuum high enough to prevent collisional heating from background gas. For some ions, such as Be^+ and Mg^+, high vacuum is also required to prevent ion loss through chemical reactions between the ion and a background neutral particle. For chemical reactions to occur, they must be energetically favorable (exothermic). If the background neutral species is a molecule, exothermic reactions almost always proceed since the internal degrees of freedom of the molecule can help satisfy energy and momentum conservation in the reaction. In the experiments discussed here, the ion can spend an appreciable amount of time in the (optically) excited state due to laser excitation (cooling and detection); in this case the extra energy due to laser excitation can make an otherwise endothermic reaction become exothermic. For example, Yb^+ and Mg^+ ions are observed to convert to their hydride forms YbH^+ [62] and MgH^+ [63], when in optically excited states and exposed to H_2. In many experiments on $^9Be^+$ ions at NIST, we observe that the ions are converted over time to ions of mass 10 u, presumably BeH^+, when resonant light was applied to the $^2S_{1/2} \rightarrow ^2P_{1/2,3/2}$ transitions [64]. This likely occurs through a similar process. Such reactions are very troublesome since they require new ions to be loaded and a significant amount of time (many minutes) may be required for the system to re-stabilize.

To achieve high vacuum, materials with minimal outgassing must be used. The trap systems are typically baked under vacuum at temperatures 200 - 350 °C for several days for outgassing, so the materials must also be compatible with these temperatures. To minimize the partial pressure of H_2, stainless steel components can be air-baked and titanium sublimation pumps can be added to the vacuum system. Typical total pressures (as measured by Bayard-Alpert gauges) are in the low 10^{-9} Pa range (low 10^{-11} Torr range). At these pressures, the lifetime of ions that react with H_2 can be several hours. For $^9Be^+$ and other ions that suffer from loss due to chemical reactions, the lifetime can be significantly improved by using cryopumping. As an example, the lifetime of $^{199}Hg^+$ ions in the NIST optical clock experiments was a few minutes in a room temperature apparatus; this time increased to several months in an apparatus cooled to 4 K [65, 66]. The offsetting penalty for cryopumping is the difficulty of dealing with laser beam and detection optics in a cryogenically cooled system.

Linear traps were first applied to atomic frequency standards where an elongated cloud of $^{199}Hg^+$ ions confined near the trap axis suppressed second-order Doppler shifts from RF micromotion [67]. The limiting case of this technique is to cool a string of ions to the trap axis [68, 69]. Reference [51] describes crystallized strings of laser-cooled $^{24}Mg^+$ ions that have been observed in a racetrack-type linear Paul trap. A close approximation to the trap shown in Fig. 1 was used for trapping small numbers of Hg^+ ions in the experiments of Ref. [68]. This trap had electrodes with $R \simeq 0.77$ mm and was constructed using conventional machining techniques. In the context of quantum computing, a linear trap for Ca^+ ions where the axial confinement is provided by rings that surround the quadrupole rod structure has been described in Ref. [27].

Most gates for trapped ions have a speed that is limited to, or proportional to, the oscillation frequency of the ions in the trap [1], [3]- [16], [25]- [33]. As noted above, for fast gates, we therefore want to maximize V_0/R^2. As R becomes smaller it is more difficult to control the relative dimensions of the electrode structures. One attempt to mitigate this problem is described in Ref. [38]. The trap was constructed from a stack of two metallized 200 μm thick alumina wafers. Laser-machined slots and gold traces created the desired electrode geometry. Gold traces of 0.5 μm thickness were made with evaporated gold that was transmitted through a shadow mask and deposited on the alumina. Subsequently, an additional 3 μm of gold was electroplated onto the electrodes, resulting in electrode surfaces smooth at the 1 μm level. These lithographic techniques allow for small traps and could be expandable to larger arrays. Even smaller traps might be constructed using MEMS-based or other fabrication techniques. However, the challenge is to find construction materials that satisfy several key requirements: (1) the ability to support large RF potentials, (2) small RF loss, (3) small trap dimensions, (4) clean electrode surfaces, (5) the ability to withstand baking to 200 - 350°C, (6) the ability maintain high vacuum, and (7) scalability as, for example, outlined in Refs. [4, 26, 29, 33].

3. Ion qubits

Trapped ions can have very long internal-state coherence times; this has been demonstrated in experiments that use ions as the basis for frequency standards

[70, 71] where decoherence times $\tau_1, \tau_2 > 10$ min have been achieved. Therefore, trapped-ion qubits can be expected to have good memory properties. Both narrow optical transitions and ground-state hyperfine transitions have been employed in quantum logic experiments to date. Optical transitions [1, 72] have the simplicity of single-photon transitions, in that only one laser beam is needed to drive qubit rotations and gates. Coherence and memory duration is limited by radiative decay from the excited state and and can be quite long ($\gg 1$ s) for weakly allowed transitions. However, phase-stable, narrow-linewidth lasers are required in order to realize the full benefit of the long decay times [65, 73]. In addition, Stark shifts from coupling to non-resonant allowed transitions become more important as the qubit lifetimes become longer since the laser intensity must be increased to maintain the same transition rates [74]. Hyperfine transitions in electronic ground states have very long radiative lifetimes and can therefore make good qubit memories. We usually consider driving transitions between hyperfine levels using stimulated Raman transitions driven by two laser beams. The strong electric field gradients associated with laser beams allow efficient coupling between the motion and internal states and the use of laser beams allows individual qubit addressing by focusing. A technical advantage of using Raman transitions is that the relevant phase for the transitions is the phase difference between the Raman beams. One of the laser beams can be derived by modulating a sample of the other beam at the relevant hyperfine transition frequency. Therefore, the relevant phase is the phase of the modulation frequency, which can usually be made quite stable if it is derived from an RF generator.[5] However, as discussed below, an important limitation to using stimulated Raman transitions is decoherence from laser-induced spontaneous Raman scattering.[6] For brevity, only hyperfine-based qubits driven by stimulated Raman transitions are discussed here. The accompanying notes by Rainer Blatt will discuss the case of optical transition qubits.

4. Stimulated Raman transitions

We assume that two electronic ground-state hyperfine levels comprise the qubit. We label the states $|\downarrow\rangle$ and $|\uparrow\rangle$ in analogy with the two levels of a spin-1/2

[5] For either case, phase fluctuations can also be caused by optical path-length fluctuations; for stimulated Raman transitions, fluctuations between the optical paths of the two Raman beams can be important.

[6] It might be possible to use single-photon transitions with frequencies near the hyperfine transition frequencies [75, 76]; however, current technology may limit the attainable gate speeds. Another interesting possibility is to use Zeeman transitions in trapped electron qubits [77]. Both of these implementations would avoid decoherence from spontaneous emission.

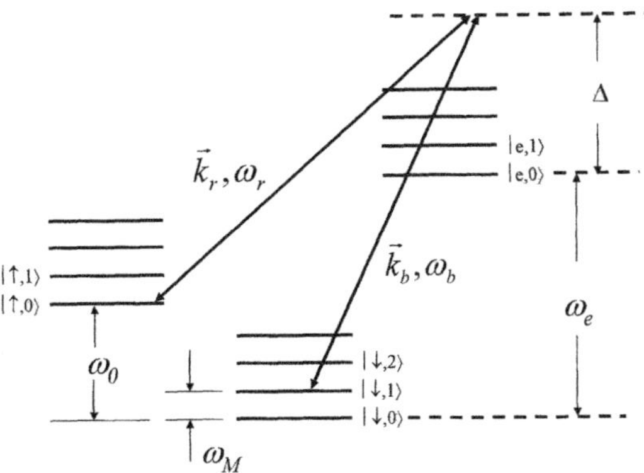

Fig. 4. Stimulated Raman transitions. We show an energy-level diagram (not to scale) for an ion bound in a one-dimensional harmonic well (frequency ω_M), and having three internal levels, designated $|\downarrow\rangle$, $|\uparrow\rangle$, and $|e\rangle$. Levels $|\downarrow\rangle$ and $|e\rangle$ are coupled by (detuned) laser beam b, and levels $|\uparrow\rangle$ and $|e\rangle$ are coupled with (detuned) laser beam r. We assume $\omega_e \gg \Delta \gg \omega_0 \gg \omega_M$. For the relative tuning of the Raman beams as shown in the figure, resonant Raman transitions are driven between states $|\downarrow, n\rangle$ and $|\uparrow, n-1\rangle$.

magnetic moment in a magnetic field. We assume the state $|\uparrow\rangle$ is at energy $\hbar\omega_0$ above the state $|\downarrow\rangle$. To understand the basic mechanisms involved we will first consider the relatively simple case of a single ion qubit bound in a harmonic one-dimensional well (frequency ω_M) in the x direction.[7] The motional mode is represented in the Fock-state basis $\{|n\rangle\}$. We assume stimulated Raman processes occur through one internal excited state $|e\rangle$, as shown in Fig. 4. Laser beam b (for "blue") couples states $|\downarrow\rangle$ and $|e\rangle$, while laser beam r (for "red") couples $|\uparrow\rangle$ and $|e\rangle$, both by electric-dipole coupling. The Hamiltonian for this system can be written

$$H = H_0 + H_I, \tag{4.1}$$

where

$$H_0 = \hbar\omega_0|\uparrow\rangle\langle\uparrow| + \hbar\omega_e|e\rangle\langle e| + \hbar\omega_M\tilde{n}, \tag{4.2}$$

and

$$H_I = -\sum_{i=b,r} e\vec{r}\cdot\vec{E}_i. \tag{4.3}$$

[7] Although we will speak in terms of a single atomic ion in this section, the discussion will apply equally well to a single neutral trapped atom.

In Eq. (4.2), $\tilde{n} \equiv a^\dagger a$, where $a^\dagger$ and a are the raising and lowering operators for the motional mode M. In Eq. (4.3), $e\vec{r}$ is the electric dipole operator for electronic transitions in the ion, and $\vec{E}_i$ is the electric field of laser beam i (assumed to be classical), which we write as

$$\vec{E}_i = \hat{\epsilon}_i E_{i0} \cos(\vec{k}_i \cdot \vec{X}_M - \omega_i t + \phi_i), \quad i \in \{b, r\}. \tag{4.4}$$

In this equation, $\hat{\epsilon}_i$, E_{i0}, $\vec{k}_i$, and ω_i are respectively the polarization, peak electric field, k-vector, and frequency of laser beam i and $\vec{X}_M = \hat{x}x_0(a + a^\dagger)$ is the position operator for deviations of the ion from its mean position ($x_0 = \sqrt{\hbar/2m\omega_M}$ is the zero-point amplitude for the motional mode). Equation (4.3) can therefore be written

$$H_I = -\sum_{i=b,r} \hbar\hat{\epsilon}_i \cdot \vec{r} \left[\frac{eE_{i0}}{2\hbar}\right] \exp[i(\vec{k}_i \cdot \vec{X}_M - \omega_i t + \phi_i)] + h.c., \tag{4.5}$$

where $h.c.$ stands for Hermitian conjugate.

Anticipating the form of the wave function, we write it as

$$\Psi = \sum_{n=0}^{\infty} \Big[C_{\downarrow,n} e^{-in\omega_M t} |\downarrow, n\rangle + C_{\uparrow,n} e^{-i[\omega_0 + n\omega_M]t} |\uparrow, n\rangle + C_{e,n} e^{-i[\omega_e + n\omega_M]t} |e, n\rangle \Big]. \tag{4.6}$$

From Schrödinger's equation, we can find an equation describing the evolution of $C_{\downarrow,n}$ by evaluating the expression $\langle\downarrow, n|(i\hbar\partial\Psi/\partial t = H\Psi)$, and find

$$\dot{C}_{\downarrow,n} = ig_b \sum_{m=0}^{\infty} e^{i(\omega_b - \omega_e + (n-m)\omega_M)t} \langle n|e^{-i\vec{k}_b \cdot \vec{X}}|m\rangle C_{e,m}. \tag{4.7}$$

In Eq. (4.7), we have used the rotating-wave approximation, neglecting terms with rapidly varying time dependence ($e^{-i(\omega_e + \omega_b)t}$ terms). The (resonant) Rabi frequency for the blue laser is

$$g_b \equiv \langle\downarrow|\hat{\epsilon}_b \cdot \vec{r}\,|e\rangle \frac{eE_{b0}e^{-i\phi_b}}{2\hbar}. \tag{4.8}$$

Similarly, we find

$$\dot{C}_{\uparrow,n} = ig_r \sum_{m=0}^{\infty} e^{i(\omega_r - (\omega_e - \omega_0) + (n-m)\omega_M)t} \langle n|e^{-i\vec{k}_r \cdot \vec{X}}|m\rangle C_{e,m} \tag{4.9}$$

and

$$\dot{C}_{e,m} = ig_b^* \sum_{p=0}^{\infty} e^{-i(\omega_b - \omega_e - (m-p)\omega_M)t} \langle m|e^{i\vec{k}_b \cdot \vec{X}}|p\rangle C_{\downarrow,p}$$

$$+ig_r^* \sum_{p=0}^{\infty} e^{-i(\omega_r - (\omega_e - \omega_0) - (m-p)\omega_M)t} \langle m|e^{i\vec{k}_r \cdot \vec{X}}|p\rangle C_{\uparrow,p}, \tag{4.10}$$

where

$$g_r \equiv \langle \uparrow |\hat{\epsilon}_r \cdot \vec{r}\ |e\rangle \frac{eE_{r0}e^{-i\phi_r}}{2\hbar}. \tag{4.11}$$

We can redefine the $|e, m\rangle$ state coefficient as

$$C_{e,m} \equiv e^{-i\Delta t} C'_{e,m}, \tag{4.12}$$

where $\Delta \equiv \omega_b - \omega_e$. This leads to

$$\dot{C}_{e,m} = e^{-i\Delta t}(\dot{C'}_{e,m} - i\Delta C'_{e,m}). \tag{4.13}$$

We now make the ansatz $\Delta C'_{e,m} \gg \dot{C'}_{e,m}$ (which we can verify later), in which case Eq. (4.13) can be written $C'_{e,m} \simeq ie^{i\Delta t}\dot{C}_{e,m}/\Delta$. From this, and Eq. (4.12), we have $C_{e,m} \simeq i\dot{C}_{e,m}/\Delta$. Using this expression in Eq. (4.10), we obtain

$$C_{e,m} = -\frac{1}{\Delta} \sum_{p=0}^{\infty} \Bigg[g_b^* e^{-i(\omega_b - \omega_e - (m-p)\omega_M)t} \langle m|e^{i\vec{k}_b \cdot \vec{X}}|p\rangle C_{\downarrow,p}$$

$$+g_r^* e^{-i(\omega_r - (\omega_e - \omega_0) - (m-p)\omega_M)t} \langle m|e^{i\vec{k}_r \cdot \vec{X}}|p\rangle C_{\uparrow,p} \Bigg], \tag{4.14}$$

and therefore $C_{e,m}$ will track the more slowly varying coefficients $C_{\downarrow,p}$ and $C_{\uparrow,p}$. We can plug this last expression into Eqs. (4.7) and (4.9) to obtain

$$\dot{C}_{\downarrow,n} = -i\frac{|g_b|^2}{\Delta} C_{\downarrow,n} - i \sum_{n'=0}^{\infty} \Omega_{n,n'}\, e^{i(\delta - (n'-n)\omega_M)t} C_{\uparrow,n'} \tag{4.15}$$

and

$$\dot{C}_{\uparrow,n'} = -i\frac{|g_r|^2}{\Delta} C_{\uparrow,n'} - i \sum_{n=0}^{\infty} \Omega^*_{n',n}\, e^{-i(\delta + (n-n')\omega_M)t} C_{\downarrow,n}, \tag{4.16}$$

where $\delta \equiv \omega_b - \omega_r - \omega_0$ is the Raman transition detuning and

$$\Omega_{n,n'} \equiv \Omega\langle n|e^{-i(\vec{k}_b-\vec{k}_r)\cdot\vec{X}}|n'\rangle = \Omega\langle n|e^{-i\eta(a+a^\dagger)}|n'\rangle = \Omega_{n',n}. \tag{4.17}$$

In Eq. (4.17),

$$\Omega \equiv g_b g_r^*/\Delta = \langle\downarrow|\hat{\epsilon}_b\cdot\vec{r}\,|e\rangle\langle e|\hat{\epsilon}_r\cdot\vec{r}\,|\uparrow\rangle\frac{e^2 E_{b0}E_{r0}}{4\hbar^2\Delta}e^{i(\phi_r-\phi_b)}, \tag{4.18}$$

$\eta \equiv (\vec{k}_b - \vec{k}_r)\cdot\hat{x}x_0$, and [78]

$$\langle n|e^{-i\eta(a+a^\dagger)}|n'\rangle = e^{-\eta^2/2}\sqrt{\frac{n_<!}{n_>!}}[-i\eta]^{|n'-n|}L_{n_<}^{|n'-n|}(\eta^2), \tag{4.19}$$

where $n_<$ ($n_>$) is the lesser (greater) of n and n'. L_n^p is the generalized Laguerre polynomial,

$$L_n^p(Y) = \sum_{m=0}^{n}(-1)^m\frac{(n+p)!}{(n-m)!(m+p)!}\frac{Y^m}{m!}. \tag{4.20}$$

We usually say the substitution step used to obtain Eqs. (4.15) and (4.16) "adiabatically eliminates" the upper states from the ground-state dynamics.

The first terms in Eqs. (4.15) and (4.16) are the AC Stark shifts of levels $|\downarrow\rangle$ and $|\uparrow\rangle$ due to the blue and red lasers respectively. The second terms give rise to the stimulated Raman coupling between the $|\downarrow\rangle$ and $|\uparrow\rangle$ states. In general, the blue laser will also couple the $|\uparrow\rangle$ and $|e\rangle$ states, and the red laser will couple the $|\downarrow\rangle$ and $|e\rangle$ states. These couplings will give negligible stimulated Raman couplings (since they will be non-resonant), but we must account for their Stark shifts, which we can do using second-order perturbation theory. Including these Stark shifts, Eqs. (4.15) and (4.16) become

$$\dot{C}_{\downarrow,n} = -i\Delta_{S\downarrow}C_{\downarrow,n} - i\sum_{p=0}^{\infty}\Omega_{n,p}\,e^{i(\delta-(p-n)\omega_M)t}C_{\uparrow,p}, \tag{4.21}$$

and

$$\dot{C}_{\uparrow,n} = -i\Delta_{S\uparrow}C_{\uparrow,n} - i\sum_{p=0}^{\infty}\Omega_{n,p}^*\,e^{-i(\delta+(p-n)\omega_M)t}C_{\downarrow,p}, \tag{4.22}$$

where, to a good approximation,

$$\Delta_{S\downarrow} \equiv |g_b|^2/\Delta + |g_{\downarrow,e,r}|^2/(\Delta-\omega_0), \quad \Delta_{S\uparrow} \equiv |g_{\uparrow,e,b}|^2/(\Delta+\omega_0) + |g_r|^2/\Delta, \tag{4.23}$$

and $g_{\downarrow,e,r} = \langle\downarrow|\hat{\epsilon}_r \cdot \vec{r}\,|e\rangle eE_{r0}e^{-i\phi_r}/2\hbar$ and $g_{\uparrow,e,b} = \langle\uparrow|\hat{\epsilon}_b \cdot \vec{r}\,|e\rangle eE_{b0}e^{-i\phi_b}/2\hbar$. We can absorb these Stark shifts into the ground-state coefficients by making the substitutions

$$C_{\downarrow,n} = C'_{\downarrow,n}e^{-i\Delta_{S\downarrow}t}, \quad C_{\uparrow,n} = C'_{\uparrow,n}e^{-i\Delta_{S\uparrow}t}. \tag{4.24}$$

Equations (4.21) and (4.22) then can be written

$$\dot{C}'_{\downarrow,n} = -i\sum_{n'=0}^{\infty} \Omega_{n,n'}e^{i\delta_{n',n}t}C'_{\uparrow,n'}, \tag{4.25}$$

and

$$\dot{C}'_{\uparrow,n'} = -i\sum_{n=0}^{\infty} \Omega^*_{n',n}e^{-i\delta_{n',n}t}C'_{\downarrow,n}, \tag{4.26}$$

where $\delta_{n',n} \equiv \delta - (\Delta_{S\uparrow} - \Delta_{S\downarrow}) - (n'-n)\omega_M$. We will usually be interested in the case where $|\Omega_{n,n'}| \ll \omega_M$ and where we are near a resonance $\delta_{n',n} \simeq 0$. Under these conditions, only one term of the sums in Eqs. (4.25) and (4.26) is relevant and we have

$$\dot{C}'_{\downarrow,n} \simeq -i\Omega_{n,n'}e^{i\delta_{n',n}t}C'_{\uparrow,n'}, \tag{4.27}$$

and

$$\dot{C}'_{\uparrow,n'} \simeq -i\Omega^*_{n',n}e^{-i\delta_{n',n}t}C'_{\downarrow,n}. \tag{4.28}$$

A general solution for these equations is given in, for example, Ref. [4]. On exact resonance $\delta_{n',n} = 0$, we find

$$\ddot{C}'_{\downarrow,n} + |\Omega_{n',n}|^2C'_{\downarrow,n} = 0, \quad \ddot{C}'_{\uparrow,n'} + |\Omega_{n',n}|^2C'_{\uparrow,n'} = 0, \tag{4.29}$$

$$C_{\downarrow,n}(t) = C_{\downarrow,n}(0)\cos|\Omega_{n,n'}t| - i\frac{\Omega_{n,n'}}{|\Omega_{n,n'}|}C_{\uparrow,n'}(0)\sin|\Omega_{n,n'}t|, \tag{4.30}$$

and

$$C_{\uparrow,n'}(t) = C_{\uparrow,n'}(0)\cos|\Omega_{n,n'}t| - i\frac{\Omega^*_{n,n'}}{|\Omega_{n,n'}|}C_{\downarrow,n}(0)\sin|\Omega_{n,n'}t|. \tag{4.31}$$

These equations describe sinusoidal "Rabi-flopping" between levels $|\downarrow\rangle$ and $|\uparrow\rangle$. For example, a "π-pulse" ($\Omega_{n',n}t = \pi/2$) completely transfers the state $|\downarrow, n\rangle$ to the state $|\uparrow, n'\rangle$. If the qubit starts in the $|\downarrow, n\rangle$ state, the probability vs. time of occupation in this state is given by $P_{\downarrow,n} = \cos^2(\Omega_{n',n}t)$.

Often, we will speak about driving transitions under the conditions that the ion is in the "Lamb-Dicke" limit, that is when the spread of the ion's wave function is much smaller than $1/2\pi$ times the wavelength or effective wavelength of the driving radiation. For two-photon stimulated Raman transitions and our 1-D harmonic well considered here, the Lamb-Dicke limit implies $\sqrt{\langle \vec{X}_M^{\,2}\rangle} \ll 1/|\vec{k}_b - \vec{k}_r|$, or equivalently, $\eta\sqrt{\langle (a+a^\dagger)^2\rangle} \ll 1$. To achieve the Lamb-Dicke limit, we must have $\eta \ll 1$ and typically use motional states with small n.

In the Lamb-Dicke limit, we will be primarily concerned with "carrier" ($n' = n$), first "red-sideband" ($n' = n-1$), and first "blue-sideband" ($n' = n+1$) transitions. Second-sideband transitions ($n' = n \pm 2$) will be suppressed relative to the first-sideband transitions by approximately another factor of η. We consider a few values of $\Omega_{n',n}$ in the Lamb-Dicke limit for small values of n and n'. In the Lamb-Dicke limit, the complete expressions given in Eqs. (4.19) and (4.20) can be simplified by noting that to second order in η,

$$\langle n|e^{-i\eta(a+a^\dagger)}|n'\rangle \simeq \langle n|\left[1 - i\eta(a+a^\dagger) - \eta^2(1+2\tilde{n}+a^2+(a^\dagger)^2)/2\right]|n'\rangle. \quad (4.32)$$

4.1. Carrier transitions

From Eq. (4.32) we have $\Omega_{n,n} = \Omega(1-\eta^2(n+1/2)) \simeq \Omega e^{-\eta^2/2}(1-n\eta^2)$. The exponential factor in this expression (and in Eq. (4.19)) is the Debye-Waller factor [78], originally coming from studies of X-ray scattering in solids [79]. Both of the factors multiplying Ω express the suppression of the interaction with the laser fields due to the spread of the ion's wavepacket; effectively, this spreading tends to average out the interaction with the laser wave.

4.2. Sidebands

For first-order sideband transitions where $n' = n \pm 1$, Eq. (4.32) gives $\Omega_{n',n} \simeq -i\Omega\eta\sqrt{n_>}$. For $n' = n-1$ (first red sideband), the resulting dynamics described by Eqs. (4.27) and (4.28) could as well have been derived by assuming $H_I = \hbar\eta\Omega|\downarrow\rangle\langle\uparrow|a^\dagger + h.c.$. This Hamiltonian is formally equivalent to the Jaynes-Cummings Hamiltonian from quantum optics [80]. In cavity-QED (for a recent review, see Ref. [81]), this Hamiltonian describes the coupling between a single atom and a single mode of the radiation field. Here, the transition of an atom from the upper to the lower state ($|\downarrow\rangle\langle\uparrow|)|\uparrow\rangle = |\downarrow\rangle$ is accompanied by the emission of a single photon to the cavity $a^\dagger|n\rangle = \sqrt{n+1}|n\rangle$. The red-sideband transitions for a trapped ion and the Jaynes-Cummings coupling both describe the coupling of single atom to a single mode of a harmonic oscillator. In cavity-QED, the relevant harmonic oscillator is a single mode of the radiation field, whereas for the trapped ion, the relevant oscillator is a single mode of the ion's motion; we can think of photons being replaced by phonons.

For $n' = n + 1$, the first blue sideband, excitation of the ion's internal state is accompanied by the creation of a phonon, the energy being supplied by the classical drive field. Sideband transitions where $n' = n \pm m$ ($|m| > 1$), are analogous to higher-order processes in optics (see Refs. [82] - [85] and references therein). Some of these processes can be simulated experimentally [86].

Eqs. (4.30) and (4.31) indicate the key entangling mechanism in an ion trap quantum processor. One way this can be seen is by noting that if the ion starts in a particular eigenstate, say $|\downarrow, n\rangle$, then after a certain evolution time, the ion evolves to an entangled state between motion and internal states of the form

$$|\downarrow, n\rangle \rightarrow \cos|\Omega_{n,n'}t||\downarrow, n\rangle - i\frac{\Omega^*_{n,n'}}{|\Omega_{n,n'}|}\sin|\Omega_{n,n'}t||\uparrow, n'\rangle. \tag{4.33}$$

Even for carrier transitions ($n = n'$), the Rabi frequency $\Omega_{n,n}$ is dependent on n so that internal state dynamics is conditioned on the state of motion. These processes can be combined to make universal logic gates as explained in the original proposal [1] and as demonstrated in various experiments.

4.3. Spontaneous emission

A fundamental limitation to the coherence of atomic ion qubits is spontaneous emission. When the qubits are formed from ground-state hyperfine levels, memory is not affected by spontaneous emission since these levels have very long radiative lifetimes. The problem arises from spontaneous emission during Raman transitions (see for example Refs. [87] - [89]). As an estimate of the decoherence rate due to spontaneous emission we calculate the total spontaneous emission rate R_{SE} from the excited state e. This rate is given by

$$R_{SE} \simeq \gamma_e \sum_{m=0}^{\infty} |C_{e,m}|^2, \tag{4.34}$$

where γ_e is the spontaneous decay rate from level e. With Eq. (4.14), we have

$$R_{SE} \simeq \gamma_e \Bigg[|C_{\downarrow,n}|^2 \left(\frac{|g_b|^2}{\Delta^2} + \frac{|g_{\downarrow,e,r}|^2}{(\Delta - \omega_0)^2} \right) + |C_{\uparrow,n'}|^2 \left(\frac{|g_{\uparrow,e,b}|^2}{(\Delta + \omega_0)^2} + \frac{|g_r|^2}{\Delta^2} \right) \Bigg]. \tag{4.35}$$

As a measure of the rate of spontaneous emission, we compare it to the Raman Rabi frequency Ω for carrier transitions in the Lamb-Dicke limit. If we assume

$|g_r| \simeq |g_b| \simeq |g_{\downarrow,e,r}| \simeq |g_{\uparrow,e,b}|$ and $|\Delta| \gg \omega_0$, then $R_{SE}/\Omega \simeq \gamma_e/|\Delta|$. Therefore we can suppress the spontaneous emission rate relative to the Raman Rabi rate by making Δ large enough (and presumably increasing the laser beam intensities to maintain the same value of Ω).

5. Multiple modes, multiple excited states

We now consider a more general situation where N ion qubits are confined in the same trap. We will still assume that one qubit in the array can be selectively addressed with a focused laser beam and the ions are cold so that the motion may be described by (harmonic) normal modes. Therefore the operator that describes the departure of an ion from its equilibrium position is written as

$$\vec{X}_M = \sum_{i=1}^{3N} \hat{u}_i q_{i0}(a_i + a_i^\dagger), \tag{5.1}$$

where $\hat{u}_i$ is a unit vector in the direction of the ion's displacement for normal mode i, and q_{i0} is the zero-point amplitude of the addressed ion for ith mode. We will now also allow the possibility of coupling to multiple excited states $\{|e\rangle\}$; however, as in the previous section, we first assume that the blue laser couples only the $|e\rangle$ states to $|\downarrow\rangle$ and the red laser couples only the $|e\rangle$ states to the $|\uparrow\rangle$ state. (Stark shifts from the other possible couplings can be added in later as a perturbation as in Eqs. (4.21) and (4.22)). We now have

$$H_0 = \hbar\omega_0|\uparrow\rangle\langle\uparrow| + \hbar\sum_e \omega_e|e\rangle\langle e| + \hbar\sum_{i=0}^{3N}\omega_i\tilde{n}_i, \tag{5.2}$$

where ω_i is now the frequency of mode i, $\tilde{n}_i = a_i^\dagger a_i$, and $a_i^\dagger$ and a_i are the raising and lowering operators for mode i. Similarly to Eq. (4.6), we now write

$$\Psi = \sum_{\{n\}}\Big[C_{\downarrow,\{n\}}e^{-i\{n\}\omega_{\{n\}}t}|\downarrow,\{n\}\rangle + C_{\uparrow,\{n\}}e^{-i[\omega_0+\{n\}\omega_{\{n\}}]t}|\uparrow,\{n\}\rangle$$
$$+ C_{e,\{n\}}e^{-i[\omega_e+\{n\}\omega_{\{n\}}]t}|e,\{n\}\rangle\Big], \tag{5.3}$$

where we have used a short hand notation

$$\sum_{\{n\}} C_{\downarrow,\{n\}}e^{-i\{n\}\omega_{\{n\}}t}|\downarrow,\{n\}\rangle \equiv \sum_{n_1=0}^{\infty}\sum_{n_2=0}^{\infty}\cdots\sum_{n_{3N}=0}^{\infty} C_{\downarrow,n_1,n_2,\ldots n_{3N}}$$
$$\times \exp(-i[n_1\omega_1 + n_2\omega_2 \ldots + n_{3N}\omega_{3N}]t)|\downarrow,n_1,n_2,\ldots n_{3N}\rangle. \tag{5.4}$$

Following the same procedure as in the previous section, we arrive at

$$\dot{C}'_{\downarrow\{n\}} = -i \sum_e \sum_{\{n\}} \Omega_{e\{n\}\{n'\}} e^{i\delta_{\{n'\}\{n\}}t} C'_{\uparrow\{n'\}} \tag{5.5}$$

and

$$\dot{C}'_{\uparrow\{n'\}} = -i \sum_e \sum_{\{n\}} \Omega^*_{e\{n'\}\{n\}} e^{-i\delta_{\{n'\}\{n\}}t} C'_{\downarrow\{n\}}, \tag{5.6}$$

where $\delta_{\{n'\}\{n\}} \equiv \delta - (\Delta_{S\uparrow} - \Delta_{S\downarrow}) - (\{n'\}\omega_{\{n'\}} - \{n\}\omega_{\{n\}})$. As in the previous section, the Stark shifts must be modified to include couplings between $|\downarrow\rangle$ and $\{|e\rangle\}$ and $|\uparrow\rangle$ and $\{|e\rangle\}$ from both laser beams, so that

$$\Delta_{S\downarrow} \equiv \sum_e \Big[|g_{\downarrow eb}|^2/\Delta_e + |g_{\downarrow er}|^2/(\Delta_e - \omega_0) \Big], \tag{5.7}$$

and

$$\Delta_{S\uparrow} \equiv \sum_e \Big[|g_{\uparrow eb}|^2/(\Delta_e + \omega_0) + |g_{\uparrow er}|^2/\Delta_e \Big], \tag{5.8}$$

where $g_{\downarrow eb} = \langle\downarrow|\hat{\epsilon}_b \cdot \vec{r}\,|e\rangle e E_{b0} e^{-i\phi_b}/2\hbar$ and $g_{\uparrow er} = \langle\uparrow|\hat{\epsilon}_r \cdot \vec{r}\,|e\rangle e E_{r0} e^{-i\phi_r}/2\hbar$. In Eqs. (5.5) and (5.6),

$$\begin{aligned} \Omega_{e\{n\}\{n'\}} &\equiv \Omega_e \langle\{n\}| e^{-i(\vec{k}_b - \vec{k}_r)\cdot\vec{X}} |\{n'\}\rangle \\ &= \Omega_e \langle\{n\}| \exp\Bigg[-i\Big(\sum_{i=1}^{3N} \eta_i (a_i + a_i^\dagger) \Big) \Bigg] |\{n'\}\rangle = \Omega_{e\{n'\}\{n\}}, \end{aligned} \tag{5.9}$$

where

$$\Omega_e \equiv g_{\downarrow eb} g^*_{\uparrow er}/\Delta_e = \langle\downarrow|\hat{\epsilon}_b \cdot \vec{r}\,|e\rangle\langle e|\hat{\epsilon}_r \cdot \vec{r}\,|\uparrow\rangle \frac{e^2 E_{b0} E_{r0}}{4\hbar^2 \Delta_e} e^{i(\phi_r - \phi_b)}, \tag{5.10}$$

and $\eta_i \equiv (\vec{k}_b - \vec{k}_r) \cdot \hat{u}_i q_{i0}$. As before, we will usually be interested in the case where $\delta_{\{n'\}\{n\}} \simeq 0$. Moreover, we will usually be interested in the case where the quantum state of at most one (spectrally-isolated) motional mode, say the kth mode, is changed. In that case, we can write

$$\Omega_{e\{n\}\{n'\}} = \Omega_{en_k,n'_k} \prod_{p \neq k} \langle n_p| e^{-i\eta_p(a_p + a_p^\dagger)} |n_p\rangle. \tag{5.11}$$

In this expression, Ω_{en_k,n'_k} is the same as $\Omega_{n,n'}$ given in Eq. (4.17), except that it expresses the stimulated Raman amplitude for coupling through a particular excited state e. According to Eqs. (5.5) and (5.6), we must sum these amplitudes over all excited states $\{e\}$ to arrive at the full expression for the Rabi frequency. The remaining product terms in Eq. (5.11) represent the Debye-Waller factors for the $3N - 1$ modes other than the kth mode that we are interested in. Motion, including zero-point motion, in all other modes reduces the interaction with the laser beam and must be taken into account, particularly if the motion in these modes fluctuates from experiment to experiment [4].

We can suppress the Debye-Waller factors and Raman-coupling from 2/3 of the motional modes by choosing $\vec{k}_b - \vec{k}_r$ to be parallel to one of the principle mode axes (and perpendicular to the other principle axes). Typically we choose $\vec{k}_b - \vec{k}_r$ to be parallel to $\hat{z}$, the direction along the trap axis. In practice, this means that Doppler cooling in the x and y directions is usually sufficient since the stimulated Raman Rabi rates are insensitive to motion in these directions. We have assumed that we can spectrally isolate the kth mode from all remaining modes of motion that give rise to Raman coupling, but if the remaining modes are not cooled to their ground states we must be careful to ensure that $(n_{k'} - n_k)\omega_k \neq n_p\omega_p + n_q\omega_q$, where n_p and n_q are integers and ω_p and ω_q are frequencies of two other modes. If $(n_{k'} - n_k)\omega_k \simeq n_p\omega_p + n_q\omega_q$, then we not only drive the desired $n_k \leftrightarrow n_{k'}$ transition, but we can also drive (unwanted) transitions that transfer quanta to or between modes p and q and lead to unwanted entanglement with those modes [4].

As noted above, to arrive at the complete expression for the Rabi frequency given in Eqs. (5.5) and (5.6), we must sum over all excited states $\{e\}$. For most ions (and neutral atoms) of interest for quantum information processing, the relevant excited states are contained in the first excited $^2P_{1/2}$ and $^2P_{3/2}$ fine-structure manifolds. Unfortunately, we find that we cannot suppress the rate of spontaneous emission relative to the Raman transition Rabi rate by making $|\Delta_e|$ arbitrarily large, as we could in the simple case of one excited state, treated in § 4; in fact we are limited to values of $|\Delta_e|$ approximately equal to the $^2P_{1/2}$, $^2P_{3/2}$ fine structure splitting. This will ultimately limit our possible choices of ion qubits that will satisfy the conditions of fault-tolerant computation if we use stimulated Raman transitions [13].

Acknowledgements

We thank M. Barrett, J. Britton, J. Chiaverini, J. Jost, C. Langer, D. Leibfried, M. Lombardi, R. Ozeri, T. Schaetz, P. Schmidt, and D. Smith for helpful comments on the manuscript. This paper is a submission from the U. S. National Institute of Standards and Technology; not subject to U. S. copyright.

References

[1] J I Cirac and P Zoller. Quantum computation with cold, trapped ions. *Phys. Rev. Lett.*, 74(20):4091–4094, May 1995.

[2] D P DiVincenzo. The physical implementation of quantum computation. In S L Braunstein, H -K Lo, and P Kok, editors, *Scalable Quantum Computers*, pages 1–13, Berlin, 2001. Wiley-VCH.

[3] A Steane. The ion trap quantum information processor. *Appl. Phys. B*, 64:623–642, 1997.

[4] D J Wineland, C Monroe, W M Itano, D Leibfried, B E King, and D M Meekhof. Experimental issues in coherent quantum-state manipulation of trapped atomic ions. *J. Res. Nat. Inst. Stand. Tech.*, 103:259–328, 1998.

[5] D J Wineland, C Monroe, W M Itano, B E King, D Leibfried, D M Meekhof, C Myatt, and C Wood. Experimental primer on the trapped ion quantum computer. *Fortschritte der Physik*, 46:363–389, 1998.

[6] D F V James. Quantum computation with hot and cold ions: an assessment of proposed schemes. In S L Braunstein, H -K Lo, and P Kok, editors, *Scalable Quantum Computers*, pages 53–68, Berlin, 2001. Wiley-VCH.

[7] G J Milburn, S Schneider, and D F V James. Ion trap quantum computing with warm ions. In S L Braunstein, H -K Lo, and P Kok, editors, *Scalable Quantum Computers*, pages 31–40, Berlin, 2001. Wiley-VCH.

[8] J F Poyatos, J I Cirac, and P Zoller. Schemes of quantum computations with trapped ions. In S L Braunstein, H -K Lo, and P Kok, editors, *Scalable Quantum Computers*, pages 15–30, Berlin, 2001. Wiley-VCH.

[9] A Sørensen and K Mølmer. Ion trap quantum computer with bichromatic light. In S L Braunstein, H -K Lo, and P Kok, editors, *Scalable Quantum Computers*, pages 41–52, Berlin, 2001. Wiley-VCH.

[10] A M Steane and D M Lucas. Quantum computing with trapped ions, atoms and light. In S L Braunstein, H -K Lo, and P Kok, editors, *Scalable Quantum Computers*, pages 69–88, Berlin, 2001. Wiley-VCH.

[11] C Monroe. Quantum information processing with atoms and photons. *Nature*, 416:238–246, 2002.

[12] D J Wineland. Trapped ions and quantum information processing. In F. De Martini and C. Monroe, editors, *Experimental Quantum Computation and Information, Proc. Int. School of Physics "Enrico Fermi"*, volume 148, pages 165–196, Amsterdam, 2002. IOS Press.

[13] D J Wineland, M Barrett, J Britton, J Chiaverini, B DeMarco, W M Itano, B. Jelenković, C Langer, D Leibfried, V Meyer, T Rosenband, and T Schätz. Quantum information processing with trapped ions. *Phil. Trans. R. Soc. Lond. A*, 361:1349–1361, 2003.

[14] S Gulde, H Häffner, M Riebe, G Lancaster, A Mundt, A Kreuter, C Russo, C Becher, J Eschner, F Schmidt-Kaler, I L Chuang, and R Blatt. Quantum information processing and cavity QED experiments with trapped Ca^+ ions. In H R Sadeghpour, E J Heller, and D E Pritchard, editors, *Proceedings of the XVIII International Conference on Atomic Physics*, pages 293–302, Singapore, 2003. World Scientific.

[15] D J Wineland, D Leibfried, B DeMarco, V Meyer, M Rowe, A Ben-Kish, M Barrett, J Britton, J Hughes, W M Itano, B M Jelenković, C Langer, D Lucas, and T Rosenband. Quantum information processing and multiplexing with trapped ions. In H R Sadeghpour, E J Heller, and D E Pritchard, editors, *Proceedings of the XVIII International Conference on Atomic Physics*, pages 263–272, Singapore, 2003. World Scientific.

[16] C Wunderlich and C Balzer. Quantum measurements and new concepts for experiments with trapped ions. *quant-ph/0305129*, 2003.

[17] W Paul. Electromagnetic traps for charged and neutral particles. *Rev. Mod. Phys.*, 62:531–540, 1990.

[18] H G Dehmelt. Radiofrequency spectroscopy of stored ions I: storage. *Adv. Atom. Mol. Phys.*, 3:53–72, 1967.

[19] H G Dehmelt. Radiofrequency spectroscopy of stored ions II: spectrocopy. *Adv. Atom. Mol. Phys.*, 5:109–153, 1967.

[20] D J Wineland, W M Itano, and R S VanDyck. High resolution spectroscopy of stored ions. *Adv. Atom. Mol. Phys.*, 19:135–186, 1983.

[21] R C Thompson. Spectroscopy of trapped ions. *Adv. Atom. Mol. Phys.*, 31:63–136, 1993.

[22] D Leibfried, R Blatt, C Monroe, and D Wineland. Quantum dynamics of single trapped ions. *Rev. Mod. Phys.*, 75:281Ű–324, 2003.

[23] P K Ghosh. *Ion Traps*. Clarendon Press, Oxford, 1995.

[24] J Drees and W Paul. Beschleunigung von elektronen in einem plasmabetatron. *Z. Phys.*, 180:340–361, 1964.

[25] C A Schrama, E Peik, W W Smith, and H Walther. Novel miniature ion traps. *Opt. Commun.*, 101:32–36, 1993.

[26] R G DeVoe. Elliptical ion traps and trap arrays for quantum computation. *Phys. Rev. A*, 58(2):910–914, 1998.

[27] H C Nägerl, W Bechter, J Eschner, F Schmidt-Kaler, and R Blatt. Ion strings for quantum gates. *Appl. Phys. B*, 66:603–608, 1998.

[28] P A Barton, C J S Donald, D M Lucas, D A Stevens, A M Steane, and D N Stacey. Measurement of the lifetime of the 3d $^2D_{5/2}$ state in $^{40}\mathrm{Ca}^+$. *Phys. Rev. A*, 62:032503–1–10, 2000.

[29] J I Cirac and P Zoller. A scalable quantum computer with ions in an array of microtraps. *Nature*, 404:579–581, 2000.

[30] G R Guthöhrlein, M Keller, K Hayasaka, W Lange, and H Walther. A single ion as a nanoscopic probe of an optial field. *Nature*, 414:49–51, 2001.

[31] L Hornekær, N Kjærgaard, A M Thommesen, and M Drewsen. Structural properties of two-component Coulomb crystals in linear Paul traps. *Phys. Rev. Lett.*, 86:1994–1997, 2001.

[32] D J Berkeland. Linear Paul trap for strontium ions. *Rev. Sci. Instrum.*, 73(8):2856–2860, 2002.

[33] D Kielpinski, C Monroe, and D J Wineland. Architecture for a large-scale ion-trap quantum computer. *Nature*, 417:709–711, 2002.

[34] N W McLachlan. *Theory and Applications of Mathieu Functions*. Clarendon, 1947.

[35] M Drewsen and A Brøner. Harmonic linear Paul trap: stability diagram and effective potentials. *Phys. Rev. A*, 62:045401–1–4, 2000.

[36] W M Itano and D J Wineland. Laser cooling of ions stored in harmonic and Penning traps. *Phys. Rev. A*, 25(1):35–54, 1982.

[37] D J Wineland, W M Itano, J C Bergquist, and R G Hulet. Laser-cooling limits and single-ion spectroscopy. *Phys. Rev. A*, 36(5):2220–2232, 1987.

[38] M A Rowe, A Ben-Kish, B DeMarco, D Leibfried, V Meyer, J Beall, J Britton, J Hughes, W M Itano, B Jelenković, C Langer, T Rosenband, and D J Wineland. Transport of quantum states and separation of ions in a dual RF ion trap. *Quant. Inform. Comp.*, 2(4):257–271, 2002.

[39] R J Cook, D G Shankland, and A L Wells. Quantum theory of particle motion in a rapidly oscillating field. *Phys. Rev. A*, 31:564–567, 1985.

[40] M Combescure. A quantum particle in a quadrupole radio-frequency trap. *Ann. Inst. Henri Poincaré*, 44(3):293–314, 1986.

[41] R J Glauber. The quantum mechanics of trapped wave packets. In E Arimondo, W D Phillips, and F Strumia, editors, *Proceedings of the International School of Physics "Enrico Fermi"*, volume CXVIII, pages 643–660, New York, 1992. North-Holland.

[42] V N Gheorghe and F Vedel. Quantum dynamics of trapped ions. *Phys. Rev. A*, 45:4828–4831, 1992.

[43] B Baseia, R Vyas, and V S Bagnato. Particle trapping by oscillating fields: squeezing effects. *Quant. Opt.*, 5:155–159, 1993.

[44] P J Bardroff, C Leichtle, G Schrade, and W P Schleich. Endoscopy in the Paul trap: measurement of the vibratory quantum state of a single ion. *Phys. Rev. Lett.*, 77(11):2198–2201, 1996.

[45] P J Bardroff, C Leichtle, G Schrade, and W P Schleich. Paul trap multi-quantum interactions. *Act. Phys. Slov.*, 46:231–240, 1996.

[46] Q A Turchette, C S Wood, B E King, C J Myatt, D Leibfried, W M Itano, C Monroe, and D J Wineland. Deterministic entanglement of two trapped ions. *Phys. Rev. Lett.*, 81:1525–1528, August 1998.

[47] D G Enzer, M M Schauer, J J Gomez, M S Gulley, M H Holzscheiter, P G Kwiat, S K Lamoreaux, C G Peterson, V D Sandberg, D Tupa, A G White, R J Hughes, and D F V James. Observation of power-law scaling for phase transitions in linear trapped ion crystals. *Phys. Rev. Lett.*, 85(12):2466–2469, 2000.

[48] R G DeVoe, J Hoffnagle, and R G Brewer. Role of laser damping in trapped ion crystals. *Phys. Rev. A*, 39(9):4362–4365, 1989.

[49] R Blümel, C Kappler, W Quint, and H Walther. Chaos and order of laser-cooled ions in a Paul trap. *Phys. Rev. A*, 40(2):808–823, 1989.

[50] J I Cirac, L J Garay, R Blatt, A S Parkins, and P Zoller. Laser cooling of trapped ions: The influence of micromotion. *Phys. Rev. A*, 49:421–432, 1994.

[51] H Walther. Phase transitions of stored laser-cooled ions. *Adv. At. Mol. Opt. Phys.*, 31:137–182, 1993.

[52] N Yu, W Nagourney, and H Dehmelt. Demonstration of new Paul-Straubel trap for trapping single ions. *Am. J. Phys.*, 69(6):3779–3781, 1991.

[53] J D Miller, D J Berkeland, F C Cruz, J C Bergquist, W M Itano, and D J Wineland. A cryogenic linear ion trap for $^{199}Hg^{+}$ frequency standards. *IEEE Int. Frequency Control Symposium*, 1996:1086–1088, 1996.

[54] D J Berkeland, J D Miller, J C Bergquist, W M Itano, and D J Wineland. Minimization of ion micromotion in a Paul trap. *J. Appl. Phys.*, 83(10):5025–5033, 1998.

[55] M D Barrett, B DeMarco, T Schaetz, V Meyer, D Leibfried, J Britton, J Chiaverini, W M Itano, B Jelenković, J D Jost, C langer, T Rosenband, and D J Wineland. Sympathetic cooling of $^{9}Be^{+}$ and $^{24}Mg^{+}$ for quantum logic. *Phys. Rev. A*, 68:042302–1–8, 2003.

[56] D Leibfried. Individual addressing and state readout of trapped ions utilizing rf micromotion. *Phys. Rev. A*, 60:R3335–R3338, 1999.

[57] D F V James. Quantum dynamics of cold trapped ions with applications to quantum computing. *Appl. Phys. B*, 66:181–190, 1998.

[58] J J García-Ripoll, P Zoller, and J I Cirac. Speed optimized two-qubit gates with laser coherent control techniques for ion trap quantum computing. *Phys. Rev. Lett.*, 91(15):157901–1–4, 2003.

[59] N Kjærgaard, L Hornekær, A M Thommesen, Z Videsen, and M Drewsen. Isotope selective loading of an ion trap using resonance-enhanced two-photon ionization. *Appl. Phys. B*, 71:207–210, 2000.

[60] S Gulde, D Rotter, P Barton, F Schmidt-Kaler, R Blatt, and W Hogervorst. Simple and efficient photo-ionization loading of ions for precision ion-trapping experiments. *Appl. Phys. B*, 73:861–863, 2001.

[61] Q A Turchette, D Kielpinski, B E King, D Leibfried, D M Meekhof, C J Myatt, M A Rowe, C A Sackett, C S Wood, W M Itano, C Monroe, and D J Wineland. Heating of trapped ions from the quantum ground state. *Phys. Rev. A*, 61:063418–1–8, 2000.

[62] K Sugiyama and J Yoda. Production of YbH^+ by chemical reaction of Yb^+ in excited states with H_2 gas. *Phys. Rev. A*, 55:R10–R13, 1997.

[63] K Mølhave and M Drewsen. Formation of translationally cold MgH^+ and MgH^+ molecules in an ion trap. *Phys. Rev. A*, 62:011401–1–4, 2000.

[64] X P Huang, J J Bollinger, T B Mitchell, W M Itano, and D H E Dubin. Precise control of the global rotation of strongly coupled ion plasmas in a Penning trap. *Phys. Plasmas*, 5(5):1656–1663, 1998.

[65] R J Rafac, B C Young, J A Beall, W M Itano, D J Wineland, and J C Bergquist. Sub-dekahertz ultraviolet spectroscopy of $^{199}Hg^+$. *Phys. Rev. Lett.*, 85(12):2462–2465, 2000.

[66] J C Bergquist, R J Rafac, B C Young, J A Beall, W M Itano, and D J Wineland. Sub-dekahertz spectroscopy of $^{199}Hg^+$. In J. L. Hall and J. Ye, editors, *Laser Frequency Stabilization, Standards, Measurement, and Applications*, volume 4269, pages 1–7, Bellingham, WA, 2001. Proc. SPIE.

[67] J D Prestage, G J Dick, and L Maleki. Linear ion trap based atomic frequency standard. *IEEE Trans. Instrum. Meas.*, 40:132–136, 1991.

[68] M G Raizen, J M Gilligan, J C Bergquist, W M Itano, and D J Wineland. Ionic crystals in a linear Paul trap. *Phys. Rev. A*, 45(9):6493–6501, May 1993.

[69] D J Berkeland, J D Miller, J C Bergquist, W M Itano, and D J Wineland. Laser-cooled mercury ion frequency standard. *Phys. Rev. Lett.*, 80(10):2089–2092, 1998.

[70] P T H Fisk. Trapped-ion and trapped-atom microwave frequency standards. *Rep. Prog. Phys.*, 60(8):761–816, 1997.

[71] P. Gill, editor. *Proceedings of the 6th Symposium on Frequency Standards and Metrology*, Singapore, 2002. World Scientific.

[72] F Schmidt-Kaler, H Häffner, M Riebe, S Gulde, G P T Lancaster, T Deuschle, C Becher, C Roos, J Eschner, and R Blatt. Realization of the Cirac-Zoller controlled-NOT quantum gate. *Nature*, 422:408–411, 2003.

[73] F Schmidt-Kaler, S Gulde, M Riebe, T Deuschle, A Kreuter, G Lancaster, C Becher, J Eschner, H Häffner, and R Blatt. The coherence of qubits based on single Ca^+ ions. *J. Phys. B: At. Mol. Opt. Phys.*, 36:623Ű–636, 2003.

[74] H Häffner, S Gulde, M Riebe, G Lancaster, C Becher, J Eschner, F Schmidt-Kaler, and R Blatt. Precision measurement and compensation of optical Stark shifts for an ion-trap quantum processor. *Phys. Rev. Lett.*, 90:143602–1–4, 2003.

[75] D J Wineland, J J Bollinger, W M Itano, F L Moore, and D J Heinzen. Spin squeezing and reduced quantum noise in spectroscopy. *Phys. Rev. A*, 46(11):R6797–R6800, December 1992.

[76] F Mintert and C Wunderlich. Ion-trap quantum logic using long-wavelength radiation. *Phys. Rev. Lett.*, 87(25):257904–1–4, 2001.

[77] G Ciaramicoli, I Marzoli, and P Tombesi. Scalable quantum processor with trapped electrons. *Phys. Rev. Lett.*, 91(1):017901–1–4, 2003.

[78] D J Wineland and W M Itano. Laser cooling of atoms. *Phys. Rev. A*, 20(4):1521–1540, October 1979.

[79] H J Lipkin. *Quantum Mechanics*. North-Holland, New York, 1973.

[80] E T Jaynes and F W Cummings. Comparison of quantum and semiclassical radiation theories with application to the beam maser. *Proceedings of the IEEE*, 51:89–109, January 1963.

[81] J M Raimond, M Brune, and S Haroche. Manipulating quantum entanglement with atoms and photons in a cavity. *Rev. Mod. Phys.*, 73:565–582, 2001.

[82] S Wallentowitz and W Vogel. Quantum-mechanical counterpart of nonlinear optics. *Phys. Rev. A*, 55(6):4438–4442, 1997.

[83] D J Wineland, C Monroe, W M Itano, B E King, D Leibfried, C Myatt, and C Wood. Trapped-ion quantum simulator. *Phys. Scr.*, T76:147–151, 1998.

[84] R L de Matos Filho and W Vogel. Engineering the Hamiltonian of a trapped atom. *Phys. Rev. A*, 58(3):R1661–1164, 1998.

[85] S Wallentowitz and W Vogel. High-order nonlinearities in the motion of a trapped atom. *Phys. Rev. A*, 59(1):531–538, 1999.

[86] D Leibfried, B DeMarco, V Meyer, M Rowe, A Ben-Kish, J Britton, W M Itano, B Jelenković, C Langer, T Rosenband, and D J Wineland. Trapped-ion quantum simulator; experimental application to nonlinear interferometers. *Phys. Rev. Lett.*, 89:247901–1–4, 2002.

[87] M B Plenio and P L Knight. Decoherence limits to quantum computation using trapped ions. *Proc. R. Soc. Lond. A*, 453:2017–2041, 1997.

[88] C Di Fidio and W Vogel. Damped Rabi oscillations of a cold trapped ion. *Phys. Rev. A*, 62:031802–1–4, 2000.

[89] A A Budini, R L de Matos Filho, and N Zagury. Localization and dispersive-like decoherence in vibronic states of a trapped ion. *Phys. Rev. A*, 65:041402–1–4, 2002.

Course 7

QUANTUM CRYPTOGRAPHY WITH AND WITHOUT ENTANGLEMENT

N. Gisin and N. Brunner

Group of Applied Physics, University of Geneva, 1211 Geneva 4, Switzerland

D. Estève, J.-M. Raimond and J. Dalibard, eds.
Les Houches, Session LXXIX, 2003
Quantum Entanglement and Information Processing
Intrication quantique et traitement de l'information
© *2004 Elsevier B.V. All rights reserved*

Contents

Abstract

Quantum cryptography is reviewed, first using entanglement both for the intuition and for the experimental realizations. Next, the implementation is simplified in several steps until it becomes practical. At this point entanglement has disappeared. This method can be seen as a lesson of Applied Physics. Finally, security issues, e.g. photon number splitting attacks, and counter-measures are discussed.

1. Introduction

Quantum cryptography is a beautiful idea! It covers aspects from fundamental quantum physics to Applied Physics via classical and quantum information theories [1]. During the last ten years, quantum cryptography progressed tremendously, in all directions: from mathematical security proofs of idealized scenarii to commercial prototypes. In these proceedings we review the intuition, the experimental progress in optical fibers implementations and some security aspects, each viewed first with entanglement, and then without. Undoubtedly, quantum cryptography is intellectually more fascinating and conceptually easier with entanglement, but much more practical without it. Hence both aspects, with and without entanglement, are equally beautiful!

The next section presents the intuition behind quantum cryptography. Section 3 can be seen as a lesson in Applied Physics: how to simplify a theorist's implementation of a nice idea until it is practical, while keeping the essential. This shows that Applied Physics requires a lot of imagination and a deep understanding of the essential physical ingredients. Finally, section 4 reviews some security issues: coherent and individual eavesdropping, Trojan horse attacks, photon number splitting attacks and means to limit their efficiency.

2. Intuitions

2.1. *Key distribution*

The general scenario for key distribution, whether classical or quantum, goes as follows. Alice and Bob, the honest parties, hold many realizations of random variables X and Y respectively. The adversary, Eve, holds realizations of a third random variable Z. Hence the scenario is described by a joint probability distribution $P(X, Y, Z)$ [2]. Intuitively it is clear that if X and Y are strongly correlated (e.g. almost identical) and furthermore, if Z is essentially uncorrelated, then Alice and Bob can use a public communication channel to distil secret bits. This intuition is made precise in the following theorem. The useful measure of correlation here is the mutual Shannon information.

Theorem [3] For a given $P(X, Y, Z)$, Alice and Bob can establish a secret key (using only error correction and classical privacy amplification) if and only if $I(X, Y) \geq \min\{I(X, Z), I(Y, Z)\}$, where $I(X, Y) = H(X) - H(X|Y)$ denotes the mutual information and H is the Shannon entropy.

Note that by definition *privacy amplification* uses only 1-way communication. If Alice and Bob use 2-way communication, the situation is more complex [4–6]. But these 2-way protocols are so inefficient that in practice they are always ignored.

2.2. *Quantum key distribution with entanglement*

Let us assume that the random variables X, Y and Z introduced above result from quantum measurements that Alice, Bob and Eve perform on a quantum state ψ_{ABE}. It is clear for the quantum physicists, that if the partial state ρ_{AB} shared by Alice and Bob is close to maximally entangled, then Eve is "factorized out", i.e. is uncorrelated. This is because a maximally entangled state is a pure state $\rho_{AB} \approx |\psi_{AB}\rangle\langle\psi_{AB}|$, hence the global state has to be close to a product state: $\psi_{ABE} \approx \psi_{AB} \otimes \psi_E$. If one understands entanglement, more precisely, if one is familiar with the algebra of tensor products, then the reason why quantum key distribution with entanglement is secure becomes very intuitive!

2.3. *Quantum key distribution without entanglement*

Assume now that Alice and Bob do not share an entangled state, but - following the original idea [7] - that Alice sends individual quanta to Bob (when the quanta are described by a 2-dimensional Hilbert space, one speaks of qubits). Alice and Bob use two (or more) incompatible bases to prepare and measure each quanta. Because of the use of incompatible bases, there is no way for Eve to make copies of the flying quanta. Indeed, the no-cloning theorem guarantees that there is no way to copy an unknown quanta without perturbing its state [8]. Thus Alice and Bob can check for the presence of an adversary, Eve, by comparing a sample of their data: if the data is perfectly correlated, then Eve did not try to copy it and the remaining data is safe. Each time Alice and Bob happen to have used the same basis, their data provides them with a secret bit.

This view of QKD without entanglement can be based on different aspects of quantum physics, like Heisenberg's uncertainty relation or that quantum measurements perturb the system. But in the end all these are based on the linearities of quantum kinematics (the Hilbert space) and dynamics (Schrödinger's equation). And this linearity is also the basis for entanglement, which appears when one introduces linear combinations of product states. Hence, intuitively one feels that both QKD schemes are closely related.

3. Experiments: a lesson in applied physics

The first choice when thinking about an experimental realization of QKD concerns the degree of freedom used for encoding the qubit. Indeed, if one goes for optical fibers, then the system is imposed: telecom photons. A first possible choice would be polarization. Unfortunately this is a quite unstable degree of freedom: actual fibers have some birefringence (different polarization modes travel at different speeds), moreover the polarization modes suffer from random polarization mode coupling [9]. And if the fiber is hanging between posts, the situation is even worse: Berry phase would be random, leading to fast (ms) random polarization fluctuations [10]. Hence, better choices should be envisaged. In Geneva, we chose time-bin qubits [11]. The idea is depicted in Fig. 1. Each photon is brought into a superposition of two time-bins, an early and a delayed one. The probability amplitudes of each time-bin and their relative phase allow one to prepare any possible qubit state. Also any possible projective measurement can be realized using a similar interferometer shown on the right hand side of Fig 1.

In the following sub-sections we review step by step simplifications of the theorist's implementation of QKD.

3.1. Basic experiment with entangled time-bin qubits

The configuration presented in Fig. 2 is close to Ekert's original proposal [12], but uses time-bin qubits instead of polarization. The source at the center contains a non-linear crystal in which a pump photon spontaneously splits into two twin photons. Energy conservation guarantees that the twins' energies (i.e the optical frequencies) add up to the well defined energy of the pump photon, although each of the twin photon has itself an uncertain energy, uncertain in the usual quantum mechanical sense. The pump photon is part of a large classical pulse, about 500 ps long. Since the probability of "splitting", i.e. of spontaneous parametric down-conversion, is low (typically 10^{-10}, up to $10^{-}6$ in PPLN waveguides [13]), the pulse energy can be adjusted such that the probability that a pair of twin photons is generated is around 10%. In order to produce entangled time-bin qubits, the pump pulse passes through an unbalanced interferometer, where the imbalance

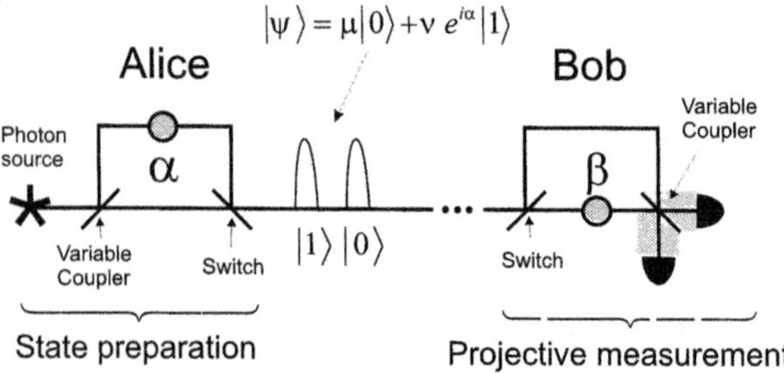

Fig. 1. Time-bin qubits. After Alice's interferometer the photon is brought into a superposition of the two time-bins (early and late) corresponding to the two arms of the interferometer (short and long). The logical values 0 (1) is attributed to early (late). Note that by tuning their respective phases and coupling ratios, Alice can prepare any qubit state and Bob can perform a measurement in any qubit basis. The switch allows in principle the state preparation and the measurement without losses. In practice however one often replaces the switch by a 50-50 coupler and uses postselection.

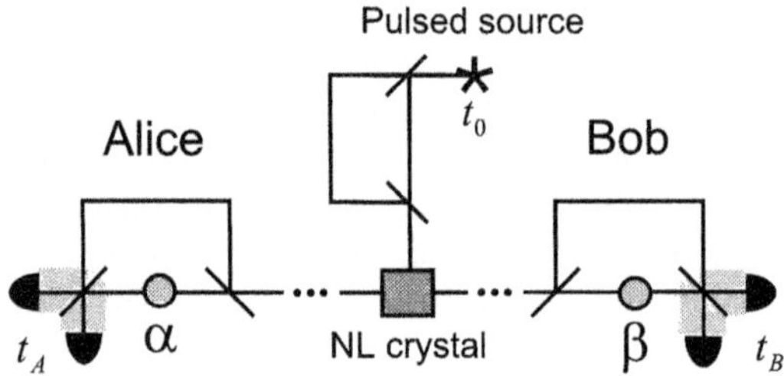

Fig. 2. Quantum cryptography with entangled time-bin qubits. The source sends a pulse at time t_0. The detection on Alice's (Bob's) side occurs at time t_A (t_B).

is much longer than the pulse duration. Alice and Bob both use the standard time-bin qubit analyzer presented in Fig. 1. They fix the phases (relative to the pump interferometer) of their interferometers such that the two twin photons always emerge at the same output port, hence their detectors clicks are perfectly correlated. For a second, incompatible, basis Alice and Bob could use different phase settings. But a first simplification can immediately be implemented. They replace the switch of their measuring interferometer by a much simpler and less lossy fiber optical coupler. Hence, Alice and Bob can also detect photons at an earlier or at a later time: earlier if the pump and their twin photons passed through the short arms of the interferometers, later if they both travelled the long way. Consequently, whenever Alice and Bob both detect their twin photon in the

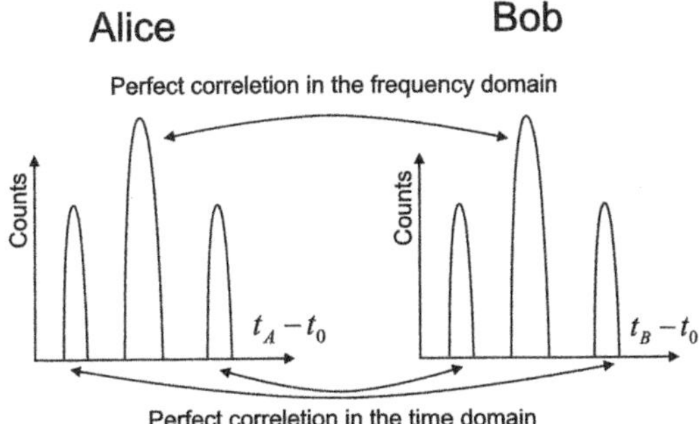

Fig. 3. Time and frequency correletions

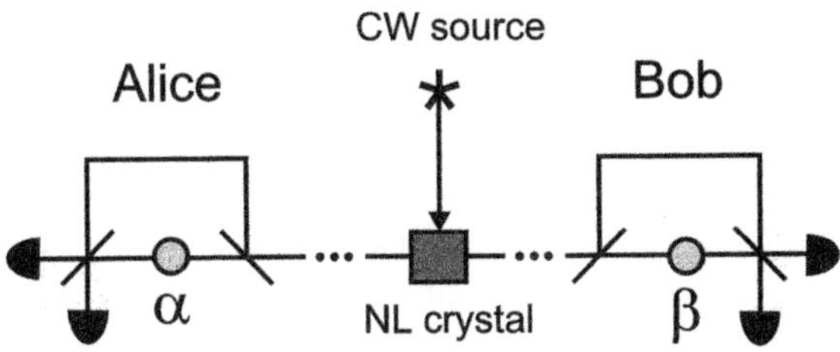

Fig. 4. First simplification. A continuous wave (cw) source can replace the pump interferometer.

lateral peaks (i.e. early or late), then they both definitely have the same detection time: either both early or both late. And if Alice and Bob both detect their twin photon in the central time-bin, then they definitely have a click in the same detector (assuming their phases are fixed at $\alpha = \beta = 0$). The first case uses the time-basis, the second the frequency-basis (see Fig. 3).

Conceptually this is quite elegant. It has even been realized in our lab [14], and recent results show that it is feasible over a significant distance. But this configuration is not very practical: there are three interferometers to align and stabilize, and the polarization of the three photons has to be kept under control. Hence, let's simplify this!

3.2. *First simplification: energy-time entanglement*

A first simplification of the previous scheme consists in suppressing the pump interferometer and replacing the pulsed laser by a continuous pump laser, see Fig. 4. If the coherence length of this cw pump laser is larger than the imbalance of

Alice and Bob's interferometers, then 2-photon interference can still be observed. Indeed, when Alice and Bob post-select coincidence detections, then there are two possibilities: either both photons passed through the short arm of both interferometers, or both passed through the long arm. Since the pump laser's coherence is large, these two possibilities are indistinguishable. Hence, according to quantum mechanics, one should add the probability amplitudes and observe interference. In this configuration Alice and Bob need to randomly choose the settings of their phase modulators: 0, 90, 180 and 270 degrees, let's say. Whenever they happen to use settings corresponding to a phase difference multiple of 180°, then their detectors always fire together.

This is a nice configuration, but admittedly more suited for tests of Bell inequality (i.e. of quantum non-locality) than for a practical quantum cryptography setup. Actually, this configuration has been proposed in 1989 by J. Franson on the context of quantum nonlocality and is called a Franson interferometer [15]. This is the configuration we used in 1997 for our long distance Bell test over 18km in optical fibers (10km in straight line) [16].

3.3. *Somewhat simpler*

The next step notices that there is no need to put the source at the middle, halfway between Alice and Bob. The middle position is merely elegant. But it is more practical to put the source on one side, let's say Alice's side. Notice that Alice doesn't become the sender of the quantum key: the key results eventually from independent random choices made by both partners and by Nature, there is nothing like a quantum key sender. But now, only one photon must travel a long distance. Hence, the photon that stays on Alice's side can be chosen at a more convenient wavelength for efficient detection, that is at a wavelength where silicium APDs are available, i.e. below 1 μ, around 800 nm. This configuration, with some additional nice tricks, was demonstrated in 2001 by G. Ribordy [17], who founded id Quantique a few years later, the first company to propose a quantum cryptography setup [18]. His experiment was the first one targeting primarily quantum cryptography with entangled photons - all other experiments, including ours, where tailored for Bell tests and merely adapted to fashion. Ribordy's experiment still holds the distance record of QKD using entangled photons. But admittedly, is not yet that practical since two photons must be detected. Hence, let's make it simpler!

3.4. *The first main step towards a practical system*

The first step towards a really practical system consists in moving the photon source to the other side of Alice's interferometer. At first this may look like a complete change, but it really isn't! Let's first use formulas. Whenever a unitary

Fig. 5. The source can be moved on the other side of Alice's interferometer.

operator U acts on one subsystem of a maximally entangled pair state $\Phi^{(+)}$, then the same effect can be obtained by acting with a related unitary operator on the other subsystems:

$$U \otimes \mathbf{1}\, \Phi^{(+)} = \mathbf{1} \otimes U^t\, \Phi^{(+)} \tag{3.1}$$

where U^t denotes the transpose.

This formula applied to our case simply tells us that for Alice's interferometer, the long arm with a central source is equivalent to the short arm with a source moved to the left of the interferometer, as shown in Fig. 5. Now the interference results from the indistinguishability of the following two paths: short-long and long-short, where the first term applies to the path in Alice's interferometer and the second to the path in Bob's interferometer. The significant simplification follows quite naturally. Since the photon travelling to the left on Fig. 5 is actually not used, or only as a trigger, one may as well use a single photon source. Well, that is even less practical, at least as long as single photon sources at telecom wavelength do not exists. But now one can also use the much more practical pseudo-single photon sources. These sources are simply very attenuated telecom laser pulses, such that the mean photon number per pulse is only of the order of 0.1. Hence the probability that a pulse contains two photons is almost negligible (in section III we come back to the issue of multi-photon pulses). Attenuating a laser pulse that low is not trivial, but still much simpler and much more stable, which is very important, than twin-photon sources.

This configuration presented in Fig. 5 was first used by Paul Townsend, then at BT, and John Rarity, then at DERA [19], and is still developed at Los Alamos National Laboratories, USA, in the group of Richard Hughes [20]. But looking at Fig. 5 one still sees two interferometers that need to be stabilized: the difference long-short has to be the same for both interferometers. And since this scheme relies on interferences, the polarization of the pseudo-single photons must be controlled. All this requires active feedback, which is not impossible to achieve, but not yet entirely practical. So let's simplify it further!

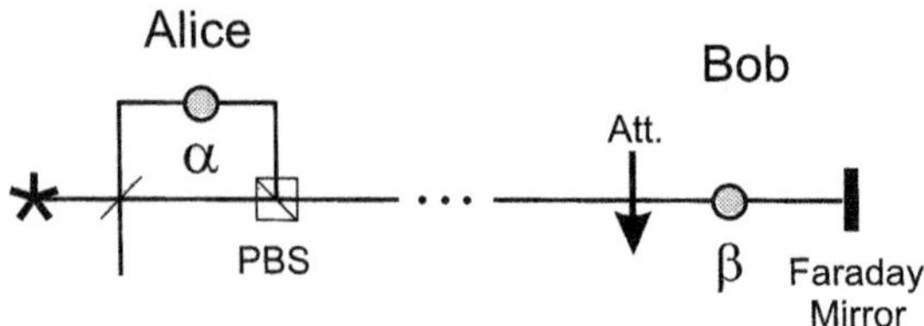

Fig. 6. Plug & Play setup [23,24].

3.5. *A practical setup: the Plug & Play configuration*

The next step realizes that there is no need for two interferometers, one is enough (see Fig. 6). But then the pulse must travel go-&-return, using a mirror as indicated on the figure. The indistinguishable paths are still short-long and long-short, but now referring to the paths during the go and the return propagations. Notice that in this scheme the role of Alice and Bob are inverted: Bob chooses one among four phase settings and Alice chooses a measurement basis. A serious drawback is that the photons must travel twice the distance, hence suffer from twice the loss. But this can be circumvented. Actually it is only on the return flight that the pulse has to be attenuated down to the pseudo-single photon level. Consequently, a bright pulse is sent out, attenuated by Bob and reflected to Alice. Notice that since there is now only a single interferometer, there is no longer any need to align it! All that is needed is that it remains stable during the time of a go-&-return, i.e. a few micro-seconds. But there remains the polarization. Here again there is an elegant solution, first suggested in a different context by Martinelli [21]. It consists of using a Faraday mirror. The details can be found in [1, 21]. Essentially such mirrors act on polarization like a phase conjugating mirror acts on phase. The net result is that when a light pulse arrives back on Alice's side, it is in a fixed polarization state, independent of all the polarization fluctuation light underwent during propagation: all the fluctuations where undone during the return journey. Faraday mirrors use the non-reciprocal Faraday effect, the same effect used in isolators and in circulators. Hence the telecom industry has developed this technology to a remarkable point and Faraday mirrors can readily be bought [22].

A further simplification comes from the fact that a Faraday mirror exchanges vertical and horizontal polarization. Hence, replacing the output coupler of Alice's interferometer by a polarization beam splitter guarantees that a photon that passed through the long arm when emitted, will return via the short arm, and vice-versa. Consequently, Alice doesn't need to post-select the cases where the photon arrives in the correct time-bin since all detected photons arrive at the correct time.

When this setup was first tested using classical light (i.e. without the attenuator), experimentalists in Geneva were very pleased to measure visibilities up to $V = 99.8\%$, without much effort, even over tens of km! Accordingly this configuration was named Plug-&-Play [23]. Using this setup for QKD, the noise (QBER) is largely dominated by the detector noise: $QBER_{optical} = \frac{1+V}{2} << 1\%$.

The Plug & Play configuration has been demonstrated in a QKD experiment between Geneva and Lausanne over a distance of 67km using the swiss telecom network, with terrestrial and with a cable under lake Geneva [24] (see also [25]). This experiment received quite a lot of attention. But actually, another experiment presented in the same paper deserves probably more attention. It used aerial cables and was done in mountains near Geneva. This clearly demonstrated the very high stability of the Plug & Play configuration. Indeed, it would be almost impossible to demonstrate QKD with any of the previously discussed configurations using aerial cables!

4. Security

4.1. Security proofs based on entanglement

The most general proofs of security, often termed à la Shor-Preskill, are quite surprising [26]. Following ideas by Mayer [27], Lo and Chau [28] and the development of quantum error codes, these proofs essentially show that from Alice and Bob's points of view everything is as if they had used close to maximally entangled states, although they actually did use a scheme without entanglement. More details can be found in I. Chuang's contribution to these proceedings. Let us simply emphasize that it is still not known whether Eve can in principle reach these bounds, or whether these bounds are sub-optimal. From a practical point of view this is a pity, since we do not know whether we do really need to sacrifice qubits to these bounds or could use the more optimistic bound summarized in the next sub-section.

4.2. Security proofs without entanglement

The proofs in this subsection do not consider the most general attack, but only what is called the individual, or incoherent attack. Actually, these proofs also treat the case of finite-coherent attacks, hence let us concentrate on the later. All the security proofs are valid only in the limit of arbitrarily long keys. If not, the statistical arguments wouldn't apply. Now, let's assume that Eve can attack

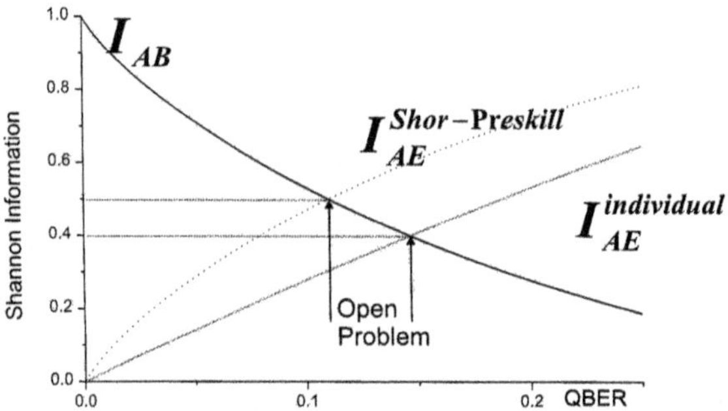

Fig. 7. The Shannon mutual information as a function of the QBER for the BB84 protocol. It is still not known if the Shor-Preskill bound (QBER≈11 %) can be saturated or not. The bound for invidual attacks (QBER≈15%) is known to be optimal.

several qubits in a coherent way, i.e. she can coherently let auxiliary systems under her control interact (unitarily of course) with the flying qubits. Assume that Eve can do this up to a maximum number of N qubits. We call this finite-coherent attacks. If Alice and Bob use key lengths much longer than N, then Eve is in the same situation as if she would be limited to individual attacks (one auxiliary system per qubit). Hence, the proofs "without entanglement" are valid for all senarii except if Eve can attack coherently an unlimited number of qubits - a conceptually interesting scenario, but hard to take seriously for the practical physicist.

In 1997 Fuchs et al. [31] presented the optimal individual attack, see also [32]. Since then it has been generalized to more than two bases [33] and to higher dimensions [34]. By now, the BB84 case is well known. The main results are summarized in Fig. 7.

4.3. Trojan horse attacks and technological loopholes

As shown in Fig. 6 the configuration opens a new possible attack for Eve, the so-called Trojan horse attack. Eve could send into Bob's apparatus a bright laser pulse to sense the phase modulator's setting. This illustrates that for every simplification step one has to carefully check the security of the configuration. In the present case there is a simple way to avoid Trojan horse attacks. Bob adds a coupler taking out a large fraction (typically 90%) of the light at his apparatus input. This coupler can be considered as part of the attenuator shown in Fig. 6. The extracted light is directed onto a standard detector that monitors the energy of each incoming pulse. Additionally this detector is very useful for the synchronization of the phase modulator. Eve could now use a different wavelength at which either the coupler or the detector is inefficient. To avoid this Bob has to use a filter which blocks all unwanted wavelengths. This discussion could be extended more or less for ever. Let us emphasize two important points. First, this is not specific to the Plug-&-Play configuration, every real optical component has some imperfection, in particular they do all reflect some light. Hence Eve could always try to send a sensing pulse and Alice and Bob should always have warning detectors and protecting filters. The second point is that this brief discussion illustrates the limit of mathematical proofs of security. Indeed, such proofs have either to assume perfect components, or components with precisely defined defects. In practice a central issue is how to make sure that an actual prototype satisfies the assumptions of a mathematical theorem? In this respect, it should be mentioned that when we made the simplification from a 2-photon to a 1-photon configurations, we lost the possibility of using the violation of Bell's inequality as a signature of quantumness (i.e. if the correlation measured by Alice and Bob violate some Bell inequality, then they definitely share an entanglement preserving quantum channel). Using Bell inequalities in this sense is a very nice idea [29]. However the detection efficiency loophole that affects all optical tests of Bell inequality renders this kind of control infeasible with near future technology [30]. Note also that a violation of a Bell inequality could not detect a Trojan horse type of attack.

4.4. *What is secure?*

Since there is some controversy on this, let us ask "what is secure in QKD?". It is clear that Eve should not have access to Alice nor to Bob's electronics. Indeed, there the information is classical and Eve could merely copy it. On the contrary, the quantum channel, i.e. the optical fiber, is secure thanks to quantum physics. But now comes an old question in a new context: where does the quantum/classical transition happen? As long as the information is quantum, the no-cloning theorem applies. As soon as it is classical, security is lost (i.e. must be guaranteed by other means). Surprisingly to us, many physicists (mainly theorists) consider the detector on the quantum side. This is of course a simple way to be on the safe side [35]. But it implies a very significant waste of qubits. It seems really hard to imagine Eve modifying Bob's detector's dark count probability from a distance. And if we give her this capability, why not also give her the power to change Alice's source from a distance? Let's say that quantum cryptography offers "only" secure key distribution over a quantum channel, assuming the hardware on both sides are secured by classical means.

There remains though an issue. Eve could modify the apparent detection efficiency of Bob's detector by sending brighter pulses. This is clearly feasible and Bob thus has to continuously monitor the coincidence rate between his detectors.

If this coincidence rate exceeds the threshold corresponding to accidentals (due mainly to dark counts), then he should interrupt the protocol.

4.5. *Multi-photon pulses: problem and solutions*

Another potential security loophole comes from the cases where the pseudo-single photon source actually produces more than one photon. These events being rare one may think that they are negligible. However, if the losses on the quantum channel are high, e.g. the fiber is long, then the cases where the desired photon makes it to Bob are also rare. Hence Eve could perform the following attack [36]. Directly at the exit of Alice's office, Eve counts the number of photons in each pulse, without perturbing the degree of freedom used to encode the qubit, i.e. Eve performs quantum nondemolition measurements on each pulse (this is total science fiction with today's technology, but if one assumes that Eve is limited only by the laws of physics, she could do so). Next, Eve blocks all single-photon pulses. Whenever a pulse contains 2 or more photons, she keeps one and sends the others to Bob through a perfect channel, or even better she teleports them to Bob. If the fraction of pulses Eve blocks balanced the fraction of pulses that would have got lost in normal operation, then Bob notices no difference. But now Eve holds a copy of the qubits. The main point is that she didn't need to make any copy, Alice unwillingly offered her some.

Of course, once the attack was performed, Eve has to conserve her photons, waiting for the basis reconciliation, when Alice publicly announces which basis she used to encode each qubit. Thus Eve clearly needs a quantum memory, which again is far from today's technology but could in principle be designed.

A first way around such PNS (Photon Number Splitting) attack consists of using sources producing sub-poissonian light. Indeed in such sources, the probability of 2-photon pulses is reduced compared to a poissonian light source like a laser, for the same probability of a 1-photon pulse. Such sources are often named single-photon sources and are an active field of research [37].

Another approach realizes that the weakness of the BB84 protocol against PNS attacks is that whenever Eve holds a copy, she has full information about the quantum state. But then, why not replace in the protocol the bases by sets of non-orthogonal states [39]. Remember that unambiguous discrimination of non orthogonal states is possible, but at the cost of some inconclusive results [38]. So even when Eve has a perfect copy of the state she cannot find out what the bit is with certainty. A particularly simple example of such new protocols uses precisely the same states and measurements as in the BB84 protocol, but the sifting procedure differs [40]. This protocol is called SARG and is described in Fig.8

Though PNS attacks seem completely unrealistic with today's technology, it is

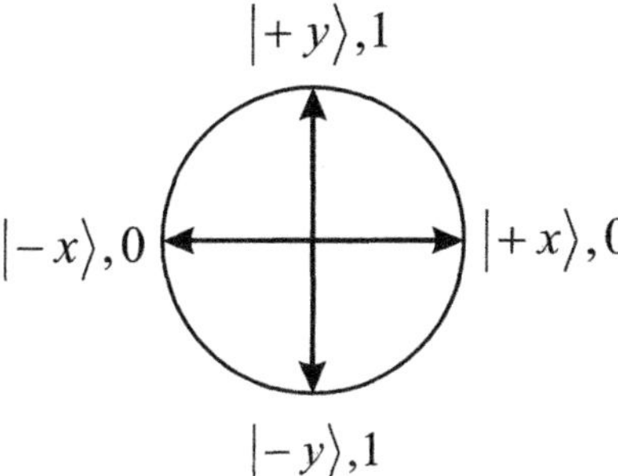

Fig. 8. The SARG protocol. The hardware is exactly the same as for the well known BB84 protocol. The only difference is in the sifting procedure. To exchange a secret bit Alice and Bob proceed as follows. Alice prepares one of the 4 states shown above, say $|+x\rangle$. Then she announces to Bob a set of two non-orthogonal states containing the state she actually prepared, for example $\{|+x\rangle, |+y\rangle\}$. Whenever Bob measures in the x basis (the *correct* basis), he always finds $|+x\rangle$ and cannot conclude anything. But when Bob measures in the y (the *wrong* basis) he finds $|-y\rangle$ half of the time. In these cases he concludes that Alice prepared the state $|+x\rangle$. The PNS attack is much less effective for this protocol than for the BB84, since Eve has to distinguish between two non orthogonal states. The probability of success of such a measurement is $p = 1 - \gamma \approx 0.29$, where $\gamma = \langle \pm x | \pm y \rangle = \frac{1}{\sqrt{2}}$ since we used two maximally conjugated basis.

nice to see that new protocols can still be devised, inspired by practical considerations. In this respect, see also Ph. Grangier's contribution to these proceedings.

5. Conclusion

Quantum cryptography is a beautiful idea! It is also an excellent teaching tool, encompassing basic quantum physics (no-cloning theorem, entanglement) and Applied Physics (telecom engineering). It also involves a significant part of classical and of quantum information theory.

Acknowledgements

Financial support by the Swiss OFES within the European project RESQ and the NCCR *Quantum Photonics* are acknowledged. Many thanks to Rob Thew for careful reading of the manuscript.

References

[1] For a recent review see N. Gisin, G. Ribordy, W. Tittel and H. Zbinden, "Quantum Cryptography", Rev. Modern Phys. (2002), 74, 145-195.

[2] U.M. Maurer, "Secret key agreement by public discussion from common information", IEEE Transactions on Information Theory **39** (1993), 733-742.

[3] I. Csiszár and J. Körner, "Broadcast channels with confidential messages", IEEE Transactions on Information Theory, Vol. IT-24 (1978), 339-348.

[4] U.M. Maurer and S. Wolf, "Unconditionnally secure key agreement and intrinsic information", IEEE Transactions on Information Theory, **45** (1999), 499-514.

[5] N. Gisin and S. Wolf, "Quantum cryptography on noisy channels: quantum versus classical key-agreement protocols", Phys. Rev. Lett. 83 (1999), 4200-4203; N. Gisin and S. Wolf, "Linking Classical and Quantum Key Agreement: Is There "Bound Information"? ", Advances in cryptology - Proceedings of Crypto 2000, Lecture Notes in Computer Science (2000), Vol. 1880, 482-500.

[6] T. Acin, N. Gisin and L. Massanes, "Equivalence between two-qubit entanglement and secure key distribution", Phys. Rev. Lett. 91 (2003), 167901, also quant-ph/0303053,

[7] Ch.H. Bennett and G. Brassard, "Quantum cryptography: public key distribution and coin tossing", Int. conf. Computers, Systems & Signal Processing, Bangalore, India, 1984, 175-179.

[8] W.K. Wooters and W.H. Zurek, "A single quanta cannot be cloned", Nature **299** (1982), 802-803; P.W. Milonni and M.L. Hardies, "Photons cannot always be replicated", Phys. Lett. A **92** (1982), 321-322; D. Dieks, "Communication by EPR devices", Phys. Lett. A **92** (1982), 271-272; E.P. Wigner, 1961, "The probability of the existence of a self-reproducing unit", in "The logic of personal knowledge" Essays presented to Michael Polanyi in his Seventieth birthday, 11 March 1961 Routledge & Kegan Paul, London, pp 231-238.

[9] N. Gisin and J.P. Pellaux, "Polarization mode dispersion: Time versus frequency domain", Optics Commun. 89 (1992), 316-323; N. Gisin, J.P. Von Der Weid and J.P. Pellaux, "Polarization mode dispersion of short and long single mode fibers", IEEE J. Lightwave Technology (1991), 9, 821-827.

[10] A. Tomita and R. Y. Chiao, "Observation of Berry's topological phase by use of an optical fiber", Phys. Rev. Lett. **57** (1986), 937-940.

[11] J. Brendel, N. Gisin, W. Tittel, and H. Zbinden, "Pulsed energy-time entangled twin-photon source for quantum communication", Phys. Rev. Lett. **82** (12) (1999), 2594-2597; W. Tittel and G. Weihs, "Photonic Entanglement for Fundamental Tests and Quantum Communication", QIC, Vol. 1, No. 2 (2001) 3-56, Rinton Press.

[12] A.K. Ekert, "Quantum cryptography based on Bell's theorem", Phys. Rev. Lett. **67** (1991), 661-663.

[13] S. Tanzilli, H. De Riedmatten, W. Tittel, H. Zbinden, P. Baldi, M. De Micheli, D.B. Ostrowsky, and N. Gisin, "Highly efficient photon-pair source using a Periodically Poled Lithium Niobate waveguide", Electr. Lett. **37** (2001), 26-28.

[14] W. Tittel, J. Brendel, H. Zbinden and N. Gisin, "Quantum Cryptography Using Entangled Photons in Energy-Time Bell States", Phys. Rev. Lett. **84** (2000), 4737-4740

[15] J.D. Franson, "Bell Inequality for Position and Time", Phys. Rev. Lett. **62** (1989), 2205-2208.

[16] W. Tittel, J. Brendel, H. Zbinden, and N. Gisin, "Violation of Bell inequalities by photons more than 10 km apart", Phys. Rev. Lett. **81** (1998), 3563-3566; H. Zbinden, N. Gisin, J. Brendel, and W. Tittel, Phys. Rev. A, 63, 022111/1-10, 2001.

[17] G. Ribordy, J. Brendel, J.D. Gautier, N. Gisin and H. Zbinden, "Long distance entanglement based quantum key distribution", Phys. Rev. A **63** (2001), 012309.

[18] www.idQuantique.com, for another company see www.MagiQtech.com.

[19] P. Townsend, J.G. Rarity and P.R. Tapster, "Single photon interference in a 10 km long optical fiber interferometer", Electron. Lett. **29** (1993), 634-639.

[20] R. Hughes, G. Morgan and C. Peterson, "Quantum key distribution over a 48km optical fibre network", J. Modern Opt. **47** (2000), 533-547.

[21] M. Martinelli, "A universal compensator for polarization changes induced by birefringence on a retracing beam", Opt. Commun. **72** (1989), 341-344; M. Martinelli, "Time reversal for the polarization state in optical systems", J. Modern Opt. **39** (1992), 451-455.

[22] see for example www.jdsu.com or www.ofr.com.

[23] A. Muller, T. Herzog, B. Huttner, W. Tittel, H. Zbinden and N. Gisin, " 'Plug and play' systems for quantum cryptography" , Applied Phys. Lett. **70** (1997), 793-795.

[24] D. Stucki, N. Gisin, O. Guinnard, G. Ribordy and H. Zbinden, "Quantum Key Distribution over 67 km with a plug & play system", New Journal of Physics, 4 (2002), 41.1-41.8.

[25] D. Bethune and W. Risk, "An Autocompensating Fiber-Optic Quantum Cryptography System Based on Polarization Splitting of Light", IEEE J. Quantum Electron. **36** (2000), 340-347; M. Bourennane, F. Gibson, A. Karlsson, A. Hening, P. Jonsson, T. Tsegaye, D. Ljunggren and E. Sundberg, "Experiments on long wavelength (1550nm) 'plug and play' quantum cryptography systems', Opt. Express **4** (1999), 383-387

[26] P.W. Shor and J. Preskill, "Simple proof of security of the BB84 Quantum key distribution protocol", Phys. Rev. Lett. **85** (2000), 441-444.

[27] D. Mayers, 1998, "Unconditional security in quantum cryptography", Journal for the Association of Computing Machinery (to be published); also quant-ph/9802025.

[28] H.-K. Lo and H.F. Chau, "Unconditional security of quantum key distribution over arbitrary long distances" Science **283** (1999), 2050-2056; also quant-ph/9803006.

[29] D. Mayers and A. Yao, "Quantum Cryptography with Imperfect Apparatus", Proceedings of the 39th IEEE Conference on Foundations of Computer Science (1998).

[30] B. Gisin and N. Gisin, "A local hidden variable model of quantum correlation exploiting the detection loophole", Phys. Lett. A **260** (1999), 323-327.

[31] C.A. Fuchs, N. Gisin, R.B. Griffiths, C.-S. Niu and A. Peres, "Optimal Eavesdropping in Quantum Cryptography. I", Phys. Rev. A **56** (1997), 1163-172.

[32] N. Lütkenhaus, "Security against individual attacks for realistic quantum key distribution", Phys. Rev. A , **61** (2000), 052304.

[33] H. Bechmann-Pasquinucci and N. Gisin, "Incoherent and Coherent Eavesdropping in the 6-state Protocol of Quantum Cryptography", Phys. Rev. A **59** (1999), 4238-4248; D. Bruss, "Optimal eavesdropping in quantum cryptography with six states", Phys. Rev. Lett. 81 (1998), 3018-3021.

[34] N. Cerf, M. Bourennane, A. Karlsson and N. Gisin, "Security of quantum key distribution using d-level systems", Phys. Rev. Lett. **88**, 127902/1-4, 2002; : D. Bruss and C. Macchiavello, "Optimal eavesdropping in cryptography with three-dimensional quantum states", Phys. Rev. Lett. **88** (2002), 127901 ; M. Bourennane, A. Karlsson, G. Björn, N. Gisin and N. Cerf, "Quantum Key Distribution using Multilevel Encoding: Security Analysis", J. Phys. A : Math. and Gen., 35 (2002), 10065-10076; D. Kaszlikowski, A. Gopinathan, Y. C. Liang, L. C. Kwek, B.-G. Englert, "Quantum and classical advantage distillation are not equivalent", quant-ph/0310144.

[35] H. Inamori, N. Lütkenhaus and D. Mayers, "Unconditional Security of Practical Quantum Key Distribution", quant-ph/0107017.

[36] G. Brassard, N. Lütkenhaus, T. Mor, and B.C. Sanders, "Limitations on Practical Quantum Cryptography", Phys. Rev. Lett. **85** (2000), 1330-1333.

[37] A. Beveratos, R. Brouri, T. Gacoin, A. Villing, J.-P. Poizat and P. Grangier, "Single photon quantum cryptography", Phys. Rev. Lett. **89** (2002), 187901; E. Waks, K. Inoue, C. Santori, D. Fattal, J. Vuckovic, G. S. Solomon and Y. Yamamoto, "Secure communication: Quantum cryptography with a photon turnstile", Nature **420** (2002), 762.

[38] A. Peres, *Quantum Theory: Concepts and Methods* (Kluwer, Dordrecht, 1998), section 9-5

[39] A. Acin, V. Scarani and N. Gisin, "Coherent pulse implementations of quantum cryptography protocols resistant to photon number splitting attacks", quant-ph/0302037, accepted in Phys. Rev. A.

[40] V. Scarani, A Acin, G. Ribordy and N. Gisin, "Quantum cryptography protocols robust against photon number splitting attacks for weak laser pulses implementations", quant-ph/0211131 , submitted to Phys. Rev. Lett.

Course 8

QUANTUM CRYPTOGRAPHY: FROM ONE TO MANY PHOTONS

Philippe Grangier

Laboratoire Charles Fabry de l'Institut d'Optique, F91403 Orsay, France

D. Estève, J.-M. Raimond and J. Dalibard, eds.
Les Houches, Session LXXIX, 2003
Quantum Entanglement and Information Processing
Intrication quantique et traitement de l'information
© 2004 Elsevier B.V. All rights reserved

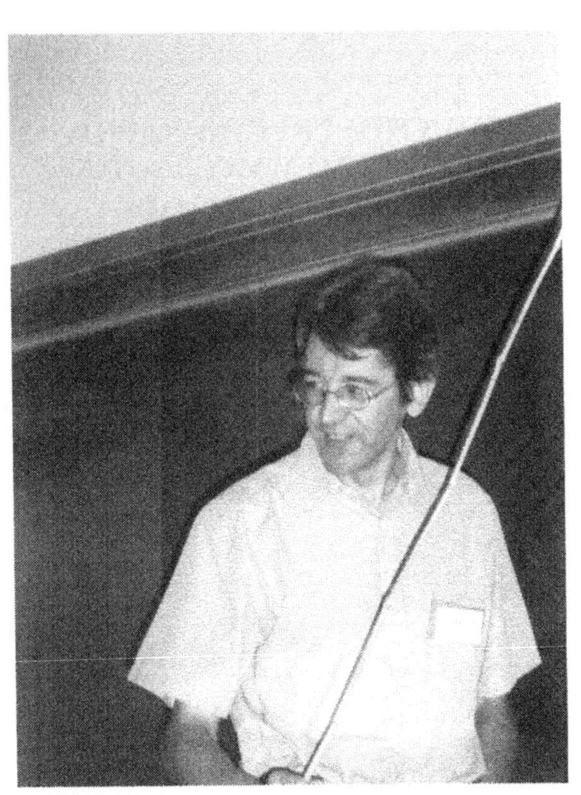

Contents

1. Introduction

During recent years the techniques for secure Quantum Key Distribution (QKD) have been progressing steadily, and they are now getting closer and closer to practical applications (see contribution by Nicolas Gisin in this volume). Here our purpose is to address two issues which have been long-standing challenges, and which have been satisfactorily addressed only very recently :

• The practical realization of light sources emitting single photons "on demand", in order to eliminate the possibility for the eavesdropper (Eve) to break the secret key transmission by using so called "photon number splitting" (PNS) attacks.

• The question whether quantum continuous variables (QCV) may provide a valid alternative to the usual QKD schemes based on single photon counting. The initial proposals to use QCV for QKD were based on the use of "non-classical" light beams, namely squeezed light or entangled light beams. It was recently shown, however, that there is actually no need for squeezed light in this context: an equivalent level of security may be obtained simply by generating and transmitting random distributions of quasi-classical (coherent) states. The basic ideas of these techniques will be reviewed and discussed in this contribution.

2. Single photons sources for quantum cryptography

2.1. Why single photons?

Since its initial proposal in 1984 [1] and first experimental demonstration in 1992 [2], Quantum Key Distribution (QKD) has reached maturity through many experimental realizations [3], and it is now commercially available [4]. However, most of the practical realizations of QKD rely on weak coherent pulses (WCP) which are only approximation of single photon pulses (SPP), that would be desirable in principle. The presence of pulses containing two photons or more in WCPs is an open door to information leakage towards an eavesdropper. In order to remain secure, the WCP schemes require to attenuate more and more the initial pulse, as the line losses become higher and higher, resulting in either a vanishingly low transmission rate - or a loss of security [5, 6]. The use of an efficient source of true single photons would therefore considerably improve the performances of existing or future QKD schemes, especially as far as high-losses schemes such as satellite QKD [7] are considered.

This motivated extensive studies of single photon sources during recent years, and many approaches have been proposed and implemented [8–13]. Here we will focus on the approach developed in our group, which is based on the fluorescence of a single Nitrogen-Vacancy (NV) color center [14] inside a diamond nanocrystal [15, 16] at room temperature. This molecular-like system has a lifetime of 23 ns when it is contained in a 40 nm nanocrystal [15]. Its zero-phonon line lies at 637 nm and its room temperature fluorescence spectrum ranges from 637 nm to 750 nm [17]. The two main advantages of the NV center when compared to other approaches is that it operates at room temperature, and is very photostable: no photobleaching has been observed over a week of continuous saturating irradiation of the same center. The nanocrystals are held by a 30 nm thick layer of polymer that has been spin coated on a dielectric mirror [15]. The mirror is initially slightly fluorescing, but this background light is reduced to a negligible value by hours of full power excitation that leads to a complete photobleaching of the dielectric coating, the NV center being unaffected.

This system was used to implement a complete QKD set-up, able to distribute a secret key over a distance of 50 m in free-space, at a typical rate of 10 kbits per second including error correction and privacy amplification [22]. Using the published criteria that warrant absolute secrecy of the key against any type of individual attacks [5, 6], this set-up reaches the region where a single photon QKD scheme takes a quantitative advantage over a similar system using WCP. Similar results have been reported by using a single photon source based on a quantum dot in a microcavity, operating at cryogenic temperatures [23].

2.2. *Experimental set-up*

The experimental set-up is shown in Fig. 1. Alice's station consists of a pulsed single photon source, a photon correlation detection to control the quality of the SPP, and a 4-state polarization encoding scheme. The single photon source is pumped by a home built pulsed laser at a wavelength of 532 nm that delivers 800 ps long pulses of energy 50 pJ with a repetition rate of 5.3 MHz, synchronized on a stable external clock [16]. The green excitation light is focused by a metallographic objective of high numerical aperture ($NA = 0.95$) onto the nanocrystals. The partially polarized fluorescence light (polarization rate of 46%) is collected by the same objective. It is then polarized horizontally by passing through a polymer achromatic half-wave plate and a polarizing cube, spectrally filtered by a long-pass filter (low cut-off 645 nm) that eliminates the reflected laser light, and spatially filtered by a confocal set-up. In order to control the quality of the

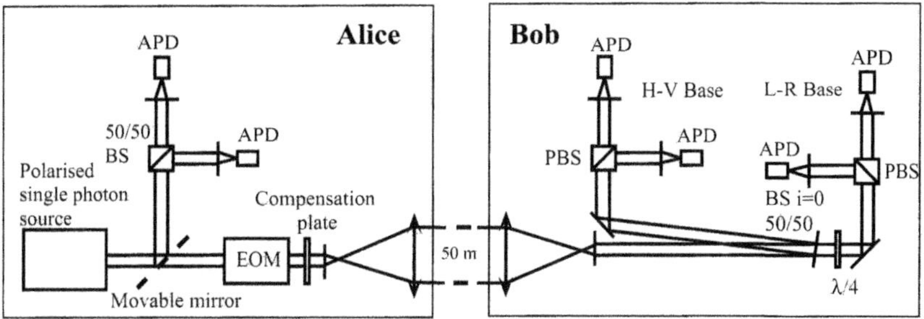

Fig. 1. Experimental set-up

SPP, the light can be sent via a movable mirror onto a photon correlation detection scheme consisting of two avalanche photodiodes (APD) in a Hanbury-Brown and Twiss set-up.

The total number of polarized photons detected by the two APDs altogether is $N_D^{(a)} = 70000\ \mathrm{s}^{-1}$ for an excitation repetition rate of 5.3 MHz. Correcting for the quantum efficiency of the control APDs ($\eta = 0.6$), the number of emitted polarized photons per pulse (before data encoding) is thus 2.2%. The autocorrelation function of the emitted light at saturation, displayed on Fig. 2, shows that the number of photon pairs within a pulse is strongly reduced with respect to Poisson statistics. The normalized area of the central peak is $C(0) = 0.07$, where this area would be unity for WCPs [16]. This means that the number of two-photon pulses of our source is reduced by a factor of $1/C(0) = 14$ compared to a WCP.

The Bennett-Brassard protocol [1] (BB84) is implemented by using two non-orthogonal basis, namely the horizontal-vertical (H-V) and circular left-circular right (L-R) basis. These four polarization states are obtained by applying four levels of high voltage on an electro-optical modulator (EOM). The EOM is driven by a home made module, that can switch 500 V in 30 ns to ensure the 5.3 MHz repetition rate of single photon source. The driving module is fed by sequences of pseudo-random numbers, that are produced using a linear feedback shift register in the Fibonacci configuration. In order to minimize polarization errors due to the broad bandwidth of the emitted photons, the EOM is operating very close to exact zero-path difference (white light fringe). In a first version of the experiment [22], the source emission rate of 2.2% was reduced down to 1.4% by the transmission of the EOM ($T_{\mathrm{EOM}} = 0.65$), and the average photon number per pulse was $\mu = 0.014$. Using an improved EOM with $T_{\mathrm{EOM}} = 0.95$ increased this number up to $\mu = 0.021$.

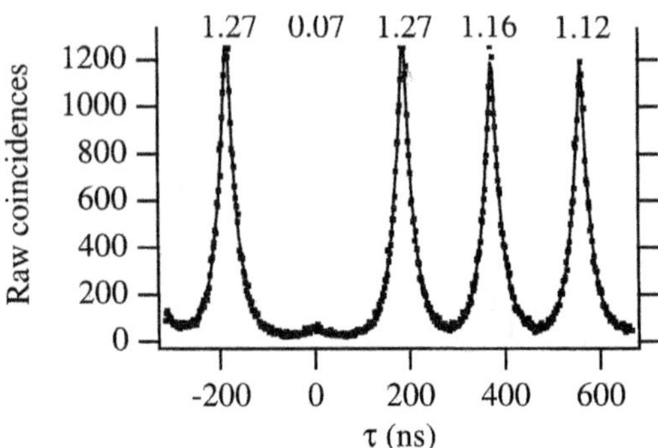

Fig. 2. Autocorrelation function of a single NV center on Alice's side. The raw coincidences are given as a function of the delay between the arrival times of the photons at Alice's correlation detection set-up. The exciting laser has a repetition period of 187.5 ns, a pulse width of 0.8 ns and an average power of 0.2 mW. The count rates are about 35000 s^{-1} on each avalanche photodiode, and the integration time is 166 s. The coincidences between peaks do not go down to zero because of the overlapping of adjacent peaks. The number above each peak represents its normalized area. The dots are experimental data. The line is an exponential fit for each peak, and takes into account the background light.

The detection at Bob's site lies 50 m away from Alice, either inside a building (first version) or between two buildings (second version). The single photons are sent via a 2 cm diameter beam so that diffraction effects are negligible. The H-V or L-R basis are passively selected by a near normal incidence 50/50 beam splitter that is polarization insensitive. A polymer achromatic quarter-wave plate is inserted in the L-R basis arm. In each basis a polarizing beam splitter sends the two polarizations on two APDs. In the first version of the experiment, the data was sent in bursts with a duration 10 ms, using a coaxial cable to transmit the synchronization signal. In the improved second version, the synchronization was provided by sendind also the green laser pulses, and the data bursts were extended to 200 ms, making high-speed processing of the data much easier.

The typical total number of photons detected by Bob is $N_D^{(b)} = 40000\ s^{-1}$. The dark counts on Bob's APDs with no signal at the input are $(d_H, d_V, d_L, d_R) = (150, 180, 380, 160)\ s^{-1}$. This includes APD's dark counts and background noise due to ambient light, that is carefully shielded using dark screens and pin-holes. Considering the 23 ns lifetime of the NV center, a post selection of pulses within a 50 ns gate selects approximatively $\eta_g = 90\%$ of all single photons, and keeps only $\beta_g = 27\%$ of the background counts. Taking into account the detection gate, the fraction of dark counts versus useful photons during a detection gate is therefore $p_{\text{dark}} = 0.7\%$. Other imperfections are due to the the achromatic wave plate, and to non-ideal white-fringe compensation and electronics driving the EOM. By comparing the full key that Bob received to the one that Alice sent, the measured QBER was found to be $e = 4.6\% \pm 1\%$ (first version), and improved to $e = 1.7\% \pm 0.3\%$ (second version).

The complete secret key transmission was achieved by carrying out error correction and privacy amplification using the public domain sofware "QUCRYPT" designed by Louis Salvail [18]. This leads to an average of 77 secret bits shared by Alice and Bob in a 10 ms sequence (first version), and of 3200 secret bits in a 200 ms sequence (improved second version).

2.3. *Single photons vs attenuated light*

We now compare the performance of our single photon BB84 set-up with QKD schemes using WCPs [7, 19]. The comparison is carried out by taking the detection efficiency and the dark counts of Bob in the present set-up. For WCP we assume a detection gate of 2 ns that is typical for recent experiments [7, 19]. The quantities that are compared are the maximum allowed on-line losses and the secret bit rate. Since QKD is supposed to offer unconditional security, it is assumed that a potentiel eavesdropper (Eve) has an unlimited technological power to carry out individual attacks within the rules of quantum mechanics. Eve can then access all the information leakage caused by the quantum bit error rate e and by the multiphoton pulses [5]. In the case of WCP with an average number μ of photons per pulse at Alice's station ($\mu \ll 1$), the probability of a multiphoton pulse is given by $S_m^{\mathrm{WCP}} = \mu^2/2$. The only way to reduce the fraction of multiphoton pulses in WCP is therefore to reduce the bit rate by working with smaller μ. To the contrary, SPP offers the possibility of achieving a vanishing ratio of multiphoton pulses without any trade-off on the filling of the pulses. In the present experiment the probability of a multiphoton pulse is reduced to $S_m^{\mathrm{SPP}} = C(0)\mu^2/2$ with $C(0) = 0.07$. The important figure to be evaluated is the number of secure bits per pulse (G) after error correction and privacy amplification, given by [6]

$$G = \frac{p_{\mathrm{exp}}}{2} \left\{ \frac{p_{\mathrm{exp}} - S_m}{p_{\mathrm{exp}}} \left(1 - \log_2 \left[1 + \frac{4\,e\,p_{\mathrm{exp}}}{p_{\mathrm{exp}} - S_m} - \left(\frac{2\,e\,p_{\mathrm{exp}}}{p_{\mathrm{exp}} - S_m}\right)^2\right]\right) + f[e]\left[e \log_2 e + (1-e)\log_2(1-e)\right]\right\} \tag{2.1}$$

The quantity p_{exp} is the probability that Bob has a click on his detectors (including possible dark counts) during a detection gate, and S_m is the probability of a multiphoton photon pulse just at the output of Alice's station. The function $f[e]$ depends on the algorithm that is used for the error correction. The Shannon limit gives $f[e] = 1$ for any e, which is the value taken in Fig. 3. For the best known algorithm, $f[e] = 1.16$ for $e \leq 5\%$.

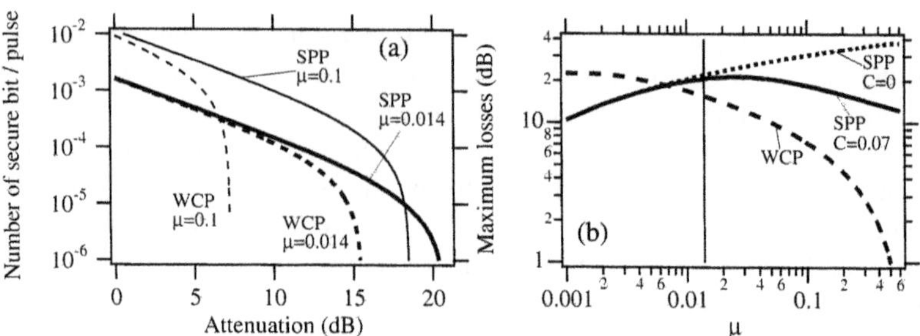

Fig. 3. These plots give theoretical evaluations obtained by using Eq. (2.1), together with the experimental parameters for our single-photon source (first version). (a) Calculated number of secure bit per time detection time gate G as a function of the on-line losses for SPPs and WCPs, for different average photon number per pulse μ. The SPP traces correspond to the measured value of the zero time autocorrelation $C = 0.07$. (b) The maximum allowed on-line losses for secure communication is deduced from (a) and corresponds to the attenuation for which $G = 10^{-6}$. This value is plotted as a function of the mean photon number per pulse μ, for a WCP system, for a SPP system with our value of the zero time autocorrelation ($C = 0.07$), and for an ideal SPP system with $C = 0$. The vertical line corresponds to $\mu = 0.014$.

In the first version of our set-up, the parameters were $(p_{\text{exp}}, S_m, e) = (7.4 \times 10^{-3}, 7 \times 10^{-6}, 4.6 \times 10^{-2})$ so that $G = 1.68 \times 10^{-3}$. The number of secure bits per second given by eq. (2.1) is thus $N_{\text{QKD}} = 8900\ \text{s}^{-1}$, which is reasonably close to the experimental value of $7700\ \text{s}^{-1}$.

As can be seen in Fig. 3, the SPP quantum cryptographic system has a quantitative advantage over the best existing WCP systems. When any type of individual attacks, without any technological limitations, are taken into account, the SPP system can deliver absolutely secure secret key at higher bit rate and offers the possibility of transmitting this key over longer distances.

Our quantum cryptographic set-up compares also favorably with QKD experiments using pairs of entangled photons [20], with a significantly higher secure bit rate in our case (though this may depend on various other practical limitations). Moreover, several relatively simple improvements could give SPP-QKD protocols an even greater advantage. In particular, inserting the emitter in a microcavity [21] is within experimental reach. This may be helpful to increase the collection efficiency, and therefore the secret bit rate, and also to narrow the emission spectrum, and thus to reduce polarisation errors.

This experiment demonstrates a complete single photon quantum key distribution set-up by using a reliable room temperature single photon source. Despite the fairly broad spectrum of the single photons, a 4-states polarization encoding and decoding can be implemented with a low error rate. In a more recent and improved version of the experiment, the quantum bit error rate was decreased down to 1.7%, and free-space transmission between two buildings was successfully achieved with a final secure bit rate of $N_{\text{QKD}} = 16000\ \text{s}^{-1}$. These results show that single photon QKD is a realistic candidate for long distance quantum cryptography, such as surface-to-satellite QKD.

3. Quantum key distribution using gaussian-modulated coherent states

3.1. Continuous variables QKD vs discrete variables QKD

Much interest has arisen recently in using the electromagnetic field amplitudes to obtain possibly more efficient quantum continuous variable (QCV) alternatives [25–36] to the usual photon-counting QKD techniques [37]. Most of the initial proposals [25–33] for achieving this goal were relying on "non- classical" light beams, involving explicitly squeezing or entanglement. In fact, it was shown in ref. [35] that secure key distribution can be obtained by transmitting "quasi-classical" coherent states. When the line transmission is larger than 50% (line loss $\leq$ 3 dB), the physical limits on QCV cloning [38–40] ensure that this protocol is secure against individual attacks. There are in principle various ways for the partners Alice and Bob to distribute keys beyond this 3 dB limit, for instance by using entanglement purification [41] or postselection [34].

Actually, we will show here that a suitably designed coherent-state QKD protocol remains, in principle, secure for any value of the line transmission [42]. As in ref. [35], this protocol relies on the distribution of a gaussian key [29] obtained by continuously modulating the phase and amplitude of coherent light pulses [35] at Alice's side, and subsequently performing homodyne detection at Bob's side. The continuous data are then converted into a common binary key via a specifically designed reconciliation algorithm [30, 32].

The security against arbitrarily high transmission losses is achieved by reversing the reconciliation protocol, that is, Alice attempts to guess what was received by Bob rather than Bob guessing what was sent by Alice. Such a "Reverse Reconciliation" protocol [36] gives Alice an advantage over a potential eavesdropper Eve, regardless of the line loss. In practice, there will be a maximum value of the tolerable losses, but these limitations will appear to be essentially technical, and are due mostly to the limited efficiency of the reconciliation software.

3.2. The coherent state protocol

The protocol runs as follows [35,42]. First, Alice draws two random numbers x_A and p_A from a gaussian distribution of mean zero and variance $V_A N_0$, where N_0 denotes the shot-noise variance. Then, she sends the coherent state $|x_A + ip_A\rangle$ to Bob, who randomly chooses to measure either quadrature x or p. Later, using a public authenticated channel, he informs Alice about which quadrature he measured, so she may discard the irrelevant data. After many similar exchanges, Alice and Bob (and possibly the eavesdropper Eve) share a set of correlated gaussian variables, which we call "key elements".

Classical data processing is then necessary for Alice and Bob to obtain a fully secret binary key. First, Alice and Bob publicly compare a random sample of their key elements to evaluate the error rate and transmission efficiency of the quantum channel. From the observed correlations, Alice and Bob evaluate the amount of information they share ($I_{AB} = I_{BA}$) and the maximum information Eve may have obtained (by eavesdropping) about their values (I_{AE} and I_{BE}).

It is known that Alice and Bob can, in principle, distill from their data a common secret key of size $S > \sup(I_{AB} - I_{AE}, I_{BA} - I_{BE})$ bits per key element [43,44]. This requires classical communication over an authenticated public channel, and may be divided into two steps : reconciliation (that is, correcting the errors while minimizing the information revealed to Eve) and privacy amplification (that is, making the key secret).

As we deal here with continuous data, we developed a "sliced" reconciliation algorithm [30,32] to extract common bit strings from the correlated key elements. In order to reconcile Bob's measured data with Alice's sent data, the most natural way to proceed is that Bob gets R extra bits of information from Alice in order to correct the transmission errors. The corresponding direct reconciliation (DR) protocols [29, 35] allow the generation of a common string of $I_{AB} + R$ bits, of which Eve may know up to $I_{AE} + R$ bits. Here we rather consider "reverse reconciliation" (RR) protocols [36, 42], where Bob sends R bits of information to Alice so that she incorporates the transmission errors in her initial data. These RR protocols allow the generation of a common string of $I_{BA} + R$ bits, of which Eve may know $I_{BE} + R$ bits. This turns out to be particularly well suited to QCV QKD, because it is more difficult for Eve to control the errors at Bob's side than to read Alice's modulation.

The last step of key extraction, namely privacy amplification, consists of filtering out Eve's information by properly mixing the reconciled bits to spread Eve's uncertainty over the entire final key. This procedure requires an estimate of Eve's information on the reconciled key, so we need a bound on I_{AE} for DR, or I_{BE} for RR. In addition, Alice and Bob must keep track of the information publicly revealed during reconciliation. This knowledge is destroyed at the end of the privacy amplification procedure, reducing the key length by the same amount. The DR bound [35] on I_{AE} implies that the security cannot be warranted if the line transmission G is below 50%. We will now establish the RR bound on I_{BE}, and show that it is not associated with a minimum value of G.

3.3. *Bounding Eve's information in a reverse reconciliation protocol*

In a RR scheme, Eve needs to guess Bob's measurement outcome without adding too much noise on his data. This can be done via an "entangling cloner", which creates two quantum-correlated copies of Alice's quantum state, so Eve simply keeps one of them while sending the other to Bob. Let $(x_{\text{in}}, p_{\text{in}})$ be the input field quadratures of the entangling cloner, and (x_B, p_B), (x_E, p_E) the quadratures of Bob's and Eve's output fields. To be safe, we must assume Eve uses the best possible entangling cloner compatible with the parameters of the Alice- Bob channel: Eve's cloner should minimize the conditional variances [45,46] $V(x_B|x_E)$ and $V(p_B|p_E)$, that is, the variances of Eve's estimates of Bob's field quadratures (x_B, p_B). These variances are constrained by Heisenberg-type relations, which limit what can be obtained by Eve:

$$\begin{aligned} &V(x_B|x_A)V(p_B|p_E) \geq N_0^2 \\ &V(p_B|p_A)V(x_B|x_E) \geq N_0^2 \end{aligned} \tag{3.1}$$

where $V(x_B|x_A)$ and $V(p_B|p_A)$ denote Alice's conditional variances.

Demonstration : In a RR protocol, Alice's estimator for x_B and Eve's estimator for p_B can be denoted respectively as αx_A and βp_E, where α, β are real numbers. The corresponding errors are $x_{B|A,\alpha} = x_B - \alpha x_A$, and $p_{B|E,\beta} = p_B - \beta p_E$. Because Alice's, Bob's, and Eve's operators commute, we have $[x_{B|A,\alpha}, p_{B|E,\beta}] = [x_B, p_B]$, and thus $\Delta x^2_{B|A,\alpha} \Delta p^2_{B|E,\beta} \geq N_0^2$ (Heisenberg relation). Defining the conditional variances as $V(x_B|x_A) = \min_\alpha\{\Delta x^2_{B|A,\alpha}\}$ and $V(p_B|p_E) = \min_\beta\{\Delta p^2_{B|E,\beta}\}$, we obtain $V(x_B|x_A)V(p_B|p_E) \geq N_0^2$, or, by exchanging x and p, $V(p_B|p_A)V(x_B|x_E) \geq N_0^2$.

The inequalies 3.1 mean that Alice and Eve cannot jointly know more about Bob's conjugate quadratures than is allowed by the uncertainty principle. Now, Alice's variances can be bounded by using the measured parameters of the quantum channel, which in turn makes it possible to bound Eve's variances.

The channel is described by the linear relations $x_B = G_x^{1/2}(x_{\text{in}} + B_x)$ and $p_B = G_p^{1/2}(p_{\text{in}} + B_p)$, with $\langle x_{\text{in}}^2\rangle = \langle p_{\text{in}}^2\rangle = VN_0 \geq N_0$, $\langle B^2_{x,p}\rangle = \chi_{x,p}N_0$, and $\langle x_{\text{in}}B_x\rangle = \langle p_{\text{in}}B_p\rangle = 0$. Here χ_x, χ_p represent the channel noises referred to its input, called equivalent input noises [45,46], while G_x, G_p are the channel gains in x and p, and V is the variance of Alice's field quadratures in shot-noise units $(V = V_A + 1)$.

A crucial observation is that the output-output correlations of the entangling cloner, described by $V(x_B|x_E)$ and $V(p_B|p_E)$, depend only on the density matrix D_{in} of the input field $(x_{\text{in}}, p_{\text{in}})$, and not on the way it is produced, namely whether it is a gaussian mixture of coherent states or one of two entangled beams. Inequalities 3.1 thus have to be fulfilled for all physically allowed values of $V(x_B|x_A)$

and $V(p_B|p_A)$, given D_{in}. Therefore, the values of $V(x_B|x_A)$ and $V(p_B|p_A)$ that should be used in inequalities 3.1 to limit Eve's knowledge are the minimum values Alice might achieve by using the maximal entanglement compatible with V, namely

$$V(x_B|x_A)_{\min} = G_x(\chi_x + V^{-1})N_0$$
$$V(p_B|p_A)_{\min} = G_p(\chi_p + V^{-1})N_0 \tag{3.2}$$

Demonstration : Alice has the estimators (x_A, p_A) for the field $(x_{\text{in}}, p_{\text{in}}) = (x_A + A_x, p_A + A_p)$ that she sends, with $\langle A_x^2\rangle = \langle A_p^2\rangle = sN_0$. Here s measures the amount of squeezing possibly used by Alice in her state preparation [36], with $s \geq V^{-1}$ for consistency with Heisenberg's relations. By calculating $\langle p_A^2\rangle = (V-s)N_0$, $\langle p_B^2\rangle = G_p(V+\chi_p)N_0$, $\langle p_A p_B\rangle = G_p^{1/2}\langle p_A^2\rangle$, we obtain the conditional variance $V(p_B|p_A) = \langle p_B^2\rangle - |\langle p_A p_B\rangle|^2/\langle p_A^2\rangle = G_p(s + \chi_p)N_0$. This equation and the constraint $s \geq V^{-1}$ gives $V(p_B|p_A) \geq G_p(V^{-1} + \chi_p)N_0$, and similarly $V(x_B|x_A) \geq G_x(V^{-1} + \chi_x)N_0$.

These lower bounds are thus directly connected with entanglement, even though Alice does not use it in practice. They may be compared with the actual values when Alice sends coherent states, that is, $V(x_B|x_A)_{\text{coh}} = G_x(\chi_x + 1)N_0$ and $V(p_B|p_A)_{\text{coh}} = G_p(\chi_p + 1)N_0$. The lower bounds on Eve's conditional variances are then obtained from equations 3.1 and 3.2, as:

$$V(p_B|p_E) \geq N_0/\{G_x(\chi_x + V^{-1})\}$$
$$V(x_B|x_E) \geq N_0/\{G_p(\chi_p + V^{-1})\} \tag{3.3}$$

Let us emphasize that the bound on $V_{B|A}$ is obtained by assuming that Alice may use squeezed or entangled beams, while the bound on $V_{B|E}$ can only be achieved if Eve uses an "entangling cloner attack", which is sketched in ref. [36]. This reflects the fact that squeezing or entanglement play a crucial role in our security demonstration, even though the protocol implies coherent states. A more detailed discussion of the role of this "virtual entanglement" is presented in ref. [47]. Though the security proof given here addresses individual gaussian attacks only, it can be expected that it encompasses all incoherent (non-collective) eavesdropping strategies. Security against more general attacks (non-gaussian and/or collective but finite-size) were actually demonstrated very recently by using the entropic version of Heisenberg's principle [48].

To assess quantitatively the security of the RR scheme, one assumes that Eve's ability to infer Bob's measurement can reach the limit put by inequalities 3.3. For simplicity, we consider the channel gains and noises and the signal variances to be the same for x and p (in practice, deviations should be estimated by statistical tests). The information rates can be derived using Shannon's theory for gaussian additive-noise channels [49], giving

$$
\begin{aligned}
I_{BA} &= (1/2)\log_2[V_B/(V_{B|A})_{\text{coh}}] \\
&= (1/2)\log_2[(V+\chi)/(1+\chi)] && (3.4) \\
I_{BE} &= (1/2)\log_2[V_B/(V_{B|E})_{\min}] \\
&= (1/2)\log_2[G^2(V+\chi)(V^{-1}+\chi)] && (3.5)
\end{aligned}
$$

expressed in bits per symbol (or per key element). Here $V_B = \langle x_B^2 \rangle = \langle p_B^2 \rangle = G(V+\chi)N_0$ is Bob's variance, $(V_{B|E})_{\min} = V(x_B|x_E)_{\min} = V(p_B|p_E)_{\min} = N_0/\{G(\chi + V^{-1})\}$ is Eve's minimum conditional variance, and $(V_{B|A})_{\text{coh}} = V(x_B|x_A)_{\text{coh}} = V(p_B|p_A)_{\text{coh}} = G(\chi+1)N_0$ is Alice's conditional variance for a coherent-state protocol. The secret bit rate of a RR protocol is thus

$$
\begin{aligned}
\Delta I_{RR} &= I_{BA} - I_{BE} \\
&= -(1/2)\log_2[G^2(1+\chi)(V^{-1}+\chi)] && (3.6)
\end{aligned}
$$

and the security is guaranteed if $\Delta I_{RR} > 0$. The equivalent input noise χ can be split into a "vacuum noise" component due to the line losses, given by $\chi_{\text{vac}} = (1-G)/G$, and an "excess noise" component defined as $\epsilon = \chi - \chi_{\text{vac}}$. In the high-loss limit ($G \ll 1$), the RR protocol remains secure if $\epsilon < (V-1)/(2V) \approx 1/2$, that is, if the amount of excess noise ϵ is not too large. In contrast, a DR protocol requires low-loss lines, as the security is warranted only if $\chi < 1$, that is, if $G > 1/(2-\epsilon)$. Note that DR tolerates an excess noise up to $\epsilon \approx 1$, so it might be preferred to RR for low-loss but noisy channels.

3.4. An experimental implementation

Our experimental implementation of the protocol descibed above uses 120-ns coherent pulses at a 800-kHz repetition rate. These pulses are chopped from a continuous-wave, grating-stabilized laser diode at 780 nm, by using an acousto-optical modulator (Fig. 4). Data bursts of 60,000 pulses have been analysed (Fig. 5). For each burst, a subset of the values are disclosed to evaluate the transmission G and the total added noise variance. The output noise has four contributions: the shot noise N_0, the channel noise $\chi_{\text{line}}\,N_0$, the electronics noise of Bob's detector ($N_{\text{el}} = 0.33\ N_0$), and the noise due to imperfect homodyne detection efficiency ($N_{\text{hom}} = 0.27\ N_0$). The two detection noises N_{el} and N_{hom} originate from Bob's detection system, so they must be taken into account when estimating I_{BA}. In contrast, we may reasonably assume that Eve cannot know or control the corresponding fluctuations, so her attack is inferred on the basis of the line noise χ_{line} only. Figure 6 shows explicitly the mutual information between all parties, which makes straightforward the comparison between the DR and RR protocols.

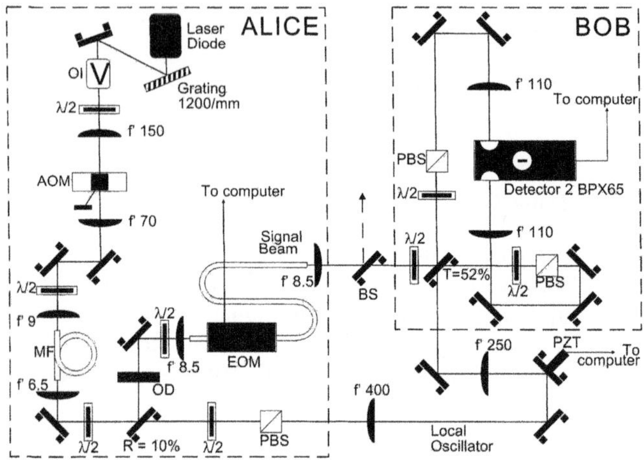

Fig. 4. Experimental set-up. Laser diode, SDL 5412 (780 nm); OI, optical isolator; $\lambda/2$, half-wave plate; AOM, acousto-optic modulator; MF, polarization maintaining single-mode fibre; OA, optical attenuator; EOM, electro-optic amplitude modulator; PBS, polarizer; BS, beam splitter; PZT, piezoelectric transducer. Focal lengths (f') are given in millimetres. R and T are reflection and transmission coefficients.

We wrote a computer program that implements the reconciliation algorithm followed by privacy amplification. Although Alice and Bob are not spatially separated in the present set-up, the analysed data have the same structure as in a realistic cryptographic exchange. Table 1 shows the ideal and practical net key rates for the reverse QKD protocol, as well as the DR values for comparison.

In principle, the RR scheme is efficient for any value of G provided that the reconciliation protocol achieves the limit given by I_{BA}. However, unavoidable deviations of the algorithm from Shannon's limit reduce the actual reconciled information I_{rec} between Alice and Bob, while I_{BE} is of course assumed unaffected. For high modulation ($V \approx 40$) and low losses, the reconciliation efficiency lies around 80%, which makes it possible to distribute a secret key at a rate of several hundreds of kilobits per second. However, the achievable reconciliation efficiency drops when the signal-to-noise ratio decreases, but this can be improved by reducing the modulation variance, which increases the ratio I_{BA}/I_{BE}. Although the ideal secret key rate is then lower, we could process the data with a reconciliation efficiency of 78% for $G = 0.49$ (3.1 dB) and $V = 27$, resulting in a net key rate of 75 kbits s^{-1}. This clearly demonstrates that RR continuous-variable protocols operate efficiently at and beyond the 3 dB loss limit of DR protocols. We emphasize that this result is obtained despite the fact that the evaluated reconciliation cost is higher for RR than for DR : the better result for RR

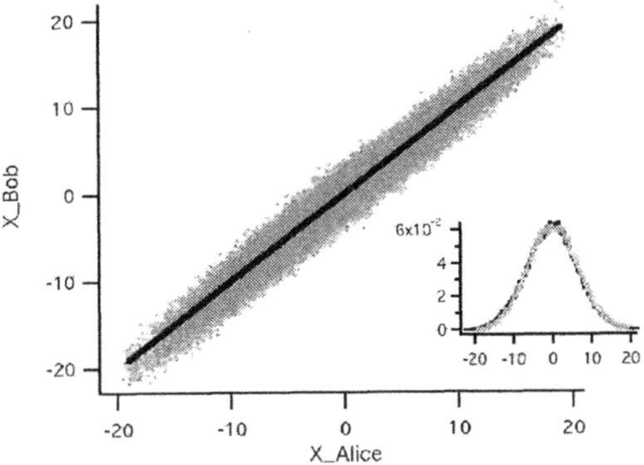

Fig. 5. Bob's measured quadrature as a function of the amplitude sent by Alice (in Bob's measurement basis) for a burst of 60,000 pulses. The line transmission is 100% and the modulation variance is V = 41.7. The solid line represents the expected unity slope. Inset, the corresponding histograms of Alice's (grey curve) and Bob's (black curve) data.

is essentially due to its initial "quantum advantage".

4. Conclusion

In photon-counting QKD, the key rate is limited by the single-photon detectors, in which the avalanche processes are difficult to control reliably at very high counting rates. In contrast, homodyne detection may run at frequencies up to tens of MHz. In addition, a specific advantage of the high dimensionality of the QCV phase space is that the field quadratures can be modulated with a large dynamics, allowing the encoding of several key bits per pulse (see Table 1). Very high secret bit rates are therefore attainable with the coherent-state protocol on low-loss lines. For high-loss lines, the protocol is at present limited by the reconciliation efficiency. However, most of the limitations of the present proof-of-principle experiment appear to be of a technical nature, and there is still a considerable margin for improvement, both in the hardware (increased detection bandwidth, better homodyne efficiency, lower electronic noise), and in the software (better reconciliation algorithms [50]). In conclusion, the way seems open for implementing the present proposal at telecommunications wavelengths as a practical, high bit- rate, quantum key distribution scheme over long distances.

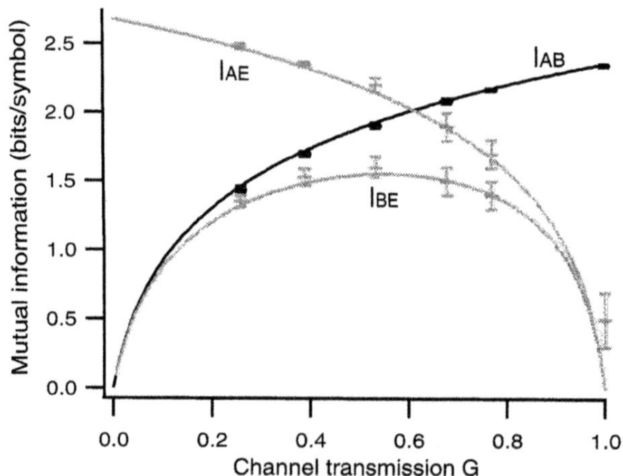

Fig. 6. Values of I_{BA}, I_{BE}, and I_{AE} as a function of the line transmission G for $V \approx 40$. Here, I_{BA} is given by eq. (3.4), including all transmission and detection noises for evaluating V_B and $(V_{B|A})_{\rm coh}$. The expression for I_{BE} is given by eq. (3.5), using the same V_B and $(V_{B|E})_{\min} = N_0/\{G(\chi_{\rm line} + V^{-1})\} + N_{\rm el} + N_{\rm hom}$. This expression realistically assumes that Eve cannot know the noises $N_{\rm el}$ and $N_{\rm hom}$, which are internal to Bob's detection set-up. For comparison with DR, the value of I_{AE} is also plotted (the theoretical value of I_{AE} is obtained from ref. [35]).

Table 1

Ideal and practical net secret key rates for the Reverse Reconciliation protocol. The parameters of the quantum key exchange are measured for several values of the channel transmission G (the corresponding losses are also given in decibels). The variations of the Alice's variance V are due to different experimental adjustments. The information I_{BA} is given in bits per time slot. Also shown are the maximum information gained by Eve (I_{BE}) and the extracted information by reverse reconciliation ($I_{\rm rec}$). The ideal secret key bit rates would be obtained from our measured data with perfect key distillation that yields exactly $I_{BA} - I_{BE}$ bits (RR), whereas the practical secret key bit rates are the one achieved with our current key distillation procedure ("–" means that no secret key is generated). Both bit rates are calculated over bursts of about 60000 pulses at 800 kHz, not taking into account the duty cycle ($\sim$ 5%) in the present set-up.

V	$G_{\rm line}$	Losses (dB)	I_{BA} (bit)	I_{BE} (% I_{BA})	$I_{\rm rec}$ (% I_{BA})	Ideal rate (kbit s^{-1})	Practical rate (kbit s^{-1})
41.7	1	0	2.39	0	88	1920	1690
38.6	0.79	1.0	2.17	58	85	730	470
32.3	0.68	1.7	1.93	67	79	510	185
27	0.49	3.1	1.66	72	78	370	75
43.7	0.26	5.9	1.48	93	71	85	–

Acknowledgements

The work presented in this contribution was carried out in collaboration with Alexios Beveratos, Jean-Philippe Poizat, Rosa Tualle-Brouri, André Villing, Thierry Gacoin, Romain Alleaume, Francois Treussart, Jean-Francois Roch (single-photon QKD), and Frédéric Grosshans, Jerôme Wenger, Rosa Tualle-Brouri, Gilles Van Assche, Nicolas Cerf (coherent states QKD).

References

[1] C.H. Bennett and G. Brassard, Int. conf. Computers, Systems and Signal Processing, Bangalore, India, 175 (1984).

[2] C.H. Bennett, F. Bessette, G. Brassard, L. Salvail, and J. Smolin, Journal of Cryptology **5**, 3 (1992).

[3] N. Gisin, G. Ribordy, W. Tittel, and H. Zbinden, Rev. Mod. Phys. **74**, 145 (2002) and references therein.

[4] id Quantique SA, http://www.idquantique.com

[5] G. Brassard, N. Lütkenhaus, T. Mor, and B. C. Sanders, Phys. Rev. Lett. **85**, 1330 (2000).

[6] N. Lütkenhaus, Phys. Rev. A **61**, 052304 (2000).

[7] W.T. Buttler, R.J. Hughes, S.K. Lamoreaux, G.L. Morgan, J.E. Nordholt, and C.G. Peterson, Phys. Rev. Lett. **84**, 5652 (2000).

[8] F. de Martini, G. di Guiseppe, and M. Marrocco, Phys . Rev. Lett. **76**, 900 (1996).

[9] R. Brouri, A. Beveratos, J.-Ph. Poizat, and P. Grangier, Phys. Rev. A **62**, 063817 (2000).

[10] Th. Basché, W. E. Moerner, M. Orrit, and H. Talon, Phys. Rev. Lett. **69**, 1516 (1992); C. Brunel, B. Lounis, Ph. Tamarat, and M. Orrit, Phys. Rev. Lett. **83**, 2722 (1999); L. Fleury, J.M. Segura, G. Zumofen, B. Hecht, and U.P. Wild, Phys. Rev. Lett. **84**, 1148 (2000); B. Lounis and W.E. Moerner, Nature (London) **407**, 491 (2000); F. Treussart, A. Clouqueur, C. Grossman, and J.-F. Roch, Opt. Lett. **26**, 1504 (2001).

[11] P. Michler, A. Imamoğlu, M.D. Mason, P.J. Carson, G.F. Strouse, and S.K. Buratto, Nature (London) **406**, 968 (2000).

[12] P. Michler, A. Kiraz, C. Becher, W. V. Schoenfeld, P. M. Petroff, L. Zhang, E. Hu, and A. Imamoğlu, Science **290**, 2282 (2000); C. Santori, M. Pelton, G. Solomon, Y. Dale, and Y. Yamamoto, Phys. Rev. Lett. **86** 1502 (2001); V. Zwiller, H. Blom, P. Jonsson, N. Panev, S. Jeppesen T. Tsegaye, E. Goobar, M.-E. Pistol, L. Samuelson, and G. Björk, Appl. Phys. Lett. **78**, 2476 (2001); E. Moreau, I. Robert, J.-M. Gérard, I. Abram, L. Manin, and V. Thierry-Mieg, Appl. Phys. Lett. **79**, 2865 (2001).

[13] J. Kim, O. Benson, H. Kan, and Y. Yamamoto, Nature (London) **397**, 500 (1999); Z. Yuan, B.E. Kardynal, R.M. Stevenson, A.J. Shields, C.J. Lobo, K. Cooper, N.S. Beattie, D.A. Ritchie, and M. Pepper, Science **295**, 102 (2002).

[14] R. Brouri, A. Beveratos, J.-Ph. Poizat, and P. Grangier, Opt. Lett. **25**, 1294 (2000); C. Kurtsiefer, S. Mayer, P. Zarda, and H. Weinfurter, Phys. Rev. Lett. **85**, 290 (2000).

[15] A. Beveratos, R. Brouri, T. Gacoin, J-Ph. Poizat, and Ph. Grangier, Phys. Rev. A. **64**, 061802(R) (2001).

[16] A. Beveratos, S. Kühn, R. Brouri, T. Gacoin, J.-Ph. Poizat, and Ph. Grangier, Eur. Phys. J. D **18**, 191 (2002).

[17] A. Gruber, A. Dräbenstedt, C. Tietz, L. Fleury, J. Wrachtrup, and C. von Borczyskowki, Science **276**, 2012 (1997).

[18] P.M. Nielsen, C. Schori, J.L. Sorensen, L. Salvail, I. Damgard, and E. Polzik, J. Mod. Opt. **48**, 1921 (2001); http://www.cki.au.dk/experiment/qrypto/doc/

[19] D. Stucki, N. Gisin, O. Guinnard, G. Ribordy, and H. Zbinden, quant-ph/0203118 (2002).

[20] T. Jennewein, C. Simon, G. Weihs, H. Weinfurter, and A. Zeilinger, Phys. Rev. Lett. **84**, 4729 (2000); D.S. Naik, C.G. Peterson, A.G. White, A.J. Berglund, and P.G. Kwiat, *ibid*, **84**, 4733 (2000); W. Tittel, J. Brendel, H. Zbinden, and N. Gisin, *ibid*, **84**, 4737 (2000); G. Ribordy, J. Brendel, J.D. Gautier, N. Gisin, H. Zbinden, Phys. Rev. A. **63**, 012309 (2001).

[21] J.M. Gérard, B. Sermage, B. Gayral, B. Legrand, E. Costard, and V. Thierry-Mieg, Phys. Rev. Lett. **81**, 1110 (1998).

[22] A. Beveratos, R. Brouri, T. Gacoin, A. Villing, J.-Ph. Poizat and P. Grangier, Phys. Rev. Lett. **89**, 187901 (28 october 2002)

[23] E. Waks, K. Inoue, C. Santori, D. Fattal, J. Vukovic, G.S. Solomon, and Y. Yamamoto, Nature (London) **420**, 762 (19 december 2002).

[24] S.L. Braunstein and A.K. Pati, *Quantum information theory with continuous variables* (Kluwer Academic, Dordrecht, 2003).

[25] M. Hillery, Phys. Rev. A **61**, 022309 (2000).

[26] T.C. Ralph, Phys. Rev. A **61**, 010303(R) (2000); *ibid*, Phys. Rev. A **62**, 062306 (2000).

[27] M.D. Reid, Phys. Rev. A **62**, 062308 (2000).

[28] D. Gottesman and J. Preskill, Phys. Rev. A **63**, 022309 (2001).

[29] N.J. Cerf, M. Lévy, G. Van Assche, Phys. Rev. A **63**, 052311 (2001).

[30] G. Van Assche, J. Cardinal, and N.J. Cerf, E-print arXiv:cs.CR/0107030. To appear in IEEE Trans.Inform. Theory.

[31] K. Bencheikh, Th. Symul, A. Jankovic and J.A. Levenson, J. Mod. Optics **48**, 1903-1920 (2001).

[32] N.J. Cerf, S. Iblisdir, G. Van Assche, Eur. Phys. J. D **18**, 211-218 (2002).

[33] Ch. Silberhorn, N. Korolkova, G. Leuchs, Phys. Rev. Lett. **88**, 167902 (2002).

[34] Ch. Silberhorn, T.C. Ralph, N. Lütkenhaus, G. Leuchs, Phys. Rev. Lett. **89**, 167901 (2002).

[35] F. Grosshans and P. Grangier, Phys. Rev. Lett. **88**, 057902 (2002).

[36] F. Grosshans and P. Grangier, E-print arXiv:quant-ph/0204127. Proc. 6th Int. Conf. on Quantum Communications, Measurement, and Computing, (Rinton Press, Princeton, 2003).

[37] N. Gisin, G. Ribordy, W. Tittel, H. Zbinden, Rev. Mod. Phys. **74**, 145-195 (2002).

[38] N.J. Cerf, A. Ipe, X. Rottenberg, Phys. Rev. Lett. **85**, 1754-1757 (2000).

[39] N.J. Cerfand S. Iblisdir, Phys. Rev. A **62**, 040301(R) (2000).

[40] F. Grosshans and Ph. Grangier, Phys. Rev. A **64**, 010301(R) (2001).

[41] L.M. Duan, G. Giedke, J.I. Cirac, P. Zoller, Phys. Rev. Lett. **84**, 4002-4005 (2000).

[42] F. Grosshans, G. Van Assche, J. Wenger, R. Brouri, N.J. Cerf and P. Grangier, Nature (London) **421**, 238-241 (16 January 2003).

[43] I. Csiszár and J. Körner, IEEE Trans. Inform. Theory **24**, 339-348 (1978).

[44] U.M. Maurer, IEEE Trans. Inform. Theory **39**, 733-742 (1993).

[45] J.-Ph Poizat, J.-F. Roch and Ph. Grangier, Ann. Phys. (Paris) **19**, 265-ă297 (1994).

[46] Ph. Grangier, J.A. Levenson and J.-Ph. Poizat, Nature (London) **396**, 537-542 (1998).

[47] F. Grosshans, N. J. Cerf, J. Wenger, R. Tualle-Brouri and Ph. Grangier, Quant. Inf. Comput. **3**, 535 (2003); see also e-print quant-ph/0306141.

[48] F. Grosshans and N.J. Cerf, e-print quant-ph/0311006.

[49] C.E. Shannon, Bell Syst. Tech. J. **27**, 623-656 (1948).

[50] W.T. Buttler, S.K. Lamoreaux, J.R. Torgerson, G.H. Nickel and C.G. Peterson, E-print arXiv:quant-ph/0203096.

[51] G. Brassard and L. Salvail, *Advances in Cryptology - Eurocrypt'93*, Lecture Notes in Computer Science, 411-423, (ed. T. Helleseth, Springer, New York, 1993).

[52] K. Nguyen, Thesis, Université Libre de Bruxelles (2002).

[53] H.-K. Lo, Eprint arXiv: quant-ph/ 0201030.

[54] C.H. Bennett, G. Brassard, C. Crépeau and U.M. Maurer, IEEE Trans. Inform. Theory **41**, 1915-1935 (1995).

Course 9

ENTANGLED PHOTONS AND QUANTUM COMMUNICATION

M. Aspelmeyer, C. Brukner, A. Zeilinger

Institut für Experimentalphysik, Universität Wien, Vienna, Austria
Institut für Quantenoptik und Quanteninformation, Österreichische Akademie der Wissenschaften
Vienna, Austria

D. Estève, J.-M. Raimond and J. Dalibard, eds.
Les Houches, Session LXXIX, 2003
Quantum Entanglement and Information Processing
Intrication quantique et traitement de l'information
© 2004 Elsevier B.V. All rights reserved

Contents

1. Introduction

Quantum entanglement is at the heart of quantum physics [1]. It offers unique insights into the fundamental principles of our physical world and it provides at the same time the basis of novel communication protocols, which allow efficient communication and computation beyond the capabilities of their classical counterparts. Prominent examples are quantum cryptography [2–5], the simultaneous distribution of a cryptographic key that is ultimately secured by the laws of quantum physics, quantum dense coding [6, 7], a protocol to double the classically allowed capacity of a communication channel by encoding 2 bits of information per bit sent, or quantum teleportation [8, 9], the remote transfer of an arbitrary quantum state between distant locations. Further examples include entanglement-assisted classical communication [10, 11] to enhance the communication capacity in a noisy environment or methods to exploit the computational advantages provided by quantum entanglement for communication complexity problems [12–14]. These quantum communication protocols utilize entanglement as a resource and form the basis for a new emerging quantum information technology. To achieve quantum communication within a network it is a central task to be able to distribute and manipulate quantum entanglement, in principle up to a global scale. At present, the only suitable system for transmitting information in long-distance quantum communication are photons.

2. Distributing quantum entanglement

The generation of entangled photons is nowadays routine work in the laboratory. One of the methods to produce polarization-entangled photons is spontaneous parametric downconversion, in which pairs of single photons are emitted into separated spatial modes. The photons can then be distributed between different receivers, say Alice (A) and Bob (B). The overall two-photon state shared by the two can be any of the four maximally entangled Bell-states [15], for example

$$|\Psi^-\rangle_{AB} = \frac{1}{\sqrt{2}}(|H\rangle_A|V\rangle_B - |V\rangle_A|H\rangle_B). \tag{2.1}$$

The Bell states have the unique feature that all information on polarization properties is completely contained in the (polarization-)correlations between the separate photons, while the individual particle does not have any polarization prior to measurement. In other words, all of the information is distributed among two particles, and none of the individual systems carries any information. This is the essence of entanglement. At the same time, these (polarization-)correlations are stronger than classically allowed since they violate bounds imposed by local realistic theories via the Bell-inequality [16] or they lead to a maximal contradiction between such theories and quantum mechanics as signified by the Greenberger-Horne-Zeilinger theorem [17, 18]. Distributed entanglement thus allows to establish non-classical correlations between distant parties and can therefore be considered the quantum analogue to a classical communication channel, a quantum communication channel [1]. This has, for example, immediate influence on the security of quantum key distribution protocols [19] and it is the basis for the quantum advantage of entanglement-based quantum communication and quantum computation schemes[2].

The possibility to establish such quantum communication channels over large distances offers the fascinating perspective to eventually take advantage of these novel communication capabilities in networks of increasing size. Naturally, non-trivial problems emerge in scenarios involving long distances or multiple parties. Experiments based on present fiber technology have demonstrated that entangled photon pairs can be separated by distances ranging from several hundreds of meters up to about 10 km [28–30], but no improvements by orders of magnitude are to be expected. Optical free-space links could provide a solution to this problem since they allow in principle for much larger propagation distances of photons because of the low absorption of the atmosphere in certain wavelength ranges. Single optical free-space links have been studied and successfully implemented already for several years for their application in quantum cryptography based on

[1] Note, however, that in contrast to classical communication no information is physically transferred via a quantum channel. Instead, non-local correlations are established via local measurements and are subsequently utilized for communication or computation purposes.

[2] A further step is the utilization of higher-dimensional entanglement. Higher-dimensional quantum systems have several interesting properties, for example fundamental tests of quantum mechanics provide more striking divergence from classical theory with respect to tests of non-locality [20]. Further, the increased complexity provides fertile ground for the development of new and more efficient protocols not possible classically or even with qubits [14, 21, 22]. We have realized several experiments exploring quantum entanglement in higher-dimensional Hilbert spaces using orbital angular momentum of photons [23–25]. In a recent experiment we demonstrated quantum tomography of a qutrit, a three dimensional orbital angular momentum state of photons. This is the first experiment which provides full control over the quantum state of a triggered qutrit and is thus of importance for the implementation of more advanced quantum communication protocols [26, 27].

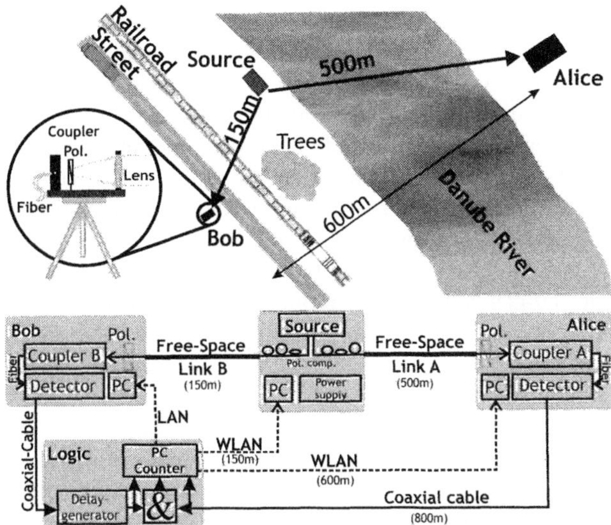

Fig. 1. Free-space distribution of polarization-entangled photons (from [33]). The entangled-photon source was positioned on the bank of the Danube River. The two receivers, Alice and Bob, were located on rooftops and separated by approx. 600m, without a direct line of sight between each other. The inset shows the schematics of the telescopes consisting of a single-mode fibre coupler and a 5cm diameter lens. At the receiver telescopes, polarizers (Pol.) were attached to determine the polarization correlations and eventually violate a Bell-inequality. The lower figure shows a functional block diagram of the experiment. Detection signals from Alice were relayed to Bob using a long BNC cable. Singles and coincidence counting was performed locally at Bob and the results were shared between all three stations using LAN and Wave-LAN connections.

faint classical laser pulses [31, 32]. We have recently demonstrated a next crucial step, namely the distribution of quantum entanglement via two simultaneous optical free-space links in an outdoor environment [33]. Polarization-entangled photon pairs have been transmitted across the Danube River in the city of Vienna via optical free-space links to independent receivers separated by 600m and without a line of sight between them (see Figure 1). A Bell inequality between those receivers was violated by more than 4 standard deviations confirming the quality of the entanglement. In this experiment, the setup for the source generating the entangled photon pairs has been miniaturized to fit on a small optical breadboard and it could easily be operated completely independent from an ideal laboratory environment.

Obviously, terrestrial free-space links are limited to rather short distances because they suffer from possible obstruction of objects in the line of sight, from atmospheric attenuation and, eventually, from the Earth's curvature. To fully exploit the advantages of free-space links, it will eventually be necessary to use space and satellite technology. By transmitting and/or receiving either photons or entangled photon pairs to and/or from a satellite, entanglement can be distributed over truly large distances. This would allow quantum communication applications on a global scale. From a fundamental point of view, satellite-based distribution of quantum entanglement is also the first step towards exploiting quantum correlations on a scale larger by orders of magnitude than achievable in laboratory and even ground-based experimental environments. State of the art photon sources and detectors would already suffice to achieve a satellite-based quantum communication link over some thousands of kilometers [34–36]. The outdoor experiments described above represent a step towards satellite-based distributed quantum entanglement.

One principle limitation of photonic quantum communication schemes is the loss of photons in the quantum channel. This limits the bridgeable distance for single photons to the order of 100 km in present silica fibers [37, 38]. Recent experiments already achieve such distances [39, 40]. In principle, this drawback can eventually be overcome by subdividing the larger distance to be bridged into smaller sections over which entanglement can be teleported. The consequent application of so-called "entanglement swapping" [41] may result in transporting entanglement over larger distances. Additionally, to diminish decoherence effects possibly induced by the quantum channel, quantum purification might be applied to eventually implement a full quantum repeater [42]. In fact, the experimental building blocks for a full-scale quantum repeater based on linear optics have been successfully demonstrated over the last years by the realization of teleportation and entanglement swapping [9,43,44] and, only recently, by the implementation of quantum entanglement purification and distillation protocols [45–47].

3. Quantum teleportation and entanglement swapping

Teleportation of quantum states [8] is an intriguing concept within quantum physics and a striking application of quantum entanglement. Besides its importance for quantum computation [48, 49], teleportation is at the heart of the quantum repeater [42], a concept eventually allowing the distribution of quantum entanglement over arbitrary distances and thus enabling quantum communication over large distances and even networking on a global scale.

The purpose of quantum teleportation is to transfer an arbitrary quantum state to a distant location, say from Alice to Bob, without transmitting the actual physical object carrying the state. Classically this is an impossible task, since Alice cannot obtain the full information of the state to be teleported without previous knowledge about its preparation. Quantum physics, however, provides a working strategy. Suppose, Alice and Bob share an ancilla entangled pair in advance. Alice then performs a Bell-state measurement between the teleportee particle and her shared ancilla, i.e. she projects the two particles into the basis of Bell-states. The four possible outcomes of this measurement provide her with two bits of classical information. Thus the quantum state to be teleported can be reconstructed at Bob's side. After communicating the classical result to Bob, he can perform one out of four unitary operations to obtain the original state to be teleported. Note, that in one of the cases the original state is transferred instantaneously and no active transformation is required at Bob's side. If Bob would have used his state beforehand as an input state for a quantum computational operation, the teleportation would define his input only *after* the actual computation has taken place. One thus achieves for these cases instantaneous quantum computation, which can provide additional computational advantage (although with exponentially small probability)for certain schemes [50].

An important feature of teleportation is that it provides no information whatsoever about the state being teleported. This means that an arbitrary unknown quantum state can be teleported. In fact, the quantum state of a teleportee particle does not have to be well defined and it could thus even be entangled with another photon. This means that a Bell-state measurement of two of the photons - one each from two pairs of entangled photons - results in the remaining two photons becoming entangled, even though they have never interacted in the past. Recently, we could demonstrate that by violating a Bell inequality between particles that never interacted with each other [44]. A repeated application of this so-called "entanglement-swapping" [41] can in principle be used to transfer quantum entanglement between distant sites (see Figure 2). In the context of Bell experiments to test for violations of local realism, entanglement-swapping can also be seen as a method to achieve event-ready detection of entangled particles, since a successful Bell-state measurement unambiguously indicates the completed preparation of an entangled pair [41].

Two recent results, both of relevance for long-distance quantum communication, are the demonstration of quantum state teleportation over a distance of several tens of meters [51] and the first realization of freely propagating teleported qubits [52], which eventually will allow the subsequent use of teleported states. In previous realizations of teleportation the teleported qubit had to be

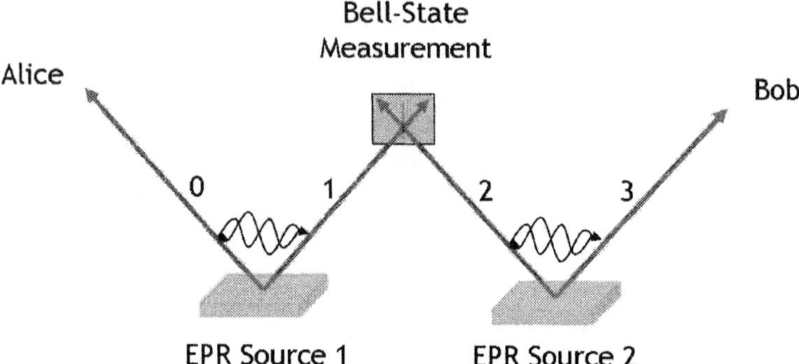

Fig. 2. Scheme for Entanglement Swapping, i.e. the teleportation of entanglement. Two entangled pairs $0-1$ and $2-3$ are produced by two entangled-photon sources (EPR). One particle from each of the pairs is sent to two separated observers, say 0 is sent to Alice and 3 to Bob. 1 and 2 become entangled through a Bell-state measurement, by which 0 and 3 also become entangled. This requires the entangled qubits 0 and 3 neither to come from a common source nor to have interacted in the past.

detected (and thus destroyed) to verify the success of the procedure. This can be avoided by providing, in average, more entangled ancilla pairs than states to be teleported. In the new teleportation scheme, a successful Bell-state analysis results in freely propagating individual qubits, which can be used for further cascaded teleportation.

4. Purification of entanglement

Owing to unavoidable decoherence in the quantum communication channel, the quality of entangled states generally decreases with the channel length. Entanglement purification is a way to extract a subset of states of high entanglement and high purity from a larger set of less entangled states - and is thus needed to overcome the decoherence of noisy quantum channels. We were able for the first time to experimentally demonstrate a general quantum purification scheme for mixed polarization-entangled two-particle states [47]. The crucial operation for a successful purification step is a bilateral conditional NOT (CNOT) gate, which effectively detects single bit-flip errors in the channel by performing local CNOT operations at Alice's and Bob's side between particles of shared entangled states. The outcome of these measurements can be used to correct for such errors and eventually end up in a less noisy quantum channel [53]. For the case of polarization entanglement, such a "parity-check" on the correlations can be performed in a straight forward way by using polarizing beamsplitters (PBS) [54] that transmit horizontally polarized photons and reflect vertically polarized ones.

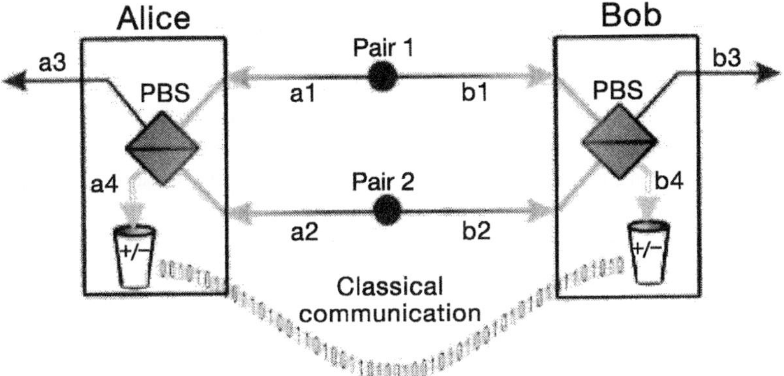

Fig. 3. Scheme for entanglement purification of polarization-entangled qubits (from [47]). Two shared pairs of an ensemble of equally mixed, entangled states ρ_{AB} are fed in to the input ports of polarizing beamsplitters that substitute the bilateral CNOT operation necessary for a successful purification step. Alice and Bob keep only those cases where there is exactly one photon in each output mode. This can only happen if no bit-flip error occurs over the channel. Finally, to obtain a larger fraction of the desired pure (Bell-)state they perform a polarization measurement in the $|\pm\rangle$ basis in modes a4 and b4. Depending on the results, Alice performs a specific operation on the photon in mode a3. After this procedure, the remaining pair in modes a3 and b3 will have a higher degree of entanglement than the two original pairs.

Consider the situation in which Alice and Bob have established a noisy quantum channel, i.e. they share a set of equally mixed, entangled states ρ_{AB}. At both sides the two particles of two shared pairs are directed into the input ports a_1, a_2 and b_1, b_2 of a PBS (see Figure 3). Only if the entangled input states have the same correlations, i.e. they have the same parity with respect to their polarization correlations, the four photons will exit in four different outputs (four-mode case) and a projection of one of the photons at each side will result in a shared two-photon state with a higher degree of entanglement. All single bit-flip errors are effectively suppressed.

For example, they might start off with the mixed state $\rho_{AB} = F \cdot |\Phi^+\rangle\langle\Phi^+|_{AB} + (1-F) \cdot |\Psi^-\rangle\langle\Psi^-|_{AB}$, where $|\Phi^+\rangle = (|HH\rangle + |VV\rangle)$ is another Bell state. Then, only the combinations $|\Phi^+\rangle_{a_1,a_2} \otimes |\Phi^+\rangle_{b_1,b_2}$ and $|\Psi^-\rangle_{a_1,a_2} \otimes |\Psi^-\rangle_{b_1,b_2}$ will lead to a four-mode case, while $|\Phi^+\rangle_{a_1,a_2} \otimes |\Psi^-\rangle_{b_1,b_2}$ and $|\Psi^-\rangle_{a_1,a_2} \otimes |\Phi^+\rangle_{b_1,b_2}$ will be rejected. Finally, a projection of the output modes a_4, b_4 into the basis $|\pm\rangle = \frac{1}{\sqrt{2}}(|H\rangle \pm |V\rangle)$ is needed to create the pure states $|\Phi^+\rangle_{a_3,b_3}$ with probability $F' = F^2/[F^2 + (1-F)^2]$ and $|\Psi^+\rangle_{a_3,b_3}$ with probability $1 - F'$, respectively. The fraction F' of the desired state $|\Phi^+\rangle$ becomes larger for $F > \frac{1}{2}$. In other words, the new state ρ'_{AB} shared by Alice and Bob after the bilateral parity operation demonstrates an increased fidelity with respect to a pure, maximally entangled state. This is the purification of entanglement.

Typically, in the experiment, one photon pair of fidelity 92% could be obtained from two pairs, each of fidelity 75%. Also, although only bit-flip errors in the channel have been discussed, the scheme works for any general mixed state, since any phase-flip error can be transformed to a bit-flip by a rotation in a complementary basis. In our experiments, decoherence is overcome to the extent that the technique would achieve tolerable error rates for quantum repeaters in long-distance quantum communication based only on linear optics and polarization entanglement.

5. Quantum entanglement and information

Quantum communication utilizes the information content of entangled systems as an additional resource. We suggest that the relation between quantum entanglement and information is a much more fundamental one and even inherent to quantum physics.

The various debates about the conceptual significance of quantum mechanics can to a large part be seen as a debate of what quantum physics refers to. Does it refer to reality directly or does it refer to (our) knowledge, and therefore to information? If quantum physics refers to reality, which reality is it? Is it the reality which appears to us, or is it a more complicated reality, like the one alluded to in the Many-Worlds interpretation?

Significant inspiration can be obtained from Niels Bohr, who, for example, according to Aage Petersen liked to say [55]: "There is no quantum world. There is only an abstract quantum physical description. It is wrong to think that the task of physics is to find out how Nature is. Physics concerns what we can say about Nature."

To us, it is thus suggestive that knowledge is *the* central concept of quantum physics. In modern language, knowledge can be equated with information. Therefore, one firstly needs a proper measure of information. One might be tempted to use Shannon's measure

$$I = -\sum_i p_i \log p_i, \tag{5.1}$$

where, p_i is the probability of sign i to occur in a sequence.

Yet it turns out that Shannon's measure is not adequate in order to describe the knowledge gained in an individual quantum experiment [56]. This feature can be understood in various ways. One most central one is that the logarithmic dependence of the Shannon measure of information is related to the postulate that the information gained in a series of observations of different properties must be independent from the specific sequence in which the properties are read out. Clearly, such a requirement is not valid in quantum mechanics any more, unless the properties are commuting, which is in general an exception. So, what is desired is a measure of information which accounts for complementarity and describes the total information obtainable in a complete set of quantum experiments. We have suggested elsewhere [57] that the most appropriate measure of information is

$$I = \sum_i p_i p_i \tag{5.2}$$

which may be viewed as the sum of the individual probabilities weighed by these probabilities themselves. The total information content I_{total} of a quantum system is then obtained as a sum of individual measures of information I_j (of the type given in (5.2)) over a complete set of maximally mutually complementary observables (indexed by j)

$$I_{total} = \sum_j I_j = \sum_j \sum_i p_{ij} p_{ij}. \tag{5.3}$$

Here p_{ij} denotes the probability to observe the i-th outcome of the j-th observable.

Maybe the notion of mutually complementary observables might need to be explained further. A famous example is given by the three components of a spin-1/2 particle taken along three directions orthogonal in space (not to be confused with orthogonal quantum states). From an operational point of view, two variables A and B are maximal mutually complementary if the knowledge of one completely precludes any knowledge of the other. In the case of spin, if A represents the spin along the z-direction, then B might represent the spin along any direction orthogonal to z in space. It is a well known feature that if the spin along z is well defined, the spin along these other directions is maximally undefined. To come back to our example, the sum $\sum_j$ in (5.3) in the case of spin has to be taken along any three spatially orthogonal directions, i.e., $j = x, y, z$.

It has not escaped our attention that equation (5.3) can be put onto a nicely visualizable foundation if one defines an information space spanned by mutually complementary states [58]. Then I_{total} just represents the square of the length of a vector in that information space when the square of the length of individual components is just given by I_j.

If, as we have suggested above, quantum physics is about information then we have to ask ourselves what we mean by a quantum system. It is then imperative to avoid assigning any variant of naive classical objectivity to quantum states [59].

Rather it is then natural to assume that the quantum system is just that to which the probabilities in equations (5.2) and (5.3) refer, and no more. So, the notion of an independently existing reality becomes void.

We might therefore ask how much information a quantum system might carry stressing again that by "carry" we just refer to the total amount of information and not to the objective existence of any subject actually carrying the information.

It is obvious that a large system, being our mental representative of the information characterizing it, carries a lot of information, very many bits. Then, how does that amount of information scale with the size of the object? It is very suggestive to assume that the smaller a system, the less information it carries. One may even consider the amount of information carried by a system as defining its "size". So, basically we postulate (1) that the amount of information carried by any system is finite and (2) that the amount of information is lesser the smaller the system. These assumptions may be supported by referring to Feynman [60]: "It always bothers me that, according to the laws as we understand them today, it takes a computing machine an infinite number of logical operations to figure out what goes on in no matter how tiny a region of space, and no matter how tiny a region of time. How can all that be going on in that tiny space? Why should it take an infinite amount of logic to figure out what one tiny piece of space/time is going to do?" Evidently, Feynman's problem is solved if the "tiny piece of space-time" only contains a finite amount of information, and the less, the smaller the piece is.

We arrive at a natural limit when a system only represents one bit of information. Once that is achieved, the system can only represent the yes/no answer to one question. If the system is asked another question, the answer by necessity has to be random. Thus, randomness is a fundamental feature of our world [61, 62]. This, we suggest, also provides a natural foundation for complementarity. Consider, for example, a simple two-path interferometer as shown in Figure 4.

As is well known, in such an interferometer, we can either define the path $|a\rangle$ or $|b\rangle$ taken by a particle. In that case, the trajectory after the semi-reflecting beam-splitter is completely random, or, in other words, detectors I or II each will register particles with the same probability of 50%. On the other hand, we can prepare the state in a coherent superposition of $|a\rangle$ and $|b\rangle$ in such a way that, by adjusting the relative phase, all particles end up in detector I and no particles in detector II. In other words, in the outgoing beam leading to detector I, constructive interference happens, and destructive interference happens in the outgoing beam II. The observations of the path the particle takes inside the interferometer and of the interference are an example of two mutually maximally complementary observations.

This behaviour can be understood very simply on the basis of our most elementary quantum system carrying just one bit of information. It is then up to the

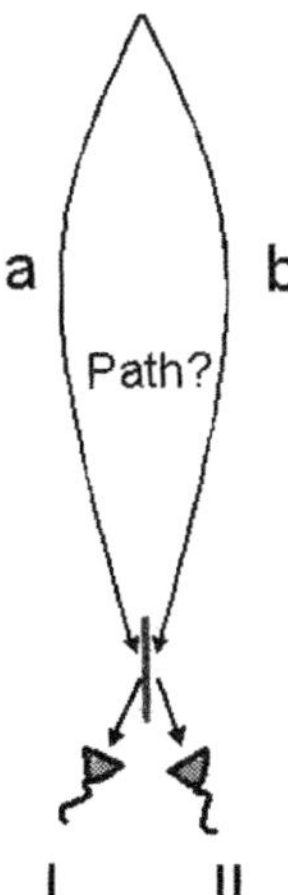

Fig. 4. Two-Path Interferometer. The source emits coherent waves of which two beams are selected and incident on the beam-splitter. Each of the two beams has the same amplitude for being transmitted or reflected at the beam-splitter, and thus the outgoing beams are coherent superpositions of the incoming ones.

experimentalist to decide whether she wants to prepare the system such that the one bit of information is used to completely determine the path, a binary variable, in which case no information is left to determine the fate of the particle after the beam splitter. Then the outcome, which detector, I or II, fires, must be completely random. Alternatively, the experimentalist prepares the system such that the one bit of information defines which detector fires, i.e. the interference, in which case the path is completely undefined. Evidently, intermediate cases are possible, where both, path information and interference, each are partly defined.

We note that we are thus led to a natural explanation of quantum complementarity. Our measure of information defined in equation (5.3) provides a perfect measure also for the intermediate cases, where path is partially defined and also interference is obtained with partial visibility only such that their information contents totally sum up to one bit [63].

Concluding this chapter, we note that using our approach, we were able to explain some other important features of quantum mechanics, most notably, Malus' law [58] which describes the cosine dependence of the probability upon the angle between the measurement direction and the direction along which the spin is well defined. We were also able to obtain a natural understanding of entanglement [64]. For example, if one considers entanglement of two spin-1/2 particles, one has two elementary systems in our sense, and thus two bits of information available. These two bits can be used in principle to encode properties of the individual particles themselves, which is basically classical coding. On the other hand, the two bits can be completely used up to fully define joint information only, that is, information about how possible measurement results on the particles relate to each other. If done in that way, one automatically obtains the four Bell states. That way one obtains a natural basis for Schrödinger's definition of entanglement [1]. Finally, we note that using our approach we were able to derive the Liouville equation describing the quantum evolution in time of a two-state system [58].

Clearly, a number of important questions remain open. Of these, we mention here two. The first one refers to continuous variables. The problem there is that with continuous variables, one in principle has an infinite number of complementary observables. One might attack this question by generalizing the definition of equation (5.2) to infinite sets. This, while mathematically possible, leads to conceptually difficult situations. The conceptual problem in our eyes is related to the fact that we wish to define all notions on operationally verifiable bases or foundations, that is, on foundations which can be verified directly in experiment. It is obvious that an infinite number of complementary observables can never be realized in experiment. In our opinion, it is therefore suggestive that the concept of an infinite number of complementary observables and therefore, indirectly, the assumption of continuous variables, are just mathematical constructions which might not have a place in a final formulation of quantum mechanics.

This leads to the second question, namely, how to derive the Schrödinger equation. If our assumption just expressed is correct, namely that continuous variables are devoid of operational and therefore physical meaning in quantum mechanics, there is no need to express the Schrödinger equation based on continuous variables in our new language. Indeed, one then should refer to situations where one always has a finite number of complementary observables only. In our opinion such a point of view is experimentally well founded, as any experiment will always lead only to a finite number of bits and a finite number of the experimental results on the basis on which only a finite number of observables can be operationally defined.

It has not escaped our attention that our way of reasoning also leads to new possibilities of understanding why we have quantum physics, i.e. Wheeler's famous question "Why the quantum?". Identifying systems with the information they carry, we note that information necessarily is quantized. One can have one proposition, two propositions, three propositions etc., but obviously the concept of, say, $\sqrt{2}$ propositions is devoid of any meaning. Therefore, since information is quantized that way, our description of information, which is quantum mechanics, also has to be quantized.

References

[1] E. Schrödinger. Die gegenwärtige Situation in der Quantenmechanik. *Naturwissenschaften*, 23:807–812; 823–828; 844–849, 1935.

[2] A. K. Ekert. Quantum cryptography based on bell's theorem. *Phys. Rev. Lett.*, 67:661–663, 1991.

[3] T. Jennewein, C. Simon, G. Weihs, H. Weinfurter, and A. Zeilinger. Quantum cryptography with entangled photons. *Phys. Rev. Lett.*, 84:4729, 2000.

[4] W. Tittel, J. Brendel, H. Zbinden, and N. Gisin. Quantum cryptography using entangled photons in energy-time bell states. *Phys. Rev. Lett.*, 84:4737, 2000.

[5] D. S. Naik, C. G. Peterson, A. G. White, A. J. Berglund, and P. G. Kwiat. Entangled state quantum cryptography: Eavesdropping on the ekert protocol. *Phys. Rev. Lett.*, 84:4733, 2000.

[6] C. H. Bennett and S. J. Wiesner. Communication via one- and two-particle operators on Einstein-Podolsky-Rosen states. *Phys. Rev. Lett.*, 69:2881, 1992.

[7] Klaus Mattle, Harald Weinfurter, Paul G. Kwiat, and Anton Zeilinger. Dense coding in experimental quantum communication. *Phys. Rev. Lett.*, 76(25):4656–4659, 1996.

[8] C. H. Bennett, G. Brassard, C. Crépeau, R. Jozsa, A. Peres, and W. K. Wootters. Teleporting an unknown quantum state via dual classical and einstein-podolsky-rosen channels. *Phys. Rev. Lett.*, 70:1895, 1993.

[9] D. Bouwmeester, J.-W. Pan, K. Mattle, M. Eibl, H. Weinfurter, and A. Zeilinger. Experimental quantum teleportation. *Nature*, 390:575, 1997.

[10] C. H. Bennett, C. A. Fuchs, and J. A. Smolin. Entanglement-enhanced classical communication on a noisy quantum channel. *quant-ph/9611006 (1996).*

[11] K. Banaszek, D. W. Wasilewski, and C. Radzewicz. Experimental demonstration of entanglement-enhanced classical communication over a quantum channel with correlated noise. *quant-ph/0403024 (2004).*

[12] G. Brassard. Quantum communication complexity (a survey). *quant-ph/0101005*, 2001.

[13] H. Buhrman, W. Van Dam, P. Høyer, and A. Tapp. Multiparty quantum communication complexity. *Phys. Rev. A*, 60:2737–2741, 1999.

[14] C. Brukner, M. Zukowski, and A. Zeilinger. Quantum communication complexity protocol with two entangled qutrits. *Phys. Rev. Lett.*, 89:197901, 2002.

[15] D. Bouwmeester, A. Ekert, and A. Zeilinger, editors. *The Physics of Quantum Information.* Springer-Verlag, Berlin, 2000.

[16] J. Bell. On the Einstein Podolsky Rosen paradox. *Physics*, 1:195, 1964.

[17] D. M. Greenberger, M. A. Horne, and A. Zeilinger. Going beyond bell's theorem. In M. Kafatos, editor, *Bell's Theorem, Quantum Theory, and Conceptions of the Universe*, page 69. Kluwer, Dordrecht, 1989.

[18] D. M. Greenberger, M. A. Horne, A. Shimony, and A. Zeilinger. Bell's theorem without inequalities. *Am. J. Phys.*, 58:1131–1143, 1990.

[19] M. Curty, M. Lewenstein, and N. Lütkenhaus. Entanglement as precondition for secure quantum key distribution. *quant-ph/0307151 (2003).*

[20] D. Kaszlikowski, P. Gnacinski, M. Żukowski, W. Miklaszewski, and A. Zeilinger. Violations of local realism by two entangled n-dimensional systems are stronger than for two qubits. *Phys. Rev. Lett.*, 85:4418, 2000.

[21] H. Bechmann-Pasquinucci and A. Peres. Quantum cryptography with 3-state systems. *Phys. Rev. Lett.*, 85:3313, 2000.

[22] M. Bourennane, A. Karlsson, and G. Björk. Quantum key distribution using multilevel encoding. *Phys. Rev. A*, 64:12306, 2001.

[23] A. Mair, A. Vaziri, G. Weihs, and A. Zeilinger. Entanglement of the orbital angular momentum states of photons. *Nature*, 412:313, 2001.

[24] A. Vaziri, G. Weihs, and A. Zeilinger. Experimental two-photon, three-dimensional entanglement for quantum communication. *Phys. Rev. Lett.*, 89:240401, 2002.

[25] A. Vaziri, J.-W. Pan, T. Jennewein, G. Weihs, and A. Zeilinger. Concentration of higher dimensional entanglement: Qutrits of photon orbital angular momentum. *Phys. Rev. Lett.*, 91:227902, 2003.

[26] G. Molina-Terriza, A. Vaziri, J. Rehacek, Z. Hradil, and A. Zeilinger. Triggered qutrits for quantum communication protocols. *quant-ph/0401183 (2004).*

[27] R.T.Thew, A. Acin, H. Zbinden, and N. Gisin. Experimental realization of entangled qutrits for quantum communication. *quant-ph/0307122 (2003).*

[28] P. R. Tapster, J. G. Rarity, and P. C. M. Owens. Violation of bell's inequality over 4 km of optical fiber. *Phys. Rev. Lett.*, 73:1923, 1994.

[29] W. Tittel, J. Brendel, H. Zbinden, and N. Gisin. Violation of Bell inequalities by photons more than 10 km apart. *Phys. Rev. Lett.*, 81:3563, 1998.

[30] G. Weihs, T. Jennewein, C. Simon, H. Weinfurter, and A. Zeilinger. Violation of Bell's inequality under strict Einstein locality condition. *Phys. Rev. Lett.*, 81:5039, 1998.

[31] W. T. Buttler, R. J. Hughes, P. G. Kwiat, S. K. Lamoreaux, G. G. Luther, G. L. Morgan, J. E. Nordholt, C. G. Peterson, and C. M. Simmons. Practical free-space quantum key distribution over 1 km. *Phys. Rev. Lett.*, 81:3283–3286, 1998.

[32] C. Kurtsiefer, P. Zarda, M. Halder, H. Weinfurter, P.M. Gorman, P.R. Tapster, and J.G. Rarity. A step towards global key distribution. *Nature*, 419:450, 2002.

[33] M. Aspelmeyer, H. R. Böhm, T. Gyatso, T. Jennewein, R. Kaltenbaek, M.Lindenthal, G. Molina-Terriza, A. Poppe, K. Resch, M. Taraba, R. Ursin, P. Walther, and A. Zeilinger. Long-distance free-space distribution of quantum entanglement. *Science*, 301:621, 2003.

[34] M. Aspelmeyer, T. Jennewein, R. Kaltenbaek, M. Lindenthal, H. R. Böhm, J. Petschinka, R. Ursin, C. Brukner, A. Zeilinger, M. Pfennigbauer, and W. Leeb. Quantum communications in space. Technical Report 16358/02, European Space Agency (ESA), 2003.

[35] M. Aspelmeyer, T. Jennewein, M. Pfennigbauer, W. Leeb, and A. Zeilinger. Long-distance quantum communication with entangled photons using satellites. *quant-ph/0305105 (2003).*

[36] R. Kaltenbaek, M. Aspelmeyer, T. Jennewein, C. Brukner, M. Pfennigbauer, W. R. Leeb, and A. Zeilinger. Proof-of-concept experiments for quantum physics in space. *quant-ph/0308174 (2003).*

[37] E. Waks, A. Zeevi, and Y. Yamamoto. Security of quantum key distribution with entangled photons against individual attacks. *Phys. Rev. A*, 65:52310, 2002.

[38] N. Gisin, G. Ribordy, W. Tittel, and H. Zbinden. Quantum cryptography. *Rev. Mod. Phys.*, 74:145–195, 2002.

[39] T. Kimura, Y. Nambu, T. Hatanaka, A. Tomita, H. Kosaka, and K. Nakamura. Single-photon interference over 150-km transmission using silica-based integrated-optic interferometers for quantum cryptography. *quant-ph/0403104*, 2004.

[40] Z. Yuan, C. Gobby, and A. J. Shields. Quantum key distribution over 101 km telecom fibre. *presented at: CLEO-Europe, Munich*, 2003.

[41] M. Żukowski, A. Zeilinger, M. A. Horne, and A. K. Ekert. "Event-ready-detectors" Bell experiment via entanglement swapping. *Phys. Rev. Lett.*, 71(26):4287–4290, 1993.

[42] H.-J. Briegel, W. Dür, J. I. Cirac, and P. Zoller. Quantum repeaters: The role of imperfect local operations in quantum communication. *Phys. Rev. Lett.*, 81:5932–5935, 1998.

[43] Jian-Wei Pan, Dik Bouwmeester, Harald Weinfurter, and Anton Zeilinger. Experimental entanglement swapping: Entangling photons that never interacted. *Phys. Rev. Lett.*, 80:3891–3894, 1998.

[44] T. Jennewein, G. Weihs, J.-W. Pan, and A. Zeilinger. Experimental nonlocality proof of quantum teleportation and entanglement swapping. *Phys. Rev. Lett.*, 88:17903, 2002.

[45] P. G. Kwiat, S. Barraza-Lopez, A. Stefanov, and N. Gisin. Experimental entanglement distillation and 'hidden' non-locality. *Nature*, 409:1014, 2001.

[46] T. Yamamoto, M. Koashi, S. K. Özdemir, and N. Imoto. Experimental extraction of an entangled photon pair from two identically decohered pairs. *Nature*, 421, 343.

[47] J.-W. Pan, S. Gasparoni, R. Ursin, G. Weihs, and A. Zeilinger. Experimental entanglement purification. *Nature*, 423:417–422, 2003.

[48] D. Gottesman and I. L. Chuang. Demonstrating the viability of universal quantum computation using teleportation and single-qubit operations. *Nature*, 402:390, 1999.

[49] E. Knill, R. Laflamme, and G. Milburn. A scheme for efficient quantum computation with linear optics. *Nature*, 409:46–52, 2000.

[50] C. Brukner, J.-W. Pan, C. Simon, G. Weihs, and A. Zeilinger. Probabilistic instantaneous quantum computation. *Phys. Rev. A*, 67:34304, 2003.

[51] I. Marcikic, H. de Riedmatten, W. Tittel, H. Zbinden, and N. Gisin. Long-distance teleportation of qubits at telecommunication wavelengths. *Nature*, 421:509, 2003.

[52] J.-W. Pan, S. Gasparoni, M. Aspelmeyer, T. Jennewein, and A. Zeilinger. Freely propagating teleported qubits. *Nature*, 421:721, 2003.

[53] C. H. Bennett, G. Brassard, S. Popescu, B. Schumacher, J. A. Smolin, and W. K. Wootters. Purification of noisy entanglement and faithful teleportation via noisy channels. 76:722, 1996.

[54] J.-W. Pan, C. Simon, C. Brukner, and A. Zeilinger. Entanglement purification for quantum communication. *Nature*, 410:1067, 2001.

[55] A. Petersen. The philosophy of Niels Bohr. In A.P. French and P.I.Kennedy, editors, *Niels Bohr, A Centenary Volume*, page 299. Harvard University Press, Cambridge, 1985.

[56] C. Brukner and A. Zeilinger. Conceptual inadequacy of the shannon information in quantum measurements. *Phys. Rev. A*, 63:22113, 2001.

[57] C. Brukner and A. Zeilinger. Operationally invariant information in quantum measurements. *Phys. Rev. Lett.*, 83:3354, 1999.

[58] C. Brukner and A. Zeilinger. In L. Castell and O. Ischebek, editors, *Information and Fundamental Elements of the Structure of Quantum Theory*, Heidelberg, 2003. Springer.

[59] E. Fox Keller. Cognitive repression in contemporary physics. *Am. J. Phys.*, 47:718, 1979.

[60] R. P. Feynman. *The Character of Physical Law*. MIT Press, Cambridge, Mass., 1965.

[61] A. Zeilinger. A foundational principle for quantum mechanics. *Found. Phys.*, 29:631, 1999.

[62] A. Bohr, B.R. Mottelson, and O. Ulfbeck. The principle underlying quantum mechanics. *Found. Phys.*, 34:405, 2004.

[63] C. Brukner and A. Zeilinger. Young's experiment and the finiteness of information. *Phil. Trans. R. Soc. Lond. A*, 360:1061, 2002.

[64] C. Brukner, M. Zukowski, and A. Zeilinger. The essence of entanglement. *quant-ph/0106119*.

Course 10

NUCLEAR MAGNETIC RESONANCE QUANTUM COMPUTATION

J. A. Jones

Centre for Quantum Computation, Clarendon Laboratory, Parks Road, Oxford OX1 3PU, UK

D. Estève, J.-M. Raimond and J. Dalibard, eds.
Les Houches, Session LXXIX, 2003
Quantum Entanglement and Information Processing
Intrication quantique et traitement de l'information
© 2004 Elsevier B.V. All rights reserved

Contents

Abstract

Nuclear Magnetic Resonance (NMR) is arguably both the best and the worst technology we have for the implementation of small quantum computers. Its strengths lie in the ease with which arbitrary unitary transformations can be implemented, and the great experimental simplicity arising from the low energy scale and long time scale of radio frequency transitions; its weaknesses lie in the difficulty of implementing essential non-unitary operations, most notably initialisation and measurement. This course will explore both the strengths and weaknesses of NMR as a quantum technology, and describe some topics of current interest.

1. Nuclear magnetic resonance

Before describing how Nuclear Magnetic Resonance (NMR) techniques can be used to implement quantum computation I will begin by outlining the basics of NMR.

1.1. Introduction

Nuclear Magnetic Resonance (NMR) is the study of the direct transitions between the Zeeman levels of an atomic nucleus in a magnetic field [1, 2, 3, 4, 5, 6, 7]. Put so simply it is hard to see why NMR would be of any interest[1], and the field has been largely neglected by physicists for many years. It has, however, been adopted by chemists, who have turned NMR into one of the most important branches of chemical spectroscopy [8].

Some of the importance of NMR can be traced to the close relationship between the information which can be obtained from NMR spectra and the information about molecular structures which chemists wish to determine, but an equally important factor is the enormous sophistication of modern NMR experiments [3], which go far beyond simple spectroscopy. The techniques developed to implement these modern NMR experiments are essentially the techniques of coherent quantum control, an area in which NMR exhibits unparalleled abilities.

[1] The interest in and importance of NMR is hinted at by the fact that research into NMR has led to Nobel prizes in Physics (Bloch and Purcell, 1952), Chemistry (Ernst, 1991, and Wütrich, 2002) and Medicine (Lauterbur and Mansfield, 2003).

It is, of course, this underlying sophistication which has led to the rapid progress of NMR implementations of quantum computing.

The basis of NMR quantum computing will be described in subsequent lectures, but I will begin by outlining the ideas and techniques underlying conventional NMR experiments. This is important, not only to gain an understanding of the key physics behind NMR quantum computing, but also to understand the language used in this field. Throughout these lectures I will use the Product Operator notation, which is almost universally used in conventional NMR [2, 9, 6, 10, 11]. Although ultimately based on traditional treatments of spin physics this notation differs from the usual physics notation in a number of subtle ways.

1.2. The Zeeman interaction and chemical shifts

Most atomic nuclei possess an intrinsic angular momentum, called spin, and thus an intrinsic magnetic moment. If the nucleus is placed in a magnetic field the spin will be quantised, with a small number of allowed orientations with respect to the field. For both conventional NMR and NMR quantum computing the most important nuclei are those with a spin of one half: these have two spin states, which are separated by the Zeeman splitting

$$\Delta E = \hbar\gamma B \tag{1.1}$$

where B is the magnetic field strength at the nucleus and γ, the gyromagnetic ratio, is a constant which depends on the nuclear species. Among these spin-half nuclei the most important species [5] are ^{1}H, ^{13}C, ^{15}N, ^{19}F and ^{31}P.

Transitions between the Zeeman levels can be induced by an oscillating magnetic field with a resonance frequency $\nu = \Delta E/h$ (the Larmor frequency). As the Larmor frequency depends linearly on the magnetic field strength it is usually desirable to use the strongest magnetic fields conveniently available. This is achieved using superconducting magnets, giving rise to fields in the range of 10 to 20 Tesla. For ^{1}H nuclei, which are the most widely studied by conventional NMR, the corresponding resonance frequencies are in the range of 400 to 800 MHz, lying in the radiofrequency (RF) region of the spectrum, and the field strengths of NMR magnets are usually described by stating the ^{1}H resonance frequency.

The relatively low frequency of NMR transitions has great significance for NMR experiments. The energy of a radio frequency photon (about 1 μeV) is so low that it is essentially impossible to detect single photons, and it is necessary to use fairly large samples (around 1 mg) containing an ensemble of about 10^{19} identical molecules. Even then the signal is weaker than one might hope, as the nuclei are distributed between the upper and lower energy levels according to the

Boltzmann distribution, and the population excess in the lower level is less than 1 in 10^4.

From the description above one would expect all the ^{1}H nuclei in a sample to have the same resonance frequency, but in fact variations are seen. These arise from the *chemical shift* interaction [6], which causes the magnetic field strength experienced by the nucleus to differ from that of the applied field. Atomic nuclei do not occur in isolation, but are surrounded by electrons, and the applied field will induce circulating currents in the electron cloud; these circulating currents cause local fields which will combine with the applied field to give a total field which determines the NMR frequency. Clearly the local fields will depend on the nature of the surrounding electrons, and thus on the chemical environment of the nucleus. Chemical shifts can in principle be calculated using quantum mechanics, but in practice it is more useful to interpret them using semi-empirical methods developed by chemists [5].

Three further points about chemical shifts should be considered. Firstly the strength of the induced fields depends linearly on the strength of the applied field, and so chemical shifts measured as frequencies increase linearly with field strength. For this reason it is more useful to measure chemical shifts as fractions, usually stated in *parts per million* (ppm). Secondly it is usually impractical to define chemical shifts with respect to the applied field, and so they are usually defined by the shift from some conventional reference system. Thirdly the induced fields depend on the relative orientation of the magnetic field and the molecular axes, and so chemical shift is a tensor, not a scalar [6, 12]. In spectra from solid powder samples [12] one observes the whole range of the tensor, and so very broad lines, but in liquids and solutions molecular tumbling causes rapid modulation of the chemical shift tensor. This averages the chemical shift interaction to its isotropic value.

1.3. Spin–spin coupling

When NMR spectra are acquired with better resolution, peaks split into groups called multiplets. Patterns in these splittings clearly indicate that they must come from some sort of coupling between spins. The most obvious explanation is direct coupling between pairs of magnetic dipoles, but it is easily seen that this cannot be the case. The dipole–dipole coupling strength is given by

$$D_{ij} \propto \frac{3\cos^2\theta_{ij} - 1}{r_{ij}^3} \tag{1.2}$$

where r_{ij} is the separation of nuclei i and j and θ_{ij} is the angle between the internuclear vector and the main magnetic field. In solid samples the dipolar coupling is clearly visible [12], but in liquids and solutions the coupling is modulated by molecular tumbling and averages to its isotropic value, which is zero.

In fact the splittings arise from the the so-called J-coupling interaction, also called scalar coupling [5, 6]. This additional coupling is related to the electron-nuclear hyperfine interaction. It is mediated by valence electrons, and thus only occurs between "nearby" spins; in particular it does not occur between nuclei in different molecules. Like dipolar coupling J-coupling is anisotropic, but unlike dipolar coupling it has a non-zero average (the isotropic value) which survives the molecular motion.

J-coupling has the form of a Heisenberg interaction, but in practice it is often truncated to an Ising form. For two coupled spins the total spin Hamiltonian is given by

$$\begin{aligned} \mathcal{H} &= \tfrac{1}{2}\omega_1\sigma_{1z} + \tfrac{1}{2}\omega_2\sigma_{2z} + \tfrac{1}{4}\omega_{J_{12}}\sigma_1 \cdot \sigma_2 \\ &\approx \tfrac{1}{2}\omega_1\sigma_{1z} + \tfrac{1}{2}\omega_2\sigma_{2z} + \tfrac{1}{4}\omega_{J_{12}}\sigma_{1z}\sigma_{2z} \end{aligned} \tag{1.3}$$

where all energies have been written in angular frequency units. Replacing the Heisenberg coupling by an Ising coupling corresponds to first-order perturbation theory, and is usually called the weak coupling approximation.

1.4. The vector model and product operators

NMR spectroscopy appears quite different from conventional optical spectroscopy, as NMR experiments are essentially always in the coherent control regime. This is because it is trivial to make intense coherent RF fields and because NMR relaxation times are extremely long. For these reasons incoherent NMR spectroscopy is essentially unknown: all modern NMR spectroscopy is built round Rabi flopping and Ramsey fringes.

Simple NMR experiments are usually described using the *vector model*, which is based on the Bloch sphere [5, 3, 9]. A single isolated spin in a pure state $|\psi\rangle$ can be described by a density matrix

$$|\psi\rangle\langle\psi| = \tfrac{1}{2}\left(\mathbf{1} + r_x\sigma_x + r_y\sigma_y + r_z\sigma_z\right) \tag{1.4}$$

and for a pure state $r_x^2 + r_y^2 + r_z^2 = 1$ so $\mathbf{r}$, the nuclear spin vector, lies on the surface of the Bloch sphere. For a mixed state the situation is similar but the Bloch vector is not of unit length.

The behaviour of a single isolated spin is exactly described by its Bloch vector, and the behaviour of the Bloch vector is identical to that of a classical magnetisation vector. Thus the average behaviour of a single isolated spin can be described using the classical vector model. This is not true of coupled spin systems, where it is essential to use quantum mechanics.

The behaviour of coupled spin systems in NMR experiments is usually described using *product operators* [2, 9, 6, 10, 11]. These are very closely related to conventional angular momentum operators, but differ in normalisation and other conventions. While they can seem strange is is essential to get used to them! The state of a single spin is described as a combination of four one-spin operators: $\frac{1}{2}E$, I_x, I_y and I_z. (In NMR experiments the first spin is traditionally called I, while later spins are usually called S, R and T in that order.) The last three operators are simply related to the conventional Pauli matrices by $I_x = \frac{1}{2}\sigma_x$, and so on, while $\frac{1}{2}E = \mathbf{1}/2$ is the identity matrix normalised to have trace one (the maximally mixed state). In this notation nuclear spin Hamiltonians will be subtly different from their traditional forms: for a single spin $\mathcal{H} = \omega_I I_z$.

The initial state of an isolated nuclear spin at thermal equilibrium is given by the usual Boltzmann formula

$$\begin{aligned} \rho &= \exp(-\hbar\omega_I I_z/kT)/\operatorname{tr}\left[\exp(-\hbar\omega_I I_z/kT)\right] \\ &\approx \tfrac{1}{2}E - \hbar\omega_I I_z/kT \end{aligned} \tag{1.5}$$

The first term (the maximally mixed state) is not affected by subsequent unitary evolutions and so is of little interest; for this reason it is usually dropped. Similarly the factors in front of the I_z term simply determine the size of the NMR signal, and are also usually neglected. Thus the thermal state of a single state is usually described as I_z.

Clearly this approach must be used with caution as I_z is not a proper density matrix: it corresponds to negative populations of some spin states! These apparent negative populations arise simply because the maximally mixed component has been neglected. The traditional NMR approach of concentrating on the traceless part of the density matrix is usually not a problem; in particular the evolution of an improper density matrix under a Hamiltonian can be calculated using the standard Liouville–von Neumann equation [2, 4], as unitary evolutions are linear. For simple Hamiltonians the evolution can be calculated algebraically

$$I_x \xrightarrow{\omega t I_z} e^{-i\omega t I_z} I_x e^{i\omega t I_z} = I_x \cos\omega t + I_y \sin\omega t \tag{1.6}$$

and the product operator notation has been developed to enable this algebraic approach to be used as far as possible.

The success of this approach relies on the properties of commutators [9, 6, 10, 11]. Consider an initial density matrix $\rho(0) = A$ evolving under a Hamiltonian $\mathcal{H} = bB$ for a time t. Suppose that $[A,\ B] = iC$ and that $[C,\ B] = -iA$; in this case the three operators A, B and C form a triple, and in general

$$\rho(t) = A\cos bt - C\sin bt \tag{1.7}$$

which can be summarised as

$$A \xrightarrow{B} -C \xrightarrow{B} -A \xrightarrow{B} C \xrightarrow{B} A. \tag{1.8}$$

Clearly I_x, I_z and $-I_y$ form such a triple, but many analogous triples exist, allowing many quantum mechanical calculations to be performed using nothing more than elementary trigonometry and a table of commutators!

1.5. Experimental practicalities

Before proceeding to more sophisticated experiments it is useful to consider the elementary experimental phenomena of excitation, detection, and relaxation.

At thermal equilibrium the Bloch vector lies along the z-axis, and we must begin by exciting the spins. This can be achieved by a magnetic field of strength B_1 which rotates around the z-axis at the Larmor frequency. The situation is most simply viewed in a rotating frame which also rotates around the z-axis at the Larmor frequency: thus the excitation field appears static, along the y-axis for example. The Bloch vector will precess around this excitation field at a rate $\omega_1 = \gamma B_1$ towards the x-axis of the rotating frame. After a time t the Bloch vector has precessed through an angle $\theta = \omega_1 t$, and particularly important cases are the $\pi/2$ and π pulses. The magnetic field is obtained by applying RF radiation, and we can choose the axis (in the rotating frame) about which the precession occurs by choosing the RF phase. Thus we can talk about, for example, x and y pulses [3].

NMR signal detection is best described using a classical view [8]. The ensemble average of the spins behaves like a classical magnetisation rotating at the Larmor frequency, and the NMR detector is a coil of wire wrapped around the sample. As the magnetisation cuts across the wires it induces an EMF in the coil which can be detected. This detection method corresponds to a weak ensemble measurement, rather than the hard projective measurements more usually considered in quantum systems. This fact, which can be ultimately traced back to the low energy of NMR transitions, has considerable significance for both conventional NMR experiments and for NMR quantum computing.

Another consequence of the low energy scale of NMR transitions is that spontaneous emission is essentially negligible, and only stimulated processes occur. Because of this NMR relaxation times can be very long (several seconds). Stimulated emission requires a magnetic field oscillating at the Larmor frequency, and modulation of the chemical shift and dipole–dipole Hamiltonians by molecular motion is the main source of relaxation for spin-half nuclei in liquids.

NMR relaxation of a single isolated spin is well described by two time constants: T_2 (the transverse relaxation time) is the time scale of the loss of xy-magnetisation, that is the decoherence time, while T_1 (the longitudinal relaxation time) is the time scale of recovery of the Boltzmann equilibrium population difference, and determines the repetition delay between experiments. For more complex spin systems the behaviour is broadly similar but more complex. Relaxation effects (especially short T_2 times) can be a hindrance, but detailed studies of relaxation properties can provide useful information on molecular motions.

1.6. Spin echoes and two-spin systems

Spin echoes [3, 9, 13] play a central role in almost all NMR pulse sequences. In the one-spin case they are easily understood using the vector model. Start off with magnetisation along the x-axis and allow it to undergo free precession at the Larmor frequency ω for a time t: the magnetisation will rotates towards the y-axis through an angle ωt. Now apply a π_x pulse, giving a 180° rotation around the x-axis, so that the magnetisation appears to have rotated by $-\omega t$. Allow the magnetisation to precess for a further time t; it will now return back to the x-axis whatever the value of ω! This behaviour can be easily calculated using product operators

$$\begin{aligned} I_x &\xrightarrow{\omega t\, I_z} I_x \cos\omega t + I_y \sin\omega t \\ &\xrightarrow{\pi\, I_x} I_x \cos\omega t - I_y \sin\omega t \\ &\xrightarrow{\omega t\, I_z} I_x \cos\omega t \cos\omega t + I_y \cos\omega t \sin\omega t \\ &\qquad - I_y \sin\omega t \cos\omega t + I_x \sin\omega t \sin\omega t \\ &= I_x \left[\cos^2\omega t + \sin^2\omega t\right] = I_x \end{aligned} \tag{1.9}$$

to get exactly the same result.

The situation is similar but slightly more complex in two spin systems. These are described using 16 basic operators, formed by taking products of the four I spin and S spin operators and multiplying by two:

$$\begin{array}{cccc} \frac{1}{2}E & S_x & S_y & S_z \\ I_x & 2I_xS_x & 2I_xS_y & 2I_xS_z \\ I_y & 2I_yS_x & 2I_yS_y & 2I_yS_z \\ I_z & 2I_zS_x & 2I_zS_y & 2I_zS_z \end{array} \tag{1.10}$$

The (weak coupling) Hamiltonian for a two spin system is then

$$\mathcal{H} = \omega_I I_z + \omega_S S_z + \pi J 2I_zS_z. \tag{1.11}$$

Product operators have the extremely useful property that all pairs of operators either commute or form triples, just like I_x, I_y and I_z; this means that the method of commutators, described in section (1.4) can also be used in two spin systems.

For a table of the main commutators see Appendix A; a more complete list is available in [11].

Spin echoes can easily be performed in two-spin systems, but the result depends on whether the system is *heteronuclear* (the two spins are of different nuclear species, with very different Larmor frequencies) or *homonuclear* (the two spins are of the same nuclear species, with very similar Larmor frequencies). In a heteronuclear spin system only one spin (say I) will be excited by the π pulse. In this case the I spin Zeeman interaction and the spin–spin coupling are refocused by the spin echo but the S spin Zeeman interaction is retained:

$$I_x + S_x \xrightarrow{\mathcal{H}} \xrightarrow{\pi I_x} \xrightarrow{\mathcal{H}} I_x + S_x \cos \omega_S t + S_y \sin \omega_S t. \tag{1.12}$$

In a homonuclear spin system, by contrast, both spins will normally be excited by the π pulse. In this case both Zeeman interactions are refocused but the spin–sin coupling is retained:

$$\begin{aligned} I_x + S_x \xrightarrow{\mathcal{H}} \xrightarrow{\pi(I_x+S_x)} \xrightarrow{\mathcal{H}} & I_x \cos \pi J t + 2 I_y S_z \sin \pi J t \\ & + S_x \cos \pi J t + 2 I_z S_y \sin \pi J t. \end{aligned} \tag{1.13}$$

It is of course possible to perform a "homonuclear" spin echo in a heteronuclear spin system, by simply applying separate π pulses to spins I and S at the same time. It also possible to perform a "heteronuclear" spin echo in a homonuclear spin system by using low power *selective* pulses, which will excite one spin while leaving the other untouched. A high power pulse which excites all the spins of one nuclear species is usually called a *hard* pulse.

More complex pulse sequences can be built up by combining spin echoes and selective and hard pulses. This is a highly developed NMR technique which has led to a host of conventional NMR experiments with whimiscal names such as COSY, NOESY and INEPT [6, 11, 8]. Using this approach one can create a pulse sequence whose total propagator corresponds to all sorts of unitary transformations—including quantum logic gates!

2. NMR and quantum logic gates

In this section I will describe how NMR techniques can be used to implement the basic gates required for quantum computation.

2.1. Introduction

Quantum logic gates [14] are simply unitary transformations which implement some desired logic operation. It has long been know by the NMR community that NMR techniques in principle provide a universal set of Hamiltonians, that is they can be used to implement *any* desired unitary evolution, including quantum logic gates. Building NMR quantum logic gates is very similar to designing conventional NMR pulse sequences, and progress in this field has been very rapid. Furthermore many of the pulse sequences used to implement quantum logic are in fact very similar to common NMR pulse sequences, and it could be argued that many conventional NMR experiments are in fact quantum computations!

Although NMR techniques could be used to directly implement any desired quantum logic gate, this is not a particularly sensible approach. Instead it is usually more convenient to implement a *universal* set of quantum logic gates, and then obtain other gates by joining these basic gates together to form networks [14]. However one should be careful not to take this process too far. Theoreticians are often interested in implementing networks using the smallest possible set of basic resources, and it is known that in principle only *one* basic logic gate is required for quantum computation [15, 16, 17, 18]. For experimentalists gates usually come in families, such that the ability to implement any one member of a family implies the ability to implement any other member of the family in much the same way, and it is more sensible to develop a fairly small set of simple but useful families of logic gates. For NMR quantum computing [19, 20, 21, 22] the best set seems to be a set containing many (but not all) single qubit gates and the family of Ising coupling gates.

2.2. *Single qubit gates*

Single qubit gates correspond to rotations of a spin about some axis. The simplest gates are rotations about axes in the xy-plane, as these can be implemented using resonant RF pulses. The *flip angle* of the pulse (the angle through which the spin is rotated) depends on the length and the power of the RF pulse, while the *phase angle* of the pulse (and hence the azimuthal angle made by the rotation axis in the xy-plane) can be controlled by choosing the initial phase angle of the RF. Rotations about the z-axis can be implemented using periods of precession under the Zeeman Hamiltonian, while rotations around tilted axes can be achieved using off-resonance RF excitation. It is, however, usually simpler not to use these last two approaches: instead all single qubit gates are built out of rotations in the xy-plane.

A simple example is provide by the composite z-pulse [23], which implements a z-rotation using x and y-rotations,

$$\theta_z \equiv 90_{-x}\theta_y 90_x \equiv 90_y \theta_x 90_{-y} \tag{2.1}$$

where the pulse sequence has been written using NMR notation, with time running from left to right, rather than using operator notation, in which operators are applied sequentially from right to left. A similar approach can be used to implement tilted rotations, such as the Hadamard gate

$$H \equiv 180_z 90_y \equiv 90_y 180_x 90_{-y} 90_y \equiv 90_y 180_x \tag{2.2}$$

Any desired single qubit gate can be built in this fashion.

Even this approach, however, is over complex, as there is a particularly simple method of implementing z-rotations. Rather than rotating the spin, it is simpler to rotate its reference frame. This can be achieved by passing z-rotations forwards or backwards through a pulse sequence

$$\psi_z \theta_\phi \equiv \theta_{\phi-\psi} \psi_z \tag{2.3}$$

and altering pulse phase to reflect the new reference frame. This technique, often called *abstract reference frames* [24, 22] has the advantage that z-rotations can be implemented without using any time or resources! Many modern implementations of NMR quantum logic gates use only rotations in the xy-plane and changes in reference frames to implement all single qubit gates.

2.3. *Two qubit gates*

In addition to single qubit gates a design for a quantum computer must include at least one non-trivial two qubit gate. The most commonly discussed two qubit gate is the controlled-NOT gate, but this is not the most natural two qubit gate for NMR quantum computing. A controlled-NOT gate can be replaced by a pair of Hadamard gates and a controlled-phase-shift gate [22]

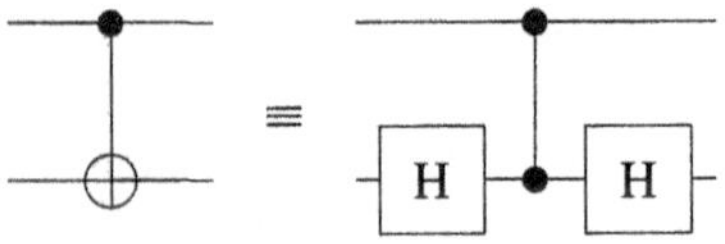

≡ H H (2.4)

where the controlled-phase-shift gate acts to negate the state $|11\rangle$ while leaving other states unchanged. Note that this gate acts symmetrically on the two qubits; it does not have control and target bits. The asymmetry in the controlled-NOT gate arises from the asymmetry in the placement of the Hadamard gates.

The controlled-phase-shift gate is itself equivalent (up to single qubit z-rotations, which can be adsorbed into abstract reference frames) to the Ising coupling gate

$$e^{i(\phi/2)2I_zS_z} = \begin{pmatrix} e^{-i\phi/4} & 0 & 0 & 0 \\ 0 & e^{-i\phi/4} & 0 & 0 \\ 0 & 0 & e^{i\phi/4} & 0 \\ 0 & 0 & 0 & e^{i\phi/4} \end{pmatrix} \tag{2.5}$$

where the case $\phi = \pi$ forms the basis of the controlled-NOT gate. This "gate" is nothing more than a period of evolution under the Ising coupling Hamiltonian, which can be achieved using a homonuclear spin echo.

2.4. *Practicalities*

The description above is adequate for simple two qubit systems, but subtleties arise in larger spin systems. Foremost among these is the so-called "do-nothing" problem. In a traditional quantum computer gates are implemented by applying additional interactions when necessary, but in an NMR quantum computer J-coupling is part of the background Hamiltonian. Thus J-couplings are always active unless they are specifically disabled. This can be done using heteronuclear spin echoes, but this means that a great deal of effort is spent in a large NMR quantum computer ensuring that spins which are not involved in a logic gate do not evolve while a gate is being implemented.

In a fully coupled N-spin system there are roughly $\frac{1}{2}N^2$ coupling interactions, and the simplest method for turning off these interactions requires $O(2^N)$ pulses. Consider a two spin system with spins called I^0 and I^1; the coupling between these spins can be eliminated using the sequence

I^0

I^1

(2.6)

where boxes correspond to 180° pulses. This sequence retains the Hamiltonian corresponding to the chemical shift of spin 0 (I_z^0), but this can be dealt with later. Similarly the final 180° pulse (shown as a dashed box), which is needed to restore spin 1 to its initial state, can often be omitted. In larger systems it is not sufficient simply to place simultaneous 180° pulse on all the spins except spin 0: while this will remove all couplings to spin 0 couplings *between* the remaining spins will survive. An obvious solution is simply to nest spin echoes within one another

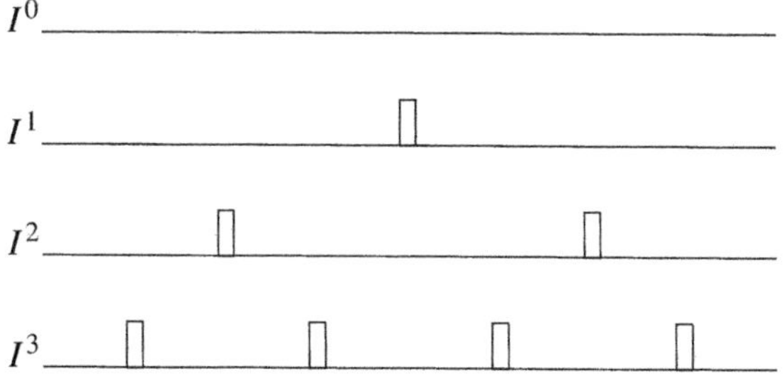

(2.7)

(once again the term I_z^0 survives) but this soon becomes unwieldy. Fortunately more efficient schemes can be designed based on Hadamard matrices [25, 26].

The effect of a 180° pulse on a spin system is, in effect, to negate the sign of the Zeeman and spin coupling terms involving that spin; simultaneous 180° pulses on two spins will negate the coupling between these spins *twice*, thus leaving it unchanged. This gives a simple way of analysing the effect of spin echo sequences. Each interaction term in the Hamiltonian begins the sequence with a relative strength of $+1$, and each 180° pulse on a spin negates every term involving that spin. The effect of a spin echo sequence on a Zeeman interaction can be determined by writing down a vector of $+1$ and -1 terms, and then summing along the components of the vector. The effect on a J-coupling between two spins can be determined by multiplying corresponding elements in the two vector and then summing them, that is by taking the dot product of two vectors. A spin echo sequence refocuses Zeeman interactions if vectors sum to zero, and refocuses J-couplings if vectors are orthogonal.

This approach can be used to analyse existing spin echo schemes, but it can also be used to design new ones: a set of vectors with the desired properties is constructed, and then a pulse sequence is designed by applying a 180° pulse to a spin every time to vector changes sign. Suitable vectors can easily be obtained by taking rows from Hadamard matrices to obtain efficient refocusing schemes. For example the four by four Hadamard matrix

$$H_4 = \begin{pmatrix} 1 & 1 & 1 & 1 \\ 1 & -1 & 1 & -1 \\ 1 & 1 & -1 & -1 \\ 1 & -1 & -1 & 1 \end{pmatrix} \tag{2.8}$$

can be used to derive an efficient scheme for four spins:

(2.9)

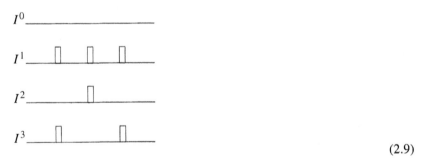

The gain is not huge for small spin systems, but becomes dramatic in large systems: Hadamard based schemes [25, 26] require only $O(N^2)$ pulses to refocus all couplings in an N-spin system.

When building NMR quantum computers with more than three spins, it may be easier to use "linear" spin systems, in which each spin is only coupled to its immediate neighbours, or other partially coupled systems. A linear spin system can be used to implement any logic gate by using SWAP gates to move qubits around the system; this imposes an overhead but this is only linear in the number of spins in the system.

Whatever refocusing scheme is adopted, large NMR quantum computers will require the use of selective pulses in homonuclear spin systems (it is not possible to build a large fully heteronuclear spin system as there are not enough spin half nuclei). One can selectively excite a single nuclear spin in a homonuclear spin system, while leaving the others untouched, by using long low-power pulses. The excitation bandwidth of a pulse is given approximately by the inverse of its duration, and selective pulses are usually shaped, that is amplitude and phase modulated, to give them better excitation profiles. Many complicated shaped pulses have been designed [3], which rely on sophisticated NMR hardware for their implementation, but for NMR quantum computing some of the simplest types (Gaussian and Hermite pulses) seem to be best.

An alternative scheme is to implement selective pulses using sequences of hard pulses and delays [27, 22]. During delay periods spins will evolve under the background Hamiltonian, which is dominated by Zeeman interactions, and so different spins will experience different z-rotations. Sandwiching these z-rotations between $90^\circ_{\pm y}$ pulses converts the varying z-rotations into corresponding x-rotations, in effect implementing selective pulses [27, 22].

The opposite approach, using selective pulses to implement two qubit gates has also been demonstrated [28]. In this case it is necessary to use extremely long low power pulses which excite one line in a multiplet while leaving other lines untouched. This provides a simple method for implementing multiply-controlled-NOT gates, such as TOFFOLI gates, but it seems unlikely that this approach will be generally useful.

Finally when considering quantum logic gates it is essential to remember that writing down a Hamiltonian is not the same as implementing a gate! Real experimental gates are vulnerable to both random and systematic errors, and the effects of these must be considered. This point will be treated in some depth in lecture 4.

2.5. Non-unitary gates

Although quantum computations are usually thought of as a sequence of unitary gates, non-unitary gates also play a key role in quantum information processing. The most obvious examples are projective measurements and the initialisation of qubits, but as discussed in lecture 3 these processes are difficult or impossible to implement in NMR systems. It is, however, possible to implement other non-unitary gates, and these are extremely important.

In general a non-unitary gate can be implemented by using a unitary gate to entangle the system with some aspect of the environment and then tracing out this environmental information. The two basic non-unitary gates in NMR use the position of spins in the spatial ensemble or the time at which an experiment was performed as the environmental label.

In modern NMR experiments the most common non-unitary is a *gradient pulse* [11]. For a short time the magnetic field is made highly inhomogeneous, so that the Larmor frequency varies strongly over the sample. This causes off-diagonal terms in the density matrix to dephase over the sample, and thus to disappear when the final NMR signal is detected. The situation is not, however, as simple as is sometimes described, as some off-diagonal terms (known in NMR notation as homonuclear zero quantum coherences) will survive the dephasing: these dephasing free subspaces are analogous to the decoherence free subspaces [29, 30] suggested for building robust quantum bits.

Gradient pulses are most commonly used as *crush* pulses; these convert visible NMR terms, such as I_x and I_y, into the maximally mixed state, in effect destroying them. Crush pulses are automatically applied to all the spins in a spin system, but some spins may be unaffected because of their initial state. The action of projecting spins onto the z-axis can be used, for example, to render error terms invisible or to change the relative polarisations of two spins

$$I_z + S_z \xrightarrow{\pi/3 I_y} \tfrac{1}{2} I_z + \tfrac{\sqrt{3}}{2} I_x + S_z \xrightarrow{\text{crush}} \tfrac{1}{2} I_z + S_z. \tag{2.10}$$

It is important to realise that crush pulses are only apparently non-unitary: the dephasing retains its spatial label and can be refocused. In particular crush pulses will interact with spin echoes; this can be a problem in sequences with many gradients, as it can lead to accidental refocusing of supposedly crushed terms. One solution to this is to use gradients along different spatial axes, and well equipped spectrometers will have three orthogonal gradients (x, y, and z); similar effects can be achieved by dephasing the spin system with inhomogeneous RF fields. More usefully the combination of gradients and spin echoes gives a route to selective crush pulses

$$I_x + S_x \xrightarrow{\text{crush}} \xrightarrow{\pi I_y} \xrightarrow{\text{crush}} I_x \tag{2.11}$$

which only affect one spin in a mult-spin system.

If necessary it is possible to obtain a true non-unitary gate by destroying the spatial label. This can be achieved by spatial diffusion of the spin system within the ensemble, either during the crush pulse or between two crush pulses. This approach is sometimes called *engineered decoherence* [30].

A second route to non-unitary processes in NMR is to use temporal rather than spatial labels. This can be acheived by repeating the same basic pulse sequence several times, making subtle changes each time, and then taking linear combinations of the resulting NMR signals, so that some terms add together while other terms cancel out. The simplest approach is to alter the relative phase of pulses, in which case it is known as phase cycling [11]. Phase cycling techniques were very widely used in conventional NMR experiments, but in recent years have been largely superseded by gradients. They have, however, found new applications in NMR quantum computing where they form the basis of the popular *temporal averaging* schemes for initialisation.

3. NMR quantum computers

In this section I will describe how NMR quantum computers overcome the difficulties inherent in NMR to perform initialisation and readout. In particular I will describe the use of pseudo-pure states, and the implications of this approach for the efficiency of NMR quantum computing. Finally I will briefly describe the implementation of a quantum cloning on an NMR quantum computer.

3.1. Introduction

From the description given in the previous lecture it would seem that NMR was very well suited to the task of implementing quantum computers. There are, however, substantial problems with NMR as a quantum information processing technology [31], which stem from difficulties in initialising nuclear spin states and in reading out the final result.

Conventional designs for quantum computers [32] use single quantum systems which start in a well defined initial state. While details may vary, this initialisation is usually achieved by cooling the system to its thermodynamic ground state. NMR quantum computers [19, 20, 21, 22], by contrast, use an ensemble of molecules which start in a hot thermal state, because even for the very large fields used in NMR spectrometers the Zeeman energy gap between the two spin states is tiny compared to kT. One could imagine lowering the temperature so that NMR enters the low temperature regime, but this would require cooling the system well below 1 mK; although this is possible the sample would certainly not remain in the liquid state. A potentially better approach is to use non-Boltzmann initial populations, as discussed in lecture 5. Almost all implementation of NMR quantum computing, however, simply sidestep this issue by forming a "pseudo-pure" initial state from the thermal state as discussed below.

Similar problems also occur with methods for reading out the final result. Conventional quantum computers achieve read out by hard (projective) measurements, while NMR quantum computers use weak ensemble measurements, which do not project the spin system. This can be seen by realising that a conventional NMR measurement (observation of the free induction decay) can be described quantum mechanically as the continuous and simultaneous observation of two non-commuting observables, I_x and I_y. This is also the approach used for readout in NMR quantum computers, and a simple example is shown in Fig. 1

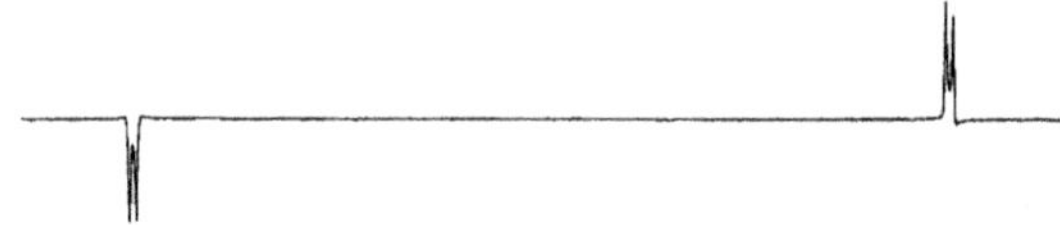

Fig. 1. NMR spectrum showing readout from a two qubit NMR quantum computer based on the two ^{1}H nuclei in cytosine [33]; the negative intensity on the left hand multiplet indicates that the corresponding qubit was in state $|1\rangle$, while the positive intensity indicates that this qubit was in state $|0\rangle$.

These weak measurements might seem more powerful than conventional projective measurements, but in fact they are less useful for two reasons. Firstly the use of projective measurements permits the use of measurements followed by classical control; by contrast NMR quantum computers can only use quantum control methods. More importantly, projective measurements provide an excellent initialisation method: just measure a bit, and then flip it if it has the wrong value! In particular reinitialisation of ancilla qubits through the use of projective measurements plays a key role in quantum error-correction protocols [34].

3.2. Pseudo-pure states

The history of NMR quantum computing in effect begins with the realisation by David Cory and coworkers [19, 20] that while it is difficult to form a pure initial state it is easy to form states whose behavior is almost identical. Such states, known a pseudo-pure states or effective pure states, take the form

$$\rho = (1-\epsilon)\frac{\mathbf{1}}{2^n} + \epsilon|0\rangle\langle 0|, \tag{3.1}$$

that is mixtures of the maximally mixed state and the desired initial state with purity ϵ. As the maximally mixed state does not evolve under any unitary transformation it will be unchanged by any quantum computation. Furthermore, all NMR observables are traceless [11], and so the maximally mixed state gives no observable signal. For this reason the presence of the maximally mixed state can, in effect, be ignored, and the behaviour of a pseudo-pure state is identical to that of the corresponding pure state up to a scaling factor [22].

As an example, consider a homonuclear spin system of two spin-half nuclei. This has four energy levels with nearly equal populations, but the population of the lowest level will of course be slightly greater than that of any other level. This excess population provides the basis of pseudo-pure state formation, but the state as described is *not* a pseudo-pure state, as the upper levels do not all have the same population. Suppose, however, that some non-unitary process is applied which equalises the populations of the upper levels, while leaving the lowest level untouched: the result will be the desired pseudo-pure state [35]. To understand the behaviour of this state imagine going through the ensemble, taking out molecules in groups of four (one in each spin state) and placing them in a box; eventually there will be a large box containing equal populations of all four spin states and a small excess of the $|00\rangle$ spin state remaining. The NMR signals from the molecules in the box will all cancel out, leaving only the signal from the small excess: the pseudo-pure state.

Pseudo-pure states can also be described more accurately using the product operator approach [36, 22]. The Boltzmann equilibrium state of a homonuclear two-spin system is approximately

$$\rho_B \approx \tfrac{1}{2}E + \delta(I_z + S_z) = \tfrac{1}{2}E + \delta\{1, 0, 0, -1\} \tag{3.2}$$

where the braces indicate a *diagonal* density matrix described by listing its diagonal elements. The ideal pure ground state takes the form

$$\rho_0 = \tfrac{1}{2}(\tfrac{1}{2}E + I_z + S_z + 2I_zS_z) = \{1, 0, 0, 0\} \tag{3.3}$$

and so forming a pseudo-pure ground state will require the creation of a $2I_zS_z$ component and the rescaling of other terms so that each term is present in the correct *relative* quantity. Clearly this will require a combination of unitary and non-unitary processes, and three main approaches have been described.

The original *spatial avaeraging* method of Cory *et al.* [19, 20] for creating a pseudo-pure state in a two spin system used a sequence of (unitary) pulses and delays combined with (non-unitary) crush gradients. The method is easily understood using product operators:

$$\begin{aligned}
I_z + S_z &\xrightarrow{60^\circ S_x} I_z + \tfrac{1}{2}S_z - \tfrac{\sqrt{3}}{2}S_y \\
&\xrightarrow{\text{crush}} I_z + \tfrac{1}{2}S_z \\
&\xrightarrow{45^\circ I_x} \tfrac{1}{\sqrt{2}}I_z - \tfrac{1}{\sqrt{2}}I_y + \tfrac{1}{2}S_z \\
&\xrightarrow{\text{Ising}} \tfrac{1}{\sqrt{2}}I_z + \tfrac{1}{\sqrt{2}}2I_xS_z + \tfrac{1}{2}S_z
\end{aligned}$$

$$\stackrel{45^\circ I_x}{\longrightarrow} \quad \tfrac{1}{2}I_z - \tfrac{1}{2}I_x + \tfrac{1}{2}2I_xS_z + \tfrac{1}{2}S_z + \tfrac{1}{2}2I_zS_z$$
$$\stackrel{\text{crush}}{\longrightarrow} \quad \tfrac{1}{2}(I_z + S_z + 2I_zS_z). \tag{3.4}$$

An widely used alternative, *temporal averaging* [37], uses permutation operations to create different initial states

$$\begin{aligned} I_z + S_z &\stackrel{P_0}{\longrightarrow} \{1, 0, 0, -1\} \\ I_z + S_z &\stackrel{P_1}{\longrightarrow} \{1, 0, -1, 0\} \\ I_z + S_z &\stackrel{P_2}{\longrightarrow} \{1, -1, 0, 0\} \end{aligned} \tag{3.5}$$

and averaging over these three separate experiments gives an effective pure state

$$\rho_{TA} = \{1, -\tfrac{1}{3}, -\tfrac{1}{3}, -\tfrac{1}{3}\}. \tag{3.6}$$

This method has the advantage of being easy to understand and to generalise to larger spin system, but the disadvantage that several different experiments are required. Indeed if the most obvious scheme, exhaustive permutation, is implemented a very large number of experiments may be required; fortunately less profligate partial averaging schemes are known [37].

Finally the *logical labelling* approach of Gershenfeld and Chuang [21] provides a conceptually elegant method for using naturally occurring subsets of levels in larger systems as pseudo-pure states. As an example consider a three spin system

$$I_z + S_z + R_z = \tfrac{1}{2}\{3, 1, 1, -1, 1, -1, -1, -3\} \tag{3.7}$$

and pick out the subset of four levels with relative populations 3, -1, -1 and -1, that is the levels $|000\rangle$, $|011\rangle$, $|101\rangle$ and $|110\rangle$. The most direct approach is just to work in this subset, but it usually more convenient to permute populations so that the levels $|000\rangle$, $|001\rangle$, $|010\rangle$ and $|011\rangle$ can be used; this makes implementing logic gates much simpler.

Perhaps the most practical general scheme for preparing pseudo-pure states is based on the use of "cat" states [24], which are states of the form

$$\psi_\pm^n = |00\ldots0\rangle \pm |11\ldots1\rangle \tag{3.8}$$

for an n-qubit system, that is equally weighted superpositions of the state in which all n qubits are in $|0\rangle$ and the state in which all qubits are in $|1\rangle$. It is easy both to reach the state ψ_+^n starting from the ground state $|00\ldots0\rangle$, and to convert the cat state back to the ground state. This may not seem useful, but it is relatively simple to design non-unitary filter schemes, using either spatial or temporal averaging, which convert all states except $\psi_\pm^n$ into the maximally mixed state. The Boltzmann equilibrium state can thus be converted to a mixed state including a component of $\psi_{\pm}^{n}$, and after filtration the ψ_{+}^{n} state can be converted back to $|00\ldots0\rangle$. The filter schemes, however, also retain any ψ_{-}^{n} component, and this is converted into $|10\ldots0\rangle$. The overall effect is to produce the state $I_z \otimes |0\ldots0\rangle\langle0\ldots0|$, that is a pseudo-pure state of $n-1$ qubits.

3.3. *Efficiency of NMR quantum computing*

The discussion so far has neglected any consideration of the level of purity which can be achieved in a pseudo-pure state; this is most simply quantified by the value of ϵ in Eq. 3.1. At one level this is unimportant, as ϵ simply determines the intensity of the observed NMR signal, but if ϵ becomes too small this will render the NMR signal undetectable. Unfortunately for NMR quantum computing, the value of ϵ drops exponentially with the number of qubits in the system: for every additional qubit the available signal intensity approximately halves [38].

This effect is not, as is sometimes suggested, a peculiar fault of NMR quantum computers: rather it is a simple consequence of working in the high temperature limit. It does, however, mean that pseudo-pure states extracted from thermal equilibrium systems cannot provide a route to scalable NMR quantum computers.

More controversially some authors have implied that NMR quantum computers are not quantum computers at all! How this claim is assessed depends on exactly what is meant by "NMR quantum computers", and even what is meant by "quantum computing". However, while there is substantial room for philosophical debates, the underlying science is now relatively clear. On the one hand it is known that high temperature pseudo-pure states cannot lead to provably entangled states [39], and that such systems cannot give efficient implementations of Shor's quantum factoring algorithm [40]. On the other hand it so far proved impossible to develop a purely classical model of pseudo-pure state NMR quantum computing: while it is possible to describe the state of an NMR device at any point in a computation using a classical model, it appears to be impossible to develop a classical model of the transitions between these states [41].

It is also vital to remember that these arguments apply only to NMR quantum computers built using pseudo-pure states, and that there are other types of NMR quantum computing. For example some quantum algorithms only require one pure qubit: the other qubits can be in maximally mixed states [42]. Indeed, even the single "pure" qubit need not be pure: a pseudo-pure state will suffice. This type of NMR quantum computing is clearly scalable, although it can only be used for a limited range of algorithms. An alternative approach is to use a scheme described by Schulman and Vazirani, which allows a small number of nearly pure qubits to be distilled from a large number of impure qubits using only unitary operations [43]. This scheme needs $O(\epsilon^{-2})$ impure spins for each pure spin extracted: this is a constant multiplicative overhead, and so has no scaling problem. Thus high temperature states, such as those used in NMR, do allow true quantum computing! Unfortunately the overhead for NMR systems of about 10^{10} means that this method has only theoretical interest.

3.4. NMR quantum cloning

Finally I will end this lecture by briefly describing an NMR implementation of approximate quantum cloning [44]. This experiment is complicated enough to be interesting, but simple enough that the basic ideas can be described in a fairly straightforward manner.

The no-cloning theorem, which states that an unknown quantum state cannot be exactly copied [45], is one of the oldest results in quantum information theory. Approximate quantum cloning is, however, possible, and a range of different schemes have been described. If one qubit is converted into two *identical* copies, such that the fidelity of the copies is independent of the initial state, then the maximum fidelity that can be achieved is $\frac{5}{6}$, and an explicit quantum circuit which achieves this is known [46]. If a state $|\psi\rangle$ is cloned, the two copies take the form

$$\tfrac{5}{6}|\psi\rangle\langle\psi| + \tfrac{1}{6}|\psi^{\perp}\rangle\langle\psi^{\perp}| = \tfrac{2}{3}|\psi\rangle\langle\psi| + \tfrac{1}{3}(\mathbf{1}/2). \tag{3.9}$$

This circuit can also be used to clone a mixed state, ρ producing even more mixed clones of the form

$$\tfrac{2}{3}\rho + \tfrac{1}{3}(\mathbf{1}/2). \tag{3.10}$$

In the language of vectors on Bloch spheres, the two clones have Bloch vectors parallel to the original Bloch vector, but with only $\frac{2}{3}$ the length [44].

The cloning circuit comprises two stages: *preparation*, which prepares two qubits into an initial "blank paper" state, suitable for receiving a copy, and *copying*, in which the initial qubit is copied onto these qubits. As the preparation stage simply prepares two blank qubits, and is independent of the state of the unknown qubit, the preparation stage can be replaced by any other transformation which has the same effect, and the NMR implementation, which is shown in Fig. 2 does indeed use a modified preparation stage. The copying stage, however, must implement the correct unitary transformation, and the implementation used the conventional copying circuit.

The cloning circuit was implemented on a three-qubit NMR quantum computer based on the molecule based on the single ^{31}P nucleus (P) and the two ^{1}H nuclei (A and B) in E-(2-chloroethenyl)phosphonic acid (Fig. 3) dissolved in D_2O. The NMR pulse sequence used is shown in Fig. 4.

This comprises two main sections: an initial purification sequence (a), used to generate an initial pseudo-pure state corresponding to $P_z \otimes |0_A 0_B\rangle\langle 0_A 0_B|$, and

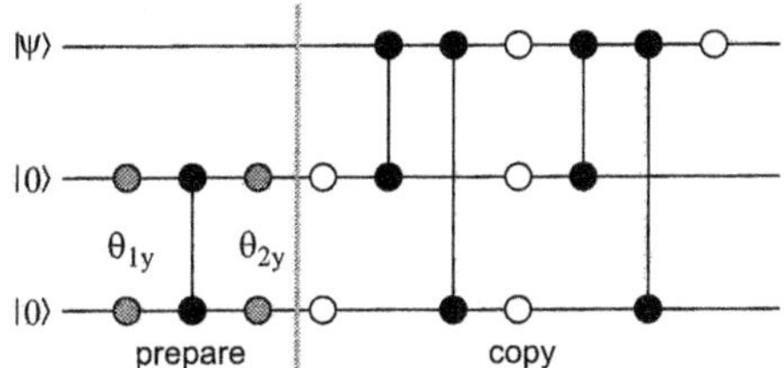

Fig. 2. A modified version of the approximate quantum cloning network: the new version is simpler to implement on the NMR system used. Filled circles connected by control lines indicate controlled phase shift gates, empty circles indicate single qubit Hadamard gates, while grey circles indicate other single qubit rotations. The two rotation angles in the preparation stage are $\theta_1 = \arcsin\left(1/\sqrt{3}\right) \approx 35°$ and $\theta_2 = \pi/12 = 15°$.

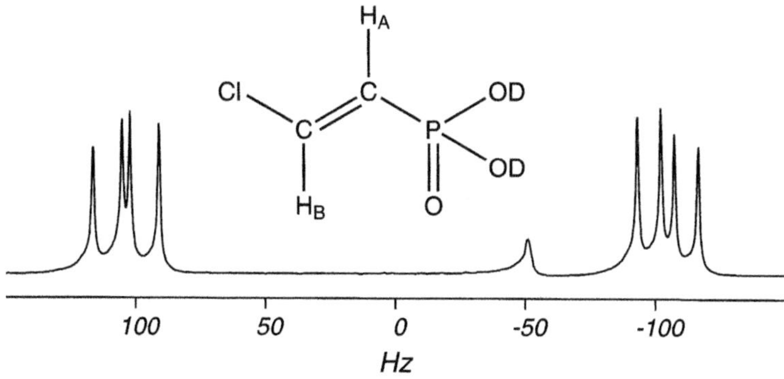

Fig. 3. The three qubit system provided by E-(2-chloroethenyl)phosphonic acid dissolved in D_2O and its ^1H NMR spectrum. Following standard NMR conventions the spectrum has been plotted with frequencies measured as offsets from the reference RF frequency, and with frequency increasing from right to left. The broad peak near -50 Hz can be ignored.

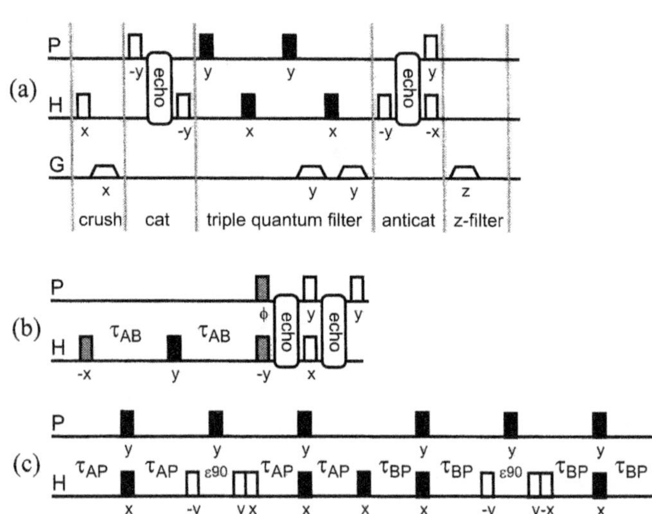

Fig. 4. The NMR pulse sequences used to implement quantum cloning. White and black boxes are 90° and 180° pulses, while grey boxes are pulses with other flip angles; pulse phases and gradient directions are shown below each pulse. All RF pulses are hard, with ^{1}H frequency selection achieved using "jump and return" methods.

a preparation and cloning sequence (b), which implements the circuit shown in Fig. 2, cloning the state of P onto A and B. Both of these are built around the "echo" sequence (c), which implements the coupling element of the PA and PB controlled phase shifts by evolution of the spin system under the weak coupling Hamiltonian with undesirable Zeeman evolutions refocused by spin echoes. This requires selective 90° pulses, which are built out of hard pulses and delays as described in my second lecture. For further details see the original paper [44].

The results of the cloning circuit can be observed by detecting the NMR signal from the two ^{1}H nuclei, A and B. The ideal spectrum should have equal intensities on the two outer lines of each multiplet, and no signal on the two central lines. Errors are seen in the experimental spectra, but the overall behaviour is clearly observed: Fig. 5 shows the result of cloning the state P_x. Results of similar quality are obtained when cloning other initial states [44].

4. Robust logic gates

In this section I will describe how techniques adapted from conventional NMR experiments can be used to develop robust logic gates for NMR quantum comput-

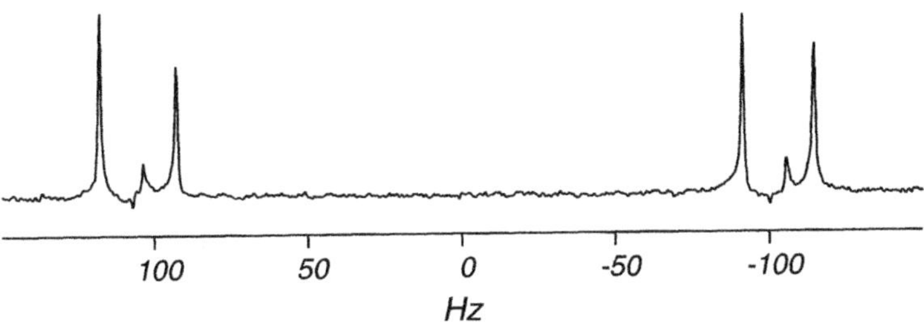

Fig. 5. The experimental result from cloning the initial state P_x; the receiver phase was set using a separate experiment so that x-magnetization appears as positive absorption mode lines.

ers. Although developed and described within the context of NMR, these robust gates could be used in other implementations of quantum computing.

4.1. Introduction

Quantum computers implement logic gates as periods of evolution under Hamiltonians which can be external (*e.g.*, RF pulses) or internal (*e.g.*, Ising couplings). Computation requires extremely accurate logic gates, and thus extremely accurate control of evolution rates. Naive estimates suggest that it may be difficult or impossible to control Hamiltonians with sufficient accuracy, but fortunately robust logic gates can be designed to tolerate small errors in these rates.

The approach described here is based on the NMR concept of *composite rotations* [9, 47, 48, 3], which have long been used to reduce the impact of systematic errors on conventional NMR experiments, but the basic idea is general and can be applied in many other fields. As usual it is not necessary to design robust versions of every conceivable logic gate: it suffices to develop a complete set of one and two qubit gates.

When considering the accuracy of logic gates it is necessary to measure the fidelity of the actual operation V in comparison with the desired operation U, and an obvious measure is provided by the propagator fidelity [49]

$$\mathcal{F} = \frac{|\operatorname{tr}(VU^{\dagger})|}{\operatorname{tr}(UU^{\dagger})} \tag{4.1}$$

where it is necessary to take the absolute value of the numerator to deal with (irrelevant) differences in global phases. The propagator fidelity works for any unitary operation, although it can be over complicated in practice and alternative measures have been suggested.

4.2. Composite rotations

The use of composite rotations to reduce the effects of systematic errors in conventional NMR experiments relies on the fact that any state of a single isolated qubit can be mapped to a point on the Bloch sphere, and any unitary operation on a single isolated qubit corresponds to a rotation on the Bloch sphere. The result of applying any series of rotations (a composite rotation) is itself a rotation, and so there are many apparently equivalent ways of performing a desired rotation. These different methods may, however, show different sensitivity to errors: composite rotations can be designed to be much less error prone than simple rotations!

A rotation can go wrong in two basic ways: the rotation angle can be wrong or the rotation axis can be wrong. In an NMR experiment (viewed in the rotating frame) ideal RF pulses cause rotation of a spin through an angle $\theta = \omega_1 t$ around an axis in the xy-plane. So called *pulse length errors* occur when the pulse power ω_1 is incorrect, so that the flip angle θ is systematically wrong by some fraction. This can be due to experimenter carelessness, but more usually arises from the inhomogeneity in the RF field over a macroscopic sample. The second type of error, *off-resonance effects* (Fig. 6), occur when the excitation frequency doesnŠt

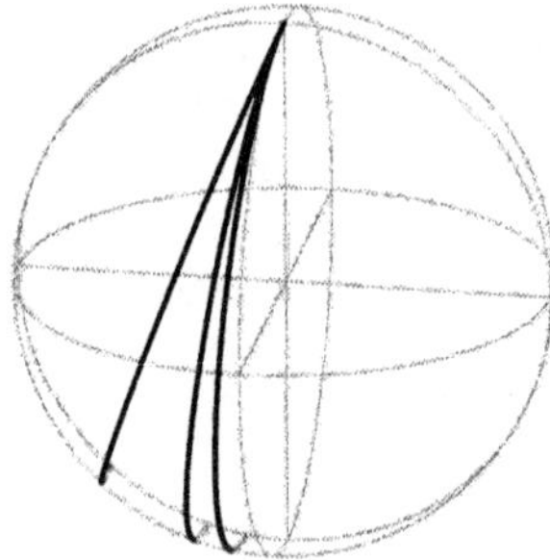

Fig. 6. Effect of applying an off-resonance 180° pulse to a spin with initial state I_z; the spin rotates around a tilted axis. Trajectories are shown for small, medium and large off-resonance effects.

match the transition frequency, so that the Hamiltonian is the sum of RF and off-resonance terms. This results in rotations around a tilted axis, and the rotation angle is also increased.

The first composite rotation [47] was designed to compensate for pulse length errors in an inversion pulse, that is a pulse which takes the state I_z to $-I_z$. This can be achieved by, for example, a simple 180°_y pulse, but this is quite sensitive to pulse length errors. The composite rotation $90^\circ_x 180^\circ_y 90^\circ_x$ has the same effect in the absence of errors, but will also partly compensate for pulse length errors. This is shown in Fig. 7 which plots the inversion efficiency of the simple and

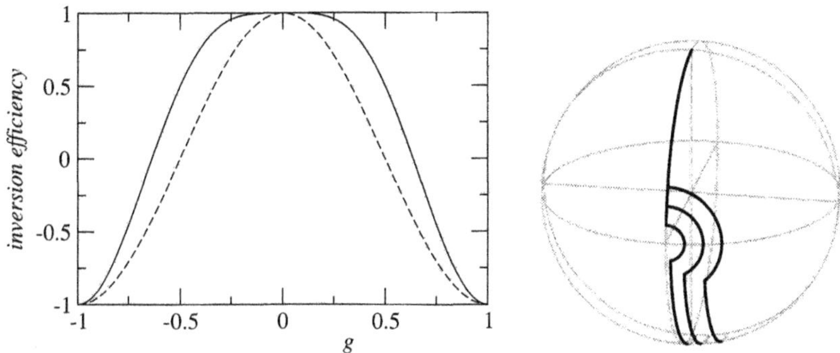

Fig. 7. The inversion efficiency of a simple 180° pulse (dashed line) and of the composite pulse $90^\circ_x 180^\circ_y 90^\circ_x$ (solid line) as a function of the fractional pulse length error g. The way in which the composite pulse works can be understood by examining trajectories on the Bloch sphere, which are shown on the right for three values of g.

composite 180° pulses as a function of the fractional pulse length error g. (The inversion efficiency of an inversion pulse measures the component of the final spin state along $-I_z$ after the pulse is applied to an initial state of I_z.)

Composite pulses of this kind are very widely used within conventional NMR, and many different pulses have been developed [48], but most of them are not directly applicable to quantum computing [50]. This is because conventional NMR pulse sequences are designed to perform specific motions on the Bloch sphere (such as inversion), in which case the initial and final spin states are known, while for quantum computing it is necessary to use general rotations, which are accurate whatever the initial state of the system. Perhaps surprisingly composite pules are known which have the desired property, of performing accurate rotations whatever the initial spin state.

4.3. *Quaternions and single qubit gates*

Quaternions provide a simple and powerful way of describing rotations (single qubit gates), as they can be easily formed, combined, and compared. The quaternion corresponding to a θ rotation around an axis at an azimuthal angle ϕ in the xy-plane is given by

$$\mathsf{q}_{\theta\phi} = \{s, \mathbf{v}\} = \{\cos(\theta/2), \sin(\theta/2)(\cos(\phi), \sin(\phi), 0)\} \tag{4.2}$$

where s is a scalar depending on the rotation angle, and $\mathbf{v}$ is a vector whose length depends on the rotation angle and which lies parallel to the rotation axis.

The result of applying two rotations is given by the quaternion product

$$\mathsf{q}_1 * \mathsf{q}_2 = \{s_1 \cdot s_2 - \mathbf{v_1} \cdot \mathbf{v_2}, s_1\mathbf{v_2} + s_2\mathbf{v_1} + \mathbf{v_1} \wedge \mathbf{v_2}\} \tag{4.3}$$

and two quaternions can be compared using the quaternion fidelity

$$\mathcal{F}(\mathsf{q}_1, \mathsf{q}_2) = |\mathsf{q}_1 \cdot \mathsf{q}_2| = |s_1 \cdot s_2 + \mathbf{v_1} \cdot \mathbf{v_2}|. \tag{4.4}$$

As a simple example consider a NOT gate, that is a 180°_x rotation. The quaternion for an ideal rotation is

$$\mathsf{q}_0 = \{0, (1, 0, 0)\} \tag{4.5}$$

while the quaternion representing this rotation in the presence of a fractional pulse length error g is

$$\mathsf{q}_1 = \{\cos[(1+g)\pi/2], (\sin[(1+g)\pi/2], 0, 0)\} \tag{4.6}$$

and so the quaternion fidelity is

$$\mathcal{F}_1 = |\sin((1+g)\pi/2)| = |\cos(g\pi/2)| \approx 1 - \frac{\pi^2 g^2}{8} \tag{4.7}$$

As an alternative consider the conventional composite pulse sequence for a 180°_x rotation, $90^\circ_y 180^\circ_x 90^\circ_y$, which has the quaternion form

$$\mathsf{q}_2 = \{\sin^2[g\pi/2], (\cos[g\pi/2], -\sin[g\pi]/2, 0)\} \tag{4.8}$$

and gives exactly the same fidelity, $\mathcal{F}_2 = |\cos(g\pi/2)| = \mathcal{F}_1$. This confirms that the conventional sequence does not actually correct for errors when considered as a general rotation: the good behaviour for certain initial states is obtained at the cost of poor behaviour for other initial states.

An example of a NOT gate which does give genuine improvement [51, 52] is provided by the sequence $90^\circ_0 180^\circ_\phi 360^\circ_{3\phi} 180^\circ_\phi 90^\circ_0$, with $\phi = \arccos(-1/4)$. The quaternion for this composite rotation in the presence of errors is complicated, but its fidelity is given by

$$\mathcal{F}_3 \approx 1 - \frac{5\pi^6 g^6}{1024} \tag{4.9}$$

showing that the second and fourth order error terms are completely cancelled. This BB1 sequence was originally developed by Wimperis for conventional NMR experiments [51], and later rederived using quaternions in the context of NMR quantum computing [52]. As shown in Fig. 8 the BB1 gate outperforms a naive NOT gate for all pulse length errors g, especially for errors in the range $\pm 25\%$. Its behaviour is essentially perfect for errors of less than 1%.

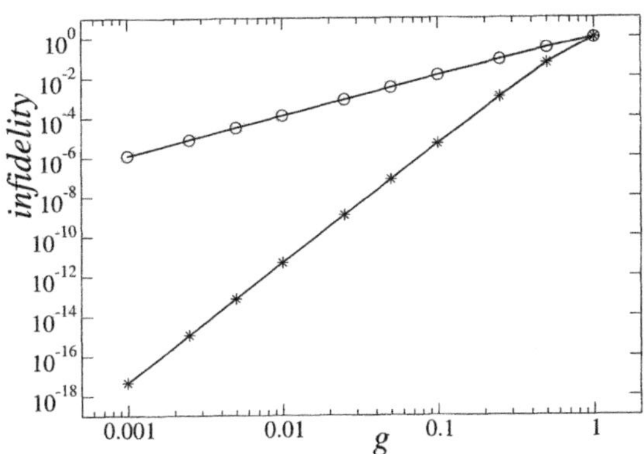

Fig. 8. The infidelity $(1 - \mathcal{F})$ of simple and BB1 composite pulses to perform a NOT gate in the presence of a fractional pulse length error g; note that both axes are plotted on log scales. The BB1 gate can achieve an infidelity of 10^{-6} with an error in ω_1 of up to 10%, in contrast with the 0.1% accuracy required for a simple gate.

Similar gates can be developed to tackle off-resonance effects. Intriguingly the sequence $90^\circ_x 180^\circ_y 90^\circ_x$ provides some compensation for off-resonance effects as long as the pulse length is correct, but as before this compensation only occurs for inversion, and so the composite pulse is not suitable for quantum computing. However suitable composite rotations are known: an early result by Tycko [53] has been refined and extended [54, 52]: a simple θ_x rotation should be replaced by the CORPSE sequence of three rotations along x, $-x$ and x with

$$\begin{aligned}\theta_1 &= 2\pi + \frac{\theta}{2} - \arcsin\left(\frac{\sin(\theta/2)}{2}\right)\\ \theta_2 &= 2\pi - 2\arcsin\left(\frac{\sin(\theta/2)}{2}\right)\\ \theta_3 &= \frac{\theta}{2} - \arcsin\left(\frac{\sin(\theta/2)}{2}\right).\end{aligned} \tag{4.10}$$

The *simultaneous* correction of pulse length errors and off-resonance effects is still being studied [52].

With any proposal for a "robust" gate it is vital to check that the errors take the form expected [55]. For a BB1 gate it is not necessary to get the absolute lengths of the pulses right, but it is essential to get the relative lengths correct. For a BB1 NOT gate ($90^\circ_0 180^\circ_\phi 360^\circ_{3\phi} 180^\circ_\phi 90^\circ_0$) this is simple as all pulses are multiples of 90 degrees, but other cases may be more tricky. The BB1 gate also requires very accurate control of pulse phases, and it is likely that phase errors will dominate in experimental implementations.

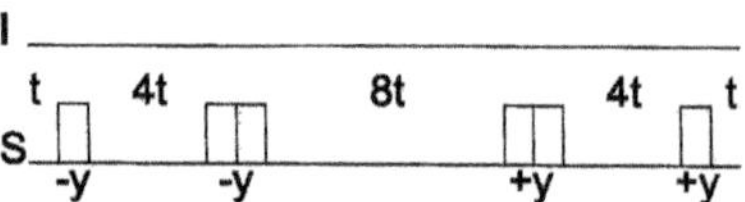

Fig. 9. Pulse sequence for an Ising gate to implement a controlled-NOT gate which is robust to small errors in J. Boxes correspond to single qubit rotations with rotation angles of $\phi = \arccos(-1/8) \approx 97.2°$ applied along the $\pm y$ axes as indicated; time periods correspond to free evolution under the Ising coupling, $\pi J\, 2I_zS_z$ for multiples of the time $t = 1/4J$. The naive Ising gate corresponds to free evolution for a time $2t$.

4.4. Two qubit gates

To obtain a complete set of robust gates it is also necessary to develop a family of robust two qubit gates, and the Ising coupling gate is the obvious choice [22]. The Ising gate is implemented by evolution under the Ising coupling Hamiltonian

$$\mathcal{H}_{IS} = \pi J\, 2I_zS_z \tag{4.11}$$

for a time $\tau = \phi/\pi J$, where J is the coupling strength and ϕ is the desired evolution angle. In order to implement accurate controlled phase-shift gates it is necessary to know J with corresponding accuracy. Remarkably a very similar approach to that used for one qubit gates can also be used to tackle systematic errors in Ising coupling gates [56]; in effect Ising coupling corresponds to rotation about the $2I_zS_z$ axis, and errors in J correspond to errors in a rotation angle about this axis. These can be parameterised by a fractional error

$$\epsilon = \frac{J_{real}}{J_{nominal}} - 1. \tag{4.12}$$

and can be compensated by rotating about a sequence of axes tilted from $2I_zS_z$ towards another axis, such as $2I_zS_x$. Defining

$$\theta_\phi \equiv \exp[-i \times \theta \times (2I_zS_z \cos\phi + 2I_zS_x \sin\phi)] \tag{4.13}$$

permits the naive sequence θ_0 to be replaced by a BB1 style sequence of the form

$$(\theta/2)_0\, \pi_\phi\, 2\pi_{3\phi}\, \pi_\phi\, (\theta/2)_0 \tag{4.14}$$

with $\phi = \arccos(-\theta/4\pi)$. Note that the BB1 NOT gate described previously is simply a special case of this with $\theta = \pi$. The tilted evolutions are implemented by sandwiching a $2I_zS_z$ rotation (evolution under the Ising Hamiltonian) between $\phi_{\mp y}$ pulses applied to spin S. After combining and cancelling pulses the final sequence for the case $\theta = \pi/2$ (which forms the basis of the controlled-NOT gate [22]) is shown in Fig. 9.

Fig. 10. Pulse sequence for a PB1 Ising gate to implement a controlled-NOT gate. Boxes are single qubit rotations with angles of $\phi = \arccos(-1/16) \approx 93.6°$.

The BB1 Ising gate outperforms the naive gate much as before: once again the error is sixth order in g. The robust gate can tolerate errors in J of around 10% with an infidelity of 10^{-6} [56, 55]. To perform a robust gate it is necessary to get the relative lengths of the coupling periods correct, but this is fairly simple as all times are multiples of $1/4J$. It is also important to use accurate pulses between the delays, but these can be achieved using robust single qubit gates.

4.5. Suppressing weak interactions

This approach can easily be adapted to tackle another problem: developing a composite rotation which suppresses the effect of weak rotations. When converted to the two qubit equivalent, this gives a controlled phase-shift gate which effectively suppresses evolution under small Ising couplings [57]. This can be achieved using the same basic sequence as before, and comparing the composite quaternion with the null quaternion

$$\mathsf{q}_0 = \{1, \{0, 0, 0\}\} \tag{4.15}$$

and then using the Maclaurin series expansion around the point $g = -1$. The first order error terms can be removed by choosing $\phi_2 = -\phi_1$ and $\phi_1 = \arccos(-\theta/4\pi)$ as before.

Although derived independently, this sequence is in fact essentially identical to the NB1 composite rotation previously described by Wimperis [51]. The NB1 sequence does effectively suppress weak interactions, but this suppression is achieved at the cost of decreased robustness to small errors in strong interactions [57]. Clearly it would be desirable to gain similar suppression effects without this increased sensitivity to errors. Amazingly another composite rotation developed by Wimperis [51], this time called PB1, provides an excellent solution. This takes the form

$$(\theta/2)_x 360_{\phi_1} 720_{\phi_2} 360_{\phi_1} (\theta/2)_x \tag{4.16}$$

with $\phi_2 = -\phi_1$ and $\phi_1 = \arccos(-\theta/8\pi)$. The pulse sequence for the two qubit version corresponding to a controlled-NOT gate is shown in Fig. 10 and its performance is compared with simple, BB1 and NB1 composite rotations in Fig. 11

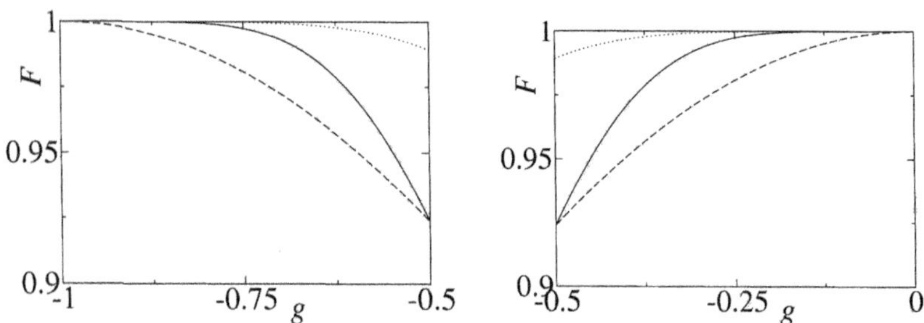

Fig. 11. Calculated fidelity of simple (dashed line), PB1 (solid line), and NB1 or BB1 (dotted line) 90° rotations as a function of the fractional error in the rotation rate g. The left hand plot shows the fidelity defined against the identity operation, with the dotted line showing the NB1 sequence; the right hand plot shows the fidelity defined against a 90° rotation, with the dotted line showing the BB1 sequence. Note that the horizontal axes differ in the two plots.

This shows that the PB1 Ising gate outperforms a simple gate both in suppressing small couplings and in robustness to small errors in coupling strengths. If, however, only one of these effects is important, even better results can be obtained by using the NB1 or BB1 gate as appropriate.

5. An NMR miscellany

In this final section I will describe a variety of miscellaneous topics relevant to NMR quantum computers. In particular I will relax my previous restriction of considering only spin-half nuclei in liquids and solutions, and only spins systems beginning at thermal equilibrium.

5.1. Introduction

In the final lecture I will address a range of topics in the general field of NMR quantum computing. I will begin by describing how geometric phases (Berry phases) can be used to implement logic gates in NMR systems [58]. I shall then consider the limits to NMR quantum computing [31], and whether these can be overcome with some of the more exotic schemes which have been suggested for performing NMR quantum computing, in particular those based on systems in the solid state or systems with nuclear spins greater than one half. Finally I shall discuss some of the non-Boltzmann methods which could in principle be used to perform initialisation of spin states, in particular those based on *para*-hydrogen.

5.2. Geometric phase gates

Classical geometric phases arise from motion of an object in a curved space [59]. For example, when an object is transported on the surface of a sphere, it can undergo a rotation arising solely from the curvature of the surface. BerryŠs phase [60], the simplest example of a geometric phase in quantum mechanics, arises in a quantum system when the Hamiltonian is varied adiabatically along a cyclic path. In NMR experiments it is usually simplest to apply a cyclic excursion to the Hamiltonian in the rotating frame. The two states of a spin-half nucleus will acquire equal and opposite geometric phases, in addition to any dynamic phases acquired during the evolution.

These geometric phases can be used to implement quantum logic gates [58]. This has the potential advantage that the Berry phase depends only on the geometry of the path, and not how it is traversed, and so should be insensitive to certain errors. Note, however, that a careful distinction should be made between *geometric* phase gates, such as those described here, and *topological* phase gates, which should exhibit extreme robustness [61]. Topological phase gates are an exciting idea, but have not yet been demonstrated experimentally.

To see how geometric phases can be implemented in NMR, recall that off-resonance excitation gives rise to a Hamiltonian which is tilted in the rotating frame. The tilt angle can be controlled by varying the off-resonance fraction, which can be achieved either by changing the frequency offset or by changing the RF intensity. The phase angle can be controlled by simply changing the phase of the RF. Thus the Hamiltonian can be moved around the Bloch sphere at will. The simplest scheme is to begin by raising the RF intensity slowly from 0 up to some maximum value, so that the Hamiltonian is tilted away from the z-axis to some final tilt angle θ, changing the phase of the RF so that the Hamiltonian rotates around a cone with cone angle θ, and finally reducing the RF intensity back to zero. The geometric phases picked up during this process are

$$\pm\gamma = \pm\Omega/2 = \pm\pi(1 - \cos\theta) \tag{5.1}$$

where the $\pm$ sign corresponds to the phase picked up by the $\pm\frac{1}{2}$ spin states, which correspond to qubits in states $|0\rangle$ and $|1\rangle$.

The geometric phase is most conveniently observed in NMR experiments by applying the adiabatic sweep to a spin in a superposition state, such as I_x; the phases are then seen as a shift 2γ in the *relative* phase of the superposition, that is as a $2\gamma I_z$ rotation. However if the experiment is carried out as described the desired geometric phase will be completely swamped by the dynamic phase which arises simply from the integrated size of the Hamiltonian. Even worse, this dynamic phase will vary over the sample, as a result of RF inhomogeneity, and so when the final signal is averaged over the macroscopic ensemble the dynamic

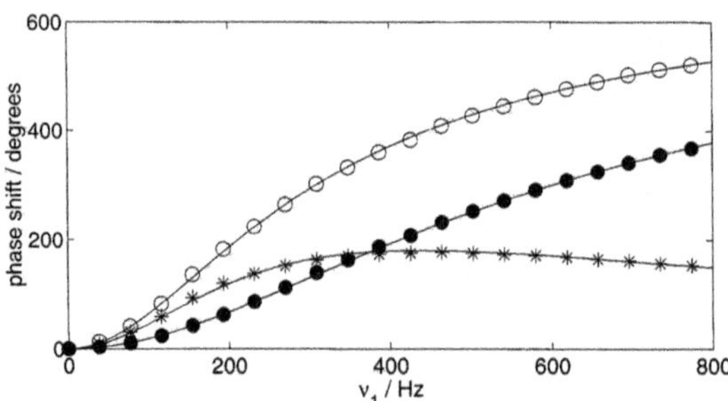

Fig. 12. Controlled geometric phases in $^{1}H^{13}CCl_3$. Filled and empty circles show the phase acquired by the ^{1}H nucleus when the ^{13}C nucleus is in $|0\rangle$ and $|1\rangle$, while stars show the difference between these (the controlled phase); solid lines show theoretical predictions.

phase will cause extensive dephasing. It is, therefore, essential to refocus the dynamic phase, and as usual this can be achieved by using spin echoes: two adiabatic sweeps are applied with the second sweep sandwiched between a pair of 180° pulses. It might seem that the geometric phase would also be refocussed by this approach, but this can be overcome by reversing the direction of the phase rotation in the second sweep: the geometric term is reversed twice, and so adds up, while the dynamic term cancels out.

The description given so far has neglected the effects of spin–spin couplings. These can be assumed to take the Ising form, and so mean that the transition frequency of a spin depends on the spin state of its coupling partners. Thus the off-resonance frequency, the tilt angle, and so the geometric phase acquired, all depend on the state of the other spin (the control spin). (Note that in a heteronuclear spin system the control spin is very far from resonance and so not directly affected by the RF field.) The results of an experiment implementing this approach [58] are shown in Fig. 12. This shows the geometric phases acquired as a function of the maximum RF intensity applied. As the maximum RF intensity is increased so are the cone angles, and thus the geometric phase acquired. More subtle behaviour is seen for the controlled geometric phase shift, $\Delta\gamma$, which first rises then falls. The extremely broad maximum in $\Delta\gamma$ indicates that this is a robust method for generating differential phase shifts; the position and height of the maximum (here chosen to be 180°) is determined by the average off-resonance frequency [58].

5.3. *Limits to NMR quantum computing*

Nuclear magnetic resonance is in many ways the leading quantum technology available to us for building small quantum computers [22]. Although it has been clear from the beginning that current NMR approaches are not scalable, and so cannot be used to build practical large scale devices, there is still substantial interest in the question of what the limits to NMR quantum computing really are.

The most commonly cited difficulty with current NMR approaches is their apparent inability to access pure initial states [38], leading to the use of non-scalable pseudo-pure state methods. This difficulty could in principle be overcome by using non-Boltzmann initial states [31], as described below. Note that schemes of this kind would also permit the creation of genuine entangled states, removing any concern about how "quantum" NMR approaches really are [39]. One serious problem might, however, remain: most of the non-Boltzmann schemes suggested would only allow the spin state to be initialised at the start of the computation; they would not allow the reinitialisation which lies at the heart of error correction schemes [34].

Another issue worth considering is the complexity of implementing pulse sequences with very large numbers of spins. This problem can be addressed mathematically by determining how the number of pulses necessary to implement a logic gate scales with the size of the spin system; the development of efficient refocussing schemes [25, 26] means that the problem scales only quadratically, which is reasonable. It is also, however, important to consider practical questions, such as how individual qubits can be addressed. NMR quantum computing uses the different Larmor frequencies of different spins to achieve this, and this approach does not scale well, as the frequency space available is quite limited [31].

Finally it is necessary to consider issues of decoherence. Although NMR decoherence times can be extremely long compared with other techniques, what matters is not the absolute length of the decoherence time, but rather the ratio of the decoherence time to the gate time. Furthermore when estimating this number it is essential to use the time needed for the *slowest* gate in the system, which in NMR systems will correspond to the smallest coupling used. Experience to date suggests that NMR quantum computations are limited to around 500 gates before the effects of decoherence become overwhelming [50, 62]. Note that this number lies well below the value required for effective error correction schemes [34].

Putting all these issues together, it seems that the limit to current NMR approaches lies around 10–15 qubits. While this is far beyond the abilities of any currently competing technology, it is not enough to make NMR a practical quantum technology.

5.4. *Exotica*

Throughout these lectures NMR has been used to refer solely to studies of spin-half nuclei in liquids and solutions. This restricted field dominates both conventional NMR studies and NMR implementations of quantum computing for a number of related reasons. There is an obvious way to map qubits onto spin-half nuclei, and the spin-half Hamiltonian in the liquid state takes a particularly nice form, which is powerful enough to be computationally universal, but simple enough to be easy to work with. Experience from conventional NMR means that the field is extremely well understood, and the available technology is highly sophisticated. It is, however, worth briefly relaxing these restrictions and seeing what the rest of NMR might have to offer.

Studying spin-half nuclei in the solid state [12] appears to have many advantages. The solid state permits access to very low temperatures, and so the preparation of pure initial states by simple Boltzmann means. This does not, however, solve the detection problem, or provide a method of reinitialisation. Solid state samples also retain the full dipolar coupling Hamiltonian, which is much larger than the isotropic scalar coupling, and so should allow much faster gates. On the down side, however, dipolar coupling networks are much more extended than scalar coupling networks, as the dipolar coupling falls off only slowly with distance. Furthermore the homonuclear dipole-dipole coupling Hamiltonian is not truncated to the (convenient) Ising form. These effects make multiplets extremely broad, rendering selective excitation of individual qubits difficult or impossible. One extreme possibility which has been suggested is to adopt techniques from magnetic resonance imaging to select spins according to their positions in space, but implementing this will not be easy.

Between liquid and solid state NMR lies the study of molecules in liquid crystal solvents. These combine some of the features of both extremes, and in particular give some access to dipole-diploe couplings in a controlled manner. This approach has been used in implementations of NMR quantum computing [63, 64], but while intriguing it is unlikely to prove important.

Another possibility which has been suggested is to use nuclei with spin greater than one half, usually in the solid state. Such nuclei have *both* a magnetic dipole moment *and* a nuclear quadrupole moment, and are often called quadrupolar nuclei [12]. The quadrupole moment will interact with electric field gradients, and the behaviour of quadrupolar nuclei is dominated by the interplay between this interaction and the interaction with magnetic fields. This can become quite complex, as the relative importance of the two interactions varies greatly for different nuclei and for different chemical environments, and the electric field gradient and magnetic field are usually attached to quite different reference frames. While this interplay could, in principle, be useful, it can also lead to many difficulties, the most obvious example being rapid dephasing.

Quadrupolar nuclei also have more than two spin states, whcih obviously suggests using one spin to store more than one qubit. It is in principle possible to build a two-qubit device in a single nucleus with spin-$\frac{3}{2}$, or even a three-qubit device in a spin-$\frac{7}{2}$ nucleus. The problem with this approach is, of course, that it does not scale beyond three qubits.

Going beyond nuclei, electrons are spin-half particles, and it should be possible to perform NMR like experiments using electron spin resonance, or ESR, studies of unpaired electrons in radicals [65]. (ESR is sometimes known as electron paramagnetic resonance, or EPR.) The electron magnetic moment is much larger than nuclear moments, and so ESR is usually a microwave technique. This also means that pure spin states could be reached by cooling too the temperature of liquid helium. The theory of ESR looks very similar to that of NMR, but the experiments are as yet much less developed. Another problem is that radicals with multiple unpaired electrons are very rare, and for this reason most proposals have concentrated on artificial nanostructures. Although this field will be challenging, it is certainly worth a look.

In the long term one of the most promising techniques is ENDOR, or electron nuclear double resonance [66], which combines NMR and ESR techniques. NMR and ESR have very different energy scales, which makes the experiments tricky, but may also prove very useful, allowing the high energy scale of ESR to be used for quantum gates, and the low energy scale of NMR to be used for storage. This is, of course, the ultimate basis of the Kane proposal for a large scale quantum computer [67].

5.5. Non-Boltzmann initial states

Although the problem of initialisation is the not the only factor limiting NMR implementations of quantum computing, it remains an important and annoying problem. The low polarisation of Boltzmann states is also an important issue in conventional NMR studies, as it results in a low signal strength, greatly limiting the sensitivity of NMR as an analytical technique. For this reason there has long been interest in developing ways to enhance NMR polarisations, and many different techniques have been developed [31]. Of these, however, only two have any realistic prospect of preparing essentially pure states: optical pumping, and the use of *para*-hydrogen.

Optical pumping is, of course, widely used to prepare atoms and ions in desired electronic states, and these techniques can be extended to prepare essentially pure nuclear spin states. Two systems dominate optically pumped NMR: ^{3}He, which can be pumped directly, and ^{129}Xe and ^{131}Xe, which are pumped indirectly via rubidium atoms. In both cases it is possible to achieve extremely high polarisations, close to unity. However these near pure spin states are almost useless for quantum computing, as each atom can only hold one qubit (or two in the case of the quadrupolar nucleus ^{131}Xe), and the atoms do not interact to form molecules. It is possible in principle to transfer these high polarisations to other nuclei [68], but so far the efficiency of such transfers has been too low to be useful.

An intriguing alternative is provided by *para*-hydrogen. This relies on the fact that the rotational and nuclear spin states of homonuclear diatomic molecules such as hydrogen are inextricably connected by the Pauli principle; in particular the even rotational states of hydrogen must have antisymmetric nuclear spin states. Cooling hydrogen to the rotational ground state would give pure *para*-hydrogen with a singlet spin state. As the interconversion of *ortho* and *para*-hydrogen is forbidden, it is necessary to use a catalyst, but this means that the enhancement will be retained on warming if the catalyst is removed. Thus it is possible to obtain a bottle of hydrogen gas at room temperature with essentially pure nuclear spin states!

It is, of course, not possible to implement quantum computing with *para*-hydrogen directly because the hydrogen molecule is too symmetric: it is essential to break the symmetry so that the two nuclei can be addressed individually. This is easily achieved by adding the *para*-hydrogen to some other molecule, such as Vaska's catalyst, to produce a complex where the two ^{1}H nuclei have different chemical shifts and so can be separately addressed [69, 70, 71]. The addition reaction occurs with retention of nuclear spin state, and so the purity is conserved. In most studies to date, however, the reaction occurs quite slowly in comparison with the Zeeman frequency difference of the two spins, leading to dephasing of the off-diagonal elements of the density matrix, converting the singlet state to an incoherent mixture of $|01\rangle$ and $|10\rangle$. In conventional *para*-hydrogen studies this dephasing is accepted, but for NMR quantum computing it is necessary to have fully coherent addition. One approach used to date [72] is to apply an *isotropic mixing* pulse sequence [3], such as MLEV-16, which removes the dephasing Zeeman interaction and preserves the pure singlet spin state. Early experiments using this approach have achieved a purity of around 10% in a two qubit system [72], significantly below the entanglement threshold. A more recent experiment has generated systems with a purity of around 92% [73].

6. Summary

Nuclear Magnetic Resonance (NMR) is arguably both the best and the worst technology we have for the implementation of small quantum computers. Its strengths lie in the ease with which arbitrary unitary transformations can be implemented, and the great experimental simplicity arising from the low energy scale and long time scale of radio frequency transitions; its weaknesses lie in the difficulty of implementing essential non-unitary operations, most notably initialisation and measurement.

	I_x	I_y	I_z	S_x	S_y	Sz	$2I_zS_z$
I_x	0	$+iI_z$	$-iI_y$	0	0	0	$-i2I_yS_z$
I_y	$-iI_z$	0	$+iI_x$	0	0	0	$+i2I_xS_z$
I_z	$+iI_y$	$-iI_x$	0	0	0	0	0
S_x	0	0	0	0	$+iS_z$	$-iS_y$	$-i2I_zS_y$
S_y	0	0	0	$-iS_z$	0	$+iS_x$	$+i2I_zS_x$
S_z	0	0	0	$+iS_y$	$-iS_x$	0	0
$2I_xS_x$	0	$+i2I_zS_x$	$-i2I_yS_x$	0	$+i2I_xS_z$	$-i2I_xS_y$	0
$2I_xS_y$	0	$+i2I_zS_y$	$-i2I_yS_y$	$-i2I_xS_z$	0	$+i2I_xS_x$	0
$2I_xS_z$	0	$+i2I_zS_z$	$-i2I_yS_z$	$-i2I_xS_y$	$+i2I_xS_x$	0	$-iI_y$
$2I_yS_x$	$-i2I_zS_x$	0	$+i2I_xS_x$	0	$+2iI_yS_z$	$-i2I_yS_y$	0
$2I_yS_y$	$-i2I_zS_y$	0	$+i2I_xS_y$	$-i2I_yS_z$	0	$+i2I_yS_z$	0
$2I_yS_z$	$-i2I_zS_z$	0	$+i2I_xS_z$	$+i2I_yS_y$	$-i2I_yS_x$	0	$+iI_x$
$2I_zS_x$	$+i2I_yS_x$	$-i2I_xS_x$	0	0	$+i2I_zS_z$	$-i2I_zS_y$	$-iS_y$
$2I_zS_y$	$+i2I_yS_y$	$-i2I_xS_y$	0	$-i2I_zS_z$	0	$+i2I_zS_x$	$+iS_x$
$2I_zS_z$	$+i2I_yS_z$	$-i2I_xS_z$	0	$+i2I_zS_y$	$-i2I_zS_x$	0	0

The debate over whether NMR quantum computers are "real" quantum computers has generated much heat, but also some useful light. It is now clear that NMR is indeed quantum mechanical, and can in principle be used to build quantum computers, but that the current approach based on pseudo-pure states will never lead to true quantum computing.

Current NMR techniques are not a serious candidate for real quantum computing, but NMR remains a great technique for playing around with small numbers of qubits. The unparalleled sophistication of NMR will almost certainly prove a rich source of insights which will find their ultimate applications in other fields.

Appendix A. Commutators and product operators

The product operator formalism allows the behaviour of spin systems undergoing complicated NMR pulse sequences to be calculated using nothing more that elementary trigonometry and a table of commutators. A table of the most important commutators in a two spin system is given below. As an example consider the element in the row labelled I_x and the column labelled I_z; the commutator $[I_x, I_z] = -iI_y$ and this is entered in the table. From this element one can immediately deduce that

$$I_x \xrightarrow{\theta I_z} I_x \cos\theta + I_y \sin\theta.$$

In the same way the next element in the table can be used to deduce that I_x commutes with S_x, and so

$$I_x \xrightarrow{\theta S_x} I_x.$$

These rules permit easy calculation of the evolution of any state of a two spin system under any one of the product operators found in simple Hamiltonians, but real Hamiltonians will contain several terms at once: for example the weak coupling Hamiltonian of a two spin system contains terms proportional to I_z, S_z and $2I_zS_z$. When, as in this case, the terms all commute the situation is simple, and the total evolution can be calculated by applying the terms sequentially in any desired order. A similar situation occurs when pulses are applied simultaneously to two or more spins: as one-spin operators on different spins all commute the pulses can be applied separately. Note, however, that pulse Hamiltonians do *not* commute with the background Hamiltonian, and it is necessary to neglect this during pulses: this is a good approximation for short high power hard pulses. In the same way the three components contributing to a Heisenberg coupling ($2I_xS_x$, $2I_yS_y$, and $2I_zS_z$) do not commute, and so the product operator approach can only be used in the weak coupling regime where the Heisenberg coupling is truncated to the Ising ($2I_zS_z$) form.

References

[1] A. Abragam. *Principles of Nuclear Magnetism.* Clarendon Press, Oxford, UK, 1961.

[2] R. R. Ernst, G. Bodenhausen, and A. Wokaun. *Principles of Nuclear Magnetic Resonance in One and Two Dimensions.* Clarendon Press, Oxford, UK, 1987.

[3] R. Freeman. *Spin Choreography: Basic Steps in High Resolution NMR.* Spektrum, Oxford, UK, 1997.

[4] M. Goldman. *Quantum Description of High-Resolution NMR in Liquids.* Clarendon Press, Oxford, UK, 1988.

[5] P. J. Hore. *Nuclear Magnetic Resonance.* Oxford Chemistry Primers, Oxford, UK, 1995.

[6] M. H. Levitt. *Spin Dynamics: Basics of Nuclear Magnetic Resonance.* Wiley, Chichester, UK, 2001.

[7] R. Freeman. *Magnetic Resonance in Chemistry and Medicine.* Oxford University Press, Oxford, UK, 2003.

[8] T. Claridge. *High-resolution NMR Techniques in Organic Chemistry.* Pergamon, 1999.

[9] R. Freeman. *A Handbook of Nuclear Magnetic Resonance.* Longman, Harlow, UK, 1987.

[10] O. W. Sørensen, G. W. Eich, M. H. Levitt, G. Bodenhausen, and R. R. Ernst. *Prog. Nucl. Magn. Reson. Spectrosc.*, 16:163–192, 1983.

[11] P. J. Hore, J. A. Jones, and S. Wimperis. *NMR: The Toolkit.* Oxford Chemistry Primers, Oxford, UK, 2000.

[12] K. Schmidt-Rohr and H. W. Spiess. *Multidimensional Solid-State NMR and Polymers.* Academic Press, San Diego, CA, 1994.

[13] E. L. Hahn. *Phys. Rev. Lett.*, 80:580, 1950.

[14] D. P. DiVincenzo. *Proc. Roy. Soc. Lond. A*, 454:261–276, 1998.

[15] D. Deutsch. *Proc. Roy. Soc. Lond. A*, 425:73–90, 1989.

[16] D. Deutsch, A. Barenco, and A. Ekert. *Proc. Roy. Soc. Lond. A*, 449:669–677, 1995.

[17] A. Barenco, C. H. Bennett, R. Cleve, D. P. DiVincenzo, N. Margolus, P. Shor, T. Sleator, J. A. Smolin, and H. Weinfurter. *Phys. Rev. A*, 52:3457–3467, 1995.

[18] A. Barenco. *Proc. Roy. Soc. Lond. A*, 449:679–683, 1995.

[19] D. G. Cory, A. F. Fahmy, and T. F. Havel. In T. Toffoli, M. Biafore, and Leão, editors, *Proceedings of Phys Comp '96*, pages 87–91. New England Complex Systems Institute, 1996.

[20] D. G. Cory, A. F. Fahmy, and T. F. Havel. *Proc. Natl. Acad. Sci. USA*, 64:1634–1639, 1997.

[21] N. A. Gershenfeld and I. L. Chuang. *Science*, 275:350–356, 1997.

[22] J. A. Jones. *Prog. Nucl. Magn. Reson. Spectrosc.*, 38:325–360, 2001.

[23] R. Freeman, T. A. Frenkiel, and M. H. Levitt. *J. Magn. Reson.*, 44:409–412, 1981.

[24] E. Knill, R. Laflamme, R. Martinez, and C. H. Tseng. *Nature*, 404:368–370, 2000.

[25] J. A. Jones and E. Knill. *J. Magn. Reson.*, 141:322–325, 1999.

[26] D. W. Leung, I. L. Chuang, F. Yamaguchi, and Y. Yamamoto. *Phys. Rev. A*, 61:042310, 2000.

[27] J. A. Jones and M. Mosca. *Phys. Rev. Lett.*, 83:1050–1053, 1999.

[28] N. Linden, H. Barjat, and R. Freeman. *Chem. Phys. Lett.*, 296:61–67, 1998.

[29] P. Zanardi and M. Rasetti. *Phys. Rev. Lett.*, 79:3306–3309, 1997.

[30] L. Viola, E. M. Fortunato, M. A. Pravia, E. Knill, R. Laflamme, and D. G. Cory. *Science*, 93:2059–2063, 2001.

[31] J. A. Jones. *Fort. der Physik*, 48:909–924, 2000.

[32] S. L. Braunstein and H. K. Lo, editors. *Scalable Quantum Computers: Paving the Way to Realisation*. Wiley, 2001.

[33] J. A. Jones and M. Mosca. *J. Chem. Phys.*, 109:1648–1653, 1998.

[34] A. M. Steane. *Phil. Trans. Roy. Soc. Lond. A*, 356:1739–1758, 1998.

[35] J. A. Jones. *PhysChemComm*, 11, 2001.

[36] J. A. Jones, R. H. Hansen, and M. Mosca. *J. Magn. Reson.*, 135:353–360, 1998.

[37] E. Knill, I. Chuang, and R. Laflamme. *Phys. Rev. A*, 57:3348–3363, 1998.

[38] W. S. Warren. *Science*, 277:1688–1689, 1997.

[39] S. L. Braunstein, C. M. Caves, R. Jozsa, Lindenn N., S. Popescu, and R. Schack. *Phys. Rev. Lett.*, 83:1054–1057, 1999.

[40] N. Linden and S. Popescu. *Phys. Rev. Lett.*, 87:047901, 2001.

[41] R. Schack and C. M. Caves. *Phys. Rev. A*, 60:4354–4362, 1999.

[42] E. Knill and R. Laflamme. *Phys. Rev. Lett.*, 81:5672–5675, 1998.

[43] L. J. Schulman and U. Vazirani. In *Proc. 31st ACM STOC*, pages 322–329, 1999. E-print quant-ph/9804060.

[44] H. K. Cummins, C. Jones, A. Furze, N. Soffe, M. Mosca, J. M. Peach, and J. A. Jones. *Phys. Rev. Lett.*, 88:187901, 2002.

[45] W. K. Wooters and W. H. Zurek. *Nature*, 299:802–803, 1982.

[46] V. Bužek and M. Hillery. *Phys. Rev. A*, 54:1844–1852, 1996.

[47] M. H. Levitt and R. Freeman. *J. Magn. Reson.*, 33:473, 1979.

[48] M. H. Levitt. *Prog. Nucl. Magn. Reson. Spectrosc.*, 18:61–122, 1986.

[49] M. D. Bowdrey, D. K. L. Oi, A. J. Short, K. Banaszek, and J. A. Jones. *Phys. Lett. A*, 294:258–260, 2002.

[50] H. K. Cummins and J. A. Jones. *New J. Phys.*, 2:1–12, 2000.

[51] S. Wimperis. *J. Magn. Reson. A*, 109:221–231, 1994.

[52] H. K. Cummins, G. Llewellyn, and J. A. Jones. *Phys. Rev. A*, 67:042308, 2003.

[53] R. Tycko. *Phys. Rev. Lett.*, 51:775, 1983.

[54] H. K. Cummins and J. A. Jones. *Contemp. Phys.*, 41:383–399, 2000.

[55] J. A. Jones. *Phil. Trans. Roy. Soc. Lond. A*, 361:1429–1440, 2003.

[56] J. A. Jones. *Phys. Rev. A*, 67:012317, 2003.

[57] J. A. Jones. *Phys. Lett. A*, 316:24–28, 2003.

[58] J. A. Jones, V. Vedral, G. Ekert, and G. Castagnoli. *Nature*, 403:869–871, 2000.

[59] A. Shapere and F. Wilczek, editors. *Geometric Phases in Physics*. World Scientific, Singapore, 1989.

[60] M. V. Berry. *Proc. Roy. Soc. Lond. A*, 392:45, 1984.

[61] A. Y. Kitaev. *Annals Phys.*, 303:2–30, 2003. E-print quant-ph/9707021.

[62] L. M. K. Vandersypen, M. Steffen, G. Breyta, C. S. Yannoni, M. H. Sherwood, and I. L. Chuang. *Nature*, 414:883–887, 2001.

[63] M. Marjanska, I. L Chuang, and M. G. Kubinec. *J. Chem. Phys.*, 112:5095–5099, 2000.

[64] B. M. Fung. *J. Chem. Phys.*, 115:8044–8048, 2001.

[65] A. Schweiger and G. Jeschke. Oxford University Press, Oxford, UK, 2001.

[66] M. Mehring, J. Mende, and W. Scherer. *Phys. Rev. Lett.*, 90:153001, 2003.

[67] B. E. Kane. *Nature*, 393:133–137, 1998.

[68] A. S. Verhulst, O. Liivak, M. H. Sherwood, H. Vieth, and I. L. Chuang. *Appl. Phys. Lett.*, 79:2480–2482, 2001.

[69] C. R Bowers and D. P. Weitekamp. *Phys. Rev. Lett.*, 57:2645–2648, 1986.

[70] J. Natterer and J. Bargon. *Prog. Nucl. Magn. Reson. Spectrosc.*, 31:293–315, 1997.

[71] S. B. Duckett and C. J. Sleigh. *Prog. Nucl. Magn. Reson. Spectrosc.*, 34:71–92, 1999.

[72] P. Hübler, J. Bargon, and S. J. Glaser. *J. Chem. Phys.*, 113:2056–2059, 2000.

[73] M. S. Anwar *et al.* quant-ph/0312014, 2003.

Course 11

INTRODUCTION TO QUANTUM CONDUCTORS

D. C. Glattli

Service de Physique de l'Etat Condensé, CEA Saclay
Gif-sur-Yvette France
Laboratoire Pierre Aigrain, Département de Physique
ENS, 24 rue Lhomond, Paris

D. Estève, J.-M. Raimond and J. Dalibard, eds.
Les Houches, Session LXXIX, 2003
Quantum Entanglement and Information Processing
Intrication quantique et traitement de l'information
© 2004 Elsevier B.V. All rights reserved

Contents

1. Introduction

The aim of this lecture is to provide an introduction to the physics of phase coherent conductors using the quantum scattering approach. We will describe the electronic interference effects observable in the conductance, and the quantum noise associated with the granularity of the quasi-particles. We will also discuss some recent proposals to produce electronic entanglement using ballistic quantum conductors and to test Bell's inequality. We will show the importance of the Fermi statistics which leads to remarkable properties: quantum conductors show sub-Poissonian current fluctuations and can even be noiseless; unlike photons, electrons can be entangled without the need of interactions but only with equilibrium sources and beam splitters. The importance of the statistics will be emphasized using analogies between photons and electrons each time it is possible. Examples will be taken from experiments on quantum ballistic systems (mostly 2D electron systems) rather than metallic diffusive systems as scattering problems in these systems are very close to scattering problems in optics.

We hope the reader not familiar with condensed matter will find here the basic elements to understand further readings on the use of electronic mesoscopic systems to realize quantum information devices. More on the properties of quantum conductors can be find in references [1, 2, 3].

1.1. The mesoscopic scale and the mesoscopic conductor

We are interested in electron quasi-particles propagating in quantum conductors. The quasi-particles reveal many interesting quantum effects and their manipulation can be envisaged in the near future for quantum information.

We first briefly describe the frame in which the quasi-particles live : the phase coherent conductor. The resistor schematically shown in Fig.1 is usually considered classical. No quantum mechanics is apparently needed to describe this conductor in a circuit at low frequency and not too low temperature T: it obeys Ohm's law and is characterized by a resistance relating the voltage drop V to the current I. At the most microscopic level however the resistor is a metal. A metal cannot be understood without requiring quantum mechanics and particularly the following quantum effects: -first, the Fermi statistics forces the electrons to spread their kinetic energy up to the so-called Fermi energy. - second, the elec-

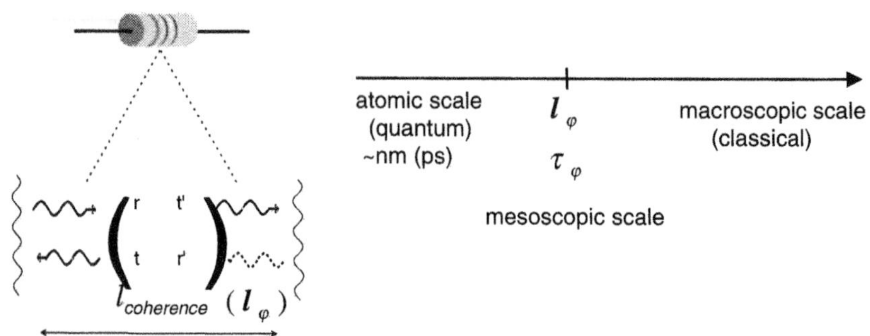

Fig. 1. A macroscopic resistor can be view as an ensemble of quantum conductors of size l_ϕ, the electron coherence length. Below this length the conduction reduces to a problem of electronic wave scattering.

tron wave diffracts on the host periodic lattice and free propagation occurs only for certain energy range; in the good metals considered here, the Fermi energy corresponds to propagating states with renormalization of some electronic parameters as the effective mass m^*. - third, the Fermi statistics reduces the number of electrons participating to conduction to a narrow energy range around the Fermi energy; by the way it strongly reduces the possible number of channels for electron-electron interaction; the electron becomes a compound object : the Landau quasi-particle, a weakly interacting screened electron with long coherence time.

The description of the correct scenario interpolating from the quantum description at short-scale to a classical picture at macroscopic scale has been the subject of a recent active field of quantum physics: Mesoscopic Physics. The key word is decoherence. After diffracting on the regular lattice and eventually on static localized impurities or boundaries, the electronic wave looses coherence by interaction with the environment : the phonon bath, electrons, magnetic impurities and electromagnetic fields. The coherence time τ_φ and length l_φ define the mesocopic scales separating the classical world from the quantum world experienced by the quasi-particle. At macroscopic scale, the resistor of Fig.1 can be considered as a huge collection of mesoscopic conductors of size l_φ. One can add classically the resistance of each mesocopic domain, while inside a domain Ohm's addition law fails as the properties are non longer local.

In the following, we will focus on a single mesoscopic domain. Practically, this is possible because progress in lithography and low temperature allow to study conductors whose size is smaller than the mesoscopic scale l_φ. In a well designed experiment, the narrow region defining the mesoscopic domain is immediately surrounded by much wider regions called contacts or reservoirs. The

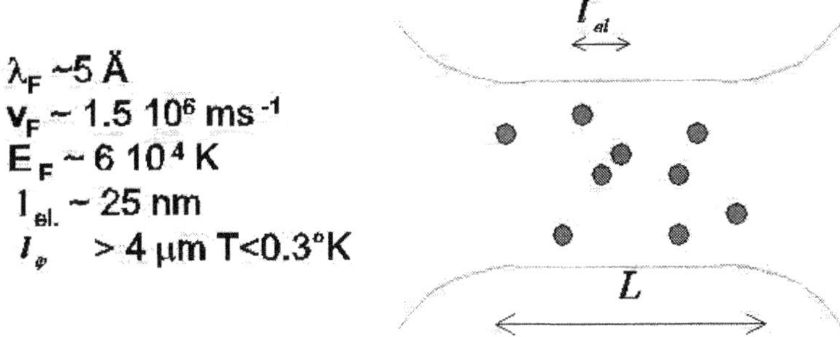

Fig. 2. Schematic representation of the multiple scattering of electrons in a diffusive mesoscopic conductor like a metallic films. Typical elastic mean free path l_{el}, Fermi wavelength λ_F and phase coherence length are indicated.

resistance to electrical flow is imposed by the narrow region, not by the resistance of the wide contacts assumed to be negligible. As the narrow region is pase-coherent, there is no inelastic scattering and the rate of energy dissipation IV occurs *in the reservoirs,* not in the mesoscopic region. In the quantum scattering approach used in part II and III, it is assumed that the voltage V remains constant any time. This means that the resistance of the external circuit resistance is infinitely small at all frequency and no power is taken by the circuit when measuring current: the ammeter is ideal.

Example of quantum conductors are given in Fig. 2 and 3 with values of the coherence length and of the elastic mean free path.

1.2. Short overview of the properties of quantum conductors discussed here

In part II, we will describe the conduction in term of scattering of waves emitted by contacts. The scattering region can be characterize by its scattering matrix S which relates outgoing waves to incoming waves. The left reservoir injects in the phase-coherent region electronic waves occupied by electrons up to the chemical potential μ_L. The injected waves, incoming on the scattering region, return to the reservoirs as transmitted or reflected outgoing waves. If the wave is occupied, an electron will appear in the reservoir where decoherence occurs with a probability given by the reflection or the transmission coefficients. To satisfy Fermi statistics, an electron will also "have to check" wether another electron emitted by the right reservoir at the same energy (up to chemical potential μ_R) is not going to fill the same outgoing state.

It sometime helps to use optical analogies, both when considering the reflection/transmission of waves but also when considering the reservoirs. The key idea

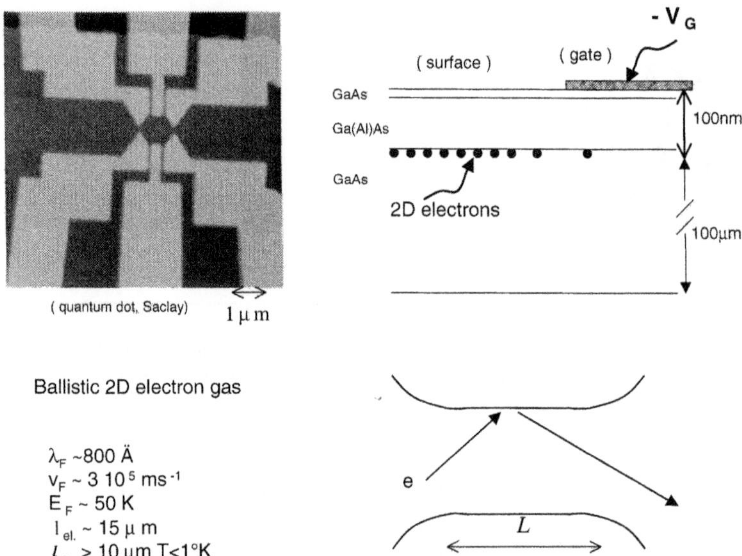

Fig. 3. Ballistic quantum dot realized in a 2D electron gas. The electrons are confined in a plane at the interface between the GaAs and Ga(Al)As semiconductors. Gates negatively voltage biased can deplete electrons underneath and so realize 0D or 1D structures. Lower right: schematic representation of ballistic electron trajectories

here is that reservoirs are black-body sources of electrons (perfectly absorbing all electrons while emitting electrons by filling states according to the Fermi-Dirac distribution). The intensity I of the electrical current measures the transmission D_n of the electronic waves as do the intensity of a light beam for an optical medium. At $T = 0$, for chemical potential difference $\mu_L - \mu_R = eV$ across the reservoirs, the current is:

$$I = \frac{e^2}{h} V \sum_n D_n$$

where $\frac{e^2}{h}$ =1/25.812kOhm is the quantum of the conductance. Eventually, the D_ns reveal interferences or resonances of electron waves in the conductor (Aharonov-Bohm effect, resonant tunnelling, ...).

While the intensity I is sensitive to the wave nature of the electrons, its fluctuations reveal new informations. When repeating similar finite time measurements and comparing the results, one finds current fluctuations. They arise from the granularity of the current and are a manifestation of the electronic shot noise. An enlightening comparison is given by the fluctuations of light beams. Their study has given direct evidence of the granularity of the light and marked the foundation of quantum optics. Similarly, a new field of mesoscopic physics has recently appeared: electrical quantum shot noise. In a shot noise experiment, the reservoirs play the role of the photo-detectors. Before detection, the electron quasi-particle emitted by a reservoir is in a superposition of reflected and transmitted states. Decoherence reduces the electron wave-packet and makes the electron appear in one of the reservoirs. In part III, we will show that, unlike photons, the noise is naturally sub-Poissonian. Thanks to the Fermi statistics, electrons are emitted by the reservoir as a regular noiseless flow toward the conductor. The only noise mechanism is binomial partition noise: is the electron transmitted or reflected. The resulting quantum shot noise power S_I (at low frequency) is [4, 5, 6]

$$S_I = 2e\frac{e^2}{h}V\sum_n D_n(1-D_n) = 2eI\frac{\sum_n D_n(1-D_n)}{\sum_n D_n} \tag{1.1}$$

where the $D_n(1-D_n)$ factor comes from the binomial law. One can realize situations where $D_n = 0, 1$ and the quantum conductor can even be noiseless [7, 8]. References and review on quantum shot noise can be found in Ref.[9, 10, 11].

Similarly to optics [12], Hanbury-Brown Twiss correlations with electron quasiparticles can be made by correlating current fluctuations of different leads [13, 14]. They can be a useful tool to indicate entanglement. But, what is entanglement in a quantum conductor made of so many electrons? Electrons interact easily and must satisfy the anti-symmetrization postulate. This immediately implies that the ground state of the many electron wavefunction is not a product of elementary quantum states but a non-separable state with a high degree of entanglement. However this entanglement is not useful. In a real experiments one manipulates only few elementary excitations (the quasi-particles) above the ground state which is the analog of the vacuum in optics. By definition, the quasi-particles are not entangled for time smaller than their lifetime which identifies to the coherence time in the absence of external perturbations. Physicists have to do some work to realize controlled entanglement. In part IV we first consider an example of sources of pre-entangled electrons: the superconducting - normal junction where the super-conducting part injects pairs of electrons in a spin singlet state in a normal reservoir [15, 16, 17, 18]. Then, we briefly review recent proposals to generate quasi-particle entanglement using equilibrium contacts combined with beam-splitters [19, 20, 27]. We will see that current noise measurements can replace coincidence measurements to test the violation of Bell's inequalities and demonstrate entanglement [17, 18, 19, 20, 27].

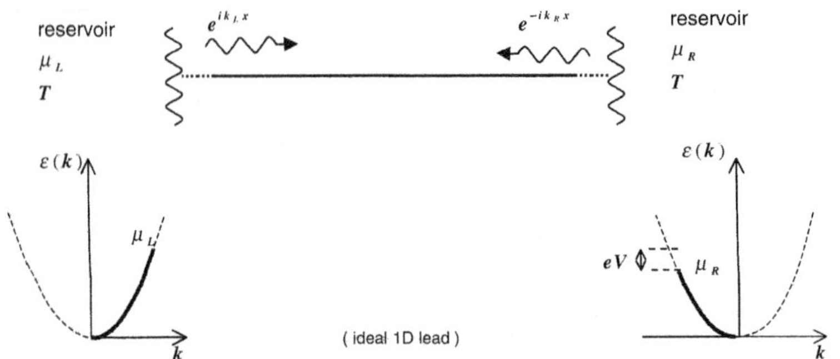

Fig. 4. Perfect 1D conductor connected between two electronic reservoirs. The bold regions of the left and right parabola, representing energy versus wavenumber, correspond to states occupied by electrons for the left and right reservoirs respectively.

2. The scattering approach to quantum conduction

2.1. The quantum of conductance

2.1.1. The ideal one dimensional conductor

To introduce notations we first consider an ideal single mode conductor (or equivalently a 1D conductor). Spin is disregarded here. In zero magnetic field the spin degeneracy can be accounted for by double occupation of the orbital modes. Each end of the conductor is connected to a reservoir (Fig.4). The left (right) reservoir injects electrons to the right (left) according to an equilibrium distribution given by temperature T and chemical potential $\mu_{L(R)}$. A reservoir is assumed to perfectly absorb electrons at all energy.

Assuming non interacting electrons with kinetic energy $\varepsilon_{\pm k} = \hbar^2k^2/2m$ and momentum $\pm k$ ($k \geq 0$), the normalized wavefunction are $\varphi_{\pm k}(x,t) = e^{i(\pm kx-\epsilon t)}/\sqrt{2\pi}$ with $\langle\varphi_{k'} \mid \varphi_k\rangle = \delta\left(k'-k\right)$ and $\left\langle\varphi_{-k'} \mid \varphi_k\right\rangle = 0$. The sign $+$ corresponds to right moving electrons emitted by the left reservoir, the sign $-$ to left moving electrons emitted by the right reservoir. The fermion field operator describing right moving electrons is thus :

$$\widehat{\Psi}_L(x,t) = \frac{1}{\sqrt{2\pi}} \int dk \widehat{a}_L(k) e^{i(kx-\epsilon t)}$$

where $\widehat{a}_L(k)$ are the fermion annihilation operators acting on the reservoir's Fock space, with:

$$\left\{\widehat{a}_L(k), \widehat{a}_L^{\dagger}(k')\right\} = \delta(k-k') \quad , \quad \left\langle\widehat{a}_L^{\dagger}(k')\widehat{a}_L(k)\right\rangle = f_L(\varepsilon_k)\delta(k-k')$$

and $f_L(\varepsilon_k) = \frac{1}{e^{(\varepsilon_K - \mu_L)/k_B T} + 1}$ is the Fermi-Dirac distribution for the left reservoir in equilibrium at temperature T and electrochemical potential μ_L. Similar definitions with $L \to R$ and $k \to -k$ holds to describe left moving electrons emitted by the right reservoir.

As k is a single valued function of $\varepsilon_k = \hbar^2 k^2/2m$ one can rewrite the previous expressions as a function of the energy

$$\widehat{\Psi}_L(x,t) = \frac{1}{\sqrt{2\pi}} \int d\varepsilon \frac{\widehat{a}_L(\varepsilon)}{\sqrt{\hbar v(\varepsilon)}} e^{i(kx - \epsilon t)} \tag{2.1}$$

where $\widehat{a}_L(\varepsilon) = \widehat{a}_l(k(\varepsilon))/\sqrt{\hbar v(\varepsilon)}$ and $v(\varepsilon) = \partial\varepsilon/\hbar\partial k$ is the electron velocity. The new fermion operators satisfy:

$$\left\{\widehat{a}_L(\varepsilon), \widehat{a}_L^\dagger(\varepsilon')\right\} = \delta(\varepsilon - \varepsilon') \quad , \quad \left\langle\widehat{a}_L^\dagger(\varepsilon')\widehat{a}_L(\varepsilon)\right\rangle = f_L(\varepsilon)\delta(\varepsilon - \varepsilon') \tag{2.2}$$

The advantage of using wavefunctions normalized with respect to energy is double: first we will exclusively consider elastic scattering which preserves energy but not necessarily the modulus of the momentum $|\hbar k|$; second, in this units, the amplitude $a(\varepsilon)$ of the normalized wave $\frac{a_L(\varepsilon)}{\sqrt{2\pi\hbar v(\varepsilon)}} e^{i(kx - \epsilon t)}$ is the amplitude of the current probability which is conserved in elastic scattering. Finally, the density of state $\mathrm{D}(\varepsilon) = \frac{1}{2\pi\hbar v(\varepsilon)}$ for right moving electrons at energy ε enters as a square root in the wavefunction.

In 1D, the current density is the electrical current $I(x,t)$ (the charge per unit time passing through point x at time t). It satisfies the conservation equation $e\frac{\partial(\widehat{\Psi}^\dagger\widehat{\Psi})}{\partial t} + \frac{\partial}{\partial x}(\widehat{I}) = 0$ where $e\widehat{\Psi}^\dagger\widehat{\Psi}$ is the local charge density. For right moving electrons:

$$\begin{aligned}\widehat{I}_L(x,t) &= e\frac{\hbar}{i2m}\left(\widehat{\Psi}_L^\dagger(x,t)\frac{\partial\widehat{\Psi}_L(x,t)}{\partial x} - \frac{\partial\widehat{\Psi}_L^\dagger(x,t)}{\partial x}\widehat{\Psi}_L(x,t)\right) \\ &= \frac{e}{h}\int d\varepsilon d\varepsilon' \widehat{a}_L^\dagger(\varepsilon')\widehat{a}_L(\varepsilon)\frac{v(\varepsilon) + v(\varepsilon')}{2\sqrt{v(\varepsilon)}\sqrt{v(\varepsilon')}} e^{i(k(\varepsilon) - k(\varepsilon'))x} e^{i(\varepsilon' - \varepsilon)t}\end{aligned} \tag{2.3}$$

This expression, and in particular the limit when energy differences $|\varepsilon' - \varepsilon| \ll \varepsilon$ are small will be useful when considering time dependent transport of current fluctuations. For the moment we are interested in the average value of the current which is (with no approximations):

$$I_L = \frac{e}{h}\int d\varepsilon f_L(\varepsilon) \tag{2.4}$$

For fully occupied states ($f(\varepsilon) = 1$) within the energy range $d\varepsilon$ the current carried by right moving electrons takes the universal value

$$dI_L = \frac{e}{h} d\varepsilon \tag{2.5}$$

Relation 2.5 is very general and holds for any source of fermions. With no electrical units it tells us that the maximum rate of fermions emitted by a source within the energy interval $d\varepsilon$ is $d\dot{N} = \frac{1}{h} d\varepsilon$. A similar relation holds for bosons. The rate of photons within the frequency interval $d\nu$ is $d\dot{N}_{Ph.} = \frac{N_{Ph.}}{h} d(h\nu)$ but the population $N_{Ph.}$ of photon states emitted by the source is not quantized to 1 and can be much larger. For thermal photons f_L has to be replaced in Eq.2.4 by the Bose-Einstein distribution . The corresponding intensity of the photon light beam is $dI_{Ph.} = N_{Ph.} \frac{(h\nu)}{h} d(h\nu)$.

2.1.2. *The quantum of conductance*

With similar definitions left moving electrons emitted by the right reservoir give a current $I_R = -\frac{e}{h} \int d\varepsilon f_R(\varepsilon)$. The total current is:

$$I = \frac{e}{h} \int d\varepsilon \, (f_L(\varepsilon) - f_R(\varepsilon))$$

For electrochemical potential difference $eV = \mu_L - \mu_R$ applied between the reservoirs, the conductance, defined as $G = I/V = \frac{e^2}{h} \int d\varepsilon \partial f(\varepsilon)/\partial \varepsilon$ for $V \to 0$, takes the universal value:

$$G = \frac{e^2}{h} \tag{2.6}$$

Eq.(2.6) holds at all temperature and defines the quantum of conductance for an ideal lead. Using the quantum Hall effect which occurs in a 2D electron metal in high perpendicular magnetic field, a metrological determination of $h/4e^2$ has been possible giving a new definition of the resistance.

2.2. *The conductance as a measure of electron transmission*

We now consider the one dimensional lead with a localized point scatter in the middle. A wave emitted by the left reservoir and incoming on the scatter gives a reflected and a transmitted outgoing wave. We expect that the incoming current e^2V/h is reduced. We show here that the current is proportional to the transmission.

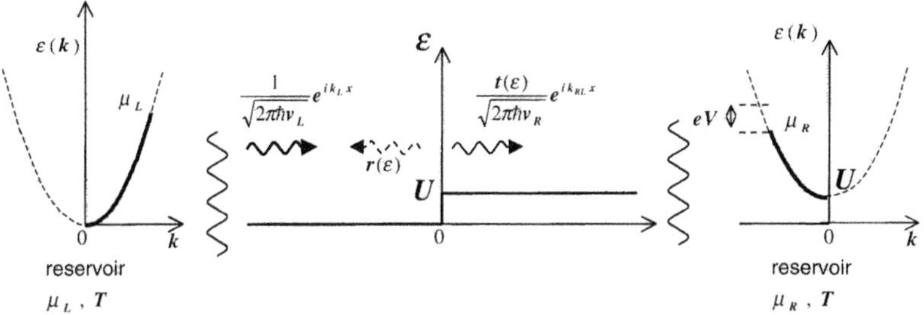

Fig. 5. 1D conductor with a point scatter at $x = 0$ induced by a potential discontinuity.

2.2.1. Scattering states

To introduce the definitions let us consider the following 1D conductor made of two distinct semi-infinite ideal leads connected at point $x = 0$ (see Fig. 5). The electron kinetic energy is $\varepsilon(k_L) = \frac{\hbar^2 k_L^2}{2m}$ in the left lead and $\varepsilon(k_R) = U + \frac{\hbar^2 k_R^2}{2m}$ in the right lead. An electron emitted at energy $\varepsilon \geq U$ by the left lead experiences a reduction of its velocity from $v_L(\varepsilon)$ to $v_R(\varepsilon)$ in the right region. This discontinuity induces a scattering of the incoming wave into two outgoing waves: a reflected wave in the left region and a transmitted wave in the right region. The wave incoming from the left and normalized with respect to energy is:

$$\varphi_L(\varepsilon; x) = \frac{1}{\sqrt{2\pi}} \frac{1}{\sqrt{\hbar v_L(\varepsilon)}} \left(e^{ik_L(\varepsilon)x} + r(\varepsilon) e^{-ik_L(\varepsilon)x} \right) \qquad x < 0 \tag{2.7}$$

$$= \frac{1}{\sqrt{2\pi}} \frac{1}{\sqrt{\hbar v_R}} t(\varepsilon) e^{i(k_R(\varepsilon)x} \qquad x > 0$$

Similarly a wave incoming from the right is:

$$\varphi_R(\varepsilon; x) = \frac{1}{\sqrt{2\pi}} \frac{1}{\sqrt{\hbar v_L(\varepsilon)}} t'(\varepsilon) e^{-ik_L(\varepsilon)x} \qquad x < 0 \tag{2.8}$$

$$= \frac{1}{\sqrt{2\pi}} \frac{1}{\sqrt{\hbar v_R(\varepsilon)}} \left(e^{-ik_R(\varepsilon)x} + r'(\varepsilon) e^{ik_R(\varepsilon)x} \right) \qquad x > 0$$

The scattering matrix S relates the amplitude of the outgoing waves to the amplitude of the incoming waves. From 2.7 and 2.8 we immediately get:

$$S = \begin{pmatrix} r(\varepsilon) & t'(\varepsilon) \\ t(\varepsilon) & r'(\varepsilon) \end{pmatrix}$$

The diagonal elements are the reflection amplitudes of the current, the off-diagonal elements the transmission amplitudes. Current conservation implies unitarity of S:

$$S^{\dagger}S = SS^{\dagger} = 1$$

This gives useful relations between reflection and transmission coefficients:

$$|r|^2 + |t|^2 = \left|r'\right|^2 + \left|t'\right|^2 = |r|^2 + \left|t'\right|^2 = \left|r'\right|^2 + |t|^2 = 1 \tag{2.9}$$

$$t'^{*}(\varepsilon)r(\varepsilon) + r'^{*}(\varepsilon)t(\varepsilon) = t^{*}(\varepsilon)r(\varepsilon) + r'^{*}(\varepsilon)t'(\varepsilon) = 0 \tag{2.10}$$

All these relations can be verified in the particular example. By matching the continuity of the wavefunction and its derivative, one finds: $t = t' = \frac{2\sqrt{\eta}}{1+\eta}$ and $r = -r' = \frac{1-\eta}{1+\eta}$ with $\eta = \sqrt{1 - U/\varepsilon}$. This particular example with *different* velocities between right and left leads show that the amplitudes of the reflected of transmitted waves $\frac{1}{\sqrt{2\pi}}\frac{1}{\sqrt{\hbar v_R}}t(\varepsilon)$ and $\frac{1}{\sqrt{2\pi}}\frac{1}{\sqrt{\hbar v_R}}r(\varepsilon)$ do not obey simple unitarity relations unlike the amplitude of the current $t(\varepsilon)$, $r(\varepsilon)$. Finally, using 2.10 and 2.9, it is straightforward to show that the wavefunctions 2.7 and 2.8 are orthogonal:

$$\left\langle \varphi_{L,R}(\varepsilon') \mid \varphi_{L,R}(\varepsilon) \right\rangle = \delta(\varepsilon' - \varepsilon) \quad \text{and} \quad \left\langle \varphi_L(\varepsilon') \mid \varphi_R(\varepsilon) \right\rangle = 0$$

To calculate the current, and later current noise, a rigorous way to incorporate the filling of the states by the reservoirs is to introduce the second quantized Fermions fields in presence of scattering:

$$\widehat{\Psi}(x,t) = \int d\varepsilon \left\{ \widehat{a}_L(\varepsilon)\varphi_L(\varepsilon) + \widehat{a}_R(\varepsilon)\varphi_R(\varepsilon) \right\} \tag{2.11}$$

where the $\widehat{a}_{L,R}$ act on the two separate Fock spaces which define the occupation number of the states emitted by the left or by the right reservoirs. If we call $\widehat{b}_{L,R}$ the operators associated with the outgoing sates:

$$\begin{pmatrix} \widehat{b}_L \\ \widehat{b}_R \end{pmatrix} = \begin{pmatrix} r & t' \\ t & r' \end{pmatrix} \begin{pmatrix} \widehat{a}_L \\ \widehat{a}_R \end{pmatrix}$$

the Fermion field operator takes the explicit form:

$$\widehat{\Psi}(x,t) = \int d\varepsilon \frac{1}{\sqrt{2\pi\hbar v_L(\varepsilon)}} \left\{ \widehat{a}_L(\varepsilon) e^{i(k_L x - \epsilon t)} + \widehat{b}_L(\varepsilon) e^{i(-k_L x - \epsilon t/\hbar)} \right\} \quad \text{x<0} \tag{2.12}$$

$$= \int d\varepsilon \frac{1}{\sqrt{2\pi\hbar v_L(\varepsilon)}} \left\{ \widehat{b}_R(\varepsilon) e^{i(k_R x - \epsilon t)} + \widehat{a}_R(\varepsilon) e^{i(-k_R x - \epsilon t/\hbar)} \right\} \quad \text{x>0} \tag{2.13}$$

The current operator calculated at $x < 0$ is:

$$\widehat{I}(x,t) = \frac{e}{h}\int d\varepsilon d\varepsilon' \left\{\widehat{a}_L^\dagger(\varepsilon')\widehat{a}_L(\varepsilon) - \widehat{b}_L^\dagger(\varepsilon')\widehat{b}_L(\varepsilon)\right\}$$

$$\times \frac{v(\varepsilon)+v(\varepsilon')}{2\sqrt{v(\varepsilon)}\sqrt{v(\varepsilon')}} e^{i(k_L(\varepsilon)-k_L(\varepsilon'))x} e^{i(\varepsilon'-\varepsilon)t/\hbar} \tag{2.14}$$

$$\simeq \frac{e}{h}\int d\varepsilon d\varepsilon' \left\{\widehat{a}_L^\dagger(\varepsilon')\widehat{a}_L(\varepsilon) - \widehat{b}_L^\dagger(\varepsilon')\widehat{b}_L(\varepsilon)\right\} e^{i(\varepsilon'-\varepsilon)(t-x/v_F)/\hbar} \tag{2.15}$$

Similarly for $x > 0$:

$$\widehat{I}(x,t) \simeq -\frac{e}{h}\int d\varepsilon d\varepsilon' \left\{a_R^\dagger(\varepsilon')a_R(\varepsilon) - b_R^\dagger(\varepsilon')b_R(\varepsilon)\right\} e^{i(\varepsilon'-\varepsilon)(t+x/v_F)/\hbar} \tag{2.16}$$

In the last two expressions we neglect the energy dependence of the velocity. This is an excellent approximation for most experiments where the energy range probed $|\varepsilon' - \varepsilon|$ is small compared with the Fermi energy E_F.

2.2.2. The Landauer formula

The expectation value of the current calculated on the right side is:

$$\begin{aligned} I &= \frac{e}{h}\int d\varepsilon \left\{\left\langle a_R^\dagger(\varepsilon)a_R(\varepsilon)\right\rangle - \left\langle b_R^\dagger(\varepsilon)b_R(\varepsilon)\right\rangle\right\} \\ &= \frac{e}{h}\int d\varepsilon \left\{-f_R(\varepsilon) + \left(f_L(\varepsilon)|t|^2 + f_R(\varepsilon)|r'|^2\right)\right\} \end{aligned}$$

Using 2.9 we find the important relation:

$$I = e/h \int d\varepsilon\, (f_L(\varepsilon) - f_R(\varepsilon))\, |t(\varepsilon)|^2 \tag{2.17}$$

A calculation of the current on the right side gives the same result as expected from current conservation. For vanishing bias voltage V we obtain the Landauer formula of the conductance:

$$G = e^2/h \int d\varepsilon D(\varepsilon)\partial f(\varepsilon)/\partial\varepsilon \equiv \frac{e^2}{h} D(\varepsilon = E_F) \quad T = 0$$

where we have defined the transmission probability $D = |t|^2 = 1 - R$ and $f(\varepsilon) = f_{L,R}$ when $eV \to 0$ ($\mu_{R,L} = E_F$). The conductance *measures the transmission probability of the electron waves.*

The analogy between the intensity of the current in a conductor and the intensity of a light beam scattered by an optical medium is straightforward and is

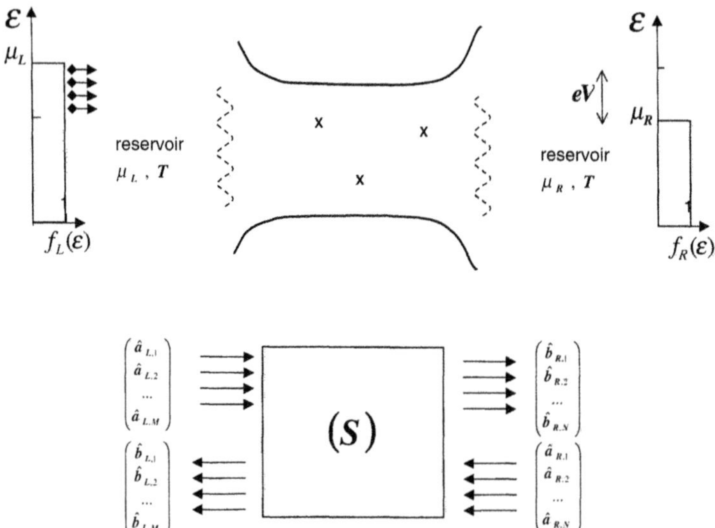

Fig. 6. Multimode quantum conductor.

often useful to get intuition on electron quantum transport. For both cases, the intensity is proportional to the transmission. Here, the spectral width of the photon source $\delta\nu$ is replaced by eV/h at $T = 0$ or k_BT/h at $V = 0$. If $D(\varepsilon)$ contains an oscillatory part resulting from electron wave interferences, the oscillations are smeared at large voltage or temperature. Similarly, Young fringes are smeared by a non monochromatic light source.

2.3. *The two-terminal Landauer formula*

2.3.1. *Modelling of the reservoirs*

Fig.6 shows a quantum conductor connected to two reservoirs. The conductor has a length much shorter than the coherence length. The contacts are wider than the quantum region such that an electron entering a reservoir is diluted and neither come back. This realizes the assumption of perfectly absorbing reservoirs. Also, the internal resistance of the wide contact is negligible with respect to that of the conductor. Assuming that the specific nature of the reservoirs is not critical in the determination of the conductance, we can replace the reservoirs by ideal leads. The length of each ideal lead is infinite in order to ensures irreversibility. The ideal leads are connected at infinity to a thermodynamic bath with chemical potential $\mu_{L,R}$ and temperature T.

The transverse quantization in the left lead of width W_L gives rise to a set of eigenmodes labelled by the integer m. A right moving wave, mode m, with unit amplitude at energy $\varepsilon = \frac{\hbar^2}{2m^*}\left[\left(\frac{m\pi}{w_L}\right)^2 + k_{L,m}(\varepsilon)^2\right]$ and incoming on the scattering region is given by

$$\chi_{L,m}(y_L)\frac{1}{\sqrt{2\pi}}\frac{1}{\sqrt{\hbar v_{L,m}(\varepsilon)}}e^{ik_{L,m}(\varepsilon)x_L} \tag{2.18}$$

where $\chi_{L,m}(y) = \left(\frac{2}{W_L}\right)^{1/2}\sin\left(\frac{\pi m}{W_L}y\right)$ and $v_{L,m}(\varepsilon) = \partial\varepsilon/\partial\hbar k_{L,m}$. Here only propagating modes with $\frac{\hbar^2}{2m^*}\left(\frac{m\pi}{w_L}\right)^2 < \varepsilon$ are considered. Evanescent waves in the leads are excluded.

2.3.2. *Multimode scattering states*

For simplicity, we will consider the conductor and the leads as 2D conductors (mode labelling for 3D conductors give more cumbersome expressions and generalization is straightforward).

The incoming wave emitted by the left reservoir can be reflected by the scattering region into outgoing modes m' in the left lead:

$$\sum_{m'} s_{LL,m'm}\chi_{L,m'}(y_L)\frac{1}{\sqrt{2\pi}}\frac{1}{\sqrt{\hbar v_{L,m'}(\varepsilon)}}e^{-ik_{L,m'}(\varepsilon)x_L} \tag{2.19}$$

or transmitted into outgoing modes n' in the right lead

$$\sum_{n'} s_{RL,n'm}\chi_{L,n'}(y_R)\frac{1}{\sqrt{2\pi}}\frac{1}{\sqrt{\hbar v_{R,n'}(\varepsilon)}}e^{-ik_{L,n'}(\varepsilon)x_R} \tag{2.20}$$

where $\chi_{R,n'}(y) = \left(\frac{2}{W_R}\right)^{1/2}\sin\left(\frac{\pi n'}{W_R}y_R\right)$, $v_{R,n'}(\varepsilon) = \partial\varepsilon/\partial\hbar k_{R,n'}$ and $\varepsilon = \frac{\hbar^2}{2m^*}\left[\left(\frac{n'\pi}{w_R}\right)^2 + k_{R,n'}(\varepsilon)^2\right]$.

The combination of 2.18, 2.19 and 2.20 gives the wave function $\varphi_{L,m}(\varepsilon)$ as in 2.7. Note the sign in the argument of the exponentials according to the choice of the axis x_α for lead $\alpha = L, R$.

Similarly a left moving wave of type n emitted by the right reservoir with unit amplitude will be scattered into reflected and transmitted outgoing waves with amplitude $s_{RR,n''n}$ and $s_{LR,m''n}$ respectively. This defines the wavefunctions $\varphi_{R,n}(\varepsilon)$ as in 2.11.

$$\widehat{\Psi}(x,t) = \int d\varepsilon\left\{\sum_{m=1}^{M(\varepsilon)}\widehat{a}_{L,m}(\varepsilon)\varphi_{L,m}(\varepsilon) + \sum_{n=1}^{N(\varepsilon)}\widehat{a}_{R,n}(\varepsilon)\varphi_{R,n}(\varepsilon)\right\}$$

where $M(\varepsilon) = Int\left(\frac{W_L}{\pi}\sqrt{\frac{2m^*\varepsilon}{\hbar^2}}\right)$ (and similarly $N(\varepsilon)$) denotes the highest index of the modes propagating in the leads at energy ε and

$$\begin{aligned}
\left\{\widehat{a}_{\alpha,k}(\varepsilon), \widehat{a}^{\dagger}_{\beta,l}(\varepsilon')\right\} &= \delta_{\alpha,\beta}\delta_{k,l}\delta(\varepsilon-\varepsilon') \;; \\
\left\langle\widehat{a}^{\dagger}_{\alpha,k}(\varepsilon')\widehat{a}_{\beta,l}(\varepsilon)\right\rangle &= f_{\alpha}(\varepsilon)\delta_{\alpha,\beta}\delta_{k,l}\delta(\varepsilon-\varepsilon')
\end{aligned}$$

The Fermion operators $\widehat{b}_{\alpha,k}(\varepsilon)$ for the outgoing waves can be defined as a linear superposition of the $\widehat{a}_{\alpha,k}(\varepsilon)$ via the multi-mode scattering matrix $S(\varepsilon)$:

$$\begin{pmatrix} \widehat{b}_{L,1} \\ \cdots \\ \widehat{b}_{L,M(\varepsilon)} \\ \widehat{b}_{R,1} \\ \cdots \\ \widehat{b}_{R,N(\varepsilon)} \end{pmatrix} = \left(\begin{array}{c|c} (S_{L,L}(\varepsilon)) & (S_{L,R}(\varepsilon)) \\ \hline (S_{R,L}(\varepsilon)) & (S_{R,R}(\varepsilon)) \end{array}\right) \begin{pmatrix} \widehat{a}_{L,1} \\ \cdots \\ \widehat{a}_{L,M(\varepsilon)} \\ \widehat{a}_{R,1} \\ \cdots \\ \widehat{a}_{R,N(\varepsilon)} \end{pmatrix}$$

where the $M \times N$ matrix $S_{R,L}(\varepsilon) = \left(s_{RL,n'm}\right)$ and so on. Although the expressions may look complicated, the signification is very simple: $s_{\beta\alpha,lk}$ is the scattering amplitude of a wave emitted by lead α in mode k and absorbed in lead β in mode l.

2.3.3. The multimode Landauer formula

The current at point x in a lead $\alpha = 1, 2$ is given by the current density integrated over the transverse width w_α. For a given energy, the current in lead α is given by the sum of probabilities $\left|s_{\alpha\beta,lk}(\varepsilon)\right|^2$ with the right sign and with the weight $f_\beta(\varepsilon)$. The average value of the current operator calculated in the right lead is:

$$I = e/h\int d\varepsilon \sum_n \left\{\left\langle a^{\dagger}_{R,n}(\varepsilon)a_{R,n}(\varepsilon)\right\rangle - \left\langle b^{\dagger}_{R,n}(\varepsilon)b_{R,n}(\varepsilon)\right\rangle\right\} \tag{2.21}$$

$$\begin{aligned}
&= e/h\int d\varepsilon \sum_n \Bigg\{-f_R(\varepsilon) + \Bigg(f_L(\varepsilon)\sum_m \left|s_{RL,nm}\right|^2 \\
&\quad + \sum_{n'} f_R(\varepsilon)\left|s_{RR,n'n}\right|^2\Bigg)\Bigg\}
\end{aligned} \tag{2.22}$$

Current conservation implies the unitarity of the S matrix : $S^{\dagger}S = SS^{\dagger} = \mathbf{1}$. This gives useful relations analogous to relations 2.9 and 2.10 which can be used to simplify expressions of the current. For example $SS^{\dagger} = \mathbf{1}$ gives the relation:

$$S_{RL}S^{\dagger}_{RL} + S_{RR}S^{\dagger}_{RR} = \mathbf{1}$$

and taking the trace leads to a simple expression of the current:

$$I = e/h \int d\varepsilon \left(f_L(\varepsilon) - f_R(\varepsilon)\right) \sum_{m,n} \left|s_{RL,nm}(\varepsilon)\right|^2 \tag{2.23}$$

$$= \frac{e}{h} \int d\varepsilon \left(f_L(\varepsilon) - f_R(\varepsilon)\right) \mathrm{Tr}\left(S_{RL} S_{RL}^{\dagger}\right) \tag{2.24}$$

$$= \frac{e}{h} \int d\varepsilon \left(f_L(\varepsilon) - f_R(\varepsilon)\right) \sum_{\lambda} D_{\lambda}(\varepsilon) \tag{2.25}$$

where the transmission probabilities $D_{\lambda}(\varepsilon)$ are the eigenvalues of the transmission matrix $S_{RL} S_{RL}^{\dagger}$.

The three expressions of the current are equally used in the literature. In the last expression, when transmission matrix S_{RL} is nearly diagonal (i.e. the scattering region do not mixes the modes) the D_{λ} identifies to physical eigenmodes of the quantum conductor. This is the case for ballistic conductors such as Quantum Point Contacts realized in 2D systems or Atomic Point Contacts realized with metallic break junctions. When the conductor is diffusive (many random scatters) or ballistic but chaotic (i.e. the equivalent classical trajectories would show deterministic chaos) the random matrix theory predicts particular forms of the probability distribution $P(\{D_{\lambda}\})$ of the transmissions for each case and for various fundamental symmetries.

Finally, the conductance is obtained for $eV = \mu_R - \mu_L \rightarrow 0$ and the zero temperature expressions are :

$$G = \frac{e^2}{h} \sum_{\lambda} D_{\lambda}(E_F) = \frac{e^2}{h} \mathrm{Tr}\left(S_{RL} S_{RL}^{\dagger}\right)_{E_F} = \frac{e^2}{h} \sum_{m,n} \left|s_{RL,nm}(E_F)\right|^2 \tag{2.26}$$

These expressions are various forms of the Landauer formula.

2.4. The Landauer formula at work

2.4.1. The quantum point contact (QPC)

A quantum point contact (QPC) is a constriction of size comparable to the Fermi wavelength of the electrons. Although point contacts have been used in 3D metals in order to understand properties of the Fermi surface, its 2D realization in the mid-80's provided direct support to the Landauer Formula [22]. 2D electron gas confined at the interface of two semiconductors has provided the physicists with very clean 2D metals with electron mean-free paths larger than 10μm. The very dilute electrons (density $n_s = 10^{15}\mathrm{m}^{-2}$) have a Fermi wavelength $\simeq$ 800Ä much larger than that of metals and comparable to electron beam lithography resolution. The planar geometry allows to evaporate gates on the sample surface.

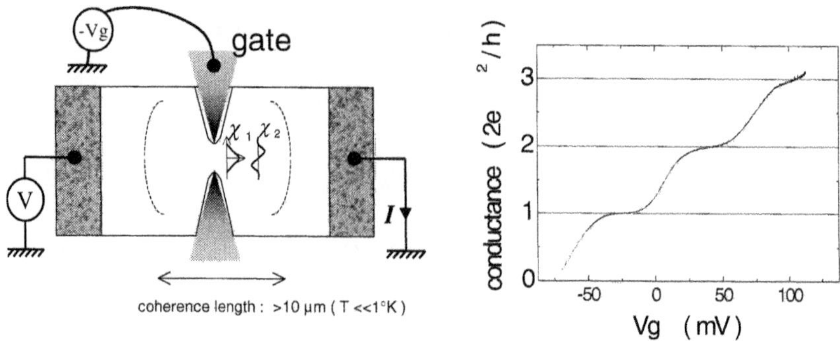

Fig. 7. Quantum point contact in a 2D electron gas. Left: metallic gates depositied above the 2D electron gas are used to create a constriction comparable to the Fermi wavelength. Right: the conductance versus gate voltage shows quantized plateaus each time a new electronic mode is fully transmitted.

With suitable geometry (see fig.7) the field effect of negatively voltage biased gates can be used to deplete electrons and to vary at will the width of a narrow undepleted regions. This realizes a QPC connecting two wide semi-infinite 2D regions. If we consider the QPC as a short waveguide, the ratio of the width to the Fermi wavelength controls the number of transmitted modes (this is similar to electromagnetic waveguides). Increasing the width from zero will allow to go from the tunnelling regime (all $D_\lambda \ll 1$, $G \ll 2\frac{e^2}{h}$) to single mode transmission regime ($D_1 \simeq 1$ and $D_{\lambda>1} \ll 1$, $G \simeq 2\frac{e^2}{h}$) and so on. The conductance shows plateaus quantized in multiple of $2\frac{e^2}{h}$ when varying the gate voltage acting on the width. Here a factor 2 appears because spin degeneracy in zero magnetic field doubles the occupancy of each orbital mode.

If we approximate the electrostatic potential at the center of the constriction in the form of a saddle potential:

$$U(x, y) = U_0 - \frac{1}{2}m^*\Omega_x^2 x^2 + \frac{1}{2}m^*\Omega_y^2 y^2$$

the transmission probabilities are given by: $D_\lambda(\varepsilon) = \frac{1}{1+\exp\left(-2\pi\frac{(\varepsilon - U_0 - \hbar\Omega_y(\lambda+1/2))}{\hbar\Omega_x}\right)}$

U_0 is controlled by the gate potential, Ω_x and Ω_y are characteristic frequencies related to the negative and positive curvature of the potential. The smaller Ω_x (i.e. the longer the QPC) the sharper are the plateaus.

2.4.2. *The mesoscopic quantum Hall effect*

In the following parts of the lecture, we will provide several examples where the chiral edge states of a 2D electron gas in quantizing magnetic field (integer quantum Hall regime) are used to realize nice electron analog of quantum optical systems. Here we briefly introduce the physics of electrons in this regime and show the relation between the quantization of the Hall conductance and the Landauer formula. We will consider a mesoscopic system and will not discuss the physics of the quantum Hall effect which exists in macroscopic samples.

The Hamiltonian of free electrons in magnetic field $B\vec{z}$ and confined in a strip in the plane (x, y) along the direction x by the potential $U(y)$ is:

$$H = \frac{1}{2m^*}\left(\vec{p} + e\vec{A}\right)^2 + U(y)$$

The symmetry of the problem suggests the choice of the Landau gauge for the vector potential $\vec{A} = (-yB, 0, 0)$. The x component of the momentum is a good quantum number $p_x = \hbar k$ and the wavefunction writes :$\sim e^{ikx}\phi(y)$. The hamiltonian now acting only on $\phi(y)$ is:

$$H = \frac{p_y^2}{2m^*} + \frac{1}{2}m^*\omega_C^2\left(y - kl_C^2\right)^2 + U(y)$$

where we have introduced the cyclotron frequency $\omega_C = eB/m^*$and the cyclotron length $l_C = (\hbar/eB)^{\frac{1}{2}}$. Well away from the edges, in the center of the strip, the confining potential $U(y)$ is zero. The wavefunctions are those of a harmonic oscillator centered in $y_k = kl_C^2$ and the energies correspond to Landau levels : $(n + \frac{1}{2})\hbar\omega_C$ (n =0, 1, ...). The energies do not depend on k as long as $U = 0$. If L is the length of the strip, periodic boundary conditions allow to estimate the degeneracy. The distance between two consecutives y_k is $\frac{2\pi}{L}\frac{\hbar}{eB}$. The area separating 2 states is $\frac{h}{eB}$: the area of a flux quantum Φ_0. Thus the *degeneracy* for a given Landau level is equal to the *number of flux quanta in the plane*.

Edge states form in the region where U is rising (see Fig.8). Assuming smooth variation over the scale l_C such that no different Landau levels are coupled, the energies can be obtained by replacing y by y_k in U :

$$\varepsilon_n \simeq (n + \frac{1}{2})\hbar\omega_C + U(y_k)$$

The velocity is only along the x direction and is given by:

$$v_x^{(n)}(k) = \frac{1}{\hbar}\frac{\partial\varepsilon_n}{\partial k} = \frac{1}{eB}\frac{\partial U}{\partial y}$$

This expresses that the Lorentz force compensates the confining electric field resulting in a drift of the electrons. Electrons move along equipotential lines and in opposite direction for opposite edges. Simultaneously the Landau levels are

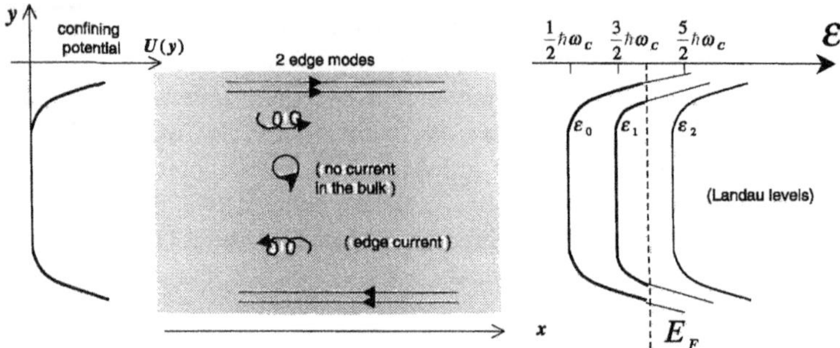

Fig. 8. 2D electrons in high magnetic field. At the edges, the confining potential induces permanent chiral currents. If p Landau levels are filled (here $p = 2$), p edge channels form, each associated with a quantized conductance.

bent by U. The crossing of each filled Landau level with the Fermi energy defines a line with gapless excitations: a 1D chiral edge mode. This is where the net edge current can be modified and conduction occurs. The number of chiral modes is equal to the number of filled Landau levels in the bulk. If $\nu = p$ Landau levels are filled, the Landauer formula gives $G = p\frac{e^2}{h}$

2.4.3. Interference using the edge states of the quantum Hall effect

Experiments showing electron interference effects have given strong support to the idea that conductance measures the square of the amplitude probability and that phase coherent conductors are realizable. The most famous are based on the Aharonov-Bohm effect. Here we present a more modern experiment. Fig.9 shows the electronic analog of the optical Mach-Zehnder interferometer from Ref. [23]. The chiral edge channels of a 2D electron gas in quantizing magnetic field (integer quantum Hall regime) are used as ideal 1D conductor. A contact placed at potential eV with respect to the other is used as the electron source. Two contacts D1 and D2 absorb the current.

The electron analog of semi-transparent mirrors is provided by two QPCs. According to the Landauer formula the current measures the transmission. The current at contact D1 (the 'detector') is proportional to $\left|t_1 t_2 + r_1 r_2 e^{i2\pi\frac{\Phi}{\Phi_0}}\right|^2$ where $T_{1,2} = \left|t_{1,2}\right|^2 = 1 - \left|r_{1,2}\right|^2$ are the transmission of the QPCs. The magnetic flux Φ is the flux enclose by the two trajectories which separate at QPC1 and recombine at QPC2. For half transmission the fringe visibility is maximum and reaches $\simeq 60\%$ for an electron temperature of 20mK (the area of the loop is $45\mu m^2$). The advantage of the Mach-Zehnder over the electronic Fabry-Pérot

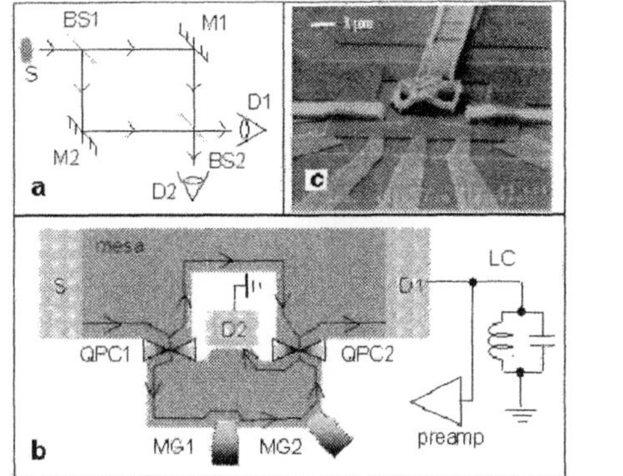

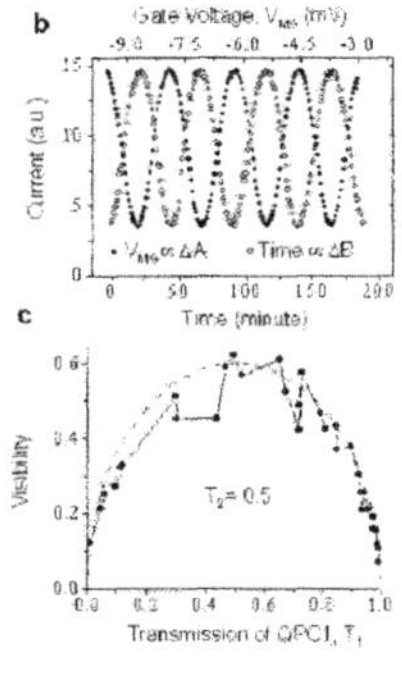

Fig. 9. Electron Mach-Zehnder interferometer after Yang Ji *et al*[23]

resonator below is that thermal smearing is reduced at a minimum for equal interfering path length.

2.4.4. *Resonant tunnelling*

For simplicity will we consider a 1D quantum dot coupled to the left and right reservoirs by barriers of transmission $T_{L,R} = \left|t_{L,R}\right|^2 = 1 - R_{L,R} \ll 1$. This can be realized using the quantum dot of fig.3 connected to leads by QPCs of transmission $T_{R,L}$ and in the Quantum Hall effect edge states regime. The phase accumulated by an electron inside the quantum dot of length l is ϕ. The amplitude of the transmission from L to R is:

$$s_{LR} = t_L e^{i\phi} t_R + t_L e^{i\phi} r_R e^{i\phi} r_L e^{i\phi} t_R + t_L e^{i\phi}\left(r_R e^{i\phi} r_L e^{i\phi}\right)^2 t_R + ...$$

$$|s_{LR}|^2 = \frac{T_L T_R}{(1 + R_L R_R - 2\sqrt{R_L R_R}\cos 2\phi)}$$

Here we have used the reflection amplitudes $r_{L,R} = \sqrt{R_{L,R}}$.

A peak occurs in the conductance $G = \frac{e^2}{h}|s_{LR}|^2$ each time $\phi = kl = n\pi$. The resonant condition corresponds to energies $\varepsilon = \frac{\hbar^2 k^2}{2m^*} = \varepsilon_n$ where the ε_n are the energy level expected for an isolated quantum dot and allows for a conductance spectroscopy of the dot. Remarkably, when $R_L = R_R$ the transmission is unity and $G_{peak} = \frac{e^2}{h}$. When $T_{L,R} \ll 1$, the conductance peak is

$$G_{peak} \simeq 4\frac{T_L T_R}{(T_L + T_R)^2}.$$

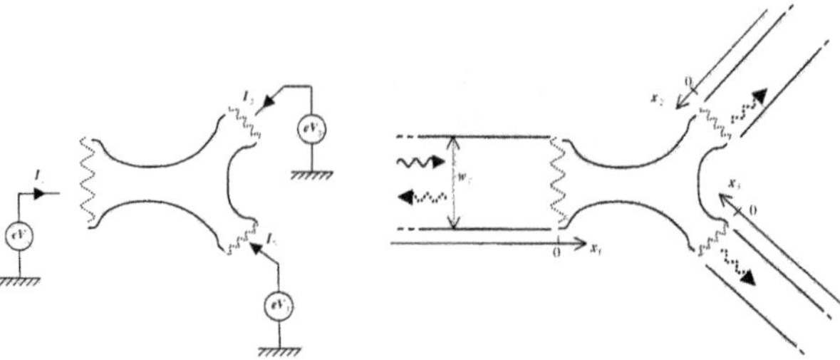

Fig. 10. Multiterminal conductor.

This example is the electron analog of the Fabry-Pérot resonator in optics. Here, we have considered for simplicity 1D resonant tunnelling. The results also apply to micron size quantum dots realized using 2D electron systems or to a 3D quantum dots made of a nm size metallic grain. For weak tunnel coupling to external leads, resonant transmission conductance peaks develop each time the Fermi energy of the leads is in the vicinity of an energy level of the dot (Coulomb blockade effects will not be discussed here). The resonance is also of the Breit-Wigner form:

$$G_{peak} \simeq 4\frac{T_L T_R}{(T_L + T_R)^2}\frac{1}{1 + \left(\frac{\varepsilon - \varepsilon_n}{\hbar\Gamma}\right)^2}$$

where $\hbar\Gamma \simeq \Delta(T_R + T_L)$ is the intrinsic quantum level width in absence of decoherence and Δ is the typical level spacing.

2.5. *Multiterminal conductors: Landauer-Büttiker formula*

Multiterminal conductors will be used in part III when considering current-current correlations as in the case of Hanbury-Brown Twiss experiments with electrons. At scale shorter than the coherence length, local quantities such as conductivity or resistivity are meaningless as the quantum system has to be considered as a whole. Ohm's addition law of resistances fails. In a network made of several leads connected to different reservoirs one have to find what are the quantum Kirchhof laws relating currents to potentials. Fig.10 shows a multibranch quantum conductor connected to three reservoirs $\alpha = 1$ to 3. Each reservoirs in equilibrium at temperature T is at the chemical potential $\mu_\alpha = E_F + eV_\alpha$. I_α is the current which enters in the contact α. Each contact α is replaced by an ideal lead of witdth w_α. The number of modes at energy ε in lead α is M_α. The x-axis of each ideal lead is chosen pointing toward the scattering region.

The procedure is the same as the one followed to derive the multimode two-terminal Landauer formula. The scattering matrix is now made of 3 × 3 block matrices:

$$S = \begin{pmatrix} S_{11} & S_{12} & S_{13} \\ S_{21} & S_{22} & S_{23} \\ S_{31} & S_{32} & S_{33} \end{pmatrix}$$

where the $M_\alpha \times M_\alpha$ diagonal sub-matrix $S_{\alpha\alpha} = (s_{\alpha\alpha,m'm})$ describes the reflection amplitudes (mode m to mode m') in lead α and the $M_\beta \times M_\alpha$ off-diagonal sub-matrix $S_{\beta\alpha} = (s_{\beta\alpha,n'm})$ describes the transmission amplitudes of mode m in lead α to mode n' in lead β.

2.5.1. The quantum Kirchhof law

Calculation of the current in reservoir α for a given energy is obtained by summing the probabilities $\left|s_{\beta\alpha,lk}(\varepsilon)\right|^2$ with the right sign and with the weight $f_\alpha(\varepsilon)$. The sign of the current is taken positive for the contribution of incoming waves and negative for outgoing waves. The average value of the current operator calculated in the right lead is:

$$I_\alpha = \frac{e}{h}\int d\varepsilon \sum_{m=1}^{M_\alpha} \left\{ \left\langle a^\dagger_{\alpha,m}(\varepsilon) a_{\alpha,m}(\varepsilon)\right\rangle - \left\langle b^\dagger_{\alpha,m}(\varepsilon) b_{\alpha,m}(\varepsilon)\right\rangle \right\} \quad (2.27)$$

$$= \frac{e}{h}\int d\varepsilon \sum_{m=1}^{M_\alpha} \left\{ f_\alpha(\varepsilon) - \left(f_\beta(\varepsilon) \sum_{n=1}^{M_\beta} \left|s_{\alpha\beta,mn}\right|^2 + f_\alpha(\varepsilon) \sum_{m'=1}^{M_\alpha} \left|s_{\alpha\alpha,m'm}\right|^2 \right) \right\} \quad (2.28)$$

It is convenient to define the total reflections $R_{\alpha\alpha}$ =Tr$(S_{\alpha\alpha} S^\dagger_{\alpha\alpha})$ and transmissions $T_{\alpha\beta}$ =Tr$(S_{\alpha\beta} S^\dagger_{\alpha\beta})$ to simplify the current expression:

$$I_\alpha = \frac{e}{h}\int d\varepsilon \left\{ f_\alpha(\varepsilon)\left(M_\alpha - R_\alpha\alpha\right) - \sum_\beta T_\alpha \beta f_\beta(\varepsilon) \right\}$$

At zero temperature an even more compact form can be obtained by defining:

$$G_{\alpha a} = \frac{e^2}{h}\mathrm{Tr}(\mathbf{1}_\alpha - S_{\alpha\alpha} S^\dagger_{\alpha\alpha}) = \frac{e^2}{h}\left(M_\alpha - R_{\alpha\alpha}\right) \quad (2.29)$$

$$G_{\alpha\beta} = -\frac{e^2}{h}\mathrm{Tr}(S_{\alpha\beta} S^\dagger_{\alpha\beta}) = -\frac{e^2}{h} T_{\alpha\beta} \quad (2.30)$$

This gives the quantum Kirchhoff law for the node α of the circuit:

$$I_\alpha = \sum_\beta G_{\alpha\beta} V_\beta \tag{2.31}$$

3. Electronic quantum noise

3.1. *Electrons versus photons: importance of the statistics*

In the previous chapter we have emphasized the similarities between the intensity of the current in a quantum conductor and the intensity of a light beam in an optical medium. Expressions of the time dependent current operator are similar:

$$I = e/h \int d\varepsilon \left\{ \left\langle \widehat{a}^\dagger(\varepsilon)\widehat{a}(\varepsilon) \right\rangle - \left\langle \widehat{b}^\dagger(\varepsilon)\widehat{b}(\varepsilon) \right\rangle \right\}$$

for electrons, $\widehat{a}, \widehat{b}$ fermion operators, and for photons:

$$I = \frac{1}{h} \int \varepsilon d\varepsilon \left\{ \left\langle \widehat{a}^\dagger(\varepsilon)\widehat{a}(\varepsilon) \right\rangle - \left\langle \widehat{b}^\dagger(\varepsilon)\widehat{b}(\varepsilon) \right\rangle \right\}$$

$\widehat{a}, \widehat{b}$ boson operators, and $\varepsilon = h\nu$.

Quantitative differences however exist due to the quantum statistics. The field of photon can be very intense as one can put several photons in the same wavefunction. This makes observations of interferences easy. The single occupancy of electron wavepackets limits the range of current to very small values. At low temperature, single occupancy, gives rise to a remarkable regular flow of single electrons at a rate eV/h. Single photon sources are not naturally available and difficult to realize. This is a good motivation to study electron quantum noise. Understanding light beam correlations have led to very fundamental concepts such as entanglement, squeezing and quantum non-demolition measurements. The emergence of electron quantum noise in mesoscopic physics about ten years ago have stimulated many new type of experiments and theoretical proposals in a parallel way.

The Fermi-Dirac and Bose-Einstein quantum statistics contribute very differently to fluctuations. For bosons, the thermal fluctuations of the number of particles of a quantum state with average population $\langle N \rangle$ has a variance $\left\langle (\Delta N)^2 \right\rangle = f(1+f) = \langle N \rangle (1 + \langle N \rangle)$ characteristic of a super-Poissonian statistics. Here, $f = \langle N \rangle = \frac{1}{\exp \frac{h\nu}{k_B T} - 1}$ is the Bose-Einstein distribution which can take large values $\simeq \frac{k_B T}{h\nu}$ when $k_B T > h\nu$. For fermions the variance of the fluctuations of the particle number n of a quantum sate is sub-Poissonian: $\left\langle (\Delta n)^2 \right\rangle = f(1 - f) =$

$\langle n\rangle(1-\langle n\rangle)$, where $f=\langle n\rangle=\frac{1}{\exp\frac{\varepsilon-\mu}{k_BT}+1}$. The differences between the two quantum statistics are important not only in the fluctuation of the population. If one considers quantum states with a well defined number of particle, the quantum exchange between particle gives very different results for fermion and bosons. Consider the semi-transparent mirror were two sources (a) and (b) each inject a single particle at the same time. The probability to have i transmitted and j reflected particles detected at the two outputs is $P(i,j)$. D and $R=1-D$ are the single particle transmission and reflection probabilities. For classical particles one have: $P(2,0)=DR$, $P(0,2)=DR$, and $P(1,1)=D^2+R^2$. Performing a series of identical experiments gives for the number $N_{t(r)}$ of transmitted (reflected) particles the following average values are: $\langle N_t\rangle=\langle N_r\rangle=1$ and the variance and cross-correlations of the fluctuations are::

$$\left\langle(\Delta N_t)^2\right\rangle=\left\langle(\Delta N_r)^2\right\rangle=2RD=-\langle\Delta N_t\Delta N_r\rangle$$

Consider now quantum particles. As in the previous chapter, it is convenient to built annihilation operators $\widehat{b}_t$ and $\widehat{b}_r$ which will be used to measure the results in the outgoing states from the operators $\widehat{a}_a$ and $\widehat{a}_b$ via the scattering matrix:

$$\begin{pmatrix}\widehat{b}_t\\ \widehat{b}_r\end{pmatrix}=\begin{pmatrix}ir & t\\ t & ir\end{pmatrix}\begin{pmatrix}\widehat{a}_a\\ \widehat{a}_b\end{pmatrix}$$

where $r=\sqrt{R}$ and $t=\sqrt{D}$. With these definitions: $P(2,0)=\langle\Psi|\widehat{b}_t^{\dagger}\widehat{b}_t\widehat{b}_t^{\dagger}\widehat{b}_t|\Psi\rangle$, $P(0,2)=\langle\Psi|\widehat{b}_r^{\dagger}\widehat{b}_r\widehat{b}_r^{\dagger}\widehat{b}_r|\Psi\rangle$ are both equal to 0 for fermions and to $2RD$ for bosons. $P(1,1)=\langle\Psi|\widehat{b}_r^{\dagger}\widehat{b}_r\widehat{b}_t^{\dagger}\widehat{b}_t|\Psi\rangle$ is 1 for fermions and $(R-D)^2$ for bosons. The particle number fluctuations for a series of experiments are:

$$\left\langle(\Delta N_t)^2\right\rangle=\left\langle(\Delta N_r)^2\right\rangle=-\langle\Delta N_t\Delta N_r\rangle=\left\{\begin{array}{l}0\ \text{fermions}\\ 4RD\ \text{bosons}\end{array}\right.$$

In the last expression, fermions give zero cross-correlation because of the Pauli principle. For bosons the correlation is negative because of bunching.

3.2. *Quantum shot noise*

3.2.1. *Schottky formula*

Electrical shot noise associated with poissonian transfer of charges between two conductors has been derived by W. Schottky in 1918. He considered the thermionic emission of a vacuum diode but the result is also valid for tunnel barrier with weak transmission. Consider a series of experiments in which the number of charges N is measured during a time $\tau\gg e/I$. The average current is given by

$$I=e\langle N\rangle/\tau$$

and the fluctuations of the current is

$$\left\langle (\Delta I)^2 \right\rangle = e^2 \left\langle (\Delta N)^2 \right\rangle / \tau^2 = 2eI\Delta f$$

were we have used $\left\langle (\Delta N)^2 \right\rangle = \langle N \rangle$ and the effective bandwidth Δf of the measurement for a time slot τ. The low frequency power spectrum of the current fluctuations $S_I = \left\langle (\Delta I)^2 \right\rangle / \Delta f$ is white and given by the Schottky formula:

$$S_I = 2eI \tag{3.1}$$

3.2.2. The quantum partition noise of a single mode quantum conductor
We consider a 1D conductor with a scattering region defined by a transmission D. For example, this can be a Quantum Point Contact for which the first mode contribute to the transmission while higher modes have a vanishing transmission. At zero temperature if the left reservoir is at a chemical potential eV higher than the chemical potential of the right reservoir, there is a net current of right moving incoming electrons $I_0 = e(eV/h)$. The remarkable rate eV/h of arrival of the electrons is acompanied by another remarkable property: the incoming charges are time correlated. This can be easily shown by building a complete orthogonal wavepackets basis from a superposition of incoming plane waves whose energy span in the range eV [6]. The wavepackets are found separated in time by h/eV. Pauli principle requires occupying each of them by only one electron, resulting in a *noiseless* flow of charges e. The transmitted current is $I = DI_0 = D\frac{e^2}{h}V$. The only source of current noise is the binomial partition noise of electrons being either transmitted with probability D or reflected with probability $1 - D$. For $N_0 = eV\tau/h$ incoming electrons, the variance of transmitted electrons is $\left\langle (\Delta N)^2 \right\rangle = N_0 D(1 - D)$ and the current noise power becomes:

$$S_I = F.2eI \tag{3.2}$$

Here $F = 1 - D$ is the Fano factor (a similar definition is used in optics). It is always lower than 1 (sub-Poissonian noise) and even can vanish. A unit transmission conductor is noiseless.

3.2.3. Derivation of the shot noise formula for a single mode conductor
We will consider the following spectral density of the current fluctuations:

$$S(\omega) = 2 \int dt \left\langle \widehat{I}(0)\widehat{I}(t) \right\rangle e^{i\omega t}$$

To simplify the expressions, we make the non-restrictive assumption that $\hbar\omega \ll E_F$ and neglect the energy dependence of the velocities and spatial variations of the current and use the expression 2.15. Calculating the current fluctuation at the left side, after integration of time, the expression contains terms like : $\frac{e^2}{h}$ $\int d\varepsilon''' d\varepsilon'' d\varepsilon \left\langle \widehat{a}_i^\dagger(\varepsilon''')\widehat{a}_j(\varepsilon'')\widehat{a}_k^\dagger(\varepsilon - \hbar\omega)\widehat{a}_l(\varepsilon)\right\rangle$ which are non zero for $\omega \neq 0$ only when $\iota = l$ and $j = k$ and if $\varepsilon''' = \varepsilon$ and $\varepsilon'' = \varepsilon - \hbar\omega$. The last condition tells us that current fluctuations results from *exchange* between electrons [24]. When $i = j = k = l$ fluctuations originate from the same reservoir and contribute to the "emission noise", while when $i = l \neq j = k$ fluctuations originate from the beating between transmitted electrons emitted from one reservoir and reflected electrons coming from the other reservoir: the partition noise.

$$S(\omega) = 2\frac{e^2}{h}\int d\varepsilon\{(1 - \sqrt{R_\varepsilon R_{\varepsilon-\hbar\omega}})^2 f_1(\varepsilon)(1 - f_1(\varepsilon - \hbar\omega))$$
$$+D_\varepsilon D_{\varepsilon-\hbar\omega} f_2(\varepsilon)(1 - f_2(\varepsilon - \hbar\omega)) + D_\varepsilon D_{\varepsilon-\hbar\omega} f_1(\varepsilon)(1 - f_2(\varepsilon - \hbar\omega))$$
$$+D_\varepsilon R_{\varepsilon-\hbar\omega} f_2(\varepsilon)(1 - f_1(\varepsilon - \hbar\omega))\} \quad (3.3)$$

where $R(\varepsilon) = 1 - D(\varepsilon)$. $S(\omega)$ contains fluctuations due to transition from an occupied state at ε to an unoccupied state at $\varepsilon - \hbar\omega$. In a real measurement, this corresponds to a positive energy flow from the noisy conductor to the external circuit with an ammeter in the ground state. If we neglect the energy dependence of the transmission, the single mode finite frequency current noise power becomes

$$S(\omega) = 2\frac{e^2}{h}\Big[2D^2\hbar\omega N(\hbar\omega) + D(1 - D)\big\{(eV + \hbar\omega)N(eV + \hbar\omega)$$
$$+(eV - \hbar\omega)N(eV - \hbar\omega)\big\}\Big] \quad (3.4)$$

where $N(\hbar\omega) = \frac{1}{\exp\frac{\hbar\omega}{k_BT} - 1} = \frac{1}{\hbar\omega}\int d\varepsilon f_1(\varepsilon)(1 - f_1(\varepsilon - \hbar\omega))$

At zero voltage, only thermal noise survives, and we recover the non-symmetrized form of the Johnson-Nyquist formula:

$$S(\omega) = 4G\hbar\omega\frac{1}{\exp\frac{\hbar\omega}{k_BT} - 1} \simeq 4Gk_BT \quad \hbar\omega < k_BT$$

with $G = D\frac{e^2}{h}$. At zero temperature, only partition noise survives and we recover the finite frequency shot-noise formula:

$$S(\omega) = 2D(1 - D)\frac{e^2}{h}(eV - \hbar\omega) \quad eV > \hbar\omega; \quad S(\omega) = 0 \quad eV < \hbar\omega$$

The conductor can generate noise fluctuations at frequency ω only if there are electrons in the left reservoir with energy at least $\hbar\omega$ above the electrons of the right reservoir. At zero frequency we recover the binomial result $2eIF$ with the Fano factor $F = 1 - D$ discussed previously.

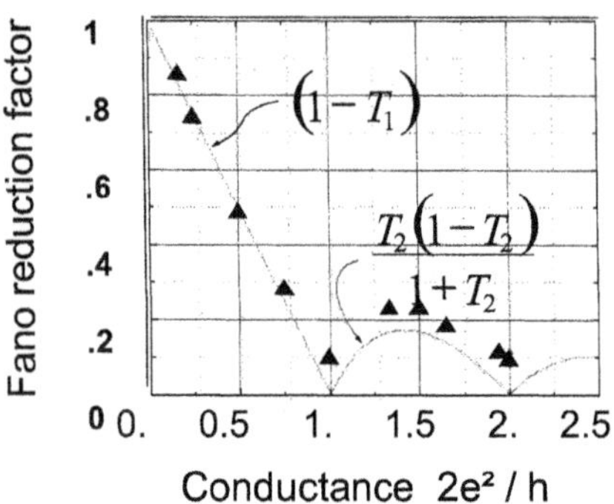

Fig. 11. Fano factor measured for a quantum point contact when the first and the second electronic mode is transmitted, from Ref.[8]. When the transmission approaches exactly 1 or 2, suppression of the noise is better than 90%. The conductor is noiseless.

3.2.4. *Shot noise for a general conductor (low frequency)*

For an arbitrary scattering matrix of a multimode conductor, with transmissions $D_n(\varepsilon)$ being the eigenvalues of the off-diagonal block matrix $S_{12}(\varepsilon)S_{12}^{\dagger}(\varepsilon)$ of the scattering matrix S, the low frequency shot noise is

$$S(\omega) = 4k_BT\frac{e^2}{h}\sum_n D_n^2 + 2eIF\frac{e^2}{h}V\coth\frac{eV}{2k_BT} \qquad F = \frac{\sum_n D_n(1-D_n)}{\sum_n D_n}$$

the first term is the thermal noise of the reservoir, while the second represents the shot noise associated with the quantum partitioning of electrons between reflected and transmitted states. The cross-over between thermal and shot noise occurs at the characteristic voltage $eV = 2k_BT$. Note the Fano factor F, always smaller than one: shot noise is sub-Poissonian. If the D_n are either 0 or 1 the conductor is even noiseless. Both the variation of the Fano factor with transmission and the variation of the noise with temperature have been measured quantitatively in experiments[7, 8]. Fig11 shows the result obtained using a Quantum Point Contact. Here the conductance gives a direct measure of the transmissions. Knowing temperature and voltage no free parameters are left and the Fano factor can be easily extracted from the data. The quantum scattering model of shot noise and the experiment agree to a very high degree of accuracy with no adjustable parameters.

For ballistic conductors, such as diffusive metals or chaotic quantum dots, the random matrix theory [10] gives the probability distribution $P(D)$ of the

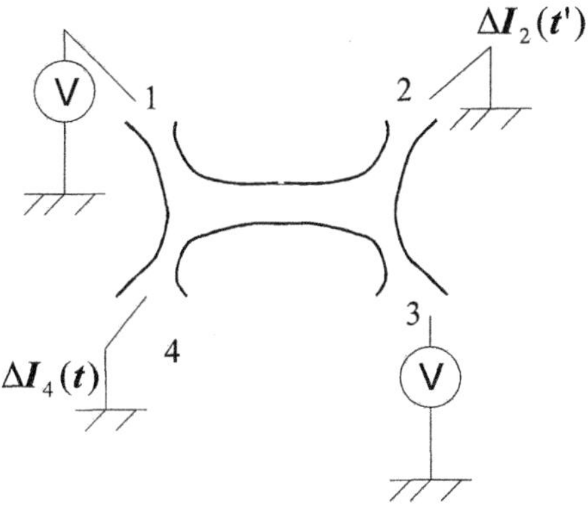

Fig. 12. Four-terminal conductor to probe non-classical current correlations.

transmissions. The Fano factor is $F = \int_0^1 dDP(D)D(1-D)/\int_0^1 dDP(D)D$. It takes the value $\frac{1}{3}$ for a diffusive metal and $\frac{1}{4}$ for a chaotic dot. In both case experiments agree with these predictions.

3.3. *Current noise correlations*

3.3.1. *Antibunching*

Experiments correlating photon beams such as Hanbury-Brown and Twiss experiments are very powerful tools in optics. Unlike photons, electrons are "naturally" noiseless. Also one expects very non-trivial results when correlating currents of different different leads for a multiterminal conductor. Let us consider the 4-terminal conductor of figure12 in which the correlations S_{24} between the current fluctuations of leads 2 and 4 are measured. As shown in ref.[24], correlations reveal strong quantum interference effects due to exchange. In a first experiment (A), contact 1 is voltage biased while the other contacts remain grounded.

The low frequency ($\omega \simeq 0$) current correlator $S_{24}^{(A)}(0) = 2\int dt \left\langle \widehat{I_2}(0)\widehat{I_4}(t)\right\rangle = 2\frac{e^2}{h}\sum_{\gamma\neq\delta}\int d\varepsilon \left[s_{2\gamma}^* s_{2\delta} s_{4\delta}^* s_{4\gamma}\right] \times \left\{f_\gamma(\varepsilon)(1-f_\delta(\varepsilon)) + f_\delta(\varepsilon)(1-f_\gamma(\varepsilon))\right\}$ gives at $T=0$:

$$S_{24}^{(A)} = -2\frac{e^2}{h}eV(s_{21}s_{21}^* s_{41}s_{41}^*) \tag{3.5}$$

It is proportional to the product of the probabilities $s_{21}s_{21}^* = D_{21}$and $s_{41}s_{41}^* = D_{41}$ to inject electrons from 1 to 2 and 1 to 4. The negative sign reflects anti-

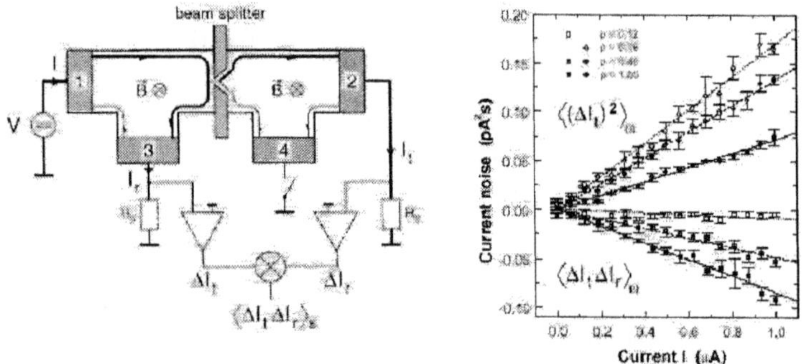

Fig. 13. Hanbury-Brown Twiss experiment with electrons in the quantum Hall edge channel regime (after Ref. [13]).

bunching. Indeed single electrons emitted at a rate eV/h by lead (1) are scattered between all four leads according to a multinomial law with probabilities D_{21}, D_{31}, D_{41}, and R_{11}. The cross-correlation between leads (2) and (4) is thus $\sim -D_{21}D_{41}$. In a second experiment (B) voltage is applied to lead (3) while all other leads (1 included) are grounded. One finds similarly: $S_{24}^{(B)} = -2\frac{e^2}{h}eV(s_{23}s_{23}^*s_{43}s_{43}^*)$. But there is more than simple anti-bunching. If, in a third experiment (A+B) the voltage is applied to leads (1) and (3) simultaneously, the current noises do not add as they classically would:

$$\begin{aligned} S_{24}^{(A+B)} &= S_{24}^{(A)} + S_{24}^{(B)} - 2\frac{e^2}{h}eV\left[(s_{21}s_{23}^*s_{43}s_{41}^*) + (s_{23}s_{21}^*s_{41}s_{43}^*)\right] \\ &\neq S_{24}^{(A)} + S_{24}^{(B)} \end{aligned}$$

An extra contribution is found which does not reduce to simple products of probabilities and results from quantum exchange [24]. This reflects entanglement of electrons coming from leads (1) to (3) and mixed in output leads (2) and (4).

3.3.2. *Hanbury-Brown and Twiss experiment with electrons*

Using the chiral edge states of the Integer Quantum Hall effect combined with a Quantum Point Contact, it is possible to realize the optical analog of the Hanbury-Brown and Twiss experiment. Fig 13 shows the results of Ref. [13] where the current correlations between leads (2) and (3) corresponding to transmitted and reflected electrons by the QPC beam splitter (transmission $D = 1/2$) are measured. The electron source is made of a contact (1), biased at potential V, followed by a second QPC whose transmission can be varied to make the statistics

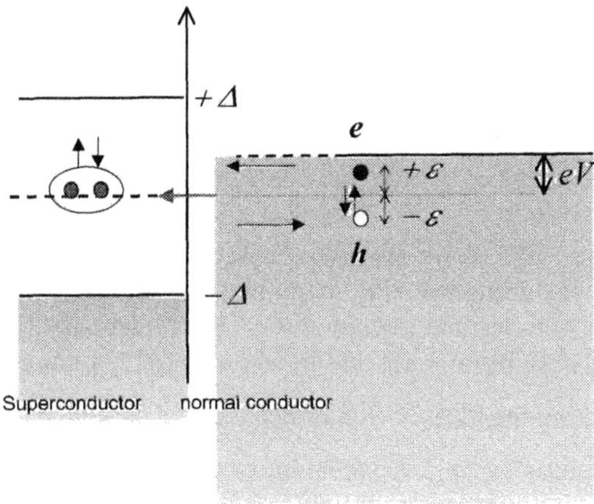

Fig. 14. Normal metal-superconductor interface. An electron incoming from the right is reflected as hole and a Cooper pair is transferred to the left.

of electrons incoming on the beam splitter fluctuationless (perfect transmission) or Poissonian (weak transmission). In the first situation, perfect anticorrelation is observed $\langle(\Delta I_t)^2\rangle = \langle(\Delta I_r)^2\rangle = -\langle\Delta I_r \Delta I_t\rangle$, while in the second case no correlation is found as expected for Poissonian statistics $\langle\Delta I_r \Delta I_t\rangle = 0$.

4. Quasi-particle entanglement in ballistics conductors

4.1. A natural source of entangled quasiparticles

A spin singlet pair of electron quasiparticles is a very simple example of entangled state. This state is often favored by the interactions. Possible physical effects leading to spin singlet states are: the Coulomb interaction exchange effect in few electron quantum dots, the Kondo scattering by a magnetic impurity, the pairing of electron in conventional superconductors. The last case is a simple and robust way to get entangled quasiparticles and is decribed in the following.

What we need is an interface between a superconductor (S) and normal metal (N) in order to extract Cooper pairs. The scattering approach describes how two convert Cooper pairs in the S region into pairs of correlated quasiparticles in the normal region. A convenient formulation is the Andreev reflection schematically shown in Fig14. Here, quasiparticle energies ε are referred to chemical potential of the superconductor μ_S. For energies $|\varepsilon| < \Delta$, the superconducting gap, a single electron can not penetrate the superconductor. A second order process is needed: an electron incoming from the normal region is reflected at the S-N interace as a hole in the normal region. From energy conservation the electron and the hole have opposite energies. A outgoing hole means the absorption of a second incoming electron: a charge $2e$ is transferred from the normal to the superconducting region. A Cooper pair being a spin singlet state, the two electrons have opposite spin $\sigma = \pm\frac{1}{2}$. More specifically, their wavefunction must be of the form $\frac{1}{2}(|\sigma; -\sigma\rangle - |-\sigma; \sigma\rangle)(|\phi_1(\varepsilon); \phi_2(-\varepsilon)\rangle + |\phi_2(-\varepsilon); \phi_1(\varepsilon)\rangle)$ where $\phi_{1,2}$ describe the incoming waves of the normal region.

For simplicity we consider a single mode normal conductor. Generalization to multimodes is straightforward. As shown in Ref. [25], the Andreev scattering matrix describing the ideal N-S interface: $S_{Andreev}(\varepsilon) = \begin{pmatrix} 0 & \gamma(\varepsilon) \\ \gamma^*(\varepsilon) & 0 \end{pmatrix}$ relates the amplitudes (a_e, a_h) of incoming electrons and holes at energy ε and $-\varepsilon$ respectively to the amplitudes (b_e, b_h) of outgoing electrons and holes at energy ε and $-\varepsilon$. Here $\gamma = \exp(-i\cos^{-1}(\varepsilon/\Delta))\exp(i\Phi)$, where Φ stands for the macroscopic phase of the superconductor. To calculate the transport and noise properties one needs the total scattering matrix $S_{NS} = \begin{pmatrix} S_{ee} & S_{eh} \\ S_{he} & S_{hh} \end{pmatrix}$ which relates the amplitudes of the waves emitted by the normal reservoir to that of the waves returning to the reservoir. It is convenient to define the scattering matrix of the normal region $S_N(\varepsilon) = \begin{pmatrix} r_{11} & t_{12} \\ t_{21} & r_{22} \end{pmatrix}$ where the index 1 stands for the normal reservoir and the index 2 corresponds to a fictive normal reservoir replacing the superconducting region. The normal reflection amplitude r_{22} is responsible for multiple Andreev reflections at the superconducting interface. Summing all amplitudes gives:

$$S_{ee}(\varepsilon) = r_{11}(\varepsilon) + \gamma^2 t_{12}(\varepsilon) r_{22}^*(-\varepsilon)\left[1 - \gamma^2 r_{22}(\varepsilon) r_{22}^*(-\varepsilon)\right]^{-1} t_{22}(\varepsilon) \quad (4.1)$$

$$S_{eh}(\varepsilon) = \gamma t_{12}(\varepsilon)\left[1 - \gamma^2 r_{22}^*(-\varepsilon) r_{22}(\varepsilon)\right]^{-1} t_{21}^*(-\varepsilon) \quad (4.2)$$

$$S_{he}(\varepsilon) = \gamma t_{12}^*(-\varepsilon)\left[1 - \gamma^2 r_{22}^*(\varepsilon) r_{22}(-\varepsilon)\right]^{-1} t_{21}(\varepsilon) \quad (4.3)$$

$$S_{hh}(\varepsilon) = r_{11}^*(-\varepsilon) + \gamma^2 t_{12}^*(-\varepsilon) r_{22}(\varepsilon)\left[1 - \gamma^2 r_{22}^*(-\varepsilon) r_{22}(\varepsilon)\right]^{-1} t_{22}(\varepsilon) \quad (4.4)$$

Replacing the incoming and outgoing waves amplitudes by corresponding fermion operators, leads to the current operator:

$$\widehat{I}_{NS}(t) \simeq 2\frac{e}{h}\int d\varepsilon d\varepsilon' \left\{\widehat{a}_e^\dagger(\varepsilon')\widehat{a}_e(\varepsilon) - \widehat{a}_h^\dagger(\varepsilon')\widehat{a}_h(\varepsilon) - \widehat{b}_e^\dagger(\varepsilon')\widehat{b}_e(\varepsilon) + \widehat{b}_h^\dagger(\varepsilon')\widehat{b}_h(\varepsilon)\right\} e^{i(\varepsilon'-\varepsilon)t/\hbar}$$

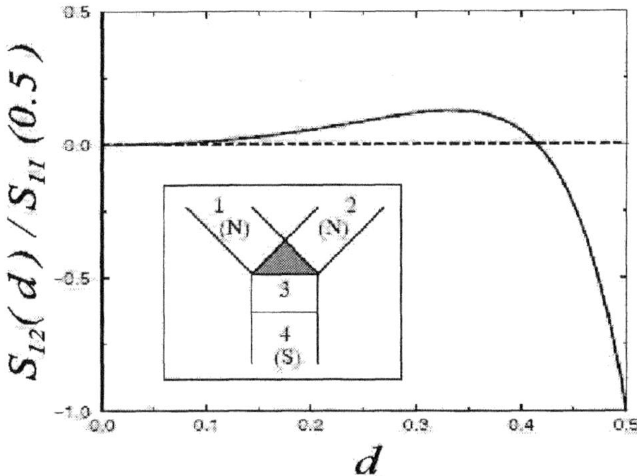

Fig. 15. Cross-correlation of the current noise of lead 1 and 2 normalized to the current noise of each lead for a symmetric normal fork connected to a superconductor. A positive correlation characterizes electron bunching (after Ref. [26]).

At zero temperature, its expectation value at small voltage bias ($eV \ll \Delta$) gives the normal-superconductor (NS) conductance:

$$G_{NS} = \frac{4e^2}{h} |S_{he}|^2 = \frac{4e^2}{h} \frac{D^2}{(1+R)^2} \tag{4.5}$$

where $D = |s_{21}|^2$ and $R = 1 - D = |r_{11}|^2$. For weak transmission, it is proportional to the square of transmission as expected for a second order process. For unit transmission $G_S = \frac{(2e)^2}{h}$ is the quantum of conductance which a doubled charge.

The shot noise can be obtained following the procedure developed in the last chapter, to give at low frequency and zero temperature:

$$S_{I_{NS}} = 2(2e)\frac{(2e)^2}{h} V |S_{he}|^2 (1 - |S_{he}|^2) = 2(2e) I_{NS}.F \tag{4.6}$$

Compared with Eq.3.2 a charge $2e$ enters in 4.6. As in the normal case the Fano factor $F = (1 - |S_{he}|^2)$ reflects the binomial law: is a Cooper pair transferred or not. The Fermi statistics makes the flow of quasi-particles emitted by the normal reservoir noiseless and only quantum partition noise remains.

The two quasi-particles are expected to be entangled in a symmetric combination of the orbital wavefunctions. How to probe this entanglement? A first,

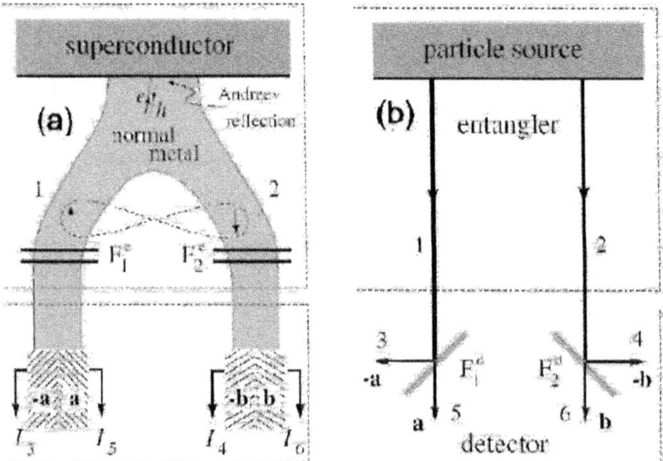

Fig. 16. (a) Proposed superconducting-normal conductor set-up to detect Bell correlations with electrons. (b) Equivalent set-up for atomic or photon detection of Bell states (after Ref. [17]).

but incomplete, way to probe the orbital entanglement is to detect the electron bunching in the normal region 4.6. Hanbury-Brown Twiss current noise correlations have been theoretically considered where a ballistic normal part divides as a fork, as shown in Fig. 15. Electron bunching should lead to positive correlations but two mechanisms compete: anticorrelation due to electron pairs transmitted either in lead (1) or in lead (2) and positive correlations due to simultaneous emission of one electron in both leads. From Ref.[26], the cross-correlation current noise spectrum for a symmetric fork with transmission $d = D_{S1} = D_{S2} \leq 0.5$ between the S part and each lead is found to be:

$$S_{12} = 2e\frac{4e^2}{h}V\frac{d^2}{2(1-d)^4}(2-(d+1)^2)$$

For $d < \sqrt{2}-1$ positive correlations reflecting bunching are found but only weak. For large transmission $d \simeq 0.5$ perfect antibunching is obtained. In order to favor positive correlations, reflecting the non-local correlation of entangled electrons in separated leads, the same authors have suggested to filter the energies at $\varepsilon = \varepsilon_0$ in one lead and $\varepsilon = -\varepsilon_0$ [16]. This could be practically achieved by using quantum dots as electronic Fabry-Pérot resonators.

HBT correlations can only indicate but not demonstrate entanglement. Placing additional beam splitters, as shown in Fig.16(a), which filters spin (polarization **a** and -**a** in lead 1, and **b** and -**b** in lead 2 could allow for a complete test. Although

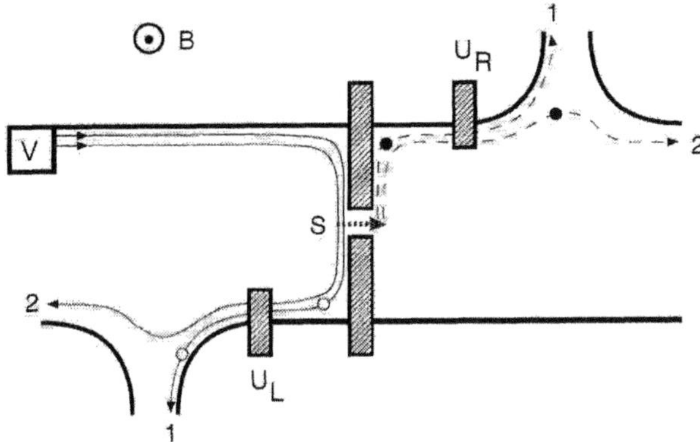

Fig. 17. Unlike photons, equilibrium electron sources (here a contact at potential V) and beam splitters can realize entanglement. A two channel electron source followed by an opaque barrier generates entangled electron-hole pairs. Unitary channel mixers followed by a spatial separation of each channel for both hole (left) and electrons (right) allow to test Bell's inequalities using current noise correlations (after Ref. [19]).

difficult to realize in practice, Ref. [17] showed that, with this set-up, tests of Bell's inequality can be done using *current noise correlations*. In Ref.[18], a simpler set-up showing entanglement, not requiring spin filters, has been proposed using two normal beam splitters and two NS contacts belonging to the same superconductor. Bell's inequality are also probed using current fluctuation correlations at the output of the beam splitters.

4.2. *Entangling electrons with thermal sources and beam-splitters*

Usual methods envisaged in quantum optics or in atomic physics to produce entanglement rely on interactions. Similarly, for electronic systems the use of interactions have been first considered. However, it is known that another way to get particle entangled is to use beam-splitters eventually combined with post-selection. For photons, it has been established that at least one source at the input of the beam-splitter must be non-classical. As a result classical Gaussian sources such as thermal sources cannot be used. Surprisingly, a completely different result is found for electrons.

During the last year, elegant proposals have been made which showed that *equilibrium* (thermal) sources of electrons combined with beam-splitters are able to produce entanglement or can be used for teleportation of quantum states. The fundamental reason is again the Fermi statistics which limits the electron occupation of the states to exactly one or zero (at $T = 0$) while Bose-Einstein statistics for photons favors large occupation of the states.

Entanglement using beam-splitters fundamentally exploits the indistinguishability of the particles ("which particle" rather than "which path") and is closely related to two-particle interferometry. In the proposal of Ref. [19] a two-channel ballistic conductor is considered. This can be realized using the edge states of a 2D conductor in the quantum Hall regime, see fig.17, but other equivalent systems can be envisaged. The left contact biased at a voltage V emits electrons toward an opaque beam-splitter while the right contact, not shown, is grounded. The 4X4 beam-splitter scattering matrix $\begin{pmatrix} r & t' \\ t & r' \end{pmatrix}$ is characterized by very small eigenvalues D_1 and D_2 of the transmission matrix product $t^\dagger t$. In this regime, nearly all left outgoing states are filled up to the energy $E_F + eV$ while the right outgoing states are nearly empty above E_F and the concept of electron-hole pair is meaningful. An electron (rarely) transmitted to the right with energy $\varepsilon \in [E_F, E_F + eV]$ above the right channel Fermi sea, is accompanied by a missing electron, called a hole, in the left channel Fermi sea. Here, the hole has a status similar to the Andreev hole of the previous chapter. There is no way to know whether the electron or the hole comes from channel 1 or channel 2 (the channel play the role of the spin) and the state describing the electron-hole pairs is non separable. To probe entanglement, interchannel mixing is realized at the left and at the right outputs and are characterized by the unitary scattering matrices U_L and U_R. Similar to the spin filters of the above S-N set-up, varying U_R and U_R can maximize a Bell parameter to experimentally probe entanglement as shown in the next part. Theoretically, it can be useful to define the so-called concurrence C, which quantitatively characterizes entanglement. In the present case it is given by $C = 2\sqrt{D_1(1-D_2)D_2(1-D_1)}/(D_1+D_2+D_1D_2)$. It approaches 1, maximal entanglement, when $D_1 = D_2 \ll 1$. Using, the current-current noise correlator $C_{i,j}$, where i, j stand for the contact 1 and 2 at the left and at the right, one can define a Bell-CHSH (Clauser-Horne-Shimony-Holt) parameter E whose maximal value $E_{\max} = 2(1+C^2)^{\frac{1}{2}} > 2$ is obtained for special values of U_R and U_L.

In [20], a variant of this proposal makes use of a chaotic quantum dot rather than an opaque beam splitter to entangle electron hole pairs. The dot is connected to leads via two separate contacts, each partially transmitting two channels. Here, the result of the random matrix theory for the transmission distribution can be used to determine the statistical distribution of the concurrence. The relation between the Bell parameter and the concurrence has been extended to the case of non-opaque regime. During the redaction of these notes, another promising set-up has been proposed by the authors of Ref. [27]. A double Hanburry-Brown Twiss configuration realized with edge states in a 2D electron gas in the QHE regime is used. Here instead of a single contact emitting electrons in two channel leads, two separate contacts (2 and 3) emit electrons in a single channel. Each contact is immediately followed by an opaque beam-splitter (C and D) which generates electron-hole pairs, see Fig. 18. The electrons (rarely) transmitted through C and D are directed toward a third beam-splitter B where they mix. Similarly, the associated holes coming from C and D are mixed by a fourth beam-splitter A. B and A are equivalent to the channel mixers $U_{L,R}$ of the previous proposal. Varying the transmissions $T_A = \sin^2\theta_A$ and $T_B = \sin^2\theta_B$ and measuring the correlations of lead 5 or 6 with lead 7 or 8 allows to probe entanglement.

4.3. *Probing entanglement using current-current correlators*

All the theoretical proposal discussed in the previous chapter as well as the N-S proposals, use current correlations between the four leads at the output of the two beam-splitter (or spin filter) to analyze entanglement via violation of Bell inequalities. Historically, Bell's inequalities are based on coincidence measurement of particles, i.e. here the joint probability $P_{\alpha,\beta}$ to find a particle at contact α and contact β. For contact emitting electrons and biased at voltage V, the correlation time of the current-current fluctuations $\langle\Delta I_a(0)\Delta I_\beta(t)\rangle$ has finite value only for time $t < \tau_c = h/eV$. This corresponds to the correlation time of electron-hole pairs (eV/h is the characteristic frequency of wavepackets incoming on the first beam-splitter). As transmissions are weak, the average time of arrival between pairs is long compare to τ_c. Thus the zero frequency current-current correlator $S_{\alpha,\beta} = 2\int_{-\infty}^{\infty} dt\,\langle\Delta I_a(0)\Delta I_\beta(t)\rangle$ is a coincidence counting measurement average over a large number events and $S_{\alpha,\beta} \propto P_{\alpha,\beta}$ [18].

Using the notations of Ref.[19], Fig.17, $\alpha \equiv L, i$ and $\beta = R, j$ where i and $j = 1, 2$, and according to equation 3.5, $S_{\alpha,\beta} = -(e^3V/h)\left|\left(rt^\dagger\right)_{i,j}\right|^2$. The effect of the local channel mixers is equivalent to replace $r \to U_L r$ and $t \to U_R t$ in the expressions of the $S_{\alpha,\beta}$'s. For current noise measurement, the Bell parameter is:

$$E = e(U_L, U_R) + e(U'_L, U_R) + e(U_L, U'_R) - e(U'_L, U'_R)$$

where $e(U_L, U_R) = \frac{S_{L1,R1}+S_{L2,R2}-S_{L1,R2}-S_{L2,R1}}{S_{L1,R1}+S_{L2,R2}+S_{L1,R2}+S_{L2,R1}}$. Measuring the four current-current correlators between leads of left and right channel mixers while trying different combinations U_R and U_L to maximize the Bell parameter E allows to test entanglement. U_R and U_L have to be replaced by the polarization orientation **a** and **b** in the S-N experiment, and by θ_A and θ_B in the set-up of Fig.18.

Experimentally, high accuracy measurement of such current-current correlators is routine in laboratories. However better results will be obtained in the "low" frequency regime where $\omega \simeq I/e$. Thus, measurements of Bell's inequal-

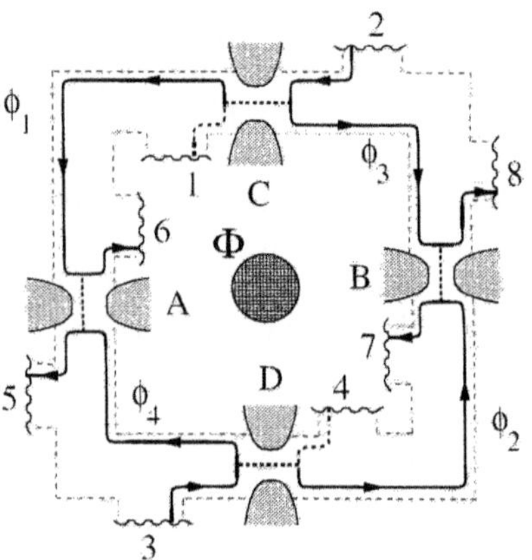

Fig. 18. Electron entangler using two equilibrium sources and four beam splitter which exploits the chirality of edge channels in the quantum Hall regime (after Ref. [27]).

ities remain challenging but not out of reach. It is of course tempting to try to benefit from the beautiful properties of the Fermi statistics to realize quantum information processing. This means going beyond statistical measurements. Instead of having contacts injecting electrons regularly at a frequency eV/h, single electron injection and fast single electron readout will be needed.

5. Conclusion

The Fermi statistics gives beautiful quantum properties to electron quasi-particles in quantum conductors. The magic value of the conductance: $\frac{e^2}{h}$ and the associated noiseless electron flow for unit transmission. The sub-Poissonian statistics and the anti-correlated current fluctuations. It provides easy way to entangle electron quasiparticles just using equilibrium contacts and beam-splitters without the need of interactions. Single occupancy of wavepackets makes coincidence measurements accessible by current noise correlations. Testing Bell's inequalities with electrons seem now experimentally accessible. All these features are best observed in ballistic quantum conductors (mostly very clean 2D electron gas at present). Photons or atoms obeying Bose-Einstein statistics do not share these remarkable properties. One can imagine that in a near future these results, obtained in the frame of the Mesoscopic Physics for degenerate electrons in coherent conductors, can also be applied to Fermionic cold atoms which offer different possibilities of control and measurements.

References

[1] S. Data *Electronic transport in mesoscopic systems* (Cambridge University Press, 1995)., and references therein.

[2] Y. Imry *Introduction to mesoscopic physics* (Oxford University Press, 1997)., and references therein.

[3] C.W.J. Beenakker and H. van Houten*Quantum transport in semiconductor nanostructures* in Solid State Physics, ED. Ehrenreich and D. Turnbull (Academic Press, San Diego 1997)., and references therein

[4] G.B. Lesovik, JETP Lett. **49** (1989) 592

[5] M. Buttiker, Phys. Rev. Lett. **65** (1990) 2901

[6] R. Landauer and Th. Martin Physica B **175** (1991) 167

[7] M. Reznikov, M. Heiblum, H. Shtrikman, and D. Mahalu, Phys. Rev. Lett **75** (1995) 3340

[8] A. Kumar, L. Saminadayar, D.C. Glattli, Y. Jinand B. Etienne, Phys. Rev. Lett **76** (1996) 2778

[9] Th. Martin in *Coulomb and Interference Effects in Small Electronic Structures* ED. D. C. Glattli, M. Sanquer, and J. Tran Thanh Van (Editions Frontières, Gifsur- Yvette, 1994) p405., and references therein

[10] M.J.M. de Jong and C.W.J. Beenakker in *Mesoscopic electron transport* ED. L.L. Sohn, L.P. Kouwenhoven, and G. Schön (Kluwer Academic Publishing, Dordrecht, 1997) Vol. 345, 225.

[11] Y.M. Blanter and M. Buttiker, *Shot noise in mesoscopic conductors* Phys. Rep. **336** (2000) 2, and references therein

[12] R. Hanbury Brown and R.Q. Twiss Nature. **177** (1956) 27

[13] M. Henny, S. Oberholzer, C. Strunk, T. Hollandand C. Schonenberger, Science **284** (1999) 296

[14] W.D. Oliver, K. Kim, R.C. Liu, and Y. Yamamoto, Science **284** (1999) 299

[15] P. Recher, E.V. Sukhorukov, and D. Loss Phys. Rev. B **63** (2001) 165314

[16] G.B. Lesovik, Th. Martin, and G. Blatter Eur. Phys. J. B **24** (2001) 287

[17] N.M. Chtchelkatchev, G. Blatter, G.B. Lesovik and Th. Martin Phys. Rev. B **66** (2002) 161320

[18] P. Samuelsson, E.V. Sukhorukov, and M. Buttiker Phys. Rev. Lett **91** (2003) 157002

[19] C.W.J. Beenakker, C. Emary, M. Kindermann and J.L. Velsen Phys. Rev. Lett. **91** (2003) 147901

[20] C.W.J. Beenakker, M. Kindermann, C.M. Marcus and A. Yacoby Cond-mat/0310199

[21] P. Samuelsson, E.V. Sukhorukov, and M. Buttiker Phys. Rev. Lett **92** (2004) 026805

[22] B.J. van Wees, H. van Houten, C.W.J. Beenakker, J.G. Williamson, L.P. Kouwenhoven, D. van de Marel, and C.T. Foxon Phys. Rev. Lett. **60** (1988) 888 ; D.A. Wharam, T.J. Thorton, R. Newbury, M. Pepper, H. Ahmed, J.E. Frost, D.G. Hasko, D.C. Peacock, D.A. Ritchie, ans G.A.C. Jones J. Phys. C **21** (1988) L209

[23] Y. Ji et al. Nature **422** (2003) 415

[24] M. Buttiker Phys. Rev. B **46** (1992) 12485

[25] M.J.M. de Jong,and C.W.J. Beenakker Phys. Rev. B **49** (1994) 16070

[26] J. Torres and Th. Martin Eur. Phys. J. B **12** (1999) 319

[27] P. Samuelsson, E.V. Sukhorukov, and M. Buttiker Phys. Rev. Lett **92** (2004) 026805

References

[illegible]

Course 12

SUPERCONDUCTING QUBITS

Michel H. Devoret[†] and John M. Martinis[*]

† *Applied Physics Department, Yale University, New Haven CT 06520*
* *National Institute of Standards and Technology, Boulder CO 80305*

D. Estève, J.-M. Raimond and J. Dalibard, eds.
Les Houches, Session LXXIX, 2003
Quantum Entanglement and Information Processing
Intrication quantique et traitement de l'information
© *2004 Elsevier B.V. All rights reserved*

Contents

1. Introduction

1.1. The problem of implementing a quantum computer

The theory of information has been revolutionized by the discovery that quantum algorithms can run exponentially faster than their classical counterparts, and by the invention of quantum error-correction protocols [1]. These fundamental breakthroughs have lead scientists and engineers to imagine building entirely novel types of information processors. However, the construction of a computer exploiting quantum – rather than classical – principles represents a formidable scientific and technological challenge. While quantum bits must be strongly inter-coupled by gates to perform quantum computation, they must at the same time be completely decoupled from external influences, except during the write, control and readout phases when information must flow in and out of the quantum computer. This difficulty does not exist for conventional (classical) bits, which follow irreversible dynamics that damp the noise of the environment.

Most proposals for implementing a quantum computer have been based on qubits constructed from microscopic degrees of freedom: electron or nuclear spin, atomic transition dipoles and so on (see other lectures in this book). These degrees of freedom are naturally very well isolated from their environment, and hence decohere very slowly. The main challenge of these implementations is enhancing the inter-qubit coupling to the level required for fast gate operations without introducing decoherence from parasitic environmental modes and noise.

In this review, we will discuss a radically different experimental approach based on "quantum integrated circuits," where qubits are constructed from *collective* electrodynamic modes of macroscopic electrical elements, rather than microscopic degrees of freedom. An advantage of this approach is that these qubits have an intrinsically large electromagnetic cross-section, which implies they may be easily coupled together in complex topologies via simple linear electrical elements like capacitors, inductors, and transmission lines. However, strong coupling also presents a related challenge: is it possible to isolate these electrodynamic qubits from ambient parasitic noise while retaining open communication channels for the write, control, and read operations? The main purpose of this article is to review the considerable progress that has been made in the past few years towards this goal, and to explain how new ideas about methodology and materials are likely to improve coherence to the threshold needed for quantum error correction.

1.2. Scope of this review

Before starting our discussion, we must warn the reader that this review is atypical in that it is neither historical nor exhaustive. Some important works have not been included or are only partially covered. We have on purpose narrowed our focus: we adopt the point of view of an engineer trying to determine the best strategy for building a reliable machine with given performances. This approach obviously runs the risk of presenting a biased and even incorrect account of recent scientific results, since the optimization of a complex system is always a intricate process with many hidden passageways and dead-ends. We hope nevertheless that the following sections will at least stimulate discussions on how to harness the physics of quantum integrated circuits into a mature quantum information processing technology.

After ending this introduction with a general presentation of quantum integrated circuits, we will first treat the simplest example of circuits, the superconducting linear LC oscillator. Although it cannot lead to a useful qubit, this circuit allows the presentation of the circuit variables and parameters with minimal mathematical complications. We will then introduce the Josephson junction as the crucial non-linear, non-dissipative element. The problem of dealing with the fluctuations in the offset charge of the junction will lead us to the three basic types of superconducting qubits. After showing how their coherence is affected by the intrinsic noise of the junction we will embark on the discussion of how to design a faithful and fast readout without compromising the coherence. Issues associated with quantum gates will be finally dealt with.

2. Basic features of quantum integrated circuits

2.1. Ultra-low dissipation: superconductivity

For an integrated circuit to behave quantum mechanically, the first requirement is very low dissipation. More specifically, all metallic parts need to be made out of a material that has negligible resistance at the qubit operating temperature and at the qubit transition frequency. The loss of only one energy quantum completely spoils quantum coherence. Low temperature superconductors [2] such as aluminium or niobium are therefore ideal for the task of carrying quantum signals. For this reason, quantum integrated circuit implementations have been nicknamed "superconducting qubits"[1].

2.2. *Ultra-low noise : low temperature*

The degrees of freedom of the quantum integrated circuit must be cooled to temperatures where the typical energy kT of thermal fluctuations is much less that the energy quantum $\hbar\omega_{01}$ associated with the transition between the states |qubit=0> and |qubit=1>. For reasons which will become clear in subsequent sections, this frequency for superconducting qubits is in the 5-20GHz range and therefore, the operating temperature temperature T must be around 20mK (Recall that 1K corresponds to about 20 GHz). These temperatures may be readily obtained by cooling the chip with a dilution refrigerator. Perhaps more importantly though, the "electromagnetic temperature" of the wires of the control and readout ports connected to the chip must also be cooled to these low temperatures, which requires careful electromagnetic filtering. Note that electromagnetic damping mechanisms are usually stronger at low temperatures than those originating from electron-phonon coupling. The techniques [3] and requirements [4] for ultra-low noise filtering have been known for about 20 years. From the requirements $kT \ll \hbar\omega_{01}$ and $\hbar\omega_{01} \ll \Delta$, where Δ is the energy gap of the superconducting material, one must use superconducting materials with a transition temperature greater than about 1K.

2.3. *Non-linear, non-dissipative elements: tunnel junctions*

Quantum signal processing cannot be performed using only purely linear components. In quantum circuits, however, the non-linear elements must obey the additional requirement of being non-dissipative. Elements like PIN diodes or CMOS transistors are thus forbidden, even if they could be operated at ultra-low temperatures.

There is only one electronic element that is both non-linear and non-dissipative at arbitrarily low temperatures: the superconducting tunnel junction (also known as a Josephson tunnel junction [5]). As illustrated in Fig. 1, this circuit element consists of a sandwich of two superconducting thin films separated by an insulating layer that is thin enough (typically ~1nm) to allow tunneling of discrete charges through the barrier. In later sections we will describe how the tunneling of Cooper pairs creates a strong non-linear inductance, thus yielding viable

[1] In principle, other condensed phases of electrons, such as high-Tc superconductivity or the quantum Hall effect, both integer and fractional, are possible and would also lead to quantum integrated circuits of the general type discussed here. We do not pursue this subject further than this note, however, because dissipation in these new phases is, by far, not as well understood as in low-Tc superconductivity.

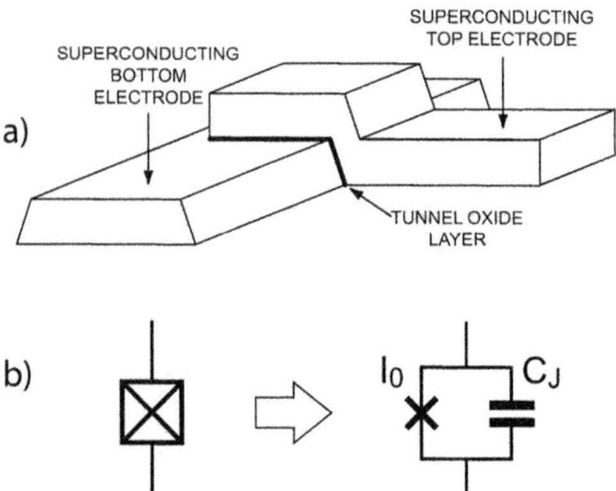

Fig. 1. a) Josephson tunnel junction made with two superconducting thin films; b) Schematic representation of a Josephson tunnel junction. The irreducible Josephson element is represented by a cross.

qubit energy levels. The tunnel barrier is typically fabricated from oxidation of the superconducting metal, which results in a reliable barrier since the oxidation process is self-terminating [6]. The materials properties of amorphous aluminum oxide (alumina) make it an attractive tunnel insulating layer. In part because of its well-behaved oxide, aluminim is the material from which good quality tunnel junctions are most easily fabricated, it is often said that aluminium is to superconducting quantum circuits what silicon is to conventional MOSFET circuits. Although the Josephson effect is a subtle physical effect involving a combination of tunneling and superconductivity, the junction fabrication process is relatively straightforward.

2.4. *Design and fabrication of quantum integrated circuits*

Superconducting junctions and wires are fabricated using techniques borrowed from conventional integrated circuits[2]. Quantum circuits are typically made on silicon wafers using optical or electron-beam lithography and thin film deposition. They present themselves as a set of micron-size or sub-micron-size circuit elements (tunnel junctions, capacitors, and inductors) connected by wires or transmission lines. The size of the chip and elements are such that, to a large extent, the electrodynamics of the circuit can be analyzed using simple transmission line equations or even a lumped element approximation. Contact to the chip

[2] It is worth mentioning that chips with tens of thousands of junctions have been successfully fabricated for the voltage standard and for the Josephson signal processors, which are only exploiting the speed of Josephson elements, not their quantum properties.

is made by wires bonded to mm-size metallic pads. The circuit can be designed using conventional layout and classical simulation programs.

Thus, many of the design concepts and tools of conventional electronics can be directly applied to quantum circuits. Nevertheless, there are still important differences between conventional and quantum circuits at the conceptual level.

2.5. Integrated circuits that obey macroscopic quantum mechanics

At the conceptual level, conventional and quantum circuits differ in that, in the former, the collective electronic degrees of freedom such as currents and voltages are classical variables, whereas in the latter, these degrees of freedom must be treated by quantum operators which do not necessarily commute. A more concrete way of presenting this rather abstract difference is to say that a typical electrical quantity, such as the charge on the plates of a capacitor, can be thought of as a simple number is conventional circuits, whereas in quantum circuits, the charge on the capacitor must be represented by a wave function giving the probability amplitude of all charge configurations. For example, the charge on the capacitor can be in a superposition of states where the charge is both positive and negative at the same time. Similarly the current in a loop might be flowing in two opposite directions at the same time. These situations have originally been nicknamed "macroscopic quantum effects" by Tony Leggett [7] to emphasize that quantum integrated circuits are displaying phenomena involving the collective behavior of many particles, which are in contrast to the usual quantum effects associated with microscopic particles such as electrons, nuclei or molecules[3].

2.6. DiVicenzo criteria

We conclude this section by briefly mentioning how quantum integrated circuits satisfy the so-called DiVicenzo criteria for the implementation of quantum computation [8]. The non-linearity of tunnel junctions is the key property ensuring that non-equidistant level subsystems can be implemented (criterion # 1: qubit existence). As in many other implementations, initialization is made possible (criterion #2: qubit reset) by the use of low temperature. Absence of dissipation in superconductors is one of the key factors in the quantum coherence of the system (criterion # 3: qubit coherence). Finally, gate operation and readout (criteria #4 and #5) are easily implemented here since electrical signals confined to and traveling along wires constitute very efficient coupling methods.

[3]These microscopic effects determine also the properties of materials, and explain phenomena such as superconductivity and the Josephson effect itself. Both classical and quantum circuits share this bottom layer of microscopic quantum mechanics.

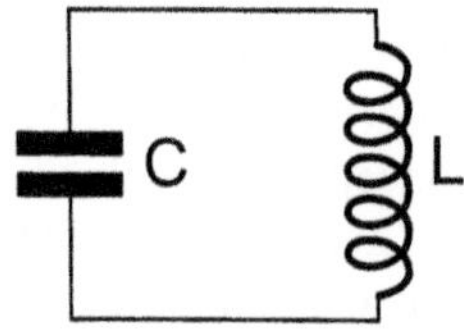

Fig. 2. Lumped element model for an electromagnetic resonator: LC oscillator.

3. The simplest quantum circuit

3.1. *Quantum LC oscillator*

We consider first the simplest example of a quantum integrated circuit, the LC oscillator. Although it cannot lead to a useful qubit, this circuit allows us to describe general key circuit variables and parameters with minimal mathematical complications. As shown in Fig. 2, it consists of an inductor L connected to a capacitor C, all metallic parts being superconducting. This simple circuit is the lumped-element version of a superconducting cavity or a transmission line resonator (for instance, the link between cavity resonators and LC circuits is elegantly discussed by Feynman [9]). The equations of motion of the LC circuit are those of an harmonic oscillator. It is convenient to take the position coordinate as being the flux Φ in the inductor, while the role of the conjugate momentum is played by the charge Q on the capacitor. The variables Φ and Q have to be treated as canonically conjugate quantum operators that obey $[\Phi, Q] = i\hbar$. The Hamiltonian of the circuit is $H = \frac{1}{2}\Phi^2/L + \frac{1}{2}Q^2/C$, which can be equivalently written as $H = \hbar\omega_0(n + \frac{1}{2})$ where n is the number operator for photons in the resonator and $\omega_0 = 1/\sqrt{LC}$ is the resonance frequency of the oscillator. It is important to note that the parameters of the circuit Hamiltonian are not fundamental constants of Nature. They are engineered quantities with a large range of possible values which can be modified easily by changing the dimensions of elements, a standard lithography operation. It is in this sense, in our opinion, that the system is unambiguously "macroscopic". The other important combination of the parameters L and C is the characteristic impedance $Z = \sqrt{L/C}$ of the circuit. Along with the residual resistance of the circuit and/or its radiation loss, both of which we can model as a series resistance R, this impedance determines the quality factor of the oscillation: $\mathcal{Q} = Z/R$. The theory of the harmonic oscillator shows that a quantum superposition of ground state and first excited state decays on a time scale precisely given by $1/RC$, yielding a quality factor for quantum coherence limited by $\mathcal{Q}$. These considerations illustrate the very useful general link between the classical measure of dissipation and the upper limit of the quantum coherence time.

3.2. Practical considerations

In practice, the circuit shown in Fig. 2 may be fabricated using planar components with lateral dimensions around 10μm, giving values of L and C approximately 0.1nH and 1pF, respectively, and yielding $\omega_0/2\pi \simeq$ 16GHz and $Z_0 = 10\Omega$. If we use aluminium, a good BCS superconductor with transition temperature of 1.1K and a gap $\Delta/e \simeq 200\mu V$, dissipation from the breaking of Cooper pairs will begin at frequencies greater than $2\Delta/h \simeq$ 100GHz. The residual resistivity of a BCS superconductor decreases exponentially with the inverse of temperature and linearly with frequency, as shown by the Mattis-Bardeen (MB) formula $\rho(\omega) \sim \rho_0 \frac{\hbar\omega}{k_B T} \exp(-\Delta/k_B T)$ [10], where ρ_0 is the resistivity of the metal in the normal state (we are treating here the case of the so-called "dirty" superconductor [11], which is well adapted to thin film systems). According to MB, the intrinsic losses of the superconductor, at the temperature and frequency (typically 20mK and 20GHz) characterizing the qubit dynamics, can be safely neglected. However, we must warn the reader that the intrisinsic losses in the superconducting material do not exhaust, by far, the causes of dissipation, even if very high quality factors have been demonstrated in superconducting cavity experiments [12].

3.3. Matching to the vacuum impedance: a useful feature, not a bug

Although the intrisinsic dissipation of superconducting circuits can be made very small, losses are in general governed by the coupling of the circuit with the electromagnetic environment that is present in the form of write, control and readout lines. These lines (which we also refer to as ports) have a characteristic propagation impedance $Z_c \simeq 50\Omega$, which is constrained to be a fraction of the impedance of the vacuum $Z_{\text{vac}} = 377\Omega$. It is thus easy to see that our LC circuit, with a characteristic impedance of $Z_0 = 10\Omega$, tends to be rather well impedance-matched to any pair of leads. This circumstance occurs very frequently in circuits, and almost never in microscopic systems such as atoms which interact very weakly with electromagnetic radiation[4]. Matching to Z_{vac} is a useful feature because it allows strong coupling for writing, reading, and logic operations. As we mentioned earlier, the challenge with quantum circuits is to isolate them from parasitic degrees of freedom. **The major task of this review is to explain how this has been achieved so far and what level of isolation is attainable.**

---

[4]The impedance of an atom can be crudely seen as being given by the impedance quantum $R_K = h/e^2$. We live in a universe where the ratio $Z_{vac}/2R_K$, also known as the fine structure constant 1/137.0, is a small number.

3.4. The consequences of being macroscopic

While our example shows that quantum circuits can be mass-produced by standard microfabrication techniques and that their parameters can be easily engineered to reach some optimal condition, it also points out evident drawbacks of being macroscopic for qubits.

The engineered quantities L and C can be written as

$$\begin{aligned} L &= L^{stat} + \Delta L(t) \\ C &= C^{stat} + \Delta C(t) \end{aligned} \tag{3.1}$$

a) The first term on the right-handside denotes the static part of the parameter. It has **statistical variations**: unlike atoms whose transition frequencies in isolation are so reproducible that they are the basis of atomic clocks, circuits will always be subject to parameter variations from one fabrication batch to another. Thus prior to any operation using the circuit, the transition frequencies and coupling strength will have to be determined by "diagnostic" sequences and then taken into account in the algorithms.

b) The second term on the right-handside denotes the time-dependent fluctuations of the parameter. It describes **noise** due to residual material defects moving in the material of the substrate or in the material of the circuit elements themselves. This noise can affect for instance the dielectric constant of a capacitor. The low frequency components of the noise will make the resonance frequency wobble and contribute to the dephasing of the oscillation. Furthermore, the frequency component of the noise at the transition frequency of the resonator will induce transitions between states and will therefore contribute to energy relaxation.

Let us stress that statistical variations and noise are not problems affecting superconducting qubit parameters only. For instance when several atoms or ions are put together in microcavities for gate operation, patch potential effects lead to expressions similar in form to Eq. 3.1 for the parameters of the hamiltonian, even if the isolated single qubit parameters are fluctuation-free.

3.5. The need for non-linear elements

Not all aspects of quantum information processing using quantum integrated circuits can be discussed within the framwork of the LC circuit which lacks an important ingredient: non-linearity. In the harmonic oscillator, all transitions between neighbouring states are degenerate as a result of the parabolic shape of the potential. In order to have a qubit, the transition frequency between states |qubit=0> and |qubit=1> must be sufficiently different from the transition between higher-lying eigenstates, in particular 1 and 2. Indeed, the maximum number of 1-qubit operations that can be performed coherently scales as $\mathcal{Q}_{01}\left|\omega_{01}-\omega_{12}\right|/\omega_{01}$ where $\mathcal{Q}_{01}$ is the quality factor of the $0 \rightarrow 1$ transition. Josephson tunnel junctions are crucial for quantum circuits since they have a strongly non-parabolic, inductive potential energy.

4. The Josephson non-linear inductance

At low temperatures, and at the low voltages and low frequencies corresponding to quantum information manipulation, the Josephson tunnel junction behaves as a pure non-linear inductance (Josephson element) in parallel with the capacitance corresponding to the parallel plate capacitor formed by the two overlapping films of the junction (Fig. 1b). This minimal, yet precise model, allows arbitrary complex quantum circuits to be analysed by a quantum version of conventional circuit theory. Even though the tunnel barrier is a layer of order ten atoms thick, the value of the Josephson non-linear inductance is very robust against static disorder, just like an ordinary inductance – such as the one considered in section 3 – is very insensitive to the position of each atom in the wire. We refer to [13] for a detailed discussion of this point.

4.1. Constitutive equation

Let us recall that a linear inductor, like any electrical element, can be fully characterized by its constitutive equation. Introducing a generalization of the ordinary magnetic flux, which is only defined for a loop, we define the **branch flux of an electric element** by $\Phi(t)=\int_{-\infty}^{t}V(t_1)dt_1$, where $V(t)$ is the space integral of the electric field along a current line inside the element. In this language, the current $I(t)$ flowing through the inductor is proportional to its branch flux $\Phi(t)$:

$$I(t)=\frac{1}{L}\Phi(t) \tag{4.1}$$

Note that the generalized flux $\Phi(t)$ can be defined for any electric element with two leads (dipole element), and in particular for the Josephson junction, even though it does not resemble a coil. The Josephson element behaves inductively, as its branch flux-current relationship [5] is:

$$I(t)=I_0\sin\left[2\pi\Phi(t)/\Phi_0\right] \tag{4.2}$$

This inductive behavior is the manifestation, at the level of collective electrical variables, of the inertia of Cooper pairs tunneling across the insulator (kinetic inductance). The discreteness of Cooper pair tunneling causes the periodic flux dependence of the current, with a period given by a universal quantum constant Φ_0, the superconducting flux quantum $h/2e$. The junction parameter I_0 is called the critical current of the tunnel element. It scales proportionally to the area of the tunnel layer and diminishes exponentially with the tunnel layer thickness. Note that the constitutive relation Eq. 4.2 expresses in only one equation the two Josephson relations [5]. This compact formulation is made possible by the introduction of the branch flux.

The purely sinusoidal form of the constitutive relation Eq. 4.2 can be traced to the perturbative nature of Cooper pair tunneling in a tunnel junction. Higher harmonics can appear if the tunnel layer becomes very thin, though their presence would not fundamentally change the discussion presented in this review. The quantity $2\pi\Phi(t)/\Phi_0 = \delta$ is called the gauge-invariant phase difference accross the junction (often abridged into "phase"). It is important to realize that at the level of the constitutive relation of the Josephson element, this variable is nothing else than an electromagnetic flux in dimensionless units. In general, we have

$$\theta = \delta \bmod 2\pi$$

where θ is the phase difference between the two superconducting condensates on both sides of the junction. This last relation expresses how the superconducting ground state and electromagnetism are tied together.

4.2. *Other forms of the parameter describing the Josephson non-linear inductance*

The Josephson element is also often described by two other parameters, each of which carry exactly the same information as the critical current. The first one is the Josephson effective inductance $L_{J0} = \varphi_0/I_0$, where $\varphi_0 = \Phi_0/2\pi$ is the reduced flux quantum. The name of this other form becomes obvious if we expand the sine function in Eq. 4.2 in powers of Φ around $\Phi = 0$. Keeping the leading term, we have $I = \Phi/L_{J0}$. Note that the junction behaves for small signals almost as a point-like kinetic inductance: a 100nm×100nm area junction will have a typical inductance of 100nH, whereas the same inductance is only obtained magnetically with a loop of about 1cm in diameter. More generally, it is convenient to define the phase-dependent Josephson inductance

$$L_J(\delta) = \left(\frac{\partial I}{\partial \Phi}\right)^{-1} = \frac{L_{J0}}{\cos\delta}$$

Note that the Josephson inductance not only depends on δ, it can actually become infinite or negative! Thus, under the proper conditions, the Josephson

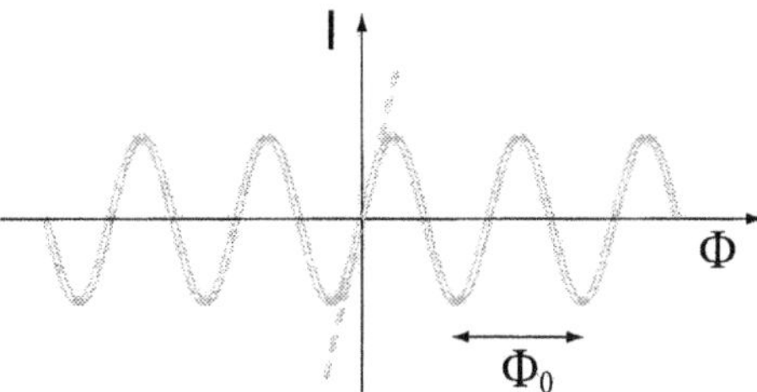

Fig. 3. Sinusoidal current-flux relationship of a Josephson tunnel junction, the simplest non-linear, non-dissipative electrical element (solid line). Dashed line represents current-flux relationship for a linear inductance equal to the junction effective inductance.

element can become a switch and even an active circuit element, as we will see below.

The other useful parameter is the Josephson energy $E_J = \varphi_0 I_0$. If we compute the energy stored in the junction as $E(t) = \int_{-\infty}^{t} I(t_1)\, V(t_1)\, dt_1$, we find $E(t) = -E_J \cos\left[2\pi\, \Phi(t)/\Phi_0\right]$. In contrast with the parabolic dependence on flux of the energy of an inductor, the potential associated with a Josephson element has the shape of a cosine washboard. The total height of the corrugation of the washboard is $2E_J$.

4.3. *Tuning the Josephson element*

A direct application of the non-linear inductance of the Josephson element is obtained by splitting a junction and its leads into 2 equal junctions, such that the resulting loop has an inductance much smaller the Josephson inductance. The two smaller junctions in parallel then behave as an effective junction [14] whose Josephson energy varies with Φ_{ext}, the magnetic flux externally imposed through the loop:

$$E_J\left(\Phi_{ext}\right) = E_J \cos\left(\pi\, \Phi_{ext}/\Phi_0\right) \tag{4.3}$$

Here, E_J the total Josephson energy of the two junctions. The Josephson energy can be modulated in a similar fashion by applying a magnetic field in the plane parallel to the tunnel layer.

5. The quantum isolated Josephson junction

5.1. *Form of the hamiltonian*

If we leave the leads of a Josephson junction unconnected, we obtain the simplest example of an non-linear electrical resonator. In order to analyse its quantum dynamics, we apply the prescriptions of quantum circuit theory briefly summarized in Appendix 1. Choosing a representation privileging the branch variables of the Josephson element, the momentum corresponds to the charge $Q = 2eN$ having tunneled through the element and the canonically conjugate position is the flux $\Phi = \varphi_0\theta$ associated with the superconducting phase difference across the tunnel layer. Here, N and θ are treated as operators that obey $[\theta, N] = i$. It is important to note that the operator N has integer eigenvalues whereas the phase θ is an operator corresponding to the position of a point on the unit circle (an angle modulo 2π).

By eliminating the branch charge of the capacitor, we obtain the hamiltonian

$$H = E_{CJ}\,(N - Q_r/2e)^2 - E_J \cos\theta \tag{5.1}$$

where $E_{CJ} = \frac{(2e)^2}{2C_J}$ is the Coulomb charging energy corresponding to one Cooper pair on the junction capacitance C_J and where Q_r is the residual offset charge on the capacitor.

One may wonder how the constant Q_r got into the hamiltonian, since no such term appeared in the corresponding LC circuit in section 3. The continuous charge Q_r is equal to the charge that pre-existed on the capacitor when it was wired with the inductor. Such offset charge is not some nit-picking theoretical construct. Its physical origin is a slight difference in work function between the two electrodes of the capacitor and/or an excess of charged impurities in the vicinity of one of the capacitor plates relative to the other. The value of Q_r is in practice very large compared to the Cooper pair charge $2e$, and since the hamiltonian 5.1 is invariant under the transformation $N \to N \pm 1$, its value can be considered completely random.

Such residual offset charge also exists in the LC circuit. However, we did not include it in our description of section 3 since a time-independent Q_r does not appear in the dynamical behavior of the circuit: it can be removed from the hamiltonian by performing a trivial canonical transformation leaving the form of the hamiltonian unchanged.

It is not possible, however, to iron this constant out of the isolated junction hamiltonian 5.1 because the potential is not quadratic in θ. The parameter Q_r plays a role here similar to the vector potential appearing in the hamiltonian of an electron in a magnetic field.

5.2. Fluctuations of the parameters of the hamiltonian

The hamiltonian 5.1 thus depends on three parameters which, following our discussion of the LC oscillator, we write as

$$\begin{aligned} Q_r &= Q_r^{stat} + \Delta Q_r(t) \\ E_C &= E_C^{stat} + \Delta E_C(t) \\ E_J &= E_J^{stat} + \Delta E_J(t) \end{aligned} \tag{5.2}$$

in order to distinguish the static sample-to-sample variation resulting from fabrication irreproducibility from the time-dependent fluctuations. While Q_r^{stat} can be considered fully random (see above discussion), E_C^{stat} and E_J^{stat} can generally be adjusted to a precision better than 20%. The relative fluctuations $\Delta Q_r(t)/2e$ and $\Delta E_J(t)/E_J$ are found to have a $1/f$ power spectral density with a typical standard deviations at 1Hz roughly of order $10^{-3}\text{Hz}^{-1/2}$ and $10^{-5}\text{Hz}^{-1/2}$, respectively, for a junction with a typical area of $0.01\mu\text{m}^2$ [15]. The noise appears to be produced by independent two-level fluctuators [16]. The relative fluctuations $\Delta E_C(t)/E_C$ are much less known, but the behavior of some glassy insulators at low temperatures might lead us to expect also a $1/f$ power spectral density, but probably with a weaker intensity than those of $\Delta E_J(t)/E_J$. We refer to the 3 noise terms in Eq.5.2 as offset charge, dielectric and critical current noises, respectively.

6. Why three basic types of Josephson qubits?

The first-order problem in realizing a Josephson qubit is to suppress as much as possible the detrimental effect of the fluctuations of Q_r, while retaining the non-linearity of the circuit. There are three main stategies for solving this problem and they lead to three fundamental basic type of qubits involving only one Josephson element.

6.1. The Cooper pair box

The simplest circuit is called the "Cooper pair box" and was first described theoretically, albeit in a slightly different version than presented here, by M. Büttiker [17]. It was first realized experimentally by the Saclay group in 1997 [18]. Quantum dynamic in the time domain was first seen by the NEC group in 1999 [19]. In the Cooper pair box, the variations of the residual offset charge Q_r are compensated by biasing the Josephson tunnel junction with a voltage source U in series with a "gate" capacitor C_g (see Fig. 4a). One can easily show that the hamiltonian of the Cooper pair box is

$$H = E_C\left(N - N_g\right)^2 - E_J \cos\theta \tag{6.1}$$

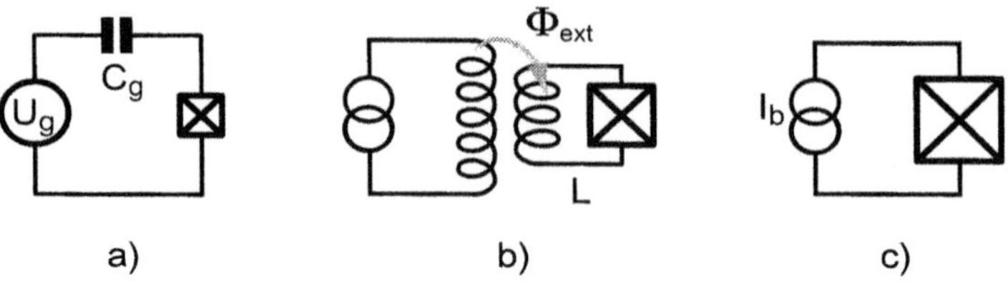

Fig. 4. a) Cooper pair box (prototypal charge qubit), b) RF-SQUID (prototypal flux qubit) and c) current-biased junction (prototypal phase qubit). The charge qubit and the flux qubit requires small junctions fabricated with e-beam lithography while the phase qubit can be fabricated with conventional optical lithography.

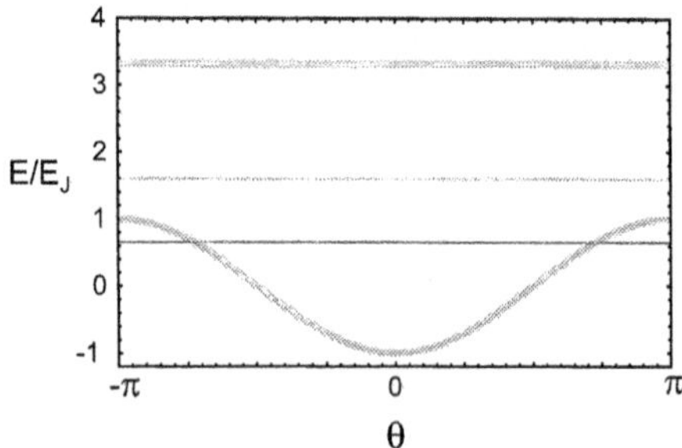

Fig. 5. Potential landscape for the phase in a Cooper pair box (thick solid line). The first few levels for $E_J/E_C = 1$ and $N_g = 1/2$ are indicated by thin horizontal solid lines.

Here $E_C = \frac{(2e)^2}{2(C_J+C_g)}$ is the charging energy of the island of the box and $N_g = Q_r + C_g U/2e$. Note that this hamiltonian has the same form as hamiltonian 5.1. Often N_g is simply written as $C_g U/2e$ since U at the chip level will deviate substantially from the generator value at high-temperature due to stray emf's in the low-temperature cryogenic wiring.

In Fig. 5 we show the potential in the θ representation as well as the first few energy levels for $E_J/E_C = 1$ and $N_g = 0$. As shown in Appendix 2, the Cooper pair box eigenenergies and eigenfunctions can be calculated with special functions known with arbitrary precision, and in Fig 6 we plot the first few eigenenergies as a function of N_g for $E_J/E_C = 0.1$ and $E_J/E_C = 1$. Thus, the Cooper box is to quantum circuit physics what the hydrogen atom is to atomic physics. We can modify the spectrum with the action of two externally controllable electrodynamic parameters: N_g, which is directly proportional to U, and E_J, which can be varied by applying a field through the junction or by using a split junction and applying a flux through the loop, as discussed in section 3. These parameters

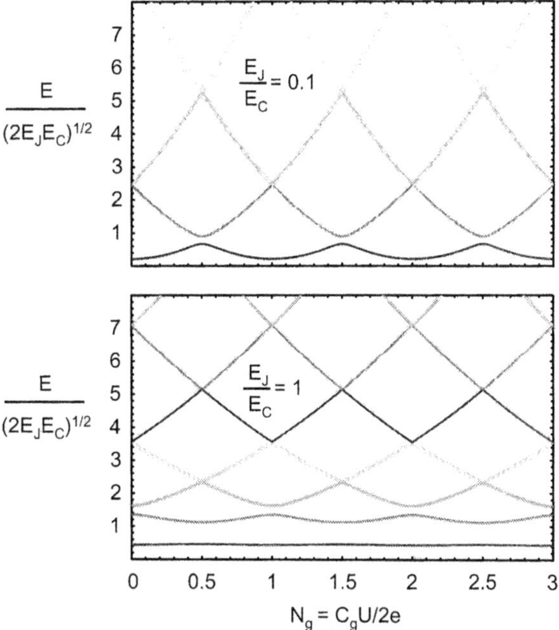

Fig. 6. Energy levels of the Cooper pair box as a function of N_g, for two values of E_J/E_C. As E_J/E_C increases, the sensitivity of the box to variations of offset charge diminishes, but so does the non-linearity. However, the non-linearity is the slowest function of E_J/E_C and a compromise advantageous for coherence can be found.

bear some resemblance to the Stark and Zeeman fields in atomic physics. For the box, however much smaller values of the fields are required to change the spectrum entirely.

We now limit ourselves to the two lowest levels of the box. Near the degeneracy point $N_g = 1/2$ where the electrostatic energy of the of the two charge states $|N = 0\rangle$ and $|N = 1\rangle$ are equal, we get the reduced hamiltonian [18, 20]

$$H_{qubit} = -E_z\left(\sigma_Z + X_{control}\sigma_X\right) \tag{6.2}$$

where, in the limit $E_J/E_C \ll 1$, $E_z = \frac{E_J}{2}$and $X_{control} = 2\frac{E_C}{E_J}\left(\frac{1}{2} - N_g\right)$. In Eq. 6.2, σ_Z and σ_X refer to the Pauli spin operators, with the X direction being chosen along the charge operator, the variable of the box we can naturally couple to.

If we plot the energy of the eigenstates of 6.2 as a function of the control parameter $X_{control}$, we obtain the universal level repulsion diagram shown in

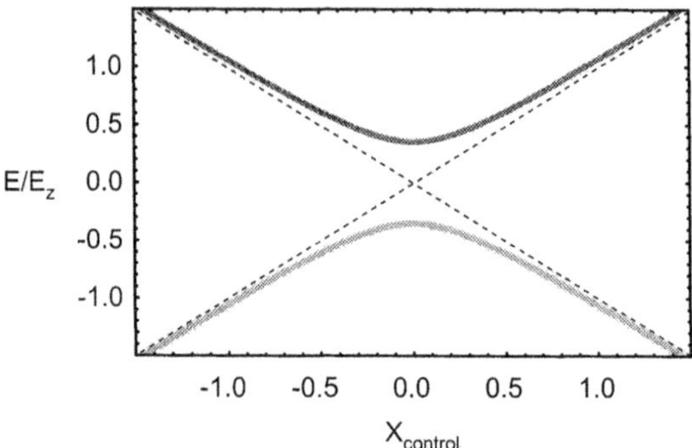

Fig. 7. Universal level anticrossing found both for the Cooper pair box and the RF-SQUID at their "sweet spot".

Fig. 7. Note that the minimum energy splitting is given by E_J. Comparing Eq. 6.2 with the spin hamiltonian in NMR, we see that E_J plays the role of the Zeeman field while the electrostatic energy plays the role of the transverse field. Indeed we can send on the control port corresponding to U time-varying voltage signals in the form of NMR-type pulses and prepare arbitrary superpositions of states [21].

The expression 6.2 shows that at the "sweet spot" $X_{control} = 0$, i.e. the degeneracy point $N_g = \frac{1}{2}$, the qubit transition frequency is to first order insentive to the offset charge noise ΔQ_r. We will discuss in the next section how an extension of the Cooper pair box circuit can display quantum coherence properties on long time scales by using this property.

In general, circuits derived from the Cooper pair box have been nicknamed "charge qubits". One should not think, however, that in charge qubits, quantum information is *encoded* with charge. Both the charge N and phase θ are quantum variables and they are both uncertain for a generic quantum state. Charge in "charge qubits" should be understood as refering to the "controlled variable", i.e. the qubit variable that couples to the control line we use to write or manipulate quantum information. In the following, for better comparison between the three qubits, we will be faithful to the convention used in Eq. 6.2, namely that σ_X represents the *controlled variable*.

6.2. The RF-SQUID

The second circuit – the so-called RF-SQUID [22] – can be considered in several ways the dual of the Cooper pair box (see Fig. 4b). It employs a superconducting transformer rather than a gate capacitor to adjust the hamiltonian. The two sides

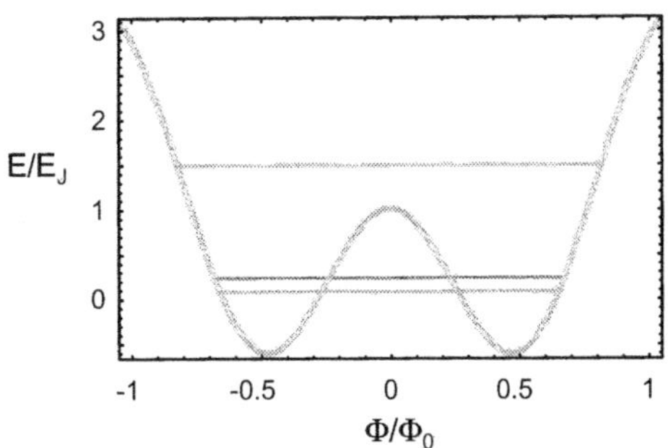

Fig. 8. Schematic potential energy landcape for the RF-SQUID.

of the junction with capacitance C_J are connected by a superconducting loop with inductance L. An external flux Φ_{ext} is imposed through the loop by an auxiliary coil. Using the methods of Appendix 1, we obtain the hamiltonian [7]

$$H = \frac{q^2}{2C_J} + \frac{\phi^2}{2L} - E_J \cos\left[\frac{2e}{\hbar}(\phi - \Phi_{ext})\right] \tag{6.3}$$

We are taking here as degree of freedom the integral ϕ of the voltage across the inductance L, i.e. the flux through the superconducting loop, and its conjugate variable, the charge q on the capacitance C_J; $[\phi, q] = i\hbar$. Note that in this representation, the phase θ, corresponding to the branch flux across the Josephson element, has been eliminated. Note also that the flux ϕ, in contrast to the phase θ, takes its values on a line and not on a circle. Likewise, its conjugate variable q, the charge on the capacitance, has continuous eigenvalues and not integer ones like N. Note that we now have three adjustable energy scales: E_J, $E_{CJ} = \frac{(2e)^2}{2C_J}$ and $E_L = \frac{\Phi_0^2}{2L}$.

The potential in the flux representation is schematically shown in Fig. 8 together with the first few levels, which have been seen experimentally for the first time by the SUNY group [23]. Here, no analytical expressions exist for the eigenvalues and the eigenfunctions of the problem, which has two aspect ratios: E_J/E_{CJ} and $\lambda = L_J/L - 1$.

Whereas in the Cooper box the potential is cosine-shaped and has only one well since the variable θ is 2π-periodic, we have now in general a parabolic potential with a cosine corrugation. The idea here for curing the detrimental effect of the offset charge fluctuations is very different than in the box. First of all Q_r^{stat} has been neutralized by shunting the 2 metallic electrodes of the junction by the superconducting wire of the loop. Then, the ratio E_J/E_{CJ} is chosen to be much larger than unity. This tends to increase the relative strength of quantum fluctuations of q, making offset charge fluctuations ΔQ_r small in comparison. The resulting loss in the non-linearity of the first levels is compensated by taking λ close to zero and by flux-biasing the device at the half-flux quantum value $\Phi_{ext} = \Phi_0/2$. Under these conditions, the potential has two degenerate wells separated by a shallow barrier with height $E_B = \frac{3\lambda^2}{2} E_J$. This corresponds to the degeneracy value $N_g = 1/2$ in the Cooper box, with the inductance energy in place of the capacitance energy. At $\Phi_{ext} = \Phi_0/2$, the two lowest energy levels are then the symmetric and antisymmetric combinations of the two wavefunctions localized in each well, and the energy splitting between the two states can be seen as the tunnel splitting associated with the quantum motion through the potential barrier between the two wells, bearing close resemblance to the dynamics of the ammonia molecule. This splitting E_S depends exponentially on the barrier height, which itself depends strongly on E_J. We have $E_S = \eta\sqrt{E_B E_{CJ}} \exp\left(-\xi\sqrt{E_B/E_{CJ}}\right)$ where the numbers η and ξ have to be determined numerically in most practical cases. The non-linearity of the first levels results thus from a subtle cancellation between two inductances: the superconducting loop inductance L and the junction effective inductance -L_{J0} which is opposed to L near $\Phi_{ext} = \Phi_0/2$. However, as we move away from the degeneracy point $\Phi_{ext} = \Phi_0/2$, the splitting $2E_\Phi$ between the first two energy levels varies linearly with the applied flux $E_\Phi = \zeta \frac{\Phi_0^2}{2L} (N_\Phi - 1/2)$. Here the parameter $N_\Phi = \Phi_{ext}/\Phi_0$, also called the flux frustration, plays the role of the reduced gate charge N_g. The coefficient ζ has also to be determined numerically. We are therefore again, in the vicinity of the flux degeneracy point $\Phi_{ext} = \Phi_0/2$ and for $E_J/E_{CJ} \gg 1$, in presence of the universal level repulsion behavior (see Fig. 7) and the qubit hamiltonian is again given by

$$H_{qubit} = -E_z\left(\sigma_Z + X_{control}\sigma_X\right) \tag{6.4}$$

where now $E_z = E_S/2$ and $X_{control} = 2\frac{E_\Phi}{E_S}\left(\frac{1}{2} - N_\Phi\right)$. The qubits derived from this basic circuit [24,32] have been nicknamed "flux qubits". Again, quantum information is not directly represented here by the flux ϕ, which is as uncertain for a general qubit state as the charge q on the capacitor plates of the junction. The flux ϕ is the system variable to which we couple when we write or control information in the qubit, which is done by sending current pulses on the primary of the RF-SQUID transformer, thereby modulating N_Φ, which itself determines the strength of the pseudo-field in the X direction in the hamiltonian 6.4. Note that the parameters E_S, E_Φ, and N_Φ are all influenced to some degree by the critical current noise, the dielectric noise and the charge noise. Another independent noise can also be present, the noise of the flux in the loop, which is not found in the box and which will affect only N_Φ. Experiments on DC-SQUIDS [14] have shown

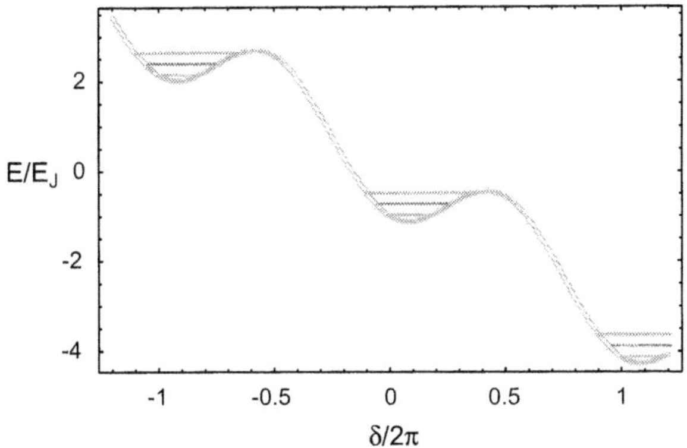

Fig. 9. Tilted washboard potential of the current-biased Josephson junction.

that this noise, in adequate conditions, can be as low as $10^{-8}(h/2e)/\mathrm{Hz}^{-1/2}$ at a few KHz. However, experimental results on flux qubits (see below) seem to indicate that larger apparent flux fluctuations are present, either as a result of flux trapping or critical current fluctuations in junctions implementing inductances.

6.3. Current-biased junction

The third basic quantum circuit biases the junction with a fixed DC-current source (Fig. 7c). Like the flux qubit, this circuit is also insensitive to the effect of offset charge and reduces the effect of charge fluctuations by using large ratios of E_J/E_{CJ}. A large non-linearity in the Josephson inductance is obtained by biasing the junction at a current I very close to the critical current. A current bias source can be understood as arising from a loop inductance with $L \to \infty$ biased by a flux $\Phi \to \infty$ such that $I = \Phi/L$. The Hamiltonian is given by

$$H = E_{CJ}p^2 - I\varphi_0\delta - I_0\varphi_0 \cos\delta\,, \tag{6.5}$$

where the gauge invariant phase difference operator δ is, apart from the scale factor φ_0, precisely the branch flux across C_J. Its conjugate variable is the charge $2ep$ on that capacitance, a continuous operator. We have thus $[\delta, p] = i$. The variable δ, like the variable ϕ of the RF-SQUID, takes its value on the whole real axis and its relation with the phase θ is $\delta \bmod 2\pi = \theta$ as in our classical analysis of section 4.

The potential in the δ representation is shown in Fig. 9. It has the shape of a tilted washboard, with the tilt given by the ratio I/I_0. When I approaches I_0, the phase is $\delta \approx \pi/2$, and in its vicinity, the potential is very well approximated by the cubic form

$$U(\delta) = \varphi_0 (I_0 - I)(\delta - \pi/2) - \frac{I_0 \varphi_0}{6} (\delta - \pi/2)^3 \tag{6.6}$$

Note that its shape depends critically on the difference $I_0 - I$. For $I \lesssim I_0$, there is a well with a barrier height $\Delta U = (2\sqrt{2}/3) I_0 \varphi_0 (1 - I/I_0)^{3/2}$ and the classical oscillation frequency at the bottom of the well (so-called plasma oscillation) is given by

$$\begin{aligned} \omega_p &= \frac{1}{\sqrt{L_J(I) C_J}} \\ &= \frac{1}{\sqrt{L_{J0} C_J}} \left[1 - (I/I_0)^2\right]^{1/4} \end{aligned}$$

Quantum-mechanically, energy levels are found in the well (see Fig. 11) [3] with non-degenerate spacings. The first two levels can be used for qubit states [25], and have a transition frequency $\omega_{01} \simeq 0.95\omega_p$.

A feature of this qubit circuit is built-in readout, a property missing from the two previous cases. It is based on the possibility that states in the cubic potential can tunnel through the cubic potential barrier into the continuum outside the barrier. Because the tunneling rate increases by a factor of approximately 500 each time we go from one energy level to the next, the population of the $|1\rangle$ qubit state can be reliably measured by sending a probe signal inducing a transition from the 1 state to a higher energy state with large tunneling probability. After tunneling, the particle representing the phase accelerates down the washboard, a convenient self-amplification process leading to a voltage $2\Delta/e$ across the junction. Therefore, a finite voltage $V \neq 0$ suddenly appearing across the junction just after the probe signal implies that the qubit was in state $|1\rangle$, whereas $V = 0$ implies that the qubit was in state $|0\rangle$.

In practice, like in the two previous cases, the transition frequency $\omega_{01}/2\pi$ falls in the 5-20GHz range. This frequency is only determined by material properties of the barrier, since the product $C_J\ L_J$ does not depend on junction area. The number of levels in the well is typically $\Delta U/\hbar\omega_p \approx 4$.

Setting the bias current at a value I and calling ΔI the variations of the difference $I - I_0$ (originating either in variations of I or I_0), the qubit Hamiltonian is given by

$$H_{qubit} = \hbar\omega_{01}\sigma_Z + \sqrt{\frac{\hbar}{2\omega_{01} C_J}} \Delta I (\sigma_X + \chi\sigma_Z), \tag{6.7}$$

where $\chi = \sqrt{\hbar\omega_{01}/3\Delta U} \simeq 1/4$ for typical operating parameters. In contrast with the flux and phase qubit circuits, the current-biased Josephson junction does not have a bias point where the $0 \rightarrow 1$ transition frequency has a local minimum. The hamiltonian cannot be cast into the NMR-type form of Eq. 6.2. However, a sinusoidal current signal $\Delta I\ (t) \sim \sin\omega_{01}t$ can still produce σ_X rotations, whereas a low-frequency signal produces σ_Z operations [26].

In analogy with the preceding circuits, qubits derived from this circuit and/or having the same phase potential shape and qubit properties have been nicknamed "phase qubits" since the controlled variable is the phase (the X pseudo-spin direction in hamiltonian 6.7).

6.4. *Tunability versus sensitivity to noise in control parameters*

The reduced two-level hamiltonians Eqs. 6.2,6.4 and 6.7 have been tested thoroughly and are now well-established. They contain the very important parametric dependence of the coefficient of σ_X, which can be viewed on one hand as how much the qubit can be tuned by an external control parameter, and on the other hand as how much it can be dephased by uncontrolled variations in that parameter. It is often important to realize that even if the control parameter has a very stable value at the level of room-temperature electronics, the noise in the electrical components relaying its value at the qubit level might be inducing detrimental fluctuations. An example is the flux through a superconducting loop, which in principle could be set very precisely by a stable current in a coil, and which in practice often fluctuates because of trapped flux motion in the wire of the loop or in nearby superconducting films. Note that, on the other hand, the two-level hamiltonian does not contain the non-linear properties of the qubit, and how they conflict with its intrinsic noise, a problem which we discuss in the next subsection.

6.5. *Non-linearity versus sensitivity to intrinsic noise*

The three basic quantum circuit types discussed above illustrate a general tendency of Josephson qubits. If we try to make the level structure very non-linear, i.e. $|\omega_{01} - \omega_{12}| \gg \omega_{01}$, we necessarily expose the system sensitively to at least one type of intrinsic noise. The flux qubit is contructed to reach a very large non-linearity, but is also maximally exposed, relatively speaking, to critical current noise and flux noise. On the other hand, the phase qubit starts with a relatively small non-linearity and acquires it at the expense of a precise tuning of the difference between the bias current and the critical current, and therefore exposes itself also to the noise in the latter. The Cooper box, finally, acquires non-linearity at the expense of its sensitivity to offset charge noise. The search for the optimal qubit circuit involves therefore a detailed knowledge of the relative intensities of the various sources of noise, and their variations with all the construction parameters of the qubit, and in particular – this point is crucial – the properties of the materials involved in the tunnel junction fabrication. No such in-depth of knowledge exists at the time of this writing and one can only make educated guesses.

The qubit optimization problem is also further complicated by the necessity to readout quantum information, which we address just after reviewing the relationships between the intensity of noise and the decay rates of quantum information.

7. Qubit relaxation and decoherence

A generic quantum state of a qubit can be represented as a unit vector $\overrightarrow{S}$ pointing on a sphere, the so-called Bloch sphere. One distinguishes two broad classes of errors. The first one corresponds to the tip of the Bloch vector diffusing along a meridian, i.e. a great circle passing through the poles (latitude fluctuations). This process is called energy relaxation or state-mixing. The second class corresponds to the tip of the Bloch vector diffusing along a parallel, i.e. a circle perpendicular to the line joining the two poles (longitude fluctuations). This process is called dephasing or decoherence.

In Appendix 2 we define precisely these rates and show that they are directly proportional to the power spectral densities of the noises entering in the parameters of the hamiltonian of the qubit. More precisely, we find that the decoherence rate is proportional to the total spectral density of the quasi-zero-frequency noise in the qubit frequency. The relaxation rate, on the other hand, is proportional to the total spectral density near the qubit frequency of the noise in the field perpendicular to the eigenaxis of the qubit.

In principle, the expressions for the relaxation and decoherence rate could lead to a ranking of the various qubit circuits: from their reduced spin hamiltonian, one can find with what coefficient each basic noise source contributes to the various spectral densities entering in the rates. One could then optimize the various parameters of the qubit to greatly reduce its sensitivity to noise. However, before discussing this question further, we must realize that the readout itself can provide substantial additional noise sources for the qubit. Therefore, the design of a qubit circuit that maximizes the number of coherent gate operations is a subtle optimization problem which must treat in parallel both the intrinsic noises of the qubit and the back-action noise of the readout.

8. Readout of superconducting qubits

8.1. Formulation of the readout problem

We have examined so far the various basic circuits for qubit implementation and their associated methods to write and manipulate quantum information. Another important task quantum circuits must perform is the readout of that information. As we mentioned earlier, the difficulty of the readout problem is to open a coupling channel to the qubit for extracting information without at the same time submitting it to noise.

Ideally, the readout part of the circuit – referred to in the following simply as "readout" – should include both a switch, which defines an "OFF" and an "ON" phase, and a state measurement device. During the OFF phase, where reset and gate operations take place, the measurement device should be completely decoupled from the qubit degrees of freedom. During the ON phase, the measurement device should be maximally coupled to a qubit variable that distinguishes the 0 and the 1 state. However, this condition is not sufficient. The back-action of the measurement device during the ON phase should be weak enough not to relax the qubit [27].

The readout can be characterized by 4 parameters. The first one describes the sensitivity of the measuring device while the next two describes its back-action, factoring in the quality of the switch (see Appendix 3 for their definition):

i) the measurement time τ_m defined as the time taken by the measuring device to reach a signal-to-noise ratio of 1 in the determination of the state.

ii) the energy relaxation time Γ_1^{ON} of the qubit in the ON state.

iii) the coherence decay rate Γ_2^{OFF} of the qubit information in the OFF state.

iv) the dead time t_d needed to reset the measuring device after a qubit measurement. The readout is usually perturbed by the energy expenditure associated with producing a signal strong enough for external detection.

Simultaneously minimizing these parameters to improve readout performance cannot be done without running into conflicts. An important quantity to optimize is the readout fidelity. By construction, at the end of the ON phase, the readout should have reached one of two classical states: 0_c and 1_c, the outcomes of the measurement process. The latter can be described by 2 probabilities: the probability $p_{00_c}(p_{11_c})$ that starting from the qubit state $|0\rangle$ ($|1\rangle$) the measurement yields $0_c(1_c)$. The readout fidelity (or discriminating power) is defined as $F = p_{00c} + p_{11_c} - 1$. For a measuring device with a signal-to-noise ratio increasing like the square of measurement duration τ, we would have, if back-action could be neglected, $F = \operatorname{erf}\left(2^{-1/2}\tau/\tau_m\right)$.

8.2. Requirements and general strategies

The fidelity and speed of the readout, usually not discussed in the context of quantum algorithms because they enter marginally in the evaluation of their complexity, are actually key to experiments studying the coherence properties of qubits and gates. A very fast and sensitive readout will gather at a rapid pace information on the imperfections and drifts of qubit parameters, thereby allowing the experimenter to design fabrication strategies to fight them or even correct them in real time.

We are thus mostly interested in "single-shot" readouts [27], for which F is order unity, as opposed to schemes in which a weak measurement is performed continuously [28]. If $F \ll 1$, of order F^{-2} identical preparation and readout cycles need to be performed to access the state of the qubit. The condition for "single-shot" operation is

$$\Gamma_1^{ON} \tau_m < 1$$

The speed of the readout, determined both by τ_m and t_d, should be sufficiently fast to allow a complete characterization of all the properties of the qubit before any drift in parameters occurs. With sufficient speed, the automatic correction of these drifts in real time using feedback will be possible.

Rapidly pulsing the readout on and off with a large decoupling amplitude such that

$$\Gamma_2^{OFF} T_2 - 1 \ll 1$$

requires a fast, strongly non-linear element, which is provided by one or more auxiliary Josephson junctions. Decoupling the qubit from the readout in the OFF phase requires balancing the circuit in the manner of a Wheatstone bridge, with the readout input variable and the qubit variable corresponding to 2 orthogonal electrical degrees of freedom. Finally, to be as complete as possible even in presence of small asymmetries, the decoupling also requires an impedance mismatch between the qubit and the dissipative degrees of freedom of the readout. In the next subsection, we discuss how these general ideas have been implemented in 2nd generation quantum circuits. The examples we have chosen all involve a readout circuit which is built-in the qubit itself to provide maximal coupling during the ON phase, as well as a decoupling scheme which has proven effective for obtaining long decoherence times.

8.3. Phase qubit: tunneling readout with a DC-SQUID on-chip amplifier.

The simplest example of a readout is provided by a system derived from the phase qubit (See Fig. 10). In the phase qubit, the levels in the cubic potential

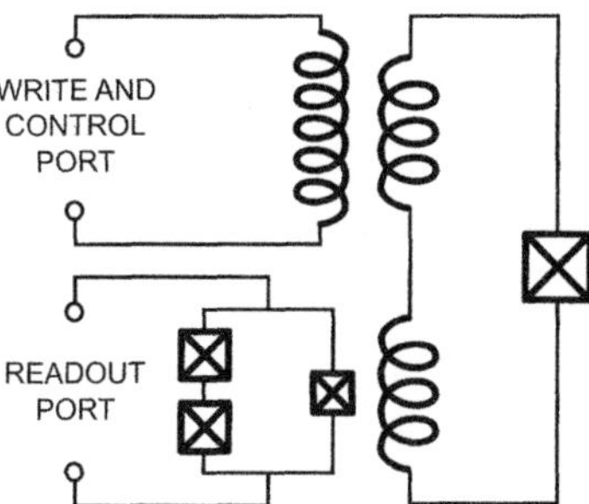

Fig. 10. Phase qubit implemented with a Josephson junction in a high-inductance superconducting loop biased with a flux sufficiently large that the phase across the junction sees a potential analogous to that found for the current-biased junction. The readout part of the circuit is an asymmetric hysteretic SQUID which is completely decoupled from the qubit in the OFF phase. Isolation of the qubit both from the readout and control port is obtained through impedance mismatch of transformers.

are metastable and decay in the continuum, with level $n+1$ having roughly a decay rate Γ_{n+1} 500 times faster than the decay Γ_n of level n. This strong level number dependence of the decay rate leads naturally to the following readout scheme: when readout needs to be performed, a microwave pulse at the transition frequency ω_{12} (or better at ω_{13}) transfers the eventual population of level 1 into level 2, the latter decaying rapidly into the continuum, where it subsequently loses energy by friction and falls into the bottom state of the next corrugation of the potential (because the qubit junction is actually in a superconducting loop of large but finite inductance, the bottom of this next corrugation is in fact the absolute minimum of the potential and the particle representing the system can stay an infinitely long time there). Thus, at the end of the readout pulse, the sytem has either decayed out of the cubic well (readout state 1_c) if the qubit was in the $|1\rangle$ state or remained in the cubic well (readout state 0_c) if the qubit was in the $|0\rangle$ state. The DC-SQUID amplifier is sensitive enough to detect the change in flux accompanying the exit of the cubic well, but the problem is to avoid sending the back-action noise of its stabilizing resistor into the qubit circuit. The solution to this problem involves balancing the SQUID loop in such a way, that for readout state 0_c, the small signal gain of the SQUID is zero, whereas for readout state 1_c, the small signal gain is non-zero [16]. This signal dependent gain is obtained by having 2 junctions in one arm of the SQUID whose total Josephson inductance equals that of the unique junction in the other arm. Finally, a large impedance mismatch between the SQUID and the qubit is obtained by a transformer. The fidelity of such readout is remarkable: 95% has been demonstrated. In Fig. 11, we show the result of a measurement of Rabi oscillations with such qubit+readout.

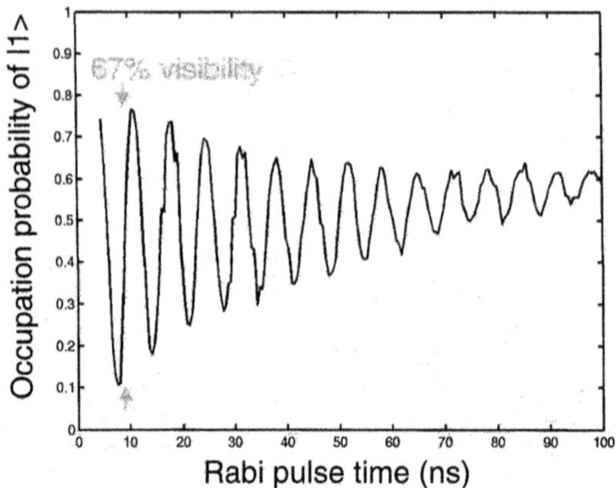

Fig. 11. Rabi oscillations observed for the qubit of Fig. 10.

8.4. Cooper-pair box with non-linear inductive readout: the "Quantronium" circuit

The Cooper-pair box needs to be operated at its "sweet spot" (degeneracy point) where the transition frequency is to first order insensitive to offset charge fluctuations. The "Quantronium" circuit presented in Fig. 12 is a 3-junction bridge configuration with two small junctions defining a Cooper box island, and thus a charge-like qubit which is coupled capacitively to the write and control port (high-impedance port). There is also a large third junction, which provides a non-linear inductive coupling to the read port. When the read port current I is zero, and the flux through the qubit loop is zero, noise coming from the read port is decoupled from the qubit, provided that the two small junctions are identical both in critical current and capacitance. When I is non-zero, the junction bridge is out of balance and the state of the qubit influences the effective non-linear inductance seen from the read port. A further protection of the impedance mismatch type is obtained by a shunt capacitor across the large junction: at the resonance frequency of the non-linear resonator formed by the large junction and the external capacitance C, the differential mode of the circuit involved in the readout presents an impedance of the order of an ohm, a substantial decoupling from the 50Ω transmission line carrying information to the amplifier stage. The readout protocol involves a DC pulse [21, 29] or an RF pulse [30] stimulation of the readout mode. The response is bimodal, each mode corresponding to a state of the qubit. Although the theoretical fidelity of the DC readout can attain 95%, only a maximum of 40% has been obtained so far. The cause of this discrepancy

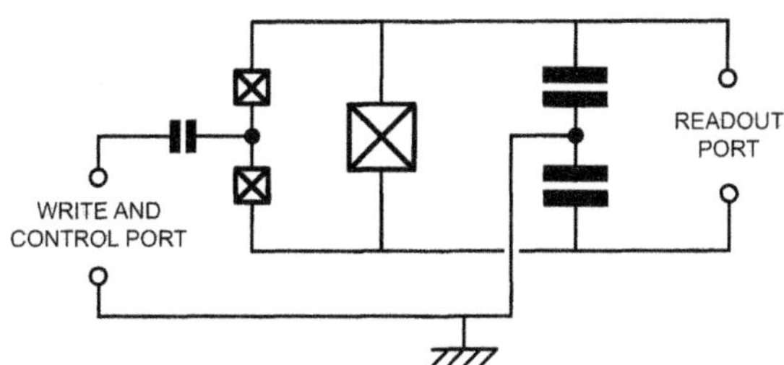

Fig. 12. "Quantronium" circuit consisting of a Cooper pair box with a non-linear inductive readout. A Wheatstone bridge configuration decouples qubit and readout variables when readout is OFF. Impedance mismatch isolation is also provided by additional capacitance in parallel with readout junction.

is still under investigation.

In Fig. 13 we show the result of a Ramsey fringe experiment demonstrating that the coherence quality factor of the quantronium can reach 25 000 at the sweet spot [21]. By studying the degradation of the qubit absorption line and of the Ramsey fringes as one moves away from the sweet spot, it has been possible to show that the residual decoherence is limited by offset charge noise and by flux noise [31]. In principle, the influence of these noises could be further reduced by a better optimization of the qubit design and parameters. In particular, the operation of the box can tolerate ratios of E_J/E_C around 4 where the sensitivity to offset charge is exponentially reduced and where the non-linearity is still of order 15%. The quantronium circuit has so far the best coherence quality factor. We believe this is due to the fact that critical current noise, one dominant intrinsic source of noise, affects this qubit far less than the others, relatively speaking, as can be deduced from the qubit hamiltonians of section 6.

8.5. 3-junction flux qubit with built-in readout

Fig. 14 shows a third example of buit-in readout, this time for a flux-like qubit. The qubit by itself involves 3 junctions in a loop, the larger two of the junctions playing the role of the loop inductance in the basic RF-SQUID [32]. The advantage of this configuration is to reduce the sensitivity of the qubit to external flux variations. The readout part of the circuit involves 2 other junctions forming a hysteretic DC-SQUID whose offset flux depends on the qubit flux state. The critical current of this DC-SQUID has been probed by a DC pulse, but an RF pulse could be applied as in another flux readout. Similarly to the two previous cases, the readout states 1_c and 0_c, which here correspond to the DC-SQUID having switched or not, map very well the qubit states $|1\rangle$ and $|0\rangle$, with a fidelity better than 60%. Here also, a bridge technique orthogonalizes the readout mode, which

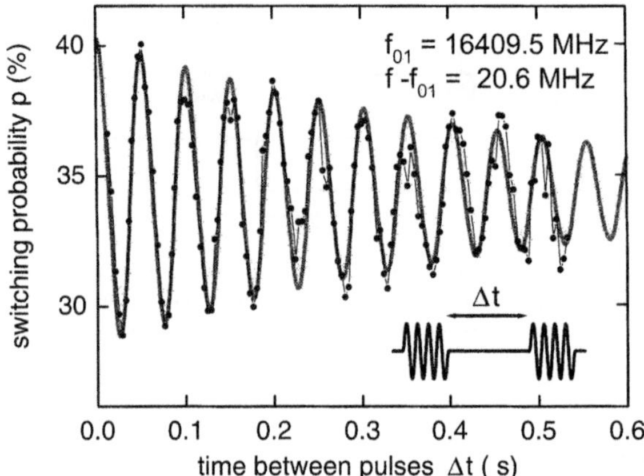

Fig. 13. Measurement of Ramsey fringes for the Quantronium. Two $\pi/2$ pulses separated by a variable delay are applied to the qubit before measurement. The frequency of the pulse is slightly detuned from the transition frequency to provide a stroboscopic measurement of the Larmor precession of the qubit.

is the common mode of the DC-SQUID, and the qubit mode, which is coupled to the loop of the DC-SQUID. External capacitors provide additional protection through impedance mismatch. Fig. 15 shows Ramsey fringes obtained with this system.

8.6. Too much on-chip dissipation can be bad: Do not stir up the dirt!

All the circuits above include an on-chip amplification scheme producing high-level signals which can be read directly by high-temperature low-noise electronics. In the second and third examples, these signals lead to non-equilibrium quasi-particle excitations being produced in the near vicinity of the qubit junctions. An elegant experiment has recently demonstrated that the presence of these excitations increases the offset charge noise [33]. More generally, one can legitimately worry that large energy dissipation on the chip itself will lead to an increase of the noises discussed in section 5.2. A broad class a new readout schemes addresses this question [30, 34, 35]. They are based on a purely dispersive measurement of a qubit susceptibility (capacitive or inductive). A probe signal is sent to the qubit. The signal is coupled to a qubit variable whose average value is identical in the 2 qubit states (for instance, in the capacitive susceptibility, the variable is the island charge in the charge qubit at the degeneracy point). The state-dependent phase

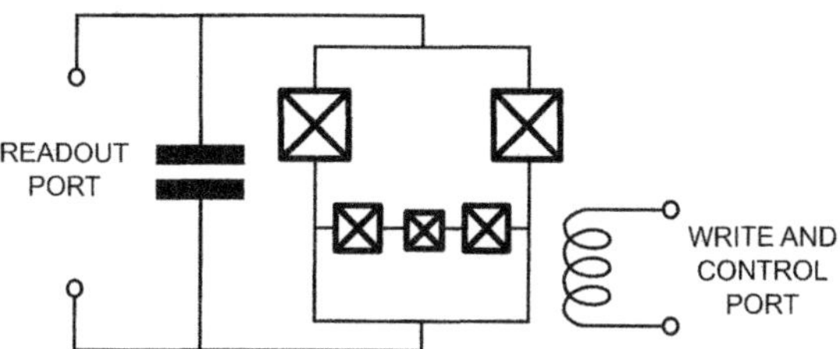

Fig. 14. Three-junction flux qubit with a non-linear inductive readout. The medium-size junctions play the role of an inductor. Bridge configuration for nulling out back-action of readout is also employed here, as well as impedance mismatch provided by additional capacitance.

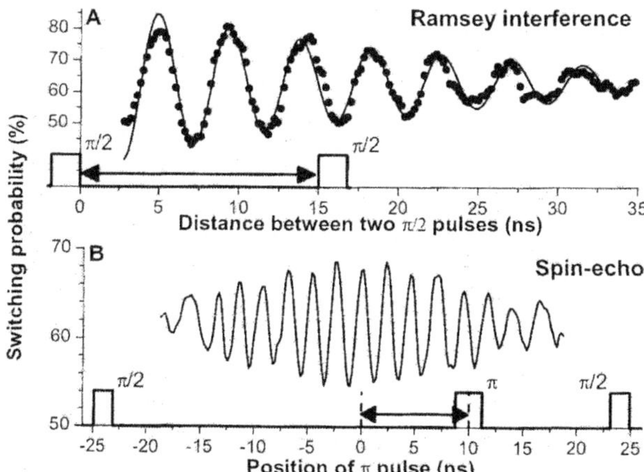

Fig. 15. Panel A: Ramsey fringes obtained for qubit of Fig. 14. Panel B: echo showing the fast dynamics of decoherence processes.

shift of the reflected signal is then amplified by a linear low-temperature amplifier and finally discriminated at high temperature against an adequately chosen threshold. In addition to being very thrifty in terms of energy being dissipated on chip, these new schemes also provide a further natural decoupling action: when the probe signal is off, the back-action of the amplifier is also completely shut off.

9. Coupling superconducting qubits

A priori, 3 types of coupling scheme can be envisioned:

a) In the first type, the transition frequency of the qubits are all equal and the coupling between any pair is switched on using one or several junctions as non-linear elements [36, 37].

b) In the second type, the couplings are fixed, but the transition frequencies of a pair of qubits, originally detuned, are brought on resonance when the coupling between them needs to be turned on [38, 39].

c) In the third type, which bears close resemblance to the methods used in NMR [1], the couplings and the resonance frequencies of the qubits remain fixed, the qubits being always detuned. Being off-diagonal, the coupling elements have negligible action on the qubits. However, when a strong microwave field is applied to the target and control qubits at their mean frequency, they become in "speaking terms" for the exchange of energy quanta and gate action can take place [40].

So far only scheme b) has been tested experimentally.

The advantage of schemes b) and c) is that they work with purely passive reactive elements like capacitors and inductors which should remain very stable as a function of time and which also should present very little high-frequency noise. In a way, we must design quantum integrated circuits in the manner that vacuum tube radios were designed in the 50's: only 6 tubes were used for a complete heterodyne radio set, including the power supply. Nowadays several hundreds of transistors are used in a radio or any hi-fi system. In that ancient era of classical electronics, linear elements like capacitors, inductors or resistors were "free" because they were relatively reliable whereas tubes could break down easily. We have to follow a similar path in quantum integrated circuit, the reliability issues having become noise minimization issues.

10. Can coherence be improved with better materials?

Up to now, we have discussed how, given the power spectral densities of the noises ΔQ_r, ΔE_C and ΔE_J, we could design a qubit equipped with control, readout and coupling circuits. It is worthwhile to ask at this point if we could improve the material properties to gain in the coherence of the qubit, assuming all other problems like noise in the control channels and the back-action of the readout have been solved. A model put forward by one of us (JMM) and collaborators shed some light on the direction one would follow to answer this question. The $1/f$ spectrum of the materials noises suggests that they all originate from 2-level fluctuators in the amorphous alumina tunnel layer of the junction itself, or its close vicinity. The substrate or the surface of the superconducting films are also suspect in the case of ΔQ_r and ΔE_C but their influence would be relatively weaker and we ignore them for simplicity. These two-level systems are supposed to be randomly distributed positional degrees of freedom ξ_i with effective spin-1/2 properties, for instance an impurity atom tunneling between two adjacent potential wells. Each two-level system is in principle characterized by 3 parameters: the energy splitting $\hbar\omega_i$, and the two coefficients α_i and β_i of the Pauli matrix representation of $\xi_i = \alpha_i \sigma_{iz} + \beta_i \sigma_{ix}$. The random nature of the problem leads us to suppose that α_i and β_i are both Gaussian random variables with the same standard deviation ρ_i. By carrying a charge, the thermal and quantum motion of ξ_i can contribute to $\Delta Q_r = \sum_i q_i \xi_i$ and $\Delta E_C = \sum_i c_i \frac{\beta_i^2}{\omega_i} \sigma_{iz}$. Likewise, by modifying the transmission of a tunneling channel in its vicinity, the motion of ξ_i can contribute to $\Delta E_J = \sum_i g_i \xi_i$. We can further suppose that the quality of the material of the junction is simply characterized by a few numbers. The essential one is the density ν of the transition frequencies ω_i in frequency space and in real space, assuming a ω^{-1} distribution (this is necessary to explain the $1/f$ behavior) and a uniform spatial distribution on the surface of the junction. Recent experiments indicate that the parameter ν is of order$10^5 \mu\mathrm{m}^{-2}(\mathrm{decade})^{-1}$. Then, assuming a universal ρ independent of frequency, only one coefficient is needed per noise, namely, the average modulation efficiency of each fluctuator. Such analysis provides a common language for describing various experiments probing the dependence of decoherence on the material of the junction. Once the influence of the junction fabrication parameters (oxydation pressure and temperature, impurity contents, and so on) on these noise intensities will be known, it will be possible to devise optimized fabrication procedures, in the same way perhaps as the $1/f$ noise in C-MOS transistors has been reduced by careful material studies.

11. Concluding remarks and perspectives

The logical thread through this review of superconducting qubits has been the question "What is the best qubit design?". We unfortunately cannot, at present, conclude by giving a definitive answer to this complex optimisation problem.

Yet, a lot has already been achieved, and superconducting qubits are becoming serious competitors of trapped ions and atoms. The following properties of quantum circuits have been demonstrated:

a) Coherence quality factors $Q_\varphi = T_\varphi \omega_{01}$ can attain at least 2.10^4
b) Readout and reset fidelity can be greater than 95%
c) All states on the Bloch sphere can be addressed
d) Spin echo techniques can null out low frequency drift of offset charges
e) Two qubits can be coupled and RF pulses can implement gate operation
f) A qubit can be fabricated using only optical lithography techniques

The major problem we are facing is that these various results have not been obtained at the same time IN THE SAME CIRCUIT, although succesful design elements in one have often been incorporated into the next generation of others. The complete optimization of the single qubit+readout has not been achieved yet. However, we have presented in this review the elements of a systematic methodology resolving the various conflicts that are generated by all the different requirements. Our opinion is that, once noise sources are better characterized, an appropriate combination of all the known circuit design strategies for improving coherence, as well as the understanding of optimal tunnel layer growth conditions for lowering the intrinsic noise of Josephson junctions, should lead us to reach the 1-qubit and 2-qubit coherence levels needed for error correction [43]. Along the way, good medium term targets to test overall progress on the simultaneous fronts of qubit coherence, readout and gate coupling are the measurement of Bell 's inequality violation or the implementation of the Deutsch-Josza algorithm, both of which requiring the simultaneous satisfaction of properties a)-e).

Acknowledgements

The authors have greatly benefited from discussions with I. Chuang, D. Esteve, S. Girvin, S. Lloyd, H. Mooij, R. Schoelkopf, I. Siddiqi, C. Urbina and D. Vion. They would like also to thank the participants of the Les Houches Summer School on Quantum Information Processing and Entanglement held in 2003 for useful exchanges. Finally, funding from ARDA/ARO and the Keck Fundation is gratefully acknowledged.

12. Appendix1: Quantum circuit theory

The problem we are addressing in this section is, given a superconducting circuit made up of capacitors, inductors and Josephson junctions, how to systematically write its quantum hamiltonian, the generating function from which the quantum dynamics of the circuit can be obtained. This problem has been considered first by Yurke and Denker [44] in a seminal paper and analyzed in further details by Devoret [45]. We will only summarize here the results needed for this review.

The circuit is given as a set of branches, which can be capacitors, inductors or Josephson tunnel elements, connected at nodes. Several independent paths formed by a succession of branches can be found between nodes. The circuit can therefore contain one or several loops. It is important to note that a circuit has not one hamiltonian but many, each one depending on a particular representation. We are describing here one particular type of representation, which is usually well adapted to circuits containing Josephson junctions. Like in classical circuit theory, a set of independent current and voltages has to be found for a particular representation. We start by associating to each branch b of the circuit, the current i_b flowing through it and the voltage v_b across it (a convention has to be made first on the direction of the branches). Kirchhoff's laws impose relations among branch variables and some of them are redundant. The following procedure is used to eliminate redundant branches: one node of the circuit is first chosen as ground. Then from the ground, a loop-free set of branches called spanning tree is selected. The rule behind the selection of the spanning tree is the following: each node of the circuit must be linked to the ground by one and only one path belonging to the tree. In general, inductors (linear or non-linear) are preferred as branches of the tree but this is not necessary. Once the spanning tree is chosen (note that we still have many possibilities for this tree), we can associate to each node a "node voltage" v_n which is the algebraic sum of the voltages along the branches between ground and the node. The conjugate "node current" i_n is the algebraic sum of all currents flowing to the node through capacitors ONLY. The dynamical variables appearing in the hamiltonian of the circuit are the node fluxes and node charges defined as

$$\phi_n = \int_{-\infty}^{t} v(t_1)\, dt_1$$

$$q_n = \int_{-\infty}^{t} i(t_1)\, dt_1$$

Using Kirchhoff's laws, it is possible to express the flux and the charge of each branch as a linear combination of all the node fluxes and charges, respectively. In this inversion procedure, the total flux through loops imposed by external flux bias sources and polarisation charges of nodes imposed by charge bias sources, appear.

If we now sum the energies of all branches of the circuit expressed in terms of node flux and charges, we will obtain the hamiltonian of the circuit corresponding to the representation associated with the particular spanning tree. In this hamiltonian, capacitor energies behave like kinetic terms while the inductor energies behave as potential terms. The hamiltonian of the LC circuit written in section 2 is an elementary example of this procedure.

Once the hamiltonian is obtained it is easy get its quantum version by replacing all the node fluxes and charges by their quantum operator equivalent. The flux and charge of a node have a commutator given by $i\hbar$, like the position and momentum of a particle:

$$\begin{aligned} \phi &\rightarrow \hat{\phi} \\ q &\rightarrow \hat{q} \\ \left[\hat{\phi}, \hat{q}\right] &= i\hbar \end{aligned}$$

One can also show that the flux and charge operators corresponding to a branch share the same commutation relation. Note that for the special case of the Josephson element, the phase $\hat{\theta}$ and Cooper pair number $\hat{N}$, which are its dimensionless electric variables, have the property:

$$\left[\hat{\theta}, \hat{N}\right] = i$$

In the so-called charge basis, we have

$$\begin{aligned} \hat{N} &= \sum_N N \left|N\right\rangle \left\langle N\right| \\ \cos\hat{\theta} &= \frac{1}{2}\sum_N \left(\left|N\right\rangle \left\langle N+1\right| + \left|N+\right\rangle \left\langle N\right|\right) \end{aligned}$$

while in the so-called phase basis, we have

$$\hat{N} = \left|\theta\right\rangle \frac{\partial}{i\partial} \left\langle \theta\right|$$

Note that since the Cooper pair number $\hat{N}$ is an operator with integer eigenvalues, its conjugate variable $\hat{\theta}$, has eigenvalues behaving like angles, i.e. they are defined only modulo 2π.

In this review, outside this appendix, we have dropped the hat on operators for simplicity.

13. Appendix 2: Eigenenergies and eigenfunctions of the Cooper pair box

From Appendix 1, it easy to see that the hamiltonian of the Cooper pair box leads to the Schrodinger equation

$$\left[E_C\left(\frac{\partial}{i\partial\theta}-N_g\right)^2-E_J\cos\theta\right]\Psi_k(\theta)=E_k\Psi_k(\theta)$$

The functions $\Psi_k(\theta)\,\mathrm{e}^{-iN_g}$ and energies E_k are solutions of the Mathieu equation and can be found with arbitrary precision for all values of the parameters N_g and E_J/E_C [46]. For instance, using the program Mathematica, we find

$$\begin{aligned}E_k &= E_C\mathcal{M}_A\left[k+1-(k+1)\bmod 2+2N_g(-1)^k,-2E_J/E_C\right]\\ \Psi_k(\theta) &= \frac{e^{iN_g\theta}}{\sqrt{2\pi}}\left\{\mathcal{M}_C\left[\frac{4E_k}{E_C},\frac{-2E_J}{E_C},\frac{\theta}{2}\right]\right.\\ &\quad\left.+i(-1)^{k+1}\mathcal{M}_S\left[\frac{4E_k}{E_C},\frac{-2E_J}{E_C},\frac{\theta}{2}\right]\right\}\end{aligned}$$

where $\mathcal{M}_A(r,q)=$ `MathieuCharacteristicA[r,q]`,
$\mathcal{M}_C(a,q,z)=$ `MathieuC[a,q,z]`,
$\mathcal{M}_S(a,q,z)=$ `MathieuS[a,q,z]`.

14. Appendix 3: Relaxation and decoherence rates for a qubit

Definition of the rates

We start by introducing the spin eigenreference frame $\hat{z}$, $\hat{x}$ and $\hat{y}$ consisting of the unit vector along the eigenaxis and the associated orthogonal unit vectors ($\hat{x}$ is in the XZ plane). For instance, for the Cooper pair box, we find that $\hat{z}=\cos\alpha\hat{Z}+\sin\alpha\hat{X}$, with $\tan\alpha=2E_C\left(N_g-1/2\right)/E_J$, while $\hat{x}=-\sin\alpha\hat{Z}+\cos\alpha\hat{X}$.

Starting with $\overrightarrow{S}$ pointing along $\hat{x}$ at time $t=0$, the dynamics of the Bloch vector in absence of relaxation or decoherence is

$$\overrightarrow{S}_0(t)=\cos(\omega_{01})\,\hat{x}+\sin(\omega_{01})\,\hat{y}$$

In presence of relaxation and decoherence, the Bloch vector will deviate from $\overrightarrow{S}_0(t)$ and will reach eventually the equilibrium value $S_z^{eq}\hat{z}$, where $S_z^{eq}=\tanh\frac{\hbar\omega_{01}}{2k_BT}$. We define the relaxation and decoherence rates as

$$\Gamma_1 = \lim_{t\to\infty} \frac{\ln\left\langle S_z(t) - S_z^{eq}\right\rangle}{t}$$

$$\Gamma_\phi = \lim_{t\to\infty} \frac{\ln\left[\frac{\left\langle \vec{S}(t).\vec{S}_0(t)\right\rangle}{\left|\vec{S}(t) - S_z^{eq}\vec{z}\right|}\right]}{t}$$

Note that these rates have both a useful and rigorous meaning only if the evolution of the components of the average Bloch vector follows, after a negligibly short settling time, an exponential decay. The Γ_1 and Γ_ϕ rates are related to the NMR spin relaxation times T_1 and T_2 [47] by

$$T_1 = \Gamma_1^{-1}$$
$$T_2 = \left(\Gamma_\phi + \Gamma_1/2\right)^{-1}$$

The T_2 time can be seen as the net decay time of quantum information, including the influence of both relaxation and dephasing processes. In our discussion of superconducting qubits, we must separate the contribution of the two type of processes since their physical origin is in general very different and cannot rely on the T_2 time alone.

Expressions for the rates

The relaxation process can be seen as resulting from unwanted transitions between the two eigenstate of the qubit induced by fluctuations in the effective fields along the x and y axes. Introducing the power spectral density of this field, one can demonstrate from Fermi's Golden Rule that, for perturbative fluctuations,

$$\Gamma_1 = \frac{S_x(\omega_{01}) + S_y(\omega_{01})}{\hbar^2}$$

Taking the case of the Cooper pair box as an example, we find that $S_y(\omega_{01}) = 0$ and that

$$S_x(\omega) = \int_{-\infty}^{+\infty} dt e^{i\omega t} \left\langle A(t)\,A(0)\right\rangle + \left\langle B(t)\,B(0)\right\rangle$$

where

$$
\begin{aligned}
A(t) &= \frac{\Delta E_J(t)\, E_{el}}{2\sqrt{E_J^2 + E_{el}^2}} \\
B(t) &= \frac{E_J \Delta E_{el}(t)}{2\sqrt{E_J^2 + E_{el}^2}} \\
E_{el} &= 2E_C\left(N_g - 1/2\right)
\end{aligned}
$$

Since the fluctuations $\Delta E_{el}(t)$ can be related to the impedance of the environment of the box [18,20,48], an order of magnitude estimate of the relaxation rate can be performed, and is in rough agreement with observations [21,49].

The decoherence process, on the other hand, is induced by fluctuations in the effective field along the eigenaxis z. If these fluctuations are Gaussian, with a white noise spectral density up to frequencies of order several Γ_ϕ (which is often not the case because of the presence of 1/f noise) we have

$$\Gamma_\phi = \frac{S_z(\omega \simeq 0)}{\hbar^2}$$

In presence of a low frequency noise with an 1/f behavior, the formula is more complicated [50]. If the environment producing the low frequency noise consists of many degrees of freedom, each of which is very weakly coupled to the qubit, then one is in presence of classical dephasing which, if slow enough, can in principle be fought using echo techniques. If, one the other hand, only a few degrees of freedom like magnetic spins or glassy two-level systems are dominating the low frequency dynamics, dephasing is quantum and not correctable, unless the transition frequencies of these few perturbing degrees of freedom is itself very stable.

References

[1] M. A. Nielsen and I. L. Chuang, *Quantum Computation and Quantum Information* (Cambridge, 2000).

[2] M. Tinkham, *Introduction to Superconductivity* (Krieger, Malabar, 1985).

[3] J.M. Martinis, M.H. Devoret, J. Clarke, Phys. Rev. Lett. 55, 1543-1546 (1985); M.H. Devoret, J.M. Martinis, J. Clarke, Phys. Rev. Lett. 55, 1908-1911 (1985); J. M. Martinis, M.H. Devoret and J. Clarke, Phys. Rev. 35, 4682 (1987).

[4] J.M. Martinis and M. Nahum, Phys Rev. B48, 18316-19 (1993).

[5] B.D. Josephson, in *Superconductivity*, R.D. Parks, ed. (Marcel Dekker, New York, 1969).

[6] I. Giaever, Phys. Rev. Lett. 5, 147, 464 (1960)

[7] A.O. Caldeira and A.J. Leggett, Ann. Phys. (NY) 149, 347-456 (1983); A.J. Leggett, J. Phys. CM 14, R415-451 (2002).

[8] D. P. DiVincenzo, arXiv:quant-ph/0002077,

[9] R. P. Feynman, *Lectures on Physics* (Addison-Wesley, Reading, 1964) Vol. 2, Chap. 23.

[10] D.C. Mattis and J. Bardeen, Phys. Rev. 111, 412 (1958)

[11] P.G. de Gennes, *Superconductivity of Metals and Alloys* (Benjamin, New York, 1966)

[12] J.M. Raimond, M. Brune and S. Haroche, Rev. Mod. Phys. 73, 565 (2001).

[13] J.M. Martinis and K. Osborne, in *Quantum Information and Entanglement*, Eds. J.M. Raimond, D. Esteve and J. Dalibard, Les Houches Summer School Series, arXiv:cond-mat/0402430

[14] J. Clarke, Proc. IEEE 77, 1208 (1989)

[15] D. J. Van Harlingen, B.L.T. Plourde, T.L. Robertson, P.A. Reichardt, and John Clarke, preprint

[16] R. W. Simmonds, K. M. Lang, D. A. Hite, D. P. Pappas, and J. M. Martinis, submitted to Phys. Rev. Lett.

[17] M. Büttiker, Phys. Rev. B36, 3548 (1987).

[18] V. Bouchiat, D. Vion, P. Joyez, D. Esteve, M.H. Devoret, Physica Scripta T76 (1998) p.165-70.

[19] Y. Nakamura , Yu. A Pashkin, and J. S. Tsai, Nature 398, 786 (1999).

[20] Yu. Makhlin, G. Schön, and A. Shnirman, Rev. Mod. Phys. 73, 357-400 (2001).

[21] D. Vion, A. Aassime, A. Cottet, P. Joyez, H. Pothier, C. Urbina, D. Esteve, and M.H. Devoret, Science 296 (2002), p. 286-9.

[22] A. Barone and G. Paternò, *Physics and Applications of the Josephson Effect* (Wiley, New York, 1992)

[23] S. Han, R. Rouse and J.E. Lukens, Phys. Rev. Lett. 84, 1300 (2000); J.R. Friedman, V. Patel, W. Chen, S.K. Tolpygo and J.E. Lukens, Nature 406, 43 (2000).

[24] J.E. Mooij, T.P. Orlando, L. Levitov, Lin Tian, C.H. van der Wal and S. Lloyd, Science 285, 1036 (1999); C.H. van der Wal, A.C.J. ter Haar, F.K. Wilhem, R.N. Schouten, C. Harmans, T.P. Orlando, S. Loyd and J.E. Mooij, Science 290, 773 (2000).

[25] J. M. Martinis, S. Nam, J. Aumentado, and C. Urbina, Phys. Rev. Lett. 89, 117901 (2002)

[26] M. Steffen, J. Martinis and I.L. Chuang, PRB 68, 224518 (2003).

[27] M.H. Devoret and R.J. Schoelkopf, Nature, 406, 1039 (2002).

[28] A.N. Korotkov, D.V. Averin, arXiv:cond-mat/0002203

[29] A. Cottet, D. Vion, A. Aassime, P. Joyez, D. Esteve, and M.H Devoret, Physica C 367, 197 (2002)

[30] I. Siddiqi, R. Vijay, F. Pierre, C.M. Wilson, M. Metcalfe, C. Rigetti, L. Frunzio and M. Devoret, submitted.

[31] D. Vion, A. Aassime, A. Cottet, P. Joyez, H. Pothier, C. Urbina, D. Esteve, and M.H. Devoret, Fortschritte der Physik, 51,

[32] I. Chiorescu, Y. Nakamura, C. J. P. M. Harmans, and J. E. Mooij, Science 299, p. 1869 (2003).

[33] J. Mannik and J.E. Lukens, arXiv:cond-mat/0305190 v2

[34] A. Blais, Ren-Shou Huang, A. Wallraff, S. M. Girvin, R. J. Schoelkopf, arXiv:cond-mat/0402216

[35] A. Lupascu, C. J. M. Verwijs, R. N. Schouten, C. J. P. M. Harmans, J. E. Mooij, submitted.

[36] Variable electrostatic transformer: controllable coupling of two charge qubits, D.V. Averin, C. Bruder, Phys. Rev. Lett. 91, 057003 (2003).

[37] A. Blais, A. Maassen van den Brink, A. M. Zagoskin, Phys. Rev. Lett. 90, 127901 (2003)

[38] Pashkin Yu. A., Yamamoto T., Astafiev O., Nakamura Y., Averin D.V., and Tsai J. S.:"Quantum oscillation in two coupled charge qubit", Nature 421 (2003).

[39] J. B. Majer, F.G. Paauw, A.C.J. ter Haar, C.J.P.M. Harmans, J.E. Mooij, arXiv:cond-mat/0308192.

[40] C. Rigetti and M. Devoret, unpublished

[41] Nakamura Y., Pashkin Yu. A., and Tsai J. S., Phys. Rev. Lett. 88, 047901 (2002).

[42] D. Vion, A. Aassime, A. Cottet, P. Joyez, H. Pothier, C. Urbina, D. Esteve, and M.H. Devoret, Forts. der Physik, 51, 462 (2003).

[43] J. Preskill, J. Proc. R. Soc. London Ser. A454, 385 (1998).

[44] B. Yurke and J.S. Denker, Phys. Rev. A 29 (1984) 1419.

[45] M.H. Devoret in "*Quantum Fluctuations*", S. Reynaud, E. Giacobino, J. Zinn-Justin, eds. (Elsevier, Amsterdam, 1996) p. 351

[46] A. Cottet, *Implementation of a quantum bit in a superconducting circuit,* PhD Thesis, Université Paris 6, 2002

[47] A. Abragam, Principles of Nuclear Magnetic Resonance (Oxford University Press, Oxford, 1985).

[48] R. J. Schoelkopf, A. A. Clerk, S. M. Girvin, K. W. Lehnert, M. H. Devoret. arXiv:cond-mat/0210247.

[49] K.W. Lehnert, K. Bladh, L.F. Spietz, D. Gunnarsson, D.I. Schuster, P. Delsing, and R.J. Schoelkopf, Phys. Rev.Lett., 90, 027002 (2002).

[50] J.M. Martinis, S. Nam, J. Aumentado, K.M. Lang, and C. Urbina, Phys. Rev. B67, 462 (2003).

Course 13

SUPERCONDUCTING QUBITS AND THE PHYSICS OF JOSEPHSON JUNCTIONS

John M. Martinis

National Institute of Standards and Technology, 325 Broadway, Boulder, CO 80305-3328, USA

D. Estève, J.-M. Raimond and J. Dalibard, eds.
Les Houches, Session LXXIX, 2003
Quantum Entanglement and Information Processing
Intrication quantique et traitement de l'information
© *2004 Elsevier B.V. All rights reserved*

Contents

1. Introduction

Josephson junctions are good candidates for the construction of quantum bits (qubits) for a quantum computer[1]. This system is attractive because the low dissipation inherent to superconductors make possible, in principle, long coherence times. In addition, because complex superconducting circuits can be microfabricated using integrated-circuit processing techniques, scaling to a large number of qubits should be relatively straightforward. Given the initial success of several types of Josephson qubits[2, 3, 4, 5, 6, 7, 9, 8, 10], a question naturally arises: what are the essential components that must be tested, understood, and improved for eventual construction of a Josephson quantum computer?

In this paper we focus on the physics of the Josephson junction because, being nonlinear, it is the fundamental circuit element that is needed for the appearance of usable qubit states. In contrast, *linear* circuit elements such as capacitors and inductors can form low-dissipation superconducting resonators, but are unusable for qubits because the energy-level spacings are degenerate. The nonlinearity of the Josephson inductance breaks the degeneracy of the energy level spacings, allowing dynamics of the system to be restricted to only the two qubit states. The Josephson junction is a remarkable nonlinear element because it combines negligible dissipation with extremely large nonlinearity - the change of the qubit state by only one photon in energy can modify the junction inductance by order unity!

Most theoretical and experimental investigations with Josephson qubits assume perfect junction behavior. Is such an assumption valid? Recent experiments by our group indicate that coherence is limited by microwave-frequency fluctuations in the critical current of the junction[10]. A deeper understanding of the junction physics is thus needed so that nonideal behavior can be more readily identified, understood, and eliminated. Although we will not discuss specific imperfections of junctions in this paper, we want to describe a clear and precise model of the Josephson junction that can give an intuitive understanding of the Josephson effect. This is especially needed since textbooks do not typically derive the Josephson effect from a microscopic viewpoint. As standard calculations use only perturbation theory, we will also need to introduce an exact description of the Josephson effect via the mesoscopic theory of quasiparticle bound-states.

The outline of the paper is as follows. We first describe in Sec. 2 the nonlinear

Josephson inductance. In Sec. 3 we discuss the three types of qubit circuits, and show how these circuits use this nonlinearity in unique manners. We then give a brief derivation of the BCS theory in Sec. 4, highlighting the appearance of the macroscopic phase parameter. The Josephson equations are derived in Sec. 5 using standard first and second order perturbation theory that describe quasiparticle and Cooper-pair tunneling. An exact calculation of the Josephson effect then follows in Sec. 6 using the quasiparticle bound-state theory. Section 7 expands upon this theory and describes quasiparticle excitations as transitions from the ground to excited bound states from nonadiabatic changes in the bias. Although quasiparticle current is typically calculated only for a constant DC voltage, the advantage to this approach is seen in Sec. 8, where we qualitatively describe quasiparticle tunneling with AC voltage excitations, as appropriate for the qubit state. This section describes how the Josephson qubit is typically insensitive to quasiparticle damping, even to the extent that a phase qubit can be constructed from microbridge junctions.

2. The nonlinear Josephson inductance

A Josephson tunnel junction is formed by separating two superconducting electrodes with an insulator thin enough so that electrons can quantum-mechanically tunnel through the barrier, as illustrated in Fig. 1 . The Josephson effect describes the supercurrent I_J that flows through the junction according to the classical equations

$$I_J = I_0 \sin\delta \tag{2.1a}$$

$$V = \frac{\Phi_0}{2\pi}\frac{d\delta}{dt}, \tag{2.1b}$$

where $\Phi_0 = h/2e$ is the superconducting flux quantum, I_0 is the critical-current parameter of the junction, and $\delta = \phi_L - \phi_R$ and V are respectively the superconducting phase difference and voltage across the junction. The dynamical behavior of these two equations can be understood by first differentiating Eq. 2.1a and replacing $d\delta/dt$ with V according to Eq. 2.1b

$$\frac{dI_J}{dt} = I_0 \cos\delta \frac{2\pi}{\Phi_0} V. \tag{2.2}$$

With dI_J/dt proportional to V, this equation describes an inductor. By defining a Josephson inductance L_J according to the conventional definition $V = L_J dI_J/dt$, one finds

$$L_J = \frac{\Phi_0}{2\pi I_0 \cos\delta}. \tag{2.3a}$$

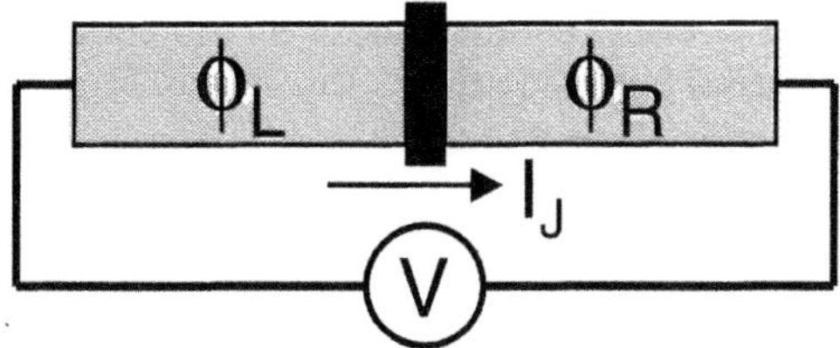

Fig. 1. Schematic diagram of a Josephson junction connected to a bias voltage V. The Josephson current is given by $I_J = I_0 \sin\delta$, where $\delta = \phi_L - \phi_R$ is the difference in the superconducting phase across the junction.

The $1/\cos\delta$ term reveals that this inductance is nonlinear. It becomes large as $\delta \to \pi/2$, and is negative for $\pi/2 < \delta < 3\pi/2$. The inductance at zero bias is $L_{J0} = \Phi_0/2\pi I_0$.

An inductance describes an energy-conserving circuit element. The energy stored in the junction is given by

$$\begin{aligned} U_J &= \int I_J V dt &\quad (2.4a)\\ &= \int I_0 \sin\delta \frac{\Phi_0}{2\pi}\frac{d\delta}{dt} dt &\quad (2.4b)\\ &= \frac{I_0\Phi_0}{2\pi}\int \sin\delta\, d\delta &\quad (2.4c)\\ &= -\frac{I_0\Phi_0}{2\pi}\cos\delta\,. &\quad (2.4d) \end{aligned}$$

This calculation of energy can be generalized for other nondissipative circuit elements. For example, a similar calculation for a current bias gives $U_{\text{bias}} = -(I\Phi_0/2\pi)\delta$. Conversely, if a circuit element has an energy $U(\delta)$, then the current-phase relationship of the element, analogous to Eq. 2.1a, is

$$I_J(\delta) = \frac{2\pi}{\Phi_0}\frac{\partial U(\delta)}{\partial\delta}\,. \tag{2.5}$$

A generalized Josephson inductance can be also be found from the second derivative of U ,

$$\frac{1}{L_J} = \left(\frac{2\pi}{\Phi_0}\right)^2 \frac{\partial^2 U(\delta)}{\partial\delta^2}\,. \tag{2.6}$$

The classical and quantum behavior of a particular circuit is described by a Hamiltonian, which of course depends on the exact circuit configuration. The procedure for writing down a Hamiltonian for an arbitrary circuit has been described in detail in a prior publication[11]. The general form of the Hamiltonian for the Josephson effect is $H_J = U_J$.

3. Phase, flux, and charge qubits

A Josephson qubit can be understood as a nonlinear resonator formed from the Josephson inductance and its junction capacitance. nonlinearity is crucial because the system has many energy levels, but the operating space of the qubit must be restricted to only the two lowest states. The system is effectively a two-state system[12] only if the frequency ω_{10} that drives transitions between the qubit states $0 \longleftrightarrow 1$ is different from the frequency ω_{21} for transitions $1 \longleftrightarrow 2$.

We review here three different ways that these nonlinear resonators can be made, and which are named as phase, flux, or charge qubits.

The circuit for the phase-qubit circuit is drawn in Fig. 2(a). Its Hamiltonian is

$$H = \frac{1}{2C}\widehat{Q}^2 - \frac{I_0\Phi_0}{2\pi}\cos\widehat{\delta} - \frac{I\Phi_0}{2\pi}\widehat{\delta}\,, \tag{3.1}$$

where C is the capacitance of the tunnel junction. A similar circuit is drawn for the flux-qubit circuit in Fig. 2(b), and its Hamiltonian is

$$H = \frac{1}{2C}\widehat{Q}^2 - \frac{I_0\Phi_0}{2\pi}\cos\widehat{\delta} + \frac{1}{2L}(\Phi - \frac{\Phi_0}{2\pi}\widehat{\delta})^2\ . \tag{3.2}$$

The charge qubit has a Hamiltonian similar to that in Eq. 3.1, and is described elsewhere in this publication. Here we have explicitly used notation appropriate for a quantum description, with operators charge $\widehat{Q}$ and phase difference $\widehat{\delta}$ that obey a commutation relationship $[\widehat{\delta}, \widehat{Q}] = 2ei$. Note that the phase and flux qubit Hamiltonians are equivalent for $L \to \infty$ and $I = \Phi/L$, which corresponds to a current bias created from an inductor with infinite impedance.

The commutation relationship between $\widehat{\delta}$ and $\widehat{Q}$ imply that these quantities must be described by a wavefunction. The characteristic widths of this wavefunction are controlled by the energy scales of the system, the charging energy of the junction $E_C = e^2/2C$ and the Josephson energy $E_J = I_0\Phi_0/2\pi$. When the energy of the junction dominates, $E_J \gg E_C$, then $\widehat{\delta}$ can almost be described classically and the width of its wavefunction is small $\langle\widehat{\delta}^2 - \langle\widehat{\delta}\rangle^2\rangle \ll 1$. In contrast, the uncertainty in charge is large $\langle\widehat{Q}^2 - \langle\widehat{Q}\rangle^2\rangle \gg (2e)^2$.

If the Josephson inductance is constant over the width of the $\widehat{\delta}$ wavefunction, then a circuit is well described as a L_J-C harmonic oscillator, and the qubit states

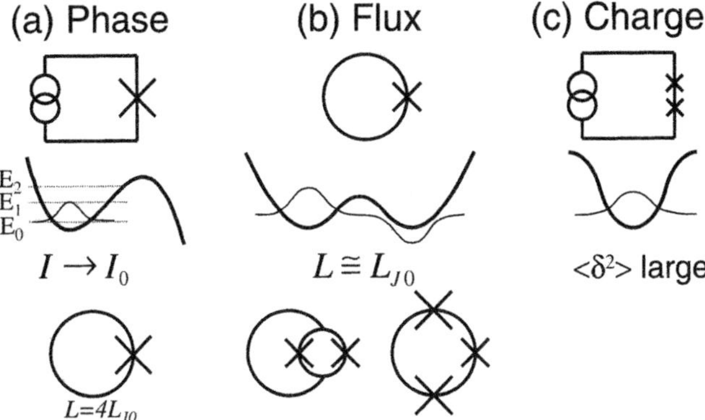

Fig. 2. Comparison of the phase (a), flux (b), and charge (c) qubits. Top row illustrates the circuits, with each "X" symbol representing a Josephson juncton. Middle row has a plot of the Hamiltonian potential (thick line), showing qualitatively different shapes for three qubit types. Ground-state wavefunction is also indicated (thin line). Key circuit parameters are listed in next row. Lowest row indicates variations on the basic circuit, as discussed in text. The lowest three energy levels are illustrated for the phase qubit (dotted lines).

are degenerate and not usable. Usable states are created only when the Josephson inductance changes over the δ-wavefunction.

The most straightforward way for the wavefunction to be affected by the Josephson nonlinearity is for $\widehat{\delta}$ to have a large width , which occurs when $E_J \sim E_C$. A practical implementation of this circuit is illustrated in Fig 2(c), where a double-junction Coulomb blockade device is used instead of a single junction to isolate dissipation from the leads[2, 4]. Because the wavefunction extends over most of the -cos $\widehat{\delta}$ Hamiltonian, the transition frequency ω_{10} can differ from ω_{21} by more than 10 %, creating usable qubit states[13].

Josephson qubits are possible even when $E_J \gg E_C$, provided that the junction is biased to take advantage of its strong nonlinearity. A good example is the phase qubit[6], where typically $E_J \sim 10^4 E_C$, but which is biased near $\delta \lessapprox \pi/2$ so that the inductance changes rapidly with δ (see Eq. 2.3a). Under these conditions the potential can be accurately described by a cubic potential, with the barrier height $\Delta U \to 0$ as $I \to I_0$. Typically the bias current is adjusted so that the number of energy levels in the well is $\sim 3 - 5$, which causes ω_{10} to differ from ω_{21} by an acceptably large amount ~ 5 %.

Implementing the phase qubit is challenging because a current bias is required with large impedance. This impedance requirement can be met by biasing the junction with flux through a superconducting loop with a large loop inductance L, as discussed previously and drawn in Fig. 2(a). To form multiple stable flux states and a cubic potential, the loop inductance L must be chosen such that $L \gtrsim 2L_{J0}$. We have found that a design with $L \simeq 4.5L_{J0}$ is a good choice since the potential well then contains the desired cubic potential and only one flux state into which the system can tunnel, simplifying operation.

The flux qubit is designed with $L \lesssim L_{J0}$ and biased in flux so that $\langle\widehat{\delta}\rangle = \pi$. Under these conditions the Josephson inductance is negative and is almost canceled out by L. The small net negative inductance near $\widehat{\delta} = \pi$ turns positive away from this value because of the $1/\cos\delta$ nonlinearity, so that the final potential shape is quartic, as shown in Fig. 2(b). An advantage of the flux qubit is a large net nonlinearity, so that ω_{10} can differ from ω_{21} by over 100 %.

The need to closely tune L with L_{J0} has inspired the invention of several variations to the simple flux-qubit circuit, as illustrated in Fig. 2(b). One method is to use small area junctions[7] with $E_J \sim 10E_C$, producing a large width in the $\widehat{\delta}$ wavefunction and relaxing the requirement of close tuning of L with L_{J0}. Another method is to make the qubit junction a two-junction SQUID, whose critical current can then be tuned via a second flux-bias circuit[14, 15]. Larger junctions are then permissible, with $E_J \sim 10^2E_C$ to 10^3E_C. A third method is to fabricate the loop inductance from two or more larger critical-current junctions[16]. These junctions are biased with phase less than $\pi/2$, and thus act as positive inductors. The advantage to this approach is that junction inductors are smaller than physical inductors, and fabrication imperfections in the critical currents of the junctions tend to cancel out and make the tuning of L with L_{J0} easier.

In summary, the major difference between the phase, flux, and charge qubits is the shape of their nonlinear potentials, which are respectively cubic, quartic, and cosine. It is impossible at this time to predict which qubit type is best because their limitations are not precisely known, especially concerning decoherence mechanisms and their scaling. However, some general observations can be made.

First, the flux qubit has the largest nonlinearity. This implies faster logic gates since suppressing transitions from the qubit states 0 and 1 to state 2 requires long pulses whose time duration scales as $1/|\omega_{10} - \omega_{21}|$[12]. The flux qubit allows operation times less than $\sim$ 1 ns, whereas for the phase qubit 10 ns is more typical. We note, however, that this increase in speed may not be usable. Generating precise shaped pulses is much more difficult on a 1 ns time scale, and transmitting these short pulses to the qubit with high fidelity will be more problematic due to reflections or other imperfections in the microwave lines.

Second, the choice between large and small junctions involve tradeoffs. Large junctions ($E_J \gg E_C$) require precise tuning of parameters (L/L_{J0} for the flux qubit) or biases (I/I_0 for the phase qubit) to produce the required nonlinearity. Small junctions ($E_J \sim E_C$) do not require such careful tuning, but become sensitve to $1/f$ charge fluctuations because E_C has relatively larger magnitude.

Along these lines, the coherence of qubits have been compared considering the effect of low-frequency $1/f$ fluctuations of the critical current[17]. These calculations include the known scaling of the fluctuations with junction size and the sensitivity to parameter fluctuations. It is interesting that the calculated coherence times for the flux and phase qubits are similar. With parameters choosen to give an oscillation frequency of ~ 1 GHz for the flux qubit and ~ 10 GHz for the phase qubit, the number of coherent logic-gate operations is even approximately the same.

4. BCS theory and the superconducting state

A more complete understanding of the Josephson effect will require a derivation of Eqs. 2.1a and 2.1b. In order to calculate this microscopically, we will first review the BCS theory of superconductivity[18] using a "pair spin" derivation that we believe is more physically clear than the standard energy-variational method. Although the calculation follows closely that of Anderson[19] and Kittel[20], we have expanded it slightly to describe the physics of the superconducting phase, as appropriate for understanding Josephson qubits.

In a conventional superconductor, the attractive interaction that produces superconductivity comes from the scattering of electrons and phonons. As illustrated in Fig. 3(a), to first order the phonon interaction scatters an electron from one momentum state to another. When taken to second order (Fig. 3(b)), the scattering of a virtual phonon produces a net attractive interaction between two pairs of electrons. The first-order phonon scattering rates are generally small, not because of the phonon matrix element, but because phase space is small for the final electron state. This implies that the energy of the second order interaction can be significant if there are large phase-space factors.

The electron pairs have the largest net interaction if every pair is allowed by phase space factors to interact with every other pair. This is explicitly created in the BCS wavefunction by including only pair states (Cooper pairs) with zero net momentum. Under this assumption and using a second quantized notation where $c_k^\dagger$ is the usual creation operator for an electron state of wavevector k, the most general form for the electronic wavefunction is

$$\Psi = \prod_k (u_k + v_k e^{i\phi_k} c_k^\dagger c_{-k}^\dagger)\, |0\rangle \ , \tag{4.1}$$

where u_k and v_k are real and correspond respectively to the probability amplitude for a pair state to be empty or filled, and are normalized by $u_k^2 + v_k^2 = 1$. For

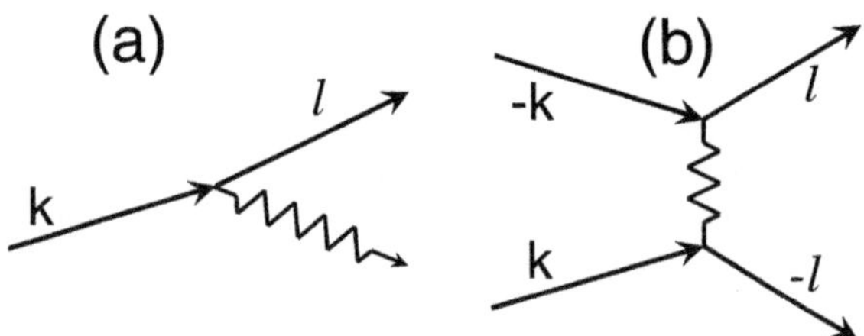

Fig. 3. Feynman diagram of electron-phonon interaction showing (a) first- and (b) second-order processes.

generality we have included a separate phase factor ϕ_k for each pair. Because each pair state is described as a two state system, the wavefunction may also be described equivalently with a "pair-spin" tensor product

$$\Psi = \prod_k \begin{pmatrix} u_k \\ v_k e^{i\phi_k} \end{pmatrix} \otimes \ , \tag{4.2}$$

and the Hamiltonian given with Pauli matrices σ_{xk}, σ_{yk}, and σ_{zk}.

The kinetic part of the Hamiltonian must give Ψ in the ground state with pairs occupied only for $|k| < k_f$, where k_f is the Fermi momentum. If we define the kinetic energy of a single electron, relative to the Fermi energy, as ξ_k, then the kinetic Hamiltonian for the pair state is

$$H_K = -\sum \xi_k \sigma_{zk} \ . \tag{4.3}$$

The solution of $H_K \Psi = E_{k\pm} \Psi$ gives for the lowest energy, E_{k-}, the values $v_k = 1$ for $|k| < k_f$, and $v_k = 0$ for $|k| > k_f$, as required. An energy $E_{k+} - E_{k-} = 2|\xi_k|$ is needed for the excitation of pairs above the Fermi energy or the excitation of holes (removal of pairs) below the Fermi energy.

The potential part of the pair-spin Hamiltonian comes from the second-order phonon interaction that both creates and destroys a pair, as illustrated in Fig. 3(b). The Hamiltonian for this interaction is given by

$$H_\Delta = -\frac{V}{2} \sum_{k,l} (\sigma_{xk}\sigma_{xl} + \sigma_{yk}\sigma_{yl}) \ , \tag{4.4}$$

and can be checked to correspond to the second-quantization Hamiltonian $H_\Delta = -V \sum c_k^\dagger c_{-k}^\dagger c_k c_{-k}$ by using the translation $\sigma_{xk} \to c_k c_{-k} + c_k^\dagger c_{-k}^\dagger$ and $\sigma_{yk} \to i(c_k c_{-k} - c_k^\dagger c_{-k}^\dagger)$.

We will first understand the solution to the Hamiltonian $H_K + H_\Delta$ for the phase variables ϕ_k. This Hamiltonian describes a bath of spins that are all coupled to each other in the x-y plane (H_Δ) and have a distribution of magnetic fields in the z-direction (H_K). Because H_Δ is negative, each pair of spins becomes aligned with each other in the x-y plane, which implies that every spin in the bath has the same phase ϕ_k. This condition explains why the BCS wavefunction has only one phase $\phi = \phi_k$ for all Cooper pairs[21]. Because there is no preferred direction in the x-y plane, the solution to the Hamiltonian is degenerate with respect to ϕ and the wavefunction for ϕ is separable from the rest of the wavefunction. Normally, this means that ϕ can be treated as a classical variable, as is done for the conventional understanding of superconductivity and the Josephson effects. For Josephson qubits, where ϕ must be treated quantum mechanically, then the behavior of ϕ is described by an external-circuit Hamiltonian, as was done in Sec. 3.

For a superconducting circuit, where one electrode is biased with a voltage V, the voltage can be accounted for with a gauge transformation on each electron state $c_k^\dagger \to e^{i(e/\hbar)\int V dt} c_k^\dagger$. The change in the superconducting state is thus given by

$$\begin{aligned}\Psi &\to \prod_k (u_k + v_k e^{i\phi} e^{i(e/\hbar)\int V dt} c_k^\dagger e^{i(e/\hbar)\int V dt} c_{-k}^\dagger)\,|0\rangle && (4.5)\\ &= \prod_k (u_k + v_k e^{i[\phi + i(2e/\hbar)\int V dt]} c_k^\dagger c_{-k}^\dagger)\,|0\rangle\ . && (4.6)\end{aligned}$$

The change in ϕ can be written equivalently as

$$\frac{d\phi}{dt} = \frac{2eV}{\hbar}\ , \qquad (4.7)$$

which leads to the AC Josephson effect.

The solution for u_k and v_k proceeds using the standard method of mean-field theory, with

$$\langle H_\Delta\rangle = -\frac{V}{2}\sum_{k,l}(\sigma_{xk}\langle\sigma_{xl}\rangle + \sigma_{yk}\langle\sigma_{yl}\rangle)\ , \qquad (4.8)$$

$$\begin{aligned}\langle\sigma_{xl}\rangle &= \left(u_l,\, v_l e^{-i\phi}\right)\cdot\sigma_x\cdot\begin{pmatrix}u_l\\ v_l e^{i\phi}\end{pmatrix} && (4.9)\\ &= 2u_l v_l\cos\phi\ , && (4.10)\\ \langle\sigma_{yl}\rangle &= 2u_l v_l\sin\phi\ . && (4.11)\end{aligned}$$

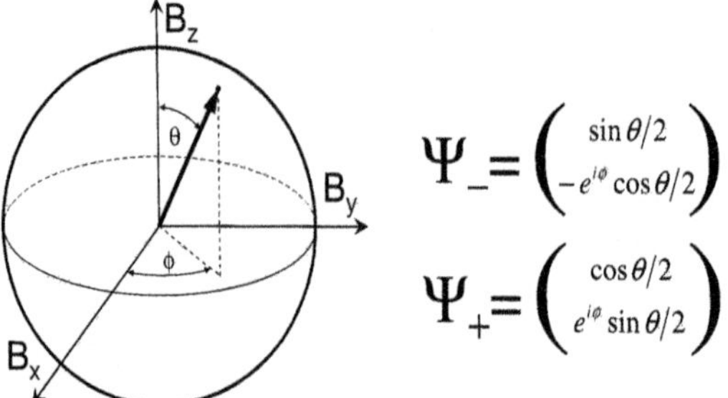

Fig. 4. Bloch sphere solution of the Hamiltonian $(\sigma_x, \sigma_y, \sigma_z) \bullet (B_x, B_y, B_z)$. The vector $\vec{B}$ gives the direction of the positive energy eigenstate.

Using the standard definition of the gap potential, one finds

$$\Delta = V \sum_l u_l v_l \,, \tag{4.12a}$$

$$H = H_K + \langle H_\Delta \rangle \tag{4.12b}$$

$$= -\sum_k (\sigma_{xk}, \sigma_{yk}, \sigma_{zk}) \cdot (\Delta \cos\phi, \Delta \sin\phi, \xi_k) \,. \tag{4.12c}$$

This Hamiltonian is equivalent to a spin 1/2 particle in a magnetic field, and its solution is well known. The energy eigenvalues of $H\Psi = E_{k\pm}\Psi$ are given by the total length of the field vector,

$$E_{k\pm} = \pm(\Delta^2 + \xi_k^2)^{1/2} \,, \tag{4.13}$$

and the directions of the Bloch vectors describing the E_{k+} and E_{k-} eigenstates are respectively parallel and antiparallel to the direction of the field vector, as illustrated in Fig. 4. The ground state solution Ψ_{k-} is given by

$$u_k = \sqrt{\frac{1}{2}\left(1 + \frac{\xi_k}{E_k}\right)} \,, \tag{4.14}$$

$$v_k = \sqrt{\frac{1}{2}\left(1 - \frac{\xi_k}{E_k}\right)} \,, \tag{4.15}$$

$$\phi_k = \phi \,, \tag{4.16}$$

with the last equation required for consistency. The excited state Ψ_{k+} is similarly described, but with u_k and v_k interchanged and $\phi \to \phi + \pi$.

At temperature $T = 0$ the energy gap Δ may be solved by inserting the solutions for u_k and v_k into Eq. 4.12a

$$\Delta = V \sum_l \frac{\Delta}{2(\Delta^2 + \xi_k^2)^{1/2}} . \tag{4.17}$$

Converting to an integral by defining a density of states N_0 at the Fermi energy, and introducing a cutoff of the interaction V at the Debye energy θ_D, one finds the standard BCS result,

$$\Delta = 2\theta_D e^{-1/N_0 V} . \tag{4.18}$$

Two eigenstates E_{k-} and E_{k+} have been determined for the pair Hamiltonian. Two additional "quasiparticle" eigenstates must exist, which clearly have to be single-particle states. These states may be solved for using diagonalization techniques, giving

$$\Psi_{k0} = c_k^\dagger |0\rangle , \tag{4.19a}$$

$$\Psi_{k1} = c_{-k}^\dagger |0\rangle . \tag{4.19b}$$

Fortunately, these states may be easily checked by inspection. The kinetic part of the Hamiltonian gives $H_K \Psi_{k0,1} = 0$ since $\Psi_{k0,1}$ corresponds to the creation of an electron and a hole, and the electron-pair and hole-pair states have opposite kinetic energy. The potential part of the energy also gives $\langle H_\Delta \rangle \Psi_{k0,1} = 0$ since the interaction Hamiltonian scatters pair states. Thus the eigenenergies of $\Psi_{k0,1}$ are zero, and these states have an energy $E_k = |E_{k-}|$ above the ground state.

The quasiparticle operators that take the ground-state wavefunction to the excited states are

$$\gamma_{k0}^\dagger = u_k c_k^\dagger - v_k e^{-i\phi} c_{-k} , \tag{4.20}$$

$$\gamma_{k1}^\dagger = u_k c_{-k}^\dagger + v_k e^{-i\phi} c_k , \tag{4.21}$$

which can be easily checked to give

$$\gamma_{k0}^\dagger (u_k + v_k e^{i\phi} c_k^\dagger c_{-k}^\dagger) |0\rangle = c_k^\dagger |0\rangle , \tag{4.22}$$

$$\gamma_{k1}^\dagger (u_k + v_k e^{i\phi} c_k^\dagger c_{-k}^\dagger) |0\rangle = c_{-k}^\dagger |0\rangle . \tag{4.23}$$

A summary of these results is illustrated in Fig. 5, where we show the energy levels, wavefunctions, and operators for transitions between the four states. The quasiparticle raising and lowering operators $\gamma_{k0}^\dagger$, $\gamma_{k1}^\dagger$, γ_{k0}, and γ_{k1} produce transitions between the states and have orthogonality relationships similar to those of the electron operators.

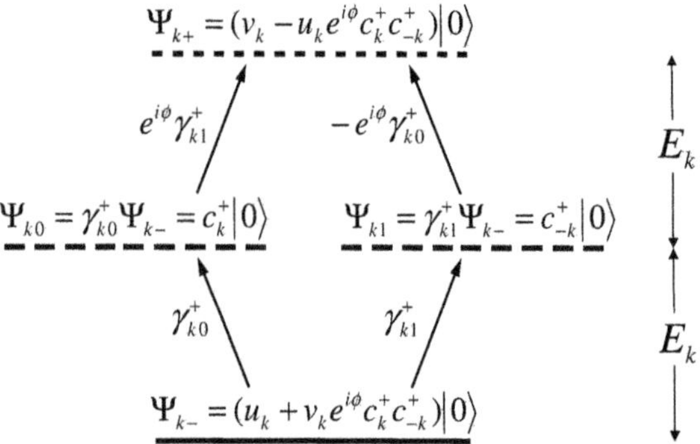

Fig. 5. Energy-level diagram for the ground-pair state (solid line), two quasiparticle states (dashed lines), and the excited-pair state (short dashed line).

It is interesting to note that the ground and excited pair states are connected by the two quasiparticle operators $e^{-i\phi}\gamma_{k1}^{\dagger}\gamma_{k0}^{\dagger}\Psi_{k-} = \Psi_{k+}$. Because the value $u_l v_l$ changes sign between Ψ_{k-} and Ψ_{k+}, and is zero for $\Psi_{k0,1}$, the gap equation 4.12a including the effect of quasiparticles is proportional to $\langle 1 - \gamma_{k0}^{\dagger}\gamma_{k0} - \gamma_{k1}^{\dagger}\gamma_{k1}\rangle$. Along with the energy levels, these results imply that the two types of quasiparticles are independent excitations.

5. The Josephson effect, derived from perturbation theory

We will now calculate the quasiparticle and Josephson current for a tunnel junction using first and second order perturbation theory, respectively. We note that our prior calculations have not been concerned with electrical transport. In fact, the electron operators describing the superconducting state have not been influenced by charge, and thus they correspond to the occupation of an effectively neutral state. Because a tunneling event involves a real transfer of an electron, charge must now be accounted for properly. We will continue to use electron operators for describing the states, but will keep track of the charge transfer separately.

When an electron tunnels through the barrier, an electron and hole state is created on the opposite (left and right) side of the barrier. The tunneling Hamiltonian for this process can be written as

$$H_T = \overrightarrow{H}_{T+} + \overrightarrow{H}_{T-} + \overleftarrow{H}_{T+} + \overleftarrow{H}_{T-} \tag{5.1a}$$

$$= \sum_{L,R} \Big(t_{LR} c_L c_R^\dagger + t_{-L-R} c_{-L} c_{-R}^\dagger + t_{LR}^* c_L^\dagger c_R + t_{-L-R}^* c_{-L}^\dagger c_{-R} \Big) , \tag{5.1b}$$

where t_{LR} is the tunneling matrix element, and the L and R indices refer respectively to momentum states k on the left and right superconductor. The first two terms $\overrightarrow{H}_{T+}$ and $\overrightarrow{H}_{T-}$ correspond to the tunneling of one electron from left to the right, whereas $\overleftarrow{H}_{T+}$ and $\overleftarrow{H}_{T-}$ are for tunneling to the left. The Hamiltonian is explicitly broken up into $\overrightarrow{H}_{T+}$ and $\overrightarrow{H}_{T-}$ to account for the different electron operators $c_k^\dagger$ and $c_{-k}^\dagger$ for positive and negative momentum.

The electron operators must first be expressed in terms of the quasiparticle operators γ because these produce transitions between eigenstates of the superconducting Hamiltonian. Equations 4.20, 4.21, and their adjoints are used to solve for the four electron operators

$$\begin{array}{ll} c_k = u_k \gamma_{k0} + v_k e^{i\phi} \gamma_{k1}^\dagger & c_{-k} = u_k \gamma_{k1} - v_k e^{i\phi} \gamma_{k0}^\dagger \\ c_k^\dagger = u_k \gamma_{k0}^\dagger + v_k e^{-i\phi} \gamma_{k1} & c_{-k}^\dagger = u_k \gamma_{k1}^\dagger - v_k e^{-i\phi} \gamma_{k0} \, . \end{array} \tag{5.2}$$

Substituting Eqs. 5.2 into 5.1b, one sees that all four terms of the Hamiltonian have operators $\gamma^\dagger$ that produce quasiparticles. We calculate here to first order the quasiparticle current from L to R given by $\overrightarrow{H}_{T+} + \overrightarrow{H}_{T-}$. The Feynman diagrams (a) and (b) in Fig. 6 respectively describe the tunneling Hamiltonian for the $\overrightarrow{H}_{T+}$ and $\overrightarrow{H}_{T-}$ terms. In this diagram a solid line represents a Cooper pair state in the ground state, whereas a quasiparticle state is given by a dashed line. Only one pair participates in the tunneling interaction, so only one of the three solid lines is converted to a dashed line. The line of triangles represents the tunneling event and is labeled with its corresponding H_T Hamiltonian, with the direction of the triangles indicating the direction of the electron tunneling. The $c_k^\dagger$ operators, acting on the L or R lead, is rewritten in terms of the γ operators and placed above or below the vertices. Since only $\gamma^\dagger$ operators give a nonzero term when acting on the ground state, the effect of the interaction is to produce final states $\Psi_f^{L,R}$ with total energy $E_R + E_L$, and with amplitudes given at the right of the figure.

The two final states in Fig 6(a) and (b) are orthogonal, as well as states involving different values of L and R. The total current is calculated as an incoherent

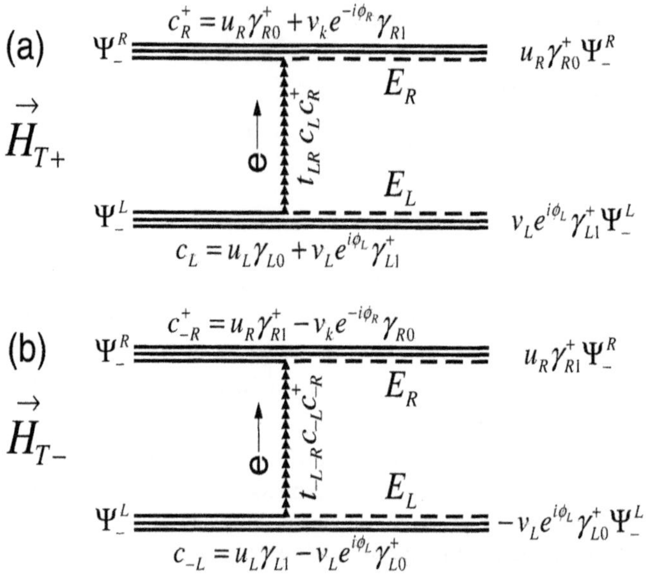

Fig. 6. First-order Feynmann diagrams for interaction $\vec{H}_{T+}$ (a) and $\vec{H}_{T-}$ (b). Solid lines are Cooper-pair states, dashed lines are quasiparticle excitations, and arrow-lines represents tunneling interaction. Electron operators arising from interaction are displayed next to vertices.

sum over all possible final quasiparticle states, under the condition that the total quasiparticle energy for the final state is equal to the energy gained by the tunneling of the electron

$$E_R + E_L = eV \; . \tag{5.3}$$

The total current from L to R is given by e multiplied by the transition rate

$$\begin{aligned}
\overrightarrow{I_{qp}} &= e\frac{2\pi}{\hbar} \sum_{L,R}^{(E_R+E_L=eV)} \left| \left\langle \Psi_f^L \right| \left\langle \Psi_f^R \right| \vec{H}_{T+} + \vec{H}_{T-} \left| \Psi_-^R \right\rangle \left| \Psi_-^L \right\rangle \right|^2 && (5.4a)\\
&= \frac{2\pi e}{\hbar} \sum_{L,R}^{(E_R+E_L=eV)} \left[\left| t_{LR} \right|^2 + \left| t_{-L-R} \right|^2 \right] (u_R v_L)^2 && (5.4b)\\
&= \frac{4\pi e}{\hbar} |t|^2 N_{0R} N_{0L} \int_{-\infty}^{\infty} v_L^2 \, d\xi_L \int_{-\infty}^{\infty} u_R^2 \, d\xi_R \; \delta(eV - E_L - E_R) \; , && (5.4c)
\end{aligned}$$

where in the last equation we have expressed the conservation of energy with a Dirac δ-function, and have assumed matrix elements $|t|^2$ of constant strength. Because $E(\xi_k) = E(-\xi_k)$ and $u_k(\xi_k) = v_k(-\xi_k)$, one finds

$$\begin{aligned}\overrightarrow{I_{qp}} &= \frac{4\pi e}{\hbar}|t|^2 N_{0R} N_{0L} \int_0^\infty (v_L^2 + u_L^2) d\xi_L \\ &\quad \int_0^\infty (u_R^2 + v_R^2)\, d\xi_R\, \delta(eV - E_L - E_R) \\ &= \frac{4\pi e}{\hbar}|t|^2 N_{0R} N_{0L} \int_0^\infty d\xi_L \int_0^\infty d\xi_R\, \delta(eV - E_L - E_R)\,. \qquad (5.5a)\end{aligned}$$

This result is equivalent to the standard "semiconductor model" of the quasiparticle current, which predicts no current for $V < 2\Delta/e$, a rapid rise of current at $2\Delta/e$, and then a current proportional to V at large voltages. Note that Eq. 5.4c has a sum over the occupation probability v_L^2 of the pair state and the occupation probability u_R^2 of a hole-pair state, as is expected given the operators $c_L c_R^\dagger$ in the tunneling Hamiltonian. The final result of Eq. 5.5a does not have these factors because the occupation probability is unity when summed over the $\pm\xi_k$ states.

It is convenient to express the tunneling matrix element in terms of the normal-state resistance of the junction, obtained by setting $\Delta = 0$, with the equation

$$\begin{aligned}1/R_N &\equiv \overrightarrow{I_{qp}}/V \qquad (5.6a)\\ &= \frac{4\pi e}{\hbar}|t|^2 N_{0R} N_{0L} \int_0^\infty d\xi_L \int_0^\infty d\xi_R\, \delta(eV - \xi_L - \xi_R)/V \qquad (5.6b)\\ &= \frac{4\pi e^2}{\hbar}|t|^2 N_{0R} N_{0L}\,. \qquad (5.6c)\end{aligned}$$

We now calculate the tunneling current with second-order perturbation theory. The tunneling Hamiltonian, taken to second order, gives

$$H_T^{(2)} = \sum_i H_T \frac{1}{\epsilon_i} H_T\,, \qquad (5.7)$$

where ϵ_i is the energy of the intermediate state i. Because the terms in H_T have both $\gamma^\dagger$ and γ operators, the second-order Hamiltonian gives a nonzero expectation value for the ground state. This is unlike the first-order theory, which produces current only through the real creation of quasiparticles.

Because H_T has terms that transfer charge in both directions, $H_T H_T$ will produce terms which transfer two electrons to the right, two to the left, and with no net transfer. With no transfer, a calculation of the second-order energy gives a constant value, which has no physical effect. We first calculate terms for transfer to the right from $(\overrightarrow{H}_{T+} + \overrightarrow{H}_{T-})(\overrightarrow{H}_{T+} + \overrightarrow{H}_{T-})$, which gives nonzero expectation values only for $\overrightarrow{H}_{T+}\overrightarrow{H}_{T-} + \overrightarrow{H}_{T-}\overrightarrow{H}_{T+}$. The Feynman diagrams for these two terms are given in Fig. 7(a) and (b), where we have displayed only the amplitudes from the nonzero operators. The expectation value of these two Hamiltonian terms is given by

$$\begin{aligned}
\left\langle \overrightarrow{H_T^{(2)}} \right\rangle &= -\sum_{L,R} \left\langle \Psi_-^R \right| \left\langle \Psi_-^L \right| (v_R e^{-i\phi_R} u_L) \\
&\quad \times \frac{\gamma_{R0}\gamma_{L1}\gamma_{R0}^\dagger\gamma_{L1}^\dagger t_{LR} t_{-L-R} + \gamma_{R1}\gamma_{L0}\gamma_{R1}^\dagger\gamma_{L0}^\dagger t_{-L-R} t_{LR}}{E_R + E_L} \\
&\quad \times (u_R v_L e^{i\phi_L}) \left|\Psi_-^L\right\rangle \left|\Psi_-^R\right\rangle && (5.8a) \\
&= -2\,|t|^2\, e^{i(\phi_L-\phi_R)} \sum_{L,R} (v_R u_R)(u_L v_L) \frac{1}{E_R + E_L} && (5.8b) \\
&= -2\,|t|^2\, e^{i\delta} N_{0R} N_{0L} \int_{-\infty}^{\infty} d\xi_R \int_{-\infty}^{\infty} d\xi_L \frac{\Delta}{E_R}\frac{\Delta}{E_L}\frac{1}{E_R + E_L} && (5.8c) \\
&= -\frac{\hbar\Delta}{2\pi e^2 R_N} e^{i\delta} \int_{-\infty}^{\infty} d\theta_R \int_{-\infty}^{\infty} d\theta_L \frac{1}{\cosh\theta_R + \cosh\theta_L} && (5.8d) \\
&= -\frac{\hbar\Delta}{2\pi e^2 R_N} e^{i\delta} \left(\frac{\pi}{2}\right)^2 , && (5.8e)
\end{aligned}$$

where we have used $t_{LR}^* = t_{-L-R}$ and assumed the same gap Δ for both superconductors. A similar calculation for transfer to the left gives the complex conjugate of Eq. 5.8e. The sum of these two energies gives the Josephson energy U_J, and using Eq. 2.5, the Josephson current I_J,

$$U_J = -\frac{1}{8}\frac{R_K}{R_N}\Delta\cos\delta , \qquad (5.9)$$

$$I_J = \frac{\pi}{2}\frac{\Delta}{eR_N}\sin\delta , \qquad (5.10)$$

where $R_K = h/e^2$ is the resistance quantum. Equation 5.10 is the standard

Ambegaokar-Baratoff formula[22] for the Josephson current at zero temperature.

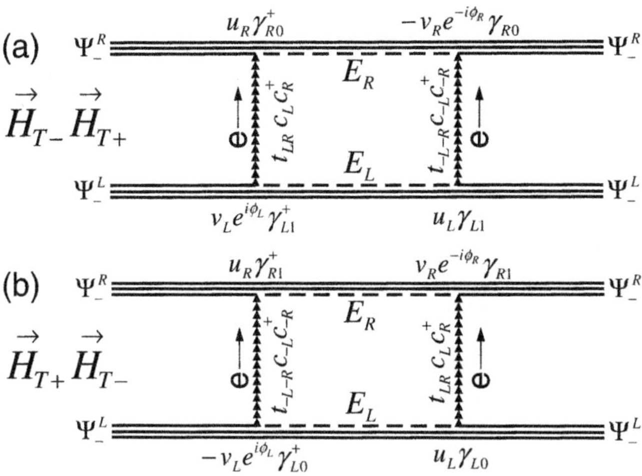

Fig. 7. Second-order Feynman diagrams for the transfer of two electrons across the junction. Only nonzero operators are displayed next to vertices.

The Josephson current is a dissipationless current because it arises from a new ground state of the two superconductors produced by the tunneling interaction. This behavior is in contrast with quasiparticle tunneling, which is dissipative because it produces excitations. It is perhaps surprising that a new ground state can produce charge transfer through the junction. This is possible only because the virtual quasiparticle excitations are both electrons and holes: the electron-part tunnels first through the junction, then the hole-part tunnels back. Only states of energy Δ around the Fermi energy are both electron- and hole-like, as weighted by the $(v_R u_R)(u_L v_L)$ term in the integral.

The form of the Josephson Hamiltonian can be understood readily by noting that the second-order Hamiltonian,

$$\vec{H}_{T+}\vec{H}_{T-} \sim |t|^2 \sum_{L,R} c_L c_{-L} c_R^\dagger c_{-R}^\dagger \tag{5.11}$$

$$= \frac{|t|^2}{2} \sum_{L,R} (\sigma_{xL}\sigma_{xR} + \sigma_{yL}\sigma_{yR}) , \tag{5.12}$$

corresponds to the pair-scattering Hamiltonian of Eq. 4.4. Comparing with the gap-equation solution, one expects $U_J \sim |t|^2 \Delta \cos\delta$, where the $\cos\delta$ term arises from the spin-spin interaction in the x-y plane.

We would like to make a final comment on a similarity between the BCS theory and the Josephson effect. In both of these derivations we see that a dissipative process that is described in first-order perturbation theory, such as phonon scattering or quasiparticle tunneling, produces in second order a new collective superfluid behavior. This collective behavior emerges from a virtual excitation of the dissipative process. Dissipation is normally considered undesirable, but by designing systems to *maximize* dissipation, it may be possible to discover new quantum collective behavior.

With this understanding of the Josephson effect and quasiparticle tunneling, how accurate is the description of the Josephson junction with the Hamiltonian corresponding to Eq. 5.9? There are several issues that need to be considered.

First, quasiparticle tunneling is a dissipative mechanism that produces decoherence. Although it is predicted to be absent for $V < 2\Delta/e$, measurements of real junctions show a small subgap current. This current is understood to arise from multiple Andreev reflections, which are described as higher-order tunneling processes. We thus need a description of the tunnel junction that easily predicts these processes for arbitrary tunneling matrix elements. This is especially needed as real tunnel junctions do not have constant matrix elements, as assumed above. Additionally, we would like to know whether a small number of major imperfections, such as "pinhole" defects, will strongly degrade the coherence of the qubit.

Second, quasiparticle tunneling has been predicted for an arbitrary DC voltage across the junction. However, the qubit state has $\langle V \rangle = 0$, but may excite quasiparticles with AC voltage fluctuations. This situation is difficult to calculate with perturbation theory. In addition, is it valid to estimate decoherence from quasiparticles at zero voltage simply from the junction resistance at subgap voltages?

Third, how will the Josephson effect and the qubit Hamiltonian be modified under this more realistic description of the tunnel junction?

All of these questions and difficulties arise because perturbation theory has been used to describe the ground state of the Josephson junction. The BCS theory gives basis states that best describe quasiparticle tunneling for large voltages, not for $V \to 0$. A theory is needed that solves for the Josephson effect *exactly*, with this solution then providing the basis states for understanding quasiparticle tunneling around $V = 0$. This goal is fulfilled by the theory of quasiparticle bound states, which we will describe next.

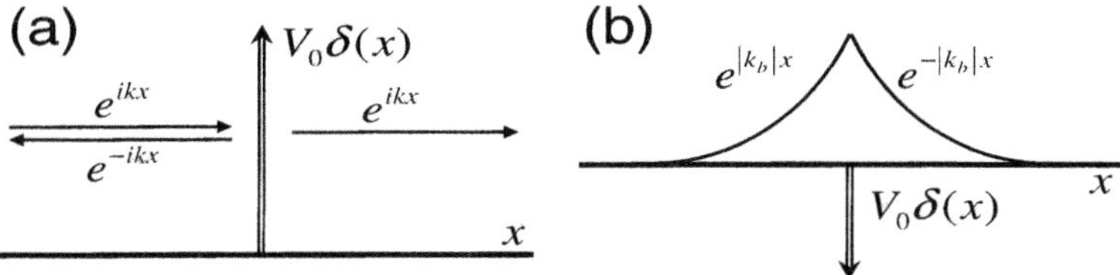

Fig. 8. Plot of potential *vs.* coordinate x with a positive delta-function tunnel barrier $V_0\delta(x)$. Scattering of plane wave states is shown in (a), whereas (b) is a plot of the bound-state wavefunction. The delta-function barrier is negative in (b), as required for producing a bound state.

6. The Josephson effect, derived from quasiparticle bound states

We begin our derivation of an exact solution for the Josephson effect with an extremely powerful idea from mesoscopic physics: electrical transport can be calculated under very general conditions by summing the current from a number of *independent* "conduction channels", with the transport physics of each conduction channel determined only by its channel transmission probability τ_i [23, 24]. For a Josephson junction, the total junction current I_j can be written as a sum over all channels i

$$I_j = \sum_i I_j(\tau_i) \,, \tag{6.1}$$

where $I_j(\tau)$ is the current for a single channel of transmission τ, which may be solved for theoretically. For a tunnel junction, the number of channels is estimated as the junction area divided by the channel area $(\lambda_f/2)^2$, where λ_f is the Fermi wavelength of the electrons. Of course, the difficulty of determining the distribution of the channel transmissions still remains. This often may be estimated from transport properties, and under some situations can be predicted from theory[25, 26, 27].

Because transport physics is determined *only* by scattering parameterized by τ, we may make two simplifying assumptions: the transport can be solved for using plane waves, and the scattering from the tunnel junction can be described by a delta function. The general theory has thus been transformed into the problem of one-dimensional scattering from a delta function, and an exact solution can be found by using a simple and clear physical picture.

Central to understanding the Josephson effect will be the quasiparticle bound state. To understand how to calculate a bound state[28, 29], we will first consider a normal-metal tunnel junction and with a δ-function barrier $V_0\overline{\delta}(x)$, as illustrated in Fig. 8. For an electron of mass m and wavevector k, the wavefunctions on the left and right side of the barrier are

$$\Psi_L = Ae^{ikx} + Be^{-ikx}, \tag{6.2}$$
$$\Psi_R = Ce^{ikx}, \tag{6.3}$$

where A, B, and C are respectively the incident, reflected, and transmitted electron amplitudes. From the continuity equations

$$\Psi_L = \Psi_R \tag{6.4}$$
$$\frac{d\Psi_L}{dx} = \frac{d\Psi_R}{dx} - \frac{2mV_0}{\hbar^2}\Psi_R \tag{6.5}$$

evaluated at $x = 0$, the amplitudes are related by

$$A + B = C \tag{6.6}$$
$$ikA - ikB = ikC - \frac{2mV_0}{\hbar^2}C. \tag{6.7}$$

The transmission amplitude and the probability are

$$\frac{C}{A} = \frac{1}{1+i\eta}, \tag{6.8}$$
$$\tau = \left|\frac{C}{A}\right|^2 = \frac{1}{1+\eta^2}, \tag{6.9}$$

where $\eta = mV_0/\hbar^2 k$. The bound state can be determined by finding the pole in the transmission amplitude. A pole describes how a state of finite amplitude may be formed around the scattering site with zero amplitude of the incident wavefunction, which is the definition of a bound state. The pole at $\eta = i$ gives $k_b = -imV_0/\hbar^2$, and a wavefunction around the scattering site $\Psi_R = Ce^{(mV_0/\hbar^2)x}$. This describes a bound state only when V_0 is negative, as expected.

A superconducting tunnel junction will also have bound states of quasiparticles excitations. These bound states describe the Josephson effect since virtual quasiparticle tunneling was necessary for the perturbation calculation in the last section. The Bogoliubov-deGennes equations describe the spatial wavefunctions, whose eigenstates are given by the solution of the Hamiltonian

$$H\varphi^{\pm}e^{i\kappa x} = \left(\pm k_f \frac{\hbar^2\kappa}{m}\sigma_z + \Delta\sigma_x\right)\varphi^{\pm}e^{i\kappa x}, \tag{6.10}$$

where $\varphi^{\pm}e^{i\kappa x}$ are the slowly varying spatial amplitudes of the exact wavefunction $\varphi^{\pm}e^{i\kappa x}e^{\pm ik_f x}$. As illustrated in Fig. 9, the $\pm k_f\kappa$ term corresponds to the kinetic

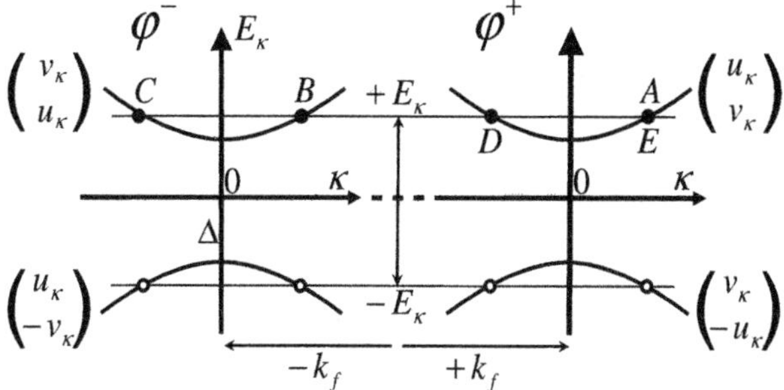

Fig. 9. Plot of quasiparticle energies E_κ verses momentum κ near the $\pm k_f$ Fermi surfaces. The two-component eigenfunctions are also displayed for each of the four energy bands. Also indicated are the quasiparticle states A-E used for the bound-state calculation.

energy at the $\pm k_f$ Fermi surfaces using the approximation $(k_f+\kappa)^2/2 \simeq$ const.+ $k_f\kappa$. As expected for a spin-type Hamiltonian, the two eigenvalues are

$$E_\kappa = \pm(\xi_\kappa^2 + \Delta^2)^{1/2} , \tag{6.11}$$

where $\xi_\kappa = \hbar^2 k_f \kappa / m$ is the kinetic energy of the quasiparticle referred to the Fermi energy. The eigenvectors are also displayed in Fig. 9, where u_κ and v_κ are given by Eqs. 4.14 and 4.15. Because the two energy bands represent quasiparticle excitations, the lower band is normally filled and its excitations correspond to the creation of hole states.

We can solve for the quasiparticle bound states by first writing down the scattering wavefunctions in the left and right superconducting electrodes. An incoming quasiparticle state, point A in Fig. 9, is reflected off the tunnel barrier to states B and C and is transmitted to states D and E [30]. The wavefunctions are then given by

$$\Psi_L = A\begin{pmatrix} u \\ ve^{i\phi_L} \end{pmatrix} e^{i\kappa x} + B\begin{pmatrix} v \\ ue^{i\phi_L} \end{pmatrix} e^{i\kappa x} + C\begin{pmatrix} u \\ ve^{i\phi_L} \end{pmatrix} e^{-i\kappa x} \tag{6.12}$$

$$\Psi_R = D\begin{pmatrix} v \\ ue^{i\phi_R} \end{pmatrix} e^{-i\kappa x} + E\begin{pmatrix} u \\ ve^{i\phi_R} \end{pmatrix} e^{i\kappa x} , \tag{6.13}$$

where we have used the relations $v \equiv v_\kappa = u_{-\kappa}$ and $u \equiv u_\kappa = v_{-\kappa}$, and we have included the phases ϕ_L and ϕ_R of the two states. The continuity conditions Eqs. 6.4 and 6.5, solved for both the components of the spin wavefunction, gives the matrix equation

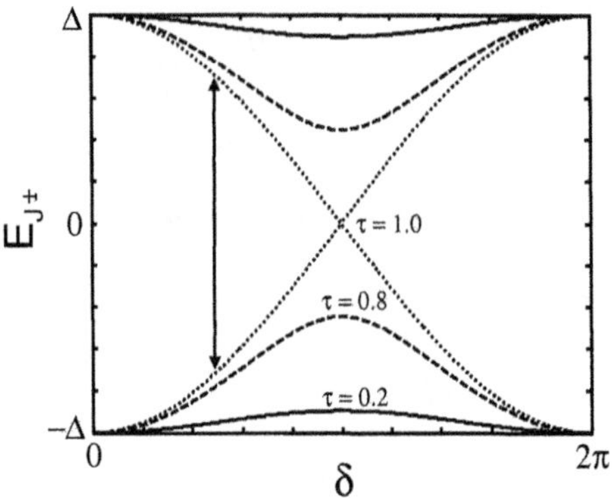

Fig. 10. Plot of quasiparticle bound-state energies E_{J-} and E_{J+} vs. the phase difference δ across the junction, for three values of tunneling transmission τ. Quasiparticles are produced by vertical transitions from the E_{J-} to E_{J+} band. As indicated by the arrow, the energy gap $E_{J+} - E_{J-}$ is always greater than $\sqrt{2}\Delta$ at $\delta = \pi/2$.

$$A\begin{pmatrix} u \\ v \\ u \\ v \end{pmatrix} = \begin{pmatrix} -v & -u & v & u \\ -u & -v & ue^{-i\delta} & ve^{-i\delta} \\ -v & u & -v(1-i2\eta) & u(1+i2\eta) \\ -u & v & -ue^{-i\delta}(1-i2\eta) & ve^{-i\delta}(1+i2\eta) \end{pmatrix} \begin{pmatrix} B \\ C \\ D \\ E \end{pmatrix} . \quad (6.14)$$

The scattering amplitudes for B-E have poles given by the solution of

$$(u^4 + v^4)(1+\eta^2) - 2(uv)^2(\eta^2 + \cos\delta) = 0 . \quad (6.15)$$

Using the relations $u^2 + v^2 = 1$, $E_J = E_k = \Delta/2uv$, and $\tau = 1/(1+\eta^2)$, the energies of the quasiparticle bound states are

$$E_{J\pm} = \pm\Delta[1 - \tau\sin^2(\delta/2)]^{1/2} . \quad (6.16)$$

Because these two states have energies less than the gap energy Δ, they are energetically "bound" to the junction and thus have wavefunctions that are localized around the junction.

The dependence of the quasiparticle bound-state energies on junction phase is plotted in Fig. 10 for several values of τ. The ground state is normally filled, similar to the filling of quasiparticle states of negative energy. The energy E_{J-} corresponds to the Josephson energy, as can be checked in the limit $\tau \to 0$ to give

$$E_{J-} \simeq -\Delta + \frac{\Delta\tau}{4} - \frac{\Delta\tau}{4}\cos\delta . \quad (6.17)$$

This result is equivalent to Eq. 5.9 after noting that the normal-state conductance of a single channel is $1/R_N = 2\tau/R_K$.

The current of each bound state is given by the derivative of its energy

$$I_{J\pm} = \frac{2\pi}{\Phi_0}\frac{\partial E_{J\pm}}{\partial\delta}\,, \tag{6.18}$$

in accord with Eq. 2.5. Since the curvature of the upper band is opposite to that of the lower band, the currents of the two bands have opposite sign $I_{J+} = -I_{J-}$. For level populations of the two states given by $f_\pm$, the average Josephson current is $\langle I_J\rangle = I_{J-}(f_- - f_+)$. For a thermal population, $f_\pm$ are given by Fermi distributions, and the Josephson current in the tunnel junction limit gives the expected Ambegaokar-Baratoff result

$$\langle I_J\rangle = \frac{\pi}{2}\frac{\Delta(T)}{eR_N}\sin\delta\cdot\left(\frac{1}{e^{-\Delta/kT}+1} - \frac{1}{e^{\Delta/kT}+1}\right) \tag{6.19}$$

$$= \frac{\pi}{2}\frac{\Delta(T)}{eR_N}\tanh(\Delta/2kT)\sin\delta\,. \tag{6.20}$$

7. Generation of quasiparticles from nonadiabatic transitions

In this description of the Josephson junction, the Josephson effect arises from a quasiparticle bound state at the junction. Two bound states exist and have energies E_{J+} and E_{J-}, with the Josephson current from the excited state being of opposite sign from that of the ground state. We will discuss here the small-voltage limit[31, 32], which can be fully understood within a semiclassical picture by considering that a linear increase in δ produces nonadiabatic transitions between the two states.

The junction creates "free quasiparticles", those with $E \geq \Delta$, via a two-step process. First, a transition is made from the ground to the excited bound state. This typically occurs because a voltage is placed across the junction, and the linear change of δ causes the ground state not to adiabatically stay in that state. For a high-transmission channel, the transition is usually made around $\delta \approx \pi$, where the energy difference between the states is the lowest and the band bending is the highest. Because this excited state initially has energy less than Δ, the state remains bound until the phase changes to 2π and the energy of the quasiparticle is large enough to become unbound and diffuse away from the junction. The quasiparticle generation rate is thus governed by $d\delta/dt$ and will increase as V increases.

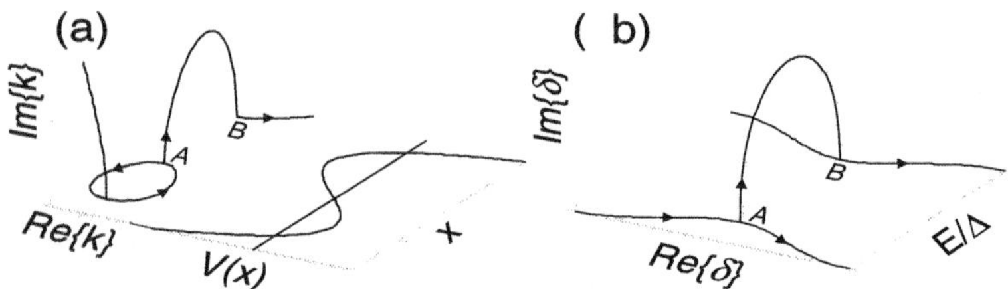

Fig. 11. Plot of semiclassical solutions for the tunneling through a barrier (a) and tunneling through an energy gap (b). Imaginary solutions to k and δ are used to calculate the tunneling rates.

The quasiparticle transition rate can be predicted using a simple semi-classical method. We will first review WKB tunneling in order to later generalize this calculation to energy tunnelling. In Fig. 11(a), we plot a cubic potential $V(x)$ versus x and its solution $k^2 = 2m[E - V(x)]/\hbar^2$. The solution for k is real or imaginary depending on whether E is greater or less than $V(x)$. A semi-classical description of the system is the particle oscillating in the well, as described by the loop in the solution of Re$\{k\}$. A solution in the imaginary part of k connects a turning point on this loop, labeled A, with the turning point of the free-running solution, labeled B. The probability of tunneling each time the trajectory passes point A is given by the standard WKB integral of the imaginary action

$$W = \exp[-2S] \tag{7.1}$$

$$S = (1/\hbar)\left|\int dx \operatorname{Im} p\right| \tag{7.2}$$

$$= \left|\int_{x_A}^{x_B} dx \operatorname{Im} k\right|. \tag{7.3}$$

The transition rate for a nonadiabatic change in a state may be calculated in a similar fashion. In Fig. 11(b) we plot the solution of Eq. 6.16 for δ versus E. In the "forbidden" region of energy $|E| < \Delta\sqrt{1-\tau}$, the solution of δ has an imaginary component. As the bias of the system changes and the system trajectory moves past point A, then this state can tunnel to point B via the connecting path in the imaginary part of δ. The probability for this event is given by Eq. 7.1 with S given by the integral of the imaginary action

$$S = (1/\hbar)\int dE\, t_{\text{imag}}\,, \tag{7.4}$$

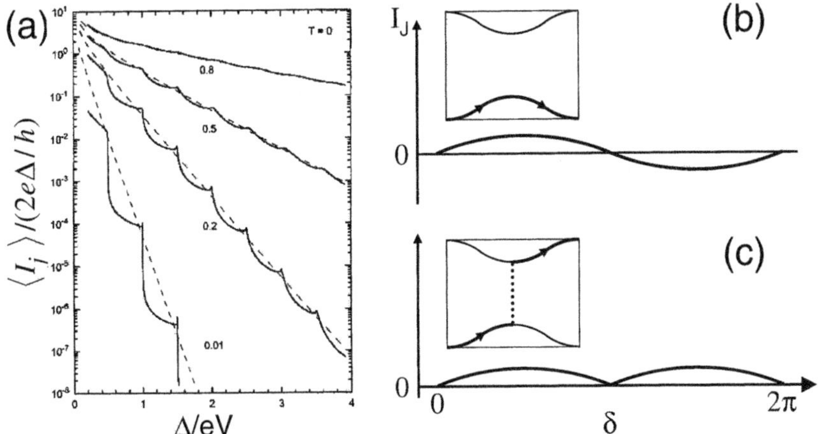

Fig. 12. (a) Plot of average junction current $\langle I_j \rangle$ versus inverse DC voltage V for transmission coefficients $\tau = 0.8, 0.5, 0.2$, and 0.01 (from Ref. [32]). Solid lines are from exact calculation, and dashed lines are from predictons of Eqs. 7.8 and 7.6. The time dependence of the Josephson current I_J is plotted for the ground state (b) and for a transition (c), where the insets show the trajectory of the bound states as E_J *vs.* δ.

where we define an imaginary time by

$$t_{\text{imag}} = \left| \frac{\operatorname{Im} \delta}{d\delta/dt} \right| . \tag{7.5}$$

Rewriting Eq. 6.16 as $(E/\Delta)^2 = 1 - \tau(1 - \cos\delta)/2$ and using $d\delta/dt = (2e/\hbar)V$, one finds the action is given by the integral

$$\begin{aligned} S &= \frac{\Delta}{2eV} \int_{-\sqrt{1-\tau}}^{\sqrt{1-\tau}} d\epsilon \ \operatorname{Im}\left\{ - \arccos\left[1 + (\varepsilon^2 - 1)2/\tau\right]\right\} \qquad (7.6) \\ &\simeq \frac{\Delta}{eV} \times \begin{cases} (1-\tau)\pi/2 & (\tau \to 1) \\ \ln(2/\tau) & (\tau \to 0) \\ (1-\tau)\left[\ln(2/\tau) + \sqrt{\tau}(\pi/2 - \ln 2)\right] , & (\text{interp.}) \end{cases} \end{aligned}$$

where the last interpolation formula approximates well a numerical solution of Eq. 7.6. The limiting expression for $\tau \to 1$ gives the standard Landau-Zener formula appropriate for a two-state system. In the tunnel-junction limit $\tau \to 0$ one finds

$$W = (\tau/2)^{2\Delta/eV} . \tag{7.7}$$

For the case of a constant DC bias voltage V, the total junction current $\langle I_j \rangle$ may be calculated with this transition rate and an attempt rate $\Gamma = (2e/h)V$

given by the frequency at which δ passes $\pi/2$. Using Eq. 7.6 and setting the power of quasiparticle generation $2\Delta\Gamma W$ to the electrical power $\langle I_j \rangle V$, one finds

$$\langle I_j \rangle = \frac{2e\Delta}{\pi\hbar} W \,. \tag{7.8}$$

This prediction is plotted in Fig. 12(a) and shows very good agreement with the results of exact calculations[32]. Only the steps in voltage are not reproduced, which are understood as arising from the quantization of energy eV from multiple Andreev reflection of the quasiparticles. The steps are not expected to be reproduced by the semiclassical theory since this theory is an expansion around small voltages, or equivalently, large quantization numbers.

The junction current may also be determined from the energies of the two bound states. For a constant voltage across the junction, we use Eq. 2.5 to calculate the charge transferred across the junction after a phase change of 2π

$$Q_j = \int_0^{2\pi/(d\delta/dt)} I_j \, dt \tag{7.9}$$

$$= \frac{2\pi}{\Phi_0} \int_0^{2\pi/(d\delta/dt)} \frac{dU_J}{d\delta} dt \tag{7.10}$$

$$= \frac{[U_J(2\pi) - U_J(0)]}{V} \,, \tag{7.11}$$

which gives the expected result that the change of energy equals $Q_j V$. When the junction remains in the ground state, the energy is constant $U_J(2\pi) - U_J(0) = 0$ and no net charge flows through the junction. Net charge is transferred, however, after a transition. The charge transfer $2\Delta/V$ multiplied by the transition rate gives an average current $Q_j \Gamma W$ that is equivalent to Eq. 7.8.

Equation 6.18 may be used to calculate the time dependence of the Josephson current, as illustrated in Fig. 12(b) and (c). When the system remains in the ground state (b), the junction current is sinusoidal and averages to zero. For the case of a transition (c), the current before the transition is the same, but the Josephson current remains positive after the transition (see Eq. 5.4 of Ref. [32]). The transition itself also produces charge transfer from multiple-Andreev reflections(MAR) [31, 33]

$$Q_{\mathrm{MAR}} = 2\Delta(1-\tau)^{1/2}/V \,. \tag{7.12}$$

This result is perhaps surprising - the junction current at finite voltage arises from transfer of charge Q_{MAR} *and* a change in the Josephson current. The relative contribution of these two currents is determined by the relative size of the gap in the bound states. For $\tau \to 1$, all of the junction current is produced by Josephson current, whereas for $\tau \to 0$ (tunnel junctions) the current comes from Q_{MAR}.

For small voltages, the transition event must transfer a large amount of charge $Q_{\rm MAR}$ in order to overcome the energy gap. In comparing this semi-classical theory with the exact MAR theory, $Q_{\rm MAR}/e$ has an integer value and represents the order of the MAR process and the number of electrons that are transferred in the transition. This description is consistent with Eq. 7.7 describing the transition probability for an n-th order MAR process, where $n = 2\Delta/eV$, and $\tau/2$ represents the matrix element for each order.

From this example it is clearly incorrect to picture the quasiparticle and Josephson current as separate entities, as suggested by the calculations of perturbation theory. To do so ignores the fact that quasiparticle tunneling, arising from a transition between the bound states, also changes the Josephson contribution to the current from $\delta = \pi$ to 2π.

8. Quasiparticle bound states and qubit coherence

The quasiparticle bound-state theory can be used to predict both the Josephson and quasiparticle current in the zero-voltage state, as appropriate for qubits. In this theory an excitation from the E_{J-} bound state to the E_{J+} state is clearly deleterious as it will change the Josephson current, fluctuating the qubit frequency and producing decoherence in the phase of the qubit state. For an excitation in one channel, the fractional change in the Josephson current is $\sim 1/N_{\rm ch}$, where $N_{\rm ch}$ is the number of conduction channels. The subgap current-voltage characteristics can be used to estimate $N_{\rm ch}$, which gives an areal density of $\sim 10^4/\mu\text{m}^2$ [10, 34]. For a charge qubit with junction area $10^{-2}\mu\text{m}^2$, the qubit frequency changes fractionally by $\sim 1/N_{\rm ch} \sim 10^{-2}$ for a single excitation, and gives strong decoherence. Although the phase qubit has a smaller change $(1/N_{\rm ch})I_0/4(I_0 - I) \sim 2 \times 10^{-5}$, the excitation of even a single bound state is clearly unwanted.

Fortunately, these quasiparticle bound states should not be excited in tunnel junctions by the dynamical behavior of the qubit. The E_{J-} to E_{J+} transition is energetically forbidden because the energy of the qubit states are typically choosen to be much less than 2Δ. Thus, the energy gap of the superconductor protects the qubit from quasiparticle decoherence.

If a junction has "pinhole" defects, where a few channels have $\tau \to 1$, then the energy gap will shrink to zero at $\delta = \pi$. However, only the flux qubit will be sensitive to quasiparticles produced at these defects since it operates near $\delta = \pi$. In contrast, the phase qubit always retains an energy gap of at least $\sqrt{2}\Delta$ around its operating point $\delta = \pi/2$ (see arrow in Fig. 10). We note this idea implies that a phase qubit can even be constructed from a microbridge junction, which has some channels[26] with $\tau = 1$. Although the phase qubit is completely insensitive to pinhole defects, this advantage is probably unimportant because Al-based tunnel junctions have oxide barriers of good quality.

Pinhole defects also change the Josephson potential away from the -cos δ form. This modification is typically unimportant because the deviation is smooth and can be accounted for by a small effective change in the critical current.

The concept that the energy gap Δ protects the junction from quasiparticle transitions suggests that superconductors with nonuniform gaps may not be suitable for qubits. Besides the obvious problem of conduction channels with zero gap, channels with a reduced gap may cause stray quasiparticles to be trapped at the junction. The high-T_c superconductors, with the gap suppressed to zero at certain crystal angles, are an obvious undesirable candidate. However, even Nb could be problematic since it has several oxides that have reduced or even zero gap. Nb based tri-layers may also be undesirable since the thin Al layer near the junction slightly reduces the gap around the junction. In contrast, Al may not have this difficulty since its gap *increases* with the incorporation of oxygen or other scattering defects. It is possible that these ideas explain why Nb-based qubits do not have coherence times as long as Al qubits[6, 10].

9. Summary

In summary, Josephson qubits are nonlinear resonators whose critical element is the nonlinear inductance of the Josephson junction. The three types of superconducting qubits, phase, flux, and charge, use this nonlinearity differently and produce qubit states from a cubic, quartic, and cosine potential, respectively.

To understand the origin and properties of the Josephson effect, we have first reviewed the BCS theory of superconductivity. The superconducting phase was explicitly shown to be a macroscopic property of the superconductor, whose classical and quantum behavior is determined by the external electrical circuit. After a review of quasiparticle and Josephson tunneling, we argued that a proper microscopic understanding of the junction could arise only from an exact solution of the Josephson effect.

This exact solution was derived by use of mesoscopic theory and quasiparticle bound states, where we showed that Josephson and quasiparticle tunneling can be understood from the energy of the bound states and their transitions, respectively. A semiclassical theory was used to calculate the transition rate for a finite DC voltage, with the predictions matching well that obtained from exact methods.

This picture of the Josephson junction allows a proper understanding of the Josephson qubit state. We argue that the gap of the superconductor strongly protects the junction from quasiparticle tunneling and its decoherence. We caution that an improper choice of materials might give decoherence from quasiparticles that are trapped at sites near the junction.

We believe a key to future success is understanding and improving this remarkable nonlinearity of the Josephson inductance. We hope that the picture given here of the Josephson effect will help researchers in their quest to make better superconducting qubits.

Acknowledgements

We thank C. Urbina, D. Esteve, M. Devoret, and V. Shumeiko for helpful discussions. This work is supported in part by the NSA under contract MOD709001.

References

[1] M. A. Nielsen and I. L. Chuang, *Quantum Computation and Quantum Information* (Cambridge University Press, Cambridge, 2000).

[2] Y. Nakamura, C. D. Chen, and J. S. Tsai, Phys. Rev. Lett. **79**, 2328 (1997).

[3] Y. Nakamura, Y. A. Pashkin, T. Yamamoto, and J. S. Tsai, Phys. Rev. Lett. **88**, 047901 (2002).

[4] D. Vion, A. Aassime, A. Cottet, P. Joyez, H. Pothier, C. Urbina, D. Esteve, and M. H. Devoret, Science **296**, 886 (2002).

[5] S. Han, Y. Yu, Xi Chu, S. Chu, and Z. Wang, Science **293**, 1457 (2001); Y. Yu, S. Han, X. Chu, S. Chu, and Z. Wang, Science **296**, 889 (2002).

[6] J. M. Martinis, S. Nam, J. Aumentado, and C. Urbina, Phys. Rev. Lett. **89**, 117901 (2002).

[7] I. Chiorescu, Y. Nakamura, C. J. P. M. Harmans, and J. E. Mooij, Science **299**, 1869 (2003).

[8] A.J. Berkley, H. Xu, R.C. Ramos, M.A. Gubrud, F.W. Strauch, P.R. Johnson, J.R. Anderson, A.J. Dragt, C.J. Lobb, and F.C.Wellstood, Science **300**, 1548 (2003).

[9] Yu. A. Pashkin, T. Yamamoto, O. Astafiev, Y. Nakamura, D. V. Averin, and J. S. Tsai, Nature **421**, 823 (2003).

[10] R. W. Simmonds, K. M. Lang, D. A. Hite, D. P. Pappas, and J. M. Martinis, submitted to Phys. Rev. Lett.

[11] M.H. Devoret, *Quantum Fluctuations in Electrical Circuits*, in "Fluctuations Quantiques", Elsevier Science (1997).

[12] M. Steffen, J. M. Martinis, and I. L. Chuang, Phys. Rev. B **68**, 2245xx (2003).

[13] Audrey Cottet, Ph.D. thesis (2002).

[14] R. Rouse, S. Han, J. Lukens, Phys. Rev. Lett. **75**, 1614 (1995).

[15] R. Koch, private communication.

[16] J. E. Mooij, T. P. Orlando, L. Levitov, L. Tian, C. H. van der Wal, S. Lloyd, Science **285**, 1036 (1999)

[17] D. J. VanHarlingen, B. L. T. Plourde, T. L. Robertson, P. A Reichardt, and J. Clarke, Proceedings of the 3rd International Workshop on Quantum Computing, to be published.

[18] J. Bardeen, L. N. Cooper, and J. R. Schrieffer, Phys. Rev. **108**, 1175 (1957).

[19] P. W. Anderson, Phys. Rev. **112**, 1900 (1958).

[20] C. Kittel, *Quantum Theory of Solids*, John Wiley (1987).

[21] U. Eckern, G. Schon, V. Ambegaokar, Phys. Rev. B **30**, 6419 (1984).

[22] V. Ambegaokar and A. Baratoff, Phys. Rev. Lett. **11**, 104 (1963).

[23] C. W. J. Beenakker, Phys. Rev. Lett. **67**,3836 (1991).

[24] C. W. J. Beenakker, Rev. Mod. Phys. **69**, 731 (1997)

[25] Y. Naveh, Vijay Patel, D. V. Averin, K. K. Likharev, and J. E. Lukens, Phys. Rev. Lett. **85**, 5404 (2000).

[26] O. N. Dorokhov, JETP Lett. **36**, 318 (1982).

[27] K. M. Schep and G. E. W. Bauer, Phys. Rev. Lett. **78**, 3015 (1997).

[28] A. Furusaki and M. Tsukada, Physica B **165-166**, 967 (1990).

[29] S. V. Kuplevakhaskii and I. I. Fal'ko, Sov. J. Low Temp. Phys. **17**, 501 (1991).

[30] Our notation for incoming and outgoing states is choosen to correspond to the boundry conditions for scatttering in Ref. [32].

[31] D. Averin and A. Bardas, Phys. Rev. Lett. **75**, 1831 (1995).

[32] E. N. Bratus', V. S. Shumeiko, E. V. Bezuglyi, and G. Wendin, Phys. Rev. B **55**, 12666 (1997).

[33] E. N. Bratus, V. S. Shumeiko, and G. A. B. Wendin, Phys. Rev. Lett. **74**, 2110 (1995).

[34] K.M. Lang, S. Nam, J. Aumentado, C. Urbina, J. M. Martinis, IEEE Trans. on Appl. Supercon. **13**, 989 (2003).

Course 14

JOSEPHSON QUANTUM BITS BASED ON A COOPER PAIR BOX

Denis Vion

CEA-Saclay, Orme des merisiers,
91191 Gif sur Yvette Cedex , France

D. Estève, J.-M. Raimond and J. Dalibard, eds.
Les Houches, Session LXXIX, 2003
Quantum Entanglement and Information Processing
Intrication quantique et traitement de l'information
© *2004 Elsevier B.V. All rights reserved*

Contents

1. Introduction

The present chapter is devoted to a particular type of electrical circuit that has been used to develop solid state quantum bit prototypes. These circuits being superconducting and involving tunneling of Cooper pairs between two superconducting electrodes, they belong to the family of Josephson qubits previously introduced in this book [1]. They are all based on the same simple device, the Cooper pair box (CPB), and are all driven by a gate electrode coupled to the charge of a small electrode. For that reason, they are often considered as forming the so-called "charge qubits" sub-family, although they essentially share the same physics with other Josephson qubits [2, 3]: their quantum state can be easily manipulated, whereas reading this state out with a high efficiency is a difficult task. Moreover, preserving their quantum coherence is a challenge (even at ultra low temperature) due to their "macroscopic" character.

This chapter is organized in six sections. After this introduction, the second section presents the Cooper pair box device in its basic version and in its improved version: the split CPB. The energy spectrum is derived as a function of the external parameters controlling the Hamiltonian and the physical properties of the corresponding eigenstates are pointed out. In the third section, we show how the two lowest energy eigenstates form a qubit, how this qubit can be manipulated with DC voltage pulses or resonant microwave pulses, and how it can be measured following various strategies. Three experiments that have demonstrated coherent control of the CPB state are also presented. Then, in section 4, we present a very simple approach to decoherence in CPBs. Considering a particular CPB device (the Quantronium) as an example, we list its different possible decoherence sources and we calculate the different physical quantities that characterize how coherence of its quantum state is lost. From these considerations, we infer design rules for Josephson qubits. Then, we present different experiments that have been used to measure the effective coherence time of a real device. Finally, we address in section 5 the problem of making a 2-qubit-gate with two capacitively coupled CPBs.

2. The Cooper pair box

2.1. The basic Cooper pair box circuit

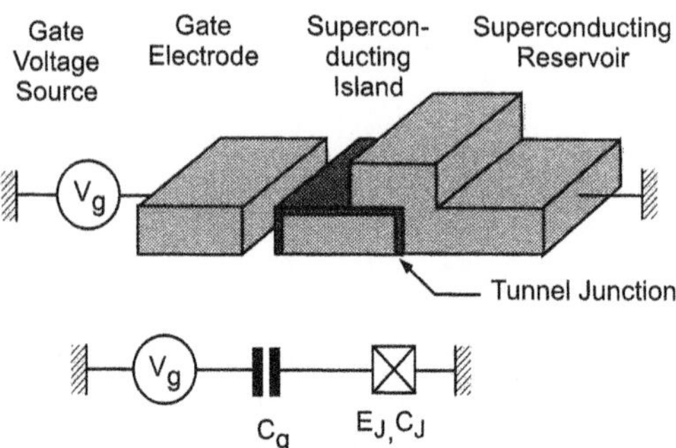

Fig. 1. *The basic Cooper pair box. Top: Schematic representation of the Cooper pair box showing the superconducting island and reservoir, the Josephson junction with energy E_J and capacitance C_J, the gate, and the voltage source V_g. Bottom: Corresponding electrical schematic drawing.*

The basic Cooper pair box (CPB) is the simplest device which combines Josephson [4] and Coulomb blockade effects [5]. It is a simplified version of a Josephson device proposed in 1987 [6], and consists [7] of a small BCS superconducting electrode, called the island, connected to a BCS superconducting reservoir by a Josephson junction with capacitance C_J and Josephson energy E_J. The island can be biased by a voltage source V_g in series with a gate capacitance C_g (see Fig. 1). In addition to E_J, the box has a second characteristic energy, the Coulomb energy E_C of a single Cooper Pair in excess in the island, with respect to electrical neutrality:

$$E_C = \frac{(2e)^2}{2C_\Sigma}, \tag{2.1}$$

where $C_\Sigma = C_g + C_J$ is the total capacitance of the island and e the electron charge. CPBs fabricated by conventional electron beam lithography having a capacitance C_Σ in the fF range (typical size of the junctions is 100nm $\times$ 100nm), E_C is typically of order of a few $k_B K$ (k_B is the Boltzmann constant). When the thermal energy $k_B T$ is reduced much below the BCS superconducting energy gap Δ of the electrodes, and when $E_C < 4\Delta$, all the electrons in the island and in the reservoirs are paired [8]. The Cooper pairs can tunnel through the Josephson junction and the only remaining degree of freedom of the system is the integer number **N** of Cooper pairs in excess or deficit on the island. Due to tunneling, **N** fluctuates quantum mechanically and has to be treated as an operator $\widehat{\mathbf{N}}$, whose eigenstates $|\mathbf{N}\rangle_c$ (index c stands for "pure charge") obey

$$\widehat{\mathbf{N}}\,|\mathbf{N}\rangle_c = \mathbf{N}\,|\mathbf{N}\rangle_c\,,\;\; \mathbf{N}\in\mathbb{Z}$$

and form a complete basis for the quantum states of the box.

Introducing the operator $\widehat{\theta}$ conjugated to $\widehat{\mathbf{N}}$ by the dimensionless relationship $[\widehat{\theta},\widehat{\mathbf{N}}]=\mathbf{i}$, one defines the variable $\theta\in[0,2\pi[$, which is the phase of the Cooper pair condensate in the island. From the conjugation relationship, one deduces the effect of the operators $\exp(\mathbf{i}\widehat{\theta})$ and $\exp(-\mathbf{i}\widehat{\theta})$ on charge states:

$$\exp(\pm\mathbf{i}\widehat{\theta})\,|\mathbf{N}\rangle_c = |\mathbf{N}\pm 1\rangle_c\,. \tag{2.2}$$

The Hamiltonian of the CPB can now be expressed as a function of the $\widehat{\mathbf{N}}$ and/or $\widehat{\theta}$ operators.

2.2. *Hamiltonian and energy spectrum*

The Hamiltonian of the whole CPB circuit (including its voltage source) is written:

$$\widehat{H}(N_g) = \widehat{H}_{el} + \widehat{H}_J = E_C(\widehat{\mathbf{N}}-N_g)^2 - E_J\cos\widehat{\theta}\,, \tag{2.3}$$

where the first term corresponds to the electrostatic energy of the circuit, $N_g = C_g V_g/(2e)$ being the reduced gate charge, and where the second term accounts for the energy cost of a phase difference θ across the Josephson junction and is responsible for the tunneling of Cooper pairs. In order to find the eigenenergies and the corresponding eigenstates of the system, (2.3) is rewritten in a form involving only $\mathbf{N}$ or only θ. Using (2.2), one finds the Hamiltonian in the charge representation,

$$\widehat{H} = \sum_{\mathbf{N}\in\mathbb{Z}}\left[E_C(\mathbf{N}-N_g)^2\,|\mathbf{N}\rangle_c\,\langle\mathbf{N}|_c - \frac{E_J}{2}(|\mathbf{N}\rangle_c\,\langle\mathbf{N}+\mathbf{1}|_c + |\mathbf{N}+\mathbf{1}\rangle_c\,\langle\mathbf{N}|_c)\right]. \tag{2.4}$$

The energy spectrum associated to this Hamiltonian is discrete and periodic in N_g with period 1. We call $|k\rangle$ the energy eigenstates and E_k their associated energies sorted in increasing order, starting from $k=0$ for the ground state:

$$\widehat{H}\,|k\rangle = E_k\,|k\rangle\,,\, k\in\mathbb{N}. \tag{2.5}$$

For a given N_g, the lowest energy eigenstates can be found in the charge representation by truncating the pure charge state basis and by diagonalizing a finite version of the matrix that corresponds to (2.4).

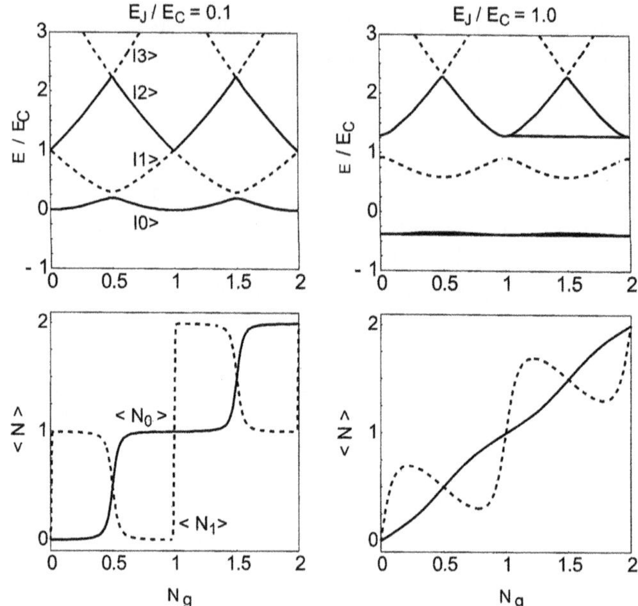

Fig. 2. Top: Energy levels of a Cooper pair box, normalized by the Cooper pair Coulomb energy E_C, as a function of the gate charge bias N_g, and for E_J/E_C ratios equal to 0.1 (left) and 1 (right). Bottom: Corresponding expectation values of the dimensionless box charges $\langle N \rangle$, for the energy eigenstates $|0\rangle$ (solid lines) and $|1\rangle$ (dotted lines).

Using $\widehat{\mathbf{N}} = (\mathbf{1}/\mathbf{i})\partial/\partial\theta$ in (2.3), one instead obtains the Hamiltonian in the phase representation and the Schrödinger equation for the $\Psi_k(\theta) = \langle\theta\,|k\rangle$ wavefunctions:

$$E_C(\frac{1}{\mathbf{i}}\frac{\partial}{\partial\theta} - N_g)^2\Psi_k(\theta) - E_J\cos(\theta)\Psi_k(\theta) = E_k\Psi_k(\theta)\,. \tag{2.6}$$

Both representations can of course be used equivalently to find the energy spectrum, which depends on N_g and on the E_J/E_C ratio, as shown on Fig. 2. When $E_J/E_C \ll 1$, the energy levels are very close to the electrostatic energies, except in the vicinity of the so-called charge degeneracy points defined by $N_g = 1/2\ (\mathrm{mod}\ 1)$, where the degeneracy between the two lowest energy charge states is lifted up by an amount E_J. With increasing E_J/E_C, the modulation by N_g of the lowest eigenenergies becomes weaker and weaker.

It is interesting to note that except for precise combinations of E_J/E_C and N_g values, the energy spectrum of a CPB is highly anharmonic. Consequently, manipulating $|0\rangle$ and $|1\rangle$ without exciting higher energy states is possible. These two states are thus regarded as defining a qubit. We now compute explicitly the $|0\rangle$ and $|1\rangle$ states in order to evaluate their physical properties, which will be used to measure the quantum bit state.

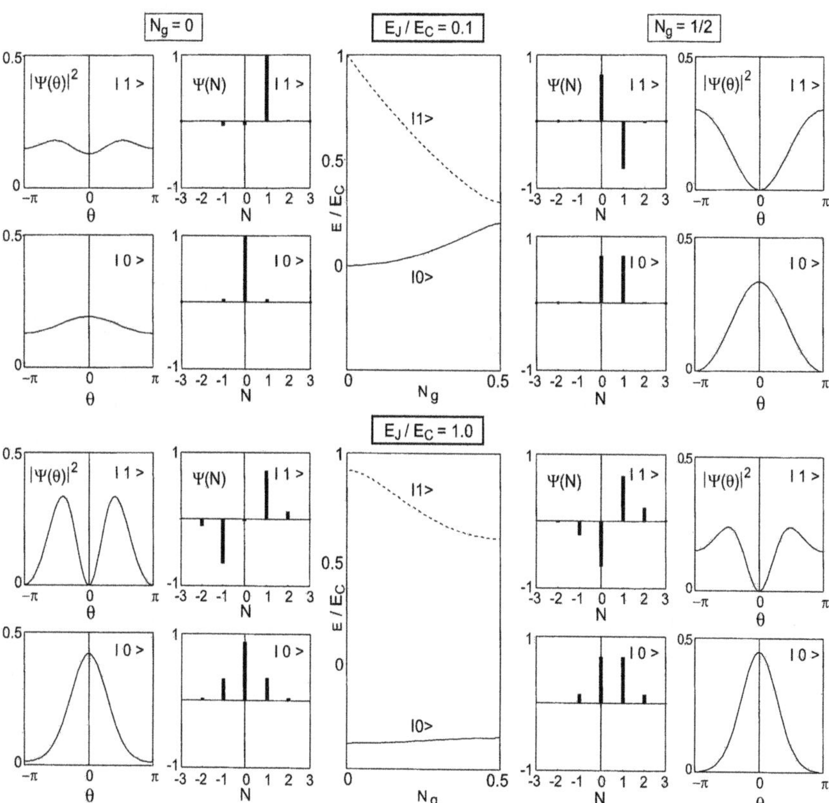

Fig. 3. Eigenenergies (middle panels) and wavefunctions of the $|0\rangle$ and $|1\rangle$ states in the charge and phase representations, for $N_g = 0$ (left panels) and for $N_g = 1/2$ (right panels), and for E_J/E_C ratios equal to 0.1 (top panels) and 1 (bottom panels). The $\Psi_k(N)$ eigenvectors are directly represented since they can be chosen real, whereas the $\Psi_k(\theta)$ wavefunctions are represented by their modulus squared.

2.3. Eigenstates in the charge and phase representations

Over an N_g period like the interval [0, 1] and for $E_J/E_C < 2$, the energy eigenstates $|k\rangle = \sum_{\mathbf{N}\in\mathbb{Z}} a_{kN} |\mathbf{N}\rangle_c$ can be found with a high accuracy by simply diagonalizing a matrix (2.4) truncated to only seven charge states. The corresponding $\Psi_k(\theta)$ functions can be found by Fourier transform of the $|k\rangle$'s expressed in the charge representation or by solving directly the Schrödinger equation (2.6). This equation is close to a Mathieu equation and its solutions are [9]

$$\begin{cases} E_k = E_C \mathcal{M}_A(k+1-(k+1)/4[\text{mod } 2] + 2n_g(-1)^k, q) \\ \Psi_k(\theta) = \frac{\exp(\mathbf{i}N_g\theta)}{\sqrt{2\pi}}[\mathcal{M}_C(a,q,\frac{\theta}{2}) + \mathbf{i}(-1)^{k+1}\mathcal{M}_S(a,q,\frac{\theta}{2})] \end{cases}, \tag{2.7}$$

where $a = 4E_k/E_C$, $q = -2E_J/E_C$, $\mathcal{M}_C$ and $\mathcal{M}_S$ are the even and odd Mathieu functions, and $\mathcal{M}_A$ is the function giving the characteristic values of $\mathcal{M}_C$. Figure 3 shows the two lowest stationary states $|0\rangle$ and $|1\rangle$ both in the charge and phase representations, for $N_g = 0$ and $N_g = 1/2$, and for two different E_J/E_C ratios. For $E_J/E_C \ll 1$, the situation is rather simple since $|0\rangle$ and $|1\rangle$ are very close to the pure charge states $|\mathbf{0}\rangle_c$ or $|\mathbf{1}\rangle_c$ at $N_g \simeq 0$, and correspond to the symmetric and antisymmetric superpositions of these charge states at $N_g = 1/2$. In this limit, it is useful to restrict the basis to $(|\mathbf{0}\rangle_c, |\mathbf{1}\rangle_c)$, so that the Hamiltonian looks like that for a spin 1/2 (like any other two-level-system [10]), after dropping out a constant term that depends on N_g only:

$$\widehat{H} = -\frac{1}{2}\vec{\sigma}.\vec{H}, \tag{2.8}$$

where $\vec{\sigma} = \hat{\sigma}_x\vec{x} + \hat{\sigma}_y\vec{y} + \hat{\sigma}_z\vec{z}$ is the vector of Pauli matrices and $\vec{H} = E_J\vec{x} + E_C(1-2N_g)\vec{z}$. Introducing the angle $\alpha = \arctan\left[E_J/\left\{E_C(1-2N_g)\right\}\right]$, the eigenenergies and the eigenstates are in this case $\mp E_J\sqrt{1+\cot^2\alpha}$ and

$$\begin{cases} |0\rangle = \cos(\alpha/2)\,|\mathbf{0}\rangle_c + \sin(\alpha/2)\,|\mathbf{1}\rangle_c \\ |1\rangle = -\sin(\alpha/2)\,|\mathbf{0}\rangle_c + \cos(\alpha/2)\,|\mathbf{1}\rangle_c \end{cases}, \tag{2.9}$$

respectively. For $E_J/E_C \sim 1$, the $|0\rangle$ and $|1\rangle$ states are, for any N_g, made up of coherent superpositions with significant contributions from at least three or four pure charge states (see Fig. 3), so that neither θ nor $\mathbf{N}$ are "good quantum numbers".

2.4. Expectation value of the box charge

The expectation value of the charge on the island or its dimensionless equivalent $\langle N_k\rangle = \langle k|\widehat{\mathbf{N}}|k\rangle$ is an interesting quantity which can be used to discriminate

$|0\rangle$ from $|1\rangle$, and thus to read out a CPB-based-qubit. It depends linearly on the derivative of the energy levels with respect to N_g:

$$\frac{\partial \hat{H}}{\partial N_g} = 2E_C(N_g - \widehat{\mathbf{N}}) \Longrightarrow \langle N_k \rangle = N_g - \frac{1}{2E_C} \frac{\partial E_k}{\partial N_g} . \tag{2.10}$$

It is plotted in Fig. 2 for the two regimes already considered. For $E_J/E_C \ll 1$ and close to half integer values of N_g, $\langle N_0 \rangle$ and $\langle N_1 \rangle$ vary as opposite rounded staircases. Within the two charge states approximation, one deduces from (2.9) the shape of the steps for $N_g \in [0, 1]$: $\langle N_0 \rangle = \sin^2 \alpha/2$ and $\langle N_1 \rangle = \cos^2 \alpha/2$. When E_J/E_C is increased, the steps are more and more rounded and have to be calculated numerically. It is important to note that the difference $\Delta N_{10} = \langle N_1 \rangle - \langle N_0 \rangle$ vanishes at the charge degeneracy points.

2.5. *The split Cooper pair box*

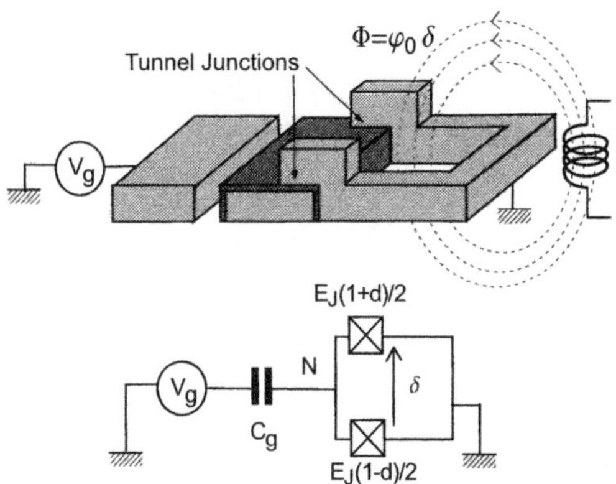

Fig. 4. *The split Cooper pair box. Top: Schematic representation showing the island, the two Josephson junctions connected to form a grounded superconducting loop, the gate circuit, and the magnetic flux bias. Bottom: Corresponding electrical drawing.*

The split Cooper pair box is an improved CPB with a tunable Josephson energy and a second access port. It is obtained by splitting its Josephson junction into two junctions with respective Josephson energies $E_J(1 + d)/2$ and $E_J(1 - d)/2$, where $d \in [0, 1]$ is an asymmetry coefficient (see Fig. 4). These two junctions are connected together to form a superconducting loop which can be biased by a magnetic flux Φ. Notice that the split CPB is similar to another Josephson device, the Bloch transistor [11] (also called the single Cooper pair transistor) that was first described in 1985. The split box has two degrees of freedom, which can be chosen either as the phase differences $\widehat{\delta_1}$ and $\widehat{\delta_2}$ across each junction, or as the linear forms $\widehat{\theta} = (\widehat{\delta_1} - \widehat{\delta_2})/2$ and $\widehat{\delta} = \widehat{\delta_1} + \widehat{\delta_2}$, which represent the phase of the island introduced previously and the phase difference across the series combination of the two junctions, respectively. The conjugate variable of $\widehat{\delta}$ is the integer number K of Cooper pairs which tunneled through both junctions.

The electrostatic "Hamiltonian" of the split CPB is that of a basic box [see (2.3)] with C_J representing now the sum of the two junction capacitances. Its Josephson "Hamiltonian" is the sum of the Josephson terms of the two junctions:

$$\widehat{H}_J^* = -E_J \frac{1+d}{2} \cos(\widehat{\delta_1}) - E_J \frac{1-d}{2} \cos(\widehat{\delta_2}) \tag{2.11}$$

$$= -E_J \cos(\frac{\widehat{\delta}}{2}) \cos(\widehat{\theta}) + d E_J \sin(\frac{\widehat{\delta}}{2}) \sin(\widehat{\theta}) \,. \tag{2.12}$$

The superconducting loop of a split CPB is designed such that its self inductance L is very small compared to the junction inductance $L_J = \varphi_0^2/E_J$, with $\varphi_0 = \hbar/2e$. Consequently, the magnetic potential energy term $\left(\varphi_0\hat{\delta} - \Phi\right)^2/2L$ attached to this inductance strongly fixes δ, which can be considered as a classical parameter $\delta = \Phi/\varphi_0$ imposed by the magnetic flux. Finally, the Hamiltonian of the split box is

$$\widehat{H}(N_g, \delta) = E_C(\widehat{\mathbf{N}} - N_g)^2 - E_J^*(d, \delta) \cos[\widehat{\theta} + \Upsilon(d, \delta)] \,, \tag{2.13}$$

with [12]:

$$\begin{aligned} E_J^*(d, \delta) &= E_J \sqrt{\tfrac{1+d^2+(1-d^2)\cos(\delta)}{2}} \\ \tan \Upsilon(d, \delta) &= -d \tan(\tfrac{\delta}{2}) \,. \end{aligned} \tag{2.14}$$

A symmetric or almost symmetric ($d \approx 0$) split CPB is thus equivalent to a basic CPB but with a magnetostatically tunable [7] Josephson energy $E_J^* = E_J \cos(\delta/2)$. Its energy spectrum (see Fig. 5) is periodic in N_g (period 1) and 2π-periodic in δ, and can now be tuned by both the electric field applied to the gate electrode and by the magnetic flux threading the superconducting loop. For that reason, the split CPB has often been presented as a kind of artificial atom showing strong Stark and Zeeman effects.

Splitting the box has also a second interest: it opens a second access port to the device, which can be used to read out its quantum state [13–15]. The quantity to be measured on this port is the persistent current in the superconducting loop, its phase equivalent across the loop inductance, or the magnetic flux it produces. This persistent current is calculated below.

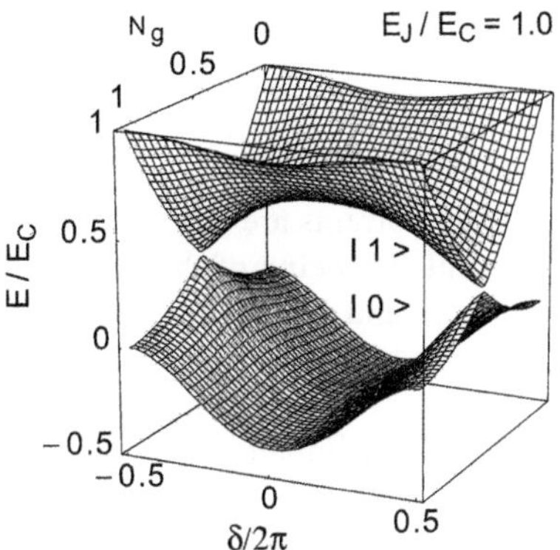

Fig. 5. Two lowest energy levels of a split Cooper pair box with $E_J/E_C = 1$, as a function of the two external parameters N_g and δ. The energy is normalized by the Cooper pair Coulomb energy E_C. The asymmetry coefficient $d = 2\%$ chosen here lifts up an energy degeneracy at $(N_g = 1/2, \delta = \pm\pi)$.

2.6. *Expectation value of the persistent current in the split box*

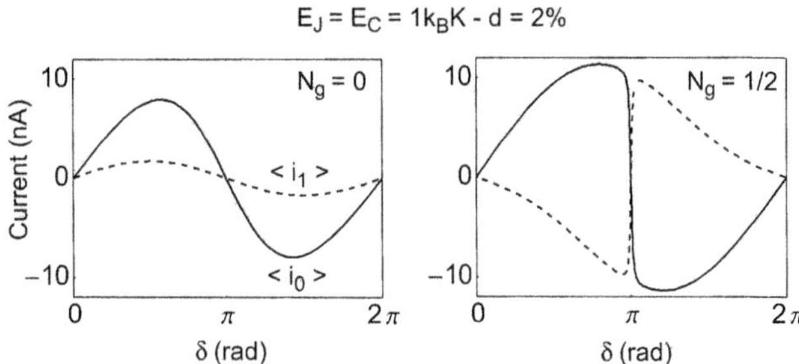

Fig. 6. Expectation value of the persistent loop currents $\langle i_0 \rangle$ for the ground state (solid line), and $\langle i_1 \rangle$ for the first excited state (dotted line), calculated for $E_C = E_J = 1$ k_BK, $d = 2\%$, $N_g = 0$ (left panel) and $N_g = 1/2$ (right panel).

The $\widehat{\delta}$ and $\widehat{K}$ operators being conjugate to each other, the operator associated to the current circulating in the loop of the split CPB is

$$\widehat{I} = (-2e)\,\frac{d\widehat{K}}{dt} = (-2e)\left(-\frac{1}{\hbar}\frac{\partial \widehat{H}}{\partial \delta}\right). \tag{2.15}$$

The average loop current $\langle i_k \rangle$ of state $|k\rangle$ follows thus the generalized Josephson relation

$$\left\langle i_k(N_g,\delta)\right\rangle = \left\langle k\left|\widehat{I}\right|k\right\rangle = \frac{1}{\varphi_0}\frac{\partial E_k(N_g,\delta)}{\partial\delta}\ . \tag{2.16}$$

Like the energy spectrum, $\langle i_k\rangle$ currents are also 2π-periodic in δ and 1-periodic in N_g, the extrema of $\langle i_0\rangle$ and $\langle i_1\rangle$ being of the order of E_J/φ_0. Also, for N_g close to 1/2 and $E_J < 3E_c$, $\langle i_0\rangle$ and $\langle i_1\rangle$ have opposite signs, as shown in Fig. 6. Note that the difference $\Delta i_{10} = \langle i_1\rangle - \langle i_0\rangle$ vanishes at $\delta = 0$ for all N_g.

Given the physical properties of the $|0\rangle$ and $|1\rangle$ CPB's states, we can now consider the different strategies for implementing, for driving and for reading a qubit based on these states.

3. The Cooper pair box as a quantum bit

As previously mentioned, the two orthogonal states chosen to define a CPB-based-qubit are its two lowest energy eigenstates $|0\rangle$ and $|1\rangle$. By varying N_g, the quantum state of the box can be manipulated within this subspace, provided that the temperature is sufficiently low, that the N_g excursion is limited, and that the anharmonicity of the energy spectrum is large enough. To implement a qubit, the CPB has also to be coupled to a readout device able to discriminate its two states at a certain measuring point (N_{gm},δ_m) in the space of the external parameters controlling its Hamiltonian. We consider here the case of a coupling between the box and its readout, weak enough so that it does not modify significantly the $|0\rangle$ and $|1\rangle$ states of the uncoupled box. When all these conditions are fulfilled, the CPB can be regarded as a fictitious dimensionless spin 1/2, $\vec{\sigma}$, with a Hamiltonian

$$\widehat{H}\left(N_g,\delta\right) = -\frac{1}{2}\vec{\sigma}.\vec{H}\left(N_g,\delta\right)\ . \tag{3.1}$$

This Hamiltonian can be expressed in any frame $R\{\vec{x},\vec{y},\vec{z}\}$ defined by

$$\left\{\begin{array}{c} \vec{H}\left(N_{g0},\delta_0\right) = h\nu_{01}\left(N_{g0},\delta_0\right)\vec{z}\left(N_{g0},\delta_0\right) \\ \partial\vec{H}/\partial N_g.\vec{y} = 0 \end{array}\right. , \tag{3.2}$$

where $\nu_{01}\left(N_g,\delta\right)$ is the transition frequency between $|0\rangle$ and $|1\rangle$ and $\left(N_{g0},\delta_0\right)$ is a particular point in the parameter space. Note that the frame introduced with (2.8) when $E_J/E_C \ll 1$ is a limit case, for which the reference states $|\mathbf{0}\rangle_c$ and $|\mathbf{1}\rangle_c$ are almost equal to $|0\rangle$ and $|1\rangle$ for N_{g0} far away from the charge degeneracy point. The time variation of the spin state can be visualized in the so-called Bloch sphere picture, where the general quantum state

$$\cos(\theta_u/2)\exp(-\mathbf{i}\varphi_u/2)\,|0\rangle + \sin(\theta_u/2)\exp(\mathbf{i}\varphi_u/2)\,|1\rangle\ , \tag{3.3}$$

is represented by a vector with polar coordinates θ_u and φ_u, precessing around $\vec{H}$ with a frequency $|\vec{H}|/h$. We now consider the different ways of modifying $\vec{H}$ in order to manipulate $\vec{\sigma}$.

3.1. Manipulation of the Cooper pair box quantum state

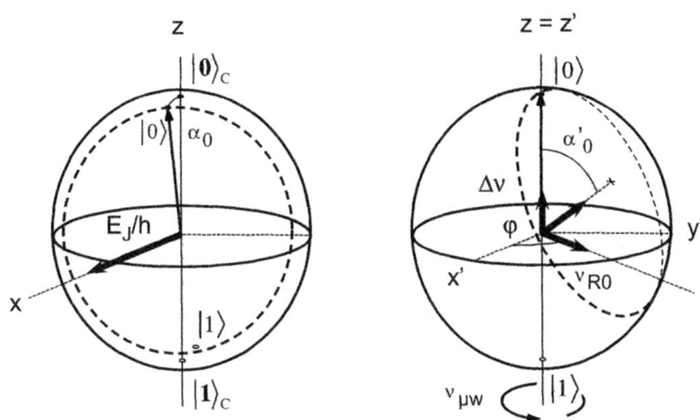

Fig. 7. Bloch sphere representations showing how the CPB state has been manipulated in two different types of experiments. The left sphere with two adjacent pure charge states at the north and south poles corresponds to a CPB with $E_J/E_C << 1$, which is driven to the charge degeneracy point with a fast dc gate pulse. The right sphere with energy eigenstates at the poles describes within the rotating wave approximation how a CPB is driven with microwave voltage pulses (see text for detailed explanation). The spin is represented by a thin arrow whereas fields are represented by bold arrows. The dotted lines show the spin trajectory, starting from the ground state.

3.1.1. Constant perturbation applied suddenly to the Hamiltonian

The first method that was used experimentally in 1999 [16] to prepare a CPB-based-qubit in a coherent quantum superposition of its 2 states consists in applying to its gate (or to a second gate specially designed for that purpose) a trapezoidal N_g pulse with rise and fall times much shorter than $1/\nu_{01}$. This method was implemented on CPBs with $E_J/E_C \ll 1$, the gate charge being initially tuned at a value N_g of the order of 0.3 during a time long enough so that the qubit has relaxed to its ground state $|0\rangle \simeq |\mathbf{0}\rangle_c$. On the Bloch sphere drawn in the pure charge state referential (see left panel of Fig. 7), the initial situation corresponds to the spin parallel to the vector $\vec{H} = E_J\vec{x} + E_C(1 - 2N_g)\vec{z}$, the latter making a small angle $\alpha_0 \sim E_J/E_C(1 - 2N_g) \ll 1$ with $\vec{z}$. Then N_g is brought suddenly to $N_{g0} = 1/2$ in a time so short that the evolution of the spin during this transition is negligible. Now $\vec{H} = E_J \vec{x}$ induces the Rabi precession of the spin around the x axis at the Rabi frequency

$$\nu_{Rabi} = E_J / h \,. \tag{3.4}$$

After a time t, the coherent superposition that is built has a weight $\cos^2 \alpha_0 \cdot \sin^2 \nu_{Rabi} t$ on $|\mathbf{1}\rangle_c$. N_g is then brought back suddenly to its initial value. The qubit precesses then around the initial $\vec{H}$ and can be measured (see section 3.2) in the $(|\mathbf{0}\rangle_c, |\mathbf{1}\rangle_c) \approx (|0\rangle, |1\rangle)$ basis. Any superposition state, i.e. any point on the Bloch sphere can thus be reached in a time shorter than $[E_J/2 + E_C(1 - 2N_g)]/h$, by applying first a single pulse and by waiting then during a precise time. Besides, it is interesting to notice (the result will be used in section 5) that if $\Delta N_g = N_{g0} - 1/2 \neq 0$, the maximum probability to measure the qubit in state $|\mathbf{1}\rangle_c$ after a single pulse is strongly reduced as

$$\frac{1}{1 + \left(2 E_C \Delta N_g / E_J\right)^2} \,. \tag{3.5}$$

The present driving method has been used successfully by two research groups [16, 17]. It has the great advantage of inducing fast Rabi oscillations that can be observed even if the coherence time is rather short. On the other hand, one needs a very fast pulse generator with rise and fall times well below 100 ps. An alternative to this method is to use a harmonic perturbation.

3.1.2. Harmonic perturbation applied to the Hamiltonian

A second way to build superposed states is to apply a small resonant or almost resonant harmonic perturbation to the spin following the techniques developed in atomic physics and in Nuclear Magnetic Resonance. More precisely, a microwave pulse $\Delta N_g \cos(2\pi \nu_{\mu w} t + \varphi)$, with $\nu_{\mu w} \approx \nu_{01}$, is added to the DC gate voltage and introduces in the Hamiltonian (2.13) a perturbation, which is written in the spin formalism as:

$$\vec{H}_{ex} = 4 E_C \Delta N_g \cos(2\pi \nu_{\mu w} t + \varphi) \left[\langle 1 | \widehat{N} | 0 \rangle \, \vec{x} \; + \Delta N_{10} \vec{z} \right]. \tag{3.6}$$

When $\nu_{\mu w}$ is close to ν_{01}, the effect of the longitudinal part $\vec{H}_{ex}.\vec{z}$ on the motion of $\vec{\sigma}$ can be neglected. Moreover the CPB is usually operated at the charge degeneracy point (see section 2) where $\Delta N_{10} = 0$, so that $\vec{H}_{ex}.\vec{z} = 0$. We are thus left with the transverse perturbation whose effect on $\vec{\sigma}$ is simpler to describe in a frame $R'\{\vec{x}', \vec{y}', \vec{z}'\}$ rotating at the frequency $\nu_{\mu w}$ around $\vec{z}' = \vec{z}$. Within the so-called rotating wave approximation [18], the free Hamiltonian and the perturbation correspond in R' to:

$$\vec{H} = h \Delta\nu \, \vec{z}' \text{ with } \Delta\nu = \nu_{01} - \nu_{\mu w} \tag{3.7}$$

$$\vec{H}_{ex} \simeq h \nu_{R0} \left[\vec{x}' \cos\varphi + \vec{y}' \sin\varphi \right] \text{ with } \nu_{R0} = 2 E_C \Delta N_g \, \langle 1 | \widehat{N} | 0 \rangle / h \,. \tag{3.8}$$

When no microwave signal is applied to the gate, $\vec{\sigma}$ precesses freely in R' around $\vec{z}'$ at the detuning frequency $\Delta\nu$, whereas during microwave pulses, it precesses around $\vec{H} + \vec{H}_{ex}$ (see right panel of Fig. 7) at the Rabi frequency

$$\nu_{Rabi} = \nu_{R0}\sqrt{1 + \left(\frac{\Delta\nu}{\nu_{R0}}\right)^2},$$

which is proportional to the dimensionless microwave amplitude ΔN_g when detuning $\Delta\nu$ is chosen well below ν_{R0}. Starting from $|0\rangle$, the probability to measure $|1\rangle$ after a single pulse with effective duration t is thus $\cos^2\alpha_0' \sin^2 \nu_{Rabi}t$, with $\tan\left(\alpha_0'\right) = \nu_{R0}/\Delta\nu$. Note that the rise and fall time of the microwave pulses do not need to be short and that the precession axis and the Rabi frequency are tunable through the three parameters ΔN_g, $\nu_{\mu w}$, and φ. Moreover, any single qubit gate (i.e. any rotator operating on the Bloch sphere) can be implemented with a sequence of resonant pulses along $\vec{x}'$ and $\vec{y}'$ only [19], and all the tricks developed in NMR like composite pulse techniques [18] are applicable. This microwave driving method has been successfully applied to a split box [20] with $E_J/E_C \sim 1$, and also to phase [2] and flux [3] Josephson qubits.

3.1.3. Adiabatic acceleration

Finally, we also mention here an alternative way to perform a rotation around $\vec{z}'$, using a technique transposed from the "Stark pulse technique" known in atomic physics [21]. It consists, starting from a freely evolving superposed state

$$a\,|0\rangle + b\exp\left[2\pi\nu_{01}\left(Ng_0, \delta_0\right)t\right]|1\rangle,$$

in applying a closed adiabatic variation of the external parameters N_g and δ away from and back to the working point $\left(N_{g0}, \delta_0\right)$ in order to decrease or increase temporarily the deterministic relative dephasing speed $2\pi\nu_{01}\left(N_g, \delta\right)$ between components $|0\rangle\left(N_g, \delta\right)$ and $|1\rangle\left(N_g, \delta\right)$, without changing their weights. This method has been successfully tested with the split box mentioned above by moving adiabatically N_g away from and back to $\left(N_{g0} = 1/2, \delta_0 = 0\right)$ along the bold line of Fig. 12.

3.2. Readout of Cooper pair box quantum states

Many different strategies [13–16,22–24,26,27] have been proposed to distinguish the $|0\rangle$ and $|1\rangle$ states of a CPB. For some of them, the readout is coupled to the box charge whereas for others, it is coupled directly or indirectly to the δ phase degree of freedom of a split box. In all cases, an important distinction is whether the readout device is designed to perform a projective measurement onto some $|0\rangle'$ and $|1\rangle'$ states close to $|0\rangle$ and $|1\rangle$, or if it designed to perform a non projective measurement involving a relaxation process of the box from $|1\rangle$ to $|0\rangle$. A second characteristic is whether the readout is designed to be switched off during the box manipulation and then switched on to measure it with a signal to noise ratio larger than 1 in a single shot, or if it is designed to measure continuously a box which is periodically prepared in the same coherent state, so that the signal becomes detectable only after many repetitions. Although almost all possibilities have been considered in theoretical proposals, we describe below only the methods that have been really implemented experimentally. A last important property that will be considered in section 4 is the back-action that the readout induces onto the qubit and that can limit its coherence time.

3.2.1. Box charge used to build a current through an additional tunnel junction
At a time when no ultrasensitive electrometer would have been fast enough to discriminate the average charges $\langle N_1\rangle$ and $\langle N_0\rangle$ of a CPB-based-qubit in a time shorter than its relaxation time, Y. Nakamura and co-workers added to a split CPB with $E_J/E_C \ll 1$ a clever readout, which demonstrated in 1999 the first Rabi oscillations of an electrical circuit (see Fig. 8). A small and very opaque additional tunnel junction is connected to the island and biased with a voltage V such that an extra Cooper pair can enter the island and be broken into two electrons which then tunnel sequentially through the junction with rates Γ_{qp1} and Γ_{qp2} [16]. In the $|0\rangle$ state, this cyclic process gives rise to a finite current through the junction, the Josephson quasiparticle current (JQP) [28]. When the box is in its $|1\rangle$ state at $N_{g0} \sim 0.2 - 0.4$ with $\langle N_1\rangle \sim 1$, it relaxes to its $|0\rangle$ state with $\langle N_0\rangle \sim 0$ in a single JQP cycle with a relaxation rate Γ_{qp1}. By repeating rapidly the preparation of the $|1\rangle$ state, the JQP current can then be made larger than in the $|0\rangle$ state, the difference being used to measure the qubit state. Note that the coupling between the box and the readout being weak, the measured states are very close to $|0\rangle$ and $|1\rangle$, although the measurement is not projective and resets automatically the qubit to $|0\rangle$. This method is by design not single shot and the voltage V is applied continuously while the same preparation of the state is repeated, using the fast trapezoidal N_g pulse technique described in section 3.1.1.

3.2.2. Capacitive coupling to an electrometer
The most natural way for discriminating the two CPB states is to measure the expectation value $\langle N\rangle$ of its island charge by coupling it capacitively to an electrometer. The basic single electron transistor (SET) [5] was the first electrometer used to characterize the $|0\rangle$ state of a CPB by measuring the 2e periodic Coulomb staircase [7] mentioned in section 2.4. This device has a maximum bandwidth of a few kHz and is far from being fast enough to measure the $|1\rangle$ state before it relaxes to $|0\rangle$. A faster version of the SET, the RFSET, was invented in 1998 [25]

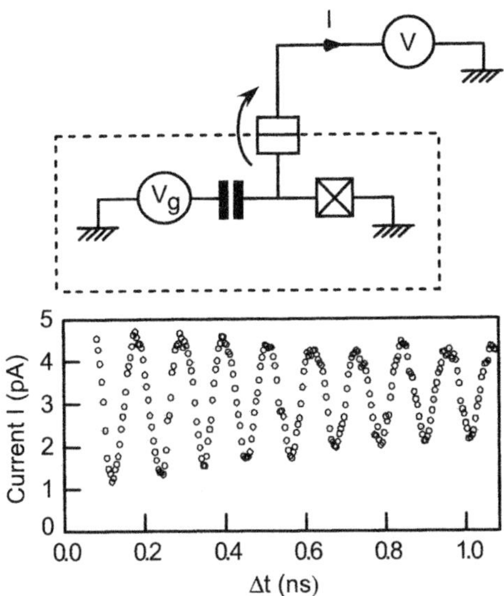

Fig. 8. First demonstration of quantum coherent control of an electrical circuit by Y. Nakamura et al. Top: simplified diagram of the setup, which includes a CPB and a readout made of a voltage biased additional opaque tunnel junction connected to the island. When fast DC pulses are applied repetitively to the CPB gate, the Josephson quasiparticle current I through the readout junction reflects the occupation probability of the CPB charge states $|\mathbf{0}\rangle_c$ and $|\mathbf{1}\rangle_c$ (see text). Bottom: Rabi oscillations of the CPB state measured by the variation of I. *Courtesy of Y. Nakamura et al., NEC, Japan.*

and was used by P. Delsing et al. [17] to measure a split CPB with $E_J/E_C \ll 1$, using the setup shown in Fig. 9. This RFSET includes a SET made up of an island defined by two tunnel junctions in series, biased with a voltage source. At DC voltages lower than or of the order of the Coulomb gap, the IV curve of the SET is modulated by the average charge of the CPB capacitively coupled to its island. By inserting the SET in parallel with the capacitance of a tank circuit that resonates in the radiofrequency domain and by applying a quasi-resonant RF signal to the ensemble, one measures a reflection coefficient that depends on the charge coupled to SET. The coupling between the RFSET and the CPB being weak, this readout is of the projective type. Besides, it can in principle be switched on and off with both the DC voltage and the RF input signal. Moreover, its sensitivity of the order of $10^{-5}e/\sqrt{Hz}$ is high enough and its back-action onto the qubit during the measurement is low enough that it can be operated in a single shot mode [27], provided that the qubit relaxation time is larger than or of order 1 μs. Unfortunately, at the time of writing (2003), it has proved difficult to measure a

CPB in this mode. Rabi oscillations [17] were demonstrated over a few nanoseconds with the fast trapezoidal N_g pulse technique and with the RFSET operated in the continuous mode (see Fig. 9).

An important point to notice is that a CPB-based-qubit measured through its average charge $\langle N \rangle$ is usually in the regime $E_J/E_C \ll 1$, for which the signal $\Delta N_{10} \sim 1$ is maximal as soon as $N_g \neq 1/2$. Consequently, its Coulomb energy E_C is rather high and its Hamiltonian is sensitive to any external charge fluctuations. It is well known that single charge devices like the CPB have always suffered from charged two-level-fluctuators, which play the role of additional noisy gate voltage sources and can thus induce decoherence of the qubit state (see section 4). It is thus interesting to increase E_J/E_C and to find an alternative to the measurement of $\langle N \rangle$.

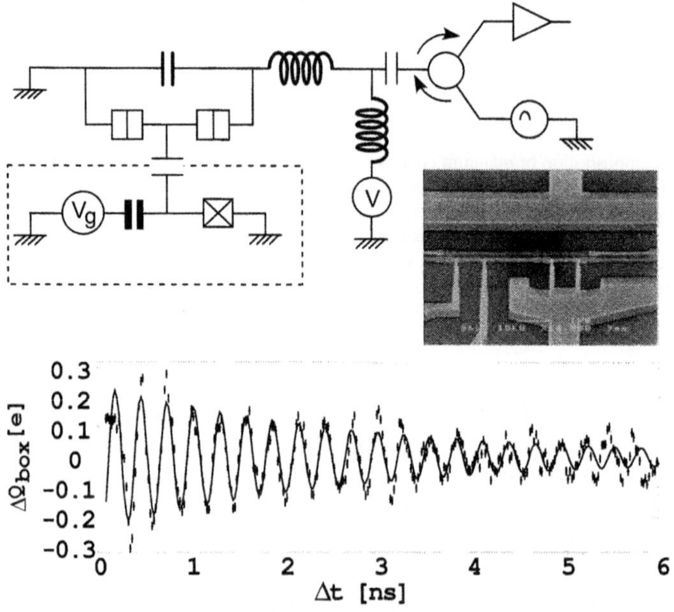

Fig. 9. Coherent control of a CPB by P. Delsing et al. Top: Simplified diagram of the setup showing that the CPB island is coupled capacitively to an RFSET, the RF reflecting power of which depends on the CPB average charge. Middle right: Scanning electron micrograph of the sample showing the SET on the left and the (split) box on the right. This sample was made by double angle shadow evaporation of aluminum through an e-beam patterned resist mask. Bottom: Rabi oscillations obtained with this readout, when applying repetitive fast DC pulses to the CPB gate. Here, ΔQ_{box} is the average charge on the box island. *Courtesy of the "Quantum Device Physics" group, Chalmers University of Technology, Göteborg, Sweden.*

3.2.3. Measurement of the split box persistent current: The Quantronium

A possible alternative to measurements of $\langle N \rangle$ consists in measuring the persistent current of a split box or the magnetic field it produces. For measuring the magnetic field, the superconducting loop of the split CPB can be coupled by mutual inductance to a SQUID or to a tank circuit whose effective inductance becomes different for the $|0\rangle$ and $|1\rangle$ CPB states, as proposed by A. Zorin et al. [14]. The latter technique is similar to that used for the RFSET and is currently under development.

We now describe the Quantronium, a setup proposed by D. Esteve [13] in order to discriminate the split box states directly through its loop persistent current. The circuit sketched in Fig. 10 consists of a split CPB with an additional current biased large Josephson junction with Josephson energy $E_{J0} \gg E_J$, inserted in the superconducting loop. During the manipulation of the qubit, the bias current I_b is kept small, so that the effective inductance of the additional junction is small and that the phase $\gamma = \arcsin(\varphi_0 I_b / E_{J0})$ across it behaves classically. The Quantronium is thus, during manipulation, a split box with $\delta = \Phi/\varphi_0 + \gamma$, the current biased junction playing only the role of an additional phase source for the split box. During the readout process, the additional junction is used to transfer adiabatically the information about the quantum state of the split box onto the phase γ, in analogy with the Stern & Gerlach experiment, where the spin state of a silver atom is entangled with its transverse position. For this transfer, a trapezoidal readout pulse $I_b(t)$ with a peak value slightly below the critical current $I_0 = E_{J0}/\varphi_0$ is applied to the circuit. Starting from $\delta = 0$, the phases $\langle \gamma \rangle$ and $\langle \delta \rangle$ grow during the current pulse and a state-dependent supercurrent develops in the loop. This current $\langle i \rangle$ adds algebraically to I_b in the large junction and modifies its switching rate Γ. By precisely adjusting the amplitude and duration of the $I_b(t)$ pulse, the large junction switches during the pulse to a finite voltage state with a large probability p_1 for state $|1\rangle$ and with a small probability p_0 for state $|0\rangle$ [13]. The switching of the large junction to the voltage state is then detected by measuring the voltage across it with an amplifier at room temperature. Although this measurement scheme is projective in a first step, it is nevertheless destructive due to the large amount of quasi particles produced when the voltage develops across the readout junction. Besides, it is designed to be single shot, its efficiency being expected to exceed $\eta = p_1 - p_0 = 95\%$ for a critical current $I_0 \sim 0.5 - 1\mu A$ and for the persistent currents plotted in Fig. 6. A Quantronium sample has been operated successfully (see Fig. 10) with the microwave N_g pulse technique, although the maximum overall efficiency of its readout was only $\eta \sim 20\%$. This sample has the longest coherence time observed so far (2004) in a Josephson qubit. The reasons for this success are analyzed in the next section devoted to decoherence of CPB-based- qubits.

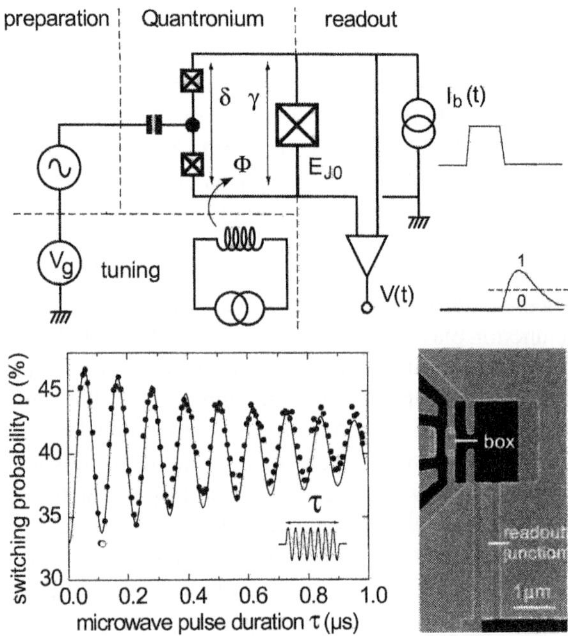

Fig. 10. Coherent control of a Quantroniun by the Quantronics group of CEA Saclay. Top: simplified diagram of the setup showing the readout Josephson junction inserted in the loop of a split box . A trapezoidal current pulse $I_b(t)$ is applied to this junction so that the latter switches out of its zero voltage state with a higher probability if the Quantronium is projected onto $|1\rangle$ than if it is projected onto $|0\rangle$. Bottom right: Scanning electron micrograph of a Quantronium made by double angle shadow evaporation of aluminum. Bottom left: Rabi oscillations obtained on a Quantronium with $E_J = 0.86k_B K$ and $E_C = 0.68k_B K$ when applying repetitively a resonant microwave pulse to the gate and a current pulse to the readout junction. Each experimental point is an average over 5×10^4 sequences.

4. Decoherence of Josephson charge qubits

4.1. Evaluation of decoherence: a simple approach

As with cany other quantum object, CPB-based-qubits are subject to decoherence due to their interaction with uncontrolled degrees of freedom present in their environment. From a general point of view, these interactions between the CPB and its environment entangle them in a complex way, which can be analyzed in principle by writing the total Hamiltonian of the system {CPB+environment} and by computing the evolution of the qubit reduced density matrix. Although this method has been used successfully for calculating decoherence induced by an RFSET [29] reading a CPB, it is in practice intractable in many cases. Moreover, it does not lead to analytical expressions showing directly the influence of each parameter and of each decoherence source, so that it is not always of great help for designing an experiment. Fortunately, decoherence during the free evolution of the qubit can be described in a much simpler way when the coupling between the qubit and its environment is weak. Indeed, an external parameter λ (such as N_g or δ) entering the Hamiltonian $\widehat{H} = -1/2\,\overrightarrow{\sigma} \cdot \overrightarrow{H}(\lambda)$ submitted to small quantum fluctuations from the external degrees of freedom can be treated as an operator of the environment. More precisely, each independent part X of the environment plays the role of an independent quantum noise source on the centered operator $\widehat{\lambda}_0 = \widehat{\lambda} - \langle\lambda\rangle$. To first order, the coupling Hamiltonian between this source and the CPB is written

$$\widehat{H}_X = -1/2\left(\overrightarrow{D}_\lambda \cdot \overrightarrow{\sigma}\right)\widehat{\lambda}_0^X\,, \tag{4.1}$$

where $\overrightarrow{D}_\lambda \cdot \overrightarrow{\sigma}$ is the restriction of $-2\widehat{\partial H/\partial\lambda}$ to the $\{|0\rangle\,, |1\rangle\}$ space. Then, from the noise properties of each source X, one calculates separately three relevant quantities that characterize X-induced decoherence: the first two characterize the depolarization of the fictitious spin representing the qubit. They are the excitation $\Gamma_{\uparrow,X}$ and relaxation $\Gamma_{\downarrow,X}$ rates giving the probability per unit time of X-induced $|0\rangle \to |1\rangle$ and $|1\rangle \to |0\rangle$ transitions of the qubit, respectively. The third relevant quantity is the "dephasing function" $f_X(t) = \langle \exp[i\Delta\varphi_X(t)]\rangle$ involving the X-induced random dephasing $\Delta\varphi_X(t)$ between the two components of a superposed state $a(t)\,|0\rangle + b(t)\exp[i\Delta\varphi_X(t)]\,|1\rangle$ (note that $f_X(t)$ is not necessarily exponential and characterized by a rate). The evolution of the qubit density matrix is then easily deduced from the values of $\Gamma_{\uparrow,X}$, the values of $\Gamma_{\downarrow,X}$ and the $f_X(t)$ functions. Introducing the total dephasing function $F(t) = \prod_X f_X(t)$ and the total upward and downward rates $\Gamma_\uparrow = \sum_X \Gamma_{\uparrow,X}$ and $\Gamma_\downarrow = \sum_X \Gamma_{\downarrow,X}$, the diagonal elements evolve exponentially towards their equilibrium values $1-\epsilon$ and $\epsilon = \Gamma_\uparrow/\Gamma_1$ with the characteristic rate $\Gamma_1 = \Gamma_\uparrow + \Gamma_\downarrow$, whereas off-diagonal elements decay as $F_2(t) = \exp[-\Gamma_1 t/2]F(t)$. In the next sections, we calculate explicitly $\Gamma_{\uparrow,X}$, $\Gamma_{\downarrow,X}$ and $f_X(t)$ for the main decoherence sources X , in the case when the noise $\widehat{\lambda}_0^X$ is Gaussian and can be fully characterized by a generalized quantum spectral density function of angular frequency ω [1]:

$$S_{\lambda_0}^X(\omega) = \frac{1}{2\pi}\int_{-\infty}^{+\infty} d\tau \left\langle \widehat{\lambda}_0^X(t)\widehat{\lambda}_0^X(t+\tau)\right\rangle \exp(-\mathbf{i}\omega\tau). \tag{4.2}$$

In this expression, the prefactor is chosen so that $S_{\lambda_0}^X(\omega)$ coincides in the classical limit and at low frequency with the spectral density of the engineer. Note that $S_{\lambda_0}^X(\omega)$ is defined for positive and negative ω's, the positive and negative parts being proportional to the number of environmental modes that can absorb and emit a quantum $\hbar\omega$, respectively.

4.2. *Overview of decoherence sources in a CPB*

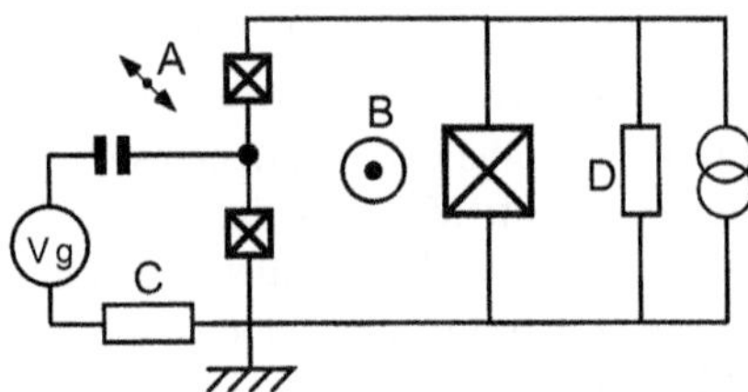

Fig. 11. Main decoherence sources in a Quantronium device. Quantum noise on N_g is generated by charged two-level-fluctuators (A) located near the CPB island and by voltage fluctuations of the series impedance (C) in the gate line. Quantum noise on δ is generated by fluctuations of the magnetic field (B) and by current fluctuations of the finite impedance (D) in parallel with the current bias source of the readout.

The uncontrolled degrees of freedom coupled to the idealized CPB of section 2 include those of the CPB substrate, those of the electrical lines of the driving and readout circuitry, and also the CPB's microscopic internal degrees of freedom which have been considered as frozen up to now. As an example, Fig. 11 shows the main decoherence sources in a Quantronium device (see section 3.2.3), which are now presented briefly.

Background charge noise First, microscopic charged two-level-fluctuators (A in Fig. 11) always present near the CPB island, either on the substrate or inside the Josephson junctions, are coupled to $\widehat{N}$ and act on the box as additional uncontrolled N_g sources. Although this background charge noise (BCN) is out of equilibrium and its generalized quantum spectral density is unknown, its classical spectral density $\widetilde{S}_{N_g,}^{BCN}(\omega) \sim B/|\omega|$ has been measured up to the MHz region, the values found for B being of order $10^{-7} - 10^{-9}$.

Impedance of the gate line The finite series impedance Z_g in the gate line (C in Fig. 11) can be regarded as an infinite collection of harmonic oscillators [1] also coupled to $\widehat{N}$ and inducing quantum noise on N_g. The circuit as seen from the pure Josephson element of the CPB (junction capacitance not included) is equivalent [9] to an effective gate capacitance C_Σ in series with a voltage source $\kappa_g V_g$ having an internal impedance Z_{eq}. In the weak coupling limit defined by $\kappa_g = C_g/C_\Sigma \ll 1$ and for all relevant frequencies, the real part of Z_{eq} is written:

$$\mathrm{Re}[Z_{eq}] \simeq \kappa_g^2 \,\mathrm{Re}[Z_g]\,. \tag{4.3}$$

At thermal equilibrium at temperature T, Z_{eq} generates voltage fluctuations whose spectral density S_u corresponds to N_g fluctuations with spectral density $S_{N_g}^{Z_g}$:

$$S_u = \frac{\hbar\omega}{2\pi}\left[1+\coth\left(\frac{\hbar\omega}{2k_B T}\right)\right]\mathrm{Re}[Z_{eq}], \tag{4.4}$$

$$S_{N_g}^{Z_g} = \left(\frac{C_\Sigma}{2e}\right)^2 S_u \simeq \kappa_g^2\frac{\hbar^2\omega}{E_C^2}\left[1+\coth\left(\frac{\hbar\omega}{2k_B T}\right)\right]\frac{\mathrm{Re}[Z_g]}{R_k}\,, \tag{4.5}$$

where $R_k = h/e^2 \simeq 26k\Omega$.

Magnetic flux noise Fluctuations of the magnetic field threading the loop of a split CPB (B in Fig. 11), either due to the motion of vortices in the vicinity of the loop or more macroscopically due to the current noise in the wires producing the field, generate directly δ noise. When the noise source is a circuit inductively coupled to the loop, its spectral density can be easily derived following the same method as we follow below for calculating the back-action of a Quantronium readout.

Readout back-action For a CPB measured by an RFSET (see section 3.2.2), the stochastic tunneling of electrons into and out of the SET island generates quantum noise on N_g. The reader can refer to [27, 29] for a characterization of this noise. For the Quantronium, the back-action of the readout circuit is the quantum noise on δ induced by the finite admittance Y_R in parallel with its current source (D in Fig. 11). More precisely, when a bias current $I_b < I_0$ is applied to the Quantronium, small oscillations of the phase δ are centered on $\delta_0 \simeq \arcsin(I_b/I_0)$ and the readout junction behaves as an inductance $L_{J0} \simeq$

$\varphi_0/[I_0 \cos\delta_0]$ much smaller than the inductance L_J of the box junction. Y_R and L_{J0} form together an effective admittance $Y_{R,eq} = Y_R // L_{J0}$ that generates current fluctuations characterized by the spectral density

$$S_I = \frac{\hbar\omega}{2\pi}\left[1 + \coth\left(\frac{\hbar\omega}{2k_B T}\right)\right] \mathrm{Re}[Y_R]. \tag{4.6}$$

$|Y_R|$ being much smaller than the effective inductance of the series combination of the two CPB Josephson junctions, this current I goes through $Y_{R,eq}$ and is converted into noise on voltage $v = \varphi_0 d\delta/dt = I/Y_{R,eq}$, or equivalently into a δ noise with spectral density

$$S_\delta^{Y_R} \simeq \left(\frac{1}{\varphi_0\omega}\right)^2 \frac{S_I}{\left|Y_R + \frac{1}{iL_{J0}\omega}\right|^2}. \tag{4.7}$$

Internal decoherence sources Finally, as examples of internal decoherence sources, one can think of out-of-equilibrium quasiparticles tunneling across the Josephson junctions or of an atom in the CPB Josephson junction jumping back and forth between two atomic sites so that a tunneling channel of the junction is open and closed randomly, such that it induces noise on E_J. Note that part of the decoherence of Josephson phase qubits has been attributed to this latter phenomenon [2].

4.3. *Depolarization of a Cooper pair box*

Relaxation and excitation proceed by exchange of an energy quantum $\hbar\Omega_{01}$ between the qubit and an oscillating field of the environment with pulsation $\omega = \Omega_{01} = 2\pi\nu_{01}$. Applying the Fermi golden rule to such processes gives:

$$\Gamma_{\downarrow,X} = \frac{\pi}{2}\left(\frac{D_{\lambda,\perp}}{\hbar}\right)^2 S_{\lambda_0,X}(\Omega_{01}), \tag{4.8}$$

$$\Gamma_{\uparrow,X} = \frac{\pi}{2}\left(\frac{D_{\lambda,\perp}}{\hbar}\right)^2 S_{\lambda_0,X}(-\Omega_{01}), \tag{4.9}$$

where the transverse part of $\vec{D}_\lambda$, $D_{\lambda,\perp} = 2|\langle 1|\widehat{\partial H/\partial\lambda}|0\rangle|$, is equal to $4E_C|\langle 0|\widehat{\mathbf{N}}|1\rangle|$ for all N_g noise sources and equal to $2\varphi_0|\langle 0|\widehat{\mathbf{I}}|1\rangle|$ for all δ noise sources, according to (2.13) and (2.15), respectively. Going further requires then specifying the origin of the noise. For the background charge noise, the spectral density is unfortunately unknown in the GHz range that corresponds to Ω_{01} so that no serious prediction of $\Gamma_{\downarrow,X}$ and $\Gamma_{\uparrow,X}$ can be made. For the gate line impedance $Z_g(\omega)$ at a temperature $T \ll \hbar\Omega_{01}/k_B$, $S_{N_g,Z_g}(-\Omega_{01})$ and $S_{N_g,Z_g}(\Omega_{01})$ obey detailed balance and

$$\frac{\Gamma_{\uparrow,Z_g}}{\Gamma_{\downarrow,Z_g}} = \exp(-\frac{\hbar\Omega_{01}}{k_B T}) \ll 1 .$$

Then, after substituting $D_{\lambda,\perp}$ and (4.5) at zero temperature in (4.8), one gets the relaxation rate

$$\Gamma_{\downarrow,Z_g} = 16\pi\kappa_g^2 \left|\langle 0|\widehat{\mathbf{N}}|1\rangle\right|^2 \frac{\mathrm{Re}[Z_g(\Omega_{01})]}{R_k}\Omega_{01} , \tag{4.10}$$

which takes the simpler form $\Gamma_{\downarrow,Z_g} = 4\pi\kappa_g^2\Omega_{01}\sin^2\alpha\,\mathrm{Re}[Z_g(\Omega_{01})]/R_k$ in the limit $E_J/E_C \ll 1$. In conclusion, a $\mathrm{Re}[Z_g(20\mathrm{GHz})]$ as large as 10Ω coupled with $\kappa_g \sim 1-2\%$ would induce relaxation of a CPB having a $1k_BK$ transition energy with a rate of only $0.1\sin^2\alpha$ MHz.

We evaluate now the relaxation induced by a resistance $R = 1/Y_R$ in parallel with the Quantronium readout junction. Substituting $D_{\lambda,\perp}$ and (4.7) at zero temperature in (4.8), one obtains after simple algebraic transformations:

$$\Gamma_{\downarrow,R} = \frac{2\left|\langle 0|\widehat{\mathbf{I}}|1\rangle\right|^2}{\hbar\Omega_{01}} \frac{R}{\left|1+\frac{R}{iL_{J0}\Omega_{01}}\right|^2} , \tag{4.11}$$

which simplifies to $\Gamma_{\downarrow,R} = 2|\langle 0|\widehat{\mathbf{I}}|1\rangle|^2 R/(\hbar\Omega_{01})$ for $R \ll L_{J0}\Omega_{01}$. At $N_g = 1/2$, $\langle 0|\widehat{\mathbf{I}}|1\rangle$ increases linearly with the asymmetry d between the box junctions. $\Gamma_{\downarrow,R}$ varies as d^2 and a Quantronium with a $1k_BK$ transition energy and an asymmetry $d = 5\%$ would relax with a rate of order 1 MHz under the influence of a readout resistance $R = 2\Omega$ at 20 GHz. Obtaining balanced junctions during the fabrication of a Quantronium is thus an important point.

4.4. *Random dephasing of a Cooper pair box*

In a semi-classical approach, the random phase $\Delta\varphi_X(t)$ between the two components of a superposed state is obtained by integration of the longitudinal fluctuations of $\widehat{H}_X$:

$$\Delta\varphi_X(t) = \frac{D_{\lambda,z}}{\hbar}\int_0^t \lambda_0^X(t')dt' , \tag{4.12}$$

where the longitudinal part of $\overrightarrow{D}_\lambda$, $D_{\lambda,z} = \langle 0|\,\widehat{\partial H/\partial\lambda}\,|0\rangle - \langle 1|\,\widehat{\partial H/\partial\lambda}\,|1\rangle \approx h\partial\nu_{01}/\partial\lambda$, is equal to $-2E_C\,\Delta N_{10}$ for all N_g noise sources and equal to $-\varphi_0\Delta i_{10}$

for all δ noises sources, according to sections 2.4 and 2.6. An important point to notice is that the coefficients of sensitivity to charge and phase noise, $D_{Ng,z}$ and $D_{\delta,z}$, vanish when ΔN_{10} and Δi_{10} are equal to zero, i.e. at $N_g = 0$ and $\delta = 0$, where the transition frequency is stationary. Then, $\lambda_0^X(t)$ being a Gaussian noise in most cases, the ensemble average $f_X(t) = \langle \exp[i\Delta\varphi_X(t)]\rangle$ is written

$$f_X(t) = \exp[-\left\langle \Delta\varphi_X^2(t)\right\rangle/2] \tag{4.13}$$

and depends only on the variance of the random phase, which can be calculated from the classical spectral density $\widetilde{S}^X_{\lambda_0}(\omega)$ of λ_0^X:

$$\left\langle \Delta\varphi_X^2(t)\right\rangle = \left(\frac{D_{\lambda,z}}{\hbar}t\right)^2 \int_{-\infty}^{+\infty} d\omega\, \widetilde{S}^X_{\lambda_0}(\omega)\mathrm{sinc}^2(\frac{\omega t}{2}), \tag{4.14}$$

with $\mathrm{sinc}(x) = \sin(x)/x$. A full quantum calculation [9] of $f_X(t)$, based on a thermal average over a bath of harmonic oscillators linearly coupled to $\widehat{\lambda}$, gives the same result:

$$f_X(t) = \exp\left[-\frac{t^2}{2}\left(\frac{D_{\lambda,z}}{\hbar}\right)^2 \int_{-\infty}^{+\infty} d\omega\, S^X_{\lambda_0}(\omega)\mathrm{sinc}^2(\frac{\omega t}{2})\right], \tag{4.15}$$

but with $\widetilde{S}^X_{\lambda_0}$ being replaced by its quantum analogue. Applied to the background charge noise, (4.15) becomes

$$f_{BCN}(t) = \exp\left[-B\left(\frac{2E_C\,\Delta N_{10}}{\hbar}\right)^2 t^2 \ln\frac{\tau}{t}\right], \tag{4.16}$$

where τ is the time taken experimentally to define the average transition frequency, this time introducing a low frequency cutoff $1/\tau$ in the integral of (4.15). The function $f_{BCN}(t)$ decays almost as a Gaussian with an effective characteristic time $T_2^{*BCN} = [2\sqrt{B\ln(\tau/t)}E_C\,\Delta N_{10}/\hbar]^{-1}$ that decreases almost as $1/(N_g - 1/2)$ close to the charge degeneracy (see the ΔN_{10} variations of Fig. 2). Assuming $B \sim 10^{-8}$ and $\tau \sim 10^2 s$, one gets $T_2^{*BCN} \sim 50ns$ for a CPB with $E_J \sim E_C \sim 1k_BK$ operated at $N_g = 0.55$.

In contrast to $\widetilde{S}^{BCN}_{N_g}(\omega)$, spectral densities $S^X_{\lambda_0}(\omega)$ of other noise sources are often rather flat below a cut-off frequency ω_c, so that for $t > 1/\omega_c$, $S^X_{\lambda_0}(\omega) \approx S^X_{\lambda_0}(0)$ in the ω range where the sinc square term gives its main contribution to the integral in (4.15). Consequently, $f_X(t) \approx \exp[-\Gamma^X_\varphi t]$ with a decay rate

$$\Gamma^X_\varphi \approx \pi\left(\frac{D_{\lambda,z}}{\hbar}\right)^2 S^X_{\lambda_0}(\omega \approx 0)\ . \tag{4.17}$$

At 50 mK, spectral densities (4.5) and (4.7) vary only by a factor two below 200 MHz so that (4.17) holds for $t > 5ns$. Substituting the correct $D_{\lambda,\perp}$ and (4.5) [resp. (4.7)] at zero frequency in (4.17), one finds the contribution to random dephasing of the gate line circuit (resp. of the Quantronium readout):

$$\Gamma_{\varphi}^{Z_g} \approx 8\pi\kappa_g^2 \Delta N_{10}^2(N_{g0}) \frac{k_B T}{\hbar} \frac{R}{R_k} , \tag{4.18}$$

$$\Gamma_{\varphi}^{R} \approx \frac{1}{8\pi} \left(\frac{\Delta i_{10}(\delta_0)}{I_0 \cos\delta_0} \right)^2 \frac{k_B T}{\hbar} \frac{R_k}{R}, \tag{4.19}$$

where R is the resistance of the gate voltage source (4.18) and the resistance in parallel with the Quantronium readout junction in (4.19), in the 0.1-100 MHz frequency range. Taking the same CPB parameters as in the previous section and assuming an electronic temperature of 50 mK, one gets $\Gamma_{\varphi}^{Z_g} < 25$kHz for $R = 10\Omega$, showing that the gate line is not an important dephasing channel. Assuming then typical values $I_0 \sim 500nA$ and $R = 100\Omega$ for a Quantronium readout circuit, one obtains $\Gamma_{\varphi}^{R} \sim 2\delta_0^2(rad)/\cos^2\delta_0$MHz. Therefore, Γ_{φ}^{R} is negligible close to $\delta_0 = 0$ and increases very rapidly up to about 100 MHz at the top of the readout current pulse, where δ approaches $\pi/2$.

4.5. *Design rules and optimal working points*

We now focus on the requirements that an experimental setup has to fulfill in order to demonstrate an operational CPB-based-qubit. First, the CPB has to be reset in its reference stable state $|0\rangle$ with a small probability of error $e \geq \epsilon = \Gamma_{\uparrow}/\Gamma_1$. This takes a reset time $t_r \sim e/\Gamma_1$ that defines the maximum repetition rate in an experiment. Secondly, during the manipulation of the state, the characteristic decay time T_2^* of $F_2(t)$ must be as long as possible in order to perform as many coherent single qubit or two qubit gate operations as possible. Consequently, $T_1 = 1/\Gamma_1$ and the characteristic decay time T_{φ}^* of the F function have to be maximized. For that purpose, a first action is of course to minimize all the noise spectral densities of section 4.2. A complementary approach is to choose a working point where the sensitivity to noise is minimal. For a split box, according to sections 4.3 and 4.4, T_1 and the T_{φ}^*'s can be maximized by choosing a manipulation point such that $\langle 1|\, I\, |0\rangle \simeq 0$ and such that the transition frequency is stationary with respect to both N_g and δ fluctuations. Figure 12 shows that the point ($N_g = 1/2, \delta = 0$) is such an "optimal manipulation point". But since both charge signal ΔN_{10} and current signal Δi_{10} vanish at this point, one has to apply a shift to N_g or δ at the end of the manipulation to measure the qubit state through the charge or phase port. Then, the elementary measuring time t_{m0}, defined as the time during which the readout interacts with the qubit after the preparation of a particular quantum state, is constrained in a way that depends on the readout strategy. When the readout absorbs the transition energy of the qubit as described in section 3.2.1, the readout port must be the main relaxation channel in order to avoid errors. Therefore, the values of $\Gamma_{\downarrow,X}$ and t_{m0} must obey $1/\Gamma_1 \simeq 1/\Gamma_\downarrow \simeq 1/\Gamma_{\downarrow,readout} \leq t_{m0}$. In all other cases, a readout error probability smaller than e_r requires $e_r T_1 > t_{m0}$. Consequently, we wish to maximize T_1 during readout by choosing a measurement point where $\langle 1| I |0\rangle \simeq 0$ as well. Moreover, t_{m0} has to be longer than T_φ^* since the qubit density matrix has to become diagonal before a projective measurement is completed. Since dephasing is required only during measurement, it is thus a good design rule to implement a switchable readout device such that T_φ^* decreases by several orders of magnitude when the readout is switched on, whereas T_1 remains long. Moving away from an "optimal manipulation point" along a "slow relaxation line" is just such a switch. Finally, it is convenient to have a single shot readout, able to distinguish the two qubit states with a small error rate e_r within the time t_{m0}. Otherwise, repeating several times the preparation and the measurement of the same quantum state is required to reach the same target error rate. The Quantronium has been designed to fulfill all the requirements mentioned here and has demonstrated experimentally good quantum coherence properties, which are presented in the next section.

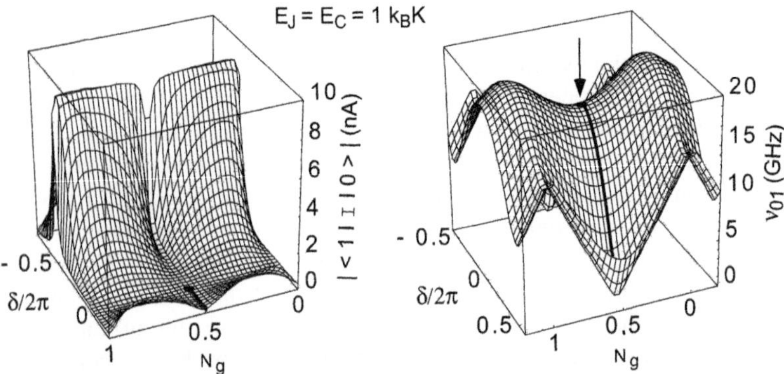

Fig. 12. Sensitivity to decoherence of a Quantronium with $E_J = E_C = 1k_B K$ and $d = 2\%$, as a function of its reduced external parameters N_g and $\delta/2\pi$. Left box: Loop current matrix element between states $|0\rangle$ and $|1\rangle$. This matrix element and consequently the relaxation rate of the qubit are minimal along the lines $\delta = 0$ and $N_g = 1/2$. Right box: Transition frequency of the Quantronium. The arrow indicates the stationary point ($N_g = 1/2$, $\delta = 0$) where pure dephasing vanishes to first order. Consequently, this is the optimal point for coherent manipulation of the Quantronium. For reading out the state, the working point is adiabatically moved along the bold solid line, where relaxation of the qubit induced by quantum noise on δ is minimal.

4.6. Experimental characterization of decoherence

The results presented below have been obtained with a Quantronium sample similar to that shown in Fig. 10, with $E_J = 0.86 k_B K$, $E_C = 0.68 k_B K$, and an asymmetry d between the CPB junctions not precisely known, but lower than 5%. We first measured the relaxation time at the optimal working point by switching on the readout at some variable time t_d after applying a microwave π pulse that prepares the qubit in state $|1\rangle$ (see Fig.13). A rough estimation of the readout resistance of the setup giving $R(20GHz) = 1\Omega - 5\Omega$, the experimental $T_1 = 1.8\mu s$ could be explained by an asymmetry coefficient $d \sim 5\% - 2\%$.

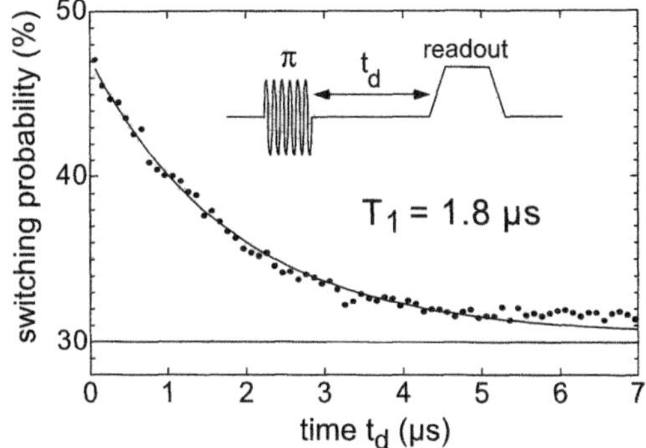

Fig. 13. Decay of the switching probability of the Quantronium's readout junction as a function of the delay between a microwave π pulse and the readout current pulse. The solid line is a fit of the data (dots) by an exponential shifted by the signal measured when no microwave is applied (horizontal bottom line).

Then, spectroscopic measurements (see Fig. 14) of ν_{01} were performed by applying to the gate a weak continuous microwave irradiation suppressed just before the readout current pulse. The variations of the switching probability as a function of the microwave frequency, display a resonance peak whose position ν_{01} as a function of N_g and δ leads to a precise determination of E_J and E_C. The resonance line shape being the Fourier transform of $F_2(t)$, the full line width at half-maximum $\Delta\nu_{01}$ leads to an effective coherence time $T_2^* = c/\Delta\nu_{01}$ with $c \sim 1/\pi$ depending on the exact line shape. As expected, $\Delta\nu_{01}$ was found to be minimal at the optimal point ($N_g = 1/2, \delta = 0$), where $\Delta\nu_{01} = 0.8$ MHz. Consequently, $2T_1 \gg T_2^* \simeq 0.4\mu s \simeq T_\varphi^*$ and decoherence is dominated by random dephasing. When departing from the optimal point, the line broadens very rapidly. For $N_g \neq 1/2$, it also becomes structured and not reproducible (see top right panel of Fig. 14) due to individual charged two-level-fluctuators. Nevertheless, the general trend (see bottom panel of Fig. 14) is that $\Delta\nu_{01}$ varies more or less linearly with $N_g - 1/2$ and δ, the proportionality coefficients $\partial \Delta\nu_{01}/\ \partial(\delta/2\pi)$ and $\partial \Delta\nu_{01}/\ \partial N_g$ being of order 0.3GHz. Noticing that ν_{01} varies quadratically in the vicinity of the optimal point so that $D_{\lambda,z} \approx h\partial\nu_{01}/\partial\lambda \propto \lambda \propto \Delta\nu_{01}$, one deduces from section 4.4, that both charge and phase noises are peaked at low frequencies and that the random dephasing functions should decay as Gaussians. This effect is well understood for the charge noise, which is dominated by the 1/f contribution of microscopic origin. Using the actual parameters of the sample in (4.16), the experimental $\partial \Delta\nu_{01}/\ \partial N_g$ leads to an amplitude $B \simeq 10^{-7}$ for the BCN, a value in agreement with previous measurements on similar Josephson devices. By contrast, the origin of the low frequency phase noise is not understood. An important point to mention here is that the experimental value of T_2^* at the optimal working point corresponds to that estimated by taking into account the second order contribution of the charge and phase noises.

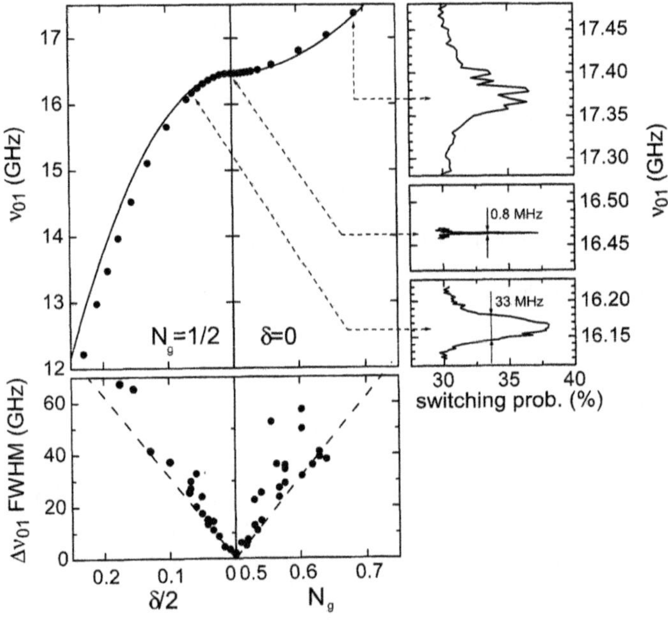

Fig. 14. Spectroscopy of a Quantronium. Top left panel: transition frequency as a function of δ at $N_g = 1/2$ and as a function of N_g at $\delta = 0$. The solid line is a fit that gives E_J and E_C. Right panels: resonance lines recorded with a small microwave power at three different working points (same scale for all lines). Bottom left panel: Full width at half-maximum $\Delta\nu_{01}$ of the resonance lines. Due to a slow and large charged two-level-fluctuator (TLF), data points can vary by a factor 2. The dotted lines indicate that $\Delta\nu_{01}$ varies linearly with the external parameters when this TLF is stable.

The direct measurement of the coherence time T_2^* during free evolution was obtained by performing a Ramsey-fringe-like experiment (see also [21]). One applies to the gate two slightly off-resonance $\pi/2$ microwave pulses separated by a delay Δt during which the spin representing the qubit state precesses freely at frequency $\Delta\nu$ in the equatorial plane of the Bloch sphere. Whereas the first pulse simply sends the spin onto the equator, the second one converts the phase accumulated during Δt into a longitudinal component of the spin. The probability to measure $|1\rangle$ at the end of the sequence oscillates as $\cos^2(\pi\Delta\nu\Delta t)$ with an amplitude that decays as $F_2(t)$. Figure 15 shows the result of such an experiment performed at the optimal manipulation point. Although the low signal to noise ratio and the long term drift due to 1/f noise prevents determination of wether the decay of the oscillations is more Gaussian than exponential, a fit of the data leads to $T_2^* \simeq 0.5\mu s$, a value that corresponds to that previously deduced from the resonance line width. Given the transition period $1/\nu_{01} \sim 60$ps, the coherence time T_2^* corresponds to about 8000 free precession turns around the z axis. Assuming that a bit flip can be performed with a microwave pulse of only 30 oscillations, i.e. in a time $\sim$ 2ns, T_2^* corresponds also to the time required for about 250 bit flips.

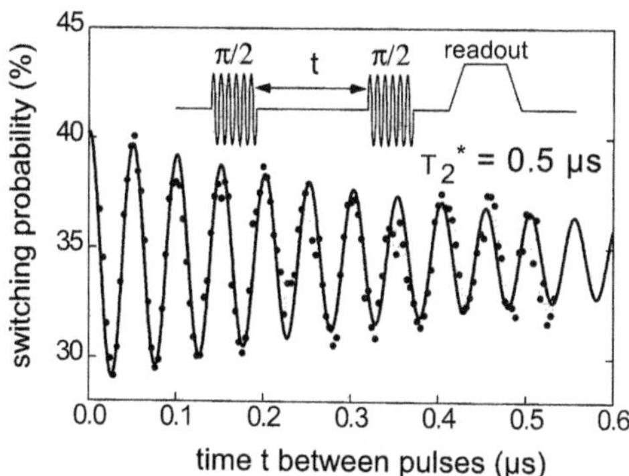

Fig. 15. Ramsey fringe experiment on a Quantronium: When two $\pi/2$ microwave pulses detuned with $\Delta\nu = 20.6$ MHz and separated by Δt are applied to the gate, the switching probability of the readout junction oscillates as a function of Δt with frequency $\Delta\nu$. Each experimental point (dot) is an average over 50000 sequences. The solid line is a fit by an exponentially decaying cosine, the decay time constant of which corresponds to the coherence time T_2^*.

The coherence can also be maintained artificially during a time longer than T_2^* using NMR-like echo sequences [18]. An intermediate π pulse is inserted in a Ramsey sequence, a time $\Delta t_1 < \Delta t$ after the first $\pi/2$ pulse (see Fig. 16). Assuming for instance that all rotations are performed around the y' axis of the

Bloch sphere, the effect of this π pulse is to change rapidly the phase of the spin from $\varphi = 2\pi\Delta\nu\Delta t_1$ to $\varphi = \pi - 2\pi\Delta\nu\Delta t_1$. Then, the phase grows again by an amount $2\pi\Delta\nu(\Delta t - \Delta t_1)$ before the last $\pi/2$ pulse. The probability to measure $|1\rangle$ at the end of the sequence is $\sin^2[\pi\Delta\nu(\Delta t - 2\Delta t_1)]$. When $\Delta t_1 = \Delta t/2$, this probability is thus less sensitive to $\Delta\nu$ fluctuations than the Ramsey function $\cos^2(\pi\Delta\nu\Delta t)$. In other words, a π pulse in the middle of an echo sequence makes the spin go a longer (resp. shorter) path along the equator when the precession speed is faster (resp. slower), so that the ending point is the same from sequence to sequence, provided that $\Delta\nu$ is constant within a sequence. Figure 16 compares a Ramsey sequence and an echo sequence with variable Δt_1 performed at $N_g = 0.52$ and $\delta = 0$, where T_2^* is reduced to 30ns. For $\Delta t = 2\Delta t_1 \sim 20T_2^*$, the amplitude of the echo is still 20% of the maximum amplitude whereas the Ramsey signal is of course zero. This result confirms that decoherence is essentially due to charge noise at frequencies lower than $1/\Delta t \sim 1$ MHz. Although mapping the amplitude of the echo as a function of Δt for different working points can give much information on the shape of noise spectral densities, no complete study could be made on the sample presented here.

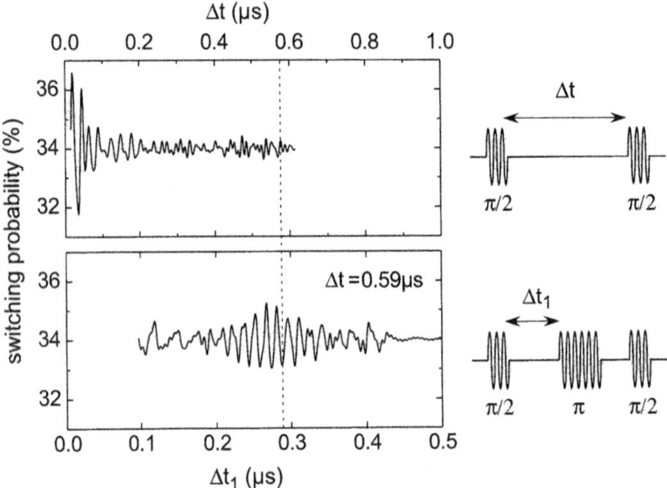

Fig. 16. Comparison between a Ramsey and an echo sequence (top right and bottom right pictograms, respectively) with a Quantronium circuit operated at the working point ($N_g = 0.52$, $\delta = 0$) with a detuning $\Delta\nu = 41$MHz. Top panel: Ramsey fringes of the Quantronium's switching probability indicating a coherence time of only 30 ns. Bottom panel: Echo signal taken at fixed $\Delta t = 0.59\mu s$ as a function of Δt_1. Very close to $\Delta t_1 = \Delta t/2$, the amplitude of the echo is maximum and equal to about 20% of the signal at $\Delta t = 0$. The dashed vertical line indicates Δt_1 and points out that the Ramsey signal has completely disappeared for the same Δt.

5. Two-qubit-gates with capacitively coupled Cooper pair boxes

Being able to implement any quantum algorithms requires adding to all possible single qubit operations a two-qubit gate such that the ensemble forms a so-called universal set of gates [30]. Many coupling schemes of two (or more) CPBs have been proposed to realize such 2-qubit gates. The nature of the coupling can be either capacitive or inductive. In the first case, two CPB islands are coupled by a capacitor, whereas in the second case, the loops of two split CPBs are coupled by a mutual inductance or galvanically by an inductor or an additional Josephson junction. Although ideally the coupling should be switchable and tunable, using a constant coupling is of course simpler. In this paper, we restrict ourselves to the constant capacitive coupling between two CPBs, which is the only scheme that has been implemented at the time of writing.

The Hamiltonian of two CPBs indexed 1 and 2, the islands of which are coupled by a capacitor C_C, is the sum of two terms (2.3) with Josephson energies including eventually a phase term if the box is split and with Cooper pair Coulomb energies involving now $C_\Sigma \approx C_g + C_J + C_C$, and of a coupling term

$$E_{CC}(\widehat{\mathbf{N}}_1 - N_{g1})(\widehat{\mathbf{N}}_2 - N_{g2}) \text{ with } E_{CC} \approx E_{C1}E_{C2}\frac{C_C}{(2e)^2}\,. \tag{5.1}$$

Within the spin formalism and when $E_J/E_C \ll 1$, Hamiltonian (5.1) is rewritten in the pure charge state basis $(|\mathbf{0}\rangle_{c1}, |\mathbf{1}\rangle_{c1}) \otimes (|\mathbf{0}\rangle_{c2}, |\mathbf{1}\rangle_{c2})$:

$$\begin{aligned}\widehat{H}(N_{g1}, N_{g2}) = &-\frac{1}{2}\left\{E_{J1}\widehat{\sigma}_{X1} + \left[(1-2N_{g1})E'_{C1} + (1-2N_{g2})\frac{E_{CC}}{2}\right]\widehat{\sigma}_{Z1}\right\}\\ &-\frac{1}{2}\left\{E_{J2}\widehat{\sigma}_{X2} + \left[(1-2N_{g2})E'_{C2} + (1-2N_{g1})\frac{E_{CC}}{2}\right]\widehat{\sigma}_{Z2}\right\}\\ &+\frac{E_{CC}}{4}\widehat{\sigma}_{Z1}\widehat{\sigma}_{Z2}\,,\end{aligned} \tag{5.2}$$

where constant terms that depend only on N_{g1} and N_{g2} have been dropped. The coherent evolution induced by this Hamiltonian has been experimentally demonstrated [31] with two strongly coupled ($E_{CC} \sim E_{J1,2}$) CPBs, using the fast DC gate pulses (see section 3.1.1) to the charge co-degeneracy point $N_{g1} = N_{g2} = 1/2$, and using the readout technique described in section 3.2.1. A conditional operation close to the Controlled-NOT gate has also been demonstrated [32] with the same system. The main idea behind this experiment is to regard the $\widehat{\sigma}_{Z1}\widehat{\sigma}_{Z2}$ coupling as shifting the CPB2 charge degeneracy point by a quantity that depends on the state of CPB1. Indeed, this degeneracy point is defined by $(1-2N_{g2})E'_{C2}+(1-2N_{g1}\mp 1/2)E_{CC}/2=0$. A gate 2 pulse bringing CPB2 to the charge degeneracy when CPB1 is in state $|\mathbf{0}\rangle_{c1}$, brings it $\Delta N_{g2}=1/2-E_{CC}/4E'_{C2}$ away from the degeneracy when CPB1 is in state $|\mathbf{1}\rangle_{c1}$. According to section 3.1.1, the maximum probability of $|\mathbf{1}\rangle_{c2}$ after the pulse drops rapidly with ΔN_{g2} as $1/\{1+[2E_C\Delta N_{g2}/E_J]^2\}$, so that bit 2 can be flipped when bit 1 is in state $|\mathbf{0}\rangle_{c1}$, whereas it is almost unchanged when bit 1 is in state $|\mathbf{1}\rangle_{c1}$. Figure 17 shows the experimental results obtained by the NEC group with such a C-NOT.

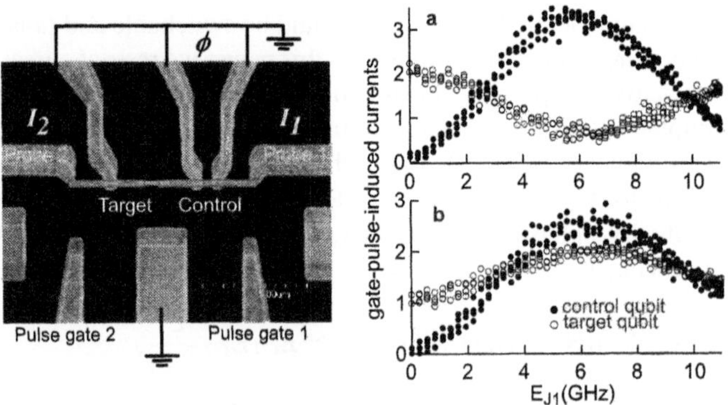

Fig. 17. Demonstration of a C-NOT-type quantum gate with two capacitively coupled CPBs by T. Yamamoto and co-workers. Left: Scanning electron micrography of the device. The qubits are manipulated using the fast dc pulse technique. Here, the target qubit is prepared in the pure charge state $|\mathbf{0}\rangle_c$ (a) or $|\mathbf{1}\rangle_c$ (b) whereas the control qubit is prepared with a dc pulse of constant width in a superposition state that depends on the Josephson energy E_{J1} of qubit 1. Finally, a gate 2 pulse performs the CNOT as explained in the text. Right panels: Anticorrelation (a) and correlation (b) between the two probe currents as a function of E_{J1}. *Courtesy of T. Yamamoto et al., NEC, Japan.*

Another type of gate can be developed by working in the uncoupled energy eigenbasis $(|0\rangle_1, |1\rangle_1)\otimes(|0\rangle_2, |1\rangle_2)$ at fixed $N_{g1}=N_{g2}=1/2$, where $\langle 0|\widehat{\mathbf{N}}|0\rangle = \langle 1|\widehat{\mathbf{N}}|1\rangle = 1/2$. The Hamitonian (5.1) is now rewritten as

$$\widehat{H}(N_{g1},N_{g2}) = -\frac{1}{2}h\nu_1\widehat{\sigma}_{Z1} - \frac{1}{2}h\nu_2\widehat{\sigma}_{Z2} + K\widehat{\sigma}_{X1}\widehat{\sigma}_{X2}, \tag{5.3}$$

with $K = E_{CC}\left|{}_1\langle 1|\widehat{\mathbf{N}}_1|0\rangle_1\right|\left|{}_2\langle 1|\widehat{\mathbf{N}}_2|0\rangle_2\right|$, the corresponding matrix being

$$\widehat{H} = \begin{pmatrix} \begin{bmatrix} -h\overline{\nu} & K \\ K & h\overline{\nu} \end{bmatrix} & [0] \\ [0] & \begin{bmatrix} -\frac{h\Delta\nu}{2} & K \\ K & \frac{h\Delta\nu}{2} \end{bmatrix} \end{pmatrix}_{(|00\rangle,|11\rangle,|01\rangle,|10\rangle)}, \tag{5.4}$$

with $\overline{\nu} = (\nu_1 + \nu_2)/2$ and $\Delta\nu = \nu_1 - \nu_2$. In the weak coupling limit defined by $C_C \ll C_{\Sigma 1,2}$, the coupling strength $K/h\overline{\nu}$ between $|00\rangle$ and $|11\rangle$ is always weak, whereas coupling strength $2K/h\Delta\nu$ between $|01\rangle$ and $|10\rangle$ can be varied by adjusting the difference between the two qubit transition frequencies. By equating ν_1 and ν_2, this coupling is maximized and $|01\rangle$ and $|10\rangle$ are simply swapped in a time $t_{SWAP} = \pi\hbar/2K$, whereas $|00\rangle$ and $|11\rangle$ are almost unchanged. When applying the effective coupling during $t_{SWAP}/2$, one obtains the so-called $\sqrt{iSWAP}$ gate, which entangles $|01\rangle$ and $|10\rangle$ in a simple and efficient way and which forms a universal set of gates when complemented with 1-qubit operations. Moreover, the $\widehat{\sigma}_{X1}\widehat{\sigma}_{X2}$ nature of the coupling has the great advantage of conserving the property of a vanishing random dephasing at the optimal manipulation point. An experiment aiming at testing such an $\sqrt{iSWAP}$ gate prototype with two capacitively coupled Quantroniums is currently in preparation.

6. Conclusions

As anticipated in 1995 immediately after the first successful characterization of a Cooper pair box [7], this device has been shown to have sufficiently good quantum properties to be used for building quantum bit prototypes. In less than ten years, two different schemes for driving the quantum state of a CPB and three very different readouts were developed and tested in several laboratories [16, 17, 20, 25]. Spectroscopy and coherent free and driven quantum evolution were demonstrated over times ranging from a few nanoseconds up to about a microsecond. Other Josephson qubits were also able to reach comparable results and Josephson qubits should now be considered as a single family, the sub-families having only historical justifications. The research community involved in the development of Josephson qubits has made great progress in understanding how decoherence occurs in electrical circuits and "quantum electrical engineering" was really born. The concept of optimal manipulation and measuring points of such circuits could for instance be formulated and experimentally tested. Moreover, experimental protocols for characterizing decohering effects and decoherence sources are continuously improving. With the development of more complex manipulations of Josephson qubits, methods to limit decoherence such as NMR-like echoes and spin locking have already been or are about to be tested. Preliminary experiments on two coupled CPB-based-qubits have demonstrated in 2003 a first solid-state-two-qubit-gate prototype. Nevertheless, the route towards a real quantum processor incorporating, for instance, quantum error correcting circuits is still long. A good quantum-non-demolition single-shot-readout is still lacking, the precision of qubit manipulations is still weak compared to that achieved in quantum optics, and coherence times must be increased by one or two orders of magnitude to start implementing simple algorithms. Finally, the scalability of Josephson qubits is still to be demonstrated. To conclude, although we do not think that any serious prediction can be made about the future existence or not of a (Josephson) Quantum computer, we are convinced that developing Josephson qubits is a very valuable program of research that paves the way towards a truly quantum electronics and toward machines in which quantum physics will manifest itself at a more "macroscopic" scale.

Acknowledgements

A large part of the physics presented here, including the development of the CPB and of the Quantronium is the fruit of a collective work in the "Quantronics group", which is part of the condensed matter physics division (SPEC) of CEA-Saclay, France. We warmly thank the other permanent members of the group, P. Orfila, P. Senat, P. Joyez, H. Pothier, C. Urbina, M. Devoret and D. Esteve, and also the past and present post-docs and Ph.D. students, E. Turlot, P. Lafarge, V. Bouchiat, A. Cottet, A. Aassime, and E. Collin for their direct participation in the development of Josephson qubits. We also gratefully acknowledge the research group of J.S. Tsai, Y. Nakamura, and T. Yamamoto from the "NEC Fundamental Research Lab" (Tsukuba, Japan) and the "Quantum Device Physics" group headed by Per Delsing at Chalmers University of Technology (Göteborg, Sweden) for providing material included in this chapter. We also should like to thank very much all the people we have been collaborating with over the past ten years, in particular I. Chuang from IBM, J. Martinis from NIST, O. Buisson from CRTBT, the group of H. Mooij from Delft University, the group of G. Schön from Karlsruhe, and all our colleagues from the SQUBIT european projects. Finally, I thank N. Birge and P. Meeson who helped to improve this text.

References

[1] See chapter by M. Devoret in this book.

[2] See chapter by J. Martinis in this book.

[3] See chapter by H. Mooij in this book.

[4] A. Barone and G. Paternò, *Physics and applications of the Josephson effect* (Wiley, New York, 1982).

[5] *Single Charge Tunneling*, edited by H. Grabert and M. H. Devoret (Plenum Press, New York, 1992).

[6] M. Büttiker, Phys. Rev. B **36**, 3548 (1987).

[7] V. Bouchiat et al., Physica Scripta **76**, 165 (1998).

[8] P. Lafarge et al., Nature **365**, 422 (1993).

[9] A. Cottet, *Implementation of a quantum bit in a superconducting circuit*, Thesis, University Paris VI, Paris (2002).

[10] C. Cohen-Tannoudji, B. Diu and F. Laloë, *Mécanique quantique T II*, Hermann, Paris.

[11] D. V. Averin, K. K. Likharev, in *Mesoscopic Phenomena in Solids*, edited by B. L. Altshuler, P. A. Lee, and R. A. Webb (Elsevier, Amsterdam, 1991).

[12] A.B. Zorin, Phys. Rev. Lett. **76**, 4408 (1996).

[13] A. Cottet, D. Vion, P. Joyez, P. Aassime, D. Esteve, M.H. Devoret, Physica C **367**, 197 (2002).

[14] A. B. Zorin, Physica C **368**, 284 (2002).

[15] J. R. Friedman and D. V. Averin, Phys. Rev. Lett. **88**, 50403 (2002).

[16] Y. Nakamura, Yu. A. Pashkin and J. S. Tsai, Nature **398**, 786, (1999)

[17] T. Duty et al., cond-mat / 0305433 (2003).

[18] See chapter by J. Jones in this book.

[19] See chapter by I. Chuang in this book.

[20] D. Vion et al., Science **296**, 886 (2002).

[21] See chapter by M. Brune in this book.

[22] F. W. J. Hekking et al., in *Electronic Correlations: from Meso- to Nanophysics*, T. Martin, G. Montambaux and J. Trân Thanh Vân, eds. (EDP Sciences, 2001), p. 515.

[23] I. Siddiqi et al., cond-mat/0312553 (2003).

[24] See chapter by R. Schoelkopf in this book.

[25] R.J. Schoelkopf *et al.*, Science, 280, 1238 (1998).

[26] A. Cottet et al., Workshop on "Macroscopic Quantum Coherence and Computing", Naples, Italy (2001).

[27] A. Aassime et al., Phys. Rev. Lett., **86**, 3376 (2001).

[28] T.A. Fulton, P.L. Gammel, D.J. Bishop, L.N.Dunkleberger, and G.J. Dolan, Phys. Rev. Lett. 63,1307 (1989).

[29] Y. Makhlin, G. Schön and A. Schnirman, Rev. Mod. Phy, **73**, 357 (2001)

[30] M.A. Nielsen and I.L. Chuang, Quantum Computation and Quantum Information (Cambridge University Press, Cambridge, 2000).

[31] Yu. A. Pashkin et al., Nature **421**,823 (2003).

[32] T. Yamamoto et al., Nature **425**, 941 (2003).

Course 15

QUANTUM TUNNELLING OF MAGNETIZATION IN MOLECULAR NANOMAGNETS

W. Wernsdorfer

Lab. L. Néel - CNRS, BP166,
38042 Grenoble Cedex 9, France,
e-mail : wernsdor@grenoble.cnrs.fr

D. Estève, J.-M. Raimond and J. Dalibard, eds.
Les Houches, Session LXXIX, 2003
Quantum Entanglement and Information Processing
Intrication quantique et traitement de l'information
© 2004 Elsevier B.V. All rights reserved

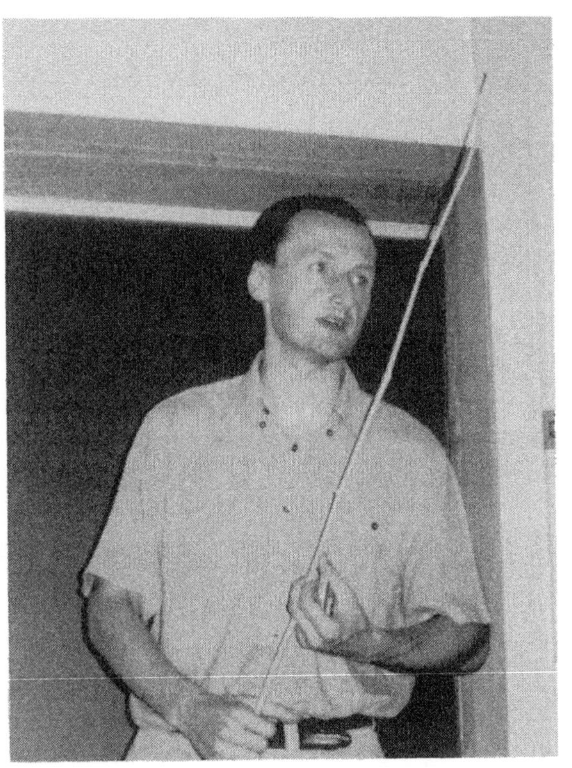

Contents

1. Introduction

The interface between classical and quantum physics has always been an interesting area, but its importance has nevertheless grown with the current explosive thrusts in nanoscience. Taking devices to the limit of miniaturization (the mesoscale and beyond) where quantum effects become important makes it essential to understand the interplay between the classical properties of the macroscale and the quantum properties of the microscale. This is particularly true in nanomagnetism, where many potential applications require monodisperse, magnetic nanoparticles.

In order to put this lecture into perspective, let us consider Fig. 1, which presents a scale of size ranging from macroscopic down to nanoscopic sizes. The unit of this scale is the number of magnetic moments in a magnetic system. At macroscopic sizes, a magnetic system is described by magnetic domains (Weiss 1907) [1] that are separated by domain walls. Magnetization reversal occurs via nucleation, propagation, and annihilation of domain walls (see the hysteresis loop on the left in Fig. 1 which was measured on an individual elliptic CoZr particle of 1 μm $\times$ 0.8 μm and a thickness of 50 nm). Shape and width of domain walls depend on the material of the magnetic system, on its size, shape and surface, and on its temperature [2, 3].

When the system size is of the order of magnitude of the domain wall width or the exchange length, the formation of domain walls requires too much energy. Therefore, the magnetization remains in the so-called single-domain state. Hence, the magnetization might reverse by uniform rotation, curling or other nonuniform modes (see hysteresis loop in the middle of Fig. 1). For system sizes well below the domain wall width or the exchange length, one must take into account explicitly the magnetic moments (spins) and their couplings. The theoretical description is complicated by the particle's boundaries.

Magnetic molecular clusters are the final point in the series of smaller and smaller units from bulk matter to atoms (Fig. 1). Up to now, they have been the most promising candidates for observing quantum phenomena because they have a well-defined structure with well-characterized spin ground state and magnetic anisotropy. These molecules can be regularly assembled in large crystals where all molecules often have the same orientation. Hence, macroscopic measurements can give direct access to single molecule properties. The most prominent

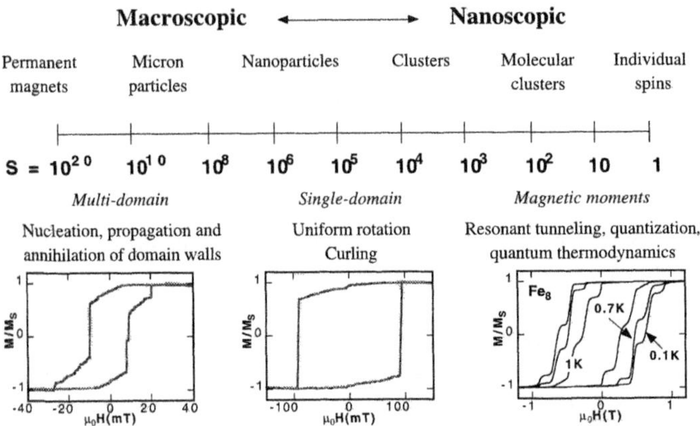

Fig. 1. Scale of size that goes from macroscopic down to nanoscopic sizes. The unit of this scale is the number of magnetic moments in a magnetic system (roughly corresponding to the number of atoms). The hysteresis loops are typical examples of magnetization reversal via nucleation, propagation, and annihilation of domain walls (*left*), via uniform rotation (*middle*), and via quantum tunneling (*right*).

examples are a dodecanuclear mixed-valence manganese-oxo cluster with acetate ligands, short Mn_{12} acetate [4], and an octanuclear iron(III) oxo-hydroxo cluster of formula $[Fe_8O_2(OH)_{12}(tacn)_6]^{8+}$ where tacn is a macrocyclic ligand, short Fe_8 (Fig. 2) [5]. Both systems have a spin ground state of $S = 10$ and an Ising-type magnetocrystalline anisotropy, which stabilizes the spin states with $m = \pm 10$ and generates an energy barrier for the reversal of the magnetization of about 67 K for Mn_{12} acetate and 25 K for Fe_8.

Thermally activated quantum tunneling of the magnetization has first been evidenced in both systems [6–9]. Theoretical discussion of this assumes that thermal processes (principally phonons) promote the molecules up to high levels with small quantum numbers $|m|$, not far below the top of the energy barrier, and the molecules then tunnel inelastically to the other side. Thus the transition is almost entirely accomplished via thermal transitions and the characteristic relaxation time is strongly temperature-dependent. An alternative explication was also presented [10]. For Fe_8, however, the relaxation time becomes temperature-independent below 0.36 K [9, 11] showing that a pure tunneling mechanism between the only populated ground states $m = \pm S = \pm 10$ is responsible for the relaxation of the magnetization. During the last years, several new molecular magnets were presented (see, for instance, Refs. [12–15]) which show also tunneling at low temperatures.

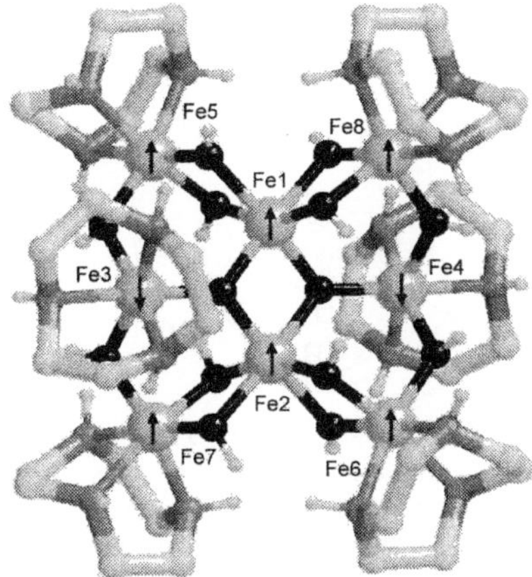

Fig. 2. Schematic view of the magnetic core of the Fe_8 cluster. The oxygen atoms are black, the nitrogen atoms are gray, and carbon atoms are white. The arrows represent the spin structure of the ground state $S = 10$.

Tunneling studies presented here were performed by magnetization measurements on single crystals using an array of micro-SQUIDs [3].

2. Giant spin model for nanomagnets

The simplest model describing the spin system of a nanomagnet like the Fe_8 molecular cluster (called the giant spin model) has the following Hamiltonian

$$\mathcal{H} = -DS_z^2 + E\left(S_x^2 - S_y^2\right) + g\mu_B\mu_0\vec{S}\cdot\vec{H} \tag{2.1}$$

S_x, S_y, and S_z are the three components of the spin operator, D and E are the anisotropy constants which were determined via HF-EPR ($D/k_B \approx 0.275$ K and $E/k_B \approx 0.046$ K [5]), and the last term of the Hamiltonian describes the Zeeman energy associated with an applied field $\vec{H}$. This Hamiltonian defines hard, medium, and easy axes of magnetization in x, y, and z directions, respectively (Fig. 3). It has an energy level spectrum with $(2S + 1) = 21$ values which, to a first approximation, can be labeled by the quantum numbers $m = -10, -9, ..., 10$ choosing the z-axis as quantization axis. The energy spectrum, shown in Fig. 4, can be obtained by using standard diagonalisation techniques of the $[21 \times 21]$ matrix describing the spin Hamiltonian $S = 10$. At

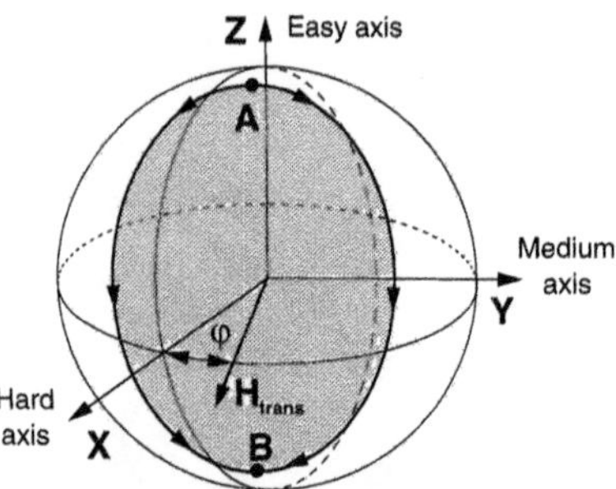

Fig. 3. Unit sphere showing degenerate minima **A** and **B** which are joined by two tunnel paths (heavy lines). The hard, medium, and easy axes are taken in x-, y-, and z-direction, respectively. The constant transverse field H_{trans} for tunnel splitting measurements is applied in the xy-plane at an azimuth angle φ. At zero applied field $\vec{H} = 0$, the giant spin reversal results from the interference of two quantum spin paths of opposite direction in the easy anisotropy yz-plane. For transverse fields in direction of the hard axis, the two quantum spin paths are in a plane which is parallel to the yz-plane, as indicated in the figure. By using Stokes'theorem it has been shown [16] that the path integrals can be converted in an area integral, yielding that destructive interference—that is a quench of the tunneling rate—occurs whenever the shaded area is $k\pi/S$, where k is an odd integer. The interference effects disappear quickly when the transverse field has a component in the y-direction because the tunneling is then dominated by only one quantum spin path.

$\vec{H} = 0$, the levels $m = \pm 10$ have the lowest energy. When a field H_z is applied, the energy levels with $m < -2$ increase, while those with $m > 2$ decrease (Fig. 4). Therefore, energy levels of positive and negative quantum numbers cross at certain fields H_z. It turns out that for Fe_8 the levels cross at fields given by $\mu_0 H_z \approx n \times 0.22$ T, with $n = 1, 2, 3, ...$ The inset of Fig. 4 displays the details at a level crossing where transverse terms containing S_x or S_y spin operators turn the crossing into an "avoided level crossing". The spin S is "in resonance" between two states when the local longitudinal field is close to an avoided level crossing. The energy gap, the so-called "tunnel spitting" Δ, can be tuned by an applied field in the xy-plane (Fig. 3) via the $S_x H_x$ and $S_y H_y$ Zeeman terms (Section 2.2).

The effect of these avoided level crossings can be seen in hysteresis loop measurements (Fig. 5). When the applied field is near an avoided level crossing, the magnetization relaxes faster, yielding steps separated by plateaus. As the temperature is lowered, there is a decrease in the transition rate due to reduced thermal-assisted tunneling.

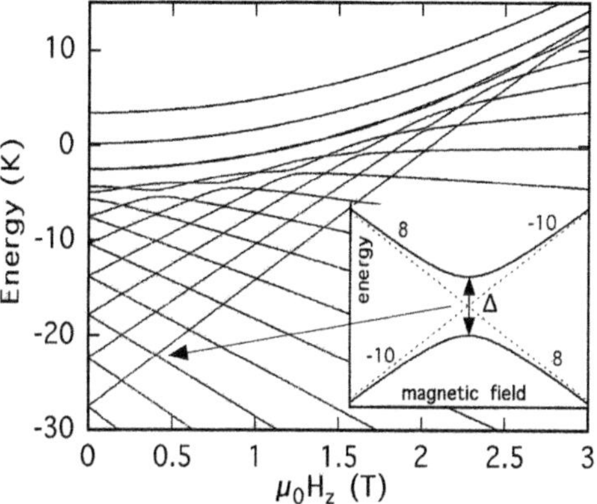

Fig. 4. Zeeman diagram of the 21 levels of the $S = 10$ manifold of Fe_8 as a function of the field applied along the easy axis [Eq. (2.1)]. >From bottom to top, the levels are labeled with quantum numbers $m = \pm 10, \pm 9, ..., 0$. The levels cross at fields given by $\mu_0 H_z \approx n \times 0.22$ T, with $n = 1, 2, 3, ...$ The *inset* displays the detail at a level crossing where the transverse terms (terms containing S_x or/and S_y spin operators) turn the crossing into an avoided level crossing. The greater the tunnel splitting Δ, the higher the tunnel rate.

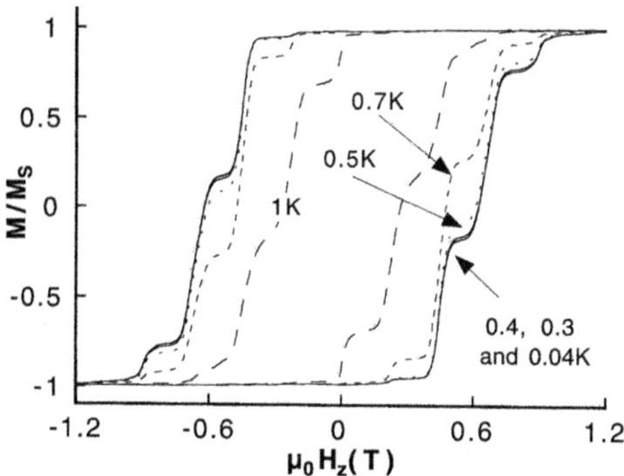

Fig. 5. Hysteresis loops of a single crystal of Fe_8 molecular clusters at different temperatures. The longitudinal field (z−direction) was swept at a constant sweeping rate of 0.014 T/s. The loops display a series of steps, separated by plateaux. As the temperature is lowered, there is a decrease in the transition rate due to reduced thermal assisted tunneling. The hysteresis loops become temperature independent below 0.35 K, demonstrating quantum tunneling at the lowest energy levels

2.1. Landau–Zener tunneling in Fe_8

The nonadiabatic transition between the two states in a two-level system has first been discussed by Landau, Zener, and Stückelberg [17–19]. The original work by Zener concentrates on the electronic states of a bi-atomic molecule, while Landau and Stückelberg considered two atoms that undergo a scattering process. Their solution of the time-dependent Schrödinger equation of a two-level system could be applied to many physical systems and it became an important tool for studying tunneling transitions. The Landau–Zener model has also been applied to spin tunneling in nanoparticles and clusters [20–24]. The tunneling probability P when sweeping the longitudinal field H_z at a constant rate over an avoided energy level crossing (Fig. 6) is given by

$$P_{m,m'} = 1 - \exp\left[-\frac{\pi\Delta^2_{m,m'}}{2\hbar g\mu_B|m - m'|\mu_0 dH_z/dt}\right] \tag{2.2}$$

Here, m and m' are the quantum numbers of the avoided level crossing, dH_z/dt is the constant field sweeping rates, $g \approx 2$, μ_B the Bohr magneton, and $\hbar$ is Planck's constant.

With the Landau–Zener model in mind, we can now start to understand qualitatively the hysteresis loops (Fig. 5). Let us start at a large negative magnetic field H_z. At very low temperature, all molecules are in the $m = -10$ ground state (Fig. 4). When the applied field H_z is ramped down to zero, all molecules will stay in the $m = -10$ ground state. When ramping the field over the $\Delta_{-10,10}$-region at $H_z \approx 0$, there is a Landau–Zener tunnel probability $P_{-10,10}$ to tunnel from the $m = -10$ to the $m = 10$ state. $P_{-10,10}$ depends on the sweeping rate [Eq. (2.2)]; that is, the slower the sweeping rate, the larger the value of $P_{-10,10}$. This is clearly demonstrated in the hysteresis loop measurements showing larger steps for slower sweeping rates [25]. When the field H_z is now further increased, there is a remaining fraction of molecules in the $m = -10$ state which became a metastable state. The next chance to escape from this state is when the field reaches the $\Delta_{-10,9}$ region. There is a Landau–Zener tunnel probability $P_{-10,9}$ to tunnel from the $m = -10$ to the $m = 9$ state. As $m = 9$ is an excited state, the molecules in this state desexcite quickly to the $m = 10$ state by emitting a phonon. An analogous procedure happens when the applied field reaches the $\Delta_{-10,10-n}$ regions ($n = 2, 3, \ldots$) until all molecules are in the $m = 10$ ground state; that is, the magnetization of all molecules is reversed. As phonon emission can only change the molecule state by $\Delta m = 1$ or 2, there is a phonon cascade for higher applied fields.

In order to apply quantitatively the Landau–Zener formula [Eq. (2.2)], we first saturated the crystal of Fe_8 clusters in a field of $H_z = -1.4$ T, yielding an initial

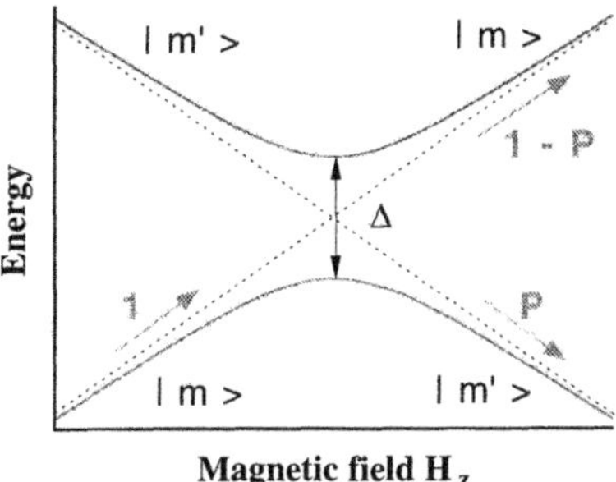

Fig. 6. Detail of the energy level diagram near an avoided level crossing. m and m' are the quantum numbers of the energy level. $P_{m,m'}$ is the Landau–Zener tunnel probability when sweeping the applied field from the left to the right over the anticrossing. The greater the gap Δ and the slower the sweeping rate, the higher is the tunnel rate [Eq. (2.2)].

magnetization $M_{\text{in}} = -M_{\text{s}}$. Then, we swept the applied field at a constant rate over one of the resonance transitions and measured the fraction of molecules which reversed their spin. This procedure yields the tunneling rate $P_{-10,10-n}$ and thus the tunnel splitting $\Delta_{-10,10-n}$ [Eq. (2.2)] with $n = 0, 1, 2, \ldots$.

We first checked the predicted Landau–Zener sweeping field dependence of the tunneling rate. We found a good agreement for sweeping rates between 10 and 0.001 T/s [25]. The deviations at lower sweeping rates are mainly due to the *hole-digging mechanism* [26] which slows down the relaxation (Section 4.2). Our measurements showed for the first time that the Landau–Zener method is particularly adapted for molecular clusters because it works even in the presence of dipolar fields which spread the resonance transition provided that the field sweeping rate is not too small.

2.2. *Oscillations of tunnel splitting*

An applied field in the xy−plane can tune the tunnel splittings $\Delta_{m,m'}$ via the S_x and S_y spin operators of the Zeeman terms that do not commute with the spin Hamiltonian. This effect can be demonstrated by using the Landau–Zener method (Section 2.1). Fig. 7 presents a detailed study of the tunnel splitting $\Delta_{\pm 10}$ at the tunnel transition between $m = \pm 10$, as a function of transverse fields applied at different angles φ, defined as the azimuth angle between the anisotropy hard axis and the transverse field (Fig. 3). For small angles φ the tunneling rate oscillates with a period of $\sim$0.4 T, whereas no oscillations showed up for large angles φ [25]. In the latter case, a much stronger increase of $\Delta_{\pm 10}$ with transverse field is observed. The transverse field dependence of the tunneling rate for

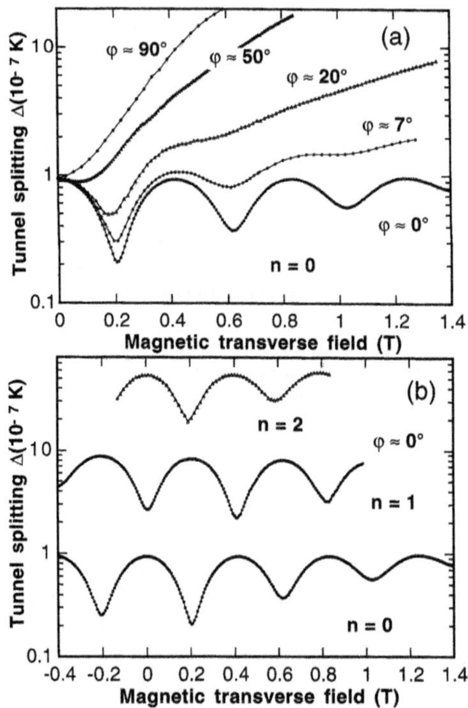

Fig. 7. Measured tunnel splitting Δ as a function of transverse field for (a) several azimuth angles φ at $m = \pm 10$ and (b) $\varphi \approx 0°$, as well as for quantum transition between $m = -10$ and $(10 - n)$. Note the parity effect that is analogous to the suppression of tunneling predicted for half-integer spins. It should also be mentioned that internal dipolar and hyperfine fields hinder a quench of Δ which is predicted for an isolated spin.

different resonance conditions between the state $m = -10$ and $(10 - n)$ can be observed by sweeping the longitudinal field around $\mu_0 H_z = n \times 0.22$ T with $n = 0, 1, 2, \ldots$. The corresponding tunnel splittings $\Delta_{-10,10-n}$ oscillate with almost the same period of ~0.4 T (Fig. 7). In addition, comparing quantum transitions between $m = -10$ and $(10-n)$, with n even or odd, revealed a parity (or symmetry) effect that is analogous to the Kramers' suppression of tunneling predicted for half-integer spins [27,28]. A similar strong dependence on the azimuth angle φ was observed for all studied resonances.

2.2.1. Semiclassical descriptions

Before showing that the above results can be derived by an exact numerical calculation using the quantum operator formalism, it is useful to discuss semiclassical models. The original prediction of oscillation of the tunnel splitting was done by using the path integral formalism [29]. Here [16], the oscillations are explained by constructive or destructive interference of quantum spin phases (Berry phases) of two tunnel paths (instanton trajectories) (Fig. 3). Since our experiments were reported, the Wentzel–Kramers–Brillouin theory has been used independently by Garg [30] and Villain and Fort [31]. The surprise is that although these models [16, 30, 31] are derived semiclassically, and should have higher-order corrections in $1/S$, they appear to be exact as written! This has first been noted in Refs. [30] and [31] and then proven in Ref. [32] Some extensions or alternative explications of Garg's result can be found in Refs. [33–36].

The period of oscillation is given by [16]

$$\Delta H = \frac{2k_{\mathrm{B}}}{g\mu_{\mathrm{B}}}\sqrt{2E(E+D)} \tag{2.3}$$

where D and E are defined in Eq. (2.1). We find a period of oscillation of $\Delta H = 0.26$ T for $D = 0.275$ K and $E = 0.046$ K as in Ref. [5]. This is somewhat smaller than the experimental value of $\sim$0.4 T. We believe that this is due to higher-order terms of the spin Hamiltonian which are neglected in Garg's calculation. These terms can easily be included in the operator formalism as shown in the next subsection.

2.2.2. Exact numerical diagonalization

In order to quantitatively reproduce the observed periodicity we included fourth-order terms in the spin Hamiltonian [Eq. (2.1)] as employed in the simulation of inelastic neutron scattering measurements [37, 38] and performed a diagonalization of the [21×21] matrix describing the $S = 10$ system. For the calculation of the tunnel splitting we used $D = 0.289$ K, $E = 0.055$ K [Eq. (2.1)] and the fourth-order terms as defined in [37] with $B_4^0 = 0.72\times 10^{-6}$ K, $B_4^2 = 1.01\times 10^{-5}$ K, $B_4^4 = -0.43\times 10^{-4}$ K, which are close to the values obtained by EPR measurements [39] and neutron scattering measurements [38].

The calculated tunnel splittings for the states involved in the tunneling process at the resonances $n = 0$, 1, and 2 are reported in Fig. 8, showing the oscillations as well as the parity effect for odd resonances.

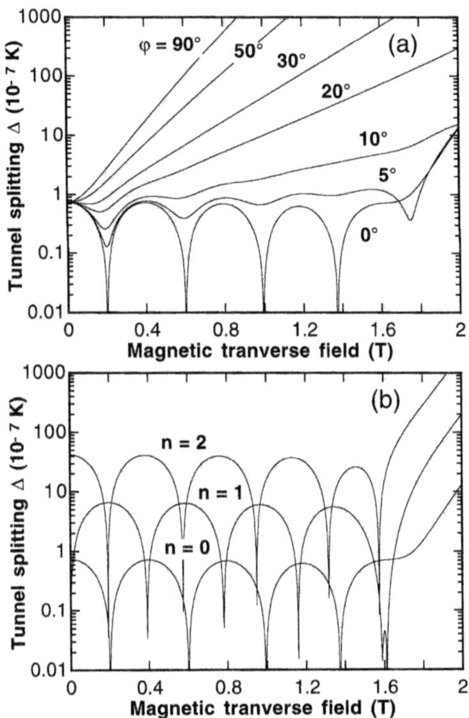

Fig. 8. Calculated tunnel splitting Δ as a function of transverse field for (a) quantum transition between $m = \pm 10$ at several azimuth angles φ and (b) quantum transition between $m = -10$ and $(10 - n)$ at $\varphi = 0°$ (Section 2.2.2). The fourth-order terms suppress the oscillations of Δ for large transverse fields $|H_x|$.

3. Quantum dynamics of a dimer of nanomagnets

Here we present a new family of dimers of nanomagnets [40, 41] in which antiferromagnetic coupling between two single-molecule magnets (SMMs) results in quantum behaviour different from that of the individual SMMs. Each SMM acts as a bias on its neighbor, shifting the quantum tunneling resonances of the individual SMMs. Hysteresis loop measurements on a single crystal of SMM-dimers established quantum tunneling of the magnetization via entangled states of the dimer. This shows that the dimer really does behave as a quantum-mechanically coupled dimer, and also allows the measurement of the longitudinal and transverse superexchange coupling constants.

The compound $[Mn_4]_2 \cdot 2C_6H_{14}$ crystallizes in the hexagonal space group

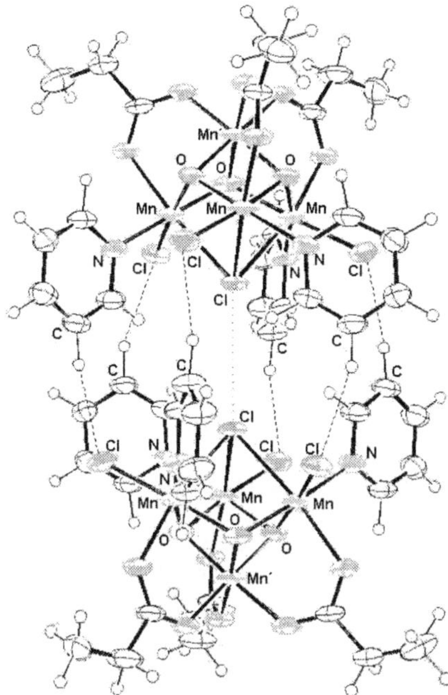

Fig. 9. The structure of the $[Mn_4]_2$ dimer of $[Mn_4O_3Cl_4(O_2CEt)_3(py)_3]$. The small circles are hydrogen atoms. The dashed lines are C-H· · ·Cl hydrogen bonds and the dotted line is the close Cl· · ·Cl approach. The labels Mn and Mn' refer to Mn^{III} and Mn^{IV} ions, respectively.

*R*3(bar) with two Mn_4 molecules per unit cell lying head-to-head on a crystallographic S_6 symmetry axis [40] (Fig. 9). Each Mn_4 monomer has a ground state spin of $S = 9/2$, well separated from the first excited state $S = 7/2$ by a gap of about 300K [42]. The Mn-Mn distances and the Mn-O-Mn angles are similar and the uniaxial anisotropy constant is expected to be the same for the two dimer systems. These dimers are held together via six C−H· · ·Cl hydrogen bonds between the pyridine (py) rings on one molecule and the Cl ions on the other, and one Cl· · ·Cl Van der Waals interaction. These interactions lead to an antiferromagnetic superexchange interaction between the two Mn_4 units of the $[Mn_4]_2$ dimer [40]. Dipolar couplings between Mn_4 molecules can be easily calculated and are more than one order of magnitude smaller than the exchange interaction.

Before presenting the measurements, we summarize a simplified spin Hamiltonian describing the $[Mn_4]_2$ dimer [40]. Each Mn_4 SMM can be modeled as a *giant spin* of $S = 9/2$ with Ising-like anisotropy [Eq. (2.1)]. The corresponding

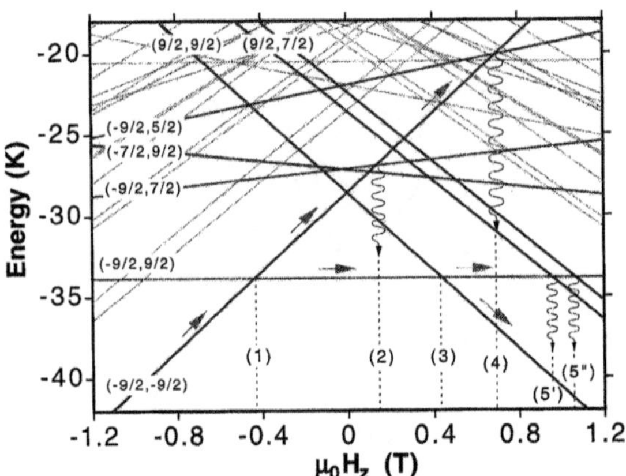

Fig. 10. Low lying spin state energies of the $[Mn_4]_2$ dimer, calculated by exact numerical diagonalization using Eq. 2 with $D = 0.77$ K and $J = 0.13$ K, as a function of applied magnetic field H_z (Zeeman diagram). The bold energy levels are labelled with two quantum numbers (M_1, M_2). Dotted lines, labelled **1** to **5**, indicate the strongest tunnel resonances: **1**: (-9/2,-9/2) to (-9/2,9/2); **2**: (-9/2,-9/2) to (-9/2,7/2), followed by relaxation to (-9/2,9/2); **3**: (-9/2,9/2) to (9/2,9/2); **4**: (-9/2,-9/2) to (-9/2,5/2), followed by relaxation to (-9/2,9/2); **5**: (-9/2,9/2) to (7/2,9/2), followed by relaxation to (9/2,9/2). For clarity, degenerate states such as (M,M') and (M',M) and lifted degenerate states such as $(M, M \pm 1)$, $(M, M \pm 2)$ ... are not both listed. For example, the $(9/2, 7/2)$ and $(7/2, 9/2)$ states are strongly split into a symmetric (labelled $5''$) and antisymmetric (labelled $5'$) combination of $(9/2, 7/2)$ and $(7/2, 9/2)$ states. This splitting is used to measure the transverse superexchange interaction constant J_{xy}. Co-tunneling and other two-body tunnel transitions have a lower probability of occurrence and are neglected [43].

Hamiltonian is given by

$$\mathcal{H}_i = -DS_{z,i}^2 + \mathcal{H}_{\text{trans},i} + g\mu_B\mu_0\vec{S}_i \cdot \vec{H} \tag{3.1}$$

where $i = 1$ or 2 (referring to the two Mn_4 SMMs of the dimer), D is the uniaxial anisotropy constant, and the other symbols have their usual meaning. Tunneling is allowed in these half-integer ($S = 9/2$) spin systems because of a small transverse anisotropy $\mathcal{H}_{\text{trans},i}$ containing $S_{x,i}$ and $S_{y,i}$ spin operators and transverse fields (H_x and H_y). The exact form of $\mathcal{H}_{\text{trans},i}$ is not important in this discussion. The last term in Eq. (3.1 is the Zeeman energy associated with an applied field. The Mn_4 units within the $[Mn_4]_2$ dimer are coupled by a weak superexchange interaction via both the six C-H· · ·Cl pathways and the Cl· · ·Cl approach. Thus, the Hamiltonian ($\mathcal{H}$) for $[Mn_4]_2$ is

$$\mathcal{H} = \mathcal{H}_1 + \mathcal{H}_2 + J_z S_{z,1} S_{z,2} + J_{xy}(S_{x,1} S_{x,2} + S_{y,1} S_{y,2}) \tag{3.2}$$

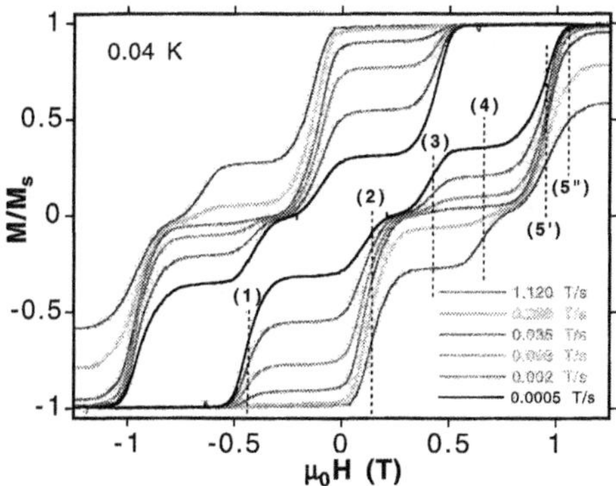

Fig. 11. Hysteresis loops for the $[Mn_4]_2$ dimer at several field sweep rates and 40 mK. The tunnel transitions (manifested by steps) are labelled from **1** to **5**, see Fig. 1.

where J_z and J_{xy} are respectively the longitudinal and transverse superexchange interactions. $J_z = J_{xy}$ is the case of isotropic superexchange. The $(2S+1)^2 = 100$ energy states of the dimer can be calculated by exact numerical diagonalization and are plotted in Fig. 10 as a function of applied field along the easy axis. Each state of $[Mn_4]_2$ can be labelled by two quantum numbers (M_1, M_2) for the two Mn_4 SMMs, with $M_1 = -9/2, -7/2, ..., 9/2$ and $M_2 = -9/2, -7/2, ..., 9/2$. The degeneracy of some of the (M_1, M_2) states is lifted by transverse anisotropy terms. For the sake of simplicity, we will discuss mainly the effect of the transverse superexchange interaction $\mathcal{J}_{trans} = J_{xy}(S_{x,1}S_{x,2} + S_{y,1}S_{y,2}) = J_{xy}(S_{+,1}S_{-,2} + S_{-,1}S_{+,2})/2$, where $S_{+,i}$ and $S_{-,i}$ are the usual spin raising and lowering operators. Because $\mathcal{J}_{trans}$ acts on $(M, M \pm 1)$ states to first order of perturbation theory, the degeneracy of those states is strongly lifted. For example, the $(9/2, 7/2)$ and $(7/2, 9/2)$ states are strongly split into a symmetric (labelled $5''$) and antisymmetric (labelled $5'$) combination of $(9/2, 7/2)$ and $(7/2, 9/2)$ states. Similarly for the (-9/2, -7/2) and (-7/2, -9/2) states. Measuring this energy splitting allows us to determine the transverse superexchange interaction constant J_{xy} because the latter is proportional to the former.

Fig. 11 shows typical hysteresis loops (magnetization versus magnetic field scans) with the field applied along the easy axis of magnetization of $[Mn_4]_2$, that is, parallel to the S_6 axis. These loops display step-like features separated by plateaus. The step heights are temperature-independent below ~0.35 K [40]. The steps are due to resonant quantum tunneling of the magnetization (QTM) between the energy states of the $[Mn_4]_2$ dimer (see figure caption 10 and 11 for a discussion of 5 tunnel transitions). QTM has been previously observed for most

SMMs, but the novelty for $[Mn_4]_2$ dimers is that the QTM is now the collective behavior of the complete $S = 0$ dimer of exchange-coupled $S = 9/2$ Mn_4 quantum systems. This coupling is manifested as an exchange bias of all tunneling transitions, and the resulting hysteresis loop consequently displays unique features, such as the absence for the first time in a SMM of a QTM step at zero field [40].

Even though the five strongest tunneling transitions are observed in Fig. 11, fine structure was not observed. For example, the hysteresis loops do not show the splitting of the $(9/2, 7/2)$ states (labelled $5'$ and $5''$), which we suspected might be due to line broadening. Usually, line broadening in SMMs is caused by dipolar and hyperfine interactions [44], and distributions of anisotropy and exchange parameters. In most SMMs, the zero-field resonance is mainly broadened by dipolar and hyperfine interactions because distributions of anisotropy parameters do not affect the zero-field resonance. For an antiferromagnetically coupled dimer, however, this resonance is shifted to negative fields. Therefore, a distribution of the exchange coupling parameter J_z can further broaden this resonance. In fact, we show in the following that the latter is the dominant source of broadening. We then use the 'quantum hole-digging' method (see Section 4.2) [26,44–47] to provide direct experimental evidence for the transitions $5'$ and $5''$, which establishes tunneling involving entangled dimer states and allows us to determine J_{xy}.

The 'quantum hole-digging' method (Section 4.2) is a relatively new method that can, among other things [48], study line broadening and its evolution during relaxation [26,44–47]. The method is based on the simple idea that after a rapid field change, the resulting magnetization relaxation at short time periods is directly related to the number of molecules in resonance at the applied field; Prokof'ev and Stamp proposed [44] that this short time relaxation should follow a $\sqrt{t}$ (t = time) relaxation law. Thus, the magnetization of the $[Mn_4]_2$ dimers in the crystal was first saturated with a large positive field, and then a 'digging field' H_{dig} was applied at 0.04 K for a chosen 'digging time' t_{dig}. Then, the fraction (and only that fraction) of the molecules that is in resonance at H_{dig} can undergo magnetization tunneling. After t_{dig}, a field H_{probe} is applied and the magnetization relaxation rate is measured for short time periods; from this is calculated the short-time relaxation rate Γ_{sqrt}, which is related to the number of $[Mn_4]_2$ dimers still available for QTM [3]. The entire procedure is then repeated at other H_{probe} fields. The resulting plot (Fig. 12a) of Γ_{sqrt} versus H_{probe} reflects the distribution of spins still available for tunneling after t_{dig}.

In the limit of very short digging times, the difference between the relaxation rate in the absence and in the presence of digging, $\Gamma_{hole} = \Gamma_{sqrt}(t_{dig} = 0) - \Gamma_{sqrt}(t_{dig})$, is approximately proportional to the number of molecules which reversed their magnetization during the time t_{dig} (Fig. 12b). Γ_{hole} is characterized

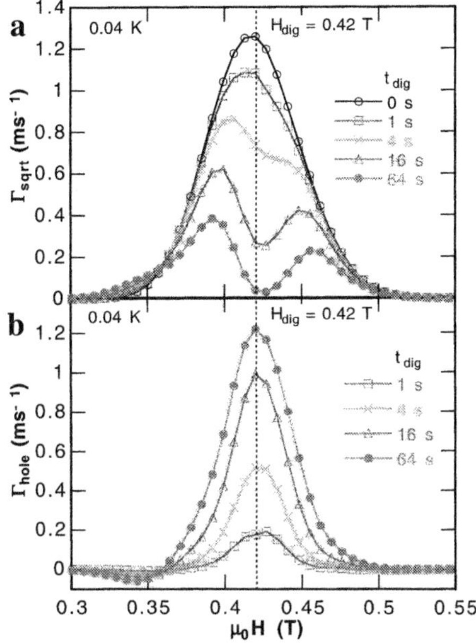

Fig. 12. (a) Field dependence of the short-time square-root relaxation rates Γ_{sqrt} are presented on a logarithmic scale showing the depletion of the molecular spin states by quantum tunneling at $H_{\text{dig}} =$ 0.42 T for various waiting times t_{dig}. (b) Difference between the relaxation rate in the absence and in the presence of digging, $\Gamma_{\text{hole}} = \Gamma_{\text{sqrt}}(t_{\text{dig}} = 0) - \Gamma_{\text{sqrt}}(t_{\text{dig}})$.

by a width that can be called the 'tunnel window'.

The width of the distribution in the absence of digging ($\sim$80 mT, Fig. 12a) is too large to be due to only dipolar ($\sim$20 mT) and hyperfine coupling ($\sim$10 mT). The following result suggests that it is due to a distribution of the exchange coupling parameter J_z.

First, the magnetization of the $[Mn_4]_2$ dimers was saturated with a large positive field, and then a 'digging field' H_{dig} was applied to reverse a fraction of the molecules that are in resonance at H_{dig} (transition **3** in Fig. 10). After the reversal of 2.5% of the molecules, the applied field is swept back to a large positive field. $5'$ and $5''$ are the first tunnel transitions that can allow the reversed molecules to tunnel back to positive saturation. Figs. 13a and 13b show the corresponding minor hysteresis loops for several 'digging fields and field sweep rates, respectively. Both figures show clearly the expected tunnel transitions $5'$ and $5''$, that were not resolved in the major hysteresis loops (Fig. 11). This suggests that the

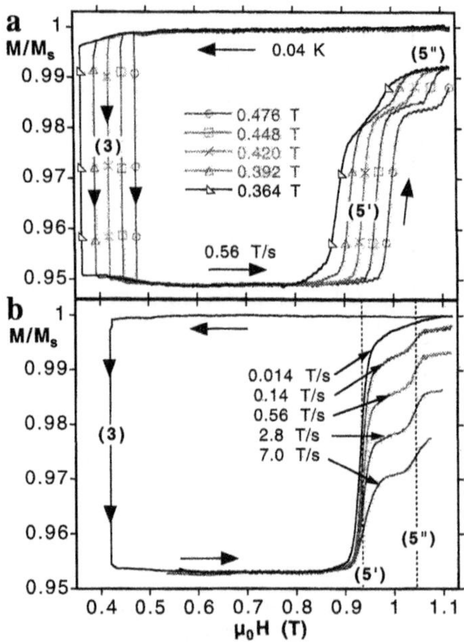

Fig. 13. Minor hysteresis loops for several (a) digging fields and (b) field sweep rates. After positive saturation, a digging field $H_{\rm dig}$ was applied to reverse ≈2.5% of the molecules that are in resonance at $H_{\rm dig}$ (transition **3** in Fig. 10). Then, the applied field is swept back to a large positive field. $5'$ and $5''$ are the first tunnel transitions allowing the reversed molecules to tunnel back to positive saturation.

broadening of tunneling transition **3** (the distribution in the absence of digging in Fig. 12a) is dominated by a distribution of the exchange coupling parameter J_z. During the application of the digging field, a subgroup of molecules is selected with an exchange coupling constant $J_z \approx g\mu_B\mu_0 H_{\rm dig}/S$, that can tunnel back at the fields of transitions $5'$ and $5''$.

This interpretation is supported by the study of the field values of $5'$ and $5''$ as a function of digging field, that is J_z, exhibiting a nearly linear variation (Fig. 14a). The field difference between transition $5'$ and $5''$ can be used to find the J_{xy}, presented in Fig. 14b. This shows that the superexchange interaction of the dimers is nearly isotropic ($J_{xy} \approx J_z$). It is important to mention that the transitions $5'$ and $5''$ are well separated, suggesting long coherence times compared to the time scale of the energy splitting.

The above results demonstrate for the first time tunneling via entangled states of a dimer of exchange coupled SMMs, showing that the dimer really does behave

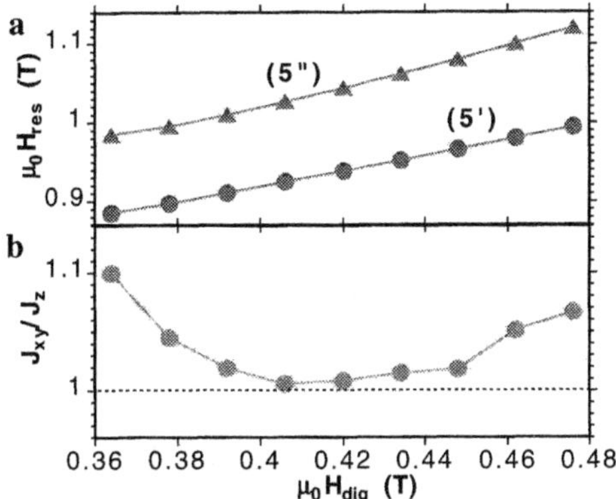

Fig. 14. (a) Resonance field positions H_{res} of $5'$ and $5''$ and (b) normalized transverse superexchange interaction J_{xy}/J_z as a function of digging field H_{dig}.

as a quantum mechanically coupled system. This result is of great importance if such systems are to be used for quantum computing.

4. Environmental decoherence effects in nanomagnets

At temperatures below 0.36 K, Fe_8 molecular clusters display a clear crossover from thermally activated relaxation to a temperature-independent quantum regime, with a pronounced resonance structure of the relaxation time as a function of the external field (Section 2). It was surprising, however, that the observed relaxation of the magnetization in the quantum regime was found to be nonexponential and the resonance width orders of magnitude too large [9, 11]. The key to understand this seemingly anomalous behavior involves the hyperfine fields as well as the evolving distribution of the weak dipole fields of the nanomagnets themselves [44]. Both effects were shown to be the main source of decoherence at very low temperature. At higher temperatures, phonons are another source of decoherence.

In the following sections, we focus on the low temperature and low field limits, where phonon-mediated relaxation is astronomically long and can be neglected. In this limit, the $m = \pm S$ spin states are coupled due to the tunneling splitting $\Delta_{\pm S}$ which is about 10^{-7} K for Fe_8 (Section 2.2) with $S = 10$. In order to tunnel between these states, the longitudinal magnetic energy bias $\xi = g\mu_B S H_{\text{local}}$ due to the local magnetic field H_{local} on a molecule must be smaller than $\Delta_{\pm S}$, implying a local field smaller than 10^{-8} T for Fe_8 clusters. Since the typical intermolecular dipole fields are of the order of 0.05 T, it seems at first that almost all molecules should be blocked from tunneling by a very large energy bias. Prokof'ev and Stamp have proposed a solution to this dilemma by proposing that fast dynamic nuclear fluctuations broaden the resonance, and the gradual adjustment of the dipole fields in the sample caused by the tunneling brings other molecules into resonance and allows continuous relaxation [44]. Some interesting predictions are briefly reviewed in the following section.

4.1. *Prokof'ev–Stamp theory*

Prokof'ev and Stamp were the first who realized that there are localized couplings of environmental modes with mesoscopic systems which cannot be modeled with an "oscillator bath" model [49] describing delocalized environmental modes such as electrons, phonons, photons, and so on. They found that these localized modes such as nuclear and paramagnetic spins are often strong and described them with a spin bath model [50]. We do not review this theory [51] but focus on one particular application which is interesting for molecular clusters [44]. Prokof'ev and Stamp showed that at a given longitudinal applied field H_z, the magnetization of a crystal of molecular clusters should relax at short times with a square-root time dependence which is due to a gradual modification of the dipole fields in the sample caused by the tunneling

$$M(H_z, t) = M_{\mathrm{in}} + (M_{\mathrm{eq}}(H_z) - M_{\mathrm{in}})\sqrt{\Gamma_{\mathrm{sqrt}}(H_z)t} \tag{4.1}$$

Here M_{in} is the initial magnetization at time $t = 0$ (after a rapid field change), and $M_{\mathrm{eq}}(H_z)$ is the equilibrium magnetization at H_z. The rate function $\Gamma_{\mathrm{sqrt}}(H_z)$ is proportional to the normalized distribution $P(H_z)$ of molecules which are in resonance at H_z

$$\Gamma_{\mathrm{sqrt}}(H_z) = c\frac{\xi_0}{E_D}\frac{\Delta_{\pm S}^2}{4\hbar}P(H_z) \tag{4.2}$$

where ξ_0 is the line width coming from the nuclear spins, E_D is the Gaussian half-width of $P(H_z)$, and c is a constant of the order of unity which depends on the sample shape. If these simple relations are exact, then measurements of the short time relaxation as a function of the applied field H_z give directly the distribution $P(H_z)$, and they allows one to measure the tunnel splitting $\Delta_{\pm S}$ which is described in the next section.

4.2. *Hole digging method to study dipolar distributions and hyperfine couplings*

Motivated by the Prokof'ev–Stamp theory [44], we developed a new technique—which we call the *hole digging method*—that can be used to observe the time

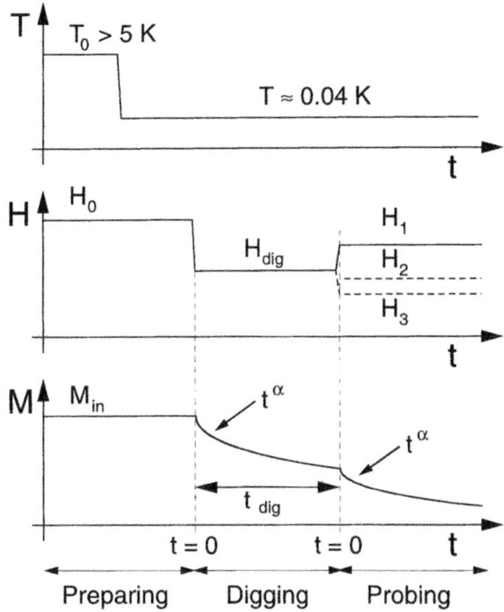

Fig. 15. Schema of the hole digging method presenting the time dependence of temperature, applied field, and magnetization of the sample.

evolution of molecular states in crystals of molecular clusters. It allowed us to measure the statistical distribution of magnetic bias fields in the Fe_8 system that arise from the weak dipole fields of the clusters themselves. A hole can be "dug" into the distribution by depleting the available spins at a given applied field. Our method is based on the simple idea that after a rapid field change, the resulting short time relaxation of the magnetization is directly related to the number of molecules which are in resonance at the given applied field. Prokof'ev and Stamp have suggested that the short time relaxation should follow a $\sqrt{t}$−relaxation law [Eq. (4.1)]. However, the hole digging method should work with any short time relaxation law—for example, a power law

$$M(H_z, t) = M_{\text{in}} + (M_{\text{eq}}(H_z) - M_{\text{in}})(\Gamma_{\text{short}}(H_z)t)^{\alpha} \tag{4.3}$$

where Γ_{short} is a characteristic short time relaxation rate that is directly related to the number of molecules which are in resonance at the applied field H_z, and $0 < \alpha < 1$ in most cases. $\alpha = 0.5$ in the Prokof'ev–Stamp theory [Eq. (4.1)] and Γ_{sqrt} is directly proportional to $P(H_z)$ [Eq. (4.2)]. The *hole digging method* can be divided into three steps (Fig. 15):

1. **Preparing the initial state.** A well-defined initial magnetization state of the crystal of molecular clusters can be achieved by rapidly cooling the sample from high down to low temperatures in a constant applied field H_z^0. For zero applied field ($H_z = 0$) or rather large applied fields ($H_z > 1$ T), one yields the demagnetized or saturated magnetization state of the entire crystal, respectively. One can also quench the sample in a small field of few milliteslas yielding any possible initial magnetization $M_{\rm in}$. When the quench is fast (< 1 s), the sample's magnetization does not have time to relax, either by thermal or by quantum transitions. This procedure yields a frozen thermal equilibrium distribution, whereas for slow cooling rates the molecule spin states in the crystal might tend to certain dipolar ordered ground state.
2. **Modifying the initial state—hole digging.** After preparing the initial state, a field $H_{\rm dig}$ is applied during a time $t_{\rm dig}$, called "digging field and digging time", respectively. During the digging time and depending on $H_{\rm dig}$, a fraction of the molecular spins tunnel (back and/or fourth); that is, they reverse the direction of magnetization. [1]
3. **Probing the final state.** Finally, a field $H_z^{\rm probe}$ is applied (Fig. 15) to measure the short time relaxation from which one yields $\Gamma_{\rm short}$ [Eq. (4.3)] which is related to the number of spins which are still free for tunneling after step (2).

The entire procedure is then repeated many times but at other fields $H_z^{\rm probe}$ yielding $\Gamma_{\rm short}(H_z, H_{\rm dig}, t_{\rm dig})$ which is related to the distribution of spins $P(H_z, H_{\rm dig}, t_{\rm dig})$ that are still free for tunneling after the hole digging. For $t_{\rm dig} = 0$, this method maps out the initial distribution.

4.3. *Intermolecular dipole interaction in Fe_8*

We applied the hole digging method to several samples of molecular clusters and quantum spin glasses. The most detailed study has been done on the Fe_8 system. We found the predicted $\sqrt{t}$ relaxation [Eq. (4.1)] in experiments on fully saturated Fe_8 crystals [11,52] and on nonsaturated samples [26]. Fig. 16 displays a detailed study of the dipolar distributions revealing a remarkable structure that is due to next-nearest-neighbor effects [26]. These results are in good agreement with simulations [47,53].

For a saturated initial state, the Prokof'ev–Stamp theory allows one to estimate the tunnel splitting $\Delta_{\pm S}$. Using Eqs. (3), (9), and (12) of Ref. [44], along with integration, we find $\int \Gamma_{\rm sqrt} d\xi = c \frac{\xi_0}{E_D} \frac{\Delta_{\pm S}^2}{4\hbar}$, where c is a constant of the order of unity which depends on the sample shape. With $E_D = 15$ mT, $\xi_0 = 0.8$ mT, $c = 1$,

[1] The field sweeping rate to apply $H_{\rm dig}$ should be fast enough to minimize the change of the initial state during the field sweep.

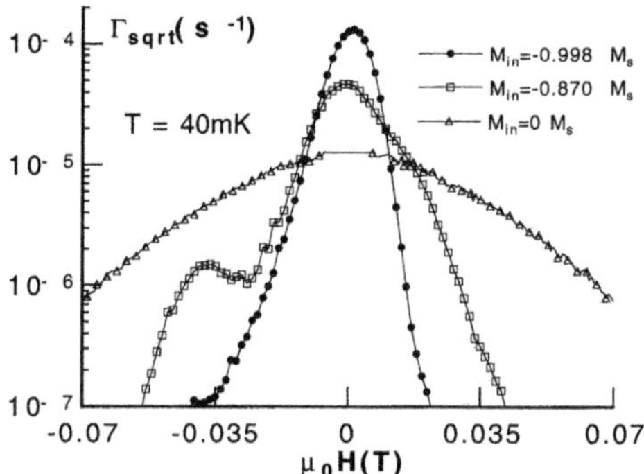

Fig. 16. Field dependence of the short time square-root relaxation rates $\Gamma_{\mathrm{sqrt}}(H_z)$ for three different values of the initial magnetization M_{in}. According to Eq. (4.2), the curves are proportional to the distribution $P(H_z)$ of magnetic energy bias due to local dipole field distributions in the sample. Note the logarithmic scale for Γ_{sqrt}. The peaked distribution labeled $M_{\mathrm{in}} = -0.998 M_{\mathrm{s}}$ was obtained by saturating the sample, whereas the other distributions were obtained by thermal annealing. $M_{\mathrm{in}} = -0.870 M_{\mathrm{s}}$ is distorted by nearest neighbor lattice effects. The peak at -0.04 T as well as the shoulder at 0.02 T and 0.04 T are originated by the clusters which have one nearest-neighbor cluster with reversed magnetization: The peak at -0.04 T corresponds to the reversal of the neighboring cluster along the **a** crystallographic axis, which almost coincides with the easy axis of magnetization, while the shoulder at 0.02 T and 0.04 T are due to the clusters along **b** and **c**.

and Γ_{sqrt} [26, 45], we find $\Delta_{\pm 10} = 1.2 \times 10^{-7}$ K which is close to the result of $\Delta_{\pm 10} = 1.0 \times 10^{-7}$ K obtained by using a Landau–Zener method (Section 2.1) [25].

Whereas the hole digging method probes the longitudinal dipolar distribution (H_z direction), the Landau–Zener method can be used to probe the transverse dipolar distribution by measuring the tunnel splittings Δ around a topological quench.

4.4. *Hyperfine interaction in* Fe_8

The strong influence of nuclear spins on resonant quantum tunneling in the molecular cluster Fe_8 was demonstrated for the first time [45] by comparing the relaxation rate of the standard Fe_8 sample with two isotopic modified samples: (i) ^{56}Fe is replaced by ^{57}Fe, and (ii) a fraction of ^{1}H is replaced by ^{2}H. By using the hole digging method, we measured an intrinsic broadening which is driven by the hyperfine fields (Fig. 17). Our measurements are in good agreement with numerical hyperfine calculations [45, 47]. For $T > 1.5$ K, the influence of nuclear spins on the relaxation rate is less important, suggesting that spin–phonon coupling dominates the relaxation rate.

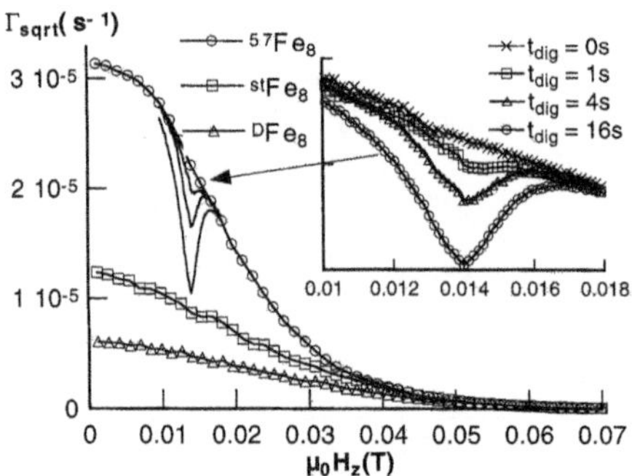

Fig. 17. Comparison of the short time relaxation rates of three different Fe_8 samples at $T = 40$ mK with $H_{trans} = 0$ and $M_{init} = 0$. The inset displays an typical example of a hole which was dug into the distribution by allowing the sample to relax for the time t_{dig} at $\mu_0 H_{dig} = 14$ mT.

5. Conclusion

In conclusion, we presented detailed measurements which demonstrated that molecular magnets offer a unique opportunity to explore the quantum dynamics of a large but finite spin. We focused our discussion on the Fe_8 molecular magnet because it is the first system where studies in the pure quantum regime were possible. The tunneling in this system is remarkable because it does not show up at the lowest order of perturbation theory.

Molecules with small spin have also been studied. For example, time-resolved magnetization measurements were performed on a spin $1/2$ molecular complex, so-called V_{15} [54]. Despite the absence of a barrier, magnetic hysteresis is observed over a time scale of several seconds. A detailed analysis in terms of a dissipative two-level model has been given, in which fluctuations and splittings are of the same energy. Spin–phonon coupling leads to long relaxation times and to a particular "butterfly" hysteresis loop [55,56].

A new family of supramolecular, antiferromagnetically exchange-coupled dimers of single-molecule magnets (SMMs) has recently been reported [40]. Each SMM acts as a bias on its neighbor, shifting the quantum tunneling resonances of the individual SMMs. Hysteresis loop measurements on a single crystal of SMM-dimers have established quantum tunneling of the magnetization via entangled states of the dimer. This showed that the dimer really does behave as a quantum-mechanically coupled dimer. The transitions are well separated, suggesting long coherence times compared to the time scale of the energy splitting [41]. This result is of great importance if such systems are to be used for quantum computing.

What remains still debated is the possibility of observing quantum coherence between states of opposite magnetization. Dipole–dipole and hyperfine interactions are source of decoherence. In other words, when a spin has tunneled through the barrier, it experiences a huge modification of its environment (hyperfine and dipolar) which prohibits the back tunneling. Prokof'ev and Stamp suggested three possible strategies to suppress the decoherence [57]. (i) Choose a system where the NMR frequencies far exceed the tunnel frequencies making any coupling impossible. (ii) Isotopically purify the sample to remove all nuclear spins. (iii) Apply a transverse field to increase the tunnel rate to frequencies much larger than hyperfine field fluctuations. Several groups are currently working on such proposals.

Concerning the perspectives of the field of single molecule magnets, we expect that chemistry is going to play a major role through the synthesis of novel larger spin clusters with strong anisotropy. We want to stress that there are already many other molecular magnets (the largest is currently a Mn_{84}) which are possible model systems. We believe that more sophisticated theories are needed which describe the dephasing effects of the environment onto the quantum system. These investigations are important for studying the quantum character of molecular clusters for applications like "quantum computers". The first implementation of Grover's algorithm with molecular magnets has been proposed [58]. However, many experimantal difficulties are still waiting for solutions.

References

[1] A. Hubert and R. Schäfer. *Magnetic Domains: The Analysis of Magnetic Microstructures*. Springer-Verlag, Berlin Heidelberg New York, 1998.

[2] Aharoni. *An Introduction to the Theory of Ferromagnetism*. Oxford University Press, London, 1996.

[3] W. Wernsdorfer. Classical and quantum magnetization reversal studied in nanometer-sized particles and clusters. *Adv. Chem. Phys.*, 118:99, 2001.

[4] R. Sessoli, H.-L. Tsai, A. R. Schake, S. Wang, J. B. Vincent, K. Folting, D. Gatteschi, G. Christou, and D. N. Hendrickson. High-spin molecules: $[Mn_{12}O_{12}(O_2CR)_{16}(H_2O)_4]$. *J. Am. Chem. Soc.*, 115:1804–1816, 1993.

[5] A.-L. Barra, P. Debrunner, D. Gatteschi, Ch. E. Schulz, and R. Sessoli. Superparamagnetic-like behavior in an octanuclear iron cluster. *EuroPhys. Lett.*, 35:133, 1996.

[6] M.A. Novak and R. Sessoli. Ac susceptibility relaxation studies on a manganese organic cluster compound: Mn_{12}-ac. In L. Gunther and B. Barbara, editors, *Quantum Tunneling of Magnetization-QTM'94*, volume 301 of *NATO ASI Series E: Applied Sciences*, pages 171–188. Kluwer Academic Publishers, London, 1995.

[7] J. R. Friedman, M. P. Sarachik, J. Tejada, and R. Ziolo. Macroscopic measurement of resonant magnetization tunneling in high-spin molecules. *Phys. Rev. Lett.*, 76:3830, 1996.

[8] L. Thomas, F. Lionti, R. Ballou, D. Gatteschi, R. Sessoli, and B. Barbara. Macroscopic quantum tunneling of magnetization in a single crystal of nanomagnets. *Nature (London)*, 383:145–147, 1996.

[9] C. Sangregorio, T. Ohm, C. Paulsen, R. Sessoli, and D. Gatteschi. Quantum tunneling of the magnetization in an iron cluster nanomagnet. *Phys. Rev. Lett.*, 78:4645, 1997.

[10] A. Garg. Lattice distortion mediated paramagnetic relaxation in high-spin high-symmetry molecular magnets. *Phys. Rev. B*, 81:1513, 1998.

[11] T. Ohm, C. Sangregorio, and C. Paulsen. Fe_8. *Euro. Phys. J. B*, 6:195, 1998.

[12] A. Caneschi, D. Gatteschi, C. Sangregorio, R. Sessoli, L. Sorace, A. Cornia, M. A. Novak, C. Paulsen, and W. Wernsdorfer. The molecular approach to nanoscale magnetism. *J. Magn. Magn. Mat.*, 200:182, 1999.

[13] S. M. J. Aubin, N. R. Dilley, M. B. Wemple, G. Christou, and D. N. Hendrickson. Resonant magnetization tunnelling in the half-integer-spin single-molecule magnet $[PPh_4][Mn_{12}O_{12}(O_2CEt)_{16}(H_2O)_4]$. *J. Am. Chem. Soc.*, 120:839, 1998.

[14] D. J. Price, F. Lionti, R. Ballou, P.T. Wood, and A. K. Powell. Large metal clusters and lattices with analogues to biology. *Phil. Trans. R. Soc. Lond. A*, 357:3099, 1999.

[15] J. Yoo, E. K. Brechin, A. Yamaguchi, M. Nakano, J. C. Huffman, A.L. Maniero, L.-C. Brunel, K. Awaga, H. Ishimoto, G. Christou, and D. N. Hendrickson. Single-molecule magnets; a new class of tetranuclear magnganese magnets. *Inorg. Chem.*, 39:3615, 2000.

[16] A. Garg. Topologically quenched tunnel splitting in spin systems without Kramers' degeneracy. *EuroPhys. Lett.*, 22:205, 1993.

[17] L. Landau. On the theory of transfer of energy at collisions II. *Phys. Z. Sowjetunion*, 2:46, 1932.

[18] C. Zener. Non-adiabatic crossing of energy levels. *Proc. R. Soc. London, Ser. A*, 137:696, 1932.

[19] E.C.G. Stückelberg. Theorie der unelastischen stoesse zwischen atomen. *Helv. Phys. Acta*, 5:369, 1932.

[20] S. Miyashita. Dynamics of the magnetization with an inversion of the magnetic field. *J. Phys. Soc. Jpn.*, 64:3207, 1995.

[21] S. Miyashita. Observation of the energy gap due to the quantum tunneling making use of the Landau-Zener mechanism [uniaxial magnets]. *J. Phys. Soc. Jpn.*, 65:2734, 1996.

[22] G.Rose and P.C.E. Stamp. Short time ac response of a system of nanomagnets. *Low Temp. Phys.*, 113:1153, 1998.

[23] M. Thorwart, M. Grifoni, and P. Hänggi. Strong coupling theory for driven tunneling and vibrational relaxation. *Phys. Rev. Lett.*, 85:860, 2000.

[24] M. N. Leuenberger and D. Loss. Incoherent Zener tunneling and its application to molecular magnets. *Phys. Rev. B*, 61:12200, 2000.

[25] W.Wernsdorfer and R. Sessoli. Quantum phase interference and parity effects in magnetic molecular clusters. *Science*, 284:133, 1999.

[26] W. Wernsdorfer, T. Ohm, C. Sangregorio, R. Sessoli, D. Mailly, and C. Paulsen. Observation of the distribution of molecular spin states by resonant quantum tunneling of the magnetization. *Phys. Rev. Lett.*, 82:3903, 1999.

[27] D. Loss, D. P. DiVincenzo, and G. Grinstein. Suppression of tunneling by interference in half-integer-spin particles. *Phys. Rev. Lett.*, 69:3232, 1992.

[28] J. von Delft and C. L. Henley. Destructive quantum interference in spin tunneling problems. *Phys. Rev. Lett.*, 69:3236, 1992.

[29] R. P. Feynman, R. B. Leighton, and M. Sand. *The Feynman lectures on physics*, volume 3. Addison-Wesley Publishing Company, London, 1970.

[30] A. Garg. Oscillatory tunnel splittings in spin systems: A discrete Wentzel-Kramers-Brillouin approach. *Phys. Rev. Lett.*, 83:4385, 1999.

[31] J. Villain and A. Fort. Magnetic tunneling told to ignorants by two ignorants. *Euro. Phys. J. B*, 17:69, 2000.

[32] E. Kececioglu and A. Garg. Diabolical points in magnetic molecules: An exactly solvable model. *Phys. Rev. B*, 63:064422, 2001.

[33] S.E. Barnes. Manifestation of intermediate spin for fe_8. *cond-mat/9907257.*

[34] J.-Q. Liang, H.J.W. Mueller-Kirsten, D.K. Park, and F.-C. Pu. Phase interference for quantum tunneling in spin systems. *Phys. Rev. B*, 61:8856, 2000.

[35] Sahng-Kyoon Yoo and Soo-Young Lee. Geometrical phase effects in biaxial nanomagentic particles. *Phys. Rev. B*, 62:3014, 2000.

[36] Hui Hu, Jia-Lin Zhu, Rong Lu, and Jia-Jiong Xiong. Effects of arbitrarily directed field on spin phase oscillations in biaxial molecular magnets. *cond-mat/0005527.*

[37] R. Caciuffo, G. Amoretti, A. Murani, R. Sessoli, A. Caneschi, and D. Gatteschi. Neutron spectroscopy for the magnetic anisotropy of molecular clusters. *Phys. Rev. Lett.*, 81:4744, 1998.

[38] G. Amoretti, R. Caciuffo, J. Combet, A. Murani, and A. Caneschi. Inelastic neutron scattering below 85 mev and zero-field splitting parameters in the Fe_8 magnetic cluster. *Phys. Rev. B*, 62:3022, 2000.

[39] A.L. Barra, D. Gatteschi, and R. Sessoli. High-frequency EPR spetra of Fe_8. a critical appraisal of the barrier for the reorientation of the magnetization in single-molecule magnets. *Chem. Eur. J.*, 6:1608, 2000.

[40] W. Wernsdorfer, N. Aliaga-Alcalde, D.N. Hendrickson, and G. Christou. Exchange-biased quantum tunnelling in a supramolecular dimer of single-molecule magnets. *Nature*, 416:406, 2002.

[41] R. Tiron, W. Wernsdorfer, D. Foguet-Albiol, N. Aliaga-Alcalde, and G. Christou. Spin quantum tunneling via entangled states in a dimer of exchange-coupled single-molecule magnets. *Phys. Rev. Lett.*, 91:227203, 2003.

[42] D. N. Hendrickson and *et al.* Photosynthetic water oxidation center: spin frustration in distorted cubane $mn^{iv}mn_3^{iv}$ model complexes. *J. Am. Chem. Soc.*, 114:2455, 1992.

[43] W. Wernsdorfer, S. Bhaduri, R. Tiron, D. N. Hendrickson, and G. Christou. Spin-spin cross relaxation in single-molecule magnets. *Phys. Rev. Lett.*, 89:197201, 2002.

[44] N.V. Prokof'ev and P.C.E. Stamp. Low-temperature quantum relaxation in a system of magnetic nanomolecules. *Phys. Rev. Lett.*, 80:5794, 1998.

[45] W. Wernsdorfer, A. Caneschi, R. Sessoli, D. Gatteschi, A. Cornia, V. Villar, and C. Paulsen. Effects of nuclear spins on the quantum relaxation of the magnetization for the molecular nanomagnet fe_8. *Phys. Rev. Lett.*, 84:2965, 2000.

[46] J. J. Alonso and J. F. Fernandez. Tunnel window's imprint on dipolar field distributions. *Phys. Rev. Lett.*, 87:097205, 2001.

[47] I. Tupitsyn and P.C.E. Stamp. 'hole-digging' in ensembles of tunneling molecular magnets. *cond-mat/0305371*, 5:5371, 2003.

[48] W. Wernsdorfer, M. Soler, D. N. Hendrickson, and G. Christou. Quantum tunneling of magnetization in a large molecular nanomagnet – approaching the mesoscale. *cond-mat/0306303.*

[49] R.P. Feynman and F.L. Vernon. The theory of the general quantum system interacting with a linear dissipative system. *Ann. Phys.*, 24:118, 1963.

[50] N.V. Prokof'ev and P.C.E. Stamp. Quantum relaxation of magnetisation in magnetic particles. *J. Low Temp. Phys.*, 104:143, 1996.

[51] N.V. Prokof'ev and P.C.E. Stamp. Theory of the spin bath. *Rep. Prog. Phys.*, 63:669–726, 2000.

[52] T. Ohm, C. Sangregorio, and C. Paulsen. Fe_8. *J. Low Temp. Phys.*, 113:1141, 1998.

[53] A. Cuccoli, A. Fort, A. Rettori, E. Adam, and J. Villain. Dipolar interaction and incoherent quantum tunneling: a Monte Carlo study of magnetic relaxation. *Euro. Phys. J. B*, 12:39, 1999.

[54] I. Chiorescu, W. Wernsdorfer, B. Barbara, A. Müller, and H. Bogge. Environmental effects on big molecule with spin 1/2. *J. Appl. Phys.*, 87:5496, 2000.

[55] W. Wernsdorfer, A. Caneschi, R. Sessoli, D. Gatteschi, A. Cornia, V. Villar, and C. Paulsen. Butterfly hysteresis loop and dissipative spin reversal in the $S = 1/2$, V_{15} molecular complex. *Phys. Rev. Lett.*, 84:3454, 2000.

[56] V. V. Dobrovitski, M. I. Katsnelson, and B. N. Harmon. Mechanisms of decoherence in weakly anisotropic molecular magnets. *Phys. Rev. Lett.*, 84:3458, 2000.

[57] N.V. Prokof'ev and P.C.E. Stamp. Spin environments and the suppression of quantum coherence. In L. Gunther and B. Barbara, editors, *Quantum Tunneling of Magnetization-QTM'94*, volume 301 of *NATO ASI Series E: Applied Sciences*, page 369. Kluwer Academic Publishers, London, 1995.

[58] M. N. Leuenberger and D. Loss. Quantum computing with molecular magnets. *Nature*, 410:789, 2001.

Course 16

PROSPECTS FOR STRONG CAVITY QUANTUM ELECTRODYNAMICS WITH SUPERCONDUCTING CIRCUITS

S. M. Girvin[1], R.-S. Huang[1,2], A. Blais[1],
A. Wallraff[3] and R. J. Schoelkopf[3]

[1]*Department of Physics, Sloane Physics Laboratory, Yale University, New Haven, CT 06520-8120*
[2]*Department of Physics, Indiana University, Bloomington, IN 47405*
[3]*Department of Applied Physics, Becton Laboratory, Yale University, New Haven, CT 06520-8284*

D. Estève, J.-M. Raimond and J. Dalibard, eds.
Les Houches, Session LXXIX, 2003
Quantum Entanglement and Information Processing
Intrication quantique et traitement de l'information
© *2004 Elsevier B.V. All rights reserved*

Contents

1. Introduction

Cavity quantum electrodynamics (cQED) studies the properties of atoms coupled to discrete photon modes in high Q cavities. Such systems are of great interest in the study of fundamental quantum mechanics of open systems, the engineering of quantum states and the study of measurement-induced decoherence [1, 2, 3], and have also been proposed as possible candidates for use in quantum information processing and transmission [1, 2, 3]. Ideas for novel cQED analogs using nano-mechanical resonators have recently been suggested by Schwab and collaborators [4, 5]. We present a realistic proposal for cQED via Cooper pair boxes coupled to a one-dimensional (1D) transmission line resonator as shown in Fig. 1, within a simple circuit that can be fabricated on a single microelectronic chip. As we discuss, 1D cavities offer a number of practical advantages in reaching the strong coupling limit of cQED over previous proposals using discrete LC circuits [6, 7], large Josephson junctions [8, 9, 10], or 3D cavities [11, 12, 13]. Besides the potential for entangling qubits to realize two-qubit gates addressed in those works, we show that the cQED approach also gives strong and controllable isolation of the qubits from the electromagnetic environment, permits high fidelity quantum non-demolition (QND) readout of multiple qubits, and can produce states of microwave photon fields suitable for quantum communication. The proposed circuits therefore provide a simple and efficient architecture for solid-state quantum computation, in addition to opening up a new avenue for the study of entanglement and quantum measurement physics with macroscopic objects. We will frame our discussion in a way that makes contact between the language of atomic physics and that of electrical engineering, and begin with a brief general overview of cQED before turning to a more specific discussion of our proposed architecture.

2. Brief review of cavity QED

In the optical version of cQED [2], one drives the cavity with a laser and monitors changes in the cavity transmission resulting from coupling to atoms falling through the cavity. One can also monitor the spontaneous emission of the atoms into transverse modes not confined by the cavity. It is not generally possible to

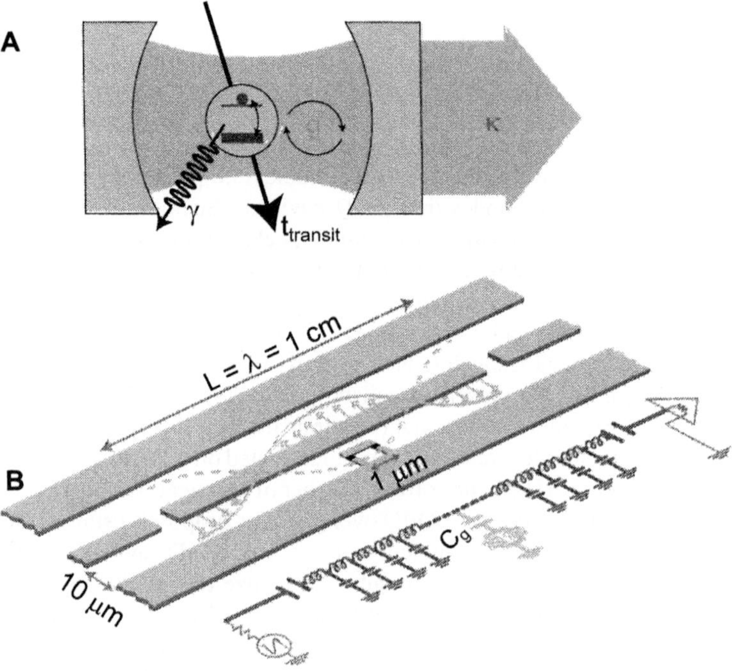

Fig. 1. a) Standard representation of cavity quantum electrodynamic system, comprising a single mode of the electromagnetic field in a cavity with decay rate κ coupled with a coupling strength $g = \mathcal{E}_{\mathrm{rms}}d/\hbar$ to a two-level system with spontaneous decay rate γ and cavity transit time t_{transit}. b) Schematic layout and effective circuit of proposed implementation of cavity QED using superconducting circuits. The 1D transmission line resonator consists of a full-wave section of superconducting coplanar waveguide, which may be lithographically fabricated using conventional optical lithography. A Cooper-pair box qubit is placed between the superconducting lines, and is capacitively coupled to the center trace at a maximum of the voltage standing wave, yielding a strong electric dipole interaction between the qubit and a single photon in the cavity. The box consists of two small ($\sim$ 100 nm $\times$ 100 nm) Josephson junctions, configured in a $\sim$ 1 μm loop to permit tuning of the effective Josephson energy by magnetic field. Input and output signals are coupled to the resonator, via the capacitive gaps in the center line, from 50 Ω transmission lines which allow measurements of the amplitude and phase of the cavity transmission, and the introduction of dc and rf pulses to manipulate the qubit states. Multiple qubits (not shown) can be similarly placed at different antinodes of the standing wave to generate entanglement and two-bit quantum gates across distances of several millimeters.

directly determine the state of the atoms after they have passed through the cavity because the spontaneous emission lifetime is on the scale of nanoseconds. One can, however, infer information about the state of the atoms inside the cavity from real-time monitoring of the cavity optical transmission.

In the microwave version of cQED [3] one uses a very high Q superconducting 3D resonator to couple photons to transitions in Rydberg atoms. Here one does not directly monitor the state of the photons, but is able to determine with high efficiency the state of the atoms after they have passed through the cavity (since the excited state lifetime is of order 30 ms). From this state-selective detection one can infer information about the state of the photons in the cavity.

The key parameters describing a cQED system (see Table I) are the cavity resonance frequency ω_r, the atomic transition frequency Ω, and the strength of the atom-photon coupling g appearing in the Jaynes-Cummings [14] Hamiltonian

$$H = \hbar\omega_r\left(a^\dagger a + \frac{1}{2}\right) + \frac{\hbar\Omega}{2}\sigma^z + \hbar g(a^\dagger\sigma^- + a\sigma^+) + H_\kappa + H_\gamma. \tag{2.1}$$

Here H_κ describes the coupling of the cavity to the continuum which produces the decay rate $\kappa = \omega_r/Q$, while H_γ describes the coupling of the atom to modes other than the cavity mode which cause the excited state to decay at rate γ (and possibly also produce additional dephasing effects). An additional important parameter in the atomic case is the transit time t_{transit} of the atom through the cavity. In the absence of damping and for the case of zero detuning $[\Delta \equiv \Omega - \omega_r = 0]$ between the atom and the cavity, an initial zero-photon excited atom state $|0, \uparrow\rangle$ flops into a photon $|1, \downarrow\rangle$ and back again at the vacuum Rabi frequency g/π. The degeneracy of the two corresponding states with n additional photons is split by $2\hbar g\sqrt{n+1}$. Equivalently, the atom's state and the photon number are entangled. The value of $g = \mathcal{E}_{\text{rms}}d/\hbar$ is determined by the transition dipole moment d and the rms zero-point electric field of the cavity mode. Strong coupling is achieved when $g \gg \kappa, \gamma$ [15].

3. Circuit implementation of cavity QED

We now consider in more specific detail the cQED setup illustrated in Fig. 1. A number of possible superconducting quantum circuits could function as the 'atom'. For definiteness we focus on the Cooper pair box [16, 6, 17, 18]. Unlike the usual cQED case, these artificial 'atoms' remain at fixed positions indefinitely and so do not suffer from the problem that the coupling g varies with position in the cavity. An additional advantage is that the zero-point energy is distributed over a very small effective volume ($\approx 10^{-5}$ cubic wavelengths) for our choice of a quasi-one-dimensional transmission line 'cavity.' This leads to significant

Table 1

Key rates and cQED parameters for optical [2] and microwave [3] atomic systems using 3D cavities, compared against the proposed approach using superconducting circuits, showing the possibility for attaining the strong cavity QED limit ($n_{\mathrm{Rabi}} \gg 1$). For the 1D superconducting system, a full-wave ($L = \lambda$) resonator, $\omega_r/2\pi = 10$ GHz, a relatively low Q of 10^4 and coupling $\beta = C_g/C_\Sigma = 0.1$ are assumed. For the 3D microwave case, the number of Rabi flops is limited by the transit time. For the 1D circuit case, the intrinsic Cooper-pair box decay rate is unknown; a conservative value equal to the current experimental upper bound $1/\gamma \geq 2\,\mu$s is assumed.

parameter	symbol	3D optical	3D microwave	1D circuit
resonance/transition frequency	$\omega_r/2\pi$, $\Omega/2\pi$	350 THz	51 GHz	10 GHz
vacuum Rabi frequency	g/π, g/ω_r	220 MHz, 3×10^{-7}	47 kHz, 1×10^{-7}	100 MHz, 5×10^{-3}
transition dipole	d/ea_0	~ 1	1×10^3	2×10^4
cavity lifetime	$1/\kappa$, Q	10 ns, 3×10^7	1 ms, 3×10^8	160 ns, 10^4
atom lifetime	$1/\gamma$	61 ns	30 ms	2 μs
atom transit time	t_{transit}	$\geq 50\,\mu$s	100 μs	∞
critical atom number	$N_0 = 2\gamma\kappa/g^2$	6×10^{-3}	3×10^{-6}	$\leq 6 \times 10^{-5}$
critical photon number	$m_0 = \gamma^2/2g^2$	3×10^{-4}	3×10^{-8}	$\leq 1 \times 10^{-6}$
# of vacuum Rabi flops	$n_{\mathrm{Rabi}} = 2g/(\kappa + \gamma)$	~ 10	~ 5	$\sim 10^2$

rms voltages $V^0_{\rm rms} \sim \sqrt{\hbar\omega_r/cL}$ between the center conductor and the adjacent ground plane at the antinodal positions, where L is the resonator length and c is the capacitance per unit length of the transmission line. At a resonant frequency of 10 GHz ($h\nu/k_B \sim 0.5$ K) and for a 10 μm gap between the center conductor and the adjacent ground plane, $V_{\rm rms} \sim 2\,\mu$V corresponding to electric fields $\mathcal{E}_{\rm rms} \sim 0.2$ V/m, some 100 times larger than achieved in the 3D cavity described in Ref. [3]. Thus, this geometry might also be useful for coupling to Rydberg atoms [19].

In addition to the small effective volume, and the fact that the on-chip realization of cQED shown in Fig. 1 can be fabricated with existing lithographic techniques, a transmission-line resonator geometry offers other practical advantages over LC circuits or large Josephson junctions. The qubit can be placed within the cavity formed by the transmission line to strongly suppress the spontaneous emission, in contrast to an LC circuit, where radiation and parasitic resonances may be induced in the wiring. Since the resonant frequency of the transmission line is determined primarily by a fixed geometry, its reproducibility and immunity to 1/f noise should be superior to Josephson junction resonators. Finally, transmission line resonances in coplanar waveguides with $Q \sim 10^6$ have already been demonstrated [20], suggesting that the internal losses can be very low. The optimal choice of the resonator Q in this approach is strongly dependent on the presently unknown intrinsic decay rates of superconducting qubits. Here we assume the conservative case of an overcoupled resonator with a $Q \sim 10^4$, which is preferable for the first experiments.

Our choice of 'atom', the Cooper pair box [16, 6] is a mesoscopic superconducting grain with a significant charging energy. The two lowest charge states having N_0 and $N_0 + 1$ Cooper pairs are coherently mixed by Josephson tunnelling between the box and a reservoir (in this case the resonator ground plane) leading to the two-level Hamiltonian [6]

$$H_Q = E_{\rm el}\sigma^x - \frac{E_{\rm J}}{2}\sigma^z. \tag{3.1}$$

Here, we have chosen the spinor basis such that the box Cooper pair number operator is [21] $\hat{N} - N_0 = (1 + \sigma^x)/2$. The electrostatic energy is given by $4E_{\rm c}(C_{\rm g}V_{\rm g}/2e - 1/2)$, where $C_{\rm g}$ is the coupling capacitance between the box and the resonator, $E_{\rm c} \equiv e^2/2C_\Sigma$ is the charging energy determined by the total box capacitance and $E_{\rm J}$ is the Josephson energy. Dc gating of the box can be conveniently achieved by applying a bias voltage to the center conductor of the transmission line. In addition to the dc part $V^{\rm dc}_{\rm g}$ the gate voltage has a quantum part $v = V_{\rm rms}^0(a^\dagger + a)$ from which we obtain

$$g = \frac{E_{\rm J}}{\sqrt{E_{\rm J}^2 + E_{\rm el}^2}} \frac{e}{\hbar} \beta \sqrt{\frac{\hbar\omega_{\rm r}}{cL}}, \tag{3.2}$$

where $\beta \equiv C_{\rm g}/C_\Sigma$. At the charge degeneracy point $E_{\rm el} = 0$ (where $n_{\rm g} = C_{\rm g}V_{\rm g}^{\rm dc}/2e = 1/2$), the two levels are split only by the Josephson energy and the 'atom' is highly polarizable, having transition dipole moment $d \equiv \hbar g/\mathcal{E}_{\rm rms} \sim 2 \times 10^4$ atomic units (ea_0), or more than an order of magnitude larger than even a typical Rydberg atom [15]. An experimentally realistic [18] coupling $\beta \sim 0.1$ leads to a vacuum Rabi rate $g/\pi \sim 100$ MHz, which is three orders of magnitude larger than in corresponding atomic microwave cQED experiments [3].

A comparison of the experimental parameters for implementations of cavity QED with optical and microwave atomic systems, and for the proposed implementation with superconducting circuits, is presented in Table I. We assume a relatively low $Q = 10^4$ and a worst case estimate, consistent with the bound set by previous experiments (discussed further below), for the intrinsic qubit lifetime of $1/\gamma \geq 2\,\mu$s. The standard figures of merit [22] for strong coupling are the critical photon number needed to saturate the atom on resonance $m_0 = \gamma^2/2g^2 \leq 1 \times 10^{-6}$ and the minimum atom number detectable by measurement of the cavity output $N_0 = 2\gamma\kappa/g^2 \leq 6 \times 10^{-5}$. These remarkably low values are clearly very favorable, and show that superconducting circuits could access the interesting regime of very strong coupling.

4. Zero detuning

For the case of zero detuning and weak coupling $g < \kappa$, the radiative decay rate of the qubit into the transmission line becomes strongly *enhanced* by a factor of Q relative to the rate in the absence of the cavity [15] because of the resonant enhancement of the density of states at the atomic transition frequency. In electrical engineering language, the $\sim 50\,\Omega$ external transmission line impedance is transformed on resonance to a high value which is better matched to extract energy from the qubit. For strong coupling, the first excited state becomes a doublet with line width $(\kappa + \gamma)/2$ since the excitation is half atom and half photon [15]. As can be seen from Table I, the coupling is so strong that, even for the low $Q = 10^4$ we have assumed, $2g/(\kappa + \gamma) \sim 100$ vacuum Rabi oscillations are possible, and the frequency splitting ($g/\pi \sim 100$ MHz) will be readily resolvable in the transmission spectrum of the resonator. This spectrum can be observed in the same manner employed in optical atomic experiments, with a continuous wave measurement at low drive, and will be of practical use to find the dc gate voltage needed to tune the box into resonance with the cavity. Of more fundamental importance than this simple avoided level crossing however, is the fact that the Rabi splitting scales with the square root of the photon number, making the level spacing anharmonic. This should cause a number of novel non-linear effects [14] to appear in the spectrum at higher drive powers when the average photon number in the cavity is large ($\langle n \rangle > 1$). A conservative estimate of the noise energy for a 10 GHz cryogenic high electron mobility (HEMT) amplifier is $n_{\text{amp}} = k_B T_N / \hbar\omega = 100$ photons, so these spectral features should be readily observable in a measurement time $t_{\text{meas}} = 2n_{\text{amp}}/\langle n \rangle \kappa$, or only $\sim 32\,\mu$s for $\langle n \rangle \sim 1$.

5. Large detuning: lifetime enhancement

For the case of strong detuning, the coupling to the continuum is substantially reduced. One can view the effect of the detuned resonator as filtering out the vacuum noise at the qubit transition frequency or, in electrical engineering terms, as providing an impedance transformation which strongly *reduces* the real part of the environmental impedance seen by the qubit. For large detuning the qubit excitation spends only a small fraction of its time as a photon [15] so that the decay rate into the transmission line is only $\gamma_\kappa = (g/\Delta)^2 \kappa \sim 1/(64\,\mu\text{s})$, much less than κ.

One of the important motivations for this cQED experiment is to determine the various contributions to the qubit decay rate γ so that we can understand their fundamental physical origins as well as engineer improvements. Besides γ_κ, there are two additional contributions to $\gamma = \gamma_\kappa + \gamma_\perp + \gamma_{\text{NR}}$. Here $\gamma_\perp$ is the decay rate into photon modes other than the cavity mode, and γ_{NR} is the rate of other (possibly non-radiative) decays. Optical cavities are relatively open and $\gamma_\perp$ is significant, but for 1D microwave cavities, $\gamma_\perp$ is expected to be negligible (despite the very large transition dipole). For Rydberg atoms the two qubit states are both highly excited levels and γ_{NR} represents (radiative) decay out of the two-level subspace. For Cooper pair boxes, γ_{NR} is completely unknown at the present time, but could have contributions from phonons, two-level systems in insulating [23] barriers and substrates, or thermally excited quasiparticles.

For Cooper box qubits *not* inside a cavity, recent experiments [18] have determined a relaxation time $1/\gamma = T_1 \sim 1.3\,\mu$s despite the back action of continuous measurement by a SET electrometer. Vion et al. [17] found $T_1 \sim 1.84\,\mu$s (without measurement back action) for their charge-phase qubit. The rate of relaxation expected from purely vacuum noise (spontaneous emission) is [18, 6]

$$\gamma_\kappa = \frac{E_\mathrm{J}^2}{E_\mathrm{J}^2 + E_\mathrm{el}^2} \left(\frac{e}{\hbar}\right)^2 \beta^2 2\hbar\Omega \,\mathrm{Re}[Z(\Omega)]. \tag{5.1}$$

It is difficult in most experiments to precisely determine the real part of the high frequency environmental impedance $Z(\Omega)$ presented by the leads connected to the qubit, but reasonable estimates [18] yield values of T_1 in the range of 1 μs. Thus in these experiments, if there are non-radiative decay channels, they are at most comparable to the vacuum radiative decay rate (and may well be much less). Experiments with a cavity will present the qubit with a simple and well controlled electromagnetic environment, in which the radiative lifetime can be enhanced with detuning to $1/\gamma_\kappa > 64\ \mu$s, allowing γ_NR to dominate and yielding valuable information about any non-radiative processes.

6. Dispersive QND readout of qubit

For large detuning, making the unitary transformation

$$U = \exp\left[\frac{g}{\Delta}(a\sigma^+ - a^\dagger\sigma^-)\right] \tag{6.1}$$

and expanding to second order in g, approximately diagonalizes the Hamiltonian (neglecting damping for the moment)

$$UHU^\dagger \approx \hbar\left[\omega_\mathrm{r} + \frac{g^2}{\Delta}\sigma^z\right]a^\dagger a + \frac{1}{2}\hbar\left[\Omega + \frac{g^2}{\Delta}\right]\sigma^z. \tag{6.2}$$

We see that there is a dispersive shift of the cavity transition by $\sigma_z g^2/\Delta$, that is the qubit pulls the cavity frequency by $\pm g^2/\kappa\Delta = \pm 2.5$ line widths for a 10% detuning. Exact diagonalization [15] shows that the pull becomes power dependent and decreases in magnitude for cavity photon numbers on the scale $n = n_\mathrm{crit} \equiv \Delta^2/4g^2 \sim 100$. In the regime of non-linear response, single-atom optical bistability [14] can be expected when the drive frequency is off resonance at low power but on resonance at high power [24].

The state-dependent pull of the cavity frequency by the qubit can be used to entangle the state of the qubit with that of the photons passing through the resonator. For $g^2/\kappa\Delta > 1$ the pull is greater than the line width and the microwave frequency can be chosen so that the transmission of the cavity is close to unity for one state of the qubit and close to zero for the other [25]. For $g^2/\kappa\Delta \ll 1$ the state of the qubit is encoded in the phase of the transmitted microwaves. An initial qubit state $|\chi\rangle = \alpha\,|\uparrow\rangle + \beta\,|\downarrow\rangle$ evolves under microwave illumination into the entangled state $|\psi\rangle = \alpha\,|\uparrow,\theta\rangle + \beta\,|\downarrow,-\theta\rangle$, where $\tan\theta = 2g^2/\kappa\Delta$, and $|\pm\theta\rangle$ are (interaction representation) coherent states with the appropriate mean photon number and opposite phases. Such an entangled state can be used to couple qubits in distant resonators and allow quantum communication [26]. If an independent measurement of the qubit state can be made, then such states can be turned into photon Schrödinger cats [15].

The phase shift of the transmitted microwaves can be measured using standard heterodyne techniques, and can therefore serve as a high efficiency quantum non-demolition dispersive readout of the state of the qubit, as described in Figure 2. Exciting the cavity to a maximal amplitude $n_{\mathrm{crit}} = 100 \sim n_{\mathrm{amp}}$ the signal-to-noise ratio, SNR = $(n_{\mathrm{crit}}/n_{\mathrm{amp}})(\kappa/2\gamma)$, can be very high if the qubit lifetime is longer than a few cavity decay times ($1/\kappa = 160$ ns). We see from Eq. (6.2) that the ac-Stark/Lamb shift of the box transition is $(2g^2/\Delta)(n + 1/2)$, so the back action of the dispersive cQED measurement is due to quantum fluctuations of the number of photons in the cavity which cause variations in the ac Stark shift, that dephase the qubit. A second possible form of back action is mixing transitions between the two qubit states induced by the microwaves. Since the coupling is so strong, large detuning $\Delta = 0.1\,\omega_{\mathrm{r}}$ can be chosen, making the mixing rate limited not by the frequency spread of the drive pulse, but rather by the width of the qubit excited state itself. The rate of driving the qubit from ground to excited state when n photons are in the cavity is $R \approx n(g/\Delta)^2\gamma$. If the measurement pulse excites the cavity to $n = n_{\mathrm{crit}}$, we see that the excitation rate is still only 1/4 of the relaxation rate, so the main limitation on the fidelity of the QND readout is the decay of the excited state of the qubit during the course of the readout. This occurs (for small γ) with probability $P_{\mathrm{relax}} \sim \gamma t_{\mathrm{meas}} \sim 15\times\gamma/\kappa \sim 3.75\,\%$ and the measurement is highly non-demolition. The numerical stochastic wave function calculations [27] shown in Fig. 2 confirm that the measurement-induced mixing is negligible and that one can determine the qubit's state in a single-shot measurement with high fidelity. The readout fidelity, including the effects of this stochastic decay, and related figures of merit of the QND readout are summarized in Table II. Since nearly all the energy used in this dispersive measurement scheme is dissipated in the remote terminations of the input and output transmission lines, it has the practical advantage of avoiding quasiparticle generation in the qubit.

Another key feature of the cavity QED readout is that it lends itself naturally to operation of the box at the charge degeneracy point ($n_g = 1/2$), where it has been shown that T_2 can be enormously enhanced [17] because the energy splitting has an extremum with respect to gate voltage and isolation of the qubit from 1/f dephasing is optimal. The derivative of the energy splitting with respect to gate voltage is the charge difference in the two qubit states. At the degeneracy point this derivative vanishes and the environment cannot distinguish the

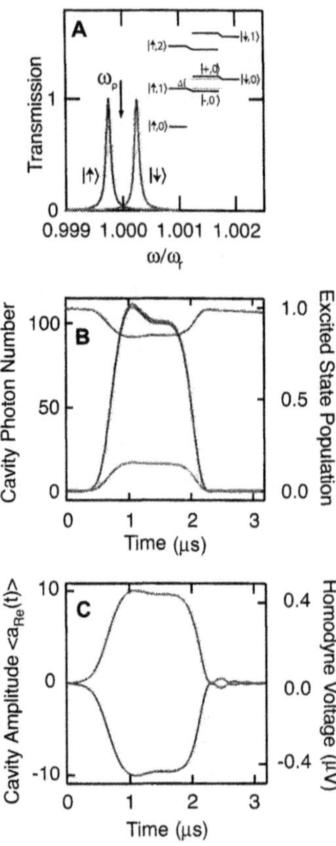

Fig. 2. Use of the coupling between a Cooper-pair box qubit and a transmission-line resonator to perform a dispersive quantum non-demolition measurement. a) Transmission spectrum of the cavity, which is "pulled" by an amount $\pm g^2/\Delta = 2.5 \times 10^{-4} \times \omega_r$, depending on the state of the qubit (red for the excited state, blue for the ground state). To perform a measurement of the qubit, a pulse of microwave photons, at a probe frequency $\omega_p = \omega_r$, is sent through the cavity. Inset shows the dressed-state picture of energy levels for the cavity-qubit system, for 10 % detuning. b) Results of numerical simulations of this QND readout using the quantum state diffusion method. A microwave pulse with duration $\sim 1.5\,\mu$s excites the cavity to an amplitude $\langle n \rangle \sim 100$. The intracavity photon number (left axis, in black), and occupation probability of the excited state (right axis), for the case in which the qubit is initially in the ground (blue) or excited (red) state, are shown as a function of time. Though the qubit states are coherently mixed during the pulse, the probability of real transitions is seen to be small. Depending on the qubit's state, the pulse is either above or below the combined cavity-qubit resonance, and so is transmitted with an large relative phase shift that can be detected with homodyne detection. c) The real component of the cavity electric field amplitude (left axis), and the transmitted voltage phasor (right axis) in the output transmission line, for the two possible qubit states. The opposing phase shifts cause a change in sign of the output, which can be measured with high signal-to-noise to realize a single-shot, QND measurement of the qubit.

Table 2

Figures of merit for readout and multi-qubit entanglement of superconducting qubits using dispersive (off-resonant) coupling to a 1D transmission line resonator. The same parameters as Table 1, and a detuning of the Cooper pair box from the resonator of 10% ($\Delta = 0.1\,\omega_r$), are assumed. Quantities involving the qubit decay γ are computed both for the theoretical lower bound $\gamma = \gamma_\kappa$ for spontaneous emission via the cavity, and (in parentheses) for the current experimental upper bound $1/\gamma \geq 2\,\mu$s. Though the signal-to-noise of the readout is very high in either case, the estimate of the readout error rate is dominated by the probability of qubit relaxation during the measurement, which has a duration of a few cavity lifetimes ($\sim 1 - 10\,\kappa^{-1}$). If the qubit non-radiative decay is low, both high efficiency readout and more than 10^3 two-bit operations could be attained.

parameter	symbol	1D circuit
dimensionless cavity pull	$g^2/\kappa\Delta$	2.5
cavity-enhanced lifetime	$\gamma_\kappa^{-1} = (\Delta/g)^2\kappa^{-1}$	64 μs
readout SNR	SNR = $(n_{\text{crit}}/n_{\text{amp}})\kappa/2\gamma$	200 (6)
readout error	$P_{\text{err}} \sim 5 \times \gamma/\kappa$	1.5 % (14 %)
1 bit operation time	$T_\pi > 1/\Delta$	> 0.16 ns
entanglement time	$t_{\sqrt{i\text{SWAP}}} = \pi\Delta/4g^2$	$\sim$ 0.05 μs
2 bit operations	$N_{\text{op}} = 1/[\gamma\, t_{\sqrt{i\text{SWAP}}}]$	> 1200 (40)

two states and thus cannot dephase the qubit. This also implies that a charge measurement cannot be used to determine the state of the system [4, 5]. While the first derivative of the energy splitting with respect to gate voltage vanishes at the degeneracy point, the second derivative, corresponding to the difference in charge *polarizability* of the two quantum states, is *maximal.* One can think of the qubit as a non-linear quantum system having a state-dependent capacitance (or in general, an admittance) which changes sign between the ground and excited states [28]. It is this change in polarizability which is measured in the dispersive QND measurement.

In contrast, standard charge measurement schemes [29, 18] require moving away from the optimal point. Simmonds et al. [23] have recently raised the possibility that there are numerous parasitic environmental resonances which can relax the qubit when its frequency Ω is changed during the course of moving the operating point. The dispersive cQED measurement is therefore highly advantageous since it operates best at the charge degeneracy point. In general, such a measurement of an ac property of the qubit is strongly desirable in the usual case where dephasing is dominated by low frequency (1/f) noise. Notice also that the proposed quantum non-demolition measurement would be the inverse of the atomic microwave cQED measurement in which the state of the photon field is inferred non-destructively from the phase shift in the state of atoms sent through the cavity [3].

7. Resonator as quantum bus: entanglement of multiple qubits

Finally, the transmission-line resonator has the advantage that it should be possible to place multiple qubits along its length ($\sim$ 1 cm) and entangle them together, which is an essential requirement for quantum computation. For the case of two qubits, they can be placed closer to the ends of the resonator but still well isolated from the environment and can be separately dc biased by capacitive coupling to the left and right center conductors of the transmission line. Any additional qubits would have to have separate gate bias lines installed. If qubits i and j are tuned in resonance with each other but detuned from the cavity, the effective Hamiltonian will contain qubit-qubit coupling due to exchange of virtual photons: $H_2 = (g^2/\Delta)(\sigma_i^+\sigma_j^- + \sigma_i^-\sigma_j^+)$. Starting with an excitation in one of the qubits, this interaction will have the pair of qubits maximally entangled after a time $t_{\sqrt{i\mathrm{SWAP}}} = \pi\Delta/4g^2 \sim 50\,\mathrm{ns}$. Making the most optimistic assumption that we can take full advantage of the lifetime enhancement inside the cavity (i.e. that γ_{NR} can be made negligible), the number of $\sqrt{i\mathrm{SWAP}}$ operations which can be carried out in one cavity decay time is $N_{\mathrm{op}} = 4\Delta/\pi\kappa \sim 1200$ for the experimental parameters assumed above. This can be further improved if the qubit's non-radiative decay is sufficiently small, and higher Q cavities are employed. When the qubits are detuned from each other, the qubit-qubit interaction in the effective Hamiltonian is turned off, hence the coupling is tunable. Numerical simulations indicate that when the qubits are strongly detuned from the cavity, single-bit gate operations can be performed with high fidelity [24]. Driving the cavity at its resonance frequency constitutes a *measurement* because the phase shift of the transmitted wave is strongly dependent on the state of the qubit and hence the photons become entangled with the qubit. On the other hand, driving the cavity at the qubit transition frequency constitutes a *rotation*. This is *not* a measurement because, for large detuning the photons are largely reflected with a phase shift which is independent of the state of the qubit. Hence there is little entanglement and the rotation fidelity is high [24].

Together with one-qubit gates, the interaction H_2 is sufficient for universal quantum computation (UQC) [30]. Alternatively, H_2 can be used to realize encoded UQC on the subspace $\mathcal{L} = \{|\uparrow\downarrow\rangle, |\downarrow\uparrow\rangle\}$ [31]. In this context, a simpler non-trivial encoded two-qubit gate can also be obtained by tuning, for a time $t = \pi\Delta/3g^2$, all four qubits in the pair of encoded logical qubits in resonance with each other but detuned from the resonator. This is closely related to the Sørensen-Mølmer scheme discussed in the context of the ion-trap proposals [32].

Interestingly, $\mathcal{L}$ is also a decoherence-free subspace with respect to global dephasing [31] and use of this encoding will provide some protection against noise.

Another advantage of the dispersive QND readout is that one may be able to determine the state of multiple qubits in a single shot without the need for additional signal ports. For example, for the case of two qubits with different detunings, the cavity pull will take on four different values $\pm g_1^2/\Delta_1 \pm g_2^2/\Delta_2$ allowing single-shot readout of the coupled system. This can in principle be extended to N qubits provided that the range of individual cavity pulls can be made large enough to distinguish all the combinations. Alternatively, one could read them out in small groups at the expense of having to electrically vary the detuning of each group to bring them into strong coupling with the resonator.

8. Summary and conclusions

In summary, we propose that the combination of one-dimensional superconducting transmission line resonators, which confine their zero point energy to extremely small volumes, and superconducting charge qubits, which are electrically controllable qubits with large electric dipole moments, constitutes an interesting system to access the strong-coupling regime of cavity quantum electrodynamics. This combined system constitutes an advantageous architecture for the coherent control, entanglement, and readout of quantum bits for quantum computation and communication. Among the practical benefits of this approach are the ability to suppress radiative decay of the qubit while still allowing one-bit operations, a simple and minimally disruptive method for readout of single and multiple qubits, and the ability to generate tunable two-qubit entanglement over centimeter-scale distances. We also note that in the structures described here, the emission or absorption of a single photon by the qubit is tagged by a sudden large change in the resonator transmission properties [24] making them potentially useful as single photon sources and detectors.

Acknowledgments

We are grateful to David DeMille, Michel Devoret, Clifford Cheung and Florian Marquardt for useful conversations. We also thank André-Marie Tremblay and the Canadian Foundation for Innovation for access to computing facilities. This work was supported in part by the National Security Agency (NSA) and Advanced Research and Development Activity (ARDA) under Army Research Office (ARO) contract number DAAD19-02-1-0045, NSF DMR-0196503, NSF

DMR-0342157, the David and Lucile Packard Foundation, the W.M. Keck Foundation and NSERC.

References

[1] H. Mabuchi, A. Doherty, *Science* **298**, 1372 (2002).
[2] C. J. Hood, T. W. Lynn, A. C. Doherty, A. S. Parkins, H. J. Kimble, *Science* **287**, 1447 (2000).
[3] J. Raimond, M. Brune, S. Haroche, *Rev. Mod. Phys.* **73**, 565 (2001).
[4] A. Armour, M. Blencowe, K. C. Schwab, *Phys. Rev. Lett.* **88**, 148301 (2002).
[5] E. K. Irish, K. Schwab, *Phys. Rev. B* **68**, 155311 (2003).
[6] Y. Makhlin, G. Schön, A. Shnirman, *Rev. Mod. Phys.* **73**, 357 (2001).
[7] O. Buisson, F. Hekking, *Macroscopic Quantum Coherence and Quantum Computing*, D. V. Averin, B. Ruggiero, P. Silvestrini, eds. (Kluwer, New York, 2001).
[8] F. Marquardt, C. Bruder, *Phys. Rev. B* **63**, 054514 (2001).
[9] F. Plastina, G. Falci, *Phys. Rev. B* **67**, 224514 (2003).
[10] A. Blais, A. Maassen van den Brink, A. Zagoskin, *Phys. Rev. Lett.* **90**, 127901 (2003).
[11] W. Al-Saidi, D. Stroud, *Phys. Rev. B* **65**, 014512 (2001).
[12] C.-P. Yang, S.-I. Chu, S. Han, *Phys. Rev. A* **67**, 042311 (2003).
[13] J. Q. You, F. Nori, *Phys. Rev. B* **68**, 064509 (2003).
[14] D. Walls, G. Milburn, *Quantum optics* (Spinger-Verlag, Berlin, 1994).
[15] S. Haroche, *Fundamental Systems in Quantum Optics*, J. Dalibard, J. Raimond, J. Zinn-Justin, eds. (Elsevier, 1992), p. 767.
[16] V. Bouchiat, D. Vion, P. Joyez, D. Esteve, M. Devoret, *Physica Scripta* **T76**, 165 (1998).
[17] D. Vion, *et al.*, *Science* **296**, 886 (2002).
[18] K. Lehnert, *et al.*, *Phys. Rev. Lett.* **90**, 027002 (2003).
[19] A.S. Sørensen, C.H. van der Wal, L. Childress, M.D. Lukin, *Phys. Rev. Lett.* **92**, 063601 (2004).
[20] P. K. Day, H. G. LeDuc, B. A. Mazin, A. Vayonakis, J. Zmuidzinas, *Nature (London)* **425**, 817 (2003).
[21] For a large ratio of E_J/E_C higher charge states will be mixed into the two lowest energy eigenstates and the coefficient of σ_x in this expression will be somewhat different from unity.
[22] H. Kimble, *Structure and dynamics in cavity quantum electrodynamics* (Academic Press, 1994).
[23] R. W. Simmonds, K. M. Lang, D. A. Hite, D. P. Pappas, J. Martinis, Decoherence in Josephson qubits from junction resonances (2003). Submitted to Phys. Rev. Lett.
[24] S. Girvin, A. Blais, R. Huang. Unpublished.
[25] We note that for the case of $Q = 10^6$, the cavity pull is a remarkable ± 250 line widths, but, depending on the non-radiative decay rate of the qubit, this may be in the regime $\kappa < \gamma$, making the state measurement too slow.
[26] S. van Enk, J. Cirac, P. Zoller, *Science* **279**, 2059 (1998).
[27] R. Schack, T. A. Brun, *Comp. Phys. Comm.* **102**, 210 (1997).
[28] D. Averin, C. Bruder, *Phys. Rev. Lett.* **91**, 057003 (2003).
[29] Y. Nakamura, Y. Pashkin, J. Tsai, *Nature (London)* **398**, 786 (1999).
[30] A. Barenco, *et al.*, *Phys. Rev. A* **52**, 3457 (1995).
[31] D. Lidar, L.-A. Wu, *Phys. Rev. Lett.* **88**, 017905 (2002).
[32] A. Sørensen, K. Mølmer, *Phys. Rev. Lett.* **82**, 1971 (1999).

Printed and bound by CPI Group (UK) Ltd, Croydon, CR0 4YY

03/10/2024

01040428-0006